complex analysis and geometry

PURE AND APPLIED MATHEMATICS

A Program of Monographs, Textbooks, and Lecture Notes

LECTURE NOTES IN PURE AND APPLIED MATHEMATICS

1. *N. Jacobson,* Exceptional Lie Algebras
2. *L.-Å. Lindahl and F. Poulsen,* Thin Sets in Harmonic Analysis
3. *I. Satake,* Classification Theory of Semi-Simple Algebraic Groups
4. *F. Hirzebruch, W. D. Newmann, and S. S. Koh, Differentiable Manifolds and Quadratic Forms*
5. *I. Chavel,* Riemannian Symmetric Spaces of Rank One
6. *R. B. Burckel,* Characterization of C(X) Among Its Subalgebras
7. *B. R. McDonald, A. R. Magid, and K. C. Smith,* Ring Theory: Proceedings of the Oklahoma Conference
8. *Y.-T. Siu,* Techniques of Extension on Analytic Objects
9. *S. R. Caradus, W. E. Pfaffenberger, and B. Yood,* Calkin Algebras and Algebras of Operators on Banach Spaces
10. *E. O. Roxin, P.-T. Liu, and R. L. Sternberg,* Differential Games and Control Theory
11. *M. Orzech and C. Small,* The Brauer Group of Commutative Rings
12. *S. Thomier,* Topology and Its Applications
13. *J. M. Lopez and K. A. Ross,* Sidon Sets
14. *W. W. Comfort and S. Negrepontis,* Continuous Pseudometrics
15. *K. McKennon and J. M. Robertson,* Locally Convex Spaces
16. *M. Carmeli and S. Malin,* Representations of the Rotation and Lorentz Groups: An Introduction
17. *G. B. Seligman,* Rational Methods in Lie Algebras
18. *D. G. de Figueiredo,* Functional Analysis: Proceedings of the Brazilian Mathematical Society Symposium
19. *L. Cesari, R. Kannan, and J. D. Schuur,* Nonlinear Functional Analysis and Differential Equations: Proceedings of the Michigan State University Conference
20. *J. J. Schäffer,* Geometry of Spheres in Normed Spaces
21. *K. Yano and M. Kon,* Anti-Invariant Submanifolds
22. *W. V. Vasconcelos,* The Rings of Dimension Two
23. *R. E. Chandler,* Hausdorff Compactifications
24. *S. P. Franklin and B. V. S. Thomas,* Topology: Proceedings of the Memphis State University Conference
25. *S. K. Jain,* Ring Theory: Proceedings of the Ohio University Conference
26. *B. R. McDonald and R. A. Morris,* Ring Theory II: Proceedings of the Second Oklahoma Conference
27. *R. B. Mura and A. Rhemtulla,* Orderable Groups
28. *J. R. Graef,* Stability of Dynamical Systems: Theory and Applications
29. *H.-C. Wang,* Homogeneous Branch Algebras
30. *E. O. Roxin, P.-T. Liu, and R. L. Sternberg,* Differential Games and Control Theory II
31. *R. D. Porter,* Introduction to Fibre Bundles
32. *M. Altman,* Contractors and Contractor Directions Theory and Applications
33. *J. S. Golan,* Decomposition and Dimension in Module Categories
34. *G. Fairweather,* Finite Element Galerkin Methods for Differential Equations
35. *J. D. Sally,* Numbers of Generators of Ideals in Local Rings
36. *S. S. Miller,* Complex Analysis: Proceedings of the S.U.N.Y. Brockport Conference
37. *R. Gordon,* Representation Theory of Algebras: Proceedings of the Philadelphia Conference
38. *M. Goto and F. D. Grosshans,* Semisimple Lie Algebras
39. *A. I. Arruda, N. C. A. da Costa, and R. Chuaqui,* Mathematical Logic: Proceedings of the First Brazilian Conference
40. *F. Van Oystaeyen,* Ring Theory: Proceedings of the 1977 Antwerp Conference
41. *F. Van Oystaeyen and A. Verschoren,* Reflectors and Localization: Application to Sheaf Theory
42. *M. Satyanarayana,* Positively Ordered Semigroups
43. *D. L Russell,* Mathematics of Finite-Dimensional Control Systems
44. *P.-T. Liu and E. Roxin,* Differential Games and Control Theory III: Proceedings of the Third Kingston Conference, Part A
45. *A. Geramita and J. Seberry,* Orthogonal Designs: Quadratic Forms and Hadamard Matrices
46. *J. Cigler, V. Losert, and P. Michor,* Banach Modules and Functors on Categories of Banach Spaces

47. *P.-T. Liu and J. G. Sutinen,* Control Theory in Mathematical Economics: Proceedings of the Third Kingston Conference, Part B
48. *C. Byrnes,* Partial Differential Equations and Geometry
49. *G. Klambauer,* Problems and Propositions in Analysis
50. *J. Knopfmacher,* Analytic Arithmetic of Algebraic Function Fields
51. *F. Van Oystaeyen,* Ring Theory: Proceedings of the 1978 Antwerp Conference
52. *B. Kadem,* Binary Time Series
53. *J. Barros-Neto and R. A. Artino,* Hypoelliptic Boundary-Value Problems
54. *R. L. Sternberg, A. J. Kalinowski, and J. S. Papadakis,* Nonlinear Partial Differential Equations in Engineering and Applied Science
55. *B. R. McDonald,* Ring Theory and Algebra III: Proceedngs of the Third Oklahoma Conference
56. *J. S. Golan,* Structure Sheaves Over a Noncommutative Ring
57. *T. V. Narayana, J. G. Williams, and R. M. Mathsen,* Combinatorics, Representation Theory and Statistical Methods in Groups: YOUNG DAY Proceedings
58. *T. A. Burton,* Modeling and Differential Equations in Biology
59. *K. H. Kim and F. W. Roush,* Introduction to Mathematical Consensus Theory
60. *J. Banas and K. Goebel,* Measures of Noncompactness in Banach Spaces
61. *O. A. Nielson,* Direct Integral Theory
62. *J. E. Smith, G. O. Kenny, and R. N. Ball,* Ordered Groups: Proceedings of the Boise State Conference
63. *J. Cronin,* Mathematics of Cell Electrophysiology
64. *J. W. Brewer,* Power Series Over Commutative Rings
65. *P. K. Kamthan and M. Gupta,* Sequence Spaces and Series
66. *T. G. McLaughlin,* Regressive Sets and the Theory of Isols
67. *T. L. Herdman, S. M. Rankin III, and H. W. Stech,* Integral and Functional Differential Equations
68. *R. Draper,* Commutative Algebra: Analytic Methods
69. *W. G. McKay and J. Patera,* Tables of Dimensions, Indices, and Branching Rules for Representations of Simple Lie Algebras
70. *R. L. Devaney and Z. H. Nitecki,* Classical Mechanics and Dynamical Systems
71. *J. Van Geel,* Places and Valuations in Noncommutative Ring Theory
72. *C. Faith,* Injective Modules and Injective Quotient Rings
73. *A. Fiacco,* Mathematical Programming with Data Perturbations I
74. *P. Schultz, C. Praeger, and R. Sullivan,* Algebraic Structures and Applications: Proceedings of the First Western Australian Conference on Algebra
75. *L Bican, T. Kepka, and P. Nemec,* Rings, Modules, and Preradicals
76. *D. C. Kay and M. Breen,* Convexity and Related Combinatorial Geometry: Proceedings of the Second University of Oklahoma Conference
77. *P. Fletcher and W. F. Lindgren,* Quasi-Uniform Spaces
78. *C.-C. Yang,* Factorization Theory of Meromorphic Functions
79. *O. Taussky,* Ternary Quadratic Forms and Norms
80. *S. P. Singh and J. H. Burry,* Nonlinear Analysis and Applications
81. *K. B. Hannsgen, T. L. Herdman, H. W. Stech, and R. L. Wheeler,* Volterra and Functional Differential Equations
82. *N. L. Johnson, M. J. Kallaher, and C. T. Long,* Finite Geometries: Proceedings of a Conference in Honor of T. G. Ostrom
83. *G. I. Zapata,* Functional Analysis, Holomorphy, and Approximation Theory
84. *S. Greco and G. Valla,* Commutative Algebra: Proceedings of the Trento Conference
85. *A. V. Fiacco,* Mathematical Programming with Data Perturbations II
86. *J.-B. Hiriart-Urruty, W. Oettli, and J. Stoer,* Optimization: Theory and Algorithms
87. *A. Figa Talamanca and M. A. Picardello,* Harmonic Analysis on Free Groups
88. *M. Harada,* Factor Categories with Applications to Direct Decomposition of Modules
89. *V. I. Istrăţescu,* Strict Convexity and Complex Strict Convexity
90. *V. Lakshmikantham,* Trends in Theory and Practice of Nonlinear Differential Equations
91. *H. L. Manocha and J. B. Srivastava,* Algebra and Its Applications
92. *D. V. Chudnovsky and G. V. Chudnovsky,* Classical and Quantum Models and Arithmetic Problems
93. *J. W. Longley,* Least Squares Computations Using Orthogonalization Methods
94. *L. P. de Alcantara,* Mathematical Logic and Formal Systems
95. *C. E. Aull,* Rings of Continuous Functions
96. *R. Chuaqui,* Analysis, Geometry, and Probability
97. *L. Fuchs and L. Salce,* Modules Over Valuation Domains

98. *P. Fischer and W. R. Smith,* Chaos, Fractals, and Dynamics
99. *W. B. Powell and C. Tsinakis,* Ordered Algebraic Structures
100. *G. M. Rassias and T. M. Rassias,* Differential Geometry, Calculus of Variations, and Their Applications
101. *R.-E. Hoffmann and K. H. Hofmann,* Continuous Lattices and Their Applications
102. *J. H. Lightbourne III and S. M. Rankin III,* Physical Mathematics and Nonlinear Partial Differential Equations
103. *C. A. Baker and L. M. Batten,* Finite Geometrics
104. *J. W. Brewer, J. W. Bunce, and F. S. Van Vleck,* Linear Systems Over Commutative Rings
105. *C. McCrory and T. Shifrin,* Geometry and Topology: Manifolds, Varieties, and Knots
106. *D. W. Kueker, E. G. K. Lopez-Escobar, and C. H. Smith,* Mathematical Logic and Theoretical Computer Science
107. *B.-L. Lin and S. Simons,* Nonlinear and Convex Analysis: Proceedings in Honor of Ky Fan
108. *S. J. Lee,* Operator Methods for Optimal Control Problems
109. *V. Lakshmikantham,* Nonlinear Analysis and Applications
110. *S. F. McCormick,* Multigrid Methods: Theory, Applications, and Supercomputing
111. *M. C. Tangora,* Computers in Algebra
112. *D. V. Chudnovsky and G. V. Chudnovsky,* Search Theory: Some Recent Developments
113. *D. V. Chudnovsky and R. D. Jenks,* Computer Algebra
114. *M. C. Tangora,* Computers in Geometry and Topology
115. *P. Nelson, V. Faber, T. A. Manteuffel, D. L. Seth, and A. B. White, Jr.,* Transport Theory, Invariant Imbedding, and Integral Equations: Proceedings in Honor of G. M. Wing's 65th Birthday
116. *P. Clément, S. Invernizzi, E. Mitidieri, and I. I. Vrabie,* Semigroup Theory and Applications
117. *J. Vinuesa,* Orthogonal Polynomials and Their Applications: Proceedings of the International Congress
118. *C. M. Dafermos, G. Ladas, and G. Papanicolaou,* Differential Equations: Proceedings of the EQUADIFF Conference
119. *E. O. Roxin,* Modern Optimal Control: A Conference in Honor of Solomon Lefschetz and Joseph P. Lasalle
120. *J. C. Díaz,* Mathematics for Large Scale Computing
121. *P. S. Milojević,* Nonlinear Functional Analysis
122. *C. Sadosky,* Analysis and Partial Differential Equations: A Collection of Papers Dedicated to Mischa Cotlar
123. *R. M. Shortt,* General Topology and Applications: Proceedings of the 1988 Northeast Conference
124. *R. Wong,* Asymptotic and Computational Analysis: Conference in Honor of Frank W. J. Olver's 65th Birthday
125. *D. V. Chudnovsky and R. D. Jenks,* Computers in Mathematics
126. *W. D. Wallis, H. Shen, W. Wei, and L. Zhu,* Combinatorial Designs and Applications
127. *S. Elaydi,* Differential Equations: Stability and Control
128. *G. Chen, E. B. Lee, W. Littman, and L. Markus,* Distributed Parameter Control Systems: New Trends and Applications
129. *W. N. Everitt,* Inequalities: Fifty Years On from Hardy, Littlewood and Pólya
130. *H. G. Kaper and M. Garbey,* Asymptotic Analysis and the Numerical Solution of Partial Differential Equations
131. *O. Arino, D. E. Axelrod, and M. Kimmel,* Mathematical Population Dynamics: Proceedings of the Second International Conference
132. *S. Coen,* Geometry and Complex Variables
133. *J. A. Goldstein, F. Kappel, and W. Schappacher,* Differential Equations with Applications in Biology, Physics, and Engineering
134. *S. J. Andima, R. Kopperman, P. R. Misra, J. Z. Reichman, and A. R. Todd,* General Topology and Applications
135. *P Clément, E. Mitidieri, B. de Pagter,* Semigroup Theory and Evolution Equations: The Second International Conference
136. *K. Jarosz,* Function Spaces
137. *J. M. Bayod, N. De Grande-De Kimpe, and J. Martínez-Maurica,* *p*-adic Functional Analysis
138. *G. A. Anastassiou,* Approximation Theory: Proceedings of the Sixth Southeastern Approximation Theorists Annual Conference
139. *R. S. Rees,* Graphs, Matrices, and Designs: Festschrift in Honor of Norman J. Pullman
140. *G. Abrams, J. Haefner, and K. M. Rangaswamy,* Methods in Module Theory

141. *G. L. Mullen and P. J.-S. Shiue*, Finite Fields, Coding Theory, and Advances in Communications and Computing
142. *M. C. Joshi and A. V. Balakrishnan*, Mathematical Theory of Control: Proceedings of the International Conference
143. *G. Komatsu and Y. Sakane*, Complex Geometry: Proceedings of the Osaka International Conference
144. *I. J. Bakelman*, Geometric Analysis and Nonlinear Partial Differential Equations
145. *T. Mabuchi and S. Mukai*, Einstein Metrics and Yang–Mills Connections: Proceedings of the 27th Taniguchi International Symposium
146. *L. Fuchs and R. Göbel*, Abelian Groups: Proceedings of the 1991 Curaçao Conference
147. *A. D. Pollington and W. Moran*, Number Theory with an Emphasis on the Markoff Spectrum
148. *G. Dore, A. Favini, E. Obrecht, and A. Venni*, Differential Equations in Banach Spaces
149. *T. West*, Continuum Theory and Dynamical Systems
150. *K. D. Bierstedt, A. Pietsch, W. Ruess, and D. Vogt*, Functional Analysis
151. *K. G. Fischer, P. Loustaunau, J. Shapiro, E. L. Green, and D. Farkas*, Computational Algebra
152. *K. D. Elworthy, W. N. Everitt, and E. B. Lee*, Differential Equations, Dynamical Systems, and Control Science
153. *P.-J. Cahen, D. L. Costa, M. Fontana, and S.-E. Kabbaj*, Commutative Ring Theory
154. *S. C. Cooper and W. J. Thron*, Continued Fractions and Orthogonal Functions: Theory and Applications
155. *P. Clément and G. Lumer*, Evolution Equations, Control Theory, and Biomathematics
156. *M. Gyllenberg and L. Persson*, Analysis, Algebra, and Computers in Mathematical Research: Proceedings of the Twenty-First Nordic Congress of Mathematicians
157. *W. O. Bray, P. S. Milojević, and Č. V. Stanojević*, Fourier Analysis: Analytic and Geometric Aspects
158. *J. Bergen and S. Montgomery*, Advances in Hopf Algebras
159. *A. R. Magid*, Rings, Extensions, and Cohomology
160. *N. H. Pavel*, Optimal Control of Differential Equations
161. *M. Ikawa*, Spectral and Scattering Theory: Proceedings of the Taniguchi International Workshop
162. *X. Liu and D. Siegel*, Comparison Methods and Stability Theory
163. *J.-P. Zolésio*, Boundary Control and Variation
164. *M. Křížek, P. Neittaanmäki, and R. Stenberg*, Finite Element Methods: Fifty Years of the Courant Element
165. *G. Da Prato and L. Tubaro*, Control of Partial Differential Equations
166. *E. Ballico*, Projective Geometry with Applications
167. *M. Costabel, M. Dauge, and S. Nicaise*, Boundary Value Problems and Integral Equations in Nonsmooth Domains
168. *G. Ferreyra, G. R. Goldstein, and F. Neubrander*, Evolution Equations
169. *S. Huggett*, Twistor Theory
170. *H. Cook, W. T. Ingram, K. T. Kuperberg, A. Lelek, and P. Minc*, Continua: With the Houston Problem Book
171. *D. F. Anderson and D. E. Dobbs*, Zero-Dimensional Commutative Rings
172. *K. Jarosz*, Function Spaces: The Second Conference
173. *V. Ancona, E. Ballico, and A. Silva*, Complex Analysis and Geometry
174. *E. Casas*, Control of Partial Differential Equations and Applications
175. *N. Kalton, E. Saab, and S. Montgomery-Smith*, Interaction Between Functional Analysis, Harmonic Analysis, and Probability

Additional Volumes in Preparation

complex analysis and geometry

proceedings of the conference at Trento

edited by

Vincenzo Ancona
Università di Firenze
Florence, Italy

Edoardo Ballico
Università di Trento
Trento, Italy

Alessandro Silva
Università La Sapienza
Rome, Italy

Marcel Dekker, Inc. **New York • Basel • Hong Kong**

Library of Congress Cataloging-in-Publication Data

Complex analysis and geometry : proceedings of the conference at Trento / edited by V. Ancona, E. Ballico, A. Silva.
p. cm. — (Lecture notes in pure and applied mathematics ; v. 173)
ISBN 0-8247-9672-1 (alk. paper)
1. Geometry, Algebraic—Congresses. 2. Functions of several complex variables—Congresses. I. Ancona, Vincenzo. II. Ballico, E. (Edoardo). III. Silva, Alessandro. IV. Series.
QA564.C65636 1996
515'.9—dc20 95-34364
CIP

The publisher offers discounts on this book when ordered in bulk quantities. For more information, write to Special Sales/Professional Marketing at the address below.

This book is printed on acid-free paper.

MARCEL DEKKER, INC.
270 Madison Avenue, New York, New York 10016

Current printing (last digit):
10 9 8 7 6 5 4 3 2 1

PRINTED IN THE UNITED STATES OF AMERICA

Preface

This volume contains the Proceedings of the C.I.R.M. Conference "Complex Analysis and Geometry" held in Trento, Italy. It contains twenty-two research papers and three (very up-to-date and very useful) survey papers. All papers (except the one of J.-P. Vigué, which is in French) are in the English language. All were refereed. All papers are in final form and no part of them will be submitted elsewhere.

In 1981 two of us thought that perhaps other complex analysts and complex geometers deserved the opportunity of having a regular forum in which a presentation of the main results of the year would be given regardless of specialties (from algebraic geometry to P.D.E. and microanalysis). The first meeting was held in 1982. This volume contains the contributions of the participants (or invited speakers not able to come) and participants (as in the case of Zampieri and Tumanov) of another C.I.R.M. meeting on related topics, whose papers fit very well in this volume.

We give a very brief preview of the papers printed in this volume. We hope that this preview (and much more the volume) will show the breadth and vitality of complex analysis.

An n-dimensional complex manifold is said to be *balanced* if it carries a hermitian metric whose Kähler form ω satisfies $d\omega^{n-1} = 0$ (hence a Kähler manifold is balanced). In the paper "The class of compact balanced manifolds is invariant under modifications" L. Alessandrini and G. Bassanelli prove the result stated in the title. It is well known that the corresponding result is false for the class of Kähler manifolds. The survey by A. Silva is concerned with the birational properties of compact complex manifolds (Kähler, Moishezon, balanced, or more generally, p-Kähler), from the classical results to the more recent ones. Two main problems are singled out: the behavior of these classes of manifolds under monoidal transformations and conditions for a suitable metric (e.g. a p-Kähler one) defined outside a suitable set to be extendable to all the manifolds.

Let $C \subset P^n$ be a smooth curve and H a hyperplane of P^n. H is said to be tangentially versal to C if the family of all $\{C \cap M\}$ with M a hyperplace near H is a versal deformation of the multigerm of singularities $C \cap H$. In the paper "On tangent hyperplanes to complex projective curves" by E. Ballico the problem is considered of the existence of embeddings $C \subset P^n$ such that every hyperplane is tangentially versal to C. The paper "Unfoldings of holomorphic maps as deformations" by J. Buchweitz and H. Flenner recasts the classical Mather-Thom theory of unfoldings of holomorphic maps from the point of view (very abstract but also very effective) of deformation theory. In this way the authors unify and generalize several results of Mather type (finite determinacy, infinitesimal stability implies stability, existence of versal deformations) and of Mather-Yau type (under suitable assumptions two germs $f,g : (C^n,0) \to (C^p,0)$ are equivalent if and only if their spaces of first order deformations are algebraically isomorphic). The authors apply their abstract theory to many particular cases and a very useful feature of this paper is that a mathematician can read or use a single application without reading the general theory.

In the first part of the paper "Infinite dimensional supergeometry" J. Bingener and T. Lehmkuhl develop a very general theory of infinite dimensional geometry. The theory is developed in a general noncommutative setting and it embraces differential calculus, analytic functions, Banach analytic spaces and Lie theory. The authors also describe possible applications of this theory in complex-analytic deformation theory.

Every real analytic manifold X admits a complexification X_C whose germ along X is uniquely determined. Recently there has been much interest in the study of specific choices of X^r. Assume X compact and with a fixed real analytic riemannian metric g. A Grauert tube X_C^r of (X,g) is a complex structure on the bundle of tangent vectors of length $< r$ (with respect to g) for some fixed $r > 0$. In the paper "On the uniqueness and characterization of Grauert tubes" D. Burns, using the Monge-Ampère equation, proves several uniqueness results for Grauert tubes. Among them there is the following theorem. Let Ω be a connected complex manifold and $\tau : \Omega \to [0, r^2)$ be a proper C^∞ strictly plurisubharmonic function such that $u = \sqrt{\tau}$ is a plurisubharmonic solution of the homogeneous Monge-Ampére equation $(\delta\delta - u)^n = 0$ on $\Omega\, \tau^{-1}(0)$; then $\tau^{-1}(0)$ is a connected real analytic manifold X and Ω is canonically biholomorphic to the Grauert tube X_C^r, where X is equipped with the metric induced from the Kähler metric on Ω with Kähler form $\omega := i\delta\delta - \tau/2$.

A *CR* manifold is said to be *pseudoconcave* if it is smooth, locally embeddable and at least 1-pseudoconcave. In "Pseudoconcave *CR* manifolds" C. Denson Hill and M. Nacinovich initiate the global study (e.g. vanishing theorems for cohomology groups) of pseudoconcave *CR* manifolds of arbitrary *CR*-dimension and *CR*-codimension. Recall that a $G_{m,h}$-bundle E on a manifold X is a fiber bundle $E \to X$ with $C^m \times R^h$ as fiber and *CR*-functions preserving the natural foliation on $C^m \times R^h$ as transition functions; let E be the sheaf of germs of smooth *CR*-sections of $E \to X$. In the paper "$G_{m,h}$-bundles over foliations with complex leaves" G. Gigante and G. Tomassini prove that such a bundle over a 1-complete foliation is a 1-complete foliation and use this result to prove that $H^j(X,E) = 0$ for every $j \geq 1$ if X and E are real analytic and X is 1-complete. This vanishing theorem is used to obtain a *CR*-tubular neighborhood and an extension theorem for smooth *CR*-functions. Let M be a smooth real generic manifold in C^N; recall that a *CR* orbit of $P \in M$ is the set of all points of M reachable from P using *CR* curves, i.e. real curves that run in complex tangential directions of M. In the paper "On the propagation of extendibility of *CR* functions" A. Tumanov gives a new simple proof of his result (which had already found an application by Merker for simultaneous wedge-extendibility of *CR* functions on globally minimal manifolds) on the *CR* extendibility of *CR* functions along real curves; the exposition is as self-contained as possible (hence understandable). The boundary of a pseudoconvex domain with Levi form of constant rank has a local foliation called the Levi foliation. In the paper "Canonical symplectic structure of a Levi foliation" G. Zampieri studies the microlocal structure of this foliation and links this work with the classical local structure. With the microlocal approach he improves all the classical results and gives two main new applications: the decomposition of the wave front set of *CR* hyperfunctions and the Edge of the Wedge Theorem for generic real analytic *CR* manifolds.

In the paper "Hölder regularity for $\square_b$ on hypersurfaces in C^n, with nondiagonalizable Levi form", M. Derridj proves the Hölder regularity for the boundary Laplacian $\square_b$ on psuedoconvex hypersurfaces of C^4. He plans to prove the general case for hypersurfaces in C^n, $n \geq 3$, in a later paper.

Consider a holomorphic vector field on C^n; the orbit of $P \in C^n$ explodes if the integral curve of P is unbounded on some finite time interval. J. E. Fornaess and S. Grellier study in "Exploding orbits of hamiltonian and contact structures" when there is a dense subclass of vector fields for each of which a dense set of orbits explodes. Their main result is for the holomorphic hamiltonian vector fields in C^2 and for the Reeb vector fields in C^3.

In "Holomorphic automorphisms of C^n: A survey" F. Forstneric surveys some very recent results on holomorphic automorphism groups of C^n, with emphasis on the approximation of biholomorphic maps and on approximation of diffeomorphisms on suitable submanifolds by automorphisms of C^n. He surveys also results on flows generated by holomorphic vector fields, stressing the cases of holomorphic hamiltonian fields and of C^2.

Let M be a compact differentiable manifold of dimension $n > 2$ with a conformal structure c, i.e. an equivalence class up to multiplication by a positive function of riemannian metrics; c is of positive (resp. negative, resp. zero) type if c contains a metric with positive (resp. negative, resp. zero) total scalar curvature. A complex structure on M (seen as an integrable almost-complex structure J) is compatible with c if J is c-antisymmetric. In the first part of the paper "Complex structures on compact conformal manifolds of negative type" P. Gauduchon gives a criterion (in terms of the scalar curvature and the minimal eigenvalue of the Weyl tensor) for the nonexistence of a compatible complex structure on a negative c. This criterion applies e.g. to all flat (M,c). Then he applies his criterion to the case in which M is a compact quotient of a symmetric riemannian space N. He shows that M has no compatible complex structure unless N is hermitian and in this case the only compatible complex structure on (M,c) is induced by the complex structure of N.

The main result of the paper "A real analytic version of Abels' theorem and complexifications of proper Lie group actions" by Heinzner, Huckleberry and Kutzschebauch is the following theorem. Let G be a Lie group with finitely many components which acts properly on the real analytic manifold M; then there exists a Stein G-complexification of M where G acts properly.

The paper "Exotic structures on C^n and C^*-actions on C^3" by S. Kaliman is a survey of the existence of exotic holomorphic structures on C^n and on suitable complex surfaces. Here are reported very recent remarkable connections discovered by P. Russell between this topic and the problem of linearizing C^*-actions on C^3.

By definition a symmetric complex manifold is a pair (M,c) with M a complex manifold and $c : M \to M$ antiholomorphic with $c^2 = Id$. In the paper "Anti-involutions, symmetric complex manifolds, and quantum spaces" J. Lawrynowicz and E. Ramirez de Arellano consider the case $\dim(M) = 1$. They prove the existence of a duality between one-dimensional complex manifolds and phase spaces related to the quantum plane.

Let X be a complex space and E be a finite, étale O_X-algebra; an E-structure is a right E-structure on the sheaf Ω^1_X of differential 1-forms on X. An active topic of research in the late 1950s was the study of integrability conditions (in terms of the Nijenhuis tensor) for an endomorphism of the (real analytic) tangent bundle satisfying polynomial identities with constant coefficients coming from geometry. An *E*-structure is essentially *n* commuting endomorphisms (instead of just one endomorphism). The paper "Newlander-Nirenberg type theorem for analytic algebras" by T. Maszczyk contains an integrability condition for an *E*-structure.

Let X be a projective manifold and L an ample line bundle on X. Suppose that the cotangent bundle Ω^1_X is not nef; then there is a unique positive real number α such that the virtual bundle $\Omega^1_X(\alpha)$ is nef but not ample (called the nef value of Ω_X). In the paper "Semi-positivity and cotangent bundles", by K. Oguiso and T. Peternell, this nef value is studied for many X: several classes of complex surfaces and Calabi-Yau hypersurfaces in CP^n, $n \geq 4$, i.e. degree $n + 1$ hypersurfaces (their nef value is always 2 because they always contain lines). The more interesting result is that every quartic surface in CP^3 with Picard number 1 has $\sqrt{2}$ as nef value (hence the nef value may be irrational).

A compact complex submanifold A of a complex manifold M is said to be exceptional if it can be contracted, i.e. if there is a complex space M'', $P \in M''$ and a holomorphic map $\pi : M \to M''$ with $\pi(A) = \{P\}$ and $\pi \mid (M \setminus A) : M \setminus A \to M'' \setminus \{P\}$ biholomorphic. Motivated by their link with the existence of a Kähler metric on M, in "On a class of (-3,1)-exceptional P^1" T. Ohsawa constructs a large class of exceptional embeddings of P^1 with normal bundle being the direct sum of a line bundle of degree –3 and a line bundle of degree 1.

It is very simple to construct deformations of a riemannian metric, but it is much more difficult to keep the curvature or some other riemannian invariants of the deformed metrics under control. The paper "Isocurved deformations of riemannian homogeneous metrics" by A. Tomassini and F. Tricerri constructs and studies deformations of riemannian homogeneous metrics which (in a suitable sense) preserve the Riemann curvature.

The paper "Completions of instantons moduli space and control theory" by G. Valli contains the construction of two compactifications of the moduli space of framed $SU(N)$-instantons over the 4-sphere or, equivalently, of the moduli space of holomorphic vector bundles over CP^2 with a fixed trivialization at the CP^1 at infinity. A nice feature of this work (seeing instantons as matrix-valued rational functions in two variables) is the connection between one of these compactifications and linear control theory.

Let D be a bounded domain of C^n and $f : D \to D$ a holomorphic map. M. Abate found (at least if D is convex) a strong relationship between the fixed points $Fix(f)$ of f and the sequence $\{f^n\}_{n>0}$ of all iterates of f (in particular if this sequence is compactly divergent). In the first part of the paper "Itérées et points fixes d'applications holomorphes" J.-P. Vigué studies the limit sets and the fixed points of a holomorphic map $f : M \times X \to X$ with M a complex manifold and X a hyperbolic taut complex manifold. He shows that the sets $Fix(f(m, \cdot))_{m\in M}$ (with $f(m, \cdot) : X \to X$ a "slice" of f) are all biholomorphic. Then he uses this theorem to study the fixed points of a holomorphic map $f : X_1 \times X_2 \to X_1 \times X_2$ with X_i a hyperbolic taut manifold, giving the connection between $Fix(f)$ and the fixed points of the "slice" maps $f(x_1, \cdot) : X_2 \to X_2$ and $f(\cdot, x_2) : X_1 \to X_1$.

Let G be a connected complex linear algebraic group and G' its commutator. Consider the following two properties: (1) G/G' is reductive; (2) $O(G)^\Gamma = C$ for every Zariski-dense subgroup Γ. It is known that (2) implies (1) and (Barth-Otte) that (1) and (2) are equivalent if

G is reductive. In the paper "Holomorphic functions on an algebraic group invariant under Zariski-dense subgroups" J. Winkelmann proves the equivalence of (1) and (2) if either G is solvable or G/G' is reductive and the semisimple elements are dense in G'.

In "Pseudoconvexity and pseudoconcavity of dihedrons of C^n" G. Zampieri generalizes (in the case of dihedrons with transversal faces) the classical vanishing theorems for local cohomology groups at a boundary point of a domain in C^n given in terms of Levi forms. He uses Hörmander L^2-estimates and algebraic analysis (hence he gets the results for microfunctions, too).

Our organizing and editorial efforts would have been to no avail without the Centro Internazionale per la Ricerca Matematica (C.I.R.M.) supported by the Istituto Trentino di Cultura ands the Italian C.N.R. C.I.R.M. is an independent center of mathematical research based in Trento, Italy, founded and headed by Mario Miranda. C.I.R.M. provides financial and logistic support for conferences in beautiful alpine surroundings. To Mario Miranda and to C.I.R.M.'s invaluable secretary, Augusto Micheletti, go our warmest thanks. The precious and friendly help we received from Augusto Micheletti went far beyond his (well done) duties: thanks a lot, again.

This volume contains one of the last papers by Franco Tricerri, who died tragically (together with all his family) in a plane accident. We would like to dedicate this volume to his memory.

Vincenzo Ancona
Edoardo Ballico
Alessandro Silva

Contents

Contributors

Lucia Alessandrini, Dipartimento di Matematica, Università di Trento, I-38050 Povo (Trento), Italy. alessandrini@itnvax.science.unitn.it

Edoardo Ballico, Dipartimento di Matematica, Università di Trento, I-38050 Povo (Trento), Italy. ballico@science.unitn.it. Fax: Italy-461-881624

Giovanni Bassanelli, Dipartimento di Matematica, Università di Trento, I-38050 Povo (Trento), Italy. bassanelli@itnvax.science.unitn.it

Jürgen Bingener, Fakultät für Mathematik der Universität, Universitätsstrasse 31, D-93040 Regensburg, Germany

Ragnar-Olaf Buchweitz, Department of Mathematics, University of Toronto, Toronto, Ont. M5S 1A1, Canada. ragnar@math.utoronto.ca

Daniel M. Burns Jr., Department of Mathematics, University of Michigan, Ann Arbor, Michigan 48109-1003, USA. dburns@umich.edu

Makhlouf Derridj, Mathématiques - Bâtiment 425, Université de Paris-Sud, F-91405 Orsay Cedex, France. mderridj@matups.matups.fr

Hubert Flenner, Mathematisches Institut der Universität Göttingen, Bunsenstr. 3-5, D-37073 Göttingen, Germany. hflenner@cfgauss.uni- math.gwdg.de

John Erik Fornaess, Department of Mathematics, University of Michigan, Ann Arbor, MI 48109-1003, USA. fornaess@math.lsa.umich.edu

Franc Forstneric, Department of Mathematics, University of Wisconsin, Madison, WI 53706, USA. forstner@math.wisc.edu

Paul Gauduchon, Centre de Mathématiques, Ecole Polytechnique, F-91128 Palaiseau Cedex, France. pg@orphee.polytechnique.fr

Giuliana Gigante, Dipartimento di Matematica, Università di Parma, via dell'Università 12, I-43100 Parma, Italy. gigante@prmat.math.unipr.it

Sandrine Grellier, Mathématiques - Bâtiment 425, Université de Paris-Sud, F-91405 Orsay Cedex, France. grellier@anh.matups.fr

Peter Heinzner, Mathematisches Institut, Ruhr-Universität Bochum, Universitätsstrasse 150, D-44780 Bochum, Germany. heinzner@ruba.rz.ruhr-uni-bochum.de

Denson C. Hill, Department of Mathematics, SUNY at Stony Brook, Stony Brook, NY 11794, USA. Dhill@math.sunysb.edu

Alan T. Huckleberry, Mathematisches Institut, Ruhr-Universität Bochum, Universitätsstrasse 150, D-44780 Bochum, Germany. huckleberry@ruba.rz.ruhr-uni-bochum.de

Shulim Kaliman, Department of Mathematics and Computer Science, University of Miami, Coral Gables, Florida 33124, USA. kaliman@paris-gw.cs.miami.edu

Frank Kutzschebauch, Mathematisches Institut, Ruhr-Universität Bochum, Universitätsstrasse 150, D-44780 Bochum, Germany. kutzschebauch@ruba.rz.ruhr-uni-bochum.de

Julian Lawrynowicz, Institute of Mathematics, Polish Academy of Sciences and the University of Lódź, Narutowicza 56, PL-90-136 Lódź, Poland. banach@impan.impan.gov.pl

Thomas Lehmkuhl, Fakultät für Mathematik der Universität, Universitätsstrasse 31, D-93040 Regensburg, Germany

Tomasz Maszczyk, Institute of Mathematics, Warsaw University, ul. Banacha 2, 02-097 Warszawa, Poland. maszczyk@mimuw.edu.pl

Mauro Nacinovich, Dipartimento di Matematica, Università di Pisa, via F. Buonarroti 2, I-56127 Pisa, Italy. nacinovi@dm.unipi.it

Keiji Oguiso, Department of Mathematics, Ochanomizu University, Otsuka Bunkyo, 112 Tokyo, Japan. oguiso@math.ocha.ac.jp

Takeo Ohsawa, Dept. of Mathematics, School of Science, Nagoya University, Chikusa-ku Nagoya 464-01, Japan. ohsawa@math.nagoya-u.ac.jp

Thomas Peternell, Mathematisches Institut der Universität Bayreuth, Postfach 101251, D-95440 Bayreuth, Germany. thomas.peternell@uni-bayreuth.d400.de

Enrique Ramirez de Arellano, Departamento de Matemáticas, Centro de Investigación y de Estudios Avanzados, 07000 México, D.F.

Alessandro Silva, Dipartimento di Matematica "G. Castelnuovo", Università di Roma "La Sapienza", P.le A. Moro 2, I-00185 Roma, Italy. silva@sci.uniroma1.it

Adriano Tomassini, Dipartimento di Matematica, Università di Firenze, Viale Morgagni 67/A, I-50134 Firenze, Italy

Giuseppe Tomassini, Scuola Normale Superiore, Piazza dei Cavalieri 7, I-56126 Pisa, Italy. tomassini@vaxsns.sns.it

Franco Tricerri, † Dipartimento di Matematica, Università di Firenze, Viale Morgagni 67/A, I-50134 Firenze, Italy

Alexander Tumanov, Department of Mathematics, University of Illinois, Urbana, Illinois 61801, USA. tumanov@math.uiuc.edu

Giorgio Valli, Dipartimento di Matematica, Università di Pavia, Via Abbiategrasso 209, I-27100 Pavia, Italy

Jean-Pierre Vigué, Mathématiques, URA CNRS D1322 Groupes de Lie et Géométrie, Université de Poitiers, 40 avenue du Recteur Pineau, F-86022 Poitiers Cedex, France. vigue@mathrs.univ-poitiers.fr

Jörg Winkelmann, Mathematisches Institut, Ruhr-Universität Bochum, Universitätstrasse 150, D-44780 Bochum, Germany. winkelmann@ruba.rz.ruhr-uni-bochum.de

Giuseppe Zampieri, Dipartimento di Matematica Pura ed Applicata, Università di Padova, Via Belzoni 7, I-35131 Padova, Italy. zampieri@pdmat1.unipd.it

complex analysis and geometry

The Class of Compact Balanced Manifolds Is Invariant Under Modifications

Lucia Alessandrini, Dipartimento di Matematica, Università di Trento, I-38050 Povo (Trento), Italy. alessandrini@itnvax.science.unitn.it

Giovanni Bassanelli, Dipartimento di Matematica, Università di Trento, I-38050 Povo (Trento), Italy. bassanelli@itnvax.science.unitn.it

0. Introduction.

Examples of compact complex manifolds, that are birationally equivalent to projective manifolds, and nevertheless are non-projective (even non-algebraic) are well-known (f.i. [HA] p. 443-444); in particular, there exist non-algebraic Moishezon manifolds.

In this context, the differential-geometric point of view cannot be neglected: for instance, a Moishezon manifold is projective if and only if it carries a Kähler metric. The examples show that if $f : \tilde{M} \to M$ is a modification, and $\tilde{M}, M$ are compact complex manifolds, the existence of a Kähler metric on one of them does not imply that the other is Kähler. Also in the simplest case: if f is a blow-up with smooth center, then the Kähler property goes back from M to $\tilde{M}$, but not always it goes down, from $\tilde{M}$ to M.

In [AB1] we took one of these examples as starting point for singling out a metric property that is weaker than the Kähler property, but is preserved via modifications.

Let us recall that an n-dimensional complex manifold is said to be balanced if it carries a balanced metric, that is, an hermitian metric whose Kähler form ω satisfies $d\omega^{n-1} = 0$ (of course, a Kähler metric is also a balanced metric). This notion, at least in the compact case, is intrinsic to the complex structure, since it can be characterized in terms of positive currents (see [MI]); more precisely, a compact manifold is balanced if and only if there exists no non-zero positive current T of degree (1,1), such that T is the (1,1)-component of a boundary (i.e., $T = \bar{\partial}S + \partial\bar{S}$, with S of degree (1,0); hence $\partial\bar{\partial}T = 0$).

As a matter of fact, we proved in [AB2] and [AB3] that if M is Kähler or, more generally, balanced, then $\tilde{M}$ is balanced.

The converse is more difficult to investigate, also in the case of a blow-up with smooth center. Our argument uses the above characterization of balanced manifolds in terms of currents; therefore we had to solve the following problem: given a positive $\partial\bar{\partial}$-closed (1,1)-current T on M, is it possible to pull-back it to $\tilde{M}$?

This work is partially supported by MURST 40%. It is in final form and no part of it will be submitted elsewhere.

A partial result in this direction (Theorem 3.9 of [AB4]) was used to prove that if $\tilde{M}$ is Kähler, then M is balanced. This means that all compact complex manifolds which are bimeromorphic to a Kähler space are balanced, but we could not conclude, in general, that if $\tilde{M}$ is balanced, then M is balanced too; this was proved only under a suitable cohomological hypothesis on $\tilde{M}$ (condition (B) in [AB4]). Let us notice that this pull-back tecnique could be applied to the study of the Lelong numbers of plurisubharmonic currents [AB5].

In this paper, first of all we study the Aeppli groups

$$V_{\mathbf{R}}^{1,1} := \frac{\partial\bar{\partial}\text{-closed real (1,1)-currents}}{\text{(1,1)-components of boundaries}}.$$

These groups can also be given in terms of smooth (1,1)-forms, so that $f : \tilde{M} \to M$ induces a natural map $f^* : V_{\mathbf{R}}^{1,1}(M) \to V_{\mathbf{R}}^{1,1}(\tilde{M})$.

Of course, one expects that the irreducible components $\{E_\alpha\}$ of the exceptional set E of the modification represent non-vanishing classes in $V_{\mathbf{R}}^{1,1}(\tilde{M})$, as well as the pull-back of non-trivial classes in $V_{\mathbf{R}}^{1,1}(M)$. But notice that, since a class in $V_{\mathbf{R}}^{1,1}(\tilde{M})$ is represented by a form α which is not closed, in general, but $\partial\bar{\partial}$-closed , if $\tilde{h}_\alpha$ is a pluriharmonic function on some E_α, then also $\tilde{h}_\alpha[E_\alpha]$ represents a class in $V_{\mathbf{R}}^{1,1}(\tilde{M})$. Since the fibers of the map f are compact, $\tilde{h}_\alpha$ is nothing but the pull-back of a pluriharmonic function on $Y_\alpha := f(E_\alpha)$. These are the only needed ingredients: in fact, we get in Theorem 5.3

$$V_{\mathbf{R}}^{1,1}(\tilde{M}) = f^* V_{\mathbf{R}}^{1,1}(M) \oplus (\oplus_\alpha f^* \mathcal{H}(Y_\alpha)\langle [E_\alpha]\rangle),$$

where $\mathcal{H}$ denotes the sheaf of germs of pluriharmonic functions.

This result implies the following theorem:

THEOREM 5.6. - *Let M and $\tilde{M}$ be complex manifolds, and $f : \tilde{M} \to M$ be a proper modification. Let T be a positive $\partial\bar{\partial}$-closed (1,1)-current on M. Then there exists a unique positive $\partial\bar{\partial}$-closed (1,1)-current $\tilde{T}$ on $\tilde{M}$ such that $f_*\tilde{T} = T$ and $\tilde{T} \in f^*\langle T\rangle \in V_{\mathbf{R}}^{1,1}(\tilde{M})$.*

As a matter of fact, the last assertion of this theorem allows to get rid of condition (B) (of [AB4]), so that:

COROLLARY 5.7 - *Let M and $\tilde{M}$ be compact complex manifolds, and $f : \tilde{M} \to M$ be a modification. Then $\tilde{M}$ is balanced if and only if M is balanced, that is, the class of compact balanced manifolds is invariant under modifications.*

1. Cohomology.

1.1 NOTATION.

Let M and $\tilde{M}$ be complex N-dimensional manifolds, not necessarily compact. A proper modification $f : \tilde{M} \to M$ is a proper holomorphic map such that, for a suitable analytic subset Y in M, called the center, $E := f^{-1}(Y)$ (the exceptional set of the modification) is a hypersurface and $f_{|\tilde{M}-E} : \tilde{M} - E \to M - Y$ is a biholomorphism. Moreover, Y has codimension ≥ 2 or f is a biholomorphism.

Let Δ_n (resp. Δ_m) be open polydiscs in $\mathbf{C}^n$ (resp. $\mathbf{C}^m$) centered at 0, with $n \geq 1$ and $m \geq 0$; denote by $\pi : \tilde{\Delta}_n \to \Delta_n$ the blow-up at the origin, and also the induced map $\pi : \tilde{\Delta}_n \times \Delta_m \to \Delta_n \times \Delta_m$.

In every kind of cohomology group, we shall denote by $\langle \alpha \rangle$ the class of α, while for an irreducible analytic subset Y, $[Y]$ denotes the current associated to Y.

1.2 DOLBEAULT COHOMOLOGY GROUPS.

Let us recall the following result (see [HI] p.153)

THEOREM. - *Let $f : X' \to X$ be a modification of complex analytic spaces which is a proper morphism. If both X and X' are non-singular, then we have the vanishing of higher direct images $\mathcal{R}^q f(\mathcal{O}_{X'})$ for all $q > 0$ ($\mathcal{R}^0 f(\mathcal{O}_{X'}) = \mathcal{O}_X$).*

Hence, in our situation, $H^{0,q}_{\bar\partial}(\tilde M) \simeq H^{0,q}_{\bar\partial}(M)$ by [GR], Satz 6 page 417.

1.3 DE RHAM COHOMOLOGY GROUPS.

It is well-known (see f.i. [GH] pages 605-608) that if $f : \tilde M \to M$ is a blow-up with smooth center, f induces an injective map $f^* : H^i(M) \to H^i(\tilde M)$ so that

$$H^*(\tilde M) \simeq f^* H^*(M) \oplus (H^*(E)/f^* H^*(Y)). \tag{1.1}$$

Moreover, the cohomology ring of E is generated, as an $H^*(Y)$-algebra, by the first Chern class $\zeta = c_1(T)$ of the tautological bundle on E, so that $H^1(E) \simeq f^* H^1(Y)$ and $H^1(\tilde M) \simeq f^* H^1(M)$. From (1.1) we get in particular that $H^i(\tilde\Delta_n) \simeq H^i(\mathbf{P}_{n-1})$ $\forall i$, and hence $H^i(\tilde\Delta_n \times \Delta_m) \simeq \mathbf{R}$ if i is even and $i < 2n$, zero otherwise.

Furthermore, call $p : \widetilde{\mathbf{P}_n} \to \mathbf{P}_n$ the blow-up at a point; let H be an hyperplane in $\mathbf{P}_n$ and denote by $\tilde H$ its strict transform (see 4.1); since $H^{0,2}_{\bar\partial}(\widetilde{\mathbf{P}_n}) \simeq H^{0,2}_{\bar\partial}(\mathbf{P}_n) = 0$, we get $H^{1,1}_{\bar\partial}(\widetilde{\mathbf{P}_n}) \simeq H^2(\widetilde{\mathbf{P}_n})$, which is generated, because of (1.1), from the classes of E and $\tilde H$.

1.4 THE SHEAF $\mathcal{H}$.

Let $\mathcal{H}$ denote the sheaf of germs of pluriharmonic functions; using the exact sequence

$$0 \to \mathbf{R} \to \mathcal{O} \to \mathcal{H} \to 0$$

we get

$$H^1(\Delta_n \times \Delta_m, \mathcal{H}) = 0,\ H^1(\tilde\Delta_n \times \Delta_m, \mathcal{H}) \simeq \mathbf{R},\ H^2(\tilde\Delta_n \times \Delta_m, \mathcal{H}) = 0,\ \mathcal{H}(\Delta_m) \simeq \mathcal{O}(\Delta_m)/\mathbf{R}.$$

1.5 THE AEPPLI GROUP $V^{1,1}_{\mathbf{R}}(M)$.

The exact sequence of sheaves that produces the Aeppli group $V^{1,1}_{\mathbf{R}}(M)$ is the following:

$$0 \to \mathcal{H} \xrightarrow{A} (\Omega^1 \oplus \bar\Omega^1 \oplus E^0)_{\mathbf{R}} \xrightarrow{B} (E^{1,0} \oplus E^{0,1})_{\mathbf{R}} \xrightarrow{\bar\partial + \partial} E^{1,1}_{\mathbf{R}} \xrightarrow{i\partial\bar\partial} E^{2,2}_{\mathbf{R}} \to \dots \tag{1.2}$$

where $A(h) := (\partial h, \bar\partial h, h)$ and $B(\alpha, \bar\alpha, g) := (\alpha - \partial g, \bar\alpha - \bar\partial g)$.

Recall that

$$V^{1,1}_{\mathbf{R}}(M) := \frac{Ker\left(i\partial\bar\partial : E^{1,1}_{\mathbf{R}}(M) \to E^{2,2}_{\mathbf{R}}(M)\right)}{\left(\bar\partial E^{1,0}(M) \oplus \partial E^{0,1}(M)\right)_{\mathbf{R}}}.$$

Consider only a part of (1.2):

$$0 \to Ker(\bar\partial + \partial) \xrightarrow{i} (E^{1,0} \oplus E^{0,1})_{\mathbf{R}} \xrightarrow{\bar\partial + \partial} Ker(i\partial\bar\partial) \to 0$$

and the associated long exact sequence

$$\ldots \to H^0(M,(E^{1,0}\oplus E^{0,1})_{\mathbf{R}}) \stackrel{(\bar\partial+\partial)_0}{\longrightarrow} H^0(M,Ker(i\partial\bar\partial)) \to H^1(M,Ker(\bar\partial+\partial)) \to 0.$$

Hence

$$H^1(M,Ker(\bar\partial+\partial)) \simeq \frac{H^0(M,Ker(i\partial\bar\partial))}{Im(\bar\partial+\partial)_0} = V^{1,1}_{\mathbf{R}}(M). \tag{1.3}$$

We can also consider the analogous sequence for real currents, instead of forms:

$$0 \to \mathcal{H} \to (\Omega^1\oplus\bar\Omega^1\oplus E'_0)_{\mathbf{R}} \to (E'_{1,0}\oplus E'_{0,1})_{\mathbf{R}} \to (E'_{1,1})_{\mathbf{R}} \to (E'_{2,2})_{\mathbf{R}} \to \ldots \tag{1.4}$$

and, as for De Rham cohomology groups, we get

$$V^{1,1}_{\mathbf{R}}(M) \simeq \frac{Ker\,(i\partial\bar\partial : E'_{1,1}(M)_{\mathbf{R}} \to E'_{2,2}(M)_{\mathbf{R}})}{(\bar\partial E'_{1,0}(M)\oplus\partial E'_{0,1}(M))_{\mathbf{R}}}.$$

1.6 Remark.
Consider the first part of (1.4):

$$0 \to \mathcal{H} \stackrel{A'}{\to} (\Omega^1\oplus\bar\Omega^1\oplus E'_0)_{\mathbf{R}} \stackrel{B'}{\to} Ker'(\bar\partial+\partial) \to 0$$

and the long exact sequence of cohomology groups:

$$\ldots \to H^0\left(\Delta_n\times\Delta_m,(\Omega^1\oplus\bar\Omega^1\oplus E'_0)_{\mathbf{R}}\right) \stackrel{B'_0}{\to}$$

$$\stackrel{B'_0}{\to} H^0\left(\Delta_n\times\Delta_m, Ker'(\bar\partial+\partial)\right) \to H^1\left(\Delta_n\times\Delta_m,\mathcal{H}\right) \to \ldots$$

Since the last group vanishes by 1.4, B'_0 is surjective.

Consider now the first part of (1.2):

$$0 \to \mathcal{H} \stackrel{A}{\to} (\Omega^1\oplus\bar\Omega^1\oplus E^0)_{\mathbf{R}} \stackrel{B}{\to} Ker(\bar\partial+\partial) \to 0$$

and

$$\ldots \to H^0\left(\tilde\Delta_n\times\Delta_m,(\Omega^1\oplus\bar\Omega^1\oplus E^0)_{\mathbf{R}}\right) \stackrel{B_0}{\to}$$

$$\stackrel{B_0}{\to} H^0\left(\tilde\Delta_n\times\Delta_m, Ker(\bar\partial+\partial)\right) \to H^1\left(\tilde\Delta_n\times\Delta_m,\mathcal{H}\right) \to \ldots.$$

Let us check that also B_0 is surjective. Let $\varphi \in H^0\left(\tilde\Delta_n\times\Delta_m, Ker(\bar\partial+\partial)\right)$, that is, $\varphi = \varphi_{1,0}+\overline{\varphi_{1,0}}$, with $\varphi_{1,0}\in E^{1,0}(\tilde\Delta_n\times\Delta_m)$ and $\bar\partial\varphi_{1,0}+\partial\overline{\varphi_{1,0}}=0$.

Since $\pi_*\varphi \in H^0\left(\Delta_n\times\Delta_m, Ker'(\bar\partial+\partial)\right)$, there exist $\omega\in\Omega^1(\Delta_n\times\Delta_m)$ and $A\in E'_0(\Delta_n\times\Delta_m)_{\mathbf{R}}$ such that $\pi_*\varphi = \omega-\partial A+\bar\omega-\bar\partial A$. Call $\psi_{1,0} := \varphi_{1,0}-\pi^*\omega$ and $\psi := \psi_{1,0}+\overline{\psi_{1,0}}$; then $\psi\in H^0\left(\tilde\Delta_n\times\Delta_m, Ker(\bar\partial+\partial)\right)$ and $\pi_*\psi = -dA$.

Hence $d\psi=0$ in $(\tilde\Delta_n\times\Delta_m)-E$, but ψ is smooth, so that $d\psi=0$ in $\tilde\Delta_n\times\Delta_m$. Since $H^1(\tilde\Delta_n\times\Delta_m)=0$ by 1.3, $\psi=-dg$, therefore

$$\varphi = \pi^*\omega+\overline{\pi^*\omega}-\partial g-\bar\partial g = B_0(\pi^*\omega,\overline{\pi^*\omega},g).$$

2. The computation of $V_{\mathbf{R}}^{1,1}(\tilde{\Delta}_n \times \Delta_m)$.

First of all, our pourpose is to use a Künneth-formula to compute some Dolbeault cohomology groups of $M \times \Delta$, where M is a compact complex manifold and Δ is a polydisc. Here, we refer to [KA] or [CA] for definitions and notation.

Recall that, if M is a complex manifold, every coherent analytic sheaf is a normal nuclear Fréchet-sheaf. Moreover, a sheaf $\mathcal{F}$ on M is called a Fréchet-m-sheaf if there exist a Leray covering $\mathcal{U}$ on M such that, $\forall p, 0 \le p \le m, H^p(\mathcal{U}, \mathcal{F})$ is a Hausdorff space. Let us state the following Künneth-formula:

THEOREM 2.1. - ([KA], THEOREM 1) *Let X and Y be topological spaces with countable topology. Let $\mathcal{F}$ (resp. $\mathcal{G}$) be a Fréchet-m-sheaf on X (resp. on Y) for every m, and $\mathcal{F}$ or $\mathcal{G}$ be nuclear. Then there exists an isomorphism of topological vector spaces*

$$H^*(X \times Y, \mathcal{F}\tilde{\otimes}\mathcal{G}) \simeq H^*(X, \mathcal{F})\tilde{\otimes}H^*(Y, \mathcal{G}),$$

which is compatible with the gradation:

$$H^m(X \times Y, \mathcal{F}\tilde{\otimes}\mathcal{G}) \simeq \bigoplus_{p+q=m} H^p(X, \mathcal{F})\tilde{\otimes}H^q(Y, \mathcal{G}).$$

REMARK 2.2. - *Let M be an n-dimensional manifold and $\Delta := \Delta_m$. The natural injective maps of sheaves*

$$\mathcal{O}_M \hookrightarrow \mathcal{O}_{M\times\Delta}, \qquad \Omega^1_M \hookrightarrow \Omega^1_{M\times\Delta}$$

$$\mathcal{O}_\Delta \hookrightarrow \mathcal{O}_{M\times\Delta}, \qquad \Omega^1_\Delta \hookrightarrow \Omega^1_{M\times\Delta}$$

induce isomorphisms:

$$\mathcal{O}_{M\times\Delta} \simeq \mathcal{O}_M\tilde{\otimes}\mathcal{O}_\Delta \quad and \quad \Omega^1_{M\times\Delta} \simeq \mathcal{O}_M\tilde{\otimes}\Omega^1_\Delta \oplus \mathcal{O}_\Delta\tilde{\otimes}\Omega^1_M.$$

PROPOSITION 2.3. - *Let M be a compact n-dimensional manifold and $\Delta := \Delta_m$. Then:*

$$H^1\left(M \times \Delta, \mathcal{O}_{M\times\Delta}\right) \simeq H^1(M, \mathcal{O}_M)\tilde{\otimes}\mathcal{O}_\Delta(\Delta)$$

and

$$H^1\left(M \times \Delta, \Omega^1_{M\times\Delta}\right) \simeq H^1(M, \mathcal{O}_M)\tilde{\otimes}\Omega^1_\Delta(\Delta) \oplus H^1(M, \Omega^1_M)\tilde{\otimes}\mathcal{O}_\Delta(\Delta).$$

PROOF. - We can use Remark 2.2 with Theorem 2.1, since M is compact, so that its cohomology groups are Hausdorff spaces, and $H^i(\Delta, \mathcal{O}_\Delta), H^i(\Delta, \Omega^1_\Delta)$ are Hausdorff spaces too.

PROPOSITION 2.4. - *The map $\mathcal{O}(\Delta_m) \ni f \to \langle \pi^* f[E] \rangle \in H^{1,1}_{\bar\partial}(\tilde{\Delta}_n \times \Delta_m)$ is an isomorphism.*

PROOF. - Call $M^* := M - \{0\}$(0 belongs to a chart in an atlas of M) and denote by p the blow-up map $p : \widetilde{\mathbf{P}_n} \times \Delta_m \to \mathbf{P}_n \times \Delta_m$.

To compute cohomology groups, let us use a Mayer-Vietoris sequence, with

$$\mathbf{P}_n = \mathbf{P}_n^* \cup \Delta_n, \qquad \widetilde{\mathbf{P}_n} = \mathbf{P}_n^* \cup \tilde{\Delta}_n,$$

and

$$\mathbf{P}_n \times \Delta_m = (\mathbf{P}_n^* \times \Delta_m) \cup (\Delta_n \times \Delta_m), \quad \widetilde{\mathbf{P}_n} \times \Delta_m = (\mathbf{P}_n^* \times \Delta_m) \cup (\tilde{\Delta}_n \times \Delta_m).$$

We get the following commutative diagram with exact rows:

$$\begin{array}{ccccc}
H_{\bar\partial}^{1,0}(\Delta_n^* \times \Delta_m) & \overset{\tilde\delta_0}{\to} & H_{\bar\partial}^{1,1}(\widetilde{\mathbf{P}_n} \times \Delta_m) & \overset{\tilde r_1}{\to} & H_{\bar\partial}^{1,1}(\mathbf{P}_n^* \times \Delta_m) \oplus H_{\bar\partial}^{1,1}(\tilde\Delta_n \times \Delta_m) \\
\uparrow \simeq & & \uparrow p^* & & \uparrow id \times \pi^* \\
H_{\bar\partial}^{1,0}(\Delta_n^* \times \Delta_m) & \overset{\delta_0}{\to} & H_{\bar\partial}^{1,1}(\mathbf{P}_n \times \Delta_m) & \overset{r_1}{\to} & H_{\bar\partial}^{1,1}(\mathbf{P}_n^* \times \Delta_m) \oplus H_{\bar\partial}^{1,1}(\Delta_n \times \Delta_m)
\end{array}$$

$$\begin{array}{ccccc}
\overset{\tilde s_1}{\to} & H_{\bar\partial}^{1,1}(\Delta_n^* \times \Delta_m) & \overset{\tilde\delta_1}{\to} & H_{\bar\partial}^{1,2}(\widetilde{\mathbf{P}_n} \times \Delta_m) & \to \dots \\
 & \uparrow \simeq & & \uparrow p^* & \\
\overset{s_1}{\to} & H_{\bar\partial}^{1,1}(\Delta_n^* \times \Delta_m) & \overset{\delta_1}{\to} & H^{1,2}(\mathbf{P}_n \times \Delta_m) & \to \dots
\end{array}$$

where the vertical maps are induced by p and π, δ_i and $\tilde\delta_i$ are the connecting homomorphisms associated to the $\bar\partial$-complex, $r_1, \tilde r_1, s_1, \tilde s_1$ are the Mayer-Vietoris homomorphisms.

Since p^* is injective, $Ker\ \tilde\delta_0 = Ker\ \delta_0$ and $Im\ s_1 \simeq Ker\ \delta_1 \simeq Ker\ \tilde\delta_1 \simeq Im\ \tilde s_1$, therefore $\tilde r_1$ induces an isomorphism

$$\frac{H_{\bar\partial}^{1,1}(\widetilde{\mathbf{P}_n} \times \Delta_m)}{p^* H_{\bar\partial}^{1,1}(\mathbf{P}_n \times \Delta_m)} \to \frac{H_{\bar\partial}^{1,1}(\tilde\Delta_n \times \Delta_m)}{\pi^* H_{\bar\partial}^{1,1}(\Delta_n \times \Delta_m)}. \tag{2.1}$$

From Proposition 2.3,

$$H_{\bar\partial}^{1,1}(\mathbf{P}_n \times \Delta_m) \simeq H^1(\mathbf{P}_n \times \Delta_m, \Omega^1) \simeq H^1(\mathbf{P}_n, \Omega^1) \tilde\otimes \mathcal{O}(\Delta_m) \simeq \mathcal{O}(\Delta_m)\langle [H] \rangle$$

where H is an hyperplane in $\mathbf{P}_n$, and also (see 1.3)

$$H_{\bar\partial}^{1,1}(\widetilde{\mathbf{P}_n} \times \Delta_m) \simeq H^1(\widetilde{\mathbf{P}_n} \times \Delta_m, \Omega^1) \simeq$$

$$\simeq H^1(\widetilde{\mathbf{P}_n}, \Omega^1) \tilde\otimes \mathcal{O}(\Delta_m) \simeq \mathcal{O}(\Delta_m)\langle [\tilde H] \rangle \oplus \mathcal{O}(\Delta_m)\langle [E] \rangle.$$

Thus, by (2.1), $H_{\bar\partial}^{1,1}(\tilde\Delta_n \times \Delta_m) \simeq \mathcal{O}(\Delta_m)\langle [E] \rangle$ and the map that gives this isomorphism is precisely $(f, \langle E \rangle) \mapsto \langle \pi^* f[E] \rangle$ since $\tilde r_1(\langle E \rangle) = \langle E \cap \tilde\Delta_n \rangle$.

In Proposition 2.5 we shall prove the key-result for our study of the Aeppli group of a modification, in fact we get the local result (for a blow-up with smooth center).

PROPOSITION 2.5. - *The map* $\mathcal{H}(\Delta_m) \ni h \to \langle \pi^* h[E] \rangle \in V_{\mathbf{R}}^{1,1}(\tilde\Delta_n \times \Delta_m)$ *is an isomorphism.*

PROOF. - Recall the exact sequence in Remark 1.6:

$$0 \to \mathcal{H} \xrightarrow{A} (\Omega^1 \oplus \bar{\Omega}^1 \oplus E^0)_{\mathbf{R}} \xrightarrow{B} Ker(\bar{\partial} + \partial) \to 0$$

and the corresponding long exact sequence:

$$0 \to H^1\left(\tilde{\Delta}_n \times \Delta_m, \mathcal{H}\right) \xrightarrow{A_1} H^1\left(\tilde{\Delta}_n \times \Delta_m, (\Omega^1 \oplus \bar{\Omega}^1 \oplus E^0)_{\mathbf{R}})\right) \xrightarrow{B_1}$$

$$\xrightarrow{B_1} H^1\left(\tilde{\Delta}_n \times \Delta_m, Ker(\bar{\partial} + \partial)\right) \to H^2\left(\tilde{\Delta}_n \times \Delta_m, \mathcal{H}\right) \to \dots$$

where A_1 is injective because B_0 is surjective (see Remark 1.6) and

$$H^1\left(\tilde{\Delta}_n \times \Delta_m, \mathcal{H}\right) = \mathbf{R}\,,\; H^2\left(\tilde{\Delta}_n \times \Delta_m, \mathcal{H}\right) = 0$$

(see 1.4). Notice that

$$(\Omega^1 \oplus \bar{\Omega}^1 \oplus E^0)_{\mathbf{R}} \simeq (\Omega^1 \oplus \bar{\Omega}^1)_{\mathbf{R}} \oplus E^0$$

and

$$\Omega^1 \simeq (\Omega^1 \oplus \bar{\Omega}^1)_{\mathbf{R}} \quad \text{via} \quad \alpha \mapsto (\alpha, \bar{\alpha})$$

so that, by Proposition 2.4 and an easy computation,

$$H^1\left(\tilde{\Delta}_n \times \Delta_m, (\Omega^1 \oplus \bar{\Omega}^1 \oplus E^0)_{\mathbf{R}})\right) \simeq H^1\left(\tilde{\Delta}_n \times \Delta_m, (\Omega^1 \oplus \bar{\Omega}^1)_{\mathbf{R}})\right) \simeq$$

$$\simeq H^1\left(\tilde{\Delta}_n \times \Delta_m, \Omega^1\right) \simeq \mathcal{O}(\Delta_m)$$

(via the map $f \mapsto \langle \pi^* f[E] \rangle$) and then, by (1.3) and 1.4,

$$V^{1,1}_{\mathbf{R}}(\tilde{\Delta}_n \times \Delta_m) \simeq H^1\left(\tilde{\Delta}_n \times \Delta_m, Ker(\bar{\partial} + \partial)\right) \simeq \mathcal{O}(\Delta_m)/\mathbf{R} \simeq \mathcal{H}(\Delta_m).$$

To check that the isomorphism is given by $h \mapsto \langle \pi^* h[E] \rangle$, it sufficies to analyze the isomorphisms in the previous formula.

3. A tie between $V^{1,1}_{\mathbf{R}}(\tilde{M})$ and $V^{1,1}_{\mathbf{R}}(M)$.

The important result here is Theorem 3.5, which gives a first relation between the Aeppli groups $V^{1,1}_{\mathbf{R}}(\tilde{M})$ an $V^{1,1}_{\mathbf{R}}(M)$, in the general case of a proper modification.

Let us recall an useful result of [AB4]:

PROPOSITION 3.1. - ([AB4], 2.5) - *Let $f : \tilde{M} \to M$ be a proper modification which is obtained as a finite sequence of blow-ups with smooth centers. Call $\{E_\alpha\}$ the set of irreducible components of the exceptional set E. Then every real $\partial\bar{\partial}$-closed (1,1)-current $\tilde{T}$ of order zero on $\tilde{M}$, supported on E, is of the form $\sum_\alpha \tilde{h}_\alpha[E_\alpha]$, where $\tilde{h}_\alpha \in \mathcal{H}(E_\alpha)$. Moreover, $\tilde{T}$ is a (weak) limit of currents which are components of boundaries if and only if every $\tilde{h}_\alpha$ vanishes.*

LEMMA 3.2. - *Let M be a complex* N*-dimensional manifold, $f : \tilde{M} \to M$ a blow-up with smooth center Y and $\tilde{T}$ a $\partial\bar{\partial}$-closed real (1,1)-current on $\tilde{M}$ such that*
(*) $\forall x \in M$, there exists an open neighborhood W of x such that $\tilde{T}_{|f^{-1}(W)}$ is a weak limit of currents which are the (1,1)-component of boundaries.
Then
(**) $\forall x \in M$, there exists an open neighborhood W of x such that $\tilde{T}_{|f^{-1}(W)}$ is the (1,1)-component of a boundary.

PROOF. - If $x \in M - Y$, we can choose W as a polydisc disjoint from Y; hence $V^{1,1}_{\mathbf{R}}(f^{-1}(W)) \simeq V^{1,1}_{\mathbf{R}}(W)$ which vanishes, so that the $\partial\bar{\partial}$-closed current $\tilde{T}_{|f^{-1}(W)}$ is in fact the (1,1)-component of a boundary.

If $x \in Y$, call m the dimension of Y and $E := f^{-1}(Y)$; let W be a neighborhood of x such that W is biholomorphic to $\Delta_n \times \Delta_m$, $W \cap Y \simeq \{0\} \times \Delta_m$ and $\tilde{T}_{|f^{-1}(W)}$ satisfies the condition in (*).

$\tilde{T}_{|f^{-1}(W)}$ represents a class in $V^{1,1}_{\mathbf{R}}(f^{-1}(W)) \simeq V^{1,1}_{\mathbf{R}}(\tilde{\Delta}_n \times \Delta_m)$, hence by Proposition 2.5 there exists a pluriharmonic map $h : W \cap Y \to \mathbf{R}$ such that

$$\tilde{T}_{|f^{-1}(W)} = f^*h[E \cap f^{-1}(W)] + \bar{\partial}R + \partial\bar{R}$$

for a suitable (1,0)-current R in $f^{-1}(W)$.

Therefore also the current $f^*h[E \cap f^{-1}(W)]$ satisfies the condition in (*), but it is supported on E, so that by Proposition 3.1 it vanishes.

Let us consider now a proper modification: the following Lemma is essentially contained in [HR]; a proof can be seen in [AB4], Lemma 2.6.

LEMMA 3.3. - *Let M and $\tilde{M}$ be complex manifolds, $f : \tilde{M} \to M$ be a proper modification. For every $x \in M$ there exist an open neighborhood V of x in M, a complex manifold Z and holomorphic maps $g : Z \to \tilde{M}$, $q : Z \to V$ such that $q = f \circ g$; moreover, $g : Z \to f^{-1}(V)$ is a blow-up and $q : Z \to V$ is obtained as a finite sequence of blow-ups with smooth centers.*

COROLLARY 3.4. - *Let M and $\tilde{M}$ be complex manifolds, and let $f : \tilde{M} \to M$ be a proper modification; let $\mathcal{W} = \{W_j\}$ be a covering of M such that every W_j is as described in Lemma 3.3. Then $f_* : H^1(f^{-1}(W_j \cap W_k)) \to H^1(W_j \cap W_k)$ is an isomorphism.*

PROOF. - Let

$$\begin{array}{ccc} Z & \xrightarrow{g} & f^{-1}(W_j \cap W_k) \\ & {\scriptstyle q}\searrow \quad \swarrow{\scriptstyle f} & \\ & W_j \cap W_k & \end{array}$$

be the diagram described in Lemma 3.3 for $W_j \cap W_k$. Since q is obtained as a finite sequence of blow-ups with smooth centers and $Z = q^{-1}(W_j \cap W_k)$, from 1.3 we get that $q_* : H^1(Z) \to H^1(W_j \cap W_k)$ is an isomorphism. But g_* and f_* are surjective, so that they are isomorphisms too.

THEOREM 3.5. - *Let M and $\tilde{M}$ be complex manifolds, and let $f : \tilde{M} \to M$ be a proper modification. Let $\tilde{T}$ be a $\partial\bar{\partial}$-closed real (1,1)-current on $\tilde{M}$ such that $\langle f_*\tilde{T}\rangle = 0$ in $V^{1,1}_{\mathbf{R}}(M)$ and*
(**) $\forall x \in M$, there exists an open neighborhood W of x such that $\tilde{T}_{|f^{-1}(W)}$ is the (1,1)-component of a boundary.

Then $\langle \tilde{T} \rangle = 0$ *in* $V^{1,1}_{\mathbf{R}}(\tilde{M})$.

PROOF. - Choose a smooth $\partial\bar{\partial}$-closed (1,1)-form φ in the class of $\tilde{T}$ in $V^{1,1}_{\mathbf{R}}(\tilde{M})$; since the hypotheses also hold for φ, we need only to prove that $\langle \varphi \rangle = 0$ in $V^{1,1}_{\mathbf{R}}(\tilde{M})$. Let us fix an open covering $\{W_j\}$ of M, such that each W_j is biholomorphic to a polydisc and, as said in (**) for φ, in $f^{-1}(W_j)$

$$\varphi = \bar{\partial} r_j + \partial \overline{r_j}$$

for suitable $r_j \in E^{1,0}(f^{-1}(W_j))$. Moreover, there exists $S \in E'_{1,0}(M)$ such that $f_*\varphi = \bar{\partial} S + \partial \bar{S}$ so that

$$f_* r_j - S + f_* \overline{r_j} - \bar{S} \in H^0(W_j, Ker(\bar{\partial} + \partial)).$$

By Remark 1.6 we get in W_j:

$$f_* r_j - S = \alpha_j - \partial A_j \quad \text{for} \quad \alpha_j \in \Omega^1(W_j), A_j \in E'_0(W_j)_{\mathbf{R}}.$$

Since

$$\varphi = \bar{\partial}(r_j - f^*\alpha_j) + \partial(\overline{r_j} - f^*\overline{\alpha_j}),$$

in $f^{-1}(W_j)$ we can consider $r_j - f^*\alpha_j$ instead of r_j, and write

$$f_* r_j - S = -\partial A_j \quad \text{in} \quad W_j$$

so that

$$f_*(r_j - r_k + \overline{r_j} - \overline{r_k}) = d(A_k - A_j) \quad \text{in} \quad W_j \cap W_k.$$

Notice that $(r_j - r_k + \overline{r_j} - \overline{r_k})$ is a smooth representative of a class in $H^1(f^{-1}(W_j \cap W_k))$, because $d(r_j - r_k + \overline{r_j} - \overline{r_k})_{|f^{-1}(W_j \cap W_k) - E} = 0$ and a smooth form cannot be supported on E.

By Corollary 3.4, f_* gives an isomorphism

$$H^1(f^{-1}(W_j \cap W_k)) \simeq H^1(W_j \cap W_k),$$

so that

$$(r_j - r_k + \overline{r_j} - \overline{r_k}) = df_{jk} \quad , \quad f_{jk} \in E^0_{\mathbf{R}}(f^{-1}(W_j \cap W_k)).$$

This implies

$$d(f_* f_{jk}) = d(A_k - A_j),$$

or

$$f_* f_{jk} = A_k - A_j + c_{jk} \quad \text{in} \quad W_j \cap W_k,$$

for suitable real constants c_{jk}. But $f_{jk} - c_{jk}$ can replace f_{jk}, so that we can suppose

$$f_* f_{jk} = A_k - A_j \quad \text{in} \quad W_j \cap W_k.$$

Since the functions f_{jk} are smooth, we get

$$f_{jk} + f_{kl} + f_{lj} = 0 \quad \text{in} \quad f^{-1}(W_j \cap W_k \cap W_l).$$

and therefore (if necessary, take a refinement of $\{W_j\}$) there exist $f_j \in E^0_{\mathbf{R}}(f^{-1}(W_j))$ such that

$$f_j - f_k = f_{jk} \quad \text{in} \quad f^{-1}(W_j \cap W_k).$$

Now the computation is over: since

$$r_j - \partial f_j = r_k - \partial f_k \quad \text{in} \quad f^{-1}(W_j \cap W_k),$$

the 1-form r , $r_{|f^{-1}(W_j)} := r_j - \partial f_j$, satisfies $\varphi = \bar\partial r + \partial \bar r$.

4. Some geometrical remarks.

4.1 NOTATION.

Let $f : \tilde M \to M$ be a proper modification with center Y and exceptional set $E = \bigcup E_\alpha$ (E_α irreducible). If V is an irreducible hypersurface of M, let $\tilde V$ denote its strict transform, i.e. $\tilde V := \overline{f^{-1}(V - V \cap Y)}$, and let f^*V denote the pull-back divisor, that is, if $\mathcal{U} = \{U_i\}$ is a covering of M and $g_i \in \mathcal{O}(U_i)$ are holomorphic functions that define the divisor V, then f^*V is given on $f^{-1}(U_i)$ by f^*g_i. We know that

$$f^*V = \tilde V + \sum c_\alpha E_\alpha$$

for some constants c_α.

In the next Proposition, we shall prove that the pull-back divisor represents the pull-back class in the Aeppli group.

PROPOSITION 4.2. - *Let $f : \tilde M \to M$ and V be as above, let $f^* : V^{1,1}_{\mathbf{R}}(M) \to V^{1,1}_{\mathbf{R}}(\tilde M)$ be the map induced by f, that is, if φ is a smooth real $\partial\bar\partial$-closed (1,1)-form that represents a class $\langle\varphi\rangle \in V^{1,1}_{\mathbf{R}}(M)$, $f^*\langle\varphi\rangle := \langle f^*\varphi\rangle$.*

Then $f^*V \in f^*\langle[V]\rangle$.

PROOF. - Let $\mathcal{U} = \{U_i\}$ be a covering of M and let $g_i \in \mathcal{O}(U_i)$ be holomorphic functions that define the divisor V; we choose a "nice" smooth representative of $\langle[V]\rangle$. Call L the line bundle associated to V, and choose an hemitian metric h on L, with curvature (1,1)-form Θ; that is, on U_i, $\Theta = -\partial\bar\partial \log h_i$.

By a routine computation,

$$[V]_{|U_i} = \frac{i}{2\pi}\partial\bar\partial \log|g_i|^2,$$

so that

$$([V] - \frac{i}{2\pi}\Theta)_{|U_i} = \frac{i}{2\pi}\partial\bar\partial \log(|g_i|^2 h_i).$$

Since $|g_i|^2 h_i = |g_j|^2 h_j$ on $U_i \cap U_j$, we get

$$[V] = \frac{i}{2\pi}\Theta + \frac{i}{2\pi}\partial\bar\partial \log s,$$

for a smooth function s on M. Hence

$$f^*\langle[V]\rangle = f^*\langle\frac{i}{2\pi}\Theta\rangle = \langle\frac{i}{2\pi}f^*\Theta\rangle.$$

But, by definition,

$$[f^*V]_{|f^{-1}(U_i)} = \frac{i}{2\pi}\partial\bar{\partial}\log|f^*g_i|^2 \quad \text{and} \quad \frac{i}{2\pi}f^*\Theta_{|f^{-1}(U_i)} = \frac{-i}{2\pi}\partial\bar{\partial}\log f^*h_i,$$

hence

$$[f^*V] - \frac{i}{2\pi}f^*\Theta = \frac{i}{2\pi}\partial\bar{\partial}\log f^*s,$$

which implies $f^*V \in f^*\langle[V]\rangle$.

Let us consider now not only divisors, but also pluriharmonic maps.

THEOREM 4.3. - *Let $f : \tilde{M} \to M$ be a proper modification, let V be an irreducible hypersurface of M, and let $h \in \mathcal{H}(V)$. Define $f^*(h[V]) := (f^*h)[f^*V]$. Then:*
a) $f_*f^*(h[V]) = h[V]$
b) $f^*(h[V]) \in f^*\langle h[V]\rangle$
c) $f^*(h[V])$ is the unique current of the form $h'[D]$, where D is a divisor in $\tilde{M}$ and $h' \in \mathcal{H}(D)$, that satisfies a) and b).

PROOF. - a) is obvious, because

$$(f^*h)[f^*V] = (f^*h)[\tilde{V}] + \sum c_\alpha(f^*h)[E_\alpha]$$

and, if ψ is a test form on M,

$$(h[V])(\psi) = \int_V h\psi = \int_{\tilde{V}}(f^*h)(f^*\psi).$$

b) Choose $x \in M$ and let U be an open neighborhood of x such that h extends from $U \cap V$ to U. We have seen in the proof of Proposition 4.2 that $[V] = \frac{i}{2\pi}\Theta + \frac{i}{2\pi}\partial\bar{\partial}\log s$, so that $\frac{i}{2\pi}\Theta$ is a smooth representative of $\langle[V]\rangle$; moreover, in U,

$$h[V] - \frac{i}{2\pi}h\Theta = \frac{i}{2\pi}h\partial\bar{\partial}\log s =$$

$$= \partial\left(\frac{i}{2}\bar{\partial}(h\log s) - i(\bar{\partial}h)\log s\right) + \bar{\partial}\left(\frac{-i}{2}\partial(h\log s) + i(\partial h)\log s\right),$$

since $\partial\bar{\partial}h = 0$.

This implies that $\langle h[V]\rangle = \langle\frac{i}{2\pi}h\Theta\rangle$, and also that $f^*\left(h[V] - \frac{i}{2\pi}h\Theta\right)$ is the (1,1)-component of a boundary in $f^{-1}(U)$. Hence

$$f^*\langle h[V]\rangle = f^*\langle\frac{i}{2\pi}h\Theta\rangle = \langle\frac{i}{2\pi}(f^*h)f^*\Theta\rangle$$

in $f^{-1}(U)$.

Let us consider now a smooth representative φ of the class $\langle h[V]\rangle$, that is,

$$h[V] = \varphi + \bar{\partial}L + \partial\bar{L};$$

call $\tilde{T} := f^*(h[V] - \varphi)$. By part a), $\langle f_* \tilde{T} \rangle = 0$ and moreover $\tilde{T}$ is the (1,1)-component of a boundary in $f^{-1}(U)$, because in U, $\langle \varphi \rangle = \langle h[V] \rangle = \langle \frac{i}{2\pi} h\Theta \rangle$ and $f^* \left(h[V] - \frac{i}{2\pi} h\Theta \right)$ is the (1,1)-component of a boundary in $f^{-1}(U)$.

Thus we can apply Theorem 3.5 to get $\langle \tilde{T} \rangle = 0$, therefore

$$f^*(h)[f^*V] = f^*\varphi + \bar{\partial} S + \partial \bar{S} \in f^* \langle h[V] \rangle.$$

To prove c), let us use the following Lemma:

LEMMA 4.4. - *Let $f : \tilde{M} \to M$ be as above, $\{E_\alpha\}$ the set of irreducible components of E, and $\tilde{h}_\alpha \in \mathcal{H}(E_\alpha)$. If $\langle \sum_\alpha \tilde{h}_\alpha [E_\alpha] \rangle = 0$, then $\tilde{h}_\alpha = 0 \quad \forall \alpha$.*

PROOF. Choose α and $x \in f(E_\alpha)$; consider the following commutative diagram given by Lemma 3.3:

$$\begin{array}{ccc} Z & \xrightarrow{g} & f^{-1}(V) \\ & q \searrow \quad \swarrow f & \\ & V & \end{array}$$

Let $E'_\beta := E_\beta \cap f^{-1}(V)$ and $T := \sum_\beta \tilde{h}_\beta [E'_\beta]$; from the hypothesis, the class of T is zero, hence also the class of g^*T vanishes, by Theorem 4.3, part b). Then we can apply Proposition 3.1 to get $g^*T = 0$, so $T = 0$.

End of the proof. Take $h'[D]$ that satisfies a) and b); then $f_*(h'[D]) = h[V]$, so that, looking at supports, we get

$$D = \gamma \tilde{V} + \sum \gamma_\alpha E_\alpha$$

for some constants γ, γ_α. On the other hand, we have

$$f^*(h[V]) = (f^*h)[\tilde{V}] + (f^*h) \sum c_\alpha E_\alpha.$$

Moreover, on $\tilde{V}, \gamma h' = f^*h$, so that

$$\langle h'[D] - f^*(h[V]) \rangle = \langle \sum_\alpha (\gamma_\alpha h' - c_\alpha f^*h)[E_\alpha] \rangle = 0$$

by b), thus by Lemma 4.4, $\gamma_\alpha h' - c_\alpha f^*h = 0 \quad \forall \alpha$, that is, $h'[D] = f^*(h[V])$.

5. The group $V_{\mathbf{R}}^{1,1}(\tilde{M})$.

Now, we can give an explicit description of the Aeppli group $V_{\mathbf{R}}^{1,1}(\tilde{M})$ in terms of the group $V_{\mathbf{R}}^{1,1}(M)$ and pluriharmonic functions on the center Y. We shall do it first of all for a blow-up in Lemma 5.1, in the general case in Theorem 5.3. The most important consequence of this description is stated in Theorem 5.6 and Corollary 5.7.

LEMMA 5.1. - *Let $f : \tilde{M} \to M$ be a blow-up with smooth center Y and exceptional set E. Then*

$$V_{\mathbf{R}}^{1,1}(\tilde{M}) = f^* V_{\mathbf{R}}^{1,1}(M) \oplus f^* \mathcal{H}(Y) \langle [E] \rangle.$$

PROOF. - Let $\langle\varphi\rangle \in V_{\mathbf{R}}^{1,1}(\tilde{M})$, where φ is a smooth form, and let ψ be a smooth (1,1)-form on M such that $\langle\psi\rangle = \langle f_*\varphi\rangle \in V_{\mathbf{R}}^{1,1}(M)$, that is, there exist a current R on M such that $f_*\varphi = \psi + \bar{\partial}R + \partial\bar{R}$. For our pourposes, we can replace φ by $\varphi - f^*\psi$, hence we can assume $f_*\varphi = \bar{\partial}R + \partial\bar{R}$.

Cover Y with open sets $\{W_j\}$ such that W_j is biholomorphic to $\Delta_n \times \Delta_m$ (m is the dimension of Y) and $W_j \cap Y \simeq \{0\} \times \Delta_m$. Then $f^{-1}(W_j) \simeq \tilde{\Delta}_n \times \Delta_m$, and by Proposition 2.5 there exist $h_j \in \mathcal{H}(Y \cap W_j)$ and currents R_j such that

$$\varphi = (f^*h_j)[E \cap f^{-1}(W_j)] + \bar{\partial}R_j + \partial\bar{R}_j \quad \text{in} \quad f^{-1}(W_j).$$

Thus, in $f^{-1}(W_j \cap W_k)$, the class of $f^*(h_j - h_k)[E \cap f^{-1}(W_j \cap W_k)]$ vanishes, and this implies $h_j = h_k$ in $W_j \cap W_k \cap Y$ by Proposition 3.1. So the functions glue together to give $h \in \mathcal{H}(Y)$, $h_{|W_j} := h_j$, such that

$$(\varphi - f^*h[E])_{|f^{-1}(W_j)} = \bar{\partial}R_j + \partial\bar{R}_j$$

and we conclude by Theorem 3.5 that $\langle\varphi - f^*h[E]\rangle = 0$. The decomposition is unique by Lemma 4.4.

COROLLARY 5.2. - *Let $f : \tilde{M} \to M$ be a proper modification, which is given by a finite sequence of blow-ups with smooth centers. Let $\{E_\alpha\}$ be the set of irreducible components of the exceptional set E, and $Y_\alpha := f(E_\alpha)$. Then*

$$V_{\mathbf{R}}^{1,1}(\tilde{M}) = f^*V_{\mathbf{R}}^{1,1}(M) \oplus (\oplus_\alpha f^*\mathcal{H}(Y_\alpha)\langle[E_\alpha]\rangle)\,.$$

PROOF. - Let us use induction on the number of blow-ups, that is, let us suppose to have the following commutative diagram:

$$\begin{array}{ccc} \tilde{M} & \xrightarrow{g} & N \\ & f\searrow \quad \swarrow q & \\ & M & \end{array}$$

where q satisfies the inductive hypothesis and g is a blow-up with smooth center X and exceptional set F. Call $\{E'_\beta\}$ the set of irreducible components of the exceptional set of q, and let $Y'_\beta := q(E'_\beta)$.

If $\langle\varphi\rangle \in V_{\mathbf{R}}^{1,1}(\tilde{M})$, where φ is a smooth form, by Lemma 5.1 there exist a smooth form ψ on N and $h \in \mathcal{H}(X)$ such that

$$\langle\varphi\rangle = \langle g^*\psi\rangle + \langle g^*h[F]\rangle,$$

and, by inductive hypothesis, there exist a smooth form χ on M and $h'_\beta \in \mathcal{H}(E'_\beta)$ such that

$$\langle\psi\rangle = \langle q^*\chi\rangle + \sum_\beta \langle h'_\beta[E'_\beta]\rangle.$$

This implies

$$\langle g^*\psi\rangle = g^*\langle\psi\rangle = g^*\langle q^*\chi\rangle + \sum_\beta g^*\langle h'_\beta[E'_\beta]\rangle = \langle f^*\chi\rangle + \sum_\beta \langle (g^*h'_\beta)[g^*E'_\beta]\rangle$$

and

$$\langle\varphi\rangle = \langle f^*\chi\rangle + \sum_\beta \langle (g^*h'_\beta)[g^*E'_\beta]\rangle + \langle g^*h[F]\rangle.$$

Notice that $\{F\} \cup \{g^*E'_\beta\}_\beta$ is the set of irreducible components of E; for each of these hypersurfaces the corresponding function g^*h or $g^*h'_\beta$ is pluriharmonic, therefore it is constant on the fibres of f. This means that it is the pull-back of a pluriharmonic function on some Y_α via the map f.

THEOREM 5.3. - *Let $f : \tilde{M} \to M$ be a proper modification, $\{E_\alpha\}$ be the set of irreducible components of the exceptional set E, and $Y_\alpha := f(E_\alpha)$. Then*

$$V^{1,1}_{\mathbf{R}}(\tilde{M}) = f^*V^{1,1}_{\mathbf{R}}(M) \oplus (\oplus_\alpha f^*\mathcal{H}(Y_\alpha)\langle [E_\alpha]\rangle).$$

PROOF. - As in the proof of Lemma 5.1, we can consider only the case of a class $\langle\varphi\rangle$ such that $\langle f_*\varphi\rangle = 0$. Locally, we have the following diagram, as explained in Lemma 3.3,

$$\begin{array}{ccc} Z & \overset{g}{\longrightarrow} & f^{-1}(V) \\ & {}_{q}\searrow \quad \swarrow_{f} & \\ & V & \end{array}$$

where g is a blow-up with exceptional set F and q is a finite sequence of blow-ups with smooth centers. Denote by $\{E_{\alpha,j}\}$ the set of irreducible components of $E \cap f^{-1}(V)$, and let $i : f^{-1}(V) \to \tilde{M}$ be the inclusion map.

Notice that $\langle g^*i^*\varphi\rangle \in V^{1,1}_{\mathbf{R}}(Z)$, so that by Corollary 5.2 there exists a smooth form ψ on V such that

$$\langle g^*i^*\varphi\rangle = \langle q^*\psi\rangle + \langle h[F]\rangle + \sum_{\alpha,j}\langle h_{\alpha,j}[g^*E_{\alpha,j}]\rangle$$

for suitable functions $h \in \mathcal{H}(F)$, $h_{\alpha,j} \in \mathcal{H}(g^*E_{\alpha,j})$. Hence

$$\langle i^*\varphi\rangle = \langle g_*q^*\psi\rangle + \sum_{\alpha,j}\langle g_*h_{\alpha,j}[E_{\alpha,j}]\rangle$$

But $0 = \langle f_*i^*\varphi\rangle = \langle f_*g_*q^*\psi\rangle = \langle\psi\rangle$, so that $\langle q^*\psi\rangle = 0$,thus

$$\langle i^*\varphi\rangle = \sum_{\alpha,j}\langle g_*h_{\alpha,j}[E_{\alpha,j}]\rangle.$$

By Lemma 4.4, for every α there exists $\tilde{h}_\alpha \in \mathcal{H}(E_\alpha)$ such that $\tilde{h}_\alpha = g_*h_{\alpha,j}$ in $E_{\alpha,j}$, so we get

$$\langle\varphi\rangle = \sum_\alpha \langle\tilde{h}_\alpha[E_\alpha]\rangle.$$

As above, to conclude it is enough to remark that the functions $\tilde{h}_\alpha$ are constant on the fibres of g, so that $\tilde{h}_\alpha = f^*k_\alpha$ for suitable $k_\alpha \in \mathcal{H}(Y_\alpha)$.

Now we can generalize Lemma 3.2 .

LEMMA 5.4. - *Lemma 3.2 also holds if $f : \tilde{M} \to M$ is given by a finite sequence of blow-ups with smooth centers.*

PROOF. - Use induction as in the proof of Corollary 5.2. Assume we have a diagram

$$\begin{array}{ccc} \tilde{M} & \xrightarrow{g} & N \\ & f\searrow \quad \swarrow q & \\ & M & \end{array}$$

where g is a blow-up with smooth center, and q satisfies the inductive hypothesis. Let $\tilde{T}$ be a $\partial\bar{\partial}$-closed real (1,1)-current on $\tilde{M}$ which satisfies

(*) $\forall x \in M$, there exists an open neighborhood W of x such that $\tilde{T}_{|f^{-1}(W)}$ is a weak limit of currents which are the (1,1)-component of boundaries.

Consider a smooth form φ on M such that $\langle\varphi\rangle = \langle f_*\tilde{T}\rangle \in V^{1,1}_{\mathbf{R}}(M)$, and let W be a polydisc in M, as given in (*). Denote still with φ and $f_*\tilde{T}$ the restriction of these objects to W; $\langle\varphi\rangle = 0$ in $V^{1,1}_{\mathbf{R}}(W)$, because this group is trivial. Obviously, also $g_*\tilde{T}$ satisfies the hypothesis on $q^{-1}(W)$, hence by the inductive hypothesis

$$\langle g_*\tilde{T}\rangle = 0 \quad \text{in} \quad V^{1,1}_{\mathbf{R}}(q^{-1}(W)).$$

But also $\langle q^*\varphi\rangle = 0$ in $V^{1,1}_{\mathbf{R}}(q^{-1}(W))$, so that we can apply Theorem 3.5 to $S := g_*\tilde{T} - q^*\varphi$ and q, and get that

$$g_*\langle\tilde{T} - f^*\varphi\rangle = \langle g_*\tilde{T} - q^*\varphi\rangle = 0 \quad \text{in} \quad V^{1,1}_{\mathbf{R}}(N).$$

We can apply now Theorem 3.5 to the current $\tilde{T} - f^*\varphi$ and the map g, since by Lemma 3.2, $\langle\tilde{T} - f^*\varphi\rangle = 0$ in $V^{1,1}_{\mathbf{R}}(g^{-1}(W))$. Hence

$$\langle\tilde{T}_{|f^{-1}(W)}\rangle = \langle f^*\varphi_{|f^{-1}(W)}\rangle = 0 \quad \text{in} \quad V^{1,1}_{\mathbf{R}}(f^{-1}(W)).$$

PROPOSITION 5.5. - *Lemma 3.2 holds in the general case, that is, when $f : \tilde{M} \to M$ is a proper modification.*

PROOF. - Use the notation of the proof of Theorem 5.3, that is, the local situation is

$$\begin{array}{ccc} Z & \xrightarrow{g} & f^{-1}(V) \\ & q\searrow \quad \swarrow f & \\ & V & \end{array}$$

and let $\tilde{T}$ be a $\partial\bar{\partial}$-closed real (1,1)-current on $\tilde{M}$ which satisfies (*) . We can prove the statement with a smooth representative φ of $\langle\tilde{T}\rangle$ in $V^{1,1}_{\mathbf{R}}(\tilde{M})$; thus assume

$$\varphi = \lim_j \bar{\partial}R_j + \partial\bar{R}_j \quad \text{in} \quad f^{-1}(V).$$

If necessary, shrink V to get, by regularisation, smooth currents ρ_j such that

$$\varphi = \lim_j \bar{\partial}\rho_j + \partial\bar{\rho}_j \quad \text{in} \quad f^{-1}(V).$$

Then $g^*\varphi$ satisfies (*) in Z with respect to the map q, and thus also (**): this implies the thesis for $\varphi = g_*(g^*\varphi)$.

Let us recall the partial result we got in [AB4], Theorem 3.9:

THEOREM. - *Let M and $\tilde{M}$ be complex manifolds, and $f : \tilde{M} \to M$ be a proper modification. Let T be a positive $\partial\bar{\partial}$-closed (1,1)-current on M. Then there exists a positive $\partial\bar{\partial}$-closed (1,1)-current $\tilde{T}$ on $\tilde{M}$ such that $f_*\tilde{T} = T$ and (*) (of Lemma 3.2) holds. Moreover, if M is compact, $\tilde{T}$ is unique.*

We can now complete it as follows.

THEOREM 5.6. - *Let M and $\tilde{M}$ be complex manifolds, and $f : \tilde{M} \to M$ be a proper modification. Let T be a positive $\partial\bar{\partial}$-closed (1,1)-current on M. Then there exists a unique positive $\partial\bar{\partial}$-closed (1,1)-current $\tilde{T}$ on $\tilde{M}$ such that $f_*\tilde{T} = T$ and $\tilde{T} \in f^*\langle T\rangle \in V^{1,1}_{\mathbf{R}}(\tilde{M})$.*

PROOF. - a) Existence: from the previous Theorem we get $\tilde{T}$, a positive $\partial\bar{\partial}$-closed (1,1)-current on $\tilde{M}$ such that $f_*\tilde{T} = T$, and $\tilde{T}$ satisfies (*) of Lemma 3.2. Let φ be a smooth representative of the class $\langle T\rangle$: then $\tilde{T} - f^*\varphi$ also satisfies (**) of Lemma 3.2, hence, by Theorem 3.5, $\langle \tilde{T} - f^*\varphi\rangle = 0$, that is, $\tilde{T} \in f^*\langle T\rangle$.

b) Uniqueness: let $\tilde{T}$ and $\tilde{T}'$ satisfy the statement; from ([B],Theorem 4.10) there exist plurisubharmonic functions h_α and h'_α on E_α such that

$$\chi_E\tilde{T} = \sum_\alpha h_\alpha[E_\alpha], \quad \chi_E\tilde{T}' = \sum_\alpha h'_\alpha[E_\alpha].$$

Since $f_*\tilde{T} = T = f_*\tilde{T}'$, we get $\tilde{T} - \tilde{T}' = \chi_E(\tilde{T} - \tilde{T}') = \sum_\alpha (h_\alpha - h'_\alpha)[E_\alpha]$, and $\forall\alpha$, $h_\alpha - h'_\alpha \in \mathcal{H}(E_\alpha)$, because $\tilde{T}$ and $\tilde{T}'$ are $\partial\bar{\partial}$-closed . But $\tilde{T}$ and $\tilde{T}'$ belong to the same class $f^*\langle T\rangle$, so that by Lemma 4.4 we get $\tilde{T} - \tilde{T}' = 0$.

Let us recall the following characterization of balanced manifolds (see also [AB4]).

THEOREM. ([MI]) - *Let M be a compact complex n-dimensional manifold. The following conditions are equivalent, and in this case the manifold is called balanced (or semi-Kähler, or (n-1)-Kähler):*

i) there exists an hermitian metric on M such that its Kähler form ω satisfies

$$d\omega^{n-1} = 0.$$

ii) if T is a positive representative of the zero class in $V^{1,1}_{\mathbf{R}}(M)$, then $T = 0$.

So we get the following result.

COROLLARY 5.7 - *Let M and $\tilde{M}$ be compact complex manifolds, and $f : \tilde{M} \to M$ be a modification. Then $\tilde{M}$ is balanced if and only if M is balanced.*

PROOF. - As said in the introduction, we need only to prove that, if $\tilde{M}$ is balanced, then M is balanced too. Let T be a positive (1,1)-current on M, such that $\langle T\rangle = 0$ in $V^{1,1}_{\mathbf{R}}(M)$. Then $\tilde{T}$, as given in Theorem 5.6, belongs to $f^*\langle T\rangle = 0$, that is, $\tilde{T} = 0$ since $\tilde{M}$ is balanced. Hence $T = f_*\tilde{T} = 0$.

REFERENCES

[AB1] L. ALESSANDRINI and G. BASSANELLI, A balanced proper modification of P_3, Comment. Math. Helv. 66 (1991) 505-511.

[AB2] L. ALESSANDRINI and G. BASSANELLI, Positive $\partial\bar{\partial}$-closed currents and non-Kähler geometry, J. Geometrical Analysis 2 (1992) 291-316.

[AB3] L. ALESSANDRINI and G. BASSANELLI, Smooth proper modifications of compact Kähler manifolds, Proc. Internat. Workshop on Complex Analysis (Wuppertal 1990), Complex Analysis, Aspects of Mathematics E, vol.17, Vieweg, Germany, 1991.

[AB4] L. ALESSANDRINI and G. BASSANELLI, Metric properties of manifolds bimeromorphic to compact Kähler spaces, J. Differential Geometry 37 (1993) 95-121.

[AB5] L. ALESSANDRINI and G. BASSANELLI, Lelong Numbers of positive plurisubharmonic currents, to appear.

[B] G. BASSANELLI, A cut-off theorem for plurisubharmonic currents, to appear in Forum Math.

[CA] A. CASSA, Formule di Künneth per la coomologia a valori in un fascio, Ann. S.N.S. 27 (II) (1973) 905-931.

[GH] P. GRIFFITHS and J. HARRIS, Principles of Algebraic Geometry, Wiley-Interscience Publ., 1978, N.Y.

[GR] H. GRAUERT and R.REMMERT, Bilder und Urbilder Analytischer Garben, Ann. of Math. 68 (1958) 393-443.

[HA] R. HARTSHORNE, Algebraic Geometry, Springer, New York, 1977.

[HI] H. HIRONAKA, Resolution of singularities of an algebraic variety over a field of characteristic zero I, Ann. of Math. 79 (1964) 109-326.

[HR] H. HIRONAKA and H. ROSSI, On the equivalence of embeddings of exceptional complex spaces, Math. Ann. 156 (1964) 313-333.

[KA] L. KAUP, Eine Künnethformel für Fréchetgarben, Math. Zeitschr. 97 (1967) 158-168.

[MI] M.L. MICHELSON, On the existence of special metrics in complex geometry, Acta Math. 143 (1983) 261-295.

On Tangent Hyperplanes to Complex Projective Curves

E. Ballico

Dept. of Mathematics, University of Trento, 38050 Povo (TN), Italy

e-mail: (bitnet) ballico@itncisca or ballico@itnvax.science.unitn.it

fax: italy + 461881624

In the report [8] of joint work with Adams, Shifrin and Varley, C. McCrory considered the following situation.

Let $C \subset \mathbf{P}^m$ be a smooth non degenerate compact complex curve and let $H \in \mathbf{P}^{m*}$; set d:= deg(C); we will say that the hyperplane H has type $(m_1,...,m_k)$ (with $m_i>0$ and $\Sigma_i m_i = d$) if there are points P_i, $1 \leq i \leq k$, of C such that $\Sigma_i m_i P_i$ is the degree d divisor $C \cap H$ of C. Set $I(C) := \{(x,H) \in C \times \mathbf{P}^{m*}: x \in H\}$ (the incidence correspondence); if there is no danger of misunderstanding we will write I instead of I(C). Let $\pi: I(C) \to \mathbf{P}^{m*}$ be the projection; the fibers of π are the hyperplane sections of C. We will say that C is *tangentially versal* if $(I(C),\pi)$ is versal.

Since the set $\mathbf{P}^{m*}$ of hyperplanes of $\mathbf{P}^m$ is smooth, to check that a smooth complex curve $C \subset \mathbf{P}^m$ is tangentially versal we need to check the following property ($):

($) For every $(P,H) \in I(C)$ (say of type $(m_1,...,m_k)$) the differential at (P,H) of the morphism from $\mathbf{P}^{m*}$ to the versal deformation space for the type $(m_1,...,m_k)$ is surjective.

In particular only the integers m_i with $m_i \geq 2$ are important; we will always assume $m_1 \geq m_2 \geq \cdots \geq m_k$ and let $x \leq k$ be the last integer (if any) with $m_x \geq 2$.

Assume that $C \subset \mathbf{P}^m$ is tangentially versal; then, by the structure of the smooth miniversal spaces involved, the following condition ($$) (which will be also called "the right codimension condition") is satisfied ([8], Corollary in the first page):

($$) The tangential behaviour occurs in the right codimension i.e. if there is $(P,H) \in I(C)$ with type $(m_1,...,m_k)$, then near H the set of hyperplanes of type $(m_1,...,m_k)$ for C has the expected codimension $\Sigma_i (m_i-1)$ (i.e. the minimal possible one); furthermore, near H there are hyperplanes with all possible types $(m_1',...,m_k')$ with $m_i' \leq m_i$ for every i and strict inequality for at least one index i; the last part of this condition is the completeness of the family near (P,H).

It is very easy to find bad embeddings of an abstract curve and in particular embeddings which are not tangentially versal. It is true (see [8]) but much more difficult to show the existence of an irreducible component, A, of the Hilbert scheme $\mathrm{Hilb}(\mathbf{P}^m)$ of curves with fixed degree and genus such that no curve parametrized by A is tangentially versal. The aim of this note is to give several examples of "sufficiently general" curves which are tangentially versal and a few examples of "general" curves which are not tangentially versal. As a sample of the results we are looking for, we state now the following Theorem 0.1. The proof of this theorem is very cheap, since it is just a collection of the relevant references; however, Theorem 0.1 is the best we are able to prove in $\mathbf{P}^3$ and will be used for the proof of Theorem 0.2.

Theorem 0.1. *Fix integers d, g, m. There is an irreducible generically smooth component M(d,g,m) of the Hilbert scheme of degree d and genus g smooth curves in* $\mathbf{P}^m$ *such that a general* $C \in M(d,g,m)$ *satisfies the right codimensional condition ($$) (in any characteristic) and is tangentially versal (in characteristic 0) if one of the following numerical conditions on (d,g,m) is satisfied:*

(a) Assume m = 3; let $\sigma(d)$ *be the unique integer such that* $\sigma(d)(\sigma(d)+1)/2 \le d < (\sigma(d)+1)(\sigma(d)+2)/2$; *set* $F_d(t) = dt+1 - (t+3)(t+2)(t+1)/6$; *assume* $0 \le g \le F_d(\sigma(d)-1)$; *there exists a component M(d,g,3) with the right number of moduli (i.e. such that the morphism* $M(d,g,3) \to M_g$ *has image of codimension* $max(0,-\rho(g,d,3)) := max(0,3g+12-4d)$.

(b) Assume $m \ge 4$; *there is a function w(g) with* $lim_{g \to \infty} w(g) = (m-2)/2$ *such that M(d,g,m) exists if* $d \ge g \cdot w(g)$.

Here is another existence result which will be proved in this paper .

Theorem 0.2. *Assume characteristic 0. Fix integers m, g, d with* $m \ge 3$ *and* $g \ge 0$ *and* $d \ge g+m$. *Then a general curve* $C \subset \mathbf{P}^m$ *with degree d, genus g and* $h^1(C, \mathcal{O}_C(1)) = 0$ *is tangentially versal.*

For all positive integers d, g, r, set $\rho(g,d,r)$: g - (r+1)(g+r-d) (the Brill - Noether number). A curve of genus g has a g^r_d if and only if $\rho(g,d,r) \ge 0$; if this condition is satisfied for some $r \ge 3$, it is known (see e.g. [3] or [5]) that such a curve has a non degenerate degree d embedding in $\mathbf{P}^r$; if $\rho(g,d,r) > 0$ the set of all g^r_d on C is irreducible; if $\rho(g,d,r) = 0$, this set is finite, but, moving C, the corresponding finite cover of a Zariski open part of M_g is integral ([6]) and hence, strectching the terminology, we will speak of "generic" g^r_d even in this case: there is even in this case a unique irreducible component (call it W(d,g;r)) of the Hilbert scheme Hilb($\mathbf{P}^r$) of $\mathbf{P}^r$ containing curves with general moduli.

For all integers d, g, m with $m \ge 3$, $d \ge m$, $d-m < g \le d-m+[(d-m-2)/(m-2)]$, in [3] (and previously in [1] for m = 3) was proved the existence of an irreducible component W(d,g;m) of the Hilbert scheme Hilb($\mathbf{P}^m$) of curves in $\mathbf{P}^m$ with degree d and genus g with very nice properties; for instance, this component is generically smooth and contains curves with the right number of moduli (hence, if $\rho(g,d,r) \ge 0$, it is the component containing curves with general moduli). By the results in [3], §2, we have W(d,g;m) = M(d,g,m) (the component claimed to exist in Theorem 0.1) for all integers d, g and m for which both components are defined. The next theorem is the more interesting result proved in this paper.

Theorem 0.3. *Assume characteristic 0. For every integer $m \geq 3$ there is an integer $a(m)$ and a function ε_m: $N \rightarrow R_+$ with $\lim_{t \rightarrow \infty} \varepsilon_m(t) = 0$, such that for all integers d, g with $g \geq 0$, $d \geq a(m)$ and $md \geq (m+2)g(1+\varepsilon_m(g))$, a general $C \in W(d,g;m)$ has no point of type $(m_1,...,m_k)$ with $\Sigma_i(m_i-1)>m$.*

To be more explicit, we give the following more precise result for the case x = 1, i.e. for the higher order ramification points ("Weierstrass points") of a general embedding; we stress that if $\rho(g,d,m) \geq 0$ the corresponding result was previously proved by D. Eisenbud and J. Harris (it is [5], Th. 2).

Theorem 0.4. *For all integers d, g, m with $m \geq 3$, $d \geq m$, $d-m<g \leq d-m+[(d-m-2)/(m-2)]$ (i.e. for all integers d, g and m for which $W(d,g;m)$ is defined), a general $C \in W(d,g;m)$ has the property that for every $P \in C$ the osculating flag to C at P has either contact order $(2,3,...,m-1,m)$ or contact order $(2,3,...,m-1,m+1)$.*

It is very cheap (see Proposition 1.2 (a)) to find non linearly normal embeddings for which ($$) fails. Here is an interesting case with linearly normal (but not general) embeddings for which ($$) fails.

Theorem 0. 5. *Fix positive integers d, g, r with $r \geq 3$ and $g \geq 2$. Assume $\rho(g,d,r+1) \geq 0$ and let C be a smooth curve of genus g with general moduli. Then there are linearly normal embeddings of C into $\mathbf{P}^r$ which are not tangentially versal; more precisely, there are linearly normal embeddings with a point, P, which is hyperosculating with $m_1>r+1$ and hence which does not satisfy condition ($$).*

All the results stated in the introduction will be proved in section 1. In section 2 we will consider the conditions "versal" or "tangential behaviour with expected codimension" with respect to lower dimensional linear subspaces, not just for hyperplanes. In the third (and last section) we will consider monodromy problems for the ramification points. Indeed, in suitable ranges (either curves of high genus g with degree $d \geq g+m$ in $\mathbf{P}^m$ or curves with degree much greater than g+m) we are able to show that the monodromy is huge for many particular subfamilies of such embeddings of all genus g curves. As an example we state and prove the case of the family of all hyperelliptic curves (see Theorem 3.1). Changing the numerical bounds the same proof works for many other families. Then we will consider (see Theorem 3.2) the existence and monodromy problem for ramification points for which the condition ($$) is not satisfied.

For work on this subject, see [9].

The author was partially supported by MURST and GNSAGA of CNR (Italy).

1. Proofs of the theorems

In this section we prove the theorems stated in the introduction and give a result (Proposition 1.2) on the non linearly normal case.

Proof of 0.1: It is explained in [Hi] the following consequence of the existence of a curve $C \subset \mathbf{P}^m$ whose normal bundle N_C satisfies the vanishing condition $h^1(C,N_C(-1)) =$

0. Fix such a curve, say of degree t, and take any hyperplane H; then the sets of all $Y \cap H$ with Y curve near C contains a Zariski open dense subset of the symmetric product $S^t(H)$; of course, since $h^1(C,N_C(-1)) = 0$ implies $h^1(C,N_C) = 0$, $Hilb(\mathbf{P}^m)$ is smooth at C. Looking at the general fiber of an appropriate incidence correspondence (and using generic smoothness in characteristic 0) we obtain that to prove the existence of M(d,g,m) it is sufficient to prove the existence of a smooth curve $C \subset \mathbf{P}^m$ with degree d, genus g and with normal bundle N_C satisfying $h^1(C,N_C(-1)) = 0$. Hence part (a) and part (b) follow respectively from [10], Th. 6.1, and [2], Th. 5, part (3). ♦

Now we give a few preliminaries for the proof of Theorem 0.2.

Remark 1.1. Assume $m \geq 4$. Let $C \subset \mathbf{P}^m$ be a smooth curve. It is easy to show that if C is not tangentially versal (resp. does not satisfy the "right codimension" condition ($$)), there is an isomorphic projection of C into $\mathbf{P}^{m-1}$ whose image is not tangentially versal (resp. does not satisfy ($$)): just take the projection from a general point of a hyperplane at which condition ($$) (or ($)) fails.

Proposition 1.2. *Fix integers n and m. Let $C \subset \mathbf{P}^m$ be a smooth non degenerate curve.*

(a) Assume $3 \leq n \leq m-2$. Then there is a smooth curve $X \subset \mathbf{P}^n$ with X linear projection of C, deg(X) = deg(C), such that X does not satisfy the right codimension condition ($$); in particular C is not tangentially versal. If C is not a rational normal curve, then the same is true if n = m-1.

(b) Assume $n \geq 3$ and that C satisfies ($$) (resp. is tangentially versal and the characteristic is 0).Then a general projection of C into $\mathbf{P}^n$ satisfies ($$) (resp. is tangentially versal).

Proof. (a1) First assume that C is not a rational normal curve. By Plücker formula we see the existence of $P \in C$ and a hyperplane $H \subset \mathbf{P}^m$ which is hyperosculating to C at P, i.e. with $C \cap H$ containing (m+1)P. Since a general point Y of H is not contained (e.g. for dimensional reasons) in the 3-dimensional secant variety Sec(C) of C, the projection of C from Y into $\mathbf{P}^{m-1}$ is biregular. Its image, X, does not satisfy ($$) since it has a point P' (the image of P) at which the osculating hyperplane intersects the curve at least in the scheme (m+1)P'. Now assume that C is a rational normal curve. The biregular image, C', of a general projection of C into $\mathbf{P}^{m-1}$ is not a rational normal curve. Hence we may apply the first part of the proof to C'.

(b) Repeating m-n times the following proof, we may assume n = m-1. Fix a type $(m_1,...,m_k)$ and let $B(\{m_i\})$ (or $B(m;\{m_i\})$ or $B(m;\{m_i\})(C)$) be the set of hyperplanes of type $(m_1,...,m_k)$ for C; set $b(\{m_i\}) := \dim(B(\{m_i\}))$ (or $b(m;\{m_i\})$ and so on). For a general $u \in \mathbf{P}^m$, $\dim\{H \in B(\{m_i\}): u \in H\} < b(\{m_i\})$. Hence the projection of C from u into $\mathbf{P}^{m-1}$ satisfies the part of condition ($$) concerning the type $(m_1,...,m_k)$. The versality condition for the projection from a general $P \in \mathbf{P}^m$ (i.e. the non vanishing of the differential of a suitable map) follows in a standard way in characteristic 0 from the corresponding versality assumption in $\mathbf{P}^m$ and the theorem on generic smoothness ♦

Proof of 0.3: We fix the integer x with $1 \le x \le m/2$ and a corresponding type $(m_1,...,m_k)$. We fix a subset $S = \{P(i), Q(j)\}$ with card(S) = m+3+x, $1 \le i \le m+3$, $1 \le j \le x$, and such that the subset $S' := \{P(i)\}_{1 \le i < m+3}$ is formed by points in linear general position. Using reducible curves as in the definition of W(d,g;m) given in [3], it is easy to check that for large d the subset, A(d,g;m;S),of W(d,g;m) formed by curves containing S has pure codimension m(m+3+x) and that how large we need d is independent of the choice of S. Then as in the sketch of proof of [2], Th. 5, part (3), we see the existence of a smooth $C \in A(d,g;m;S)$ with $h^1(C,N_C(-\Sigma_i P(i)-\Sigma_j m_j Q(j)) = 0$. Hence, as for Theorem 0.1, we win for the given type. Since using the inductive definition of W(d,g;m) we easily control all types with, say, $\Sigma_i (m_i-1) \ge 100m$, we need to use the previous proof only for finitely many types. Hence we obtain the result. ♦

Proof of 0.4: If $d \le 2m$, the result is easy, using e.g. the vanishing of $h^1(C,N_C(-1))$ or [5], Th. 2). If d>2m we will use the following induction. We fix an integer t with $t \le m+2$ such that W(d-m,g-t;m) is defined and we assume the result for the integers d-m, g-t and m. We take a general $Y \in W(d-m,g-t;m)$ and we consider $X \in W(d,g;m)$ with $X = Y \cup T$, T rational normal curve meeting X quasi-transversally and exactly at t+1 sufficiently general points; $X \in W(d,g;m)$ by [3],Lemma 2.3. The only ramification of X is at the ramification points of Y and over $Y \cap T$, since T has no ramification point. Over every $P \in Y \cap T$, the curve X has contact order at most a+1 with every subspace of dimension a, because Y and T have distinct tangents at P. ♦

Proof of 0.5: By the assumption on $\rho(g,d,r+1)$ there is a component W(d,g;r+1) of Hilb($\mathbf{P}^{r+1}$) containing degree d non degenerate embedding of C. By Proposition 1.2, part (a), there are non linearly normal degree d non degenerate embeddings of C into $\mathbf{P}^r$ which are not tangentially versal (and indeed with exactly a hyperosculating point, P, with the minimal possible, r+2, intersection with the osculating hyperplane and only ordinary osculation with the (r-2)-dimensional osculating space). Note that such a type of hyperosculation, if it occurs at some (point,hyperplane) (P,H), it occurs in codimension at most r+1. Counting dimensions we check easily that there are also linearly normal embedding of C with the same hyperosculating phenomenon, hence not tangentially versal. ♦

2. Lower dimensional linear spaces

For several reasons (see e.g. Remark 2.2) it seems worthwhile to introduce the following definition, i.e. the notion of "tangential versality with respect to higher codimensional linear subspaces".

Definition 2.1. Fix integers m, and r with $m \ge 3$ and $0<r<m$. Let $C \subset \mathbf{P}^m$ be a smooth curve; let G(r):= G(r+1,m+1) be the Grassmannian of r-dimensional linear subspaces of

$\mathbf{P}^m$; let $I(C)(r) := \{(P,V) \in C \times G: P \in V\}$ be the incidence correspondence and $\pi(C)(r)$: $I(C)(r) \to G(r)$ be the projection; write $I(r)$ and $\pi(r)$ if C is clear. We will say that C is tangentially r-versal if I(r) is versal. We define in the obvious way the equivalent condition ($)(r) and "the right codimension condition" ($$)(r) (with ($$)(m-1):= ($$)).

Note that at a point (P,V) of type $(m_1,...,m_k)$ with $m_i = 1$ if $i>1$, the condition ($$)(r) just means that $m_1 \leq r+1$, i.e. that V is at most "an ordinary hyperosculating r-dimensional linear space" (the case r = m-1 being just the notion of ordinary Weierstrass point with respect to $\mathcal{O}_C(1)$).

Remark 2.2. The proof of part (b) of proposition 1.2 works verbatim if instead of conditions ($) or ($$) we just assume conditions ($)(m) or ($$)(m).

As in the case r = m-1 we have the following result.

Proposition 2.3. *Assume characteristic 0. Let $C \subset \mathbf{P}^m$ be a smooth curve. C is tangentially r-versal if and only if the projection $\pi(r)$ is transversal to the stratification of type.*

Proof. Just use the structure of a miniversal deformation space for, say, k "curvilinear" singularities. ♦

3. Monodromy groups

In this section we consider the monodromy problem for ramification points. Indeed, in suitable ranges we are able to handle also the case of subfamilies of all curves of given degree and genus. We assume characteristic 0. As a sample of the results that can be obtained in this way we state and prove the following theorem.

Theorem 3.1. *Fix integers m, d, g with $m \geq 3$ and $d \geq g+m$. Then a general hyperelliptic curve of genus g and degree d in $\mathbf{P}^m$ has only ordinary ramification points. Assume also $g \geq m+2$; consider the subvariety of Hilb($\mathbf{P}^m$) obtained varying the hyperelliptic curve and the embedding;the family of all ramification points of such curves is irreducible, i.e. the monodromy group G for the ramification points of hyperelliptic curves is transitive. Let $t := (m+1)d+m(m+1)(g-1)$ be the number of such ramification points (given by the Plücker formula). If $g \geq 4(m+2)$ then G is either the alternating group A_t or the symmetric group S_t.*

Proof. Fix an integer $e \geq 0$ and let $F := F_e$ be a Segre - Hirzebruch surface with invariant e; as base of Pic(F) we will take smooth rational curves f and h with $h^2 = -e$, $h \cdot f = 1$ and $f^2 = 0$. First we assume m-e even and fix integers a, b such that the hyperelliptic curves in the linear system |2h+af| have arithmetic genus g and the embedding by |h+bf| of F induces a degree d embedding in a suitable projective space, $\mathbf{P}^N$, of such hyperelliptic curves, i.e. we take a = g+1+e, 2b+a-2e = d (hence we take e with the same parity of d-g-1 and we have N = 2b-e+1). By the assumption on d and g we have $N \geq m$ and we take a general projection into $\mathbf{P}^m$ of the image of F in $\mathbf{P}^N$; call J the induced smooth ruled surface in $\mathbf{P}^m$. We will prove the existence and monodromy problems for the subfamily of the hyperelliptic curves formed by the hyperelliptic curves of type $|2h+af|$ in J. Fix $P \in J$ and a hyperplane H with $P \in H$ but H not tangent to J at P; hence $T := J \cap H$ is a curve smooth at P. A curve, D, through P has P as ramification point and H as corresponding hyperosculating hyperplane if and only if D contains the length (m+1) subscheme Z of T with $Z_{red} = P$. Furthermore, the osculating hyperplane has $m_1 = m+1$ if and only if D does not contain the length (m+2) subscheme, Z', of D with support at P. Note that $a \geq (m+2)+2e$ if and only if $g \geq (m+2)+e-1$. Hence if $0 \leq e \leq 1$ by the assumption on g it is easy to check that the set A(P,H) of curves in $|2h+af|$ containing Z has codimension m+1 and the ones containing Z' has codimension m+2. Now vary P and H. Since every A(P,H) is a linear space (hence integral) and with dimension independent of P and H, we obtain easily that the scheme of ramification points is integral. Furthermore, looking at lower dimensional linear subspaces we see that in general no lower dimensional linear subspace is hyperosculating. Hence, the general such hyperelliptic curve has only ordinary ramification. If $g \geq 4(m+2)$ then we have $t \geq 25$; furthermore we may apply the same construction not just for one pair (P,H) but to 4 such pairs (obtaining something of the right codimension, 4m+4). Thus we get that the symmetric product 4 times of the variety of ramification points is integral, i.e. that the monodromy group is at least 4 transitive. Since it acts on a set with $t \geq 25$ elements, it is know (see [4], §5) that it is either A_t or S_t. ♦

The same proof shows the following existence and irreducibility theorem for hyperosculating hyperplanes.

Theorem 3.2. *Fix integers d, g, m, z, $b(j)_{1 \leq j \leq z}$, $a(j)_{1 \leq j \leq z}$ with $g \geq 0$, $m \geq 3$, $d \geq g+m$, $z > 0$, $b(j) > 0$ for every j, $a(j+1) > a(j) > 0$ for every j and such that for every integer u with $1 \leq u \leq z$ we have $g \geq 2+3b(u)m(u)+\Sigma_j b(j)m(j)$. Then there is an integral family,* ***U****, of degree d and genus g hyperelliptic curves embedded in $\mathbf{P}^m$ and such that each of these curves has, apart from ordinary ramification, exactly b(j) ramification points with osculating codimension 2 linear space with ordinary order of contact (i.e. m-1) and with osculating hyperplane with contact order $m+1+a(j)$. Furthermore, in the locus of ramification points over* ***U****, for every j with $1 \leq j \leq z$ each scheme corresponding to the ramification order $m+1+a(j)$ is integral and, if $b(u) \geq 6$ and $g \geq 2+5b(u)m(u)+\Sigma_j b(j)m(j)$ (for just one index u), the corresponding Galois group contains the alternating group $A_{b(u)}$.*

References

1. E. Ballico and Ph. Ellia, *Beyond the maximal rank for curves in $\mathbf{P}^3$*, in: Space curves, Proceedings Rocca di Papa, Lect. Notes in Math. **1266**, Springer-Verlag, 1987, pp. 1-23.
2. E. Ballico and Ph. Ellia, *Bonnes petites composantes des schemas de Hilbert de courbes lisses de $\mathbf{P}^n$*, Comptes Rend. Acad. Sc. Paris **306** (1988) 187-190.

3. E. Ballico and Ph. Ellia, *On the existence of curves with maximal rank in* $\boldsymbol{P}^n$, J. reine angew. Math. **397** (1989) 1-22.
4. P. J. Cameron, *Finite permutation groups and finite simple groups,* Bull. London Math. Soc. **13** (1981), 1-22.
5. D. Eisenbud and J. Harris, *Divisors on general curves and cuspidal rational curves,* Invent. Math. **74** (1983), 371-418.
6. D. Eisenbud and J. Harris, *Irreducibility and monodromy of some families of linear series,* Ann. scient. Ec. Norm. Sup. (4) **20** (1987), 65-87.
7. A. Hirschowitz, *Sectiones planes et multisecantes pour les courbes gauches génériques principales,* in: Spaces curves, Proceedings Rocca di Papa, Lect. Notes in Math. **1266**, Springer-Verlag, 1987, pp. 124-155.
8. C. McCrory, *Secant planes of space curves,* MSRI preprint, 1993.
9. J. McKernan, *Versality for canonical curves and complete intersections,* preprint.
10. Ch. Walther, *On the cohomology of the normal bundle of space curves I,* preprint.

Infinite Dimensional Supergeometry

Jürgen Bingener, Fakultät für Mathematik der Universität, Universitätsstrasse 31, D-93040 Regensburg, Germany

Thomas Lehmkuhl, Fakultät für Mathematik der Universität, Universitätsstrasse 31, D-93040 Regensburg, Germany

Dedicated to Prof. H. Grauert

Abstract. We develop, in a rather general noncommutative setting, an infinite dimensional geometry embracing differential calculus, analytic functions, Banach analytic spaces and Lie theory, and describe possible applications in complex-analytic deformation theory.

This paper is in final form and no part of it will be submitted elsewhere

Introduction

Some time ago we realized that an infinite dimensional superanalysis could possibly turn out to be a powerful tool to handle certain complicated formulas occuring in geometry and analysis. At that time such a new type of analysis did not yet exist, although *finite* dimensional superanalysis was already on the way to become a well established theory, thanks to the pioneering work of Berezin, the school of Manin, Kostant and many others, c.f. for instance [Be], [DW], [Ko], [Ma_1], [Ma_2], [Pe], [Rg_1] and [Rg_2]. Applications of these ideas, for example in connection with the Atiyah-Singer index theorem, were obtained more or less at the same time, cf. for instance Bismut [Bi], Getzler [Ge] and Quillen [Qu].

In this paper we present the foundations of infinite dimensional supergeometry including differential calculus, analytic functions, Banach analytic spaces and Lie theory.

This machinery definitely admits applications both to classical commutative analysis and physics. It has been known for some time that the local moduli spaces in analytic geometry can be obtained as orbit spaces associated with the adjoint representation of certain infinite dimensional $\mathbb{Z}$-graded super Lie algebras, cf. [LM]. *Using the methods developed in this paper, one can substantially simplify the presentation given in loc. cit.*. Moreover we can show that the main results of [LM] hold in the super case as well. This requires in particular in the extension of the so-called RO-analysis including the quotient theorem. Somewhat loosely, we may summarize the final result in the following form:

Theorem. *"Every" result in complex analytic deformation theory can be "superized".*

For the case of "compact" superobjects and the standard $\mathbb{Z}/2\mathbb{Z}$-grading, this was obtained by Flenner-Sundararaman [FS]. Let us mention here that the commutative $\mathbb{Z}$-graded case is partially contained in [LM]. As a possible application not included in published work, we can construct semiuniversal deformations of strictly pseudoconvex complex superspaces and in particular of germs of complex superspaces with finite dimensional T^1. More general, we determine that a recent existence theorem of V.P. Palamodov and the first-named author [BP] has a super analogue. A complete account will be given elsewhere.

Classical commutative analysis in all its variants relies heavily on the notion of a function. In contrast to this, one cannot use functions for the foundations of (anticommutative) superanalysis. In the finite dimensional case it is possible to circumvent this difficulty by basing upon the notion of a graded ringed space. This is exactly the technique used so far.

The basic notion for our approach to infinite dimensional superanalysis is the well-known notion of a cogebra. The fundamental tool is the so-called Φ-operation or induced cogebra morphism. It is only this operation which enables us to define the composition of two formal maps. In the classical set-up this merely means composition of formal power series.

The Φ-operation as presented here works only in the context of $\mathbb{Q}$-algebras. In forthcoming papers we will remove this restriction using suitable divided powers, cf. also (4.5.1). In contrast to common practice, we deal with very general gradations, namely so-called signed groups. Here by a signed group we mean an abelian group $\mathbb{J}$ endowed with a commutation factor ε, cf. (1.1). This approach has the advantage of including simultaneously the standard grading used in superanalysis ($\mathbb{J} = \mathbb{Z}/2\mathbb{Z}, \varepsilon(j,k) = (-1)^{jk}$), the classical commutative case ($\mathbb{J}$ arbitrary, $\varepsilon = 1$) and the $\mathbb{Z}$-graded super case. For the applications we have in mind, the latter is the most important one.

But as a matter of fact, our commutation factors allow an amount of "noncommutativity", which means considerably more than just "commutativity up to sign". For instance in a "commutative" ring we may have elements x, y satisfying a commutator relation $xy = qyx$ with an arbitrary unit q of degree zero. So the theory presented here can be considered as one possible approach to spaces with non-commutative structures (cf. Connes [Co] for another one).

We shall now give a brief description of the content of the various sections. In the first two sections we have gathered some basic facts and notations from the corresponding linear algebra. Not included here is the theory of (Berezin) determinants, which we will treat elsewhere, in a generalized context. The third section deals with cogebras, bigebras and the Φ-operation. In section 4 we define formal maps and their composition, and develop differential calculus. Then, in section 5, we introduce (pseudo)norms compatible with our general gradings. This requires some new ideas. We have good reasons for treating here the archimedean and ultrametric case simultaneously.

In section 6 we first define convergent maps and prove the inverse function theorem. Then we investigate the basic question, which elements can be "substituted" into a formal map with respect to a given norm on the target. Applying this concept to the Taylor series of a formal map, we arrive at the notion of an "admissible" element.

In fact, one motivation for developing the whole machinery was our belief that many nonlinear operations occuring in analysis and geometry (local or global) can be obtained canonically (more or less) by formal maps in the sense of infinite dimensional superanalysis, i.e. the given operation should arise via substitution.

The theme of section 7 is the so-called *composition map*, one of the most important formal maps occuring in analysis. Set theoretically, i.e. after substitution, this map associates with a pair of composable formal maps their composition. This property supports the previously mentioned philosophy. The composition map is not convergent with respect to any reasonable norms on the underlying spaces. But if we endow the "function" spaces of formal maps themselves with the so called canonical norm, the composition map is convergent, in the nonarchimedean sense!

Here the ground ring is endowed with the *trivial* norm. Generally speaking, it seems that in the literature on nonarchimedean analysis (cf. for instance [BGR]) the case where the given valuation is trivial, is more or less excluded. It is this "extreme" case which leads to delicate convergence questions on the boundaries of the respective convergence domains. The applications to the composition map in mind (cf. (7.3.2), (7.3.6), (7.5.2)), we overcame these difficulties in advance already in section 6. After these preparations, we develop in section 8 the calculus of analytic functions and introduce Banach analytic spaces.

In the final section 9 we show that the classical correspondence between (formal) groups and Lie algebras applies in the context of infinite dimensional supergeometry, too. Apart from the omnipresent Poincaré-Birkhoff-Witt theorem, our proof depends heavily on the Φ-operation. This approach differs considerably from the proofs which can be found in the literature, cf. for instance Lazard [La], Serre [Se], Bourbaki [Bo3] and Zink [Zi], and seems to be simpler even when applied specifically to the classical case. A new result is the characterization of the enveloping bigebras of Lie algebras among all bigebras, cf. (9.2.6). This result shows the frontiers of possible generalizations of our theory. Finally, as an important example of a formal group, based upon the composition map, we consider the formal group associated with the Lie algebra of vector fields.

In order to keep our presentation at a reasonable length, we were forced to omit proofs and calculations at many places. Those readers feeling unhappy with this situation, may later consult the monographs [B1], [B2], where one can find an extensive treatment with detailed proofs.

In a series of lectures at Regensburg starting in 1988, we presented the better part of the material contained in this paper. We would like to thank the audience for patience and helpful criticism. We are further indebted to V.P. Palamodov for stimulating our interest in noncommutative geometry. Finally, our sincere thanks go to Mrs. R. Bonn for expert typesetting using the system TEX.

1. Signed groups and commutative rings

(1.1.1) (Signed groups) Let $\mathbb{J}$ be an abelian group, written additively, and Λ a commutative ring. By a *commutation factor for* $\mathbb{J}$ *with values in* Λ we mean a map ε from $\mathbb{J} \times \mathbb{J}$ into $\Lambda^\times$ satisfying $\varepsilon(j+j',k) = \varepsilon(j,k)\cdot\varepsilon(j',k), \varepsilon(j,k) = \varepsilon(k,j)^{-1}$ and $\varepsilon(j,j) \in \{\pm 1\}$ for all elements j, j', k of $\mathbb{J}$. A group homomorphism π from $\mathbb{J}$ into $\mathbb{Z}/2\mathbb{Z}$ is called a *parity map* for ε, if for any element j of $\mathbb{J}$ the relation $\pi(j) = 0$ resp. $\pi(j) = 1$ implies $\varepsilon(j,j) = 1$ resp. $\varepsilon(j,j) = -1$. If $\operatorname{char}(\Lambda)$ is different from 2, there exists a unique parity map for ε. In case $\operatorname{char}(\Lambda) = 2$ *every* group homomorphism from $\mathbb{J}$ into $\mathbb{Z}/2\mathbb{Z}$ is a parity map for ε.

(1.1.2) By a *signed group* we understand a tuple $(\mathbb{J}, \Lambda, \varepsilon, \pi)$ consisting of an abelian group $\mathbb{J}$, a commutative ring Λ, a commutation factor ε for $\mathbb{J}$ with values in Λ and a parity map π for ε. An element j of $\mathbb{J}$ is called *even* resp. *odd*, if $\pi(j)$ is 0 resp. 1. If I is a totally ordered index set, $g = (g_i)_{i\in I}$ a tuple in $\mathbb{J}^{(I)}$ and σ a permutation of I, then

$$e(\sigma; g) = e^{\mathbb{J}}(\sigma; g) := \prod_{\substack{i>j \\ \sigma(i)<\sigma(j)}} \varepsilon(g_j, g_i) \tag{1.1.2.1}$$

is a well defined element of $\Lambda^\times$. Let $(\mathbb{J}', \Lambda', \varepsilon', \pi')$ be another signed group and $\rho: \mathbb{J} \longrightarrow \mathbb{J}'$ resp. $f: \Lambda \longrightarrow \Lambda'$ be a homomorphism of groups resp. rings. We call (ρ, f) a *morphism of signed groups*, if the relations $\varepsilon' \circ (\rho \times \rho) = f \circ \varepsilon$ and $\pi' \circ \rho = \pi$ hold. We mostly abbreviate $(\mathbb{J}, \Lambda, \varepsilon, \pi)$ (resp. (ρ, f)) by $\mathbb{J}$ (resp. ρ) and denote then Λ resp. ε resp. π (resp. f) by $\Lambda\langle\mathbb{J}\rangle$ resp. $\varepsilon^{\mathbb{J}}$ resp. $\pi^{\mathbb{J}}$ (resp. $f\langle\rho\rangle$), if there is any doubt. For $\nu = 0, 1$ let $\mathbb{J}^{\langle\nu\rangle}$ be the signed group $(\mathbb{J} \amalg \mathbb{Z}, \Lambda, \varepsilon^{\langle\nu\rangle}, \pi^{\langle\nu\rangle})$ where $\varepsilon^{\langle\nu\rangle}$ resp. $\pi^{\langle\nu\rangle}$ is given by $\varepsilon^{\langle\nu\rangle}((j,m),(k,n)) = \varepsilon(j,k)\cdot(-1)^{\nu mn}$ resp. $\pi^{\langle\nu\rangle}(j,m) = \pi(j) + [\nu m]$. Then we have natural morphisms $\mathbb{J} \longrightarrow \mathbb{J}^{\langle\nu\rangle}$, $\nu = 0, 1$, and $\mathbb{J}^{\langle 0\rangle} \longrightarrow \mathbb{J}$ of signed groups. The signed groups form, with respect to the morphisms defined above, a category denoted by (SG).

(1.2.1) (Sets of type $\mathbb{J}$) Let $\mathbb{J}$ be a signed group. By a *set of type* $\mathbb{J}$ we mean a pair (M, g) consisting of a set M and a map $g: M \to \mathbb{J}$. We mostly abbreviate (M, g) by M and denote then g by g^M. Suppose now given a morphism $\rho: \mathbb{J} \to \mathbb{J}'$ of signed groups, a set M' of type $\mathbb{J}'$ and a map $f: M \to M'$. Then (f, ρ) is called a *morphism of sets over* ρ, if we have $g^{M'} \circ f = \rho \circ g^M$. With respect to morphisms defined in this way, the sets of variable type form a fibred and cofibred category (S) over (SG).

(1.3.1) (Groups and rings of type $\mathbb{J}$) Let $\mathbb{J} = (\mathbb{J}, \Lambda, \varepsilon, \pi)$ be a signed group. By a *group of type* $\mathbb{J}$ we mean a $\mathbb{J}$-graded Λ-module $E = \coprod_{j\in\mathbb{J}} E^j$ such that the E^j are Λ-submodules of E. The set of homogeneous elements of E is denoted by ${}^{\mathrm{h}}E$. Further let $g: {}^{\mathrm{h}}E \longrightarrow \mathbb{J}$ be the degree map. An element x of ${}^{\mathrm{h}}E$ is called *even* resp. *odd*, if $g(x) \in \mathbb{J}$ is even resp. odd. Suppose now given a second signed group $\mathbb{J}' = (\mathbb{J}', \Lambda', \varepsilon', \pi')$ and a morphism $\rho = (\rho, f)$ from $\mathbb{J}$ into $\mathbb{J}'$ in (SG). Then $f^*(E) = \Lambda' \otimes_\Lambda E$ is in a natural way a group of type $\mathbb{J}'$ denoted by $\rho^*(E)$.

Let F be a group of type $\mathbb{J}'$. If we set $\rho_*(F)^j := f_*(F^{\rho(j)})$ for j in $\mathbb{J}$, then $\rho_*(F) := \coprod_{j\in\mathbb{J}} \rho_*(F)^j$ is a group of type $\mathbb{J}$. For an element j' of $\mathbb{J}'$ we denote by $\mathrm{Hom}^\rho(E, F)^{j'}$ the Λ'-module consisting of all Λ'-linear maps from $\rho^*(E)$ into F sending $\rho^*(E)^{k'}$ to $F^{k'+j'}$ for every k' in $\mathbb{J}'$. Then

$$\mathrm{Hom}^\rho(E,F) = \mathrm{Hom}^\rho_f(E,F) := \coprod_{j'\in\mathbb{J}'} \mathrm{Hom}^\rho(E,F)^{j'}$$

is a group of type $\mathbb{J}'$. If $\rho = id_\mathbb{J}$ is the identity of $\mathbb{J}$, we write $\mathrm{Hom}^\mathbb{J}(E, F) = \mathrm{Hom}^\mathbb{J}_\Lambda(E, F)$ instead of $\mathrm{Hom}^\rho(E, F)$. The groups of variable type form in a natural way a cofibred category (G′) over (SG). Let

$$(\mathrm{G}) \longrightarrow (\mathrm{SG})$$

denote the fibred and cofibred subcategory of (G′) having the same objects, but admitting only those pairs (ρ, u) as morphisms, for which u is homogeneous of degree zero. Then, for a fixed signed group $\mathbb{J}$, the fibre $(\mathrm{G}'^{\mathbb{J}})$ of (G′) in $\mathbb{J}$ is an additive category, whereas the fibre $(G^{\mathbb{J}})$ of (G) in $\mathbb{J}$ is an abelian subcategory of $(G'^{\mathbb{J}})$ admitting arbitrary inductive and inverse limits.

(1.3.2) Let $\mathbb{J} = (\mathbb{J}, \Lambda, \varepsilon, \pi)$ be a signed group. By a *ring of type* $\mathbb{J}$ we mean a (unital and associative) $\mathbb{J}$-graded Λ-algebra $A = \coprod_{j\in\mathbb{J}} A^j$ such that the A^j are Λ-submodules of A. The rings of variable type form a fibred and cofibred category (R) over (SG).

(1.4.1) (Action of the symmetric group) Let $\mathbb{J} = (\mathbb{J}, \Lambda, \varepsilon, \pi)$ be a signed group, I a finite totally ordered index set and $E_i, i \in I, F$ be groups of type $\mathbb{J}$. For an element j of $\mathbb{J}$ let $\mathrm{Mult}^{\mathbb{J}}(E_i, i \in I; F)^j$ denote the Λ-module consisting of all I-linear mappings f from $\prod_{i\in I} E_i$ into F (over Λ) such that f transforms $\prod_{i\in I} E_i^{\nu_i}$ into $F^{|\nu|+j}$ for every tuple $\nu = (\nu_i)_{i\in I}$ in $\mathbb{J}^I$. Then

$$\mathrm{Mult}\,(E_i, i \in I; F) = \mathrm{Mult}^{\mathbb{J}}(E_i, i \in I; F) := \coprod_{j\in\mathbb{J}} \mathrm{Mult}^{\mathbb{J}}(E_i, i \in I; F)^j$$

is a group of type $\mathbb{J}$. - Let now σ be a permutation of I and f a map of $\mathrm{Mult}\,(E_{\sigma^{-1}(i)}, i \in I; F)$. Moreover suppose that $x = (x_i)_{i\in I}$ is a tuple of $\prod_{i\in I} E_i$. Since ${}^{\sigma}x = (x_{\sigma^{-1}(i)})_{i\in I}$ is a tuple of $\prod_{i\in I} E_{\sigma^{-1}(i)}$,

$$e(\sigma; x) \cdot f({}^{\sigma}x) := \sum_{g=(g_i)\in\mathbb{J}^I} e(\sigma; g) \cdot f((x_{\sigma^{-1}(i)}^{g_{\sigma^{-1}(i)}})_{i\in I}) \tag{1.4.1.1}$$

is a well defined element of F, c.f. (1.1.2). If there is no doubt about x resp. σ, we will sometimes write $e(\sigma)\cdot f({}^{\sigma}x)$ resp. $e(x)\cdot f({}^{\sigma}x)$ instead of $e(\sigma; x)\cdot f({}^{\sigma}x)$. In the *practical* use of these notations, the permutation σ and the map f often will not be defined explicitly. But their definition will always be clear from the context. Sometimes it will turn out to be convenient to deviate slightly from these conventions in a more or less self-explanatory way. The correspondence sending x to $e(\sigma; x) \cdot f({}^{\sigma}x)$ is a map f^{σ} in $\mathrm{Mult}\,(E_i, i \in I; F)$. If τ is a second permutation of I and h a map of $\mathrm{Mult}\,(E_{(\tau\circ\sigma)^{-1}(i)}, i \in I; F)$, then $h^{\tau\circ\sigma} = (h^{\tau})^{\sigma}$.

(1.5.1) (Commutative rings) Let A be a ring of type $\mathbb{J}$. For two elements a, b of A, not necessarily homogeneous, $[a, b] := ab - e(a, b) \cdot ba$ is called the *commutator* of a and b, cf. (1.4.1). We say that A is $\mathbb{J}$*-commutative* or simply *commutative*, if the following two conditions are satisfied:

(1) For any two elements a, b of A we have $[a, b] = 0$, i.e. $ab = e(a, b) \cdot ba$.

(2) For any odd element a of A we have $a^2 = 0$.

Note that (2) follows from (1), if 2 is a non-zero divisor in A. If σ is a permutation of a finite totally ordered index set I and $a = (a_i)_{i \in I}$ a tuple in A^I, the condition (1) gives more generally

$$\prod_{i \in I} a_i = e(\sigma; a) \cdot \prod_{i \in I} a_{\sigma^{-1}(i)}. \tag{1.5.1.1}$$

The commutative rings in the above sense form a fibred and cofibred full subcategory $({}_{\mathrm{c}}\mathrm{R})$ of (R) over (SG). For a fixed signed group $\mathbb{J}$ we denote by $(\mathrm{R}^{\mathbb{J}})$ resp. $({}_{\mathrm{c}}\mathrm{R}^{\mathbb{J}})$ the fibre of (R) resp. $({}_{\mathrm{c}}\mathrm{R})$ in $\mathbb{J}$. Further, for a fixed commutative ring S, let (R_S) resp. $({}_{\mathrm{c}}\mathrm{R}_S)$ be the category of S-objects in (R) resp. $({}_{\mathrm{c}}\mathrm{R})$, i.e. pairs consisting of a ring A in (R) resp. $({}_{\mathrm{c}}\mathrm{R})$ and a homomorphism from S into A in (R).

2. Modules and algebras

(2.1.1) (Modules) Let A be a $\mathbb{J}$-commutative ring and E a group of type $\mathbb{J}$, endowed simultaneously with the structure of a left and right A-module such that both $A^j \cdot E^k$ and $E^j \cdot A^k$ are contained in E^{j+k} for all j, k in $\mathbb{J}$. Then E is called an *A-module*, if the left and right action of A on E are connected by the formula $ax = e(a,x) \cdot xa$ for arbitrary elements a resp. x in A resp. E. Submodules of A-modules are always supposed to be graded, i.e. they are A-modules in the above sense.

Let $\varphi \colon A \to A'$ be a homomorphism in $({}_{\mathrm{c}}\mathrm{R})$ over $\rho \colon \mathbb{J} \to \mathbb{J}'$ and E resp. F be a module over A resp. A'. Then $\rho^*(E)$ resp. $\rho_*(F)$ resp. $\mathrm{Hom}^{\rho}(E,F)$ is a module over $\rho^*(A)$ resp. $\rho_*(A')$ resp. A' in a natural way. A map u of $\mathrm{Hom}^{\rho}(E,F)$ is called *φ-linear*, if we have $u(ax) = e(u,a) \cdot \varphi(a)u(x)$ and $u(xa) = u(x)\varphi(a)$ for all elements a resp. x of $\rho^*(A)$ resp. $\rho^*(E)$. Note that one of these relations implies the other. The φ-linear maps form an A'-submodule of $\mathrm{Hom}^{\rho}(E,F)$ denoted by $\mathrm{Hom}_{\varphi}(E,F) = \mathrm{Hom}^{\rho}_{\varphi}(E,F)$. The actions of A' on this module are given by the formulas $(a'u)(x) = a'u(x)$ and $(ua')(x) = e(a',x) \cdot u(x)a'$. If $\varphi = id_A$ is the identity of A, we write $\mathrm{Hom}_A(E,F) = \mathrm{Hom}^{\mathbb{J}}_A(E,F)$ instead of $\mathrm{Hom}_{\varphi}(E,F)$. The modules over variable commutative rings form a category (M') over $({}_{\mathrm{c}}\mathrm{R})$ in a natural way. By

$$(\mathrm{M}) \xrightarrow{p} ({}_{\mathrm{c}}\mathrm{R})$$

we denote the fibred subcategory of (M') having the same objects, but admitting only those pairs (φ, u) as morphisms, for which u is homogeneous of degree zero. Finally, for a fixed $\mathbb{J}$-commutative ring A the fibre $(\mathrm{M}'_A) = (\mathrm{M}'^{\mathbb{J}}_A)$ resp. $(\mathrm{M}_A) = (\mathrm{M}^{\mathbb{J}}_A)$ of (M') resp. (M) in A is an additive category resp. an abelian category admitting arbitrary inductive and inverse limits.

(2.1.2) Let E be a group of type $\mathbb{J}$ and k an element of $\mathbb{J}$. Then the *shift of* E *with respect to* k is the group $E[k]$ of type $\mathbb{J}$ given by $E[k]^j = E^{k+j}$ for j in $\mathbb{J}$. If E is an A-module, there is an unique A-module structure on $E[k]$ such that the canonical bijection T_k from $E[k]$ onto E is A-linear.

(2.2.1) (Multilinear mappings and tensor products) Let A be a $\mathbb{J}$-commutative ring, I a finite totally ordered index set and $E_i, i \in I, F$ modules over A. Then $I = \{i_1, \ldots, i_n\}$ with $i_1 < \ldots < i_n$, and $\mathrm{Mult}\,(E_i, i \in I; F)$ is an A-module in a natural way, the respective module operations given by $(af)(x) = af(x)$ and $fa = e(f,a) \cdot af$ for elements a resp. x in A resp. $\prod_{i \in I} E_i$. By

$$\mathrm{Mult}_A(E_i, i \in I; F) = \mathrm{Mult}^{\mathbb{J}}_A(E_i, i \in I; F) \subseteq \mathrm{Mult}\,(E_i, i \in I; F)$$

we denote the A-submodule consisting of all maps f such that for every element $(x_i)_{i \in I} = (x_{i_1}, \ldots . x_{i_n})$ resp. a in $\prod_{i \in I} E_i$ resp. A we have

$$\begin{aligned}
&f(x_{i_1}, \ldots, x_{i_r}a, x_{i_{r+1}}, \ldots, x_{i_n}) = f(x_{i_1}, \ldots, x_{i_r}, ax_{i_{r+1}}, \ldots, x_{i_n}),\ 1 \leq r < n, \\
&f(x_{i_1}, \ldots, x_{i_n}a) = f(x_{i_1}, \ldots, x_{i_n})a, \\
&f(ax_{i_1}, \ldots, x_{i_n}) = e(f,a) \cdot af(x_{i_1}, \ldots, x_{i_n}).
\end{aligned}$$

Here of course the last relation is a consequence of the first ones. These mappings are called *I-linear maps from $\prod_{i \in I} E_i$ into F over A.* If σ is a permuation of I and f a map of $\mathrm{Mult}_A(E_{\sigma^{-1}(i)}, i \in I; F)$, then the map f^σ (c.f. (1.3.1)) is in $\mathrm{Mult}_A(E_i, i \in I; F)$. The tensor product

$$\bigotimes_{i \in I} {}_A E_i = E_{i_1} \otimes_A \cdots \otimes_A E_{i_n}$$

is a module (of type $\mathbb{J}$) over A in a natural way, and the canonical map from $\prod_{i \in I} E_i$ into this module is I-linear over A and of degree zero. Let σ be a permutation of I. Then there exists a unique A-isomorphism

$$\tilde{\sigma}: \bigotimes_{i \in I} {}_A E_i \xrightarrow{\sim} \bigotimes_{i \in I} {}_A E_{\sigma^{-1}(i)} \tag{2.2.1.1}$$

of degree zero such that for every tuple $x = (x_i)_{i\in I}$ of $\prod_{i\in I} E_i$ we have $\tilde{\sigma}(\otimes_{i\in I} x_i) = e(\sigma; x) \cdot \otimes_{i\in I} x_{\sigma^{-1}(i)}$. If τ is another permutation of I, then $\tilde{\tau} \circ \tilde{\sigma} = (\tau \circ \sigma)^{\sim}$. If E and F are two A-modules and if we take for σ the transposition $\langle 1,2\rangle$, we obtain the A-isomorphism $\sigma_{E,F}$ from $E \otimes_A F$ onto $F \otimes_A E$ sending $x \otimes y$ to $e(x,y) \cdot y \otimes x$, called the *symmetry* of $E \otimes_A F$. In particular $\sigma_E := \sigma_{E,E}$ is an involution of $E \otimes_A E$ called also the *symmetry* of E. - Let now $\varphi: A \to A'$ be a homomorphism in $({}_c\mathrm{R})$ and F_i resp. $u_i \in \mathrm{Hom}_{\varphi}(E_i, F_i), i \in I$, be A'-modules resp. homomorphisms. The *tensor product of the* $u_i, i \in I$, is the unique map

$$\bigotimes_{i\in I} u_i = \bigotimes_{i\in I} {}_{\varphi}u_i \in \mathrm{Hom}_{\varphi}(\bigotimes_{i\in I} {}_A E_i, \bigotimes_{i\in I} {}_{A'} F_i)$$

such that $(\otimes_{i\in I} u_i)(\otimes_{i\in I} x_i) = e(u,x) \cdot \otimes_{i\in I} u_i(x_i)$ for every tuple $x = (x_i)_{i\in I}$ in $\prod_{i\in I} E_i$; here u stands for $(u_i)_{i\in I}$. If σ is a permutation of I, we have

$$\tilde{\sigma} \circ (\bigotimes_{i\in I} u_i) \circ \tilde{\sigma}^{-1} = e(\sigma; u) \cdot \bigotimes_{i\in I} u_{\sigma^{-1}(i)}. \tag{2.2.1.2}$$

The tensor product of homomorphisms is compatible with composition in an obvious sense. After these preparations, we can extend the usual "general (non)sense" concerning tensor products (tensor products and homomorphism modules, base extension, traces of endomorphisms) to our somewhat more general situation, c.f. [BL$_1$] for details. For instance, if $\varphi: A \to A'$ is a homomorphism in $({}_c\mathrm{R})$ over $\rho: \mathbb{J} \to \mathbb{J}'$ and E an A-module, then the *base extension*

$$A' \otimes_A E = \varphi^*(E) := A' \otimes_{\rho^*(A)} \rho^*(E)$$

of E is an A'-module with respect to the actions of A' given by $a'(a'' \otimes x) = (a'a'') \otimes x$ and $(a'' \otimes x)a' = e(x, a') \cdot (a''a') \otimes x$. In particular we infer that (M) is cofibred over $({}_c\mathrm{R})$.

(2.3.1) (Algebras) Let A be a commutative ring in $({}_c\mathrm{R})$. By an *algebra* over A we mean an A-module B endowed with A-bilinear map $B \times B \to B$ of degree zero. If B is unital, then the image of the structure homomorphism $A \to B$ is contained in the (naturally defined) center of B. For instance, if E is an A-module, then $\mathrm{End}_A(E) = \mathrm{Hom}_A(E, E)$ is an associative unital A-algebra. The algebras resp. associative unital algebras over variable commutative rings form a fibred and cofibred category

$$(\mathrm{Alg}) \xrightarrow{p} ({}_c\mathrm{R}) \quad \text{resp.} \quad (\mathrm{Al}) \xrightarrow{p} ({}_c\mathrm{R})$$

over $({}_c\mathrm{R})$. By $({}_c\mathrm{Al})$ we denote the full subcategory of (Al) consisting of all those algebras which, considered as objects of (R), are commutative in the sense of (1.4.1). For a fixed commutative ring A let (Alg_A), (Al_A) and $({}_c\mathrm{Al}_A)$ be the respective fibres in A.

(2.4.1) (Graded modules and graded algebras) Let A be a ring in $({}_c\mathrm{R})$. By a *graded A-module* we mean a module E over A endowed with a family $(E_n)_{n\in\mathbb{Z}}$ of A-submodules such that $E = \coprod_{n\in\mathbb{Z}} E_n$. We say that E is *positive* if $E_n = 0$ for all $n < 0$. Let $\varphi\colon A \to A'$ be a ring homomorphism over $\rho\colon \mathbb{J} \to \mathbb{J}'$ in $({}_c\mathrm{R})$ and E resp. F a graded module over A resp. A'. For an integer n of $\mathbb{Z}$ let $\mathrm{Hom}_\varphi(E,F)_n$ denote the A'-submodule of $\mathrm{Hom}_\varphi(E,F)$ consisting of all homomorphisms u which map $\rho^*(E_p)$ into F_{p+n} for every p. Then

$$\mathrm{Hom}_\varphi(E,F)^{\mathrm{gr}} := \coprod_{n\in\mathbb{Z}} \mathrm{Hom}_\varphi(E,F)_n$$

is a graded module over A' and an A'-submodule of $\mathrm{Hom}_\varphi(E,F)$ simultaneously. If $\varphi = id_A$, we write $\mathrm{Hom}_A(E,F)^{\mathrm{gr}}$ instead of $\mathrm{Hom}_\varphi(E,F)^{\mathrm{gr}}$. Let $(\mathrm{M}^{\mathrm{gr}\,\prime})$ resp. $(\mathrm{M}^{\mathrm{gr}})$ denote the category resp. fibred category over $({}_c\mathrm{R})$ consisting of all graded modules over variable commutative rings. The morphisms are defined as in (M') resp. (M), but with $\mathrm{Hom}_\varphi(E,F)^{\mathrm{gr}}$ resp. $\mathrm{Hom}_\varphi(E,F)^0_0$ replacing $\mathrm{Hom}_\varphi(E,F)$ resp. $\mathrm{Hom}_\varphi(E,F)^0$.

(2.4.2) In order to clarify the relation between graded modules and modules, we consider a graded module E over a $\mathbb{J}$-commutative ring A. If $\tau^{\langle\nu\rangle}\colon \mathbb{J} \to \mathbb{J}^{\langle\nu\rangle}$ for $\nu = 0,1$ denotes the natural morphism, then $A^{\langle\nu\rangle} := (\tau^{\langle\nu\rangle})^*(A)$ is $\mathbb{J}^{\langle\nu\rangle}$-commutative with $\tau_*^{\langle\nu\rangle}(A^{\langle\nu\rangle}) = A$. If we set $E^{\langle\nu\rangle(j,n)} := E_n^j$ for (j,n) in $\mathbb{J}^{\langle\nu\rangle}$, then

$$E^{\langle\nu\rangle} := \coprod_{(j,n)\in\mathbb{J}^{\langle\nu\rangle}} E^{\langle\nu\rangle(j,n)}$$

is a module over $A^{\langle\nu\rangle}$ in a natural way, and the correspondence sending E to $((A^{\langle\nu\rangle}, E^{\langle\nu\rangle}), A)$ gives rise to equivalences

$$(\mathrm{M}^{\mathrm{gr}\,\prime}) \longrightarrow (\mathrm{M}') \times_{({}_c\mathrm{R})} (({}_c\mathrm{R}), \tau^{\langle\nu\rangle *}),$$
$$(\mathrm{M}^{\mathrm{gr}}) \longrightarrow (\mathrm{M}) \times_{({}_c\mathrm{R})} (({}_c\mathrm{R}), \tau^{\langle\nu\rangle *})$$

of categories over ($_c$R). - Let B be an algebra over A, endowed with the structure of a graded A-module. B is called a *graded algebra* over A, if $B_m \cdot B_n$ is contained in B_{m+n} for all m, n. Then $B^{\langle\nu\rangle}$ is an algebra over $A^{\langle\nu\rangle}$ in a natural way for $\nu = 0, 1$. We assume now that B is associative and unital. We say that B is *commutative* resp. *alternating*, if $B^{\langle 0\rangle}$ resp. $B^{\langle 1\rangle}$ is commutative. It's easy to express these conditions in a more explicit way. The graded algebras resp. graded associative unital algebras over variable commutative rings form a fibred and cofibred category (Alg^{gr}) resp. (Al^{gr}) over ($_c$R). By ($_c$Al^{gr}) resp. ($_a$Al^{gr}) we denote the full subcategory of (Al) consisting of all commutative resp. alternating graded algebras.

(2.5.1) (Universal algebras) Let A be a $\mathbb{J}$-commutative ring and E an A-module. Then there exists a unique sequence $\mathrm{F}^n(E) = \mathrm{F}^n_A(E)$, $n \in \mathbb{N}^*$, of modules over A such that $\mathrm{F}^1(E) = E$ and $\mathrm{F}^n(E) = \coprod_{p=1}^{n-1} \mathrm{F}^p(E) \otimes_A \mathrm{F}^{n-p}(E)$ for $n \geq 2$. For any pair p, q of integers in $\mathbb{N}^*$ we have a natural A-bilinear map of degree zero from $\mathrm{F}^p(E) \times \mathrm{F}^q(E)$ into $\mathrm{F}^{p+q}(E)$. These mappings make the direct sum

$$\mathrm{F}(E) = \mathrm{F}_A(E) := \coprod_{n \in \mathbb{N}^*} \mathrm{F}^n_A(E)$$

into a graded A-algebra in (Alg^{gr}), called the *free algebra* of E. For an integer n in $\mathbb{N}$ the tensor product of n copies of E is denoted as usual by $\mathrm{T}^n(E) = \mathrm{T}^n_A(E)$. The natural A-bilinear maps from $\mathrm{T}^p(E) \times \mathrm{T}^q(E)$ into $\mathrm{T}^{p+q}(E)$ endow the direct sum

$$\mathrm{T}(E) = \mathrm{T}_A(E) := \coprod_{n \in \mathbb{N}} \mathrm{T}^n_A(E)$$

with the structure of a graded A-algebra in (Al^{gr}), which is called the *tensor algebra* of E. Let J_E resp. K_E denote the two-sided graded ideal of $\mathrm{T}(E)$ generated by the elements $x \otimes y - e^{\mathbb{J}}(x,y) \cdot y \otimes x$ resp. $x \otimes y + e^{\mathbb{J}}(x,y) \cdot y \otimes x$, where x, y run in E, and by the elements $x \otimes x, x \in E$ odd resp. even. The quotient

$$\mathrm{S}(E) = \mathrm{S}_A(E) := \mathrm{T}_A(E)/J_E \quad \text{resp.} \quad \Lambda(E) = \Lambda_A(E) := \mathrm{T}_A(E)/K_E$$

is a commutative resp. alternating graded A-algebra in ($_c$Alg^{gr}) resp. ($_a$Al^{gr}), which is called the *symmetric algebra* resp. *exterior algebra* of E. For an integer n in $\mathbb{N}$ the image of $\mathrm{T}^n(E)$ under the natural surjection p_E resp. q_E from $\mathrm{T}(E)$ onto $\mathrm{S}(E)$ resp. $\Lambda(E)$ is an A-submodule called the n-th *symmetric power* resp. *exterior power* of E

and denoted by $\mathrm{S}^n(E) = \mathrm{S}^n_A(E)$ resp. $\Lambda^n(E) = \Lambda^n_A(E)$. We have natural injections from E into $\mathrm{F}_A(E)$ resp. $\mathrm{T}_A(E)$ resp. $\mathrm{S}_A(E)$ resp. $\Lambda_A(E)$ of $\mathbb{J}$-degree zero which are all denoted by i_E. Then $i_E({}^{\mathrm{h}}E)$ generates $\mathrm{F}_A(E)$ as an A-algebra, and analogous for the other algebras. The algebras $\mathrm{F}_A(E), \mathrm{T}_A(E), \mathrm{S}_A(E), \Lambda_A(E)$ enjoy the usual universal properties and thus define functors

$$\begin{aligned}\mathrm{F}\colon (\mathrm{M}) \longrightarrow (\mathrm{Alg}^{\mathrm{gr}}) \quad &, \quad \mathrm{T}\colon (\mathrm{M}) \longrightarrow (\mathrm{Al}^{\mathrm{gr}}),\\ \mathrm{S}\colon (\mathrm{M}) \longrightarrow ({}_{\mathrm{c}}\mathrm{Al}^{\mathrm{gr}}) \quad &, \quad \Lambda\colon (\mathrm{M}) \longrightarrow ({}_{\mathrm{a}}\mathrm{Al}^{\mathrm{gr}})\end{aligned}$$

of cofibred categories over $({}_{\mathrm{c}}\mathrm{R})$.

(2.6.1) (Linearly topologized rings and modules) Let E be a group of type $\mathbb{J}$, endowed with a topology compatible with the group structure, and let $\mathcal{U}(E)$ denote the set consisting of all open subgroups (of type $\mathbb{J}$) of E. We say that E is *linearly topologized* (l.t.), if $\mathcal{U}(E)$ is a neighbourhood basis of zero admitting a countable cofinal subset. Then the inverse limit

$$\hat{E} := \varprojlim_{U \in \mathcal{U}(E)} E/U$$

in $(\mathrm{G}^{\mathbb{J}})$ is a l.t. group in a natural way, called the *completion* of E. Note that $\hat{E}$ is a subgroup of the completion of E in the usual sense and that the closures $\bar{U}, U \in \mathcal{U}(E)$, in $\hat{E}$ form a neighbourhood basis of zero in $\hat{E}$, contained in $\mathcal{U}(\hat{E})$. We say that E is *complete*, if the natural homomorphism from E into $\hat{E}$ is bijective. The complete l.t. groups over variable signed groups, endowed with the continuous homomorphisms in (G') resp. (G) as morphisms, form a cofibred resp. fibred and cofibred category $(\mathcal{G}')$ resp. $(\mathcal{G})$ over (SG) in a natural way.

(2.6.2) Let A be a ring of type $\mathbb{J}$ and simultaneously a l.t. group, and let $\mathcal{J}(A)$ denote the set of all open two-sided ideals (of type $\mathbb{J}$) of A. We say that A is a *linearly topologized ring*, if $\mathcal{J}(A)$ is a neighbourhood basis of zero. Then A is in particular a topological ring, and its completion $\hat{A}$ is a complete l.t. ring. The complete l.t. rings resp. complete l.t. commutative rings, endowed with the continuous ring homomorphisms in (R) as morphisms, form a fibred and cofibred category $(\mathcal{R})$ resp. $({}_{\mathrm{c}}\mathcal{R})$ over (SG). Further, for a fixed ring S in $({}_{\mathrm{c}}\mathcal{R})$, let $(\mathcal{R}_{\mathrm{S}})$ resp. $({}_{\mathrm{c}}\mathcal{R}_{\mathrm{S}})$ be the category of S-objects in $(\mathcal{R})$ resp. $({}_{\mathrm{c}}\mathcal{R})$.

(2.6.3) Similarly we define the notion of a l.t. module resp. a l.t. algebra over a commutative l.t. ring A. For a l.t. module E resp. algebra B over A, the completion

$\hat{E}$ resp. $\hat{B}$ is a complete l.t. module resp. algebra over $\hat{A}$. If $E_i, i \in I$, is a family of l.t. modules over A such that the index set I is finite and totally ordered, then the tensor product $\bigotimes_{i \in I, A} E_i$ is a l.t. module over A in a natural way. Hence its completion

$$\widehat{\bigotimes_{i \in I}}{}_A E_i$$

is a complete l.t. module over A, called the *complete tensor product* of the $E_i, i \in I$. Thus, for a single l.t. module E over A, the A-modules $\mathrm{F}^n_A(E)$ resp. $\mathrm{T}^n_A(E), \mathrm{S}^n_A(E)$, $\Lambda^n_A(E)$, $n \in \mathbb{N}^*$ resp. $n \in \mathbb{N}$, are all l.t.. Their respective completions are denoted by $\tilde{\mathrm{F}}^n_A(E), \tilde{\mathrm{T}}^n_A(E), \tilde{\mathrm{S}}^n_A(E)$, $\tilde{\Lambda}^n_A(E)$. - The complete l.t. modules over variable rings in $({}_c\mathcal{R})$, endowed with the continuous homomorphisms in (M′) resp. (M) as morphisms, form a category resp. fibred category $(\mathcal{M}')$ resp. $(\mathcal{M})$ over $({}_c\mathcal{R})$. For a fixed ring A in $({}_c\mathcal{R})$, the respective fibres $(\mathcal{M}'_A), (\mathcal{M}_A)$ are additive, and $(\mathcal{M}_A)$ admits kernels and cokernels.

Similarly the complete l.t. algebras resp. complete l.t. associative unital algebras form a fibred and cofibred category $(\mathcal{A}lg)$ resp. $(\mathcal{A}l)$ over $({}_c\mathcal{R})$. Let $({}_c\mathcal{A}l)$ denote the full subcategory of $(\mathcal{A}l)$ consisting of all commutative algebras. For a fixed ring A in $({}_c\mathcal{R})$ let $(\mathcal{A}lg_A), (\mathcal{A}l_A)$ and $({}_c\mathcal{A}l_A)$ be the respective fibres in A.

(2.6.4) Let $E = \coprod_{n \in \mathbb{Z}} E_n$ be a graded module over a ring A in $({}_cR)$. Then the descending filtration $\mathrm{F}. = \mathrm{F}.(E)$ of E given by $\mathrm{F}_r := \coprod_{n \geq r} E_n$ for r in $\mathbb{Z}$ provides E with the structure of a l.t. module over A. Here A is endowed with the discrete topology. The associated topology on E is called *canonical.* If $E_n = 0$ for $n \ll 0$, then the completion of E is given by $\hat{E} = \prod_{n \in \mathbb{Z}} E_n$.

3. Cogebras and bigebras

(3.1.1) (Formal cogebras and cogebras) Let A be a ring in $({}_c\mathcal{R})$ of type $\mathbb{J}$. A (coassociative counital) *formal cogebra* over A is a triple $M = (M, c_M, \gamma_M)$ consisting of an A-module M in $(\mathcal{M})$ and A-homomorphisms

$$c_M : M \longrightarrow M \hat{\otimes}_A M \quad , \quad \gamma_M : M \longrightarrow A$$

in $(\mathcal{M}_A)$ such that $(id_M \hat{\otimes} c_M) \circ c_M = (c_M \hat{\otimes} id_M) \circ c_M$ and $(\gamma_M \hat{\otimes} id_M) \circ c_M = id_M$ $= (id_M \hat{\otimes} \gamma_M) \circ c_M$. Let M_+ denote the kernel of γ_M. If M is discrete, we call M a *cogebra* over A. We say that M is $\mathbb{J}$-*cocommutative* or simply *cocommutative*, if $c_M(M)$

is contained in the kernel of the natural surjection from $M\hat{\otimes}_A M$ onto $\tilde{\Lambda}^2_A(M)$. Then $\sigma_M \circ c_M = c_M$, the converse being true if 2 is a unit in A. If $M_i, i \in I$, is a family of formal cogebras over A such that I is finite and totally ordered, then the complete tensor product $\widehat{\bigotimes}_{i\in I,A} M_i$ is a formal cogebra over A in a natural way. The formal cogebras resp. cogebras over variable rings in $({}_c\mathcal{R})$ resp. $({}_c\mathrm{R})$ form a cofibred category $(\mathcal{C})$ resp. (C) over $({}_c\mathcal{R})$ resp. $({}_c\mathrm{R})$. By $({}^c\mathcal{C})$ resp. $({}^c\mathrm{C})$ we denote the full subcategory of $(\mathcal{C})$ resp. (C) consisting of all cocommutative formal cogebras resp. cogebras.

(3.1.2) Let $\varphi: A \to A'$ and $\varphi': A' \to A''$ be homomorphisms in $({}_c\mathcal{R})$, M a formal cogebra over A, and let F, G resp. H be modules in $(\mathcal{M}_{A'})$ resp. $(\mathcal{M}_{A''})$. Further suppose given a homomorphism μ of $\mathrm{Hom}_{\varphi'}(F\hat{\otimes}_{A'}G, H)^0$. For an element u resp. v of $\mathrm{Hom}_\varphi(M, F)$ resp. $\mathrm{Hom}_\varphi(M, G)$ we call

$$u \cdot v = u \cdot_\mu v := \mu \circ (u\hat{\otimes}_\varphi v) \circ c_M \in \mathrm{Hom}_{\varphi'\circ\varphi}(M, H)$$

the *product* of u and v with respect to μ. Let now C be an algebra over A' in $(\mathcal{A}lg)$. Then $\mathrm{Hom}_\varphi(M, C)$ is an A'-algebra with respect to the product $(u, v) \mapsto u \cdot v$. If C is associative and unital with structure homomorphism $\varphi_{\mathrm{C}}: A' \to C$, then $\mathrm{Hom}_\varphi(M, C)$ enjoys the same properties, the identity element being $\varphi_{\mathrm{C}} \circ \varphi \circ \gamma_M$. If M is cocommutative and C is an algebra in $({}_c\mathcal{A}l)$, then $\mathrm{Hom}_\varphi(M, C)$ is commutative.

(3.1.3) Let A be a ring in $({}_c\mathrm{R})$ and $M = \coprod_{n\in\mathbb{Z}} M_n$ a graded A-module, endowed with the structure of a cogebra over A. Then M is called a *graded cogebra* over A, if $c_M(M_n)$ is contained in $\coprod_{p+q=n} M_p \otimes_A M_q$ for every n and $\gamma_M(M_n) = 0$ if $n \neq 0$. In this case $M^{\langle\nu\rangle}$ is a cogebra over $A^{\langle\nu\rangle}$ in a natural way for $\nu = 0, 1$. If M is positive, then c_M and γ_M are continuous with respect to the canonical topology. Hence the completion $\hat{M} = \prod_{n\in\mathbb{Z}} M_n$ of M with respect to this topology is a formal cogebra over A. In an obvious way one defines graded coideals and morphisms of graded cogebras.

Let $\varphi: A \to A'$ be a homomorphism in $({}_c\mathrm{R})$ and E an A'-module in $(\mathrm{M}_{A'})$. If u is a homomorphism in $\mathrm{Hom}_\varphi(M, E)$, then u has an unique representation

$$u = \sum_{n\in\mathbb{Z}} u_n$$

with homomorphisms u_n in $\mathrm{Hom}_\varphi(M_n, E)$. If M is positive, the infimum resp. the supremum of all n such that $u_n \neq 0$ is called the *order* resp. the *degree* of u and denoted by $\mathrm{ord}\,(u)$ resp. $\deg(u)$.

(3.2.1) (Pointwise topologically nilpotent homomorphisms) We fix a homomorphism $\varphi: A \to A'$ in $({}_c\mathcal{R})$ over a morphism $\rho: \mathbb{J} \to \mathbb{J}'$ in (SG), a formal cogebra M over A and an algebra C over A' in $(\mathcal{A}l)$. In the following we identify the elements of M with their canonical images in $\rho^*(M)$. A homomorphism u of $\mathrm{Hom}_\varphi(M,C)^0$ is called *power-bounded*, if for every neighbourhood V of zero in C there exists a neighbourhood U of zero in M such that $u^n(U)$ is contained in V for all $n \in \mathbb{N}$.

Further u is called *pointwise topologically nilpotent* (p.t.n.), if u is power-bounded and if for every element x of M the sequence $(u^n(x))_{n\in\mathbb{N}}$ converges to zero in C. We say that u is *pointwise topologically unipotent* (p.t.u.) if $1_{\mathrm{Hom}(M,C)} - u$ is p.t.n.. - Now let us suppose that A' is a $\mathbb{Q}$-algebra. For a p.t.n. resp. p.t.u. homomorphism u resp. v in $\mathrm{Hom}_\varphi(M,C)^0$

$$\exp(u) := \sum_{q=0}^{\infty} \frac{1}{q!}\, u^q \in \mathrm{Hom}_\varphi(M,C)^0,$$

$$\log(v) := \sum_{q=1}^{\infty} \frac{-1}{q}\, (1-v)^q \in \mathrm{Hom}_\varphi(M,C)^0$$

are well defined and moreover p.t.u. resp. p.t.n. with $\exp(\log(v)) = v$ and $\log(\exp(u)) = u$. Clearly $\exp(0) = 1$ and $\log(1) = 0$. If u' resp. v' is a second p.t.n. resp. p.t.u. homomorphism in $\mathrm{Hom}_\varphi(M,C)^0$ such that $u \cdot u' = u' \cdot u$ resp. $v \cdot v' = v' \cdot v$, then $\exp(u+u') = \exp(u) \cdot \exp(u')$ resp. $\log(v \cdot v') = \log(v) + \log(v')$.

(3.3.1) (Formal bigebras and bigebras) Let A be a ring in $({}_c\mathcal{R})$ of type $\mathbb{J}$ and B an algebra over A in $(\mathcal{A}l)$ endowed with the structure of a formal cogebra. Then B is called a *formal bigebra* over A, if both c_B and γ_B are homomorphisms of algebras over A. If B is moreover discrete, then B is called a *bigebra* over A. We say that B is $\mathbb{J}$-*bicommutative* or simply *bicommutative*, if it is simultaneously commutative and cocommutative.

In the usual way we define the notion of a *biideal* of a formal bigebra B. An element x of B is called *primitive*, if $c_B(x) = x \otimes 1 + 1 \otimes x$. The primitive elements form a (closed) submodule $\mathrm{P}(B)$ of B, and the closure of the two-sided ideal generated by $\mathrm{P}(B)$ is a biideal I_B of B, contained in B_+. If $B_i, i \in I$, is a family of formal bigebras over A such that I is finite and totally ordered, then $\widehat{\bigotimes}_{i\in I,A} B_i$ is a formal bigebra over A in a natural way. The formal bigebras resp. bigebras over variable rings in $({}_c\mathcal{R})$ resp. $({}_c\mathrm{R})$ form a cofibred category $(\mathcal{B})$ resp. (B) over $({}_c\mathcal{R})$ resp. $({}_c\mathrm{R})$, and the correspondence sending B to $\mathrm{P}(B)$ defines a covariant functor $\mathrm{P}: (\mathcal{B}) \to (\mathcal{M})$ of categories over $({}_{\mathrm{c}}\mathcal{R})$.

(3.3.2) Let now $\varphi: A \to A'$ be a ring homomorphism in $({}_{\mathrm{c}}\mathcal{R}_{\mathbb{Q}})$, M a cocommutative formal cogebra over A, B a formal bigebra over A' and $u \in \mathrm{Hom}_{\varphi}(M, B)^0$ a p.t.n. homomorphism. With these notations, it's easy to check the following proposition.

(3.3.3) Proposition. *$\exp(u): M \to B$ is a morphism of formal cogebras over φ if and only if $u(M)$ is contained in $\mathrm{P}(B)$.*

(3.3.4) Let $B = \coprod_{n \in \mathbb{Z}} B_n$ be a graded algebra over a ring A in $({}_{\mathrm{c}}\mathrm{R})$, endowed with the structure of a bigebra over A. Then B is called a *graded bigebra* over A, if both c_B and γ_B stabilize the $\mathbb{Z}$-gradations . In this case $\mathrm{P}(B)$ is a graded submodule of B, and $B^{\langle 0 \rangle}$ is a bigebra over $A^{\langle 0 \rangle}$ in a natural way. If B is positive, the completion $\hat{B} = \prod_{n \in \mathbb{Z}} B_n$ of B with respect to the canonical topology is a formal bigebra over A. In an obvious way one defines graded biideals and morphisms of graded bigebras.

(3.3.5) Example. Suppose given a module E over a ring A in $({}_{\mathrm{c}}\mathrm{R})$ and let $c: \mathrm{T}_A(E) \to \mathrm{T}_A(E) \otimes_A \mathrm{T}_A(E)$ resp. $\gamma: \mathrm{T}_A(E) \to A$ be the unique homomorphism of A-algebras such that $c(x) = x \otimes 1 + 1 \otimes x$ resp. $\gamma(x) = 0$ for x in E. Then $\mathrm{T}_A(E) =$ $= (\mathrm{T}_A(E), c, \gamma)$ is a cocommutative graded bigebra over A with $\mathrm{T}_A(E)_+ = I_{\mathrm{T}(E)}$, and its completion $\hat{\mathrm{T}}_A(E) = \prod_{n \in \mathbb{N}} \mathrm{T}^n_A(E)$ with respect to the canonical topology is a cocommutative formal bigebra over A with $\hat{\mathrm{T}}_A(E)_+ = I_{\hat{\mathrm{T}}(E)}$.

Since the kernel of the natural surjection from $\mathrm{T}_A(E)$ onto $\mathrm{S}_A(E)$ is a graded biideal, we infer that $\mathrm{S}_A(E)$ is a bicommutative graded bigebra over A with $\mathrm{S}_A(E)_+$ $= I_{\mathrm{S}(E)}$. Moreover its completion $\hat{\mathrm{S}}_A(E) = \prod_{n \in \mathbb{N}} \mathrm{S}^n_A(E)$ with respect to the canonical topology is a bicommutative formal bigebra over A with $\hat{\mathrm{S}}_A(E)_+ = I_{\hat{\mathrm{S}}(E)}$. If $\mathrm{S}_A(E)$, considered as a $\mathbb{Z}$-module, has no torsion, we have $\mathrm{P}(\hat{\mathrm{S}}_A(E)) = \mathrm{P}(\mathrm{S}_A(E)) = E$. If pr_n denotes the natural projection from $\mathrm{S}_A(E)$ onto $\mathrm{S}^n_A(E)$, we have $pr_n \cdot pr_m$ $= \binom{n+m}{n} pr_{n+m}$ in the algebra $\mathrm{Hom}_A(\mathrm{S}_A(E), \mathrm{S}_A(E))$. In particular we see that pr_1 is p.t.n.. If A is a $\mathbb{Q}$-algebra, then $\exp(pr_1) = id_{\mathrm{S}(E)}$.

(3.4.1) (Bigebras of Lie type) Let B be a cocommutative formal bigebra over a ring A in $({}_{\mathrm{c}}\mathcal{R}_{\mathbb{Q}})$. We say that B is of *Lie type*, if id_B is p.t.u.. Then $p_B := \log(id_B)$ is a p.t.n. endomorphism of B with $\exp(p_B) = id_B$. From (3.3.3) we can deduce that p_B is a projector with image $\mathrm{P}(B)$ and $p_B(1) = 0$. For instance, if A is discrete and E is an A-module, then $\mathrm{T}_A(E), \mathrm{S}_A(E), \hat{\mathrm{T}}_A(E)$ and $\hat{\mathrm{S}}_A(E)$ are of Lie type, and $p_{\hat{\mathrm{S}}(E)} = pr_1$ is the natural projection from $\hat{\mathrm{S}}_A(E)$ onto E.

(3.5.1) (The Φ-operation) Let $\varphi: A \to A'$ be a homomorphism in $({}_c\mathcal{R}_{\mathbb{Q}})$, M a cocommutative formal cogebra over A, B a formal bigebra of Lie type over A' and $u \in \mathrm{Hom}_{\varphi}(M, B)^0$ a p.t.n. homomorphism such that $u(M)$ is contained in $\mathrm{P}(B)$. With these notations we can derive from (3.3.3) the following result, which is the basis for our foundations of infinite dimensional superanalysis.

(3.5.2) Theorem. *There exists a unique morphism $\Phi(u): M \to B$ of formal cogebras over φ such that $p_B \circ \Phi(u) = u$, namely $\Phi(u) = \exp(u)$.*

(3.5.3) We call $\Phi(u)$ the *cogebra morphism induced by u*. Clearly $\Phi(0) = 1$. If u and u' are two commuting p.t.n homomorphisms in $\mathrm{Hom}_{\varphi}(M, B)^0$ such that $u(M)$ and $u'(M)$ are contained in $\mathrm{P}(B)$, then $\Phi(u + u') = \Phi(u) \cdot \Phi(u')$. Let $\varphi': A' \to A''$ be another homomorphism in $({}_c\mathcal{R}_{\mathbb{Q}})$ and C a formal bigebra of Lie type over A''. If u resp. v is a p.t.n. homomorphism in $\mathrm{Hom}_{\varphi}(M, B)^0$ resp. $\mathrm{Hom}_{\varphi'}(B, C)^0$ such that $u(M)$ resp. $v(B)$ is contained in $\mathrm{P}(B)$ resp. $\mathrm{P}(C)$, then $v \circ \Phi(u)$ is p.t.n. and $\Phi(v \circ \Phi(u)) = \Phi(v) \circ \Phi(u)$. For an arbitrary homomorphism v of $\mathrm{Hom}_{\varphi'}(B, C)$ (not necessarily homogeneous or p.t.n.) we call

$$v \,\square\, u := v \circ \Phi(u) \in \mathrm{Hom}_{\varphi' \circ \varphi}(M, C)$$

the *composition of u with v*. This composition is associative in an obvious sense. Moreover the Φ-operation is compatible with complete tensor products; the precise fomulation is left to the reader.

(3.5.4) Let $\varphi: A \to A'$ be a homomorphism in $({}_c\mathcal{R}_{\mathbb{Q}})$ such that A is discrete, $M = \coprod_{n \in \mathbb{Z}} M_n$ a positive graded cocommutative cogebra over A and B a formal bigebra of Lie type over A'. Then, for a p.t.n. homomorphism $u = \sum_{n \in \mathbb{N}} u_n$ of $\mathrm{Hom}_{\varphi}(M, B)^0$ such that $u(M)$ is contained in $\mathrm{P}(B)$, we have

$$\Phi(u)_n = \sum_{q \in \mathbb{N}} \frac{1}{q!} \sum_{\substack{\nu \in \mathbb{N}^q \\ |\nu| = n}} u_{\nu_1} \cdots u_{\nu_q} \tag{3.5.4.1}$$

in $\mathrm{Hom}_{\varphi}(M_n, B)^0$ for every integer n in $\mathbb{N}$. If $u_0 = 0$, then $\Phi(u)_1 = u_1$. Suppose now that the topology of B is coarser than the I_B-adic topology and that $\mathrm{P}(B)$ is discrete.

Then $\Phi(u)$ is continuous with respect to the canonical topology of M if and only if $\deg(u)$ is finite. In this case, if $\hat{u}, \widehat{\Phi(u)}: \hat{M} \to B$ denotes the completion of $u, \Phi(u)$ with respect to the canonical topology, then $\hat{u}$ is p.t.n. with $\Phi(\hat{u}) = \widehat{\Phi(u)}$.

Let us apply this specifically to the case $B = \hat{\mathrm{S}}_{A'}(F)$ with an A'-module F, the ring A' being discrete, too. Assume moreover that u has no constant term, i.e. $u_0 = 0$. Then $(pr_q \circ \Phi(u))|M_r = 0$ for $q > r$, so that u is p.t.n. with respect to the *discrete* topology of $\mathrm{S}_{A'}(F)$. Thus we may regard $\Phi(u)$ as a map from M into $\mathrm{S}_{A'}(F)$.

(3.5.5) Examples. (1) Suppose that $M = \mathrm{S}_A(E)$ with an A-module E, and let u be a map of $\mathrm{Hom}_{\varphi}(E, F)^0$, considered as a submodule of $\mathrm{Hom}_{\varphi}(\mathrm{S}_A(E), F)^0$. Then $\Phi(u) = \mathrm{S}_{\varphi}(u)$.

(2) Let $a_{E|A}: E \times E \to E$ be the addition of an A-module E. Then $\Phi(a_{E|A})$, regarded as a homomorphism from $\mathrm{S}_A(E) \otimes_A \mathrm{S}_A(E)$ into $\mathrm{S}_A(E)$, is the multiplication $\mu_{\mathrm{S}(E)}$ of $\mathrm{S}_A(E)$.

(3.5.6) In an obvious manner all other notions known in the theory of cogebras and bigebra (as comodules, coderivations, etc., cf. [Ab], [Bo₁], [Sw]) extend to the $\mathbb{J}$-graded context. Since we don't need these notions in the following we omit their explicit description.

4. Formal maps and differential calculus

(4.1.1) (Formal maps and polynomials) Let $\varphi: A \to A'$ be a homomorphism in $({}_{\mathrm{c}}\mathrm{R}_{\mathbb{Q}})$ and E resp. F a module over A resp. A'. For an integer n in $\mathbb{N}$ we set $\mathrm{For}_{\varphi}(E, F)_n := \mathrm{Hom}_{\varphi}(\mathrm{S}^n_A(E), F)$. The elements of the A'-module

$$\mathrm{For}_{\varphi}(E, F) := \mathrm{Hom}_{\varphi}(\mathrm{S}_A(E), F) = \prod_{n \in \mathbb{N}} \mathrm{For}_{\varphi}(E, F)_n$$

resp.

$$\mathrm{Pol}_{\varphi}(E, F) := \mathrm{Hom}_{\varphi}(\mathrm{S}_A(E), F)^{\mathrm{gr}} = \coprod_{n \in \mathbb{N}} \mathrm{For}_{\varphi}(E, F)_n$$

are called *formal maps* resp. *polynomials from E into F (over φ)*. If $\varphi = id_A$, we denote these modules by $\mathrm{For}_A(E, F)_n$, $\mathrm{For}_A(E, F)$, $\mathrm{Pol}_A(E, F)$ respectively. A formal map u from E into F has a unique representation $u = \sum_{n \in \mathbb{N}} u_n$ as an (infinite) power series with homogeneous polynomials u_n in $\mathrm{For}_{\varphi}(E, F)_n$. The constant term u_0

resp. the tangent map u_1 of u is an element of $\mathrm{Hom}_\varphi(A,F) = F$ resp. $\mathrm{Hom}_\varphi(E,F)$. By $\mathrm{For}_\varphi(E,F)_+$ resp. $\mathrm{Pol}_\varphi(E,F)_+$ we denote the submodule of $\mathrm{For}_\varphi(E,F)$ resp. $\mathrm{Pol}_\varphi(E,F)$ which consists of all u with $u_0 = 0$. For a formal map u in $\mathrm{For}_\varphi(E,F)^0$, the induced cogebra morphism $\Phi(u)$ is the formal map of $\mathrm{For}_\varphi(E, \hat{\mathrm{S}}_{A},(F))^0$ given by the formula

$$\Phi(u) = \sum_{q\in\mathbb{N}} \frac{1}{q!} u^q = \sum_{q\in\mathbb{N}} \frac{1}{q!} \sum_{\nu\in\mathbb{N}^q} u_{\nu_1}\cdots u_{\nu_q}. \tag{4.1.1.1}$$

Let now $\varphi'\colon A' \to A''$ be a second homomorphism in $({}_{\mathrm{c}}\mathrm{R}_{\mathbb{Q}})$, G a module over A'' and $v = \sum_{n\in\mathbb{N}} v_n$ a formal map of $\mathrm{For}_{\varphi'}(F,G)$. We assume that either u has no constant term or v is a polynomial. Then the composition $v \,\square\, u$ of u with v is a well defined formal map of $\mathrm{For}_{\varphi'\circ\varphi}(E,G)$, c.f. (3.5.3). If $u_0 = 0$ resp. if v is a polynomial, then $v \,\square\, u = v \circ \Phi(u)$ resp. $v \,\square\, u = \hat{v} \circ \Phi(u)$. Explicitly we have

$$v \,\square\, u = \sum_{q\in\mathbb{N}} \frac{1}{q!} v_q \circ u^q, \tag{4.1.1.2}$$

$$(v \,\square\, u)_n = \sum_{q\in\mathbb{N}} \sum_{\substack{\nu\in\mathbb{N}^q \\ |\nu|=n}} \frac{1}{q!} v_q \circ (u_{\nu_1}\cdots u_{\nu_q}) \tag{4.1.1.3}$$

for every integer n in $\mathbb{N}$. If $E_i, i \in I$, resp. $F_i, i \in I$, is a finite family of modules over A resp. A' and $u_{(i)} \in \mathrm{For}_\varphi(E_i,F_i), i \in I$, are formal maps, then $\prod_{i\in I} u_{(i)}:$ $= (u_{(i)} \,\square\, pr_i)_{i\in I}$ is a formal map in $\mathrm{For}_\varphi(\prod_{i\in I} E_i, \prod_{i\in I} F_i)$ called the *product* of the $u_{(i)}, i \in I$; here pr_j is the natural projection from $\prod_{i\in I} E_i$ onto E_j. Let

(For) resp. (Pol) resp. $(\mathrm{Pol})_+$

denote the following category. The objects of these categories are pairs (A,E) consisting of a ring A in $({}_{\mathrm{c}}\mathrm{R}_{\mathbb{Q}})$ and a module E over A. If (A,E) and (A',F) are two such pairs, then a morphism from (A,E) into (A',F) in (For) resp. (Pol) resp. $(\mathrm{Pol})_+$ is a pair (φ,u) consisting of a homomorphism φ from A into A' in $({}_{\mathrm{c}}\mathrm{R}_{\mathbb{Q}})$ and a formal map $u \in \mathrm{For}_\varphi(E,F)^0_+$ resp. a polynomial $u \in \mathrm{Pol}_\varphi(E,F)^0$ resp. a polynomial $u \in \mathrm{Pol}_\varphi(E,F)^0_+$.

If (A'',G) is a third pair and if (φ,u) resp. (φ',v) is a morphism from (A,E) into (A',E) resp. from (A',F) into (A'',G), then their composition is given by

$(\varphi' \circ \varphi, v \,\square\, u)$. The natural functor sending (A, E) to A makes (For), (Pol) and (Pol)$_+$ into a fibred and cofibred category over $({}_c\mathrm{R}_{\mathbb{Q}})$. Note that (Pol) is not a subcategory of (For), whereas (Pol)$_+$ is a subcategory of (Pol) and (For) simultaneously. For a fixed $\mathbb{J}$-commutative ring A in $({}_c\mathrm{R}_{\mathbb{Q}})$, the fibre $(\mathrm{For}_A) = (\mathrm{For}^{\mathbb{J}}_A)$ resp. $(\mathrm{Pol}_A) = (\mathrm{Pol}^{\mathbb{J}}_A), (\mathrm{Pol}_A)_+ = (\mathrm{Pol}^{\mathbb{J}}_A)_+$ of (For) resp. (Pol), (Pol)$_+$ in A admits arbitrary products resp. finite products. Using the Banach fixed point theorem applied to the ultrametric on $\mathrm{For}_A(E, F)^0_+$ given by the order of formal maps (cf. (5.7.2), (6)), we can prove the following result.

(4.1.2) Proposition (Inverse function theorem). *Suppose given two A-modules* E, F *and a formal map* $u = \sum_{n \geq 1} u_n$ *in* $\mathrm{For}_A(E, F)^0_+$ *and let* $\tau \in \mathrm{Hom}_A(F, E)^0$ *be an A-linear map such that* $u_1 \circ \tau = id_F$. *Then there exists a formal map* $v = \sum_{n \geq 1} v_n$ *in* $\mathrm{For}_A(F, E)^0_+$ *with* $v_1 = \tau$ *and* $u \,\square\, v = id_F$.

(4.2.1) (Taylor series and derivatives of formal maps) Let T, E, F be modules over a ring A in $({}_c\mathrm{R}_{\mathbb{Q}})$. Then we have a natural A-isomorphism

$$h\colon \mathrm{For}_A(T, \mathrm{For}_A(E, F)) \longrightarrow \mathrm{For}_A(T \times E, F)$$

of degree zero sending a formal map w to the formal map $\hat{w}$ given by $\hat{w}(t \cdot e) = w(t)(e)$ for elements t resp. e of $\mathrm{S}_A(T)$ resp. $\mathrm{S}_A(E)$. Since $\mathrm{For}_A(T, \mathrm{For}_A(E, F))$ identifies canonically with the product of the A-modules $\mathrm{For}_A(T, \mathrm{For}_A(E, F)_l), l \in \mathbb{N}$, we infer that every formal map v in $\mathrm{For}_A(T \times E, F)$ admits a unique representation

$$v = \sum_{l \in \mathbb{N}} \hat{v}^{(l)} \tag{4.2.1.1}$$

with formal maps $v^{(l)}$ in $\mathrm{For}_A(T, \mathrm{For}_A(E, F)_l)$, called the *Hartogs development* of v. Explicitly we have $v^{(l)}(t)(e) = v(t \cdot e)$ for t resp. e in $\mathrm{S}_A(T)$ resp. $\mathrm{S}^l_A(E)$. Now we specialize to the case $T = E$ and consider the addition $a_{E|A}\colon E \times E \to E$ of E. For an arbitrary formal map u in $\mathrm{For}_A(E, F)$ we call

$$\mathrm{Tay}\,(u) := u \,\square\, a_{E|A} \in \mathrm{For}_A(E \times E, F)$$

the *Taylor series* of u. By the previous remark there exists a unique family of formal maps $D^l(u)$ in $\mathrm{For}_A(E, \mathrm{For}_A(E, F)_l), l \in \mathbb{N}$, such that in $\mathrm{For}_A(E \times E, F)$ the formula

$$\text{Tay}(u) = \sum_{l \in \mathbb{N}} \widehat{D^l}(u) \tag{4.2.1.2}$$

holds. $D^l(u)$ is called the *l-th derivative* of u and is given by $(D^l(u)(e))(x) = u(e \cdot x)$ for elements e resp. x in $S_A(E)$ resp. $S^l_A(E)$. In particular we have $D^0(u) = u$ and $D^{l+m}(u)$ identifies canonically with $D^l(D^m(u))$. For the derivatives of a product of formal maps the Leibniz rule applies; the precise formulation is left to the reader.

(4.2.2) Let now E, F, G be three A-modules and u resp. v a formal map in $\text{For}_A(E, F)^0$ resp. $\text{For}_A(F, G)$ such that either $u_0 = 0$ or v is a polynomial. For integers $q, n \in \mathbb{N}$ let

$$\mu: \text{For}_A(F, G)_q \otimes_A \text{For}_A(E, S^q_A(F))_n \longrightarrow \text{For}_A(E, G)_n$$

denote the A-homomorphism of degree zero given by $\mu(\beta \otimes \alpha) = \beta \circ \alpha$. Then, if $\nu = (\nu_1, \ldots, \nu_q)$ is a tuple in $\mathbb{N}^q$ with $|\nu| = n$, the product $D^{\nu_1}(u) \cdots D^{\nu_q}(u)$ is a formal map of $\text{For}_A(E, \text{For}_A(E, S^q_A(F))_n)$, so that $(D^q(v) \,\square\, u) \cdot_\mu (D^{\nu_1}(u) \cdots D^{\nu_q}(u))$ is a well defined element.

(4.2.3) Proposition (Chain rule). *In* $\text{For}_A(E, \text{For}_A(E, G)_n)$ *we have the formula*

$$D^n(v \,\square\, u) = \sum_{q \in \mathbb{N}} \sum_{\substack{\nu \in \mathbb{N}^{*q} \\ |\nu| = n}} \frac{1}{q!} (D^q(v) \,\square\, u) \cdot_\mu (D^{\nu_1}(u) \cdots D^{\nu_q}(u)).$$

The *proof* results from a careful analysis of the correspondence h from (4.2.1) and a formula expressing the Taylor series $\text{Tay}(v \,\square\, u)$ of a composition of two formal maps in terms of the Taylor series $\text{Tay}(v)$, $\text{Tay}(u)$ of v, u respectively.

(4.3.1) (Elementary formal maps) Let B be an algebra over a ring A in $({}_c\text{R}_{\mathbb{Q}})$. Then $X_{B|A} := id_B$ is an element of $\text{For}_A(B, B)^0_1$. If we set $X_A := X_{A|A}$, then X^n_A is an A-basis of $\text{For}_A(A, A)_n$ for every integer n in $\mathbb{N}$, so that an arbitrary formal map u in $\text{For}_A(A, A)$ has a unique representation $u = \sum_{n \in \mathbb{N}} a_n X^n_A$ with elements a_n in A. There exists a unique homomorphism

$$\eta = \eta_{B|A}: \text{For}_A(A, A) \longrightarrow \text{For}_A(B, B) \tag{4.3.1.1}$$

of algebras over A with $\eta(\sum_{n \in \mathbb{N}} a_n X^n_A) = \sum_{n \in \mathbb{N}} a_n \cdot X^n_{B|A}$. If $\varphi: B \to C$ is a homomorphism of A-algebras and u a formal map in $\text{For}_A(A, A)$, we have $\varphi \circ \eta_{B|A}(u)$

$= \eta_{C|A}(u) \,\square\, \varphi$. By a careful direct calculation, we can verify the following assertion, which allows to extend many known formulas in elementary analysis to our more general situation.

(4.3.2) Proposition. *Let* u, v *be two formal maps in* $\mathrm{For}_A(A, A)$ *with* $g(u) = 0$ *such that either* $u_0 = 0$ *or* v *is a polynomial. Then* $\eta_{B|A}(v \,\square\, u) = \eta_{B|A}(v) \,\square\, \eta_{B|A}(u)$.

(4.3.3) Let $m = m_{B|A} \in \mathrm{Pol}_A(B \times B, B)^0_+$ denote the sum of the addition $a_{B|A}: B \times B \to B$ and the multiplication $\mu_{B|A}: B \otimes_A B \to B$ of B. The formal map

$$e_{B|A} := \sum_{n \in \mathbb{N}^*} \frac{1}{n!} X^n_{B|A}$$

resp.

$$l_{B|A} := \sum_{n \in \mathbb{N}^*} (-1)^{n-1} \frac{1}{n} X^n_{B|A}$$

of $\mathrm{For}_A(B, B)^0_+$ is called the *exponential map* resp. the *logarithm of* B *(over* A*)*. Their derivatives are given by $D(e_{B|A}) = (1 + e_{B|A}) \cdot id_B$ and $D(l_{B|A}) = (1 + X_{B|A})^{-1} \cdot id_B$. Using (4.3.2) and the chain rule, we infer $e_{B|A} \,\square\, l_{B|A} = id_B = l_{B|A} \,\square\, e_{B|A}$. Finally, if B is commutative, we have $m_{B|A} \,\square\, (e_{B|A} \times e_{B|A}) = e_{B|A} \,\square\, a_{B|A}$.

(4.4.1) (The multicase) Let $\varphi: A \to A'$ be a homomorphism in $({}_{\mathrm{c}}\mathrm{R}_{\mathbb{Q}})$, I a finite totally ordered set, $E_i, i \in I$, a family of A-modules with product $E := \prod_{i \in I} E_i$ and F a module over A'. Then we have a natural isomorphism of graded bigebras from $\bigotimes_{i \in I, A} \mathrm{S}_A(E_i)$ onto $\mathrm{S}_A(E)$. For an arbitrary tuple $n = (n_i)_{i \in I}$ in $\mathbb{N}^I$ let $\mathrm{S}^n_A(E)$ denote the image of $\bigotimes_{i \in I, A} \mathrm{S}^{n_i}_A(E_i)$ under this isomorphism. The elements of the A'-submodule

$$\mathrm{For}_\varphi(E, F)_n := \mathrm{Hom}_\varphi(\mathrm{S}^n_A(E), F)$$

of $\mathrm{For}_\varphi(E, F)$ are called *homogeneous polynomials of multidegree* n *from* E *into* F *(over* φ*)*. Since $\mathrm{For}_\varphi(E, F)_k$ is the direct sum of the submodules $\mathrm{For}_\varphi(E, F)_n$, $|n| = k$, we infer that any formal map u of $\mathrm{For}_\varphi(E, F)$ has a unique representation $u = \sum_{n \in \mathbb{N}^I} u_n$ with polynomials u_n in $\mathrm{For}_\varphi(E, F)_n$. There is an analogue of the formula (4.1.1.3) in the "multicase", the formulation of which is left to the reader. Also we can extend differential calculus to this slightly more general situation. In particular there is a multiversion of both the Leibniz and chain rule. Using partial derivatives, we can formulate and prove the implicit function theorem and the existence and uniqueness theorem for ordinary differential equations. Further (4.3.2) admits an extension to the multicase. A detailed treatment of all these things will be given in [BL_2].

(4.5.1) Remark. The symmetric algebra $S_A(E)$ of an A-module E is not suited to define formal maps over rings A not containing the rationals, in particular over rings with a positive characteristic. Instead one has to introduce divided powers and PD-algebras over signed groups, and to take the Γ-algebra $\Gamma_A(E)$ of E, which is in fact a PD-bigebra; see also [Bo_2] for these notions in the non graded context. Then it is a fundamental fact, that for any PD-algebra (B, I, γ) over A

$$(\operatorname{Hom}_A(\Gamma_A(E), B),\ \operatorname{Hom}_A(\Gamma_A(E), I))$$

carries a canonical PD-structure. Given an arbitrary A-module F, we set $\operatorname{For}_A(E, F) := \operatorname{Hom}_A(\Gamma_A(E), F)$. With this notion of a formal map, all results of this section extend to this more general situation. A detailed treatment will be given in [B_1].

5. Pseudonorms

(5.1.1) (Valued signed groups) Let $\mathbb{J} = (\mathbb{J}, \Lambda, \varepsilon, \pi)$ be a signed group and $U = U\langle\mathbb{J}\rangle$ the subgroup of $\Lambda^\times$ consisting of the elements $\varepsilon(j, k)$, where j, k run in $\mathbb{J}$. If $S = S\langle\mathbb{J}\rangle$ is the multiplicative subset of $\mathbb{Z}$, which consists of all integers n such that $n \cdot 1_\Lambda$ is a unit in Λ, then Λ is an algebra over $\mathbb{Z}_S$ in a natural way. By an *absolute value* for $\mathbb{J}$ we mean a map

$$|\cdot|: \Lambda \longrightarrow \mathbb{R}_+$$

which satisfies the following conditions: (1) $|0| = 0$ and $|1_\Lambda| = 1$. (2) $|\alpha + \beta| \leq |\alpha| + |\beta|, |\alpha \cdot \beta| \leq |\alpha|\,|\beta|$ and $|\alpha^n| = |\alpha|^n$ for all elements α, β in Λ and n in $\mathbb{N}$. (3) $|\varepsilon(j,k)^{-1}| = |\varepsilon(j,k)|^{-1}$ for all j, k in $\mathbb{J}$. (4) $|\alpha \cdot \beta| = |\alpha|\,|\beta|$ for all α in $\mathbb{Z}_S \cdot 1_\Lambda$ and β in Λ. - Then (2) and (3) imply that $|\alpha\beta| = |\alpha|\,|\beta|$ for all elements α, β of U. We say that $|\cdot|$ is *ultrametric*, if moreover $|\alpha + \beta| \leq \sup\{|\alpha|, |\beta|\}$ for all α, β in Λ.

If Λ is reduced and non-zero, then the map $|\cdot|: \Lambda \to \mathbb{R}_+$ given by $|x| = 1$ if $x \neq 0$ and $|0| = 0$ is an ultrametric absolute value for $\mathbb{J}$, being called the *trivial* absolute value.

(5.1.2) By a *valued signed group* we understand a signed group $\mathbb{J} = (\mathbb{J}, \Lambda, \varepsilon, \pi)$ endowed with an absolute value $|\cdot|$. Let us call $\mathbb{J}$ *archimedean* resp. *ultrametric* resp. *trivial*, if $|\cdot|$ is not ultrametric resp. ultrametric resp. trivial. For any two elements j, k of $\mathbb{J}$ we set

$$n(j,k) := |\varepsilon(j,k)|^{1/2} \in \mathbb{R}_+^\times.$$

Then $n(j,k) = n(k,j)^{-1}$. If j' is another element of $\mathbb{J}$, we have $n(j+j',k) = n(j,k)\cdot n(j',k)$. Further, for a finite totally ordered index set I and a tuple $g = (g_i)_{i\in I}$

$$n(g) = n(g_i, i \in I) := \prod_{i<j} n(g_i, g_j)$$

is a well defined element of $\mathbb{R}_+^*$. If I is nonempty with greatest element k and $I' := I \setminus \{k\}$, we have $n(g) = n(g|I') \cdot n(\sum_{i\in I'} g_i, g_k)$. If E is a group of type $\mathbb{J}$ and j resp. x an element of $\mathbb{J}$ resp. ${}^{\mathrm{h}}E$, we set $n(j,x) := n(j, g(x))$ and $n(x,j) := n(g(x), j)$ for abbreviation.

(5.1.3) Now we are going to define real numbers $\sigma = \sigma\langle\mathbb{J}\rangle$ and $s = s\langle\mathbb{J}\rangle$ associated with a given valued signed group $\mathbb{J} = (\mathbb{J}, \Lambda, \varepsilon, \pi)$. For this purpose, let φ denote the composition of the given absolute value $|\cdot| = |\cdot|^{\mathbb{J}}$ for $\mathbb{J}$ with the structure homomorphism from $\mathbb{Z}_{S\langle\mathbb{J}\rangle}$ into Λ. If $\mathbb{J}$ is archimedean, then φ is of the form $|\cdot|^s$ with the standard absolute value $|\cdot|$ on $\mathbb{Q}$ and a uniquely determined real number s in $]0,1]$. In this case we set $\sigma\langle\mathbb{J}\rangle := 1$ and $s\langle\mathbb{J}\rangle := s$.

Suppose now that $\mathbb{J}$ is ultrametric. If φ is trivial or if $\varphi^{-1}(0)$ is different from $\{0\}$, we set $\sigma\langle\mathbb{J}\rangle := 1$ and $s\langle\mathbb{J}\rangle := 1$. Otherwise φ is of the form a^{v_p} with a real number a in $]0,1[$, a prime number p and the p-adic valuation v_p on $\mathbb{Q}$. We set $\sigma\langle\mathbb{J}\rangle := a^{-1/p-1}$ and $s\langle\mathbb{J}\rangle := ap$.

Then we have $\sigma \geq 1$ and $1/|n!| \leq \sigma^{n-1}$ for every integer n in $\mathbb{N}^*$, if the characteristic of Λ is zero. Let us call $\mathbb{J}$ *normalized*, if $s\langle\mathbb{J}\rangle = 1$.

(5.1.4) Let now $\mathbb{J}$ and $\mathbb{J}'$ be two valued signed groups and $(f,\rho): \mathbb{J} \to \mathbb{J}'$ a morphism of the underlying signed groups. Then $\rho = (f,\rho)$ is called a *morphism of valued signed groups* if $|f(\alpha)| = |\alpha|$ for all elements α in $\Lambda\langle\mathbb{J}\rangle$. Then $\sigma\langle\mathbb{J}\rangle = \sigma\langle\mathbb{J}'\rangle$ and $s\langle\mathbb{J}\rangle = s\langle\mathbb{J}'\rangle$, and $\mathbb{J}$ is archimedean resp. ultrametric if and only if $\mathbb{J}'$ has this property. With respect to morphisms defined in this way, the valued signed groups form a category.

By (^{v}SG) we denote the full subcategory of this category consisting of those valued signed groups $\mathbb{J}$ with $s\langle\mathbb{J}\rangle = 1$ if $\mathbb{J}$ is archimedean.

(5.2.1) (Pseudonormed groups) Let $\mathbb{J} = (\mathbb{J}, \Lambda, \varepsilon, \pi, |\cdot|)$ be an archimedean resp. ultrametric valued signed group in (^{v}SG) and $E = \coprod_{j\in\mathbb{J}} E^j$ a group of type $\mathbb{J}$. By a *pseudonorm* on E we mean a mapping

$$\|\cdot\|: E \longrightarrow \mathbb{R}_+ \cup \{\infty\}$$

satisfying the following four conditions: (1) $\|0\| = 0$. (2) $\|\alpha x\| \leq |\alpha| \cdot \|x\|$ for all elements α resp. x in Λ resp. E with $\|x\| < \infty$. (3) $\|x+y\| \leq \|x\| + \|y\|$ resp. $\|x+y\| \leq \sup\{\|x\|, \|y\|\}$ for all x, y in E. (4) if x is an element of E with homogeneous components $x^j, j \in \mathbb{J}$, then $\|x\| = \sum_{j\in\mathbb{J}} \|x^j\|$ resp. $\|x\| = \sup\{\|x^j\|: j \in \mathbb{J}\}$.

Then $\|\alpha x\| = |\alpha|\,\|x\|$ for α in $U\langle\mathbb{J}\rangle$ and x in E with $\|x\| < \infty$. Because of (3) the condition (4) implies that $\|x^j\| \leq \|x\|$. A pseudonorm $\|\cdot\|$ on E with $\|x\| < \infty$ for all x is called a *seminorm*. If moreover $\|x\| > 0$ for all $x \neq 0$, we say that $\|\cdot\|$ is a *norm* on E.

If $\mathbb{J}$ is trivial, then the map $\|\cdot\|: E \to \mathbb{R}_+ \cup \{\infty\}$ given by $\|x\| = 1$ for $x \neq 0$ and $\|0\| = 0$ is a norm on E called the *trivial norm.*

(5.2.2) By a *pseudonormed* resp. *seminormed* resp. normed group of type $\mathbb{J}$ we mean a pair $(E, \|\cdot\|)$ consisting of a group E of type $\mathbb{J}$ and a pseudonorm resp. seminorm resp. norm $\|\cdot\|$ on E. In the following we abbreviate $(E, \|\cdot\|)$ by E and denote $\|\cdot\|$ by $\|\cdot\|^E$ if there is any doubt. The elements x in E with $\|x\| < \infty$ form a subgroup $\mathrm{B}(E)$ of type $\mathbb{J}$ of E, and the restriction of $\|\cdot\|$ to $\mathrm{B}(E)$ is a seminorm. We say that E is a *Banach space* of type $\mathbb{J}$, if $\|\cdot\|$ is a norm and the $E^j, j \in \mathbb{J}$, are complete. Of course this doesn't imply in general that E itself is complete. E is called *trivially normed*, if $\|\cdot\|^E$ is trivial.

Let E be a pseudonormed group of type $\mathbb{J}$ and $F \subseteq E$ a subgroup (of type $\mathbb{J}$). Then both F and E/F are pseudonormed groups of type $\mathbb{J}$ in a natural way.

(5.3.1) (Products of pseudonormed groups) Let I be an index set and $E_i, i \in I$, a family of pseudonormed groups of type $\mathbb{J}$. Further suppose given a number p in $\mathbb{R} \cup \{\infty\}$ with $p \geq 1$ such that $p = \infty$ if $\mathbb{J}$ is ultrametric. For a tuple $x = (x_i)_{i\in I}$ in $\prod_{i\in I} E_i$ with homogeneous components $x^j = (x_i^j)_{i\in I}$ in $\prod_{i\in I} E_i^j, j \in \mathbb{J}$, we set

$$
{}_p\|x\|: = \sum_{j\in\mathbb{J}} \Big(\sum_{i\in I} \|x_i^j\|^p\Big)^{1/p} \quad \text{if} \quad p < \infty \quad \text{and} \quad \mathbb{J} \quad \text{is archimedean,}
$$

$$
{}_\infty\|x\|: = \sum_{j\in\mathbb{J}} \sup\{\|x_i^j\|: i \in I\} \quad \text{if} \quad \mathbb{J} \quad \text{is archimedean,}
$$

$$
{}_\infty\|x\|: = \sup\{\|x_i\|: i \in I\} \quad \text{if} \quad \mathbb{J} \quad \text{is ultrametric.}
$$

Then $(E,_p\|\cdot\|)$ is a pseudonormed group.

(5.4.1) (Tensor products of pseudonormed groups) In the situation of (5.3.1) we suppose in addition that I is finite and totally ordered. If $x = (x_i)_{i\in I}$ is a family of homogeneous elements x_i in E_i of degree $g_i = g(x_i)$, we set $n(x) = n(x_i, i \in I)$: $= n(g_i, i \in I)$. Suppose first that I is non-empty. Then, for an element z of $\bigotimes_{i\in I,\Lambda} E_i$, let $S(z)$ denote the set of all finite families $(x_i^{(l)})_{i\in I, l\in L}$ of homogeneous elements $x_i^{(l)}$ in $\mathrm{B}(E_i)$ with $\sum_{l\in L}\bigotimes_{i\in I} x_i^{(l)} = z$. Here L is a (variable) finite index set. We set

$$
\|(x_i^{(l)})\|: = \sum_{l\in L} n(x_i^{(l)}, i \in I)\cdot \prod_{i\in I} \|x_i^{(l)}\|
$$

resp.

$$
\|(x_i^{(l)}\|: = \sup\{n(x_i^{(l)}, i \in I)\cdot \prod_{i\in I} \|x_i^{(l)}\|\}
$$

according as $\mathbb{J}$ is archimedean resp. ultrametric. Further let $\|z\|$ be the infimum of the real numbers $\|(x_i^{(l)})\|$, where $(x_i^{(l)})$ runs in $S(z)$, so that $\|z\| = \infty$ if $S(z)$ is empty. Then the map sending z to $\|z\|$ is a pseudonorm on $\bigotimes_{i\in I,\Lambda} E_i$. If I is empty, let $\|\cdot\| = |\cdot|$ be the given absolute value on $\Lambda = \bigotimes_{i\in I,\Lambda} E_i$.Endowed with this pseudonorm, the tensor product of the $E_i, i \in I$, is a pseudonormed group of type $\mathbb{J}$ called the *π-tensor product* of the $E_i, i \in I$, and denoted also by $\bigotimes^{\pi}_{i\in I,\Lambda}, E_i$, if there is any doubt. For an arbitrary tuple $x = (x_i)_{i\in I}$ in $\prod_{i\in I}\mathrm{B}(E_i)$ we set

$$
n(x)\cdot \prod_{i\in I}\|x_i\|: = \sum_{(g_i)\in\mathbb{J}^I} n(g_i, i \in I)\cdot \prod_{i\in I}\|x_i^{g_i}\| \tag{5.4.1.1}
$$

resp.

$$
n(x)\cdot \prod_{i\in I}\|x_i\|: = \sup\{n(g_i, i \in I)\cdot \prod_{i\in I}\|x_i^{g_i}\|: (g_i) \in \mathbb{J}^I\} \tag{5.4.1.2}
$$

according as $\mathbb{J}$ is archimedean resp. ultrametric. With this convention, the estimate

(5.4.1.3) $$\|\underset{i\in I}{\otimes} x_i\| \le n(x)\cdot \prod_{i\in I}\|x_i\|$$

holds. For an arbitrary permutation σ of I the corresponding Λ-isomorphism

$$\bigotimes_{i\in I}{}^{\pi}_{\Lambda} E_i \xrightarrow{\tilde{\sigma}} \bigotimes_{i\in I}{}^{\pi}_{\Lambda} E_{\sigma^{-1}(i)}$$

conserves the respective pseudonorms. Using this fact, an elementary verification shows that the π-tensor product is associative in an obvious sense.

(5.5.1) (Pseudonorms on modules of homomorphisms) Let $\rho: \mathbb{J} \to \mathbb{J}'$ be a morphism of valued signed groups and E resp. F be a pseudonormed group of type $\mathbb{J}$ resp. $\mathbb{J}'$. Then $\rho^*(E)$ resp. $\rho_*(F)$ is canonically a pseudonormed group of type $\mathbb{J}'$ resp. $\mathbb{J}$, c.f. (5.4.1). For a homomorphism $u = \sum_{j'\in\mathbb{J}'} u^{j'}$ in $\mathrm{Hom}^{\rho}(E,F)$ we set

$$\|u^{j'}\| := \inf\{C \in \mathbb{R}_+ : \|u^{j'}(x)\| \le C\cdot n(j',x)\cdot\|x\| \text{ for all } x \in {}^{\mathrm{h}}\mathrm{B}(\rho^*(E))\}$$

and

$$\|u\| := \sum_{j'\in\mathbb{J}'}\|u^{j'}\| \quad \text{resp.} \quad \|u\| := \sup\{\|u^{j'}\| : j' \in \mathbb{J}'\}$$

if $\mathbb{J}'$ is archimedean resp. ultrametric. Then $\mathrm{Hom}^{\rho}(E,F)$, endowed with the map sending u to $\|u\|$, is a pseudonormed group of type $\mathbb{J}'$. If $\|u\|$ is finite, we have the estimate $\|u(x)\| \le n(u,x)\cdot\|u\|\cdot\|x\|$ for any element x of $\mathrm{B}(\rho^*(E))$. Let $\rho': \mathbb{J}' \longrightarrow \mathbb{J}''$ be a second morphism of valued signed groups and G a pseudonormed group of type $\mathbb{J}''$. If u resp. v is a homomorphism in $\mathrm{Hom}^{\rho}(E,F)$ resp. $\mathrm{Hom}^{\rho'}(F,G)$ with finite norm, then $\|v\circ u\| \le n(v,(\rho')^*(u))\cdot\|v\|\cdot\|(\rho')^*(u)\|$. Further we have $\|id_E\| \le 1$.

Let u be a homomorphism in $\mathrm{Hom}^{\rho}(E,F)^0$. We say that u is a *weak contraction* if $\|u\| \le 1$. In the case $\rho = id_{\mathbb{J}}$ we call u an *isometry*, if u is bijective with $\|u(x)\| = \|x\|$ for all elements x in E. Endowed with the homomorphisms of degree zero with finite norm as morphisms, the pseudonormed groups of variable type form a fibred and cofibred category $({}^{\mathrm{pn}}\mathrm{G})$ over $({}^{\mathrm{v}}\mathrm{SG})$.

(5.6.1) (Pseudonormed rings) Let $A = \coprod_{j\in\mathbb{J}} A^j$ be a ring of type $\mathbb{J}$ which is simultaneously a pseudonormed group with pseudonorm $\|\cdot\|$. If $\|1_A\| \le 1$ and

$\|a \cdot b\| \leq n(a,b) \cdot \|a\| \, \|b\|$ for all elements a, b in $\mathrm{B}(A)$, we call A a *pseudonormed ring*. If moreover $\| \cdot \|$ is a seminorm resp. norm, we say that A is a *seminormed* resp. *normed ring*. Finally A is called a *Banach ring* if A is a Banach space in the sense of (5.2.2). The pseudonormed rings graded by variable valued signed groups, endowed the weak contractions as morphisms, form in a natural way a fibred and cofibred category $({}^{\mathrm{pn}}\mathrm{R})$ over $({}^{\mathrm{v}}\mathrm{SG})$, cf. (5.5.1).

Let A be a pseudonormed ring of type $\mathbb{J}$. Then $\mathrm{B}(A)$ is a subring of A and hence a seminormed ring. Further we have $\|1_A\| < 1$ iff $\| \cdot \|$ is zero on $\mathrm{B}(A)$. If $(a_i)_{i \in I}$ is a finite family of elements of $\mathrm{B}(A)$ such that I is totally ordered, then

$$\Big\| \prod_{i \in I} a_i \Big\| \leq n(a_i, i \in I) \cdot \prod_{i \in I} \|a_i\|. \tag{5.6.1.1}$$

(5.6.2) Examples. (1) With respect to the trivial graduation and the given absolute value, $\Lambda = \Lambda\langle \mathbb{J} \rangle$ is a seminormed ring of type $\mathbb{J}$.

(2) If E is a pseudonormed group, then $\mathrm{End}_\Lambda(E)$ is a pseudonormed ring, cf. (5.5.1).

(3) If $\mathfrak{a}$ is a two-sided ideal in a pseudonormed ring A, then $A/\mathfrak{a}$ is a pseudonormed ring, too.

(4) Let $A_i, i \in I$, be a family of pseudonormed rings of type $\mathbb{J}$. Then the product $\prod_{i \in I} A_i$ is a pseudonormed ring with respect to the pseudonorm ${}_\infty\| \cdot \|$, cf. (5.3.1). If I is finite and totally ordered, then also $\bigotimes^{\pi}_{i \in I, \Lambda} A_i$ is a pseudonormed ring of type $\mathbb{J}$.

(5.7.1) (Pseudonormed modules and algebras) Let A be a pseudonormed commutative ring of type $\mathbb{J}$ and E an A-module endowed with the structure of a pseudonormed group of type $\mathbb{J}$. Then E is called a *pseudonormed A-module*, if $\|a \cdot x\| \leq n(a,x) \cdot \|a\| \, \|x\|$ and $\|x \cdot a\| \leq n(x,a) \cdot \|x\| \cdot \|a\|$ holds for all elements a in $\mathrm{B}(A)$ and x in $\mathrm{B}(E)$. If moreover $\| \cdot \|^E$ is a seminorm resp. norm, we say that E is a *seminormed* resp. *normed module* over A. Further E is called a *Banach module* over A, if E is a Banach space. If E is a pseudonormed A-module, then $\mathrm{B}(E)$ is a seminormed module over $\mathrm{B}(A)$.

Let B be an A-algebra, which is simultaneously a pseudonormed resp. seminormed resp. normed resp. Banach module over A. If moreover $\|x \cdot y\| \leq n(x,y) \cdot \|x\| \, \|y\|$

holds for all elements $x, y \in B$ of finite norm, we say that B is a *pseudonormed* resp. *seminormed* resp. *normed* resp. *Banach algebra* over A.

(5.7.2) Examples. (1) A is a pseudonormed module over itself.

(2) Let E be a pseudonormed module over A and $F \subseteq E$ a submodule. Then both F and E/F are pseudonormed A-modules. Let k be an element of $\mathbb{J}$. Then there exists a unique pseudonorm $\|\cdot\|_{[k]}$ on $E[k]$ such that $\|x\|_{[k]} = n(-k, x)\|T_k(x)\|$ for all homogeneous elements x of $E[k]$; here T_k denotes the canonical isomorphism from $E[k]$ onto E. Then $E[k] = (E[k], \|\cdot\|_{[k]})$ is a pseudonormed module over A and $\|T_k\| = 1$. One checks that $\|\cdot\|_{[k+l]} = n(k,l) \cdot (\|\cdot\|_{[k]})_{[l]}$ for $k, l \in \mathbb{J}$.

(3) Let I be a finite totally ordered index set and $E_i, i \in I$, pseudonormed A-modules. Then $\bigotimes_{i \in I, A} E_i$ is a pseudonormed module over A in a natural way, cf. (2) and (5.4.1).

(4) Let E be a pseudonormed A-module and $n \in \mathbb{N}$ an integer. Then $\mathrm{T}^n_A(E)$, $\mathrm{S}^n_A(E)$ and $\Lambda^n_A(E)$ are canonically endowed with the structure of pseudonormed A-modules, cf. (3) and (2). For an arbitrary element $w = (w_n)_{n \in \mathbb{N}}$ of $\hat{\mathrm{T}}_A(E) = \prod_{n \in \mathbb{N}} \mathrm{T}^n_A(E)$ we set $\|w\| :=_1 \|w\|$ resp. $\|w\| :=_\infty \|w\|$ if $\mathbb{J}$ is archimedean resp. ultrametric, cf. (5.3.1). Then the map sending w to $\|w\|$ makes $\hat{\mathrm{T}}_A(E), \mathrm{T}_A(E)$ and hence also $\hat{\mathrm{S}}_A(E), \mathrm{S}_A(E)$ and $\hat{\Lambda}_A(E), \Lambda_A(E)$ into a pseudonormed A-algebra.

(5) Let $\varphi: A \to A'$ be a homomorphism of pseudonormed commutative rings in $({}^{\mathrm{pn}}_{c}\mathrm{R})$ and E resp. F a pseudonormed module over A resp. A'. Then $\varphi^*(E) = A' \otimes_A E$, $\mathrm{Hom}_\varphi(E, F)$ resp. $\varphi_*(F)$ is a pseudonormed module over A' resp. A in a canonical way. In particular $\mathrm{L}_\varphi(E, F) := \mathrm{B}(\mathrm{Hom}_\varphi(E, F))$ is a seminormed $\mathrm{B}(A')$-module, being denoted by $\mathrm{L}_A(E, F)$ if $\varphi = id_A$ is the identity of A. Moreover $\mathrm{End}_A(E)$ resp. $\mathrm{L}_A(E) := \mathrm{B}(\mathrm{End}_A(E))$ is a pseudonormed resp. seminormed algebra over A resp. $\mathrm{B}(A)$. Again, the pseudonormed modules over variable pseudonormed commutative rings form a fibred and cofibred category $({}^{\mathrm{pn}}\mathrm{M})$ over $({}^{\mathrm{pn}}_{c}\mathrm{R})$, the groups of morphisms from E into F over φ being $\mathrm{L}_\varphi(E, F)^0$.

(6) Let $\varphi: A \to A'$ be a homomorphism of trivially normed commutative rings in $({}^{\mathrm{pn}}_{c}\mathrm{R}_{\mathbb{Q}})$ and let E resp. F be a module over A resp. A'. For a formal map $u = \sum_{n \in \mathbb{N}} u_n$ in $\mathrm{For}_\varphi(E, F)$ let $\mathrm{ord}(u) \in \mathbb{N} \cup \{\infty\}$ denote the order of u, i.e. the infimum of the integers n with $u_n \neq 0$, cf. (3.1.3). If we set $\|u\| := e^{-\mathrm{ord}(u)}$, then the map sending u to $\|u\|$ is a norm on $\mathrm{For}_\varphi(E, F)$, called the *canonical norm.* It's clear that $(\mathrm{For}_\varphi(E, F), \|\cdot\|)$ is in fact a Banach module over A'.

(5.7.3) Let A be a pseudonormed commutative ring of type $\mathbb{J}$, I a finite totally ordered index set and $E_i, F_i, i \in I$, pseudonormed A-modules. Then an elementary verification shows that the natural A-homomorphism

$$\bigotimes_{i\in I} {}_A \operatorname{Hom}_A(E_i, F_i) \xrightarrow{\alpha} \operatorname{Hom}_A(\bigotimes_{i\in I} {}_A E_i, \bigotimes_{i\in I} {}_A F_i)$$

is a weak contraction with respect to the canonical pseudonorms. Thus, if $\bigotimes_{i\in I} u_i$ denotes the tensor product of a family of homomorphisms u_i in $\operatorname{Hom}_A(E_i, F_i), i \in I$, the estimate $\|\bigotimes_{i\in I} u_i\| \le n(u_i, i \in I) \cdot \prod_{i\in I} \|u_i\|$ applies. Finally, for any three pseudonormed A-modules E, F and G, the natural A-homomorphism resp. A-isomorphism

$$\operatorname{Hom}_A(F,G) \otimes_A \operatorname{Hom}_A(E,F) \xrightarrow{\kappa} \operatorname{Hom}_A(E,G)$$

resp.

$$\operatorname{Hom}_A(E, \operatorname{Hom}_A(F,G)) \xrightarrow{h} \operatorname{Hom}_A(E \otimes_A F, G)$$

is a weak contraction resp. an isometry.

(5.7.4) Let E be a Banach module over a pseudonormed ring A of type $\mathbb{J}$ in $({}_c\mathrm{R}_{\mathbb{Q}})$ and $(x_i)_{i\in I}$ a family of elements of E. We say that $(x_i)_{i\in I}$ is *normally summable*, if there exists a finite subset J of $\mathbb{J}$ such that all x_i are in $\coprod_{j\in J} E^j$ and if the family $(\|x_i\|)_{i\in I}$ of real numbers is summable resp. converging to zero according as $\mathbb{J}$ is archimedean resp. ultrametric. In this case $(x_i)_{i\in I}$ is in particular summable in E with sum $\sum_{i\in I} x_i$.

(5.8.1) (Pseudonormed cogebras) Let A be a pseudonormed commutative ring of type $\mathbb{J}$ and M a cogebra over A endowed with the structure of a pseudonormed A-module. Then M is called a *pseudonormed cogebra* over A, if both the coproduct c_M and the counit γ_M are weak contractions, i.e. $\|c_M\| \le 1$ and $\|\gamma_M\| \le 1$. Obviously the quotient of a pseudonormed cogebra by a coideal is a pseudonormed cogebra. If $M_i, i \in I$, is a family of pseudonormed cogebras over A such that I is finite and totally ordered, then $\bigotimes_{i\in I,A} M_i$ is a pseudonormed cogebra, too.

(5.8.2) Let $\varphi: A \to A'$ be a homomorphism of pseudonormed commutative rings, M a pseudonormed cogebra over A and F, G, H three pseudonormed A'-modules, and suppose given a homomorphism μ in $\operatorname{Hom}_{A'}(F \otimes_{A'} G, H)^0$ with $\|\mu\| < \infty$. Then for any two maps u resp. v in $\mathrm{Hom}_\varphi(M,F)$ resp. $\mathrm{Hom}_\varphi(M,G)$ of finite norm the estimate

$$\|u \cdot_\mu v\| \leq n(u,v) \cdot \|u\| \cdot \|v\| \cdot \|\mu\| \tag{5.8.2.1}$$

applies. Hence, if C is a pseudonormed algebra over A', we infer that $\mathrm{Hom}_\varphi(M,C)$ is a pseudonormed A'-algebra, too.

(5.9.1) (Pseudonormed graded modules and algebras) Let $E = \coprod_{n\in\mathbb{Z}} E_n$ be a graded A-module in the sense of (2.4.1), endowed with the structure of a pseudonormed A-module. Then E is called a *pseudonormed graded A-module*, if for any element $x = \sum_{n\in\mathbb{Z}} x_n$ of E with homogeneous components $x_n \in E_n$ we have $\|x\| = \sum_{n\in\mathbb{Z}} \|x_n\|$ resp. $\|x\| = \sup\{\|x_n\|: n \in \mathbb{Z}\}$ if $\mathbb{J}\langle A\rangle$ is archimedean resp. ultrametric. Then the pseudonorm on E extends canonically to a pseudonorm on the completion $\hat{E}$ of E with respect to the canonical filtration, cf. (5.3.1).

Suppose now that we are given a homomorphism $\varphi: A \to A'$ of pseudonormed commutative rings and a pseudonormed graded module E resp. F over A resp. A'. Then $\mathrm{Hom}_\varphi(E,F)^{\mathrm{gr}}$ is a pseudonormed graded A'-module with respect to the pseudonorm

$$\|\cdot\|^{\mathrm{gr}}: \mathrm{Hom}_\varphi(E,F)^{\mathrm{gr}} \longrightarrow \mathbb{R}_+ \cup \{\infty\}$$

given by $\|u\|^{\mathrm{gr}} = \sum_{n\in\mathbb{Z}} \|u_n\|$ resp. $\|u\|^{\mathrm{gr}} = \sup\{\|u_n\|: n \in \mathbb{Z}\}$ if $\mathbb{J}\langle A'\rangle$ is archimedean resp. ultrametric. We have $\|u\| \leq \|u\|^{\mathrm{gr}}$ with equality in the ultrametric case. Elementary examples show that in the archimedean case, the quotient $\|u\|^{\mathrm{gr}}/\|u\|$ can be arbitrarily large.

Let $B = \coprod_{n\in\mathbb{Z}} B_n$ be a graded algebra over A and simultaneously a pseudonormed graded A-module. We say that B is a *pseudonormed graded A-algebra*, if the pseudonorm of B is compatible with the algebra structure of B, cf. (5.7.1). Then B is in particular a pseudonormed algebra over A. For instance, if E is a pseudonormed A-module, then $\mathrm{T}_A(E)$, $\mathrm{S}_A(E)$ and $\Lambda_A(E)$ are pseudonormed graded algebras over A, cf. (5.7.2) (4).

(5.10.1) (Pseudonormed graded cogebras and bigebras) Let A be a pseudonormed commutative ring of type $\mathbb{J}$ and $M = \coprod_{n\in\mathbb{Z}} M_n$ a graded cogebra over A endowed with the structure of a pseudonormed graded A-module. We call M a *pseudonormed graded cogebra* over A, if γ_M and the composition $c_M^{p,q}$ of c_M with the natural projection from $M \otimes_A M$ onto $M_p \otimes_A M_q$ are weak contractions for all integer p, q in $\mathbb{Z}$. Then M is in particular a pseudonormed cogebra over A, if $\mathbb{J}$ is ultrametric. In the archimedean case, this fails; cf. (5.10.4) for an example.

(5.10.2) Let $\varphi: A \to A'$ be a homomorphism of pseudonormed commutative rings, M a pseudonormed graded cogebra over A and F, G, H, μ as in (5.8.2). Then for any two maps u resp. v in $\mathrm{Hom}_\varphi(M, F)^{\mathrm{gr}}$ resp. $\mathrm{Hom}_\varphi(M, G)^{\mathrm{gr}}$ with $||u||^{\mathrm{gr}} < \infty$ and $||v||^{\mathrm{gr}} < \infty$, we have the estimate

$$(5.10.2.1) \qquad ||u \cdot_\mu v||^{\mathrm{gr}} \leq n(u, v) \cdot ||u||^{\mathrm{gr}} \cdot ||v||^{\mathrm{gr}} \cdot ||\mu||.$$

In particular, if C is a pseudonormed algebra over A', we infer that $\mathrm{Hom}_\varphi(M, C)^{\mathrm{gr}}$ is a pseudonormed graded algebra over A'. Thus, if M is positive, we obtain that $\mathrm{Hom}_\varphi(M, C) = (\mathrm{Hom}_\varphi(M, C)^{\mathrm{gr}})^\wedge$ is canonically endowed with the structure of a pseudonormed A'-algebra.

(5.10.3) Let $B = \coprod_{n \in \mathbb{Z}} B_n$ be a positive graded bigebra over A endowed with the structure of a pseudonormed graded A-module. If $\mathbb{J}$ is archimedean, even in the most favourable cases we meet in practice, the pseudonorm $|| \cdot ||$ of B cannot be compatible with the structures of a graded cogebra and graded algebra of B simultaneously, cf. (5.10.4) for examples. So we are led to call B a *pseudonormed graded bigebra* over A, if we are given two pseudonorms $|| \cdot ||$ and $|| \cdot ||'$ on B which satisfy the following conditions: (1) $(B, || \cdot ||)$ is a pseudonormed graded cogebra over A and $(B, || \cdot ||')$ is a pseudonormed graded algebra over A. (2) If $\mathbb{J}$ is archimedean, then for every integer n in $\mathbb{N}$ there exists a positive number C resp. C' (depending on n) such that $||x||' \leq C||x||$ resp. $||x|| \leq C'||x||'$ for all x in B_n with $||x|| < \infty$ resp. $||x||' < \infty$. (3) If $\mathbb{J}$ is ultrametric, then $|| \cdot ||$ equals $|| \cdot ||'$.

(5.10.4) Example. Let E be a pseudonormed A-module, $\sigma = \sigma(\mathbb{J})$, and let $t > 0$ be a real number. Then there exist unique pseudonorms $| \cdot |_t$ and $| \cdot |'_t$ on $\mathrm{S}_A(E)$ which satisfy the following conditions. (1) $\mathrm{S}_A(E)$ is a pseudonormed graded A-module with respect to both $| \cdot |_t$ and $| \cdot |'_t$. (2) For any $n \in \mathbb{N}$ the restriction to $\mathrm{S}^n_A(E)$ of $| \cdot |_t$ resp. $| \cdot |'_t$ equals $n! t^{-n} || \cdot ||$ resp. $t^{-n} || \cdot ||$ if $\mathbb{J}$ is archimedean. If $\mathbb{J}$ is ultrametric, then $| \cdot |_t$ and $| \cdot |'_t$ coincide and their restriction to $\mathrm{S}^n_A(E)$ equals $(\sigma t)^{-n} || \cdot ||$. Here $|| \cdot ||$ denotes the canonical pseudonorm on $\mathrm{S}^n_A(E)$, cf. (5.7.2) (4). An elementary verification yields that $(\mathrm{S}_A(E),\ | \cdot |_t,\ | \cdot |'_t)$ is a pseudonormed graded bigebra over A. If $\mathbb{J}$ is archimedean, then $| \cdot |_t$ is in general not compatible with the algebra structure of $\mathrm{S}_A(E)$.

Let $\varphi: A \to A'$ be a homomorphism of pseudonormed commutative rings and F a pseudonormed A'-module, and let $|\cdot|_t$ also denote the corresponding pseudonorm on $\operatorname{Hom}_\varphi(\mathrm{S}_A(E), F) = \prod_{n\in\mathbb{N}} \operatorname{Hom}_\varphi(\mathrm{S}_A^n(E), F)$, cf. (5.9.1) and (5.10.2). Then for a homomorphism $u = \sum_{n\in\mathbb{N}} u_n$ in $\operatorname{Hom}_\varphi(\mathrm{S}_A(E), F)$ we obtain the formula

(5.10.4.1) $$|u|_t = \sum_{n\in\mathbb{N}} \frac{1}{n!} \|u_n\| \cdot t^n$$

if $\mathbb{J}$ is archimedean and

(5.10.4.2) $$|u|_t = \sup\{\|u_n\| \cdot (\sigma t)^n : n \in \mathbb{N}\}$$

if $\mathbb{J}$ is ultrametric. Again, here $\|u_n\|$ denotes the norm of u_n formed in the pseudonormed A'-module $\operatorname{Hom}_\varphi(\mathrm{S}_A^n(E), F)$. The role of the scaling factor σ in the ultrametric case will become evident in the next section. Of course, the same remarks also apply to $\mathrm{T}_A(E)$.

6. Convergent maps and substitutable elements

(6.1.1) (Convergent formal maps) Let $\varphi: A \to A'$ be a homomorphism of pseudonormed commutative rings in $({}^{\mathrm{pn}}_{\mathrm{c}}\,\mathrm{R}_{\mathbb{Q}})$ and let E resp. F be a pseudonormed module over A resp. A'. We set $\mathbb{J} := \mathbb{J}\langle A\rangle, \mathbb{J}' := \mathbb{J}\langle A'\rangle$ and $\sigma := \sigma\langle\mathbb{J}\rangle$ for abbreviation. Then, for any given real number $t > 0$, $\operatorname{For}_\varphi(E, F) = \operatorname{Hom}_\varphi(\mathrm{S}_A(E), F)$ is a pseudonormed module over A' with respect to the pseudonorm $|\cdot|_t$ introduced in (5.10.4). Hence

$$\begin{aligned} &\mathrm{B}_{t,\varphi}(E, F) := \mathrm{B}(\operatorname{For}_\varphi(E, F), |\cdot|_t) \quad \text{resp.} \\ &\mathrm{B}_{t,\varphi}(E, F)_+ := \mathrm{B}(\operatorname{For}_\varphi(E, F)_+, |\cdot|) \quad \text{resp.} \\ &\mathrm{B}(\operatorname{Pol}_\varphi(E, F)) := \mathrm{B}(\operatorname{Pol}_\varphi(E, F), |\cdot|_1) \end{aligned}$$

is a $\mathrm{B}(A')$-submodule of $\operatorname{For}_\varphi(E, F)$ resp. $\operatorname{For}_\varphi(E, F)_+$ resp. $\operatorname{Pol}_\varphi(E, F)$. If t' is another real number such that $t \le t'$, then $\mathrm{B}_{t',\varphi}(E, F)$ is contained in $\mathrm{B}_{t,\varphi}(E, F)$. Hence the $\mathrm{B}_{t,\varphi}(E, F)$, $t > 0$, form both an inverse and inductive system of $\mathrm{B}(A')$-modules. The elements of the $\mathrm{B}(A')$-submodule

$$\mathrm{B}_{\infty,\varphi}(E, F) := \varprojlim_{t>0} \mathrm{B}_{t,\varphi}(E, F)$$

resp.

$$\operatorname{Con}_\varphi(E, F) := \varinjlim_{t>0} \mathrm{B}_{t,\varphi}(E, F)$$

of $\mathrm{For}_{\varphi}(E,F)$ are called *entire formal maps* resp. *convergent formal maps from E into F* (over φ). For any $t > 0$, $\mathrm{B}_{t,\varphi}(E,F)$ contains $\mathrm{B}_{\infty,\varphi}(E,F)$ and is contained in $\mathrm{Con}_{\varphi}(E,F)$. If $\varphi = id_A$, we write $\mathrm{Con}_A(E,F)$ instead of $\mathrm{Con}_{\varphi}(E,F)$. For an individual formal map $u = \sum_{n\in\mathbb{N}} u_n$ in $\mathrm{For}_{\varphi}(E,F)$ we define the *radius of convergence* $\rho(u) \in \mathbb{R}_+ \cup \{\infty\}$ as the supremum of all t in $\mathbb{R}_+^*$ with $|u|_t < \infty$. Then u is convergent iff $\rho(u) > 0$. Moreover, if $\|u_n\|$ is finite for every n, we have

$$\rho(u) = \frac{1}{\limsup \sqrt[n]{|\frac{1}{n!}|\ \|u_n\|}}.$$

If $\mathbb{J}$ is ultrametric, we denote by

$$\mathrm{R}_{t,\varphi}(E,F) \subseteq \mathrm{B}_{t,\varphi}(E,F)$$

the $\mathrm{B}(A')$-submodule which consists of all formal maps $u = \sum_{n\in\mathbb{N}} u_n$ such that the sequence $(\|u_n\| \cdot (\sigma t)^n)_{n\in\mathbb{N}}$ converges to zero. Then $\mathrm{B}_{t,\varphi}(E,F)$ is contained in $\mathrm{R}_{t',\varphi}(E,F)$ for $0 < t' < t$.

(6.1.2) Let $\mathbb{J}$ furthermore be ultrametric, and fix a non-negative real number δ and an integer n in $\mathbb{N}$. Then there exists a unique pseudonorm

$$\|\cdot\|^{\delta}\colon \mathrm{For}_{\varphi}(E,F)_n \longrightarrow \mathbb{R}_+ \cup \{\infty\}$$

satisfying the following conditions: (1) $\|\cdot\|^{\delta} = \|\cdot\|$ if $n = 0$. (2) Let n be positive and v a polynomial in ${}^{\mathrm{h}}\mathrm{For}_{\varphi}(E,F)_n$. Then $\|v\|^{\delta}$ is the infimum of all real numbers C in $\mathbb{R}_+$ such that $\|v(x_1\cdots x_n)\| \leq C\cdot n(v, x_1,\ldots,x_n)\cdot\|x_1\|\cdots\|x_n\|$ for all x_i in ${}^{\mathrm{h}}\mathrm{B}(\rho^*(E))$ with $\|x_i\| \geq \delta$ for $1 \leq i \leq n$. -

Observe that $\|\cdot\|^{\delta}$ coincides with $\|\cdot\|$, if the restriction of the given absolute value on $\Lambda\langle\mathbb{J}'\rangle$ to $\mathbb{Q}$ or to $U\langle\mathbb{J}'\rangle$ is non-trivial. - For a given real number $t > 0$ and a polynomial v in $\mathrm{For}_{\varphi}(E,F)_n$ we set $|v|_t^{\delta} := \|v\|^{\delta}\cdot(\sigma t)^n$ with $\sigma = \sigma\langle\mathbb{J}\rangle$, and denote by

$$\mathrm{W}_{t,\varphi}(E,F) \subseteq \mathrm{B}_{t,\varphi}(E,F)$$

the $\mathrm{B}(A')$-submodule consisting of all formal maps $u = \sum_{n\in\mathbb{N}} u_n$ such that $\lim\limits_{n\to\infty} |u_n|_t^{\delta} = 0$ for every $\delta > 0$. Clearly $\mathrm{R}_{t,\varphi}(E,F)$ is contained in $\mathrm{W}_{t,\varphi}(E,F)$. If the restriction of the given absolute value to $\mathbb{Q}$ or $U\langle\mathbb{J}'\rangle$ is non-trivial, then these two submodules coincide.

Let $\varphi': A' \to A''$ be another homomorphism in $({}^{\text{pn}}_{\text{c}}\text{R}_{\mathbb{Q}})$. If ρ resp. ρ' denotes the morphism of valued signed groups underlying φ resp. φ', then $\tilde{E} := (\rho' \circ \rho)^*(E)$ resp. $\tilde{F} := (\rho')^*(F)$ is a pseudonormed module over $\tilde{A} := (\rho' \circ \rho)^*(A)$ resp. $\tilde{A}' := (\rho')^*(A)$, and φ gives rise to a homomorphism $\tilde{\varphi}': \tilde{A} \to \tilde{A}'$. Any formal map u in $\text{For}_{\varphi}(E,F)$ induces via base change a formal map $\tilde{u}$ in $\text{For}_{\tilde{\varphi}}(\tilde{E},\tilde{F})$. If u is an element of $\text{W}_{t,\varphi}(E,F)$, then $\tilde{u}$ is not necessarily in $\text{W}_{t,\tilde{\varphi}}(\tilde{E},\tilde{F})$. We denote by

$$\text{W}^{\varphi'}_{t,\varphi}(E,F) \subseteq \text{W}_{t,\varphi}(E,F)$$

the $\text{B}(A')$-submodule consisting of those formal maps u such that $\tilde{u}$ is in $\text{W}_{t,\tilde{\varphi}}(\tilde{E},\tilde{F})$. Clearly $\text{R}_{t,\varphi}(E,F)$ is contained in this submodule.

(6.1.3) Let now $T: \text{For}_{\varphi}(E,F)^0_+ \to \text{For}_{\varphi}(E,F)^0_+$ be a contraction with respect to the metric on $\text{For}_{\varphi}(E,F)^0_+$ given by the adic norm, cf. (5.7.2) (6). By Banach's fixed point theorem, T has a unique fixed point u in $\text{For}_{\varphi}(E,F)^0_+$. If there exist real number $t, r > 0$ such that $|T(w)|_t \le r$ for all w in $\text{For}_{\varphi}(E,F)^0_+$ with $|w|_t \le r$, then also $|u|_t \le r$. In particular we see that u is convergent. This elementary observation will be of constant use.

(6.2.1) (Derivatives of convergent maps) In the situation of (6.1.1), suppose given an integer $p \in \mathbb{N}$, real numbers t, t' with $0 < t' < t$ and a formal map u in $\text{For}_{\varphi}(E,F)$. Then we can easily derive the estimate

$$|D^p(u)|_{t'} \le p!\,\frac{1}{(t-t')^p}\,|u|_t \tag{6.2.1.1}$$

if $\mathbb{J}$ is archimedean and

$$|D^p(u)|_t \le \left(\frac{1}{\sigma \cdot t}\right)^p |u|_t \tag{6.2.1.2}$$

if $\mathbb{J}$ is ultrametric. In particular we have $\rho(D^p(u)) \ge \rho(u)$. Hence, if u is convergent, then so is $D^p(u)$. We see also that D^p induces a continuous homomorphism from $\text{B}_{t,\varphi}(E,F)$ into $\text{B}_{t,\varphi}(E,\text{For}_{\varphi}(E,F)_p)$ if $\mathbb{J}$ is ultrametric, whereas in the archimedean case this becomes true only "after shrinking" t.

(6.3.1) (Products of convergent maps) Let G, H be another couple of pseudonormed A'-modules, and suppose given a homomorphism μ in $\text{Hom}_{A'}(F \otimes_{A'} G, H)^0$ with $||\mu|| < \infty$ and a real number $t > 0$. Then for any two formal maps u resp. v in $\text{For}_{\varphi}(E,F)$ resp. $\text{For}_{\varphi}(E,G)$ with $|u|_t$, $|v|_t < \infty$ we have the estimate

(6.3.1.1) $$|u \cdot_\mu v|_t \leq n(u,v) \cdot |u|_t \cdot |v|_t \cdot \|\mu\|,$$

cf. (5.8.2). Let now C be a pseudonormed algebra over A'. Then we infer that $(\mathrm{For}_\varphi(E,C), |\cdot|_t)$ resp. $\mathrm{B}_{t,\varphi}(E,C)$ is a pseudonormed resp. seminormed algebra over A' resp. $\mathrm{B}(A')$. Thus $\mathrm{B}_{\infty,\varphi}(E,C)$ and $\mathrm{Con}_\varphi(E,C)$ are algebras over $\mathrm{B}(A')$, too. A convergent formal map $u = \sum_{n\in\mathbb{N}} u_n$ in $\mathrm{Con}_\varphi(E,C)^0$ is a unit in $\mathrm{Con}_\varphi(E,C)$ iff its constant term u_0 is a unit in C^0. In this case we have

$$\rho(1/u) \geq \sup\{t \in \mathbb{R}_+^*: |1 - u \cdot (1/u_0)|_t < 1\}.$$

(6.4.1) Example. Let B be a pseudonormed algebra over A and $(b_n)_{n\in\mathbb{N}}$ a family in $B^{\mathbb{N}}$. Then $u := \sum_{n\in\mathbb{N}} b_n \cdot X^n_{B|A}$ is a well defined formal map in $\mathrm{For}_A(B,B)$, cf. (4.3.1). We have

$$|u|_t = \sum_{n\in\mathbb{N}} \|b_n\| \cdot t^n$$

resp.

$$\sup\{|n!| \cdot \|b_n\| \cdot (\sigma t)^n : n \in \mathbb{N}\} \leq |u|_t \leq \sup\{\|b_n\| \cdot (\sigma t)^n : n \in \mathbb{N}\}$$

according as $\mathbb{J} = \mathbb{J}\langle A\rangle$ is archimedean resp. ultrametric. If B is commutative, then the inequality on the left becomes an equality. If $\|b_n\|$ is finite for every n, we infer

$$\rho(u) = \frac{1}{\limsup \sqrt[n]{\|b_n\|}}$$

resp.

$$\frac{1}{\sigma \cdot \limsup \sqrt[n]{\|b_n\|}} \leq \rho(u) \leq \frac{1}{\limsup \sqrt[n]{\|b_n\|}}$$

if $\mathbb{J}$ is archimedean resp. ultrametric. Again, the inequality on the right becomes an equality, if B is commutative. Applying this to the exponential map $e_{B|A}$ resp. the logarithm $l_{B|A}$ of B, we obtain

$$\rho(e_{B|A}) = \infty \quad \text{and} \quad \rho(l_{B|A}) = 1$$

resp.

$$\sigma^{-2} \leq \rho(e_{B|A}) \leq \sigma^{-1} \quad \text{and} \quad \sigma^{-1} \leq \rho(l_{B|A}) \leq 1$$

according as $\mathbb{J}$ is archimedean resp. ultrametric. If B is commutative, we have $\rho(e_{B|A}) = \sigma^{-1}$ and $\rho(l_{B|A}) = 1$ in the ultrametric case.

(6.5.1) (Composition of convergent maps) Let $\varphi: A \to A'$ and $\varphi': A' \to A''$ be two homomorphisms of pseudonormed commutative rings in $({}^{\mathrm{pn}}_{\mathrm{c}}\mathrm{R}_{\mathbb{Q}})$ and E resp. F resp. G a pseudonormed module over A resp. A' resp. A''. Further suppose given a formal map u resp. v in $\mathrm{For}_{\varphi}(E, F)^0$ resp. $\mathrm{For}_{\varphi'}(F, G)$ such that either $u_0 = 0$ or v is a polynomial. In this situation we have the following proposition.

(6.5.2) Lemma. *Let $t, s > 0$ be two real numbers with $|u|_t \leq s$ and set $\sigma = \sigma\langle \mathbb{J}\langle A\rangle\rangle$. Then:*

(1) $|v \mathbin{\square} u|_t \leq |v|_s$. *If moreover $v_0 = 0$ and $\mathbb{J}\langle A\rangle$ is ultrametric, the sharper estimate* $|v \mathbin{\square} u|_t \leq \sigma^{-1} \cdot |v|_s$ is valid.

(2) *If v is in $\mathrm{B}_{s,\varphi'}(F, G)$, then $v \mathbin{\square} u$ is in $\mathrm{B}_{t,\varphi'\circ\varphi}(E, G)$.*

(3) *Suppose that $\mathbb{J}\langle A\rangle$ is ultrametric and that u resp. v is in $\mathrm{R}_{t,\varphi}(E, F)$ resp. $\mathrm{R}_{s,\varphi'}(F, G)$. Then $v \mathbin{\square} u$ is in $\mathrm{R}_{t,\varphi'\circ\varphi}(E, G)$.*

(4) *Suppose that $\mathbb{J}\langle A\rangle$ is ultrametric and that u resp. v is in $\mathrm{W}^{\varphi'}_{t,\varphi}(E, F)$ resp. $\mathrm{W}_{s,\varphi'}(F, G)$. Then $v \mathbin{\square} u$ is in $\mathrm{W}_{t,\varphi'\circ\varphi}(E, G)$.*

Proof. We set $\mathbb{J} := \mathbb{J}\langle A\rangle$ for abbreviation, cf. (5.1.3), and regard $\mathrm{S}_{A'}(F)$ as a pseudonormed algebra over A' as in (5.7.2) (4). Then $\mathrm{For}_{\varphi}(E, \mathrm{S}_{A'}(F))$ is a pseudonormed A'-algebra with respect to $|\cdot|_t$, cf. (6.3.1). Hence $|u^q|_t \leq |u|_t^q$ for any integer q in $\mathbb{N}$. Since $v \mathbin{\square} u = \sum_{q\in\mathbb{N}} (1/q!)\, v_q \circ u^q$ by (4.1.1.2), we infer

$$\begin{aligned} |v \mathbin{\square} u|_t &\leq \sum_{q\in\mathbb{N}} \frac{1}{|q!|}\, |v_q \circ u^q|_t \\ &\leq \sum_{q\in\mathbb{N}} \frac{1}{q!}\, \|v_q\| \cdot |u^q|_t \\ &\leq \sum_{q\in\mathbb{N}} \frac{1}{q!}\, \|v_q\| \cdot |u|_t^q \\ &\leq |v|_s \end{aligned}$$

if $\mathbb{J}$ is archimedean and

$$\begin{aligned} |v \mathbin{\square} u|_t &\leq \sup\{\frac{1}{|q!|}\, |v_q \circ u^q|_t : q \in \mathbb{N}\} \\ &\leq \sup\{\frac{1}{|q!|}\, \|v_q\| \cdot |u|_t^q : q \in \mathbb{N}\} \\ &\leq \sup\{\|v_0\|, \sigma^{q-1}\|v_q\| s^q, q \in \mathbb{N}_+\} \\ &\leq |v|_s \end{aligned}$$

if $\mathbb{J}$ is ultrametric. This yields the assertions (1) and (2). We omit the proof of (3) and (4). - As an application of (6.5.2) we prove the convergent version of the inverse function theorem.

(6.5.3) Proposition (Convergent inverse function theorem). *Let E, F be two pseudonormed A-modules, $u = \sum_{n\geq 1} u_n$ a convergent formal map in* $\mathrm{Con}_A(E,F)^0_+$ *and $\tau \in \mathrm{L}_A(F,E)$ an A-linear map such that $u_1 \circ \tau = id_F$. Then there exists a convergent formal map $v = \sum_{n\geq 1} v_n$ in* $\mathrm{Con}_A(F,E)^0_+$ *with $v_1 = \tau$ and $u \,\square\, v = \mathrm{id}_F$.*

Proof. Since the formal map $\eta := \tau \,\square\, u_1 - \tau \,\square\, u$ in $\mathrm{Con}_A(E,E)^0_+$ has order ≥ 2, the mapping T from $\mathrm{For}_A(F,E)^0_+$ into itself given by $T(w) = \tau + \eta \,\square\, w$ is a contraction with respect to the adic norm. If v denotes the unique fixed point of T, we have $u \,\square\, v = id_F$. We fix a real number q in $]0,1[$. Since $\mathrm{ord}\,(\eta) \geq 2$, there exists an $r > 0$ with $|\eta|_r \leq q \cdot r$. We set $t := (1/\sigma\|\tau\|) \cdot (1-q)r$ with $\sigma := \sigma\langle\mathbb{J}\langle A\rangle\rangle$ if $\|\tau\| \neq 0$ and $t := (1-q)r$ if $\|\tau\| = 0$. Then (6.5.2) yields for any element w of $\mathrm{For}_A(F,E)^0_+$ with $|w|_t \leq r$ the estimates

$$\begin{aligned} |T(w)|_t &\leq \sigma \cdot \|\tau\| \cdot t + |\eta \,\square\, w|_t \\ &\leq \sigma \cdot \|\tau\| \cdot t + |\eta|_r \\ &\leq (1-q)r + qr \\ &= r. \end{aligned}$$

Now the assertion follows from (6.1.3).

(6.6.1) (Substitutable elements) Let $\varphi: A \to A'$ be a homomorphism of commutative Banach rings in $({}_c\mathrm{R}_{\mathbb{Q}})$ and E resp. F a Banach module over A resp. A'. Further let u be a formal map in $\mathrm{For}_\varphi(E,F)$ and e an element of E^0. We say that e is *substitutable in* u, if the family $u((1/n!) \cdot e^n)_{n\in\mathbb{N}}$ in F is normally summable in the sense of (5.7.4). Then

$$u[e] := \sum_{n\in\mathbb{N}} u(\frac{1}{n!} \cdot e^n) \in F$$

is a well-defined element of F which is called the *value of u at (the point) e*. Note that we don't use the Banach structure of E here. For instance 0 is always substitutable in u, the corresponding value $u[0] = u_0$ being the constant term of u. If u is a polynomial, then every element e of E^0 is substitutable in u and $u[e] = u \,\square\, i_e$ is the composition of u with the A-homomorphism $i_e: A \to E$ sending 1 to e.

(6.6.2) In addition, suppose given two Banach modules G, H over A' and a homomorphism μ in $\mathrm{L}_{A'}(F \otimes_{A'} G, H)^0$. Further let u resp. v be a formal map in $\mathrm{For}_{\varphi}(E, F)$ resp. $\mathrm{For}_{\varphi}(E, G)$ and $e \in E^0$ an element, which is substitutable in u and v. Then e is substitutable in the formal map $u \cdot_{\mu} v$ of $\mathrm{For}_{\varphi}(E, H)$ with value

(6.6.2.1) $$(u \cdot_{\mu} v)[e] = \mu(u[e] \otimes v[e]).$$

(6.7.1) (Quasisubstitutable elements) In the situation of (6.6.1) let M resp. N be a seminormed module resp. Banach module over A'. For an element m of M we consider the A'-homomorphism

$$\sigma_m : \mathrm{Hom}_{A'}(M, N) \to N$$

given by $\sigma_m(\alpha) = e(m, \alpha) \cdot \alpha(m)$. Let u be a convergent formal map in $\mathrm{Con}_{\varphi}(E, \mathrm{L}_{A'}(M, N))$ and e an element of E°. We say that e is *quasisubstitutable in* u, if e is substitutable in the formal map $\sigma_m \circ u$ of $\mathrm{Con}_{\varphi}(E, N)$ for every element m of M. Then there exists a unique A'-linear map

$$u[e] \in \mathrm{Hom}_{A'}(M, N)$$

with $u[e](m) = e(u, m) \cdot (\sigma_m \circ u)[e]$ for m in M. Again, $u[e]$ is called the *value of* u *at (the point)* e. If e is substitutable in u, then e is quasisubstitutable in u, the converse being false in general.

(6.8.1) (Admissible and quasiadmissible elements) In the situation of (6.6.1), let u be a convergent formal map in $\mathrm{Con}_{\varphi}(E, F)$. Then for any integer p in $\mathbb{N}$, the p-th derivative $D^p(u)$ is a convergent formal map in $\mathrm{Con}_{\varphi}(E, \mathrm{Con}_{\varphi}(E, F)_p)$ and $\mathrm{Con}_{\varphi}(E, F)_p = \mathrm{L}_{A'}(A' \otimes_A \mathrm{S}^p_A(E), F)$. We say that an element e of E° is *admissible for* u, if e is quasisubstitutable in $D^p(u)$ for every integer p. Then $D^p(u)[e]$ is a well-defined homogeneous polynomial in $\mathrm{For}_{\varphi}(E, F)_p$. The formal map

$$u\langle e \rangle := \sum_{p \in \mathbb{N}} D^p(u)[e] \in \mathrm{For}_{\varphi}(E, F)$$

is called the *Taylor series of* u *at (the point)* e. Its constant term $u\langle e \rangle_0$ is the value $u[e]$ of u at e. For instance 0 is always admissible for u, and $u\langle 0 \rangle = u$. If u is a polynomial, then every element e of E^0 is admissible for u, and $u\langle e\rangle$ is the composition of the Taylor series of u with the affine map $t_e: E \to E \times E$ given by $t_e(x) = (e, x)$ for x in E:

$$u\langle e\rangle = \operatorname{Tay}(u) \,\square\, t_e. \tag{6.8.1.1}$$

(6.8.2) In the situation of (6.7.1) we say that e is *quasiadmissible for* u, if e is admissible for all formal maps $\sigma_m \circ u$, $m \in M$. Then there exists a unique formal map $u\langle e\rangle$ in $\operatorname{For}_\varphi(E, \operatorname{Hom}_{A'}(M, N))$ with $\sigma_m \circ (u\langle e\rangle) = (\sigma_m \circ u)\langle e\rangle$ for m in M. Again, $u\langle e\rangle$ is called the *Taylor series of* u *at (the point)* e. Its constant term is the value $u[e]$ of u at e. If e is admissible for u, then e is quasiadmissible for u, the converse being false.

(6.8.3) Let now $t > 0$ be a real number such that u is in $\mathrm{B}_{t,\varphi}(E, F)$ and $\|e\| \leq t$. Suppose moreover that $\|e\| < t$ resp. that $\mathbb{J}$ is ultrametric and $u \in \mathrm{W}_{t,\varphi}(E, F)$. Then e is admissible for u and

$$|u\langle e\rangle|_{t-\|e\|} \leq |u|_t \quad \text{resp.} \quad |u\langle e\rangle|_t \leq |u|_t \quad \text{and} \quad u\langle e\rangle \in \mathrm{W}_{t,\varphi}(E, F). \tag{6.8.3.1}$$

Moreover, in the second situation, if u is in $\mathrm{R}_{t,\varphi}(E, F)$, then $u\langle e\rangle$ is in $\mathrm{R}_{t,\varphi}(E, F)$. This results from an elementary calculation. In particular we see that $u\langle e\rangle$ is convergent, too, and that $\|u[e]\| \leq |u|_t$. Hence, if we set $r := t - \|e\|$, the map ρ_e sending u to $u\langle e\rangle$ is a continuous A'-homomorphism of degree zero from $\mathrm{B}_{t,\varphi}(E, F)$ into $\mathrm{B}_{r,\varphi}(E, F)$ resp. from $\mathrm{W}_{t,\varphi}(E, F)$ into itself with $\|\rho_e\| \leq 1$. For any integer l in $\mathbb{N}$ the estimate (6.2.1.1) shows that e is admissible for $D^l(u)$ if $\|e\| < t$, whereas in the second situation (i.e., if $\mathbb{J}$ is ultrametric and u is in $\mathrm{W}_{t,\varphi}(E, F)$) the maps $\sigma_m \circ D^l(u)$, where m runs in $\mathrm{S}^l_{A'}(\varphi^*(E))$, are in $\mathrm{W}_{t,\varphi}(E, F)$. Hence e is quasiadmissible for $D^l(u)$ in this case. In both cases one verifies easily the formula

$$D^l(u)\langle e\rangle = D^l(u\langle e\rangle). \tag{6.8.3.2}$$

Let f be another element of E^0 such that $\|e\|, \|f\| \leq t$. Suppose that $\|e\| + \|f\| < t$ or that $\mathbb{J}$ is ultrametric and u is in $\mathrm{W}_{t,\varphi}(E, F)$. Then the previous remarks show that $e, e + f$ are admissible for u and that f is admissible for $u\langle e\rangle$. Moreover we have

$$(u\langle e\rangle)\langle f\rangle = u\langle e + f\rangle. \tag{6.8.3.3}$$

(6.8.4) We fix a real number $t > 0$. If $K(0; t)$ denotes the open ball in E^0 with center 0 and radius t, then $F^{K(0;t)}$ is a pseudonormed A'-module with respect to the pseudonorm $_\infty\|\cdot\|$, cf. (5.3.1). Moreover the continuous bounded functions in $F^{K(0;t)}$ form a Banach module $\mathrm{C_b}(K(0;t),F)$ over A'. Let u be a formal map in $\mathrm{B}_{t,\varphi}(E,F)$. Then the correspondence sending an element e of $K(0;t)$ to $u[e]$ is a well defined map $\hat{u}$ in $\mathrm{C_b}(K(0;t),F)$, and we have $_\infty\|\hat{u}\| \leq |u|_t$ by (6.8.3). Thus the mapping

$$\mathrm{B}_{t,\varphi}(E,F) \xrightarrow{\wedge} \mathrm{C_b}(K(0;t),F)$$

associating $\hat{u}$ with u is a weak contraction of degree zero of normed A'-modules. If F is a Banach algebra over A', then $\wedge$ is a homomorphism of algebras over A'.

(6.8.5) In the situation of (6.6.2), let $t > 0$ be a real number with $|u|_t < \infty$, $|v|_t < \infty$, and $\|e\| \leq t$. Suppose moreover that $\|e\| < t$ or that u resp. v is in $\mathrm{W}_{t,\varphi}(E,F)$ resp. $\mathrm{W}_{t,\varphi}(E,G)$. Then e is admissible for the formal map $u \cdot_\mu v$ and we have

$$(u \cdot_\mu v)\langle e\rangle = u\langle e\rangle \cdot_\mu v\langle e\rangle. \tag{6.8.5.1}$$

(6.9.1) (Taylor series of a composition of formal maps) Let $\varphi: A \longrightarrow A'$ and $\varphi': A' \longrightarrow A''$ be morphisms of Banach rings in $({}_{\mathrm{c}}\mathrm{R}_{\mathbb{Q}})$ and E resp. F resp. G a Banach module over A resp. A' resp. A'', and set $\mathbb{J} := \mathbb{J}\langle A\rangle$ for abbreviation. Further suppose we are given real numbers $t, s > 0$, a formal map u resp. v in $\mathrm{B}_{t,\varphi}(E,F)^0_+$ resp. $\mathrm{B}_{s,\varphi'}(F,G)$ such that $|u|_t \leq s$ and an element e of E^0 with $\|e\| \leq t$.

(6.9.2) Proposition. *In the situation of* (6.9.1), *assume moreover that one of the following two conditions is satisfied:*

(a) $\|e\| < t$ *and* $|u|_t < s$.

(b) $\mathbb{J}$ *is ultrametric and* u *resp.* v *is in* $\mathrm{W}^{\varphi'}_{t,\varphi}(E,F)$ *resp.* $\mathrm{W}_{s,\varphi'}(F,G)$.

Then:

(1) e *resp.* $u[e]$ *is admissible for* u *and* $v \,\square\, u$ *resp.* v.

(2) $(v \,\square\, u)\langle e\rangle = v\langle u[e]\rangle \,\square\, (u\langle e\rangle - u[e])$.

(3) $(v \,\square\, u)[e] = v[u[e]]$.

Sketch of proof. Suppose that (a) holds. Then $\|e\| < t$ implies that $\|u[e]\| < s$. Hence by (6.5.2) and (6.8.3) only (2) must be shown. Let $t' \leq t - \|e\|$ be a positive number such that $|u\langle e\rangle - u[e]|_{t'} \leq s - \|u[e]\|$. Then we infer from (6.5.2) and (6.8.3.1) that the maps from $\mathrm{B}_{s,\varphi'}(F,G)$ into $\mathrm{B}_{t',\varphi'\circ\varphi}(E,G)$ associating with v the map $(v \,\square\, u)\langle e\rangle$ resp. $v\langle u[e]\rangle \,\square\, (u\langle e\rangle - u[e])$ are continuous. So it suffices to prove the assertion in the case that v is a homogeneous polynomial. This can be done by a direct calculation. - In the situation of (b) the proof is similar.

(6.9.3) In the situation of (6.9.2) let us drop the assumption that the constant term of u vanishes. Then $v \,\square\, u := v\langle u[0]\rangle \,\square\, (u - u[0])$ is a well defined formal map, again called the *composition of u with v*. In the situation of (6.9.2) (a) resp. (6.9.2)(b), this map is in $\mathrm{B}_{t,\varphi' \circ \varphi}(E, G)$ resp. $\mathrm{W}_{t,\varphi' \circ \varphi}(E, G)$. If u has no constant term or if both u and v are polynomials, this definition reduces to the definition given in (4.1.1). Using (6.9.2), we can easily derive that this composition is associative in an obvious sense.

(6.10.1) (The category (Con) and it's variants) Let

$$\begin{array}{llllll} (\mathrm{Con}) & \text{resp.} & {}_{\infty}(\mathrm{Con}) & \text{resp.} & {}_{\infty}(\mathrm{Con})_{+} & \text{resp.} \\ (\mathrm{B}(\mathrm{Pol})) & \text{resp.} & (\mathrm{B}(\mathrm{Pol}))_{+} & & & \end{array}$$

denote the following category. The objects of these categories are pairs (A, E) consisting of a commutative Banach ring A in $({}^{\mathrm{pn}}_{\mathrm{c}}\mathrm{R}_{\mathbb{Q}})$ and a Banach module E over A. If (A, E) and (A', F) are two such pairs, then a morphism from (A, E) into (A', F) in (Con) resp. ${}_{\infty}(\mathrm{Con})$ resp. ${}_{\infty}(\mathrm{Con})_{+}$ resp. (B(Pol)) resp. $(\mathrm{B}(\mathrm{Pol}))_{+}$ is a pair (φ, u) consisting of a homomorphism φ from A into A' in $({}^{\mathrm{pn}}_{\mathrm{c}}\mathrm{R}_{\mathbb{Q}})$ and a convergent formal map $u \in \mathrm{Con}_{\varphi}(E, F)^{0}_{+}$ resp. an entire formal map $u \in \mathrm{B}_{\infty,\varphi}(E, F)^{0}$ resp. $u \in \mathrm{B}_{\infty,\varphi}(E, F)^{0}_{+}$ resp. a convergent polynomial $u \in \mathrm{B}(\mathrm{Pol}_{\varphi}(E, F))^{0}$ resp. $u \in \mathrm{B}(\mathrm{Pol}_{\varphi}(E, F)^{0}_{+}$. The composition of morphisms is defined according to (6.9.3).

Note that ${}_{\infty}(\mathrm{Con})_{+}$ resp. $(\mathrm{B}(\mathrm{Pol}))_{+}$ is simultaneously a subcategory of ${}_{\infty}(\mathrm{Con})$ and (Con) resp. of (B(Pol)) and ${}_{\infty}(\mathrm{Con})_{+}$, whereas (B(Pol)) and ${}_{\infty}(\mathrm{Con})$ are not subcategories of (Con). The natural functor sending (A, E) to A makes each of these five categories into a fibred and cofibred category over the category of all commutative Banach rings in $({}^{\mathrm{pn}}_{\mathrm{c}}\mathrm{R}_{\mathbb{Q}})$. For a fixed commutative Banach ring A of type $\mathbb{J}$, the fibre $(\mathrm{Con}_{A}) = (\mathrm{Con}^{\mathbb{J}}_{A})$ resp. ${}_{\infty}(\mathrm{Con}_{A}) = {}_{\infty}(\mathrm{Con}^{\mathbb{J}}_{A})$ resp. ${}_{\infty}(\mathrm{Con}_{A})_{+} = {}_{\infty}(\mathrm{Con}^{\mathbb{J}}_{A})_{+}$ resp. $(\mathrm{B}(\mathrm{Pol}_{A})) = (\mathrm{B}(\mathrm{Pol}^{\mathbb{J}}_{A}))$ resp. $(\mathrm{B}(\mathrm{Pol}_{A}))_{+} = (\mathrm{B}(\mathrm{Pol}^{\mathbb{J}}_{A}))_{+}$ in A admits finite products.

(6.11.1) Remarks. (1) The map ^ of (6.8.4) is in general not injective, even if we restrict it to the module of polynomials (take any nonzero E with $E^{0} = 0$). On the other hand it makes no sense to substitute elements of degree different from zero into a formal map. For instance, the formulas (6.6.2.1) and (6.9.2) (3) would not remain valid. Moreover one can show that the polarization formula, which expresses a given homogeneous polynomial by its values (cf. for example [LM] (II 6.2)), does not extend to our general context. Altogether we see that the function concept is not suitable for superanalysis.

(2) (The multicase) In the situation of (4.4.1), let us assume in addition that φ is a homomorphism of Banach rings and that $E = \coprod_{i \in I} E_i$, resp. F is a Banach module over A resp. A'. We put $\mathbb{J} := \mathbb{J}\langle A \rangle$ and $\sigma := \sigma\langle \mathbb{J} \rangle$ for abbreviation. Let $t = (t_i)_{i \in I}$ be a tuple in $(\mathbb{R}_+^*)^I$. For a formal map $u = \sum_{n \in \mathbb{N}^I} u_n$ in $\mathrm{For}_{\varphi}(E, F)$ we set

$$|u|_t := \sum_{n \in \mathbb{N}^I} \frac{1}{n!} \|u_n\| t^n$$

resp.

$$|u|_t := \sup\{\|u_n\| \cdot (\sigma t)^n : n \in \mathbb{N}^I\}$$

according as $\mathbb{J}$ is archimedean resp. ultrametric. Then $|\cdot|_t$ is a pseudonorm and $\mathrm{B}_{t,\varphi}(E, F) := \mathrm{B}(\mathrm{For}_{\varphi}(E, F), |\cdot|_t)$ is a Banach module over A'. Now we assume that $\mathbb{J}$ is ultrametric and fix tuples $\delta = (\delta_i)_{i \in I}$ resp. $n = (n_i)_{i \in I}$ in $\mathbb{R}_+^I$ resp. $\mathbb{N}^I$. Then there exists a unique pseudonorm

$$\|\cdot\|^{\delta} : \mathrm{For}_{\varphi}(E, F)_n \longrightarrow \mathbb{R}_+ \cup \{\infty\}$$

with the following property: If v is a polynomial in ${}^{\mathrm{h}}\mathrm{For}_{\varphi}(E, F)_n$, then $\|v\|^{\delta}$ is the infimum of all real numbers C in $\mathbb{R}_+$ such that the estimate

$$\|v(\prod_{\substack{j \in [1, n_i] \\ i \in I}} x_{ij})\| \leq C \cdot n(v, x) \cdot \prod_{\substack{j \in [1, n_i] \\ i \in I}} \|x_i\|$$

is valid for all tuples $x = (x_{ij})$ in $\coprod_{i \in I} ({}^{\mathrm{h}}\mathrm{B}(E_i))^{n_i}$ with $\|x_{ij}\| \geq \delta_i$ for every j in $[1, n_i]$. Here the index set is endowed with the natural ordering. Using this pseudonorm one defines a submodule $\mathrm{W}_{t,\varphi}(E, F)$ of $\mathrm{B}_{t,\varphi}(E, F)$ analogously to (6.1.2). Then all assertions and formulas presented in this section extend immediately to the multicase. Using this we can give a convergent version both of the implicit function theorem and the existence and uniqueness theorem for ordinary differential equations. A detailed treatment will be given in [B1].

7. The composition map

(7.1.1) (The formal maps $\rho_{E,F}$ and $\chi_{E,F,G}$) Let $\varphi: A \to A'$ be a homomorphism of trivially normed commutative rings in $({}^{\mathrm{pn}}_{\mathrm{c}}\mathrm{R}_{\mathbb{Q}})$ and E resp. F a module over A resp. A'. Recall that $\mathrm{For}_{\varphi}(E,F)$ is a Banach module over A' with respect to the canonical norm given by $\|u\| = e^{-\mathrm{ord}(u)}$, cf. (5.7.2) (6). By the universal property of the symmetric algebra, there exists a unique homomorphism

$$\rho = \rho_{E,F}: \mathrm{S}_{A'}(\mathrm{For}_{\varphi}(E,F)) \longrightarrow \mathrm{For}_{\varphi}(E, \mathrm{S}_{A'}(F))$$

of A'-algebras extending the natural injection from $\mathrm{For}_{\varphi}(E,F)$ into $\mathrm{For}_{\varphi}(E, \mathrm{S}_{A'}(F))$. If we regard ρ as a formal map from $\mathrm{For}_{\varphi}(E,F)$ into $\mathrm{For}_{\varphi}(E, \mathrm{S}_{A'}(F))$, we easily infer $|\rho|_1 \leq 1$ with respect to the canonical norms. In particular

$$\rho_{E,F} \in \mathrm{Con}_{A'}(\mathrm{For}_{\varphi}(E,F), \mathrm{For}_{\varphi}(E, \mathrm{S}_{A'}(F)))^0.$$

Let u be an element of $\mathrm{For}_{\varphi}(E,F)^0_+$. Then (6.8.3) shows that u is admissible for ρ, since $\|u\| \leq 1/e$. Because ρ is a homomorphism of algebras, we infer that the value $\rho[u]$ of ρ at the point u is the cogebra morphism $\Phi(u)$ induced by u. This fact explains the meaning of ρ. For sake of simplicity, let us assume now that $\varphi = id_A$ is the identity of A. If G is another A-module, then we denote by

$$\chi = \chi_{E,F,G}: \mathrm{For}_A(F,G) \times \mathrm{For}_A(E,F) \longrightarrow \mathrm{For}_A(E, \mathrm{S}_A(G))$$

the formal map of degree zero given by $\chi(v \cdot u) = \rho_{F,G}(v) \circ \rho_{E,F}(u)$ for elements u resp. v of $\mathrm{S}_A(\mathrm{For}_A(E,F))$ resp. $\mathrm{S}_A(\mathrm{For}_A(F,G))$.

(7.2.1) (The formal map $\mathrm{For}_A(T,u)$) Let A be a trivially normed ring in $({}^{\mathrm{pn}}_{\mathrm{c}}\mathrm{R}_{\mathbb{Q}})$ and let T, E, F be three A-modules and u a formal map in $\mathrm{For}_A(E,F)$. Then the composition of $\rho_{T,E}$ with the A-homomorphism from $\mathrm{For}_A(T, \mathrm{S}_A(E))$ into $\mathrm{For}_A(T,F)$ sending w to $u \circ w$ is a formal map

$$\mathrm{For}_A(T,u) \in \mathrm{For}_A(\mathrm{For}_A(T,E), \mathrm{For}_A(T,F)).$$

With respect to the canonical norms on $\mathrm{For}_A(T,E)$ and $\mathrm{For}_A(T,F)$ (cf. (5.7.2) (6)), we easily infer the estimate $|\mathrm{For}_A(T,u)|_1 \leq 1$. Hence each element w of $\mathrm{For}_A(T,E)^0_+$ is admissible for $\mathrm{For}_A(T,u)$ with value $\mathrm{For}_A(T,u)[w] = u \,\square\, w$. For $u = id_{\mathrm{S}(E)}$ this formal map coincides with $\rho_{T,E}$. Let G be another A-module, v a formal map in $\mathrm{For}_A(F,G)$, and suppose that u is of degree zero without constant term. Then $\mathrm{For}_A(T,u)$ has the same properties, and we have

$$\mathrm{For}_A(T,v) \,\square\, \mathrm{For}_A(T,u) = \mathrm{For}_A(T, v \,\square\, u).$$

(7.3.1) (The composition map $\lambda_{E,F,G}$) Let E, F and G be three modules over a trivially normed commutative ring A in $({}^{\mathrm{pn}}_{\mathrm{c}}\mathrm{R}_{\mathbb{Q}})$. The formal map

$$\lambda = \lambda_{E,F,G}: \mathrm{For}_A(F,G) \times \mathrm{For}_A(E,F) \longrightarrow \mathrm{For}_A(E,G)$$

given by $\lambda(v \cdot u) = pr_1(v) \circ \rho_{E,F}(u)$ for elements u resp. v in $\mathrm{S}_A(\mathrm{For}_A(E,F))$ resp. $\mathrm{S}_A(\mathrm{For}_A(F,G))$ is called the *composition map.* Clearly λ is of degree zero and without constant term, i.e. a morphism in (For_A). Its bihomogeneous components $\lambda_{p,q}$ (cf. (4.4.1)) are given by $\lambda_{1,q}(v \cdot u_1 \cdots u_q) = v \circ (u_1 \cdots u_q)$ for elements $u_i, 1 \le i \le q$, resp. v in $\mathrm{For}_A(E,F)$ resp. $\mathrm{For}_A(F,G)$ and $\lambda_{p,q} = 0$ if $p \ne 1$. This yields the estimates $\|\lambda_{1,q}\| \le e^q$ and $\|\lambda_{1.q}\|^{(e^{-k},\delta)} \le e^k$ for any k in $\mathbb{N}$ and $\delta > 0$ with respect to the canonical norms on the underlying "function spaces". Using the notations introduced in (6.11.1) (2) once more, we infer that the formal map λ is an element of $\mathrm{W}_{(t,1/e)}(\mathrm{For}_A(F,G) \times \mathrm{For}_A(E,F), \mathrm{For}_A(E,G))$ for any $t > 0$. In particular we see that λ is convergent, i.e.

$$\lambda_{E,F,G} \in \mathrm{Con}_A(\mathrm{For}_A(F,G) \times \mathrm{For}_A(E,F), \mathrm{For}_A(E,G))^0_+.$$

(7.3.2) Lemma. *Let u resp. v be an element in* $\mathrm{For}_A(E,F)^0_+$ *resp.* $\mathrm{For}_A(F,G)^0$. *Then (v,u) is admissible for λ and $\lambda[v,u] = v \,\square\, u$.*

(7.3.3) This lemma follows easily from the multiversion of (6.8.3), cf. (6.11.1) (2). It explains the name for λ and shows in particular that $\tilde{\lambda}\langle v,u\rangle := \lambda\langle v,u\rangle - v \,\square\, u$ is a well-defined formal map without constant term. We remark that (7.3.2) remains valid for an arbitrary u in $\mathrm{For}_A(E,F)^0$, if v is a polynomial. - Applying to λ the inverse of the canonical isomorphism h in (4.2.1), we obtain the formal map

$$h^{-1}(\lambda_{E,F,G}) \in \mathrm{For}_A(\mathrm{For}_A(F,G), \mathrm{For}_A(\mathrm{For}_A(E,F), \mathrm{For}_A(E,G)))$$

given by $h^{-1}(\lambda)(v) = \mathrm{For}_A(E, pr_1(v))$ for v in $\mathrm{S}_A(\mathrm{For}_A(F,G))$. Using the fact that this map is linear, we can derive the following explicit formula for $\Phi(\lambda)$: If

$u_i, 1 \leq i \leq m$, resp. $v_j, 1 \leq j \leq n$, are formal maps in $\operatorname{For}_A(E,F)$ resp. $\operatorname{For}_A(F,G)$, then

$$\Phi(\lambda)(v_1 \cdots v_n \cdot u_1 \cdots u_m) = \tag{7.3.3.1}$$
$$= \sum_{f:[1,m]\to[1,n]} e(v_1,\ldots,v_n,u_1,\ldots,u_m) \cdot \prod_{j=1}^{n} \Big(v_j \circ \big(\prod_{i\in f^{-1}(j)} u_i \big)\Big).$$

Using this formula, we can see immediately that the diagram

$$\begin{array}{ccc} \mathrm{S}_A(\operatorname{For}_A(F,G)) \otimes_A \mathrm{S}_A(\operatorname{For}_A(E,F)) & \xrightarrow{\Phi(\lambda_{E,F,G})} & \mathrm{S}_A(\operatorname{For}_A(E,G)) \\ {\scriptstyle \rho_{F,G}\otimes\rho_{E,F}} \downarrow & & \downarrow {\scriptstyle \rho_{E,G}} \\ \operatorname{For}_A(F,\mathrm{S}_A(G)) \otimes_A \operatorname{For}_A(E,\mathrm{S}_A(F)) & \xrightarrow[\chi_{E,F,G}]{} & \operatorname{For}_A(E,\mathrm{S}_A(G)) \end{array} \tag{7.3.3.2}$$

commutes. Now we take another A-module H and consider the following diagram in (For_A):

$$\begin{array}{ccc} \operatorname{For}(G,H) \times \operatorname{For}(F,G) \times \operatorname{For}(E,F) & \xrightarrow{id\times\lambda} & \operatorname{For}(G,H) \times \operatorname{For}(E,G) \\ {\scriptstyle \lambda\times id} \downarrow & & \downarrow {\scriptstyle \lambda} \\ \operatorname{For}(F,H) \times \operatorname{For}(E,F) & \xrightarrow[\lambda]{} & \operatorname{For}(E,H). \end{array} \tag{7.3.3.3}$$

(7.3.4) Proposition (Associativity of the composition map). *The diagram* (7.3.3.3) *is commutative.*

Proof. Let u resp. v resp. w be an element of $\mathrm{S}(\operatorname{For}(E,F))$ resp. $\mathrm{S}(\operatorname{For}(F,G))$ resp. $\mathrm{S}(\operatorname{For}(G,H))$. Using the fact that (7.3.3.2) commutes and the properties of the Φ-operation, we obtain

$$\begin{aligned} &(\lambda \,\square\, (\lambda \times id))(w \cdot v \cdot u) = \\ &= (\lambda \circ (\Phi(\lambda) \otimes id))(w \cdot v \cdot u) \\ &= \lambda(\Phi(\lambda)(w \cdot v) \cdot u) \\ &= pr_1(\Phi(\lambda)(w \cdot v)) \circ \rho(u) \\ &= \lambda(w \cdot v) \circ \rho(u) \\ &= pr_1(w) \circ \rho(v) \circ \rho(u) \\ &= pr_1(w) \circ \rho(\Phi(\lambda)(v \cdot u)) \\ &= (\lambda \circ \Phi(id \times \lambda))(w \cdot v \cdot u) \end{aligned}$$

$$= (\lambda \,\square\, (id \times \lambda))(w \cdot v \cdot u).$$

(7.3.5) Now we are going to state a generalization of (7.3.4), which is important in Lie theory for instance, cf. (9.6.1). For this purpose, let us fix an element u resp. v resp. w of $\mathrm{For}_A(E,F)^0_+$ resp. $\mathrm{For}_A(F,G)^0_+$ resp. $\mathrm{For}_A(G,H)^0$. Then, with the notation introduced in (7.3.3), we can consider the following diagram in (For_A):

$$\begin{array}{ccc} \mathrm{For}\,(G,H) \times \mathrm{For}\,(F,G) \times \mathrm{For}\,(E,F) & \xrightarrow{id \times \tilde{\lambda}\langle v,u\rangle} & \mathrm{For}\,(G,H) \times \mathrm{For}\,(E,G) \\ \tilde{\lambda}\langle w,v\rangle \times id \downarrow & & \downarrow \tilde{\lambda}\langle w, v \,\square\, u\rangle \\ \mathrm{For}\,(F,H) \times \mathrm{For}\,(E,F) & \xrightarrow[\tilde{\lambda}\langle w \,\square\, v, u\rangle]{} & \mathrm{For}\,(E,H) \end{array} \tag{7.3.5.1}$$

(7.3.6) Proposition. *The diagram* (7.3.5.1) *commutes.*

Proof. In fact, by (6.9.2) (2), the difference of the two composite maps in the diagram (7.3.5.1) is equal to $\lambda \,\square\, (\lambda \times id)\langle w,v,u\rangle - \lambda \,\square\, (id \times \lambda)\langle w,v,u\rangle$. Hence the assertion follows from (7.3.4).

(7.4.1) (The formal map $\mathrm{For}_A(T,\lambda)$**)** Let T, E, F be modules over a trivially normed ring A in $({}^{\mathrm{pn}}_{\mathrm{c}}\mathrm{R}_{\mathbb{Q}})$. Then any formal map u in $\mathrm{For}_A(T, \mathrm{For}_A(E,F))$ has a unique representation $u = (u^{(n)})_{n\in\mathbb{N}}$ with formal maps $u^{(n)}$ in $\mathrm{For}_A(T, \mathrm{For}_A(E,F)_n)$, cf. also (4.2.1). The infimum of the integers n with $u^{(n)} \neq 0$ is called the *order of* u *relative* T and denoted by $\mathrm{ord}_T(u)$. For $T = 0$, this is in accordance with (5.7.2) (6). The map sending u to $\|u\|_T := e^{-\mathrm{ord}_T(u)}$ is a norm which is called the *canonical norm relative* T. With respect to this norm, $\mathrm{For}_A(T, \mathrm{For}_A(E,F))$ is a Banach module over A. Let now G be another A-module and $\lambda = \lambda_{E,F,G}$ the associated composition map. Then

$$\mathrm{For}\,(T,\lambda)\colon \mathrm{For}\,(T, \mathrm{For}\,(F,G)) \times \mathrm{For}\,(T, \mathrm{For}\,(E,F)) \longrightarrow \mathrm{For}\,(T, \mathrm{For}\,(E,G))$$

is a formal map of degree zero without constant term, cf. (7.2.1). Let

$$\mu\colon \mathrm{For}_A(F,G) \otimes_A \mathrm{For}_A(E, \mathrm{S}_A(F)) \longrightarrow \mathrm{For}_A(E,G)$$

denote the A-homomorphism of degree zero sending $\beta \otimes \alpha$ to $\beta \circ \alpha$. Then, if $u = (u^{(n)})_{n\in\mathbb{N}}$ resp. $v = (v^{(n)})_{n\in\mathbb{N}}$ is a formal map in $\mathrm{For}_A(T, \mathrm{For}_A(E,F))^0$

resp. $\mathrm{For}_A(T, \mathrm{For}_A(F,G))^0$, the product $v^{(q)} \cdot_\mu u^q$ is a well-defined element of $\mathrm{For}_A(T, \mathrm{For}_A(E,G))^0$ for any integer q in $\mathbb{N}$. Now we assume in addition that $\mathrm{ord}_T(u) \geq 1$, i.e. $u^{(0)} = 0$. Then, by an elementary calculation, we obtain that (v,u) is substitutable in $\mathrm{For}\,(T,\lambda)$ with value

$$\mathrm{For}\,(T,\lambda)[(v,u)] = \sum_{q \in \mathbb{N}} \frac{1}{q!}\, v^{(q)} \cdot_\mu u^q. \tag{7.4.1.1}$$

(7.5.1) (The relative composition map) Let T, E, F, G be modules over a trivially normed ring A in $({}^{\mathrm{pn}}_{\mathrm{c}}\mathrm{R}_{\mathbb{Q}})$ and let $i_T: T \to T \times E$ resp. $pr_T: T \times E \to T$ denote the natural injection resp. projection. For a formal map u in $\mathrm{For}_A(T \times E, F)$ we call $\mathrm{ord}_T(u) := \mathrm{ord}_T(h^{-1}(u))$ resp. $\|u\|_T := \|h^{-1}(u)\|_T$ the *order* resp. *canonical norm of u relative* T, cf. (7.4.1). The *relative composition map* is the formal map

$$\Lambda = \Lambda_{E,F,G}\langle T\rangle : \mathrm{For}_A(T \times F, G) \times \mathrm{For}_A(T \times E, F) \longrightarrow \mathrm{For}_A(T \times E, G),$$

which is given by $\Lambda(v \cdot u) = pr_1(v) \circ (\mathrm{S}(pr_T) \cdot \rho_{T\times E,F}(u))$ for elements u resp. v in $\mathrm{S}_A(\mathrm{For}_A(T \times E, F))$ resp. $\mathrm{S}_A(\mathrm{For}_A(T \times F, G))$. For $T = 0$, it coincides with the composition map λ of (7.3.1). As in (7.3.1) we get that Λ has the properties $\Lambda_{p,q} = 0$ for $p \neq 1, \|\Lambda_{1,q}\| \leq e^q$ and $\|\Lambda_{1,q}\|^{(e^{-k},\delta)} \leq e^k$ for any q, k in $\mathbb{N}$ and $\delta > 0$, with respect to the canonical norms relative T. Consequently Λ is a map in $\mathrm{W}_{(t,1/e)}$ for any $t > 0$, in particular it is convergent. Now the multiversion of (6.8.3) easily implies the following result.

(7.5.2) Lemma. *Let u resp. v be a formal map of* $\mathrm{For}_A(T \times E, F)^0$ *resp.* $\mathrm{For}_A(T \times F, G)^0$ *such that $u \,\square\, i_T = 0$. Then (v,u) is admissible for Λ with value* $\Lambda[(v,u)] = v \,\square\, (pr_T, u)$.

Also, the other assertions of (7.3) can be carried over m.m. to our more general situation. We leave the formulation to the reader. The relation between the (absolute) composition map and the relative composition map is given by the fact that the natural diagram

$$\begin{array}{ccc} \mathrm{For}\,(T, \mathrm{For}\,(F,G)) \times \mathrm{For}\,(T, \mathrm{For}\,(E,F)) & \xrightarrow{\mathrm{For}\,(T,\lambda)} & \mathrm{For}\,(T, \mathrm{For}\,(E,G)) \\ {\scriptstyle h\times h}\wr\downarrow & & \wr\downarrow{\scriptstyle h} \\ \mathrm{For}\,(T \times F, G) \times \mathrm{For}\,(T \times E, F) & \xrightarrow[\Lambda]{\qquad} & \mathrm{For}\,(T \times E, G) \end{array}$$

in (For $_A$) commutes. We omit the verification.

8. Analytic maps and Banach analytic spaces

(8.1.1)(The sheaves $\mathcal{F}_E\langle F\rangle$ and $\mathcal{C}_E\langle F\rangle$) Let $\varphi: A \to A'$ be a homomorphism of commutative Banach rings in $(^{\text{pn}}_{\text{c}}\text{R}_{\mathbb{Q}})$ and let E resp. F be a Banach module over A resp. A'. For an open subset U of the topological space E^0 the U-fold product $\text{For}_\varphi(E,F)^U$ of $\text{For}_\varphi(E,F)$ is an A'-module in $(\text{M}_{A'})$. Its elements are families $f = (f_x)_{x\in U}$ of formal maps f_x in $\text{For}_\varphi(E,F)$. If $f(x) \in F$ denotes the constant term of f_x, the assignment sending x to $f(x)$ is a F-valued function on U. Although this function doesn't determine f uniquely we will denote it by the same symbol. Moreover $\text{Con}_\varphi(E,F)^U$ is an A'-submodule of $\text{For}_\varphi(E,F)^U$. The correspondence sending U to $\text{For}_\varphi(E,F)^U$ resp. $\text{Con}_\varphi(E,F)$ is in a natural way a sheaf

$$\mathcal{F}_{\varphi,E}\langle F\rangle \quad \text{resp.} \quad \mathcal{C}_{\varphi,E}\langle F\rangle$$

of A'-modules on E^0. Here by a sheaf of A'-modules we mean a sheaf with values in the category $(\text{M}_{A'})$ in the sense of [EGA] (0 3.1). If $\varphi = id_A$, we denote this sheaf by $\mathcal{F}_E\langle F\rangle = \mathcal{F}_{A,E}\langle F\rangle$ and $\mathcal{C}_E\langle F\rangle = \mathcal{C}_{A,E}\langle F\rangle$ respectively. Of course $\mathcal{C}_{\varphi,E}\langle F\rangle$ is an A'-submodule of $\mathcal{F}_{\varphi,E}\langle F\rangle$. For any point a of E^0 we have natural surjective A'-homomorphisms of degree zero:

$$\mathcal{F}_{\varphi,E}\langle F\rangle_a \longrightarrow \text{For}_\varphi(E,F) \quad , \quad \mathcal{C}_{\varphi,E}\langle F\rangle_a \longrightarrow \text{Con}_\varphi(E,F). \tag{8.1.1.1}$$

(8.2.1) (Analytic maps) In the situation of (8.1.1), suppose that we are given an open subset U of E^0 and a section $f = (f_x)_{x\in U}$ in $\Gamma(U, \mathcal{C}_{\varphi,E}\langle F\rangle)$. We say that f is an *analytic map on U with values in F (over φ)*, if for any point a in U there exists a real number $t > 0$ and a formal map u in $\text{B}_{t,\varphi}(E,F)$ such that the following properties are valid:

(1) The open ball $K(a;t)$ with center a and radius t is contained in U.
(2) If x is any point in $K(a;t)$ then $f_x = u\langle x-a\rangle$ is the Taylor series of u at $x-a$.

In this case the underlying function from U into F is continuous, cf. (6.8.4). Again, it doesn't determine the section f uniquely. This effect occurs in classical analysis only at the moment when we consider singular spaces with nilpotent elements. Here it happens even in the manifold case. The analytic maps on U with values in F form an A'-submodule $\Gamma(U, \mathcal{O}_{\varphi,E}\langle F\rangle)$ of $\Gamma(U, \mathcal{C}_{\varphi,E}\langle F\rangle)$, and the correspondence associating $\Gamma(U, \mathcal{O}_{\varphi,E}\langle F\rangle)$ with U is an A'-submodule

$$\mathcal{O}_{\varphi,E}\langle F\rangle$$

of $\mathcal{C}_{\varphi,E}\langle F\rangle$. Again, if $\varphi = id_A$, we write $\mathcal{O}_E\langle F\rangle = \mathcal{O}_{A,E}\langle F\rangle$ instead of $\mathcal{O}_{\varphi,E}\langle F\rangle$. Let B be a Banach algebra over A'. Then $\mathcal{F}_{\varphi,E}\langle B\rangle$, $\mathcal{C}_{\varphi,E}\langle B\rangle$ and $\mathcal{O}_{\varphi,E}\langle B\rangle$ are sheaves of A'-algebras in a natural way. In particular the structure sheaf $\mathcal{O}_E := \mathcal{O}_E\langle A\rangle$ is a sheaf of $\mathbb{J}\langle A\rangle$-commutative algebras over A. If U is an open subset of E^0 and f an analytic map in $\Gamma(U, \mathcal{O}_{\varphi,E}\langle B\rangle^0)$ such that $f(x)$ is a unit in B^0 for every point x in U, then f is a unit in $\Gamma(U, \mathcal{O}_{\varphi,E}\langle B\rangle)$ and we have $(f^{-1})_x = (f_x)^{-1}$.

(8.2.2) Let u be a formal map in $\mathrm{For}_{\varphi}(E, F)$ with positive radius $\rho(u)$ of convergence and let $U := K(a; \rho(u))$ denote the open ball with radius $\rho(u)$ around a given point a in E^0. We set $f_x := u\langle x - a\rangle$ for x in U. Then $f := (f_x)_{x\in U}$ is an analytic map in $\Gamma(U, \mathcal{O}_{\varphi,E}\langle F\rangle)$. If u is a polynomial in $\mathrm{Pol}_{\varphi}(E, F)$ with $\rho(u) > 0$, then $\rho(u) = \infty$, so that f is a section in $\Gamma(E^0, \mathcal{O}_{\varphi,E}\langle F\rangle)$.

We mention two special cases: If $u = id_E$ and $a = 0$, we have $f_x = id_E + x$ for x in E^0. In this case f is called the *identity* of E and will be denoted again by id_E. If u is the constant polynomial with value $y \in F$, we have $f_x = y$ for all x in E^0, and f is called the *constant map* with value y and denoted for short by y. The homomorphism (8.1.1.1) induces an A'-isomorphism

$$\mathcal{O}_{\varphi,E}\langle F\rangle_a \xrightarrow{\sim} \mathrm{Con}_{\varphi}(E, F)$$

of degree zero. Let now U be an arbitrary open subset of E^0, $f = (f_x)_{x\in U}$ an analytic map in $\Gamma(U, \mathcal{O}_{\varphi,E}\langle F\rangle)$ and $a \in U$ a given point. Then there exist open neighbourhoods $W \subseteq V \subseteq U$ of a (depending on f) with the following property: If b is any point in W such that $f_b = 0$, then $f|V = 0$. This easily implies that the identity theorem is valid for the sheaf $\mathcal{O}_{\varphi,E}\langle F\rangle$.

(8.3.1) (Composition of analytic maps) Let $\varphi' : A' \to A''$ be another homomorphism of Banach rings in $(^{\mathrm{pn}}_{\mathrm{c}}\mathrm{R}_{\mathbb{Q}})$ and let G be a Banach module over A''. Further let U resp. V be an open subset of E^0 resp. F^0 and $f = (f_x)_{x\in U}$ resp. $g = (g_y)_{y\in V}$ a section in $\Gamma(U, \mathcal{F}_{\varphi,E}\langle F\rangle^0)$ resp. $\Gamma(V, \mathcal{F}_{\varphi',F}\langle G\rangle)$ with $f(U) \subseteq V$, i.e. with $f(x) \in V$ for every point x in U. If $(g \circ f)_x$ for x in U denotes the composition $g_{f(x)} \,\Box\, (f_x - f(x))$ of the formal maps $g_{f(x)}$ and $f_x - f(x)$, then

$$g \circ f := ((g \circ f)_x)_{x \in U}$$

is a section in $\Gamma(U, \mathcal{F}_{\varphi' \circ \varphi, E}\langle G\rangle)$, called of course the *composition* of f with g. Clearly we have $(g \circ f)(x) = g(f(x))$ for any x in U. If f is in $\Gamma(U, \mathcal{C}_{\varphi,E}\langle F\rangle)$ and g is in $\Gamma(V, \mathcal{C}_{\varphi',F}\langle G\rangle)$, then $g \circ f$ is an element of $\Gamma(U, \mathcal{C}_{\varphi' \circ \varphi, E}\langle G\rangle)$. An easy verification shows that $g \circ f$ is analytic if both f and g are analytic. Finally we remark that this composition is associative in an obvious sense.

(8.4.1) (Derivatives of analytic maps) Let $l \in \mathbb{N}$ be an integer. Then the bounded homogeneous polynomials of degree l form a Banach module $\mathrm{Con}_{\varphi}(E, F)_l$ over A', and there exists a unique A'-homomorphism

$$D^l : \mathcal{C}_{\varphi,E}\langle F\rangle \longrightarrow \mathcal{C}_{\varphi,E}\langle \mathrm{Con}_{\varphi}(E, F)_l\rangle$$

of degree zero such that $D^l(f)_x = D^l(f_x)$ for every section $f = (f_x)_{x \in U}$ in $\Gamma(U, \mathcal{C}_{\varphi,E}\langle F\rangle)$ on an open subset U of E^0. Of course $D^l(f)$ is called the *l-th derivative* of f. It's immediate that D^l transforms the submodule $\mathcal{O}_{\varphi,E}\langle F\rangle$ into $\mathcal{O}_{\varphi,E}\langle \mathrm{Con}_{\varphi}(E, F)_l\rangle$, i.e. that derivatives of analytic maps are analytic.

(8.5.1) (The category $(\mathcal{K})$) By $(\mathcal{K})$ we denote the following category. The objects of this category are triples (A, E, U) consisting of a Banach ring A in $({}^{\mathrm{pn}}_{\mathrm{c}}\mathrm{R}_{\mathbb{Q}})$, a Banach module E over A and an open subset U of E^0. If (A, E, U) and (A', F, V) are two such triples, then a morphism from (A, E, U) into (A', F, V) is a pair (φ, f) consisting of a homomorphism $\varphi: A \to A'$ in $({}^{\mathrm{pn}}\mathrm{R})$ and an analytic map f in $\Gamma(U, \mathcal{O}_{\varphi,E}\langle F\rangle)^0$ with $f(U) \subseteq V$. Of course, the composition of morphisms in $(\mathcal{K})$ is given by the composition of analytic maps. If A is a fixed Banach ring in $({}^{\mathrm{pn}}_{\mathrm{c}}\mathrm{R}_{\mathbb{Q}})$, then we denote by $(\mathcal{K}_A)$ the subcategory of $(\mathcal{K})$ whose objects are the triples (A, E, U) in $(\mathcal{K})$ and whose morphisms are the morphisms in $(\mathcal{K})$ over the identity of A.

Frequently we abbreviate (A, E, U) resp. (A, E, E^0) by (E, U) resp. (E, E^0) or just by U resp. E. Clearly $(\mathcal{K}_A)$ has finite products and a final object e $(= (0, \{0\}))$, but the kernel of a double arrow in $(\mathcal{K}_A)$ doesn't exist in general. Further $(\mathcal{K}_A)$ is not "local", i.e. one cannot "patch" in $(\mathcal{K}_A)$. Following a fundamental idea of Douady from [Do1], we shall construct (in (8.7.1), (8.8.1)) a "local" category (the category of Banach analytic spaces) containing $(\mathcal{K}_A)$ as a full subcategory, such that in this new category kernels of double arrows and finite fibre products exist and such that the objects from $(\mathcal{K}_A)$ are "smooth".

(8.6.1) Examples. Let B be a Banach algebra over A. We regard the addition $a_{B|A}$ as well as the multiplication $\mu_{B|A}$ of B as bounded polynomials in $\operatorname{Pol}_A(B \times B, B)^0_+$. Then, by the correspondence described in (8.2.2), $a_{B|A}$ and $\mu_{B|A}$ define analytic maps

$$\operatorname{add}_{B|A}, \operatorname{mult}_{B|A} \in \Gamma(B^0 \times B^0, \mathcal{O}_{B\times B}\langle B\rangle^0)$$

with $\operatorname{add}_{B|A}(x,y) = x + y$ and $\operatorname{mult}_{B|A}(x,y) = xy$ for x, y in B^0. Since the restriction of id_B to the (open) group $(B^0)^\times$ of units of B^0 is invertible,

$$\operatorname{inv}_{B^\times|A} := 1/id_B|(B^0)^\times \in \Gamma((B^0)^\times, \mathcal{O}_B\langle B\rangle^0)$$

is a well defined analytic map. By an elementary verification we can show that $B = (B, B^0)$ is a ring object in $(\mathcal{K}_A)$ with respect to $\operatorname{add}_{B|A}$, $\operatorname{mult}_{B|A}$, the identity element being of course 1_B. Moreover $B^\times := (B, (B^0)^\times)$ is a group object in $(\mathcal{K}_A)$ with respect to the restriction $\operatorname{mult}_{B^\times|A}$ of $\operatorname{mult}_{B|A}$ to $(B^0)^\times \times (B^0)^\times$. Using the abbreviations $\mathbb{J} := \mathbb{J}\langle A\rangle$ and $\sigma := \sigma\langle\mathbb{J}\rangle$, we set $U := B^\circ$ resp. $V := K(1_B; 1)$ if $\mathbb{J}$ is archimedean and $U := K(0_B; \sigma^{-2})$ resp. $V := K(1_B; \sigma^{-1})$ if $\mathbb{J}$ is ultrametric. Then

$$\exp_{B|A} := ((e_{B|A} + 1_B)\langle x\rangle)_{x\in U} \in \Gamma(U, \mathcal{O}_B\langle B\rangle^0)$$

resp.

$$\log_{B|A} := (l_{B|A}\langle x - 1_B\rangle)_{x\in V} \in \Gamma(V, \mathcal{O}_B\langle B\rangle^0)$$

is an analytic map from U resp. V into B, called the *exponential map* resp. the *logarithm of B over A*. The underlying functions are given by the usual formulas

$$\exp_{B|A}(x) = \sum_{n\in\mathbb{N}} \frac{1}{n!} x^n \quad , x \in U,$$

$$\log_{B|A}(y) = \sum_{n\in\mathbb{N}^*} (-1)^{n-1} \frac{1}{n} (y - 1_B)^n \quad , y \in V.$$

In particular $\exp_{B|A}(0) = 1_B$ and $\log_{B|A}(1_B) = 0$. If x, y are two commuting elements of U, we have $\exp_{B|A}(x + y) = \exp_{B|A}(x) \cdot \exp_{B|A}(y)$. Thus $\exp_{B|A}(x)$ is always invertible in B^0 with $\exp_{B|A}(x)^{-1} = \exp_{B|A}(-x)$. Moreover $\exp_{B|A}$ is a local isomorphism. If U' denotes the image of $V' := K(1_B; \sigma^{-4})$ by $\log_{B|A}$, then U'

is an open subset of U with $\exp_{B|A}(U') = V'$, and $\exp_{B|A}: U' \to V'$, $\log_{B|A}: V' \to U'$ are analytic isomorphisms being mutually inverse. Finally, if B is commutative, then

$$\exp_{B|A}: (B, U) \longrightarrow (B, (B^0)^\times)$$

is a morphism of group objects in $(\mathcal{K}_A)$.

(8.7.1) ($(\mathcal{K}_A)$-functored spaces) Let X be a topological space and $(\mathrm{Sh}(X))$ the category of sheaves of sets on X. Further let $\Phi: (\mathcal{K}_A) \to (\mathrm{Sh}(X))$ be a covariant functor. Then, for any two objects W, W' of $(\mathcal{K}_A)$, we have a canonical morphism of sheaves from $\Phi(W \times W')$ into the product sheaf $\Phi(W) \times \Phi(W')$. If these morphisms are always bijective and if $\Gamma(U, \Phi(e))$ consists of exactly one element for every open subset U of X, we say that Φ is *admissible*.

By a $(\mathcal{K}_A)$-*functored space* we mean a pair $X = (X, \Phi_X)$ consisting of a topological space X and an admissible covariant functor $\Phi_X: (\mathcal{K}_A) \to (\mathrm{Sh}(X))$. Let G be a Banach module over A. If we set $\mathcal{O}_X\langle G\rangle^j := \Phi_X(G[j])$ for j in $\mathbb{J} = \mathbb{J}\langle A\rangle$, then $\mathcal{O}_X\langle G\rangle: = \coprod_{j\in\mathbb{J}} \mathcal{O}_X\langle G\rangle^j$ is canonically endowed with the structure of a ($\mathbb{J}$-graded) module over the ($\mathbb{J}$-graded) ring $\mathcal{O}_X := \mathcal{O}_X\langle A\rangle$. A $(\mathcal{K}_A)$-functored space $(X', \Phi_{X'})$ is called a *subspace* of (X, Φ_X), if X' is a subspace of X and for every object W in $(\mathcal{K}_A)$ the sheaf $\Phi_{X'}(W)$ is a quotient sheaf of $\Phi_X(W)|X'$ such that the projections from $\Phi_X(W)|X'$ onto $\Phi_{X'}(W)$ for W in $(\mathcal{K}_A)$ define a morphism of functors from $(\mathcal{K}_A)$ with values in $(\mathrm{Sh}(X'))$.

Let X and Y be two $(\mathcal{K}_A)$-functored spaces. A *morphism* $X \to Y$ *of* $(\mathcal{K}_A)$-*functored spaces* is a pair (f, f^*) consisting of a continuous map $f: X \to Y$ of the underlying topological spaces and a morphism $f^*: f^{-1} \circ \Phi_Y \to \Phi_X$ of functors on $(\mathcal{K}_A)$ with values in $(\mathrm{Sh}(X))$. Then, for any Banach module G over A, f^* induces a homomorphism $f^*\langle G\rangle$ of degree zero from $\mathcal{O}_Y\langle G\rangle$ into $f_*(\mathcal{O}_X\langle G\rangle)$ being linear over the algebra homomorphism $f^* = f^*\langle A\rangle$ from $\mathcal{O}_Y$ into $f_*(\mathcal{O}_X)$. With respect to the morphisms defined in this way, the $(\mathcal{K}_A)$-functored spaces form a category being denoted by $(\mathcal{F})/(\mathcal{K}_A)$.

(8.7.2) Let (E, U) and (G, W) be two objects of $(\mathcal{K}_A)$ and $\Phi_U(G, W)$ the sheaf of sets on U with

$$\Gamma(U', \Phi_U(G, W)) = \mathrm{Hom}_{\mathcal{K}_A}((E, U'), (G, W))$$

for any open subset U' of U. The assignment associating $\Phi_U(G,W)$ with (G,W) is an admissible covariant functor Φ_U on $(\mathcal{K}_A)$ with values in $(\mathrm{Sh}(U))$. Hence (U,Φ_U) is a $(\mathcal{K}_A)$-functored space. Moreover, it's rather clear that the two possible interpretations of $\mathcal{O}_U\langle G\rangle = \mathcal{O}_E\langle G\rangle|U$ coincide. The correspondence sending (E,U) to (U,Φ_U) gives rise to a covariant functor

$$\alpha\colon (\mathcal{K}_A) \to (\mathcal{F})/(\mathcal{K}_A).$$

Since this functor is obviously fully faithful, we can (and do) regard $(\mathcal{K}_A)$ as being a full subcategory of $(\mathcal{F})/(\mathcal{K}_A)$.

(8.8.1) (Banach analytic spaces) Let $f\colon (E,U) \to F$ be a morphism in $(\mathcal{K}_A)$, i.e. an analytic map of degree zero from U into F in the sense of (8.2.1). Then $X\colon= f^{-1}(0)$ is a closed topological subspace of U. Let (G,W) be an arbitrary object of $(\mathcal{K}_A)$ and $\mathrm{L}_A(F,G) = \mathrm{B}(\mathrm{Hom}_A(F,G))$ the Banach module consisting of all bounded A-linear maps from F into G. Then we have a canonical sheaf homomorphism

$$\Phi_U(\mathrm{L}_A(F,G)) \xrightarrow{\rho} \Phi_U(G)$$

transforming λ into $\lambda \cdot f$. Since ρ is linear over $\Phi_U(A) = (\mathcal{O}_E|U)^0$, we infer that $\mathcal{J}(f;G)\colon= \mathrm{Im}(\rho)$ is a $\Phi_U(A)$-submodule of $\Phi_U(G)$. Let $\Phi_X(G)$ denote the topological restriction of the quotient module $\Phi_U(G)/\mathcal{J}(f;G)$ to X and $\Phi_X(G,W)$ the subsheaf of sets of $\Phi_X(G)$ which consists of all sections "taking values in W".

Then one can verify that the correspondence sending (G,W) to $\Phi_X(G,W)$ is an admissible quotient functor Φ_X of $\Phi_U|X$. The $(\mathcal{K}_A)$-functored space $X = (X,\Phi_X)$ is called the *model (of a Banach analytic space) defined by* $f\colon U \to F$. Clearly (X,Φ_X) is a closed subspace of (U,Φ_U).

(8.8.2) By a *Banach analytic space over* A we mean a $(\mathcal{K}_A)$-functored space X being locally isomorphic to a model. Let

$$(\mathrm{Ban}_A)$$

denote the full subcategory of $(\mathcal{F})/(\mathcal{K}_A)$ consisting of all Banach analytic spaces over A. One can show that (Ban_A) enoys all desired properties: (1) It is "local" and contains $(\mathcal{K}_A)$ as a full subcategory. (2) In (Ban_A) kernels of double arrows and finite fibre products exist. (3) The objects of $(\mathcal{K}_A)$ are "smooth" in (Ban_A).

After these preparations, we can extend the whole theory of Banach analytic spaces (cf. [Do$_1$], [Do$_2$]) to our considerably more general situation. A detailed exposition will be given in [B$_2$].

(8.9.1) Remarks. (1) The calculus of convergence presented in the preceding sections does not achieve the same degree of generality as the formal calculus of section 4. The reason for this is the condition (3) in the definition of a valued signed group ($|\epsilon(j,k)^{-1}| = |\epsilon(j,k)|^{-1}$), which is crucial for the norm calculus to work. It is this condition which forces our rings Λ to be subrings of valued fields in practice.

On the other hand it should be possible to deform a given Banach analytic space X over a Banach ring of type $(\mathbb{J}, \Lambda, \epsilon, \pi)$ under suitable assumptions over a (convergent) parameter space to a Banach analytic space X' over a Banach ring of type $(\mathbb{J}, \Lambda, \epsilon', \pi)$ such that the commutation factor ϵ' is "classical", i.e. ϵ' takes values in $\{\pm 1\}$.

In order to attain this, we have to admit considerably more general rings Λ, for instance the ring of holomorphic functions on an open polycylinder. Definitely, a solution of this problem exists, either using suitable families of valuations or relative versions of valuations. Since these things are technically somewhat involved and we don't have enough experience at this moment to filter out the "right" solution, we will work out the details of this generalization at another place.

(2) It would be interesting to have an intrinsic characterization of the Banach analytic spaces in our sense among all $(\mathcal{K}_A)$-functored spaces in the spirit of the paper [BRP].

9. Lie algebras and formal groups

(9.1.1) (Lie algebras) Let A be a l.t. commutative ring in $({}_c\mathcal{R})$ and L a l.t. algebra over A with multiplication $[\ ,\ ] = [\ ,\ ]_L\colon L \times L \to L$. We say that L is a *l.t. Lie algebra* over A, if for any three elements x, y, z of L the following conditions apply:

(1) $[x,y] = -e(x,y)\cdot[y,x]$.
(2) $[x,[y,z]] + e(x,y,z)\cdot[y,[z,x]] + e(x,y,z)\cdot[z,[x,y]] = 0$.
(3) If x is even then $[x,x] = 0$.
(4) If x is odd then $[x,[x,x]] = 0$.

If 2 (resp. 3) is a nonzerodivisor on L, then (1) implies (3) (resp. (2) implies (4)). If L is discrete, we call L a *Lie algebra* over A. We say that L is *abelian*, if $[\ ,\ ]$ is zero.

In a similar way we define *graded Lie algebras*, cf. also (2.4.1). The complete l.t. Lie algebras resp. Lie algebras over variable rings in $({}_{\mathrm{c}}\mathcal{R})$ resp. $({}_{\mathrm{c}}\mathrm{R})$ form a fibred and cofibred category $(\mathcal{L}al)$ resp. (Lal) over $({}_{\mathrm{c}}\mathcal{R})$ resp. $({}_{\mathrm{c}}\mathrm{R})$ in a natural way. - Let B be a l.t. associative algebra over a l.t. ring A in $({}_{\mathrm{c}}\mathrm{R})$. Then B, endowed with the bracket $[\ ,\ ]$ given by $[x,y] = xy - e(x,y)\cdot yx$, is a l.t. Lie algebra over A, denoted by $\mathrm{L}(B)$ or simply by B. The correspondence sending B to $\mathrm{L}(B)$ gives rise to a covariant functor

$$\mathrm{L}\colon (\mathcal{A}l) \longrightarrow (\mathcal{L}al) \quad \text{resp.} \quad \mathrm{L}\colon (\mathrm{Al}) \longrightarrow (\mathrm{Lal})$$

of fibred and cofibred categories over $({}_{\mathrm{c}}\mathcal{R})$ resp. $({}_{\mathrm{c}}\mathrm{R})$. If B is a formal bigebra over a ring A in $({}_{\mathrm{c}}\mathcal{R})$, then $\mathrm{P}(B)$ is a Lie subalgebra of $\mathrm{L}(B)$, and the correspondence associating $\mathrm{P}(B)$ with B defines a covariant functor of categories over $({}_{\mathrm{c}}\mathcal{R})$ resp. $({}_{\mathrm{c}}\mathrm{R})$:

$$\mathrm{P}\colon (\mathcal{B}) \longrightarrow (\mathcal{L}al) \quad \text{resp.} \quad \mathrm{P}\colon (\mathrm{B}) \longrightarrow (\mathrm{Lal}).$$

(9.1.2) Let E be a module over a ring A in $({}_{\mathrm{c}}\mathrm{R})$. Then $\mathrm{End}_A(E)$ is a Lie algebra with respect to the bracket $[\ ,\ ]$. If E is a finitely generated projective A-module, the trace map $\mathrm{Tr}_E\colon \mathrm{End}_A(E) \to A$ is a homomorphism of Lie algebras over A, if we regard A as an abelian Lie algebra. - Let now $\mathrm{F}_A(E)$ be the free algebra of E and let H be the union $\bigcup_{n=1}^{\infty} {}^{\mathrm{h}}\mathrm{F}^n_A(E)$. Further let M_E denote the two-sided graded ideal of $\mathrm{F}_A(E)$ generated by the elements

$$xy + e(x,y)\cdot yx\ ,\ \ x,y \in H,$$

$$x(yz) + e(x,y,z)\cdot y(zx) + e(x,y,z)\cdot z(xy)\ , x,y,z \in H,$$

$$xx,\ x \in H \text{ even},$$

$$x(xx),\ x \in H \text{ odd}.$$

Then the quotient

$$\mathrm{FL}(E) = \mathrm{FL}_A(E) := \mathrm{F}_A(E)/M_E$$

is (graded) Lie algebra over A, which is called the *free Lie algebra* of E. For an integer n in $\mathbb{N}_+$ let $\mathrm{FL}^n(E) = \mathrm{FL}^n_A(E)$ denote the canonical image of $\mathrm{F}^n_A(E)$ in $\mathrm{FL}_A(E)$. Clearly we have $\mathrm{FL}^1_A(E) = E$ and $\mathrm{FL}^2_A(E) = \Lambda^2_A(E)$. The injection i_E from E into $\mathrm{F}_A(E)$ induces an injection from E into $\mathrm{FL}_A(E)$, again denoted by the same symbol. The Lie algebra $\mathrm{FL}_A(E)$ factors uniquely module homomorphisms of degree zero from

E into arbitrary Lie algebras. Thus the correspondence sending E to $\mathrm{FL}_A(E)$ defines a covariant functor

$$\mathrm{FL}: (M) \longrightarrow (\mathrm{Lal})$$

of cofibred categories over $({}_c\mathrm{R})$. Let A be a ring in $({}_c\mathrm{R})$ and I a set of type $\mathbb{J}\langle A\rangle$. If $E = \coprod_{i\in I} Ae_i$ is the free A-module associated with I, then $\mathrm{FL}_A(E)$ is called the *free Lie algebra* of I and denoted by $\mathrm{FL}_A[I]$. Again the correspondence sending I to $\mathrm{FL}_A[I]$ defines a covariant functor $\mathrm{FL}[\cdot]$ of cofibred categories from $(\mathrm{S}) \times_{(\mathrm{SG})} ({}_c\mathrm{R})$ into (Lal).

(9.2.1) (The enveloping algebra of a Lie algebra) Let L be a Lie algebra over a ring A in $({}_c\mathrm{R})$ of type $\mathbb{J}$ and let $\mathrm{T}_A(L)$ denote the tensor algebra of the underlying A-module. Then the two-sided ideal N_L generated by all elements of the form $x \otimes y - e^{\mathbb{J}}(x,y) \cdot y \otimes x - [x,y]$, where x, y run in L, is homogeneous with respect to the $\mathbb{J}$-gradation, but in general not with respect to the $\mathbb{Z}$-gradation. The quotient

$$\mathrm{U}(L) = \mathrm{U}_A(L) := \mathrm{T}_A(L)/N_L$$

is an algebra over A in (Al), which is called the *enveloping algebra* of L. The natural A-homomorphism i_L from L into $\mathrm{U}_A(L)$ is a homomorphism of Lie algebras over A. If L is abelian and if either $2 \in A^\times$ or all elements of $\mathbb{J}$ are even, then $\mathrm{U}_A(L) = \mathrm{S}_A(L)$ is the symmetric algebra of L. There exists a unique structure of a cocommutative bigebra on $\mathrm{U}_A(L)$ such that the natural surjection p_L from $\mathrm{T}_A(L)$ onto $\mathrm{U}_A(L)$ is a morphism of bigebras over A. Therefore $\mathrm{U}_A(L)$ is called also the *enveloping bigebra* of L. The image of i_L is contained in $\mathrm{P}(\mathrm{U}_A(L))$.

Let $\varphi: A \to A'$ be a ring homomorphism in $({}_c\mathrm{R})$, B' an A'-algebra in (Al) and $f: L \to B'$ a homomorphism of Lie algebras over φ. Then there exists a unique homomorphism $u: \mathrm{U}_A(L) \to B'$ of algebras over φ with $u \circ i_L = f$. If B' is a bigebra over A', then u is a morphism of bigebras if and only if $f(L)$ is contained in $\mathrm{P}(B')$. In particular we infer that the correspondence sending L to $\mathrm{U}_A(L)$ defines a covariant functor of cofibred categories over $({}_c\mathrm{R})$:

$$\mathrm{U}: (\mathrm{Lal}) \longrightarrow (\mathrm{B}).$$

(9.2.2) Let E be a module over a ring A in $({}_c\mathrm{R})$. For an integer n in $\mathbb{Z}$ we set $\mathrm{F}_n = \mathrm{F}_n(E) = \coprod_{i\leq n} \mathrm{T}^i_A(E)$. Then $\mathrm{F}. = \mathrm{F}.(\mathrm{T}_A(E)) = (\mathrm{F}_n)_{n\in\mathbb{Z}}$ is an ascending exhausting filtration of $\mathrm{T}_A(E)$ with $\mathrm{F}_{-1} = 0$, $\mathrm{F}_0 = A \cdot 1$ and $\mathrm{F}_m \cdot \mathrm{F}_n \subseteq \mathrm{F}_{m+n}$. In an obvious sense the coproduct stabilizes this filtration, i.e. $\mathrm{T}_A(E)$ is a *filtered* bigebra over A. Moreover the associated graded algebra $\mathrm{gr}_{\mathrm{F.}}(\mathrm{T}_A(E))$ coincides with $\mathrm{T}_A(E)$.

Let now L be a Lie algebra over A. Then $\mathrm{U}_A(L)$ is a filtered bigebra over A with respect to the quotient filtration $\mathrm{F.} = \mathrm{F.}(\mathrm{U}_A(L)) = (\mathrm{F}_n(\mathrm{U}_A(L)))_{n\in\mathbb{Z}}$ induced by $\mathrm{F.}(\mathrm{T}_A(L))$. We now assume that $2 \in A^\times$ or that all elements of $\mathbb{J} = \mathbb{J}\langle A\rangle$ are even. Then $\mathrm{gr}_{\mathrm{F.}}(\mathrm{U}_A(L))$ is commutative, so that the homomorphism $\mathrm{gr}(p_L)$ from $\mathrm{gr}_{\mathrm{F.}}(\mathrm{T}_A(L)) = \mathrm{T}_A(L)$ onto $\mathrm{gr}_{\mathrm{F.}}(\mathrm{U}_A(L))$ induces a surjective homomorphism

$$\omega_L \colon \mathrm{S}_A(L) \longrightarrow \mathrm{gr}_{\mathrm{F.}}(\mathrm{U}_A(L))$$

of graded A-algebras, which is functorial in L. By a slight improvement of the arguments given in [Ja] (V 2), we can directly prove that ω_L is bijective if L is a free A-module.

(9.2.3) Let L be an arbitrary Lie algebra over a $\mathbb{Q}$-algebra A in $({}_c\mathrm{R})$. Then $\mathrm{U}_A(L)$ is a bigebra of Lie type over A, cf. (3.4.1). We regard the natural projection pr_1 from $\mathrm{S}_A(L)$ onto L as a homomorphism into $\mathrm{U}_A(L)$. Then pr_1 is p.t.n., so that

$$\eta_L := \Phi(pr_1) \colon \mathrm{S}_A(L) \longrightarrow \mathrm{U}_A(L)$$

is a well defined morphism of cogebras over A, c.f. (3.5.2). Clearly η_L is functorial in L. For elements $x_1, \ldots, x_n$ of L the formula

$$\eta_L(x_1 \cdots x_n) = \frac{1}{n!} \sum_{\pi \in \gamma_n} e(\pi^{-1}; x) \cdot x_{\pi(1)} \cdots x_{\pi(n)} \qquad (9.2.3.1)$$

is valid. In particular we have $\eta_L(1) = 1$. In general η_L is not a ring homomorphism. Since η_L stabilizes the respective ascending filtrations, it induces an A-homomorphism $\mathrm{gr}(\eta_L)$ from $\mathrm{S}_A(L) = \mathrm{gr}_{\mathrm{F.}}(\mathrm{S}_A(L))$ into $\mathrm{gr}_{\mathrm{F.}}(\mathrm{U}_A(L))$, c.f. (9.2.2). Clearly $\mathrm{gr}(\eta_L) = \omega_L$. Now we can state the following crucial result:

(9.2.4) Theorem (Poincaré-Birkhoff-Witt). *$\eta_L \colon \mathrm{S}_A(L) \longrightarrow \mathrm{U}_A(L)$ is an isomorphism of cogebras over A.*

In the special case $\mathbb{J} = 0$, this theorem is a Bourbaki exercise ([Bo3] Chap. II, §2, Ex. 16). In order to handle the general case, we have to use that $\omega_L = \mathrm{gr}(\eta_L)$ is bijective, if L is a free A-module (cf. (9.2.2)), and the fact that the free Lie algebra of a free A-module is a free A-module. The latter assertion follows by base change from the special case $A = \mathbb{Q}$, where it is obvious. We remark that it is possible to give explicit formulas for the inverse η_L^{-1} of η_L.

(9.2.5) Corollary. *The map* $i_L: L \longrightarrow \mathrm{U}_A(L)$ *is injective and induces an isomorphism from* L *onto the Lie algebra* $\mathrm{P}(\mathrm{U}_A(L))$ *of primitive elements of* $\mathrm{U}_A(L)$. *Moreover* $\omega_L: \mathrm{S}_A(L) \to \mathrm{gr}_{\mathrm{F}_\cdot}(\mathrm{U}_A(L))$ *is an isomorphism of algebras.*

As an application of (9.2.5), we can characterize bigebras of the form $\mathrm{U}_A(L)$ among all bigebras. Also, the result justifies our terminology.

(9.2.6) Proposition. *The functor* $\mathrm{U}: (\mathrm{Lal}) \to (\mathrm{B})$ *is fully faithful. Its essential image consists precisely of all bigebras of Lie type.*[1]

Proof (Sketch). The first assertion is a direct consequence of (9.2.5). In order to show the second assertion, we consider an arbitrary bigebra B of Lie type over a ring A in $({}_c\mathrm{R}_{\mathbb{Q}})$ and the Lie algebra $P := \mathrm{P}(B)$ of its primitive elements. The natural injection from P into B extends to a morphism $\alpha: \mathrm{U}_A(P) \to B$ of bigebras over A. Let $u \in \mathrm{Hom}_A(B, \mathrm{U}_A(P))^0$ denote the composition of i_P with the projection from B onto P given by p_B. Being power-bounded with $u(1) = 0$, the homomorphism u is in fact p.t.n.. Hence $\beta := \Phi(u)$ is a well defined morphism of cogebras over A from B into $\mathrm{U}_A(P)$. By an immediate verification we see that α and β are mutually inverse.

(9.2.7) Example. Let E be a module over a ring A in $({}_c\mathrm{R})$. Then there exists a unique homomorphism $\varphi_E: \mathrm{FL}_A(E) \to \mathrm{P}(\mathrm{T}_A(E))$ of Lie algebras over A extending the natural injection from E into $\mathrm{P}(\mathrm{T}_A(E))$. By the universal property of the enveloping algebra we get a homomorphism

$$u_E: \mathrm{U}_A(\mathrm{FL}_A(E)) \longrightarrow \mathrm{T}_A(E)$$

of bigebras over A which restricts to φ_E. It's apparent that u_E is in fact bijective. Hence u_E induces an isomorphism of Lie algebras from $\mathrm{P}(\mathrm{U}_A(\mathrm{FL}_A(E)))$ onto $\mathrm{P}(\mathrm{T}_A(E))$. If A is a $\mathbb{Q}$-algebra, then $\mathrm{P}(\mathrm{U}_A(\mathrm{FL}_A(E))) = \mathrm{FL}_A(E)$ by (9.2.5), so that we can identify $\mathrm{P}(\mathrm{T}_A(E))$ canonically with the free Lie algebra $\mathrm{FL}_A(E)$ of E in this case.

---

[1] In this proposition all rings are supposed to be $\mathbb{Q}$-algebras

(9.3.1) (Formal groups and their enveloping bigebras) Let A be a ring in $({}_c\mathrm{R}_{\mathbb{Q}})$. A *formal group* over A is a group object $G = (G, m_G, j_G)$ in the category $(\mathrm{For}\,_A)$. Then the multiplication m_G resp. the inversion mapping j_G is a formal map in $\mathrm{For}\,_A(G \times G, G)^0_+$ resp. $\mathrm{For}\,_A(G, G)^0_+$ satisfying

$$m_G \,\square\, (id_G, 0) = id_G \quad , \quad m_G \,\square\, (0, id_G) = id_G,$$

$$m_G \,\square\, (m_G \times id_G) = m_G \,\square\, (id_G \times m_G),$$

$$m_G \,\square\, (j_G, id_G) = 0 \quad , \quad m_G \,\square\, (id_G, j_G) = 0.$$

The formal groups over variable rings in $({}_c\mathrm{R}_{\mathbb{Q}})$ form in a natural way a fibred and cofibred category over $({}_c\mathrm{R}_{\mathbb{Q}})$ denoted by (FG). For a fixed ring A in $({}_c\mathrm{R}_{\mathbb{Q}})$ the fibre (FG_A) in A is the category of formal groups over A.

(9.3.2) Let G be a formal group over A with multiplication m_G. Then $\Phi(m_G)$ is a cogebra morphism from $\mathrm{S}_A(G) \otimes_A \mathrm{S}_A(G) = \mathrm{S}_A(G \times G)$ into $\mathrm{S}_A(G)$ stabilizing the respective ascending filtrations. With respect to $\Phi(m_G)$ as multiplication, the A-module $\mathrm{S}_A(G)$ is an associative algebra over A admitting $1_{\mathrm{S}(G)}$ as an identity element. This is an easy consequence of the properties of the Φ-operation, cf. (3.5.3). Thus

$$\mathrm{E}(G) = \mathrm{E}_A(G) = (\mathrm{S}_A(G), \Phi(m_G), c_{\mathrm{S}(G)}, \gamma_{\mathrm{S}(G)})$$

is a cocommutative filtered bigebra over A, called the *enveloping bigebra* of G. An elementary verification shows that the associated graded algebra $\mathrm{gr}_{\mathrm{F}.}(\mathrm{E}_A(G))$ coincides with $\mathrm{S}_A(G)$. The A-submodule $\mathrm{P}(\mathrm{E}_A(G))$ of primitive elements of $\mathrm{E}_A(G)$ is a Lie subalgebra of $\mathrm{L}(\mathrm{E}_A(G))$ with respect to the bracket $[,]_{\Phi(m_G)}$, denoted by

$$\mathrm{L}(G) = \mathrm{L}_A(G)$$

and called the *Lie algebra of the formal group* G. By (3.3.5) we can identify $\mathrm{L}(G)$ and G as A-modules in a natural way. If we regard $[\,,\,]_{\mathrm{L}(G)}$ as a homomorphism from $G \otimes_A G$ into G, we have $[\,,\,]_{\mathrm{L}(G)} = (m_G)_{1,1} - (m_G)_{1,1} \circ \sigma_G$. By the universal property of the enveloping algebra, the canonical injection from $\mathrm{L}_A(G)$ into $\mathrm{E}_A(G)$ extends to a morphism

$$\alpha_G \colon \mathrm{U}_A(\mathrm{L}_A(G)) \longrightarrow \mathrm{E}_A(G)$$

of filtered bigebras over A. From (9.2.5) we infer that α_G is bijective. Since $\mathrm{E}_A(G)$, $\mathrm{L}_A(G)$ are functorial in G, we obtain functors

$$\mathrm{E} \colon (\mathrm{FG}) \longrightarrow (\mathrm{B}) \quad , \quad \mathrm{L} \colon (\mathrm{FG}) \longrightarrow (\mathrm{Lal})$$

of cofibred categories over $({}_c\mathrm{R}_{\mathbb{Q}})$ with $\mathrm{P} \circ \mathrm{E} = \mathrm{L}$. It's immediate that E is fully faithful. Finally the α_G define an isomorphism $\alpha\colon \mathrm{U} \circ \mathrm{L} \xrightarrow{\sim} \mathrm{E}$ of functors.

(9.4.1) (The Hausdorff group of a Lie algebra) Let L be a Lie algebra over a ring A in $({}_c\mathrm{R}_{\mathbb{Q}})$ and $\mu_{\mathrm{U}(L)}$ the multiplication of the enveloping algebra $\mathrm{U}_A(L)$ of L. Then $\Phi_L := \eta_L^{-1} \circ \mu_{\mathrm{U}(L)} \circ (\eta_L \otimes \eta_L)$ is an A-linear map from $\mathrm{S}_A(L) \otimes_A \mathrm{S}_A(L)$ into $\mathrm{S}_A(L)$. Since both η_L and $\mu_{\mathrm{U}(L)}$ are morphisms of cogebras, Φ_L is a cogebra morphism, too. Hence $h_L := pr_1 \circ \Phi_L$ is a formal map in $\mathrm{For}_A(L \times L, L)^0_+$ with $\Phi(h_L) = \Phi_L$, and

$$\mathrm{H}(L) = \mathrm{H}_A(L) := (L, h_L)$$

is a formal group over A. We call h_L resp. $\mathrm{H}(L)$ the *Hausdorff law* resp. the *Hausdorff group* of the Lie algebra L. The Lie algebra of $\mathrm{H}(L)$ is nothing but L. The correspondence sending L to $\mathrm{H}_A(L)$ gives rise to a covariant functor

$$\mathrm{H}\colon (\mathrm{Lal}) \longrightarrow (\mathrm{FG})$$

of cofibred categories over $({}_c\mathrm{R}_{\mathbb{Q}})$ such that $\mathrm{L} \circ \mathrm{H} = \mathrm{id}_{(\mathrm{Lal})}$. Let now G be a formal group over A. Since we can regard $\alpha_G \circ \eta_{\mathrm{L}(G)}$ as an isomorphism of bigebras over A from $\mathrm{E}_A(\mathrm{H}_A(\mathrm{L}_A(G)))$ onto $\mathrm{E}_A(L)$,

$$\exp_G := pr_1 \circ \alpha_G \circ \eta_{\mathrm{L}(G)}\colon \mathrm{H}_A(\mathrm{L}_A(G)) \longrightarrow G$$

is an isomorphism of formal groups over A called the *exponential mapping* of G. Its tangent map $(\exp_G)_1$ is the identity. The formal map $\log_G := \exp_G^{-1}$ is again an isomorphism of formal groups from G onto $\mathrm{H}_A(\mathrm{L}_A(G))$ called the *logarithm mapping* of G. Obviously we have $h_{\mathrm{L}(G)} = \log_G \square\, m_G \,\square\, (\exp_G \times \exp_G)$. The correspondences sending G to $\exp_G$ resp. $\log_G$ are isomorphisms

$$\exp\colon \mathrm{H} \circ \mathrm{L} \longrightarrow \mathrm{id}_{(\mathrm{FG})} \quad , \quad \log\colon \mathrm{id}_{(\mathrm{FG})} \longrightarrow \mathrm{H} \circ \mathrm{L}$$

of functors on (FG) into itself, being mutually inverse. This fact immediately implies the following fundamental proposition.

(9.4.2) Theorem (Formal Lie theory). *The functor* $\mathrm{L}\colon (\mathrm{FG}) \to (\mathrm{Lal})$ *is an equivalence of categories.*

(9.5.1) (Canonical formal groups) Let G be a formal group over A. Then there exists a unique sequence $\phi^{(q)}, q \in \mathbb{N}$, of formal maps $\phi^{(q)} = \phi_G^{(q)}$ in $\mathrm{For}_A(G,G)_+^0$ such that $\phi^{(0)} = 0$ and $\phi^{(q+1)} = m_G \,\square\, (id_G, \phi^{(q)})$. Obviously we have $\phi^{(1)} = id_G$. We call $\phi^{(q)}$ the *q-power mapping* of G. Its tangent map is $q \cdot id_G$. Moreover $\phi_G^{(q)}$ is functorial in G. If $id_{\mathrm{S}(G)}^q$ denotes the q-th power of $id_{\mathrm{S}(G)}$ in the A-algebra $\mathrm{For}_A(G, \mathrm{E}_A(G))$, then an elementary calculation shows $\Phi(\phi_G^{(q)}) = id_{\mathrm{S}(G)}^q$. The formal group G is called *canonical*, if $\exp_G$ is linear, i.e. $\exp_G = id_G$. We can characterize canonical formal groups as follows.

(9.5.2) Proposition. *The following properties are equivalent:*

(1) *G is canonical.*

(2) *$G = \mathrm{H}(\mathrm{L}(G))$, i.e. the multiplication m_G of G coincides with the Hausdorff law $h_{\mathrm{L}(G)}$ of the Lie algebra $\mathrm{L}(G)$ of G.*

(3) *The q-power mapping $\phi_G^{(q)}$ of G is linear for every q in $\mathbb{N}$.*

(9.5.3) Corollary. *For any Lie algebra L over A the Hausdorff group $\mathrm{H}(L)$ is canonical.*

(9.5.4) Remark. The theory of free Lie algebras can be generalized to the super context, the same being true for the Campbell-Hausdorff formula for the Hausdorff law of a Lie algebra, cf. for instance [Bo3], Chap. II, §7.2. The latter easily implies that the convergent version of Lie theory remains true in the super case as well. We intend to give the details at another place.

(9.6.1) (The formal group associated with the Lie algebra of vector fields) Let E be a module over a trivially normed ring A in $({}^{\mathrm{pn}}_{\mathrm{c}}\mathrm{R}_{\mathbb{Q}})$ and $\lambda := \lambda_{E,E,E}$ the corresponding composition map, cf. (7.3.1). Then (id_E, id_E) is admissible for λ with $\lambda[id_E, id_E] = id_E$, cf. (7.3.2). We set $m := \lambda\langle id_E, id_E\rangle - id_E$ for abbreviation. Then

$$m: \mathrm{For}_A(E,E) \times \mathrm{For}_A(E,E) \longrightarrow \mathrm{For}_A(E,E)$$

is a morphism in (For_A), and its power series development $m = \sum_{p,q\in\mathbb{N}} m_{p,q}$ is given by $m_{0,1}(u) = u$,

$$m_{1,q}(v \cdot u_{(1)} \cdots u_{(q)}) = D^q(v) \circ (u_{(1)} \cdots u_{(q)})$$

for elements $u, u_{(1)}, \ldots, u_{(q)}, v$ of $\mathrm{For}_A(E,E)$ and q in $\mathbb{N}$. In the remaining cases the bihomogeneous polynomials $m_{p,q}$ vanish. In particular we have $m_{1,1}(v \cdot u) = D(v) \circ u$.

From (7.3.6) we infer that $(\mathrm{For}_A(E,E), m)$ is a formal group over A called the *formal group of* E and denoted by

$$\mathbb{F}(E) = \mathbb{F}_A(E).$$

If u and v are two elements of the corresponding Lie algebra $\mathrm{L}(\mathbb{F}(E)) = \mathrm{For}_A(E,E)$, then $[u,v] = D(u) \circ v - e(u,v) \cdot D(v) \circ u$. We call $\mathrm{L}(\mathbb{F}(E))$ the *Lie algebra of formal vector fields* of E. It is possible to give recursive formulas for the inversion mapping $j_{\mathbb{F}(E)}$ and the exponential mapping $\exp_{\mathbb{F}(E)}$ of $\mathbb{F}(E)$. Let C be an A-algebra in (Al). Then an elementary calculation shows that the correspondence

$$\mathrm{For}_A(E,E) \longrightarrow \mathrm{Der}_A(\mathrm{For}_A(E,C))$$

sending u to the derivation D_u given by $D_u(v) = -e(u,v) \cdot D(v) \circ u$ is a homomorphism of Lie algebras over A. This fact justifies our terminology. We remark that the formal group $\mathbb{F}(E)$ (or rather a variant of this group) is of fundamental importance in deformation theory, cf. [LM]. Also, there is a relative variant of this formal group, cf. [BL] and [B$_1$].

Remark. Bourbaki, in the recent English version of Algebra II, independently gives a composition calculus based on the bigebra structure of $\Gamma(E)$, cf. [Bo$_2$]. It is confined to the classical commutative case and does not embrace a complete foundation of analysis.

References

[Ab] Abe, E.: Hopf algebras. Cambridge - New York: Cambridge Univ. Press 1980

[BRP] Bartocci, C., Bruzzo, U., Hernández Rùipérez, D., Pestov, V.G.: Foundations of supermanifold theory: the axiomatic approach. Differential Geometry and its Applications **3**, 135 - 155 (1993)

[Be] Berezin, F.A.: Introduction to Superanalysis. Dordrecht - Boston - Lancaster - Tokyo: Reidel 1987

[B$_1$] Bingener, J.: Cogebras, Bigebras and Infinite Dimensional Superanalysis. Kluwer academic publishers (in preparation)

[B$_2$] Bingener, J.: Banach Analytic Superspaces. Manuscript 1990

[LM] Bingener, J., Kosarew, S.: Lokale Modulräume in der analytischen Geometrie. Band 1, 2. Braunschweig - Wiesbaden, Vieweg 1987

[BL] Bingener, J., Lehmkuhl, T.: Infinite Dimensional Super Lie Theory. Manuscript 1990

[BP] Bingener, J., Palamodov, V.P.: Deformations of diagrams of complex spaces. In preparation

[Bi] Bismut, J.-M.: Localization Formulas, Superconnections, and the Index Theorem for Families. Commun. Math. Phy. **103**, 127 - 166 (1986)

[BGR] Bosch, S., Güntzer, U., Remmert, R.: Non-Archimedean Analysis. Grundlehren **261**, Berlin - Heidelberg - New York: Springer 1984

[Bo$_1$] Bourbaki, N.: Algebra I, Chapters 1 - 3. Paris: Hermann 1974

[Bo$_2$] Bourbaki, N.: Algebra II, Chapters 4 - 7. Berlin - Heidelberg - New York: Springer 1990

[Bo$_3$] Bourbaki, N.: Lie Groupes and Lie Algebras , Chapters 1 - 3, Paris: Hermann 1975

[Co] Connes, A.: Non-Commutative Differential Geometry. Publ. Math. IHES **62**, 41 - 144 (1985)

[CNS] Corwin, L., Neeman, Y., Sternberg, Y.: Graded Lie algebras in mathematics and physics. Review of Modern Physics **47**, 573 - 603 (1975)

[DW] De Witt, B.: Supermanifolds. Cambridge: Cambridge University Press 1984

[Do$_1$] Douady, A.: Le problème des modules pour les sous-espaces analytiques compacts d'un espace analytique donné. Ann. Inst. Fourier **16**, 1 - 95 (1966)

[Do$_2$] Douady, A.: Le problème des modules locaux pur les espaces $\mathbb{C}$-analytiques compacts. Ann. Sci. ENS **7**, 569 - 602 (1974)

[Dr] Drinfeld, V.G.: Quantum Groups. Proc. Intern. Congress in Math. Berkeley 1986, 798 - 820 (1987)

[Fl] Flenner, H.: Über Deformationen holomorpher Abbildungen. Habilitationsschrift, Osnabrück 1978

[FS] Flenner, H., Sundararaman, D.: Analytic Geometry of Complex Superspaces. Transactions AMS **330**, 1 (1 - 40) (1992)

[Ge] Getzler, E.: Pseudodifferential Operators on Supermanifolds and the Atiyah-Singer Index Theorem. Commun. Math. Phy. **92**, 163 - 178 (1983)

[EGA] Grothendieck, A., Dieudonné, J.: Élements de geométrie algebrique. Publ. Math. IHES **4** (1960)

[Ja] Jacobson, N.: Lie Algebras. New York - London - Sydney: John Wiley & Sons 1962

[Ka] Kac, V.G.: Lie superalgebras. Advances in Math. **26**, 8 - 96 (1977)

[Ko] Kostant, B.: Graded manifolds, graded Lie theory and prequantization. In Lecture Notes in Mathematics **570**, 177 - 306 (1977)

[La] Lazard, M.: Lois de groupes et analyseurs. Ann. Sci. ENS **72**, 299 - 400 (1955)

[Ma$_1$] Manin, Y.I.: New dimensions in geometry. Lecture Notes in Mathematics **1111**, 59 - 99 (1985)

[Ma$_2$] Manin, Y.I.: Gauge Field Theory and Complex Geometry. Grundlehren der math. Wiss. **289**. Berlin - Heidelberg - New York: Springer 1988

[MM] Milnor, J.W., Moore, J.C.: On the structure of Hopf algebras. Ann. of Math. **81**, 211 - 264 (1965)

[Pe] Penkov, I.: Classical Lie supergroups and Lie superalgebras and their representations. To appear

[Qu] Quillen, D.: Superconnections and the Chern character. Topology **24**, 89 - 95 (1985)

[Ro] Roby, N.: Lois polynomes et lois formelles en theorie des modules. Ann. Sci. ENS **80**, 213 - 348 (1963)

[Rg$_1$] Rogers, A.: A global theory of supermanifolds. J. Math. Phy. **21**, 1352 - 1365 (1980)

[Rg2] Rogers, A.: Super Lie groups: global topology and local structure. J. Math. Phy. **22**, 939 - 945 (1981)

[Sc] Scheunert, M.: The Theory of Lie Superalgebras. Lecture Notes in Mathematics **716**. Berlin - Heidelberg - New York: Springer 1979

[Se] Serre, J.-P.: Lie algebras and Lie groups. New York - Amsterdam: Benjamin 1965

[Sw] Sweedler, M.E.: Hopf algebras. New York: Benjamin 1969

[Zi] Zink, Th.: Cartiertheorie kommutativer formaler Gruppen. Leipzig: Teubner 1984

Unfoldings of Holomorphic Maps as Deformations

Ragnar-Olaf Buchweitz, Department of Mathematics, University of Toronto, Toronto, Ont. M5S 1A1, Canada. ragnar@math.utoronto.ca

Hubert Flenner, Mathematisches Institut der Universität Göttingen, Bunsenstr. 3-5, D-37073 Göttingen, Germany. hflenner@cfgauss.uni- math.gwdg.de

Introduction

In the nowadays classical theory of singularities of (holomorphic) mappings $f\colon (\mathbb{C}^n,0) \to (\mathbb{C}^p,0)$ one considers various equivalence relations given by groups of automorphisms, which act on such germs. Here we recast the theory from the point of view of deformations and give a unified treatment of the so called standard theorems, see e.g. [1]. The theory of unfoldings developed independently from deformation theory in the late sixties. It became apparent soon afterwards that unfoldings can be viewed as (unobstructed) deformations, see e.g. [18].

In our approach the key tool underlying all proofs is the *relative Kodaira-Spencer map* of a deformation. Its vanishing for a 1-parameter family signifies that the deformation is trivial whereas the surjectivity for the deformation over a smooth base characterizes versality. Of course the Kodaira-Spencer map appears in various disguises already in the classical proofs, e.g. as "homological equation" in [1], Chapt.3, (1.5). Systematic use of the Kodaira-Spencer class clarifies the proofs considerably and allows for generalizations. As a typical example we give a simplified and more conceptual proof of the theorems of Mather-Yau type [12], [6]. We deduce these results for mapping germs $(X,0) \to (\mathbb{C}^p,0)$ where $(X,0)$ is an arbitrary germ of a complex space.

Classically one studies the action of various groups on the space of mapping germs $f\colon (\mathbb{C}^n,0) \to (\mathbb{C}^p,0)$ and tries to determine the structure of the orbits under the action. For example, two germs $f,g\colon (\mathbb{C}^n,0) \to (\mathbb{C}^p,0)$ are called right resp. left equivalent if $f = kg$ resp. $f = gh$ for some isomorphism k of $(\mathbb{C}^p,0)$ resp. h of $(\mathbb{C}^n,0)$. In these cases the groups are the groups $G = \mathcal{L}$ resp. $G = \mathcal{R}$ of automorphisms of the target resp. source. Combining these actions yields right-left or $\mathcal{A} = \mathcal{R} \times \mathcal{L}$-equivalence. Another important group, $G = \mathcal{K}$, is given by contact equivalences, see (1.3)(4).

For each of these groups of automorphisms G one has the following standard results.

1. *Finite determinacy*: If the space of infinitesimal deformations $\mathrm{Def}_G(f,\mathbb{C})$ of f modulo G-equivalence is a finite dimensional vectorspace then f is finitely G-determined, i.e. if $h \in \mathfrak{m}^k\mathcal{O}^p_{\mathbb{C}^n,0}$, $\mathfrak{m}$ being the maximal ideal, then f and $f+h$ are G-equivalent for $k \gg 0$.

2. *Infinitesimal stability implies stability:* If every first order deformation of f is trivial modulo G-equivalence then f is called infinitesimally G-stable. The second standard result states that an infinitesimally G-stable germ is G-stable, i.e. every deformation of f is trivial with respect to G-equivalence. Clearly, this theorem is a special case of

Supported by NSERC–grant 3-642-114-80 and the SFB 170 in Göttingen
This paper is in final form and no part of it will be submitted elsewhere

3. *Existence of versal deformations*: If the space of first order deformations $\mathrm{Def}_G(f, \mathbb{C})$ is finite dimensional then f admits a versal deformation/unfolding.

4. *Results of Mather-Yau type:* Under suitable assumptions, two mapping germs $f, g: (\mathbb{C}^n, 0) \to (\mathbb{C}^p, 0)$ are G-equivalent iff the spaces of first order deformations are algebraically isomorphic (see (5.1)).

The standard results (1) – (3) were first shown by Mather for the case of $\mathcal{A}$- and $\mathcal{K}$-equivalence in his remarkable series of papers [11]. Later on they were generalized also to other groups, see [1], Chapt. 3, (2.4) for an overview, or [2] for an axiomatization. The standard result (4) is more recent and was shown in [12] for $\mathcal{K}$-equivalence of functions, i.e. $p = 1$. It was generalized by [6] to the case of arbitrary p and also to the case that G is the group of $\mathcal{A}$-equivalences.

Instead of considering orbits of the action of G on the space of mapping germs we consider the associated deformation theory in the spirit of Schlessinger. To a parameter space $S = (S, 0)$ we associate the set $E(S)$ of all mapping germs $(X \times S, 0) \to (\mathbb{C}^p, 0)$ for a fixed germ $(X, 0)$ and a fixed p. The groups in question act functorially on $E(S)$, and their elements serve as isomorphisms between the elements of $E(S)$ whereby $E(S)$ becomes a category denoted by $\mathbf{E}_G(S)$. This category is a *groupoid* as every morphism is an isomorphism. The assignment $S \mapsto \mathbf{E}_G(S)$ defines a fibration in groupoids which satisfies Schlessinger's condition, see [15] and (1.4), and thus constitutes a *deformation theory*.

For a mapping germ $f: (X, 0) \to (\mathbb{C}^p, 0)$, the action of G gives rise to a map of tangent spaces $\gamma_f: T_e G \to \mathcal{O}^p_{X,0}$, whose image is denoted by $TG(f, \mathbb{C})$ and interpreted as the tangent space to the orbit of f. In the language of deformation theory the cokernel $\mathcal{O}^p_{X,0}/TG(f, \mathbb{C})$ represents the space of first order deformations of f in $\mathbf{E}_G$. More generally, for a deformation $F(x, s)$ of $f(x) = F(x, 0)$ over a germ $S = (S, 0)$ and a coherent $\mathcal{O}_S$-module $\mathcal{M}$ there is a submodule $TG(F, \mathcal{M})$ of $\mathcal{O}^p_{X\times S,0} \otimes \mathcal{M}$, and the quotient $\mathcal{O}^p_{X\times S,0} \otimes \mathcal{M}/TG(F, \mathcal{M})$ can be interpreted as the space of "extensions" of F to deformations over the trivial extension $S[\mathcal{M}]$ of S by $\mathcal{M}$.

The formal analogues of the standard results (2) and (3) are now consequences of general statements in deformation theory. The transition from formal to convergent solutions requires deformation theories which admit *integration of vector fields*, see (3.4). In the case of the deformation theories $\mathbf{E}_G$ it is easily seen that this requirement is met, (3.5). The key tool in the proof of the standard results is the following simple but effective criterion: *A 1-parameter deformation is trivial iff its relative Kodaira-Spencer class vanishes, see* (3.6).

For a deformation $F(x, t)$ of a mapping germ $f(x) = F(x, 0)$ the relative Kodaira-Spencer class is given by the class of $\partial F/\partial t$ in $\mathcal{O}^p_{X\times\mathbb{C},0}/TG(F, \mathcal{O}_\mathbb{C})$, and so the vanishing of $[\partial F/\partial t]$ in that module is equivalent to the triviality of F. In this explicit form the criterion is well known, see [19]. This and a generalization to the relative case is used in Sect. 3 to prove the standard results about finite determinacy (4.3) and versal deformations, see (4.6). The treatment is essentially inspired by [19].

Finally in Sect. 5, we derive results of Mather-Yau type. In (5.1) we treat the case of $\mathcal{K}$-equivalence: *If* $f, g: (X, 0) \to (\mathbb{C}^p, 0)$ *are germs with* $T\mathcal{K}(f, \mathbb{C}) = T\mathcal{K}(g, \mathbb{C}) \subseteq \mathcal{O}^p_{X,0}$ *then* f *and* g *are* $\mathcal{K}$*-equivalent.* The idea of the proof is to study the Kodaira-Spencer class $[\partial F/\partial t]$ of the deformation $F(x, t) := (1 - t)f + tg$, which connects f and g. In view of the local triviality criterion it suffices to show the vanishing of $[\partial F/\partial t]$ for all t

in a Zariski open subset of $\mathbb{C}$ containing 0 and 1. But this follows easily from the fact that this class can be viewed as a section in a coherent sheaf on a neighbourhood of $0 \times \mathbb{C}$ in $X \times \mathbb{C}$. The idea of proof for $\mathcal{A}$-equivalence (5.6) is the same, but execution is more complicated, as the sheaf, in which the Kodaira-Spencer class lives, is not a priori coherent.

During the preparation of this work the first author was visiting the SFB 170 "Geometry and Analysis" in Göttingen. He would like to thank this institution for its hospitality and support.

We explain some notations used throughout this paper.

Categories are written in boldface and categories like **Sets** or **Groups** should need no further explanation. Whenever we talk about isomorphism classes of objects from a category **C**, it will be assumed that those classes form a set. Such a category **C** is sometimes called *essentially small.* The category $\mathbf{C}^0$ is the opposite of **C**.

Germs represents the category of germs of complex spaces. Such a germ is denoted by $(S, 0)$ or often simply by S. As a rule, every germ has 0 as its basepoint, and the same symbol represents the (reduced) point. For a complex space X the sheaf of holomorphic functions is as usual denoted by $\mathcal{O}_X$, whereas for a germ $S = (S, 0)$ the symbol $\mathcal{O}_S$ indicates the local ring $\mathcal{O}_{S,0}$ and $\mathfrak{m}_S$ its maximal ideal.

If $f: (X, \mathcal{O}_X) \to (Y, \mathcal{O}_Y)$ is a morphism of ringed spaces then the induced ring homomorphism $\mathcal{O}_{Y,f(x)} \to \mathcal{O}_{X,x}$ is denoted by f^{-1}. For a sheaf $\mathcal{N}$ of $\mathcal{O}_Y$-modules $f^{-1}(\mathcal{N})$ indicates the topological inverse image of $\mathcal{N}$ on X, i.e. $f^{-1}(\mathcal{N})_x = \mathcal{N}_{f(x)}$ for $x \in X$.

If $S = (S, 0)$ is the germ of a complex space, $\mathbf{Coh}(S)$ denotes the category of coherent, i.e. finite, $\mathcal{O}_{S,0}$-modules. A closed embedding $S \hookrightarrow S'$ of S into another germ $S' = (S', 0)$ is an *extension* of S by $\mathcal{M} \in \mathbf{Coh}(S)$ if the ideal sheaf $\mathcal{I} := \mathrm{Ker}(\mathcal{O}_{S'} \to \mathcal{O}_S)$ defining S in S' is of square zero and isomorphic to $\mathcal{M}$ as $\mathcal{O}_{S,0}$-module. In particular, $S[\mathcal{M}]$ indicates the *trivial extension* whose structure sheaf $\mathcal{O}_{S[\mathcal{M}],0} := \mathcal{O}_{S,0} \times \mathcal{M}$ is the direct sum $\mathcal{O}_{S,0} \oplus \mathcal{M}$ endowed with the multiplication $(s, m)(s', m') = (ss', sm' + s'm)$.

The $\mathcal{M}$*-valued vectorfields* on S — or, equivalently, the $\mathbb{C}$-derivations from $\mathcal{O}_{S,0}$ to $\mathcal{M}$ — form the $\mathcal{O}_{S,0}$-module $T^0_S(\mathcal{M}) := \mathrm{Hom}_S(\Omega^1_{S/\mathbb{C}}, \mathcal{M})$. If $f: S \to S'$ is a morphism of germs, $T^0_{S/S'}(\mathcal{M}) \subseteq T^0_S(\mathcal{M})$ represents correspondingly those derivations that are $f^{-1}(\mathcal{O}_{S'})$-linear.

1. Unfoldings as a Deformation Theory

Modern deformation theory is based upon the notion of *fibrations in groupoids*, [15]. Recall that a groupoid is a category in which all morphisms are isomorphisms. A typical example is given by a G-set X where G is a group acting on X: The objects are the elements of X, and the morphisms $x_1 \to x_2$ are the elements $g \in G$ transporting x_1 to x_2, i.e. $g.x_1 = x_2$. The set of isomorphism classes of objects of that category is the set of orbits X/G, and the group of automorphisms of $x \in X$ is the stabilizer subgroup G_x.

In general, up to an equivalence of categories, the structure of a groupoid **G** is determined by its set of isomorphism classes $\pi_0(\mathbf{G})$ and by the family of groups G_p, $p \in \pi_0(\mathbf{G})$, where $G_p \cong \mathrm{Aut}(a)$, if p is represented by $a \in \mathbf{G}$.

Let $p: \mathbf{E} \to \mathbf{B}$ be a functor between categories. For $S \in \mathbf{B}$ the *fibre* $\mathbf{E}(S)$ is the subcategory of $\mathbf{E}$ whose objects are those $X \in \mathbf{E}$ with $p(X) = S$, and whose morphisms are the morphisms φ over id_S, that is $p(\varphi) = \mathrm{id}_S$.

(1.1) Definition. [15] A *fibration in categories* is a functor $p: \mathbf{E} \to \mathbf{B}$ with the following properties:

(FC1) For every morphism $f: S' \to S$ in $\mathbf{B}$ and every object $a \in \mathbf{E}(S)$ there is a morphism $\tilde{f}: a' \to a$ over f which is *cartesian*, i.e. $\tilde{f}$ satisfies the following universal property: Every morphism $b \to a$ over f factors uniquely into $b \xrightarrow{\tilde{g}} a' \xrightarrow{\tilde{f}} a$ with $p(\tilde{g}) = \mathrm{id}'_S$.

(FC2) Compositions of cartesian morphisms are cartesian.

A fibration in categories $p: \mathbf{E} \to \mathbf{B}$ is called a *fibration in groupoids* if all fibers $\mathbf{E}(S)$ are groupoids.

Equivalently, a fibration in categories is a fibration in groupoids if a morphism $f: a \to b$ in $\mathbf{E}$ is an isomorphism iff $p(f)$ is an isomorphism. The category $\mathbf{B}$ will often be called the *basis* of the fibration. We will denote objects of $\mathbf{B}$ by capital letters whereas the objects of $\mathbf{E}$ will be written in lower case. To indicate that a is an object of $\mathbf{E}$ over $S \in \mathbf{B}$ we write simply $a \mapsto S$ (although this is *not* a morphism !). If $\tilde{f}: a' \to a$ is a cartesian morphism over $f: S' \to S$, the source of $\tilde{f}$ is, slightly abusively, denoted by $a \times_S S' := a'$ or also $f^*(a) := a'$.
Due to (FC2), morphisms $f': S'' \to S'$ and $f: S' \to S$ in $\mathbf{B}$ give rise to unique isomorphisms

$$(ff')^*(a) = a \times_S S'' \cong (a \times_S S') \times_{S'} S'' = f'^* f^* a$$

for every a in $\mathbf{E}(S)$.

(1.2) Example. (1)[15] Every functor $E: \mathbf{B}^0 \to \mathbf{Sets}$ gives rise to a fibration in groupoids $p: \mathbf{E} \to \mathbf{B}$ in the following way. The objects of $\mathbf{E}$ are pairs (a, S) with $a \in \mathbf{F}(S)$, and the morphisms $(a, S) \to (b, T)$ correspond bi-uniquely to those morphisms $f: S \to T$ in $\mathbf{B}$ for which $\mathbf{F}(f)(b) = a$. With p the canonical projection, in this example the fibers are even *discrete* categories.

(2) Let $E: \mathbf{B}^0 \to \mathbf{Sets}$ be as above and assume that $G: \mathbf{B}^0 \to \mathbf{Groups}$ is a group-valued functor acting on E (from the left) meaning that for every object $S \in \mathbf{B}$ there is an action $G(S) \times E(S) \to E(S)$ natural in S. Such an action defines a fibration in groupoids $p: \mathbf{E}_G \to \mathbf{B}$, where the objects of $\mathbf{E}_G$ are just the objects of $\mathbf{E}$, see (1), and the morphisms $(a, S) \to (b, T)$ are pairs (g, f) with $f: S \to T$ a morphism in $\mathbf{B}$ and $g \in G(S)$ satisfying $E(f)(b) = g.a$. The reader may easily verify that $p: \mathbf{E}_G \to \mathbf{B}$ is in fact a fibration in groupoids. Two objects $a, a' \in E(S)$ are called $\mathbf{G}$-*equivalent* if $a = g.a'$ for some $g \in G(S)$.

The classical examples of unfoldings fit into this framework as follows.

(1.3) Example. Let **Germs** denote the category of germs of complex spaces. For fixed germs $(X, 0), (Y, 0) \in \mathbf{Germs}$ we consider the functor $E: \mathbf{Germs}^0 \to \mathbf{Sets}$, where $E(S) = E(S, 0)$ is the set of all morphisms $F : (X \times S, 0) \to (Y, 0)$. The morphism F, or also the corresponding S-morphism

$$(F, id_S) : (X \times S, 0) \to (Y \times S, 0),$$

is called an *unfolding* of the special fibre $f := F|X \times 0$ of F, considered as a morphism $f : (X,0) \to (Y,0)$. In the terminology of [1] (3.2.1), the mapping germ $F(x,s)$ is a *deformation* of $f(x) = F(x,0)$ depending upon the parameter s, and $(F(x,s),s)$ is the associated unfolding over S.

In this paper we only consider the case that Y is smooth, hence $(Y,0) \cong (\mathbb{C}^p,0)$ for some p. The set $E(S)$ of morphisms $F : (X \times S, 0) \to (\mathbb{C}^p, 0)$ can then be identified with $\mathfrak{m}_{X\times S}\mathcal{O}^p_{X\times S,0}$, where $\mathfrak{m}_{X\times S}$ is the maximal ideal of $\mathcal{O}_{X\times S,0}$. In the classical theory , see e.g. loc.cit. , $(X,0)$ is smooth as well.

On such a functor E of mapping germs — with arbitrary X and Y — act various groups.

(1) (*Right equivalence*). For a germ $S = (S,0)$ let $\mathcal{R}_e(S) := \mathrm{Aut}_S(X \times S, 0)$ denote the group of all S-automorphisms of $(X \times S, 0)$. This group acts naturally on $E(S)$ from the left through $g.F := F \circ g^{-1}$ for $g \in \mathrm{Aut}_S(X \times S, 0)$. We define $\mathcal{R}(S)$ as the subgroup of all those g in $\mathcal{R}_e(S) = \mathrm{Aut}_S(X \times S, 0)$ with $g|_{0\times S} = \mathrm{id}_{0\times S}$.

(2) (*Left equivalence*). For $S = (S,0) \in$ **Germs** the group of automorphisms of the target, $\mathcal{L}_e(S) := \mathrm{Aut}_S(Y \times S, 0)$, acts naturally on $E(S)$ from the left. The subgroup of $\mathcal{L}_e(S)$ preserving $0 \times S$ is denoted by $\mathcal{L}(S)$.

(3) (*Right-left equivalence or $\mathcal{A}$-equivalence*). Combining the actions of $\mathcal{R}_e$ and $\mathcal{L}_e$ on E in (1), (2) we obtain an action of $\mathcal{A}_e := \mathcal{R}_e \times \mathcal{L}_e$ and similarly of $\mathcal{A} := \mathcal{R} \times \mathcal{L}$ on E.

(4) (*Contact equivalence or V-equivalence*). For a germ $S = (S,0)$, an (extended) contact equivalence is a morphism $\Phi \in \mathrm{Aut}_S(Y \times X \times S, 0)$ of the form

$$\Phi(y,x,s) = (\Phi_1(y,x,s), \Phi_2(x,s), s)$$

such that $\Phi_{23}(x,s) := (\Phi_2(x,s),s)$ is an automorphism of $(X \times S, 0)$ and $\Phi(0 \times X \times S) = 0 \times X \times S$. The set $\mathcal{K}_e(S)$ of contact equivalences forms evidently a subgroup of $\mathrm{Aut}_S(Y \times X \times S, 0)$. It contains $\mathcal{K}(S)$, the subgroup of all Φ with $\Phi_{23} \in \mathcal{R}(S)$, i.e. $\Phi_2(0,s) = 0$.

There is a natural action of $\mathcal{K}_e(S)$ and then in particular of $\mathcal{K}(S)$ on $E(S)$ whereby for $F \in E(S)$ and $\Phi \in \mathcal{K}_e(S)$ the product $\Phi.\,F$ is the unique S-morphism $\tilde{F}: (X \times S, 0) \to (Y,0)$ such that the diagram

$$\begin{array}{ccc} (X \times S, 0) & \xrightarrow{(F,\mathrm{id}_{X\times S})} & (Y \times X \times S, 0) \\ \downarrow \Phi_{23} & & \downarrow \Phi \\ (X \times S, 0) & \xrightarrow{(\tilde{F},\mathrm{id}_{X\times S})} & (Y \times X \times S, 0) \end{array}$$

commutes. In other words,

$$\tilde{F}(x,s) := \Phi_1(F(\Phi_{23}^{-1}(x,s)), \Phi_{23}^{-1}(x,s)).$$

(5) ($\mathcal{C}$-*equivalence*). For $S = (S,0)$ a germ, let $\mathcal{C}(S) \subseteq \mathcal{K}(S)$ be the subgroup of all $\Phi = (\Phi_1, \Phi_{23})$ as above with $\Phi_{23} = \mathrm{id}_{X\times S}$. In terms of this group, $\mathcal{K}$-equivalence becomes the conjunction of $\mathcal{R}$- and $\mathcal{C}$-equivalence,

$$\mathcal{K}(S) = \mathcal{R}(S) \ltimes \mathcal{C}(S) \quad \text{and} \quad \mathcal{K}_e(S) = \mathcal{R}_e(S) \ltimes \mathcal{C}(S).$$

In case that Y is smooth, two maps F, G are $\mathcal{C}$-equivalent, that is $F = \Phi.G$ for some $\Phi \in \mathcal{C}(S)$, iff the fibres $F^{-1}(0)$ and $G^{-1}(0)$ in $X \times S$ are equal, see [1]. In particular, F and

G are $\mathcal{C}$-equivalent iff the components of F and G generate the same ideal in $\mathcal{O}_{X\times S}$, iff F can already be obtained from G by the natural action of $GL(\mathcal{O}^p_{X\times S}) \subseteq \mathcal{C}(S)$, realized as that subgroup whose elements $\Phi = (\Phi_1, \mathrm{id}_{X\times S})$ have as first component a map $\Phi_1(y, x, s)$ that is linear and homogeneous in y. Similarly, F and G are $\mathcal{K}$-equivalent iff the complex analytic germs or vanishing loci $F^{-1}(0)$ and $G^{-1}(0)$ are (abstractly) isomorphic, whence the notion *V-equivalence* in (4).

(1.4) Let $p\colon \mathbf{E} \to \mathbf{Germs}$ be a fibration in groupoids. We will call p a *deformation theory* if the following homogeneity property is satisfied.

(H) For every diagram in $\mathbf{E}$

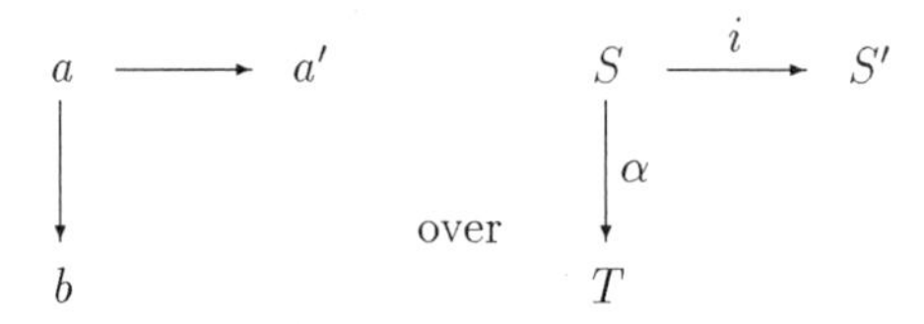

where $S \xrightarrow{i} S'$ is an extension by a coherent $\mathcal{O}_{S,0}$-module $\mathcal{M}$ and $S \xrightarrow{\alpha} T$ is finite, there exists the fibered sum $b' := a' \coprod_a b$ in $\mathbf{E}$.

Remark that, if b' exists, it lies necessarily over $S' \coprod_S T$, which exists as an analytic germ by [16].

If $a_0 \in E(0)$ is a specific object over the reduced point then a *deformation* of a_0 over a germ $S = (S, 0)$ is an object $a \in \mathbf{E}(S)$ together with a morphism $a_0 \to a$, which lies necessarily over $0 \hookrightarrow (S, 0)$.

Condition (H) guarantees that property (S1) of M.Schlessinger is satisfied, see [15]. Thus, if the space of first order deformations of a is of finite dimension, then there exists a formal versal deformation.

(1.5) Example. Assume that E: $\mathbf{Germs}^0 \to \mathbf{Sets}$ is a functor such that

$$(*) \qquad E(S' \coprod_S T) \cong E(S') \times_{E(S)} E(T),$$

whenever $(S, 0) \hookrightarrow (S', 0)$ is an extension by a coherent $\mathcal{O}_{S,0}$-module and $S \to T$ is finite. Then the fibration in groupoids $p\colon \mathbf{E} \to \mathbf{Germs}$ associated to E is a deformation theory. For the functor E of unfoldings considered in (1.3), condition $(*)$ is satisfied and so we obtain a deformation theory.

With E as in (1.4), assume that G: $\mathbf{Germs}^0 \to \mathbf{Groups}$ is a group-valued functor acting on E as in (1.2).

(1.6) Lemma. *Assume that E and G satisfy $(*)$ in (1.4) and that for every extension $S \hookrightarrow S'$ of the germ S by a coherent $\mathcal{O}_{S,0}$-module the induced map $G(S') \to G(S)$ is surjective. Then the fibration in groupoids $p\colon \mathbf{E}_G \to$ **Germs** constructed in (1.2) is a deformation theory.*

Proof. In the situation of (H), we are given elements $a \in E(S)$, $a' \in E(S')$, $b \in E(T)$ with $a \cong \alpha^*(b) \cong i^*(a')$ in $\mathbf{E}_G(S)$. This means that $\alpha^*(b) = g.i^*(a')$ for some $g \in G(S)$. We have to construct the fibered sum of a' and b over a. By assumption we can find $g' \in G(S')$ lifting g. Replacing a by $\alpha^*(b)$ and a' by $g'.a'$, we may assume that $a = \alpha^*(b) = i^*(a')$. As E satisfies $(*)$, we obtain $b' \in E(T')$ over $T' := S' \coprod_S T$ representing the fibered sum $a' \coprod_a b$ in $\mathbf{E}$. We show that b' is also a fibered sum in $\mathbf{E}_G$. For this, consider in $\mathbf{E}_G$ a diagram

$$\begin{array}{ccc} a & \longrightarrow & a' \\ \downarrow & & \downarrow \\ b & \longrightarrow & c \end{array} \quad \text{over} \quad \begin{array}{ccc} S & \xrightarrow{i} & S' \\ \downarrow{\scriptstyle \alpha} & & \downarrow{\scriptstyle \beta} \\ T & \xrightarrow{j} & T' \end{array}$$

so that $c \in E(T')$. By definition, the morphisms $a' \to c$ over β respectively $b \to c$ over j are represented by elements $g \in G(S')$ resp. $h \in G(T)$ with $g.a' = \beta^*(c)$ and $h.b = j^*(c)$. It follows that $i^*(g).a' = \alpha^*(h).b$ and so (g,h) defines an element in $G(S') \times_{G(S)} G(T)$. Using $(*)$ for G there is a unique element $h' \in G(T')$ with $j^*(h') = h$, $\beta^*(h') = g'$. Obviously $h'.b' = c$, and h' represents the unique morphism in $\mathbf{E}_G$ from b' to c that induces the given morphisms $a' \to c$ and $b \to c$. Hence b' is a coproduct in $\mathbf{E}_G$ as well, proving the lemma. □

(1.7) Corollary. *In the examples (1) – (5) of (1.3), the resulting fibrations in groupoids* $p: \mathbf{E}_G \to$ **Germs** *are deformation theories.*

Proof. We have to show that in each of the examples the group G satisfies the assumptions of (1.5). It suffices to consider the cases $G = \mathcal{R}, \mathcal{L}, \mathcal{C}$. But that then the assumptions of the lemma are fulfilled is an easy exercise left to the reader. □

2. The Kodaira-Spencer map

Let $p: \mathbf{E} \to$ **Germs** be a fibered groupoid, $S = (S, 0)$ a germ of a complex space and $a \in \mathbf{E}(S)$ an object over S. For a coherent $\mathcal{O}_S$-module $\mathcal{M}$ let **Def**$(a \mapsto S, \mathcal{M})$ be the category of all pairs $\beta = (a \to b, u)$ such that the underlying morphism $S \to T := p(b)$ is an extension of S by $\mathcal{M}$, i.e. the ideal $\mathcal{I} := \ker(\mathcal{O}_T \to \mathcal{O}_S)$ is of square zero and isomorphic to $\mathcal{M}$ under a fixed $\mathcal{O}_S$-isomorphism $u: \mathcal{M} \to \mathcal{I}$. If $\beta' = (a \to b', u')$ is another object of **Def**$(a \mapsto S, \mathcal{M})$ then a morphism $\beta \to \beta'$ consists of a morphism $b \to b'$ compatible with $a \to b$, $a \to b'$ and u, u'. We denote by $\mathrm{Def}(a \mapsto S, \mathcal{M})$ the set of isomorphism classes of **Def**$(a \mapsto S, \mathcal{M})$ and by $\mathrm{Aut}(a \mapsto S, \mathcal{M})$ the set of automorphisms of the trivial extension

$$a[\mathcal{M}] := (a \to b := a \times_S S[\mathcal{M}], \mathrm{id}_{\mathcal{M}}).$$

Let **Def**$(a, \mathcal{M})$ be the subcategory of all objects $\beta = (a \to b, u)$ with $p(b) = S[\mathcal{M}]$ and $u = \mathrm{id}_{\mathcal{M}}$, where the morphisms $\beta \to \beta' = (a \to b', \mathrm{id}_{\mathcal{M}})$ are given by all morphisms $b \to b'$ over $\mathrm{id}_{S[\mathcal{M}]}$. Again, by $\mathrm{Def}(a, \mathcal{M})$ we denote the set of isomorphism classes and by $\mathrm{Aut}(a, \mathcal{M})$ the set of automorphisms of $a[\mathcal{M}]$ in this category. If necessary for clarity, we adorn "Def" and "Aut" by the subscript $\mathbf{E}$.

Finally, let **Def**$(S, \mathcal{M})$ be the category of all infinitesimal extensions $(S \hookrightarrow T, u)$ of S by $\mathcal{M}$, i.e. **Def**$(S, \mathcal{M}) =$ **Def**$(S \mapsto S, \mathcal{M})$ where p: **Germs** $\to$ **Germs** is the identity. We let $\mathrm{Def}(S, \mathcal{M})$ be the set of isomorphism classes of **Def**$(S, \mathcal{M})$ and $\mathrm{Aut}(S, \mathcal{M})$ the set of automorphisms of the trivial extension $S[\mathcal{M}]$. It is well known that $\mathrm{Def}(S,\mathcal{M}) = T^1_S(\mathcal{M})$ and

$$\mathrm{Aut}(S,\mathcal{M}) \cong \mathrm{Def}(\mathcal{O}_S,\mathcal{M}) \cong T^0_S(\mathcal{M}),$$

where $T^i_\bullet$ are the tangent cohomology groups.

(2.1) Lemma. (1) $\mathrm{Aut}(a,\mathcal{M})$, $\mathrm{Aut}(a \mapsto S,\mathcal{M})$, $\mathrm{Def}(a,\mathcal{M})$, $\mathrm{Def}(a \mapsto S,\mathcal{M})$ *are (covariant) functors with respect to* $\mathcal{M}$.

(2) *All these functors are compatible with direct products, i.e. for coherent* $\mathcal{O}_S$-*modules* $\mathcal{M},\mathcal{N}$ *and for* F *one of these functors one has*

$$F(\mathcal{M}\times\mathcal{N}) \xrightarrow{\sim} F(\mathcal{M})\times F(\mathcal{N}).$$

The *proof* of (1) follows easily from the fact that for a morphism $\mathcal{M} \xrightarrow{\varphi} \mathcal{N}$ of coherent $\mathcal{O}_S$-modules there is a natural functor

$$\mathbf{Def}(a \mapsto S,\mathcal{M}) \to \mathrm{Def}(a \mapsto S,\mathcal{N})$$

given by pushing out extensions along φ. Moreover (2) follows easily from the fact that p is a deformation theory, see e.g. [9] or [5].

(2.2) Corollary. *The functors in* (2.1) (1) *are* $\mathcal{O}_S$-*linear, i.e. they carry natural* $\mathcal{O}_S$-*module structures.*

This follows immediately from (2.1) (2) applied to the addition map resp. scalar multiplication map of a coherent $\mathcal{O}_S$-module.

(2.3) Remark. Assume that $S = (S,0)$ is a germ and $G\colon \mathbf{Coh}(S) \to \mathbf{Sets}$ is a functor on the category of coherent $\mathcal{O}_S$-modules compatible with finite direct products and $G(0) \neq \emptyset$. Then — slightly generalizing (2.2) — $G(\mathcal{M})$ carries a natural $\mathcal{O}_S$-module structure. Moreover, if $G'\colon \mathbf{Coh}(S) \to \mathbf{Sets}$ is a second functor compatible with finite direct products and $G'(0) \neq \emptyset$, and $G \to G'$ is a natural transformation, then the maps $G(\mathcal{M}) \to G'(\mathcal{M})$ are $\mathcal{O}$-linear. If G is a group-valued functor then the addition on the $\mathcal{O}_{S,0}$-module $G(\mathcal{M})$ coincides with the given group structure; see e.g. [10], (VIII.2.Prop.4) and its proof. In particular, each group $G(\mathcal{M})$ is necessarily abelian.

(2.4) Remark. Obviously there are natural exact sequences

$$0 \to \mathrm{Aut}(a,\mathcal{M}) \to \mathrm{Aut}(a \mapsto S,\mathcal{M}) \to T^0_S(\mathcal{M})$$

and

$$\mathrm{Def}(a,\mathcal{M}) \to \mathrm{Def}(a \mapsto S,\mathcal{M}) \to \mathrm{Def}(S,\mathcal{M}) = T^1_S(\mathcal{M}).$$

(2.5) Example. Assume that $E : \mathbf{Germs}^0 \to \mathbf{Sets}$ is a functor satisfying $(*)$ in (1.4), and let $p : \mathbf{E} \to \mathbf{Germs}$ be the associated fibration in groupoids. For $a \in E(S)$ and a coherent $\mathcal{O}_S$-module $\mathcal{M}$ one has

$$\mathrm{Def}_{\mathbf{E}}(a,\mathcal{M}) = (i^*)^{-1}(a) \quad \text{and} \quad \mathrm{Aut}(a,\mathcal{M}) = \{\mathrm{id}_{a[\mathcal{M}]}\},$$

where $i^*: E(S[\mathcal{M}]) \to E(S)$ is the map induced by the natural inclusion $S \overset{i}{\hookrightarrow} S[\mathcal{M}]$. If G acts on E as in (1.5), then $G(\mathcal{M}) := \ker(G(S[\mathcal{M}]) \overset{i^*}{\to} G(S))$ acts on $\mathrm{Def}_{\mathbf{E}}(a, \mathcal{M})$ and so defines a homomorphism of $\mathcal{O}_S$-modules $\gamma_a: G(\mathcal{M}) \to \mathrm{Def}_{\mathbf{E}}(a, \mathcal{M})$. By $TG(a, \mathcal{M})$ we will denote the image of γ_a. Then $\mathrm{Def}_{\mathbf{E}_G}(a, \mathcal{M})$ is just the quotient $\mathrm{Def}_{\mathbf{E}}(a, \mathcal{M})/TG(a, \mathcal{M}) = \mathrm{Cok}(\gamma_a)$. Moreover, $\mathrm{Aut}_{\mathbf{E}_G}(a, \mathcal{M}) = \mathrm{Ker}(\gamma_a)$ is the stabilizer subgroup $G_{a[\mathcal{M}]} \subseteq G(\mathcal{M})$ of $a[\mathcal{M}] \in \mathrm{Def}_{\mathbf{E}}(a, \mathcal{M})$.

Specializing to the classical cases, all this amounts to the following

(2.6) Example. Let $(X, 0)$ be a germ of a complex space and consider the functor E: $\mathbf{Germs}^0 \to \mathbf{Sets}$ introduced in (1.3) with $(Y, 0) := (\mathbb{C}^p, 0)$, i.e. $E(S)$ is the set $\mathfrak{m}_{X\times S}(\mathcal{O}^p_{X\times S,0})$. Given a specific unfolding $F \in \mathfrak{m}_{X\times S}\mathcal{O}^p_{X\times S,0}$ and a coherent $\mathcal{O}_{S,0}$-module $\mathcal{M}$ we have

$$E(S[\mathcal{M}]) \cong \mathfrak{m}_{X\times S}\mathcal{O}^p_{X\times S,0} \oplus (\mathcal{O}^p_{X\times S,0} \otimes_{\mathcal{O}_S} \mathcal{M}),$$

and so

$$\mathrm{Def}_{\mathbf{E}}(F, \mathcal{M}) \cong \bigoplus_{i=1}^{p} \mathcal{M}_X \frac{\partial}{\partial y_i} \cong T^0_{\mathbb{C}^p\times S/S}(\mathcal{M}_X),$$

where $y_1, \ldots, y_p$ are the coordinates in $\mathbb{C}^p$ and $\mathcal{M}_X := \mathcal{O}_{X\times S} \otimes_{\mathcal{O}_S} \mathcal{M}$. In particular, this group does not depend upon the map $F \in E(S)$.

The subgroup $TG(F, \mathcal{M})$ of $\mathrm{Def}_{\mathbf{E}}(F, \mathcal{M})$ for the various groups $\mathcal{R}, \mathcal{L}, \mathcal{A}, \mathcal{C}, \mathcal{K}$ in (2.4) is given as follows.

(1) For right equivalence, $\mathcal{R}(\mathcal{M})$ is the set of all $S[\mathcal{M}]$-automorphisms $X \times S[\mathcal{M}] \to X \times S[\mathcal{M}]$ inducing the identity on $X \times S$ and on $0 \times S[\mathcal{M}]$. It is well known that this set can be naturally identified with the set of derivations $T^0_{X\times S/S}(\mathfrak{m}_X \mathcal{M}_X)$. A derivation δ in this set acts on $\mathrm{Def}_{\mathbf{E}}(F, \mathcal{M}) \cong T^0_{\mathbb{C}^p\times S/S}(\mathcal{M}_X)$ by translation along $J_F(\delta) := \sum_i \delta(F_i)\partial/\partial Y_i$, i.e. $H \mapsto H + J_F(\delta)$. Thus $T\mathcal{R}(F, \mathcal{M})$ is the image of J_F whereas $\mathrm{Aut}_{\mathbf{E}_{\mathcal{R}}}(F, \mathcal{M})$ is isomorphic to the kernel of this map. Similarly, for the extended group $\mathcal{R}_e$, we have $\mathcal{R}_e(\mathcal{M}) \cong T^0_{X\times S/S}(\mathcal{M}_X)$ where the action on $E(\mathcal{M})$ is the same as above. Comparing with the Zariski-Jacobi sequence for the T^i-functors

$$\begin{aligned} 0 \to T^0_{X\times S/Y\times S}(\mathcal{M}_X) &\to T^0_{X\times S/S}(\mathcal{M}_X) \overset{J_F}{\to} T^0_{Y\times S/S}(\mathcal{M}_X) \overset{\partial}{\to} T^1_{X\times S/Y\times S}(\mathcal{M}_X) \\ &\to T^1_{X\times S/S}(\mathcal{M}_X) \to 0 \end{aligned}$$

gives that $\mathrm{Def}_{\mathbf{E}_{\mathcal{R}_e}}(F, \mathcal{M}) \cong \mathrm{Im}(\partial)$ and $\mathrm{Aut}_{\mathbf{E}_{\mathcal{R}_e}}(F, \mathcal{M}) \cong T^0_{X\times S/Y\times S}(\mathcal{M}_X)$. In particular, if X is smooth, then $\mathrm{Def}_{E_{\mathcal{R}_e}}(F, \mathcal{M}) \cong T^1_{X\times S/Y\times S}(\mathcal{M}_X)$. Finally, the set $\mathrm{Aut}_{\mathbf{E}_{\mathcal{R}_e}}(F \mapsto S, \mathcal{M})$ is just the set of compatible pairs of derivations

$$(D, \delta) \in T^0_{X\times S}(\mathcal{M}_X) \times T^0_S(\mathcal{M}),$$

i.e. the diagram

$$\begin{array}{ccc} \mathcal{O}_{X\times S} & \xrightarrow{D} & \mathcal{M}_X = \mathcal{O}_{X\times S} \otimes_{\mathcal{O}_S} \mathcal{M} \\ \uparrow & & \uparrow \\ \mathcal{O}_S & \xrightarrow{\delta} & \mathcal{M} \end{array}$$

commutes. Such a pair (D, δ) is in $\mathrm{Aut}_{\mathbf{E}_{\mathcal{R}}}(F \mapsto S, \mathcal{M})$ iff $D(\mathfrak{m}_X \mathcal{O}_{X\times S}) \subseteq \mathfrak{m}_X \cdot \mathcal{M}_X$, that is iff D is tangent to $0 \times S$.

(2) In the case of left equivalence

$$\mathcal{L}_e(\mathcal{M}) \cong T^0_{\mathbb{C}^p\times S/S}(\mathcal{M}_Y)$$

with $\mathcal{M}_Y := \mathcal{O}_{\mathbb{C}^p\times S,0} \otimes_{\mathcal{O}_S} \mathcal{M}$. An element $\eta \in \mathcal{L}_e(\mathcal{M})$ acts on

$$H \in \mathrm{Def}_{\mathbf{E}}(F, \mathcal{M}) \cong T^0_{\mathbb{C}^p\times S/S}(\mathcal{M}_X)$$

via F^{-1}, i.e. $H \mapsto H + \eta \circ F$. Thus $T\mathcal{L}_e(F, \mathcal{M}) = F^{-1}(T^0_{\mathbb{C}^p\times S/S}(\mathcal{M}_Y))$ and

$$\begin{aligned} \mathrm{Def}_{\mathbf{E}_{\mathcal{L}_e}}(F, \mathcal{M}) &= \mathrm{Coker}(T^0_{\mathbb{C}^p\times S/S}(\mathcal{M}_Y) \xrightarrow{F^{-1}} T^0_{\mathbb{C}^p\times S/S}(\mathcal{M}_X)) \\ &\cong T^0_{\mathbb{C}^p\times S/S}(\mathcal{M}_X/F^{-1}\mathcal{M}_Y), \end{aligned}$$

whereas $\mathrm{Aut}_{\mathbf{E}_{\mathcal{L}_e}}(F, \mathcal{M})$ is the kernel of the map F^{-1} above. Observe that these are $\mathcal{O}_{\mathbb{C}^p\times S,0}$-modules. Again $\mathrm{Aut}_{\mathbf{E}_{\mathcal{L}_e}}(F \mapsto S, \mathcal{M})$ is the set of compatible derivations in $T^0_{\mathbb{C}^p\times S}(\mathcal{M}_Y) \times T^0_S(\mathcal{M})$. This is only an $\mathcal{O}_S$-module. Similarly, $\mathcal{L}(\mathcal{M}) \cong T^0_{\mathbb{C}^p\times S/S}(\mathfrak{m}_Y \mathcal{M}_Y)$ and $\mathrm{Def}_{\mathbf{E}_{\mathcal{L}}}(F, \mathcal{M})$ is the cokernel of the map F^{-1} restricted to $\mathcal{L}(\mathcal{M})$.

(3) Combining (1) and (2) yields that $\mathrm{Def}_{\mathbf{E}_{\mathcal{A}}}(F, \mathcal{M})$, $\mathrm{Aut}_{\mathbf{E}_{\mathcal{A}}}(F, \mathcal{M})$ are the cokernel resp. kernel of the map

$$T^0_{\mathbb{C}^p\times S/S}(\mathfrak{m}_Y \mathcal{M}_Y) \oplus T^0_{X\times S/S}(\mathfrak{m}_X \mathcal{M}_X) \xrightarrow{F^{-1} \oplus J_F} T^0_{\mathbb{C}^p\times S/S}(\mathcal{M}_X).$$

Similarly as above, $\mathrm{Aut}_{\mathbf{E}_{\mathcal{A}}}(F \mapsto S, \mathcal{M})$ is the set of all triples (D_X, D_Y, δ) of derivations in $T^0_{X\times S}(\mathcal{M}_X) \times T^0_{\mathbb{C}^p\times S}(\mathcal{M}_Y) \times T^0_S(\mathcal{M})$ which are compatible and satisfy $D_X(\mathfrak{m}_X \mathcal{O}_{X\times S}) \subseteq \mathfrak{m}_X \mathcal{M}_X$ and $D_Y(\mathfrak{m}_Y \mathcal{O}_{Y\times S}) \subseteq \mathfrak{m}_Y \mathcal{M}_Y$. For the extended group $\mathcal{A}_e$ one has to replace $\mathfrak{m}_Y \mathcal{M}_Y$ and $\mathfrak{m}_X \mathcal{M}_X$ by $\mathcal{M}_Y, \mathcal{M}_X$, respectively.

(4) In the case of $\mathcal{K}$-equivalence

$$\mathcal{K}(\mathcal{M}) \cong T^0_{Y\times X\times S/X\times S}(\mathfrak{m}_Y \mathcal{M}_{Y\times X}) \times T^0_{Y\times X/S}(\mathfrak{m}_X \mathcal{M}_X),$$

as follows from the description of the group $\mathcal{K}$ in (1.3)(4). For $(\delta_1, \delta_2) \in \mathcal{K}(\mathcal{M})$, the derivation δ_2 acts on $H \in \mathcal{O}^p_{X\times S,0} \otimes_{\mathcal{O}_S} \mathcal{M}$ as in the case of $\mathcal{R}$-equivalence, and $\delta_1 = \sum_{i,j} Y_i h_{ij} m_{ij} \partial/\partial Y_j$, with $h_{ij} \in \mathcal{O}_{Y\times X\times S}$ and $m_{ij} \in \mathcal{M}$, acts on H via

$$H \mapsto H + \sum_{i,j} F_i h_{ij}(F, x, s) m_{ij} \frac{\partial}{\partial Y_j}.$$

Thus $\mathrm{Def}_{\mathbf{E}_{\mathcal{K}}}(F, \mathcal{M})$ is just

$$T^0_{\mathbb{C}^p\times S/S}(\mathcal{M}_{X_0})/(J_F(T^0_{X\times S/S}(\mathfrak{m}_X \mathcal{M}_X)) \otimes_{\mathcal{O}_{X\times S}} \mathcal{O}_{X_0}),$$

where $X_0 = F^{-1}(0 \times S)$ is the preimage of the 0-section $0 \times S \subseteq \mathbb{C}^p \times S$.

The module $\mathrm{Aut}_{\mathbf{E}_{\mathcal{K}}}(F \mapsto S, \mathcal{M})$ is the set of all compatible derivations (D_1, D_2, δ) in

$$T^0_{Y\times X\times S/X}(\mathfrak{m}_Y \mathcal{M}_{Y\times X}) \times T^0_{X\times S}(\mathcal{M}_X) \times T^0_S(\mathcal{M})$$

i.e.

$$\begin{array}{ccccccc} \mathcal{O}_S & \longleftarrow & \mathcal{O}_{X\times S} & \xleftarrow{(F\times \mathrm{id})^{-1}} & \mathcal{O}_{Y\times X\times S} & \longleftarrow & \mathcal{O}_S \\ \Big\downarrow \delta & & \Big\downarrow D_2 & & \Big\downarrow D_1 & & \Big\downarrow \delta \\ \mathcal{O}_S & \longleftarrow & \mathcal{O}_{X\times S} & \xleftarrow{(F\times \mathrm{id})^{-1}} & \mathcal{O}_{Y\times X\times S} & \longleftarrow & \mathcal{O}_S \end{array}$$

commutes.

For the extended group $\mathcal{K}_e$, one has to replace $\mathfrak{m}_X\mathcal{M}_X$ by $\mathcal{M}_X$. In particular, if X is smooth and F is flat then $\mathrm{Def}_{\mathbf{E}_{\mathcal{K}_e}}(F,\mathcal{M})$ is isomorphic to $T^1_{X_0/S}(\mathcal{M}_{X_0})$.

(5) Finally, for the case of $\mathcal{C}$-equivalence

$$\mathcal{C}(\mathcal{M}) = T^0_{Y\times X\times S/X\times S}(\mathfrak{m}_Y\mathcal{M}_{Y\times X}),$$

and

$$\mathrm{Def}_{\mathbf{E}_{\mathcal{C}}}(F,\mathcal{M}) = T^0_{\mathbb{C}^p\times S/S}(\mathcal{M}_{X_0}).$$

Let $p\colon \mathbf{E} \to \mathbf{Germs}$ be a deformation theory. We recall the definition of the *Kodaira-Spencer map*

$$\delta_{KS}\colon T^0_S(\mathcal{M}) \to \mathrm{Def}(a,\mathcal{M}),$$

for $S = (S,0)$ a germ, $a \in \mathbf{E}(S)$ and $\mathcal{M}$ a coherent $\mathcal{O}_S$-Module. Let $\vartheta\colon \mathcal{O}_S \to \mathcal{M}$ be a derivation and $(1-\vartheta)\colon \mathcal{O}_S \to \mathcal{O}_S[\mathcal{M}]$ the associated homomorphism of analytic algebras. The underlying morphism $1-\vartheta\colon S[\mathcal{M}] \to S$ retracts the inclusion $S \hookrightarrow S[\mathcal{M}]$. In these terms, $\delta_{KS}(\vartheta) := [(1-\vartheta)^*(a)]$. The map δ_{KS} is functorial in $\mathcal{M}$ and so $\mathcal{O}_S$-linear, see (2.3). The Kodaira-Spencer map is the connecting homomorphism in the *Kodaira-Spencer sequence*:

$$\begin{aligned} 0 \to \mathrm{Aut}(a,\mathcal{M}) &\to \mathrm{Aut}(a \mapsto S,\mathcal{M}) \to T^0_S(\mathcal{M}) \xrightarrow{\delta_{KS}} \mathrm{Def}(a,\mathcal{M}) \\ &\to \mathrm{Def}(a \mapsto S,\mathcal{M}) \to T^1_S(\mathcal{M}). \end{aligned}$$

(2.7) Lemma. (see [5], (2.2)) *The Kodaira-Spencer sequence is exact.*

We remark that one can check the formal versality of a deformation with the modules $\mathrm{Def}(a \mapsto S,\mathcal{M})$, see [5]: *$a$ is a formally versal deformation of the special fibre a_0 iff* $\mathrm{Def}(a \mapsto S,\mathcal{O}_S/\mathfrak{m}_S) = 0$. In case S is smooth, $T^1_S(\mathcal{M})$ vanishes for every $\mathcal{M}$ and so a is formally versal iff the Kodaira-Spencer map $\delta_{KS}\colon T^0_S(\mathcal{O}_S/\mathfrak{m}_S) \to \mathrm{Def}(a,\mathcal{O}_S/\mathfrak{m}_S)$ is surjective. Conversely, we will see that for reasonable deformation theories the annihilation of a tangent vector under δ_{KS} signifies that the deformation is trivial in this direction.

(2.8) Example. Let E: $\mathbf{Germs}^0 \to \mathbf{Sets}$ be the functor of unfoldings considered in (1.3) with $Y = \mathbb{C}^p$, i.e. $E(S) = \mathfrak{m}_{X\times S}\mathcal{O}^p_{X\times S,0}$. Then the Kodaira-Spencer map

$$\delta_{KS}\colon T^0_S(\mathcal{M}) \to \mathrm{Def}_{\mathbf{E}}(F,\mathcal{M}) \cong T^0_{Y\times S/S}(\mathcal{M}_X)$$

is given by

$$\vartheta \mapsto \sum_i (\vartheta\otimes 1)(F_i)\frac{\partial}{\partial Y_i} = \sum_{i,j}\frac{\partial F_i}{\partial s_j}\vartheta(s_j)\frac{\partial}{\partial Y_i},$$

where $\vartheta \otimes 1 \in T^0_{X\times S/X}(\mathcal{M}_X) \cong T^0_S(\mathcal{M}) \otimes_{\mathcal{O}_S} \mathcal{O}_X$ is the canonical lifting and $\mathcal{O}_S \cong \mathbb{C}\{s_1, \ldots, s_k\}/J$.

If G is one of the groups considered in (1.3) (1) – (5), then

$$\mathrm{Def}_{\mathbf{E}_G}(F, \mathcal{M}) = \mathrm{Def}_{\mathbf{E}}(F, \mathcal{M})/TG(\mathcal{M}),$$

and the Kodaira-Spencer map for $\mathbf{E}_G$ is just the composition of the Kodaira-Spencer map $\delta_{KS}: T^0_S(\mathcal{M}) \to \mathrm{Def}_{\mathbf{E}}(F, \mathcal{M})$ for $\mathbf{E}$ and the canonical projection onto $\mathrm{Def}_{\mathbf{E}_G}(F, \mathcal{M})$.

3. Integration of vector fields

Nonvanishing vector fields can be integrated to exhibit locally product structures of complex spaces. This basic fact is well-known and appears in various forms in the literature, see e.g. [4], (2.12) for the complex analytic case, [13], (30.1) for the algebraic version or [19], (3.2) in the differentiable case. We begin with the local version which is apparently due to Zariski, cf. [18], p. 586. For the convenience of the reader we include a short proof.

(3.1) Proposition. (Integration of vector fields, algebraic form) *Let A be an analytic $\mathbb{C}$-algebra and $\delta: A \to A$ a $\mathbb{C}$-derivation such that $\delta(t) = 1$ for some $t \in \mathfrak{m}_A$. Then $A_0 := \ker(\delta)$ is an analytic subalgebra and the canonical map*

$$A_0\{T\} \xrightarrow{i} A$$

with $T \mapsto t$ is an isomorphism.

Proof. We consider the map

$$\begin{aligned} \varphi := \exp((T-t)\delta): A &\to A\{T\} \\ f &\mapsto \varphi(f) := \sum_{\nu=0}^{\infty} \frac{\delta^\nu(f)}{\nu!} \cdot (T-t)^\nu \end{aligned}$$

The convergence of this series follows from [7], I, §1, Satz 3. By the functional equation for the exponential function this is a morphism of analytic $\mathbb{C}$-algebras. First we show that $\varphi(A) \subseteq A_0\{T\}$. For this let $\tilde{\delta}: A\{T\} \to A\{T\}$ be the derivation with $\tilde{\delta}|A = \delta$ and $\tilde{\delta}(T) = 0$ so that $\ker(\tilde{\delta}) = A_0\{T\}$. Then

$$\tilde{\delta}\varphi(f) = \sum_{\nu=0}^{\infty} \left[\frac{\delta^{\nu+1}(f)}{\nu!}(T-t)^\nu + \frac{\delta^\nu(f)}{\nu!}\tilde{\delta}((T-t)^\nu \right] = 0$$

as $\tilde{\delta}(T-t) = -1$. So we get a map $\varphi: A \to A_0\{T\}$. We will show next that $\varphi \circ i = \mathrm{id}_{A_0\{T\}}$. Indeed, if $f = \sum a_j t^j = i(\sum a_j T^j)$ with $a_j \in A_0$ then

$$\varphi(f) = \sum_j \varphi(a_j)\varphi(t)^j,$$

and $\varphi(a_j) = a_j$, $\varphi(t) = T$ as $\delta(a_j) = 0$, $\delta(t) = 1$ and $\delta^\nu(t) = 0$ for $\nu \geq 2$. Finally we show the injectivity of φ on A. Assume that $f \in \ker\varphi$ and consider a representation $f = t^k\tilde{f}$. Then $0 = \varphi(t^k\tilde{f}) = T^k\varphi(\tilde{f})$ and so

$$\begin{aligned} 0 = \varphi(\tilde{f}) &= \sum_{\nu=0}^{\infty} \frac{\delta^\nu(\tilde{f})}{\nu!}(T-t)^\nu \\ &\equiv \tilde{f} \bmod (T, t). \end{aligned}$$

Thus $\tilde{f} \in tA$ and $f \in t^{k+1}A$. It follows that $\ker(\varphi) \subseteq \bigcap_k t^k A = 0$. □

Reformulating (3.1) in geometric terms gives the following result.

(3.2) Proposition. (Integration of vector fields) *Let X be a complex space over S and $t \in \Gamma(X, \mathcal{O}_X)$ a function. Assume that there is a S-derivation $\delta: \mathcal{O}_X \to \mathcal{O}_X$ with $\delta(t) = 1$. With $X_0 := \{t = 0\}$, there is a neighbourhood U of $X_0 \times 0$ in $X \times \mathbb{C}$ and a diagram*

$$\begin{array}{ccc} X_0 \times \mathbb{C} \supseteq U & \xrightarrow{i} & X \\ \uparrow & & \uparrow \\ X_0 \times \{0\} & \xrightarrow{\simeq} & X_0 \end{array}$$

where i is an open S-embedding and the vertical arrows are the natural inclusions. One can take $i := \exp((T - t)\delta)$ where T is the coordinate on $\mathbb{C}$.

Proof. Identifying X_0 with $X_0 \times 0$ and applying (3.1) to every point of X_0 we obtain an isomorphism of sheaves

$$\mathcal{O}_X|_{X_0} \xrightarrow{\sim} \mathcal{O}_{X_0 \times \mathbb{C}}|_{X_0 \times 0}$$

given by $\exp((T - f)\delta)$, where $\mathcal{O}_X|_{X_0}$ is the topological restriction of $\mathcal{O}_X$ to X_0, similarly for $\mathcal{O}_{X_0 \times \mathbb{C}}|_{X_0 \times 0}$. This defines an isomorphism of germs (along subspaces)

$$(X, X_0) \simeq (X_0 \times \mathbb{C}, X_0 \times 0)$$

thus proving (3.2). □

(3.3) Remark. The relative version of (3.1) holds too. Let $f: X \to Y$ be an S-morphism and $t \in \Gamma(Y, \mathcal{O}_Y)$. Assume that there is a pair of compatible vector fields

$$(D, \delta) \in T^0_{X/S}(\mathcal{O}_X) \times T^0_{Y/S}(\mathcal{O}_Y),$$

i.e.

$$\begin{array}{ccc} f^{-1}(\mathcal{O}_Y) & \xrightarrow{f^{-1}(\delta)} & f^{-1}(\mathcal{O}_Y) \\ \downarrow & & \downarrow \\ \mathcal{O}_X & \xrightarrow{D} & \mathcal{O}_X \end{array}$$

commutes. Set $X_0 := \{f^*(t) = 0\}$, $Y_0 := \{t = 0\}$ and suppose $\delta(t) = 1$. Then there are open neighbourhoods $U \subseteq X_0 \times \mathbb{C}$ of $X_0 \times 0$ and $V \subseteq Y_0 \times \mathbb{C}$ of $Y_0 \times 0$ and a commutative diagram

$$\begin{array}{ccc} U & \xrightarrow{i} & X \\ \downarrow{\scriptstyle f_0 \times 1} & & \downarrow{\scriptstyle f} \\ V & \xrightarrow{j} & Y \end{array}$$

with open embeddings i, j such that $i|X_0 \times 0$ and $j|Y_0 \times 0$ are the given inclusions.

Assume moreover that $\sigma: Y \to X$ is a section of f. Then σ induces a section $\sigma_0: Y_0 \to X_0$ of f_0, and the trivialization above is compatible with the section σ_0 iff $D(\mathcal{I}) \subseteq \mathcal{I}$, where $\mathcal{I} \subseteq \mathcal{O}_X$ is the ideal of $\sigma(Y)$, i.e. iff D is tangent to $\sigma(Y)$.

(3.4) Let $p: \mathbf{E} \to$ **Germs** be a deformation theory. For $a \in \mathbf{E}(S)$ over a germ $S = (S, 0)$, assume that there is an infinitesimal automorphism $\alpha \in \mathrm{Aut}(a \mapsto S, \mathcal{O}_S)$ such that the image, say δ, of α in $T^0_S(\mathcal{O}_S)$ is a derivation with $\delta(t) = 1$ for some $t \in \mathfrak{m}_S$. Integrating δ yields by (3.1) a decomposition $(S, 0) = (S_0 \times \mathbb{C}, 0)$. We will say that a can be *trivialized along* δ if there exists an isomorphism $a \cong \mathrm{pr}^*(a|S_0)$ over id_S, where $\mathrm{pr}: S \to S_0$ is the projection. A deformation theory admits *integration of vector fields* if every object $a \in F(S)$ for which there is an infinitesimal automorphism α as above, can be trivialized along δ.

(3.5) Example. Consider again the case of unfoldings $p: \mathbf{E}_G \to$ **Germs** as in (1.3) (1) – (5). Then p admits integration of vector fields as follows from (3.1), (3.2) and the description of the groups $\mathrm{Aut}_{\mathbf{E}_G}(a \mapsto S, \mathcal{M})$ given in (2.6) (1) – (5).

(3.6) Proposition. *Let $p: \mathbf{E} \to$ **Germs** be a deformation theory admitting integration of vector fields. Let $a \in \mathbf{E}(S \times \mathbb{C}, 0)$ be an element restricting to $a_0 \in E(S)$ over $S \cong S \times 0$. Then a is isomorphic to $\mathrm{pr}^*(a_0)$ iff the Kodaira-Spencer class $\delta_{KS}(\partial/\partial t) \in \mathrm{Def}(a, \mathcal{O}_{S \times \mathbb{C},0})$ vanishes.*

This follows immediately from the definition and the exactness of the Kodaira-Spencer sequence (2.7). Using (2.8) for unfoldings, the result becomes the following.

(3.7) Corollary. *Let E:* **Germs**$^0 \to$ **Sets** *be the functor of unfoldings considered in (1.3), with $Y = \mathbb{C}^p$, so that $E(S) = \mathfrak{m}_{X \times S} \mathcal{O}^p_{X \times S,0}$. Let G be one of the groups treated in (2.6) (1) – (5). Assume that $F = F(x, s, t) \in E(S \times \mathbb{C})$ is a 1-parameter unfolding of $f = F(x, s, 0)$. Then the following are equivalent.*

(1) F is G-equivalent to $f_{\mathbb{C}}: (X \times S \times \mathbb{C}, 0) \to (\mathbb{C}^p, 0)$, where $f_{\mathbb{C}}$ denotes the map $f_{\mathbb{C}}(x, s, t) := f(x, s)$.

(2) The residue class of $\partial F/\partial t$ in $\mathcal{O}^p_{X \times S \times \mathbb{C},0}/TG(F, \mathcal{O}_{S \times \mathbb{C},0})$ is trivial, i.e. $\partial F/\partial t \in TG(F, \mathcal{O}_{S \times \mathbb{C},0})$.

Consider in particular an unfolding of the form

$$F_h(x, s, t) := f(x, s) + h(t)g(x, s) \in E(S \times \mathbb{C}, 0)$$

with $f, g \in E(S)$ and $h \in \mathcal{O}_{\mathbb{C},0}$ a holomorphic function. By (2.7) the Kodaira-Spencer class $\delta_{KS}(\partial/\partial t)$ is given by

$$h'(t) \cdot g(x, s) \in \mathcal{O}^p_{X \times S \times \mathbb{C},0}/TG(F, \mathcal{O}_{S \times \mathbb{C}}).$$

Applying (3.7) yields the following more technical criterion.

(3.8) Lemma. *Let G be one of the groups considered in (1.3). Assume that $h(0) = 0$ and that*

(a) $TG(F_h, \mathcal{O}_{S\times\mathbb{C},0}) \supseteq TG(f_{\mathbb{C}}, \mathcal{O}_{S\times\mathbb{C},0})$ *in* $\mathcal{O}^p_{X\times S\times\mathbb{C},0}$,
(b) $g \in TG(f, \mathcal{O}_{S,0})$.
Then $f_{\mathbb{C}}$ *and* F_h *are* G*-equivalent.*

Proof. The projection $\mathrm{pr}: S \times \mathbb{C} \to S$ induces a map pr^* from $TG(f, \mathcal{O}_{S,0})$ into $TG(f_{\mathbb{C}}, \mathcal{O}_{S\times\mathbb{C},0})$. Thus from (a) and (b) we obtain that the Kodaira-Spencer class $\delta_{KS}(\partial/\partial t) = h'(t)g_{\mathbb{C}}$ of F_h vanishes in $\mathrm{Def}_{\mathbf{E}_G}(F_h, \mathcal{O}_{S\times\mathbb{C},0})$. As $F_h(x, s, 0) = f(x, s)$, the result follows from (3.7). □

(3.9) Lemma. *Let* G *be one of the groups* $\mathcal{R}, \mathcal{L}, \mathcal{A}, \mathcal{C}, \mathcal{K}$. *Assume that for every function* $h \in \mathcal{O}_{\mathbb{C},0}$ *we have*
(a) $TG(F_h, \mathcal{O}_{S\times\mathbb{C},0}) \supseteq TG(f_{\mathbb{C}}, \mathcal{O}_{S\times\mathbb{C},0})$ *in* $\mathcal{O}^p_{X\times S\times\mathbb{C},0}$,
(b) $g \in TG(f, \mathcal{O}_{S,0})$.
Then f *and* $f + g$ *are* G*-equivalent.*

Proof. Set $F_{t_0} := f + t_0 g$, $F_{t_0+t} := f + (t + t_0)g$, considered as germs in $\mathcal{O}^p_{X\times S\times\mathbb{C},0}$. It suffices to show that F_{t_0} and F_{t_0+t} are G-equivalent for every $t_0 \in \mathbb{C}$. For the (not extended) groups G under consideration, each $\gamma \in G(S \times \mathbb{C}, 0)$ satisfies $\gamma^*(\mathfrak{m}_X \mathcal{O}^p_{X\times S\times\mathbb{C},0}) \subseteq \mathfrak{m}_X \mathcal{O}^p_{X\times S\times\mathbb{C},0}$, so that the section $0 \times S \times \mathbb{C}$ in $X \times S \times \mathbb{C}$ is preserved. By our assumption (a)

$$TG(F_{t_0+t}, \mathcal{O}_{S\times\mathbb{C},0}) \supseteq TG(f_{\mathbb{C}}, \mathcal{O}_{S\times\mathbb{C},0}).$$

Using again that $\mathrm{pr}^*(TG(f, \mathcal{O}_{S,0}))$ is contained in $TG(f_{\mathbb{C}}, \mathcal{O}_{S\times\mathbb{C},0})$, assumption (b) implies the desired result as in the proof of (3.8). □

4. Applications: Finite determinacy and versality.

As a first application we will prove the following result, see e.g. [1].

(4.1) Theorem. *Let* $f, g \in \mathfrak{m}_{X\times S}\mathcal{O}^p_{X\times S,0} = E(S)$ *be* p*-tuples of functions on* $(X \times S, 0)$. *Assume that*

$$T\mathcal{K}(g, \mathcal{O}_{S,0}) \subseteq \mathfrak{m}_{X\times S} T\mathcal{K}(f, \mathcal{O}_{S,0}).$$

Then f *and* $f + g$ *are* $\mathcal{K}$*-equivalent.*

Proof. By (3.9) it is sufficient to prove that for $h \in \mathcal{O}_{\mathbb{C},0}$, $F := f + h(t)g$

(a) $T\mathcal{K}(F, \mathcal{O}_{S\times\mathbb{C},0}) = T\mathcal{K}(f_{\mathbb{C}}, \mathcal{O}_{S\times\mathbb{C},0})$, where $f_{\mathbb{C}}(x, s, t) := f(x, s)$,

(b) $g \in T\mathcal{K}(f, \mathcal{O}_{S,0})$.

Clearly (b) follows immediately from our assumption. For the proof of (a) observe that the left hand side is generated by the elements $\vartheta(F), \vartheta \in T^0_{X\times S\times\mathbb{C}/S\times\mathbb{C}}(\mathfrak{m}_X \mathcal{O}_{X\times S\times\mathbb{C},0})$, and those in $\sum F_i \mathcal{O}^p_{X\times S\times\mathbb{C},0}$. By assumption

$$\vartheta(F) - \vartheta(f) = h(t)\vartheta(g), \; F_i - f_i = h(t)g_i$$

lie in

$$\mathfrak{m}_{X\times S} T\mathcal{K}(f, \mathcal{O}_{S,0}) \subseteq T\mathcal{K}(f_{\mathbb{C}}, \mathcal{O}_{S\times\mathbb{C},0}).$$

Thus (a) follows from Nakayama's lemma. □

(4.2) Remark. (1) With the same arguments as above it is easily seen that for $f, g \in \mathfrak{m}_{X\times S}\mathcal{O}^p_{X\times S,0}$ the mappings f and $f+g$ are $\mathcal{R}$-equivalent as soon as

(i) $T\mathcal{R}(g, \mathcal{O}_{S,0}) \subseteq \mathfrak{m}_{X\times S} T\mathcal{R}(f, \mathcal{O}_{S,0})$
(ii) $g \in T\mathcal{R}(f, \mathcal{O}_{S,0})$.

(2) Similarly, f, g are $\mathcal{C}$-equivalent if $T\mathcal{C}(g, \mathcal{O}_{S,0}) \subseteq \mathfrak{m}_{X\times S} T\mathcal{C}(f, \mathcal{O}_{S,0})$, i.e. if $g_i \in \mathfrak{m}_{X\times S} \cdot \sum_j f_j \mathcal{O}_{X\times S,0}$ for all i.

Let $f: (X,0) \to (\mathbb{C}^p, 0)$ be a mapping germ. Recall that f is called r-*G-determined* if for every mapping germ $g: (X,0) \to (\mathbb{C}^p, 0)$ with $f \equiv g \bmod m_X^{r+1}$ the germs f and g are G-equivalent. Moreover, f is called *finitely G-determined* if it is r-G-determined for some $r \in \mathbb{N}$. As a second application we show the following result.

(4.3) Theorem. (see [19], (1.2)(ii)) *Let $f:(X,0) \to (\mathbb{C}^p, 0)$ be a holomorphic mapping germ and G one of the groups $\mathcal{R}, \mathcal{L}, \mathcal{A}, \mathcal{C}, \mathcal{K}$. We set $\varepsilon(G) = 1$ for $G = \mathcal{R}, \mathcal{K}, \mathcal{C}$ and $\varepsilon(G) = 2$ for $G = \mathcal{L}, \mathcal{A}$. If $m_X^{r+1}\mathcal{O}^p_{X,0} \subseteq TG(f)$ then f is $(\varepsilon r + 1)$-G-determined.*

Proof. The claim is that f and $f+g$ are G-equivalent for every $g \in m_X^{\varepsilon r+2}$. For the cases $G = \mathcal{K}$ and $G = \mathcal{R}, \mathcal{C}$ this is just (4.1) and (4.2). Now assume that $G = \mathcal{L}$. By (3.6) it is sufficient to show that for $h \in \mathcal{O}_{\mathbb{C},0}$ and $F := f + h(t)g$

(a) $T\mathcal{L}(F, \mathcal{O}_{\mathbb{C},0}) = T\mathcal{L}(f_{\mathbb{C}}, \mathcal{O}_{\mathbb{C},0})$,
(b) $g \in T\mathcal{L}(f, \mathbb{C})$.

By assumption, (b) is satisfied. For (a) consider the Taylor expansion

$$\begin{aligned} F^{-1}(\varphi) - (f_{\mathbb{C}})^{-1}(\varphi) &= \varphi(f + hg) - \varphi(f) \\ &= \sum_{\substack{i \neq 0 \\ i \in \mathbb{N}^p}} \frac{1}{i!}\frac{\partial^{|i|}\varphi}{\partial y^i}(f) \cdot (hg)^i, \end{aligned}$$

where $\varphi \in \mathfrak{m}_{\mathbb{C}^p,0}$ and $y_1, \ldots, y_p$ are the coordinates of $\mathbb{C}^p$. This expansion shows that $F^{-1}(\varphi) - (f_{\mathbb{C}})^{-1}(\varphi)$ is contained in $\mathfrak{m}_X^{2r+2}M$, where $M := \mathcal{O}^p_{X\times\mathbb{C},0}$. Thus with $I := \mathfrak{m}_X^{r+1}$ we get

$$T\mathcal{L}(F, \mathcal{O}_{\mathbb{C},0}) + I^2 M = T\mathcal{L}(f_{\mathbb{C}}, \mathcal{O}_{\mathbb{C},0}) + I^2 M,$$

and by assumption IM is contained in the right hand side. The result now follows from a variant of Nakayama's lemma due to du Plessis [14], see [19], (1.6)(ii). Finally, in the case $G = \mathcal{A}$ combine the proofs for $\mathcal{L}$ and $\mathcal{R}$. □

We now turn to versality.

(4.4) Let E: **Germs** $\to$ **Groups** be a group-valued functor and G a group acting on E satisfying the assumptions of (1.5) so that $\mathbf{E}_G \to$ **Germs** is a deformation theory. We assume that for a fixed element $a_0 \in E(0)$ the following conditions are satisfied.

1. For every closed embedding $S' \hookrightarrow S$ of germs the induced map $E(S) \to E(S')$ is surjective.

2. For every deformation $a \in E(S)$ of a_0 the module $\mathrm{Def}_{\mathbf{E}_G}(a, \mathcal{O}_S)$ is finite over $\mathcal{O}_S$.

3. $\mathbf{E}_G \to$ **Germs** admits integration of vector fields.

(4.5) Example. Let E be the functor of unfoldings and G one of the groups considered in (1.3). Assume that $f: (X,0) \to (\mathbb{C}^p, 0)$ is an element in $E(0)$ such that $\mathrm{Def}_{\mathbf{E}_G}(f, \mathbb{C}) = \mathcal{O}^p_{X,0}/TG(f, \mathbb{C})$ is a finite dimensional $\mathbb{C}$-vector space. *Then the conditions* (1) – (3) *in* (4.4) *are satisfied for E and G.* In fact, (1) is immediate, and (3) holds by (3.5). In order to prove (2) observe that

$$\mathrm{Def}_{\mathbf{E}_G}(f, \mathcal{O}_{S,0}) = \mathcal{O}^p_{X\times S,0}/TG(f, \mathcal{O}_{S,0})$$

is an $\mathcal{O}_{\mathbb{C}^p\times S,0}$-module, which for each of the groups G in question can be written as M/N, where M is a finite $\mathcal{O}_{X\times S,0}$-module and N is finite over $\mathcal{O}_{\mathbb{C}^p\times S,0}$. As by assumption $M/(N + \mathfrak{m}_S M)$ is finite dimensional over $\mathbb{C}$ it follows that $M/\mathfrak{m}_S M$ is finite over $\mathcal{O}_{\mathbb{C}^p,0}$. Thus by Weierstraß' preparation theorem M is finite over $\mathcal{O}_{\mathbb{C}^p\times S,0}$. Applying again the preparation theorem to the finite $\mathcal{O}_{\mathbb{C}^p\times S,0}$-module M/N it follows that M/N is $\mathcal{O}_{S,0}$-finite.

(4.6) Theorem. *Let $E, G, a_0 \in E(0)$ be as in (4.4). Then a_0 admits a semiuniversal deformation.*

By a semiuniversal deformation we mean here a deformation a of a_0 over some germ $S = (S, 0)$ with the following two properties:

(1) If b is any deformation of a_0 over $T = (T, 0)$, then $b \cong f^*(a)$ for some morphism of germs $f : T \to S$.

(2) The map of tangent spaces induced by f in (1) is uniquely determined by b.

The proof of (4.6) will be based on a lemma and proposition that we formulate and prove first.

(4.7) Lemma. *Let E, G and a_0 be as in (4.4) and $a \in \mathbf{E}_G(S)$ a deformation of a_0 over a germ $S = (S, 0)$. Then the following holds.*

1. *The functor $\mathcal{N} \mapsto \mathrm{Def}_{\mathbf{E}_G}(a, \mathcal{N})$ on* $\mathbf{Coh}(S)$ *is right exact.*

2. $\mathrm{Def}_{\mathbf{E}_G}(a, \mathcal{N}) \cong \mathrm{Def}_{\mathbf{E}_G}(a, \mathcal{O}_S) \otimes_{\mathcal{O}_S} \mathcal{N}$.

3. *If $S' \subseteq S$ is a closed subspace and $\mathcal{N}$ a finite $\mathcal{O}_{S'}$-module then with $a' := a|S'$*

$$\mathrm{Def}_{\mathbf{E}_G}(a, \mathcal{N}) = \mathrm{Def}_{\mathbf{E}_G}(a', \mathcal{N}).$$

Proof. That the functor considered in (1) is halfexact is contained in [5], (2.3)(3). Moreover, that it is right exact follows from our assumption (4.4)(1) applied to the closed embedding $S[\mathcal{N}''] \hookrightarrow S[\mathcal{N}]$ corresponding to a surjective morphism $\mathcal{N} \to \mathcal{N}''$ of $\mathcal{O}_S$-modules. (2) holds for any right exact functor on $\mathbf{Coh}(S)$. Finally, (3) holds generally for any deformation theory and follows easily from axiom (H) in (1.4). □

(4.8) Proposition. *Let $S = (S, 0)$ be a smooth germ, $b \in \mathbf{E}_G(S \times \mathbb{C}^r, 0)$ a deformation of a_0 and set $a := b|(S \times 0)$. Assume that the Kodaira-Spencer map*

$$\delta_{KS}: T^0_S(\mathbb{C}) \to \mathrm{Def}_{\mathbf{E}_G}(a, \mathbb{C})$$

is surjective. Then there is a map $\pi:(S \times \mathbb{C}^r, 0) \to (S, 0)$ such that b and $\pi^*(a)$ are G-equivalent by an element of $G(S \times \mathbb{C}^r, 0)$, which restricts to the identity over $S \cong S \times 0$.

Proof. It is clearly sufficient to treat the case $r = 1$. Let $t \in \mathcal{O}_{\mathbb{C},0}$ be the coordinate function and $\{\vartheta_i\}$ a basis for $T^0_S(\mathcal{O}_S)$. Then $\{\vartheta_i \otimes 1\}$ forms a basis of the $\mathcal{O}_{S\times\mathbb{C},0}$-module

$$T^0_{S\times\mathbb{C}/\mathbb{C}}(\mathcal{O}_{S\times\mathbb{C},0}) = T^0_S(\mathcal{O}_S) \otimes_{\mathcal{O}_S} \mathcal{O}_{S\times\mathbb{C},0} \subseteq T^0_{S\times\mathbb{C}}(\mathcal{O}_{S\times\mathbb{C},0}).$$

Under the Kodaira-Spencer map

$$\delta_{KS}: T^0_{S\times\mathbb{C}}(\mathcal{O}_{S\times\mathbb{C},0}) \to \mathrm{Def}_{\mathbf{E}_G}(b, \mathcal{O}_{S\times\mathbb{C},0}),$$

the elements $\vartheta_i \otimes 1$ are mapped onto a generating set of $\mathrm{Def}_{\mathbf{E}_G}(b, \mathcal{O}_{S\times\mathbb{C},0})$, since this is true modulo $\mathfrak{m}_{S\times\mathbb{C}}$ by assumption, see (4.7) (2), (3). Thus there is an equation

$$\delta_{KS}(\frac{\partial}{\partial t}) = \delta_{KS}\left(\sum_i a_i\vartheta_i \otimes 1\right),$$

for some $a_i \in \mathcal{O}_{S\times\mathbb{C},0}$. Accordingly, the derivation $\delta := \partial/\partial t - \sum_i a_i(\vartheta_i \otimes 1)$ maps to zero under δ_{KS} and satisfies $\delta(t) = 1$. Integration of vector fields yields the claim. □

Proof of (4.6). Let V be the vector space dual to $\mathrm{Def}_{\mathbf{E}_G}(a_0, \mathbb{C})$, and S_1 the trivial extension of the point 0 by V. Let $a_1 \in \mathbf{E}_G(S_1)$ be a deformation of a_0 inducing the canonical element in $\mathrm{Def}_{\mathbf{E}_G}(a_0, V)$, i.e. the element corresponding to id_V under the canonical isomorphism

$$\mathrm{Def}_{\mathbf{E}_G}(a_0, V) \cong \mathrm{Def}_{\mathbf{E}_G}(a_0, \mathbb{C}) \otimes_{\mathbb{C}} V \cong \mathrm{Hom}_{\mathbb{C}}(V, V).$$

By our assumption (1) in (4.4) the element a_1 can be lifted to an element $a \in E(S)$ if $S = (S, 0)$ is a smooth germ with first infinitesimal neighbourhood S_1. By construction the Kodaira-Spencer map

$$\delta_{KS}: T^0_S(\mathbb{C}) \to \mathrm{Def}_{\mathbf{E}_G}(a, \mathbb{C})$$

is bijective. We claim that a is a semiuniversal deformation of a_0. For this we have to verify that every deformation $b \in \mathbf{E}_G(T)$ of a_0 is induced from a over a suitable map $T \to S$. Because of (4.4) (1) we may assume that T is a smooth germ. Consider the product $S \times T$ and denote by pr_i, $i = 1, 2$ the projection onto the i-th factor and by pr the morphism to 0. Using the group stucture on E we can form $c := \mathrm{pr}_1^*(a) + \mathrm{pr}_2^*(b) - \mathrm{pr}^*(a_0)$, and c induces a resp. b on $S \cong S \times 0$ resp. $T \cong 0 \times T$. By (4.8) there is a map $\pi: (S \times T, 0) \to (S, 0)$ such that c and $\pi^*(a)$ are G-equivalent by an element of $G(S \times T, 0)$, which restricts to the identity on $S \cong S \times 0$. Thus, if $\varphi: T \to S$ is the composition of $T \cong 0 \times T \hookrightarrow S \times T$ and π, we get $\varphi^*(a) \cong b$ in $\mathbf{E}_G$ with an isomorphism inducing the identity on a_0. □

5. Theorems of Mather-Yau type

We will begin with the original theorem of Mather-Yau [12], see also [6]. It generalizes to the relative case as follows.

(5.1) Theorem. *Let $S = (S, 0)$ be a germ of a complex space and $f, g \in \mathfrak{m}_{X\times S}\mathcal{O}^p_{X\times S,0}$ mapping germs with $T\mathcal{K}(f, \mathcal{O}_S) = T\mathcal{K}(g, \mathcal{O}_S)$ in $\mathcal{O}^p_{X\times S,0}$. Then f and g are $\mathcal{K}$-equivalent.*

Proof. Consider the mapping germ $F = (1-t)f + tg$, which is defined in some neighbourhood, say U, of $0 \times 0 \times \mathbb{C}$ in $X \times S \times \mathbb{C}$. By (4.1) the mapping germ F is $\mathcal{K}$-equivalent to $f_{\mathbb{C}}(x,s,t) := f(x,s)$ resp. $g_{\mathbb{C}}$ in a neighbourhood of $(0,0,0)$ resp. $(0,0,1)$ in $X \times S \times \mathbb{C}$. Thus it suffices to show that the (open) set V of $t_0 \in \mathbb{C}$ such that F is $\mathcal{K}$-equivalent to $F_{t_0} := (1-t_0)f + t_0 g$ in a neighbourhood of $(0,0,t_0)$ in $X \times S \times \mathbb{C}$, is connected. We claim that this set is even Zariski-open. Consider the homomorphism of sheaves $J_F: \mathcal{T}^0_{U/S\times\mathbb{C}}(\mathfrak{m}_X \mathcal{O}_U) \to \mathcal{O}^p_U$ and set

$$\mathcal{D}ef_{\mathcal{K}}(F) = \mathcal{O}^p_U/(\mathrm{Im}(J_F) + \sum F_i \mathcal{O}^p_U),$$

which is a coherent sheaf on U. The Kodaira-Spencer class $\delta_{KS}(\partial/\partial t)$ may be regarded as a section in $\mathcal{D}ef_{\mathcal{K}}(F)$. The set V is just the set of points t_0 such that the germ of $\delta_{KS}(\partial/\partial t)$ at $(0,0,t_0)$ vanishes, see (3.6). As $\mathcal{D}ef_{\mathcal{K}}(F)$ is coherent, the support of $\delta_{KS}(\partial/\partial t)$ is a Zariski-closed subset and so V is Zariski-open. □

(5.2) Proposition. *Assume that $f, g \in \mathfrak{m}^2_X \mathcal{O}^p_{X\times S,0}$ are mapping germs such that $Z = f^{-1}(0)$ has singularity set* Sing $(Z) = 0 \times S$. *Then $T\mathcal{K}_e(f, \mathcal{O}_S) = T\mathcal{K}_e(g, \mathcal{O}_S)$ implies that f and g are $\mathcal{K}$-equivalent.*

Proof. Set again $F := (1-t)f + tg$. Let V be the open set of all $t_0 \in \mathbb{C}$ such that F is $\mathcal{K}$-equivalent to F_{t_0} in a neighbourhood of $(0,0,t_0)$. The argument of the proof of (5.1) shows that V is Zariski-open. We will show that it contains $0 \in \mathbb{C}$. In fact, by (3.8), applied to f and $H = f + h(t)g$ with $h(t) = t/(1-t)$, we get that H and $f_{\mathbb{C}}$ are $\mathcal{K}_e$-equivalent in a neighbourhood of $(0,0,0) \in X \times S \times \mathbb{C}$. This equivalence induces in particular an $S \times \mathbb{C}$-isomorphism α from $Z \times \mathbb{C}$ onto $\mathcal{Z} := H^{-1}(0)$ near $(0,0,0)$. As $f, g \in \mathfrak{m}^2_X \mathcal{O}^p_{X\times S,0}$, the set $0 \times S \times \mathbb{C}$ is contained in Sing $(\mathcal{Z})$. Moreover, under the isomorphism α^{-1} the set Sing $(\mathcal{Z})$ is mapped onto Sing $(Z \times \mathbb{C}) = 0 \times S \times \mathbb{C}$ so that $0 \times S \times \mathbb{C}$ is preserved, and the given equivalence is already a $\mathcal{K}$-equivalence. Similarly, applying (3.8) to g and $G = g + \tilde{h}(t)f$ with $\tilde{h} = (1-t)/t$ around $t = 1$, we get that G and $g_{\mathbb{C}}$ are $\mathcal{K}_e$-equivalent in a neighbourhood of $(0,0,1) \in X \times S \times \mathbb{C}$. As V is Zariski-open, the mapping germ $G(x,s,t_0)$ is equivalent to f for t_0 near 1, $t_0 \neq 1$. In particular Sing $G^{-1}(0,0,t_0) = 0 \times S \times \{t_0\}$ for such t_0. Moreover, Sing $(g^{-1}(0)) \supseteq 0 \times S$ as $g \in \mathfrak{m}^2_X \mathcal{O}^p_{X\times S}$. It follows that the $\mathcal{K}_e$-equivalence between G and $g_{\mathbb{C}}$ preserves $0 \times S \times \{t_0\}$, for $t_0 \neq 1$, t_0 near 1, and hence also $0 \times S \times \mathbb{C}$ in a neighbourhood of $(0,0,1)$. □

(5.3) Corollary. (see [6] ,2. (5)) *Let $(X,0)$ and $(Y,0)$ be complete intersections of codimension p with isolated singularities in $\mathbb{C}^n$. Assume that $T^1_{X,0}$ and $T^1_{Y,0}$ are isomorphic, i.e. there is an automorphism Φ of $\mathbb{C}^n$ such that $T^1_{X,0} \cong \Phi^*(T^1_{Y,0})$ as $\mathcal{O}_{\mathbb{C}^n,0}$-modules. Then $(X,0)$ and $(Y,0)$ are isomorphic.*

Proof. We may suppose that $T^1_{X,0} \cong T^1_{Y,0}$ as $\mathcal{O}_{\mathbb{C}^n,0}$ modules. Let X, Y be given by the regular sequences $f := (f_1, \ldots, f_p)$ resp. $g := (g_1, \ldots, g_p)$. Then $T^1_{X,0} = \mathcal{O}^p_{\mathbb{C}^n,0}/T\mathcal{K}_e(f, \mathbb{C})$ and $T^1_{Y,0} = \mathcal{O}^p_{\mathbb{C},0}/T\mathcal{K}_e(g, \mathbb{C})$. After passing to another minimal system of generators of the ideal of Y we may suppose that the isomorphism $T^1_{X,0} \cong T^1_{Y,0}$ is given by the identity on $\mathcal{O}^p_{\mathbb{C}^n,0}$. But then (5.2) implies the desired result. □

(5.4) Remark. The result (5.1) holds for $\mathcal{R}$–equivalence as well — by the same proof — provided that $f \in T\mathcal{R}(f, \mathcal{O}_S)$ and $g \in T\mathcal{R}(g, \mathcal{O}_S)$. For S a point and $p = 1$ this is the case if f and g are quasihomogeneous.

Moreover, if we assume that $f^{-1}(0)$ has singularity set $0 \times S$ and f, g are in $\mathfrak{m}_X^2 \mathcal{O}_{X\times S}^p$, then $\mathcal{R}$ can be replaced by $\mathcal{R}_e$. This is easily seen by the same argument as in (5.2). For similar results see also [6], 2.(8) and [17].

In the remaining part of this section we will prove a theorem of Mather-Yau type for $\mathcal{A}$-equivalence. First we show the following local result.

(5.5) Proposition. *Let $f, g \in \mathfrak{m}_{X\times S}\mathcal{O}_{X\times S,0}^p$ be mapping germs over $S = (S, 0)$ such that $\mathrm{Def}_{\mathcal{K}}(f, \mathcal{O}_S)$ is a finite $\mathcal{O}_S$-module and $T\mathcal{A}(g, \mathcal{O}_S) \subseteq T\mathcal{A}(f, \mathcal{O}_S)$. Then the mapping germs $f_{\mathbb{C}}(x, s, t) := f(x, s)$ and $F := f + tg$ in $\mathcal{O}_{X\times S\times\mathbb{C},0}^p$ are $\mathcal{A}$-equivalent.*

Proof. We set $A := \mathcal{O}_{\mathbb{C}^p\times S\times\mathbb{C},0}$, $B := \mathcal{O}_{X\times S\times\mathbb{C},0}$ and $\bar{A} := \mathcal{O}_{\mathbb{C}^p\times S,0} = A/tA$, $\bar{B} = B/tB$, where t is the coordinate function on $\mathbb{C}$. Let $J_F\colon T^0_{X\times S\times\mathbb{C}/S\times\mathbb{C}}(\mathfrak{m}_X B) \to B^p$ be the canonical map. As $T^0_{X\times S\times\mathbb{C}/S\times\mathbb{C}}(\mathfrak{m}_X B) \cong T^0_{X\times S/S}(\mathfrak{m}_X \bar{B}) \otimes_{\mathcal{O}_{X\times S,0}} B$ we have

$$\begin{aligned}\mathrm{Def}_{\mathcal{K}}(f, \mathcal{O}_S) &= \bar{B}^p/(\mathrm{Im}(J_F) \otimes_B \bar{B} + \textstyle\sum f_i \bar{B}^p) \\ &= (B^p/\mathrm{Im}(J_F)) \otimes_A (A/(\mathfrak{m}_{\mathbb{C}^p} A + tA)).\end{aligned}$$

By assumption this is a finite module over $\mathcal{O}_S = A/(\mathfrak{m}_{\mathbb{C}^p} A + tA)$. The preparation theorem gives that $B^p/\mathrm{Im}(J_F)$ is a finite $F^{-1}(A)$-module. It follows that

$$\mathrm{Def}_{\mathcal{A}}(F, \mathcal{O}_{S\times\mathbb{C},0}) = B^p/(\mathrm{Im}(J_F) + F^{-1}(\mathfrak{m}_{\mathbb{C}^p} A^p))$$

is finite over $F^{-1}(A)$. The same argument gives that $\mathrm{Def}_{\mathcal{A}}(f_{\mathbb{C}}, \mathcal{O}_{S\times\mathbb{C},0})$ is finite over $(f \times \mathrm{id}_{\mathbb{C}})^{-1}(A)$. For $\varphi \in A^p$ consider again the Taylor expansion

$$(F, \mathrm{id}_{\mathbb{C}})^{-1}(\varphi) - (f \times \mathrm{id}_{\mathbb{C}})^{-1}(\varphi) = \sum_{\substack{i\neq 0 \\ i\in\mathbb{N}^p}} \frac{1}{i!}(f \times \mathrm{id}_{\mathbb{C}})^{-1}\left(\frac{\partial^{|i|}\varphi}{\partial y^i}\right)(tg)^i.$$

Every term in this series is zero in the finite $(f \times \mathrm{id}_{\mathbb{C}})^{-1}(A)$-module $\mathrm{Def}_{\mathcal{A}}(f_{\mathbb{C}}, \mathcal{O}_{S\times\mathbb{C},0})$ by our assumption, and so $(F, \mathrm{id}_{\mathbb{C}})^{-1}(\varphi) \equiv 0 \bmod T\mathcal{A}(f_{\mathbb{C}}, \mathcal{O}_{S\times\mathbb{C},0})$. Thus

$$T\mathcal{A}(F, \mathcal{O}_{S\times\mathbb{C},0}) \subseteq T\mathcal{A}(f_{\mathbb{C}}, \mathcal{O}_{S\times\mathbb{C},0}),$$

and the quotient, say Q, satisfies $Q = tQ$ as both restrict to $T\mathcal{A}(f, \mathcal{O}_S)$ modulo t. As Q is a $\mathbb{C}\{t\}$-submodule of the finite $F^{-1}(A)$-module $\mathrm{Def}_{\mathcal{A}}(F, \mathcal{O}_{S\times\mathbb{C},0})$, it is t-adically separated and so $Q = 0$, whence

$$T\mathcal{A}(F, \mathcal{O}_{S\times\mathbb{C},0}) = T\mathcal{A}(f_{\mathbb{C}}, \mathcal{O}_{S\times\mathbb{C},0}).$$

Now the desired result follows from (3.8). □

(5.6) Theorem. *Let $f, g \in \mathfrak{m}_{X\times S}\mathcal{O}_{X\times S,0}^p$ be mapping germs over $S = (S, 0)$ such that $\mathrm{Def}_{\mathcal{K}}(f, \mathcal{O}_S)$ is a finite $\mathcal{O}_S$-module. If the subsets $T\mathcal{A}(f, \mathcal{O}_S)$ and $T\mathcal{A}(g, \mathcal{O}_S)$ of $\mathcal{O}_{X\times S,0}^p$ are equal then f and g are $\mathcal{A}$-equivalent.*

Proof. As in the proof of (5.1) we consider the unfolding $F := (1-t)f + tg$. As $T\mathcal{K}(f, \mathcal{O}_S)$ is the $\mathcal{O}_{X\times S,0}$-submodule generated by $T\mathcal{A}(f, \mathcal{O}_S)$ it follows that $\mathrm{Def}_{\mathcal{K}}(g, \mathcal{O}_S) = \mathrm{Def}_{\mathcal{K}}(f, \mathcal{O}_S)$. The proposition above shows that F and $f_{\mathbb{C}}$ are $\mathcal{A}$-equivalent in a neighbourhood of $t = 0$, whereas F and $g_{\mathbb{C}}$ are $\mathcal{A}$-equivalent near $t = 1$. Thus it is sufficient to show that the (open) set V of points $t_0 \in \mathbb{C}$, where F and $F_{t_0} = (1-t_0)f + t_0 g$ are $\mathcal{A}$-equivalent in a neighbourhood of t_0, is connected. We claim that the complement of this set is countable. Consider

$$\mathcal{D}\mathrm{ef}_{\mathcal{A}}(F) := \mathcal{O}_U^p/(\mathrm{Im}(J_F) + F^{-1}(\mathfrak{m}_{\mathbb{C}^p}\mathcal{O}^p_{\mathbb{C}^p \times S \times \mathbb{C},0})),$$

which is a sheaf in some neighbourhood U of $0 \times 0 \times \mathbb{C}$ in $X \times S \times \mathbb{C}$. The Kodaira-Spencer class $\delta_{KS}(\partial/\partial t)$ of F may be considered as a section in $\mathcal{D}\mathrm{ef}_{\mathcal{A}}(F)$. By assumption

$$\mathcal{D}\mathrm{ef}_{\mathcal{K}}(F) := \mathcal{O}_U^p/(\mathrm{Im}(J_F) + \textstyle\sum F_i \mathcal{O}_U^p)$$

and then also $\mathcal{O}_U^p/\mathrm{Im}(J_F)$ is quasifinite over $\mathbb{C}^p \times S \times \mathbb{C}$ at $(0,0,0) \in X \times S \times \mathbb{C}$, and since $\mathrm{Def}_{\mathcal{K}}(f, \mathcal{O}_S) = \mathrm{Def}_{\mathcal{K}}(g, \mathcal{O}_S)$, it is also quasifinite over $\mathbb{C}^p \times S \times \mathbb{C}$ at $(0,0,1)$. By Remmert's semicontinuity theorem [4] 3.6, the set U' of points $(x, s, t) \in U$ where $\mathcal{O}_U^p/\mathrm{Im}(J_F)$ is quasifinite over $\mathbb{C}^p \times S \times \mathbb{C}$, is Zariski open in U. Applying (5.7) to

$$\begin{array}{ccc} & Z = (0 \times 0 \times \mathbb{C}) \cap U' & \\ \swarrow & & \searrow \\ U' & \xrightarrow{\quad F|U' \quad} & Y := \mathbb{C}^p \times S \times \mathbb{C} \end{array}$$

and the natural map

$$u \colon (F|U')^{-1}(\mathcal{O}_Y^p) \to \mathcal{O}^p_{U'}/\mathrm{Im}(J_F),$$

shows that the section $\delta_{K/S}(\partial/\partial t)$ in $\mathcal{D}\mathrm{ef}_{\mathcal{A}}(F) = \mathrm{Coker}(u)$ vanishes on a nonempty set that is Zariski-open in some Zariski-open subset of Z. Thus the complement of the set $V \subseteq \mathbb{C}$ above is countable as claimed. □

(5.7) Lemma. *Let*

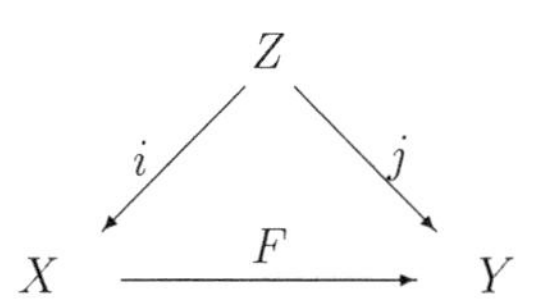

*be a diagram of complex spaces such that i, j are closed embeddings. Let $u \colon F^*(\mathcal{N}) \to \mathcal{M}$ be a morphism of $\mathcal{O}_X$-modules, where $\mathcal{N} \in$ **Coh**(Y) and $\mathcal{M} \in$ **Coh**(X). Assume that*

$$\mathrm{supp}(\mathcal{M}) \cap F^{-1}(j(Z))$$

is quasifinite over Y. Then for a section $m \in \Gamma(X, \mathcal{M})$ the set

$$V = \{z \in Z \colon m \in \mathrm{Im}(\mathcal{N}_{j(z)} \to \mathcal{M}_{i(z)})\}$$

cuts out a Zariski-open subset in some Zariski-open dense subset W of Z.

Proof. Replacing X by $\mathrm{supp}(\mathcal{M})$ equipped with the structure given by the coherent ideal sheaf $\mathrm{Ann}(\mathcal{M}) \subseteq \mathcal{O}_{X,0}$, we may assume that $Z' := F^{-1}(j(Z))$ is quasifinite over $Z \cong j(Z)$. Obviously we may assume that Z is reduced, and we equip Z' with its reduced structure. Then the ramification locus $\Gamma \subseteq Z'$ of $F : Z' \to Z \cong j(Z)$ is analytic and nowhere dense in Z'. Let $Z'' \subseteq Z'$ be the union of all irreducible components which are not contained in $i(Z)$. Replacing X by $X \setminus (\Gamma \cup Z'')$ and Z by $Z \setminus \Gamma$ we may assume that $F : Z' \xrightarrow{\sim} j(Z)$ is an isomorphism. Let $\mathcal{O}_X|Z = i^{-1}(\mathcal{O}_X)$ be the topological inverse image of $\mathcal{O}_X$, similarly for $\mathcal{O}_Y|Z, \mathcal{M}|Z, \mathcal{N}|Z$. By (5.8) below the sheaves of rings $\mathcal{O}_X|Z$, $\mathcal{O}_Y|Z$ are coherent, and $\mathcal{M}|Z$, $\mathcal{N}|Z$ are coherent modules over $\mathcal{O}_X|Z$, $\mathcal{O}_Y|Z$ respectively.

We claim that $\mathcal{O}_X|Z$ is a coherent $\mathcal{O}_Y|Z$-module. This is a local problem around a point $z \in Z$, and so after replacing X, Y by suitable neighbourhoods of z in X resp. Y we are reduced to the case that F is finite. By now $F_*(\mathcal{O}_X)$ is a coherent $\mathcal{O}_Y$-module and so $F_*(\mathcal{O}_X)|Z$ is coherent over $\mathcal{O}_Y|Z$. As $Z' = F^{-1}(j(Z)) \to j(Z)$ is bijective the canonical map $F_*(\mathcal{O}_X)|Z \to \mathcal{O}_X|Z$ is an isomorphism in every stalk and so $F_*(\mathcal{O}_X)|Z \simeq \mathcal{O}_X|Z$ is coherent over $\mathcal{O}_Y|Z$, proving the claim.

Let $\bar{m} \cdot \mathcal{O}_Y|Z \subseteq (\mathcal{M}|Z)/(\mathcal{N}|Z)$ be the subsheaf generated by the class $\bar{m}$ of the given section $m \in \Gamma(X, \mathcal{M})$. As $\mathcal{M}|Z$, $\mathcal{N}|Z$ are coherent over $\mathcal{O}_Y|Z$, this subsheaf is coherent. Thus $V = Z \setminus \mathrm{supp}(\bar{m}\mathcal{O}_Y|Z)$ is Zariski-open in Z, and we are done. □

(5.8) Lemma. *Let $f : X \to Y$ be a morphism of topological spaces and $\mathcal{O}_Y$ a sheaf of rings on Y. Then for every coherent $\mathcal{O}_Y$-module $\mathcal{N}$ the topological inverse image $f^{-1}(\mathcal{N})$ is a coherent sheaf of $f^{-1}(\mathcal{O}_Y)$-modules.*

The *proof* follows easily from the fact that f^{-1} is an exact functor. We leave the simple argument to the reader.

(5.9) Remark. Assume that $f, g \in m_X^2 \mathcal{O}^p_{X \times S, 0}$ and that $f^{-1}(0)$ has singularity set $0 \times S \subseteq X \times S$. Then (5.6) is also valid for the group $\mathcal{A}_e$. This follows with similar arguments as in (5.2).

(5.10) Remark. Let $f : X \to Y$ be a morphism of topological spaces and $\mathcal{M}, \mathcal{N}$ sheaves on X resp. Y. If $u : f^{-1}(\mathcal{N}) \to \mathcal{M}$ is a morphism of sheaves and $m \in \Gamma(X, \mathcal{M})$ is a section then the set

$$V = \{x \in X : m \in \mathrm{Im}(\mathcal{N}_{f(x)} \to \mathcal{M}_x)\}$$

is open in X. We do not know whether this set is *Zariski*-open if X, Y are complex spaces and $\mathcal{M}, \mathcal{N}$ are coherent. In the algebraic case this is tautologically true as V is open hence Zariski-open.

References

[1] Arnol'd, V.I.; Vasil'ev, V.A.; Goryunov, V.V.; Lyashko, O.V.: *Singularities, local and global theory.* In: Encyclopaedia of Mathematical Sciences, Vol. 6: Dynamical systems VI, ed. V.I. Arnol'd. Springer Verlag Berlin-Heidelberg-New York 1993

[2] Damon, J.: *The unfolding and determinacy theorems for subgroups of $\mathcal{A}$ and $\mathcal{K}$.* Memoirs AMS 50, no. **306** (1984)

[3] Dimca, A.: *Are the isolated singularities of complete intersections determined by their singular subspaces?* Math. Ann. **267**, 461–472 (1984)

[4] Fischer, G.: *Complex analytic geometry.* Lecture Notes in Math. 538, Springer Verlag Berlin-Heidelberg-New York 1976

[5] Flenner, H.: *Ein Kriterium für die Offenheit der Versalität.* Math. Z. **178**, 449 - 473 (1981)

[6] Gaffney, T.; Hauser, H.: *Characterizing singularities of varieties and mappings.* Invent. math. **81**, 427 - 447 (1985)

[7] Grauert, H.; Remmert, R.: *Analytische Stellenalgebren.* Springer Verlag Berlin-Heidelberg-New York 1971

[8] Hauser, H.: *On the singular subspace of a complex-analytic variety.* In: Complex analytic singularities, Proc. Sem., Ibaraki/Jap. 1984, Adv. Stud. Pure Math. **8**, 125 - 134 (1987).

[9] Illusie, L.: *Complexe cotangent et déformations I, II.* Lecture Notes in Math. 239, 283. Springer Verlag Berlin-Heidelberg-New York 1971, 1972

[10] MacLane, S.: *Categories for the working mathematician.* Graduate Texts in Math. 5, Springer Verlag Berlin-Heidelberg-New York 1971

[11] Mather, J.: *Stability of C^∞-mappings.*
I. The division theorem. Ann. of Math. **89**, 89 - 104 (1969)
II. Infinitesimal stability implies stability. Ann. of Math. (2) **89**, 254 - 291 (1969)
III. Finitely determined map germs. Publ. Math. IHES **36**, 127 - 156 (1968)
IV. Classification of stable germs by $\mathbb{R}$-algebras. Publ. Math. IHES **37**, 223 - 248 (1969)
V. Transversality. Advances in Math. **4**, 301 - 336 (1970)
VI. The nice dimensions. In: Liverpool Sing. Symp. I. Lecture Notes 192, 207 - 253. Springer Verlag Berlin-Heidelberg-New York 1970

[12] Mather, J.; Yau, S.S.: *Classification of isolated hypersurface singularities by their moduli-algebra.* Invent. Math. **69**, 243 - 251 (1982)

[13] Matsumura, H.: *Commutative ring theory.* Cambridge studies in advanced mathematics **8**, Cambridge University press 1989

[14] du Plessis, A.A.: *On the determinacy of smooth map-germs.* Invent. Math. **58**, 107 - 160 (1980)

[15] Rim, D.S.: *Formal deformation theory.* In: Séminaire de Géometrie Algébrique, SGA 7. Lecture Notes in Mathematics 288, Springer Verlag Berlin-Heidelberg-New York 1972

[16] Schuster, H.W.: *Infinitesimale Erweiterungen komplexer Räume.* Comment. Math. Helv. **45**, 263 - 286 (1970)

[17] Shoshitaǐshvili, A.N.: *Functions with isomorphic Jacobian ideals.* Funct. Anal. Appl. **10**, 128 - 133 (1976)

[18] Teissier, B.: *The hunting of invariants in the geometry of dicriminants.* In: Real and complex singularities, Oslo 1976 (Ed. P. Holm), Sijthoff & Noordhoff International Publishers 1977

[19] Wall, C.T.C.: *Finite determinacy of smooth map-germs.* Bull. Lond. Math. Soc. **13**, 481 – 539 (1981)

On the Uniqueness and Characterization of Grauert Tubes

DAN BURNS[1] University of Michigan, Ann Arbor, Michigan

1 Introduction

It is well-known that a real analytic manifold X admits a *complexification* X_C, a complex manifold which contains X as the *real points*, or fixed points of an anti-holomorphic involution, or complex conjugation, $\sigma : X_C \to X_C$. The germ of X_C is unique along X, as is σ, but there has been a lot of interest recently in canonical choices of a specific complexification and characterization of the resultant complex manifolds. One type of X_C is specified by the additional datum of a real analytic Riemannian metric $g = g_{ij}\, dx^i \cdot dx^j$ on X, [GS], [LS], [S1]. For compact X, these manifolds can be considered as canonical complex structures on the (real analytic) tangent bundle $T(X)$ of X, or more precisely, on the bundle of tangent vectors of length $< r$ for some *radius* $r > 0$, where length is measured with respect to the metric g. Such a manifold is sometimes called a *Grauert tube*, because of Grauert's use of Stein tubular neighborhoods of $X \subset X_C$ to settle the problem of embedding real analytic manifolds real analytically into $\mathbb{R}^N$, for some $N \gg 0$. We will adopt this nomenclature in the present paper, although in general we will follow more closely the point of view of [LS], [S1]. Thus, a Grauert tube of radius r is constructed for each $r > 0$ sufficiently small for each compact Riemannian real analytic manifold X, and the construction is functorial. In particular, the isometry group $Isom(X)$ of (X, g) acts continuously as biholomorphic transformations of X_C^r, acting as differentials of isometries, if X_C^r is viewed as $T^r(X) := \{v \in T(X) \mid \| v \| < r\}$. These Grauert tubes have many special analytic and geometric features and a very short list of characterizing properties, which we will recall below. One of the most interesting properties is that the function $\tau = \| v \|^2$ is a real analytic, strictly plurisubharmonic function on X_C^r whose square root $u := \sqrt{\tau}$ is a plurisubharmonic function on X_C^r which verifies the homogeneous complex Monge-Ampère equation

$$(\partial\bar{\partial}u)^n = 0 \tag{1.1}$$

on $X_C^r \setminus X$, where n = dimension X = complex dimension X_C^r. Our first theorem shows that the existence, regularity and boundary values of such a Monge-Ampère solution on

[1] Partially supported by NSF grant DMS-9408994.

a connected complex manifold Ω already characterize that manifold as a Grauert tube. More precisely, we prove:

Theorem 1: *Let Ω be a connected complex manifold, and let $\tau : \Omega \to [0, r^2)$ be a proper, strictly plurisubharmonic function of class C^∞ such that $u = \sqrt{\tau}$ is a plurisubharmonic solution of (1.1) on $\Omega \setminus \tau^{-1}(0)$. Then $\tau^{-1}(0)$ is a connected real analytic manifold X and Ω is canonically biholomorphic to the Grauert tube X_C^r, where X is equipped with the metric induced from the Kähler metric on Ω with Kähler form*

$$\omega := \frac{i}{2}\partial\bar{\partial}\tau. \tag{1.2}$$

This theorem should be compared to the analogous characterization result for *strictly parabolic manifolds* in [B] and [S]. The properties of Ω and τ stated above are already derived in [LS], [PW] and [S1]; our main task is to derive the existence of the manifold $X \subset \Omega$. The proof presented below for theorem 1 is differential geometric in nature, and based on the ideas in [B], [LS], [PW], [S] and [S1].

We also consider here the question of the uniqueness of the representation of a given complex manifold Ω as a Grauert tube X_C^r. That is, suppose that Ω is a Grauert tube of radius $r > 0$ for some real analytic Riemannian manifold X, g. Can it be the Grauert tube for another X and another metric? There is a trivial non-uniqueness coming from scaling: without changing Ω, we may replace the metric g on X by $\check{g} := c \cdot g$, for $c > 0$, and replacing u by $\check{u} := \sqrt{c} \cdot u$. This will only change the radius of the tube Ω, viewed as a tube for $X, \check{g}$. Thus in Theorem 2 below we will scale out this effect by assuming the radii of our representations of Ω as a Grauert tube to be equal. At any rate, for finite radius tubes, we show the following uniqueness theorem:

Theorem 2: *Let Ω be a connected complex manifold, and suppose that Ω is biholomorphic to X_C^r and $\check{X}_C^{\check{r}}$ for two real analytic Riemannian manifolds X, g and $\check{X}, \check{g}$, and radii $r = \check{r} < \infty$. Then there is a biholomorphism $\Phi : \Omega \to \Omega$, sending $X \subset \Omega$ to $\check{X} \subset \Omega$ and such that the induced map $\Phi : X \to \check{X}$ induces an isometry from X, g to $\check{X}, \check{g}$.*

The restricted case when the mapping Φ is assumed to be a biholomorphic isometry of the respective τ-metrics has already been proved by Szöke in [S2].

One could ask for a more precise form of uniqueness than that given by Theorem 2. Namely, one could ask that X and $\check{X}$ simply be identical as subsets of Ω, *i.e.*, that Φ be the identity map. Such an X is the fixed point set of an anti-holomorphic involution σ of Ω. We cannot prove at this time that $X \subset \Omega$ or its corresponding σ is unique but we can show that the number of such involutions is finite and any two are intertwined by an element $\Phi \in Aut_C(\Omega)$:= the group of biholomorphic automorphisms of Ω.

Theoerm 2bis: *If Ω is as above, with radius $r < \infty$, there exist a finite number of anti-holomorphic involutions σ_i of Ω with compact fixed point set $X_i \subset \Omega$. Any two such σ_i are conjugate by an element $\Phi \in Aut_C(\Omega)$.*

The exact uniqueness of the fixed point set $X \subset \Omega$ is equivalent to a number of different statements, for example, that every $\Phi \in Aut_C(\Omega)$ preserves X, or commutes with its involution σ, or that the natural inclusion alluded to above $Isom(X) \hookrightarrow Aut_C(\Omega)$ is in fact an isomorphism. There is one restricted class of Riemannian manifolds for which we can verify such a strong uniqueness statement, which is the point of the third theorem:

Theorem 3: *Let X be a homogeneous Riemannian manifold, i.e., $Isom(X)$ acts transitively on X, and let $\Omega \simeq X_C^r$ be an associated Grauert tube. Then X, σ are unique in Ω, and $Isom(X) \overset{\sim}{\hookrightarrow} Aut_C(\Omega)$.*

Theorems 2, 2bis and 3 are all false as stated for the case of $r = \infty$. It is unclear at present how they should be reformulated for the case of infinite radius. Theorem 4 below, due originally to Szöke [S2], is one attempt, but it is too restrictive.

Let us outline the proofs of the theorems and the remainder of the paper. Theorem 1 is proved in §2, where we first collect some geometric properties of the interplay between the leaves of the Monge-Ampère foliation associated to $u = \sqrt{\tau}$ and the Kähler geometry on Ω induced by the Kähler form ω as in Theorem 1. These properties have been treated elsewhere, so we are only collecting the statements in one convenient location. The leaves of the foliation prove to be geodesic, as in the strictly parabolic manifold case, and the uniformity of the geometry for the τ-metric enables us to stitch the various leaves together coherently along a submanifold $X \subset \Omega$. If one assumes the function τ to be real analytic from the start, the construction of X is easy, and follows from a simple dimension count. By comparison, if τ is merely smooth, then local considerations coming directly from the Monge-Ampère equation show that infinitesimally, the zero locus of τ has Zariski tangent spaces of the right dimension, but it requires a little more effort to prove that the zero locus is actually as big as an n-dimensional manifold. This is done by fitting many geodesics for the τ-metric inside the zero set. The regularity of X follows from the infinitesimal calculations already alluded to and the uniformity on compact sets of the differential injectivity of various exponential maps for the τ-metric on Ω. The real analyticity of τ follows at this point from a theorem of Lempert [L].

In §3 we prove Theorems 2 and 2bis. In his earlier study of the symmetries of Grauert tubes, Szöke had already noticed the significance of the relationship of $Isom(X)$ to $Aut_C(\Omega)$, and asked whether $Isom(X)$ could really be of smaller dimension than $Aut_C(\Omega)$. One of the main points of this paper is a simple proof that $Isom(X)$ is of finite index in $Aut_C(\Omega)$. This is based on a consideration of the geometric properties of X with respect to the Kähler-Einstein metric g_{KE} of Cheng and Yau on Ω, *c.f.*, [CY]. Since the involution σ is an anti-holomorphic isometry of g_{KE}, its fixed point set is a geodesic submanifold of Ω. It is also Lagrangian with respect to the Kähler form ω_{KE} of g_{KE}, and hence is a strictly stable minimal submanifold of Ω, g_{KE}. This controls the connected component of $Aut_C(\Omega)$ as it operates on X. The finite index property results from the compactness of $Aut_C(\Omega)$. It is very interesting to note the role played by the auxiliary symplectic structure ω_{KE} (the "obvious" one in the situation is that given by the form ω of Theorem 1). Theorems 2 and 2bis follow from some more precise considerations of detailed geometric features of Grauert tubes. The compactness of the fixed point set $X \subset \Omega$ of σ is crucial for these theorems. It is easy to construct counter-examples to Theorem 2bis if this compactness condition is dropped.

In §4 we prove Theorem 3. The point is to prove the uniqueness of the submanifold X by considering the critical points of an $Aut_C(\Omega)$-invariant exhaustion of Ω which is strictly plurisubharmonic. That the radius be finite again enters via the compactness of $Aut_C(\Omega)$. The homogeneity allows us to control the critical points of this exhaustion when restricted to the fiber of Ω over one point $x \in X$, when Ω is viewed as the tangent bundle of X. We prove that all critical points of this induced exhaustion function of a Euclidean ball must be strict local minima, using the strict plurisubharmonicity of the exhaustion. Simple topology then shows there is only one critical point. This suggests that there should be some infinite dimensional analogue of this proof in the general case, but the author has not been able to make this work. It would be interesting to know if some analogue of Hartman's uniqueness theorem for harmonic maps might be found for this case.

In a final §5 we give a second proof of a theorem of Szöke's [S2], concerning the case of infinite radius tubes. One makes the very restrictive assumption that a biholomorphism Φ of $\Omega \simeq X_C^\infty$ preserves the τ-metric or, equivalently, preserves the Kähler form ω, and is an *exact* symplectic mapping of the structure given by ω. Then we show that Φ preserves the submanifold $X \subset \Omega$. The proof given is more cumbersome than Szöke's original proof in [S2], but we choose to include it because, as in the proofs of Theorems 2 and 2bis above, considerations of symplectic geometry play a key role. Specifically, we use a theorem of Gromov [G] on the intersection of Lagrangian submanifolds to reduce the statement to an easy analogue in one complex variable.

The author would like to thank S.-J. Kan, R. Szöke, and J. Wolfson for useful communications about the questions treated in this paper.

2 A Monge-Ampère characterization of Grauert tubes

In this § we give the proof of Theorem 1 of the introduction. It is crucial to recall the result of Lempert and Szöke that any *Monge-Ampère model* is a Grauert tube. Recall first that a Monge-Ampère model is simply a triple (Ω, X, u) where Ω is a connected complex manifold of dimension n, $X \subset \Omega$ is a connected real submanifold of real dimension n, and u is a continuous function $u : \Omega \to [0, r), r \in \mathbb{R}^+ \cup \{\infty\}$ satisfying the homogeneous complex Monge-Ampère equation (1.1) on $\Omega^* := \Omega \setminus u^{-1}(0)$, and for which $\tau = u^2$ is of class $\mathcal{C}^\infty$ and strictly plurisubharmonic on all of Ω.

For Theorem 1, we suppose given the connected complex manifold Ω and the function τ of class $\mathcal{C}^\infty$ which is strictly plurisubharmonic, and we assume that $u = \sqrt{\tau}$ satisfies the Monge-Ampère equation (1.1) on Ω^*, and furthermore that $\tau : \Omega \to [0, r^2)$ is proper. To prove Theorem 1 it will suffice therefore to show that $X := \tau^{-1}(0)$ is a connected, smooth manifold of real dimension n.

We will first have to consider some geometric properties of the Kähler manifold Ω with the τ-metric, which we gather mainly from [LS], [PW] (also [B], [S]). In local holomorphic coordinates $(z^1, \ldots, z^n)$ on Ω, for a scalar valued function τ, we define

$$\tau_i = \frac{\partial \tau}{\partial z^i}$$

and

$$\tau_{\bar{\imath}} = \frac{\partial \tau}{\partial z^{\bar{\imath}}}.$$

Similarly,

$$\tau_{i\bar{\jmath}} = \frac{\partial^2 \tau}{\partial z^i \partial z^{\bar{\jmath}}},$$

and so forth. In our case the function τ is strictly plurisubharmonic. The τ-metric g_τ is the metric with Kähler form ω as in (1.2) above, that is, in our local coordinates, we have

$$g_\tau = \tau_{i\bar{\jmath}}\, dz^i \cdot dz^{\bar{\jmath}}.$$

Using the usual summation convention for repeated indices, one above and one below, we define $\tau^{i\bar{\jmath}}$ by the equation

$$\tau^{i\bar{\jmath}} \tau_{k\bar{\jmath}} = \delta^i_k$$

where δ^i_k is the usual Kronecker delta. We record the following lemma, which is well-known, and is an easy computation.

Lemma 2.1: *$u = \sqrt{\tau}$ verifies the homogeneous complex Monge-Ampère equation (1.1) where $u > 0$ if and only if*

$$n\,\partial\tau \wedge \bar\partial\tau \wedge (\partial\bar\partial\tau)^{n-1} = 2\,\tau\,(\partial\bar\partial\tau)^n. \tag{2.1}$$

One may rewrite (2.1) as

$$\tau^{i\bar{j}}\tau_i\tau_{\bar{j}} = 2\tau. \tag{2.2}$$

Note, as a corollary of Lemma 2.1 that the differential $d\tau$ of τ is non-zero wherever $\tau \neq 0$. This follows from (2.1) and the strict plurisubharmonicity of τ. Since τ is strictly plurisubharmonic on Ω, it is certainly non-constant. Since τ is proper, tending to r^2 as one passes outside any compact set in Ω, we see that τ must achieve a minimum value on Ω. At such a point, $d\tau = 0$, so we conclude that $\tau^{-1}(0)$ is non-empty.

We would like to know something about the size of this zero locus $\tau^{-1}(0)$. We know $d\tau = 0$ along $\tau^{-1}(0)$, so the next thing to check is the rank of the real Hessian of τ.

Lemma 2.2: *At every $z_0 \in \tau^{-1}(0)$, the rank of the real Hessian of τ is n.*

Proof: One differentiates equation (2.2) twice at z_0. First pick local coordinates so that $z_0 = 0$, and $\tau_{i\bar{j}}(z_0) = \delta_{i\bar{j}}$. It is well known that by a possible further unitary linear change of coordinates at 0, one can arrange that $\tau_{ij}(0) = -\lambda_i\,\delta_{ij}$, for suitable $\lambda_i \in \mathbb{R}^+$. Noting that $\tau_i(0) = \tau_{\bar{i}}(0) = 0$, for $i = 1, \ldots, n$, taking the $z^k, \bar{z}^l$ derivative of (2.2), one gets

$$\tau^{i\bar{j}}\tau_{ik}\tau_{\bar{j}\bar{l}} = \tau_{k\bar{l}},$$

at 0, or more explicitly,

$$\sum_{j=1}^{n} |\lambda_j|^2\delta_{jk}\delta_{jl} = \delta_{kl}.$$

One concludes that in such a coordinate system $\tau_{ij}(0) = -\delta_{ij}$, for all $0 \leq i, j \leq n$. Then the Taylor expansion of τ at 0 becomes

$$\tau(z) = 2\sum_{j=1}^{n}(\Im\, z^j)^2 + \text{h.o.t.},$$

and the lemma is proved.

Before proceeding let us note that one can further normalize τ. By altering the coordinates z at 0 by addition of homogeneous quadratic terms, we may assume, for $i, j, k = 1, \ldots, n$,

$$\tau_{ij\bar{k}}(0) = 0. \tag{2.3}$$

We can exploit Lemma 2.2 to show that the zero set $\tau^{-1}(0)$ is no bigger than an n-dimensional manifold locally. Specifically, set

$$\tau_{y^i} = \frac{\partial\tau}{\partial y^i},$$

and so on, for higher real derivatives, and where $y^i = \Im z^i$. Define, on a sufficiently small ball U about $0 \in \Omega$, $M = \{z \in U \mid \tau_{y^i} = 0, i = 1, \ldots, n\}$. Since $\tau_{y^iy^j}(0)$ has rank

n, for U small enough M is a closed connected submanifold of U. Since $d\tau = 0$ on all of $\tau^{-1}(0)$, $\tau^{-1}(0) \cap U \subset M$. We conclude that the Hausdorff dimension of $\tau^{-1}(0)$ is $\leq n$. Hence, a standard result tells us that Ω^* is connected, if $n \geq 2$. Since u is proper on Ω^* and has non-vanishing gradient there, it follows that, for any $c \in (0, r)$, Ω^* is homeomorphic to $u^{-1}(c) \times (0,1)$. Hence, the level sets $u^{-1}(c)$ are all connected, and hence, so is $\tau^{-1}(0) = u^{-1}(0)$. We record this as the following corollary. The reader should note that the argument given here covers only the case $n \geq 2$. Slightly special considerations are required for the case $n = 1$, and we leave this case to the reader to verify, using the map Exp_c introduced below.

Corollary 2.3: *The set $\tau^{-1}(0) \subset \Omega$ is a compact connected set.*

Let us now examine the geometric properties of the Monge-Ampère foliation $\mathcal{F}$ associated to u. This foliation is tangent to the null directions of the (1,1)-form $\partial\bar{\partial}u$. In our case we have a convenient explicit choice of complex vector field of type (1,0) tangent everywhere to the foliation. Define, in local coordinates $z = (z^1, ..., z^n)$, the vector field

$$Z := \frac{1}{2i}\tau^{-1/2}\tau^{i\bar{j}}\tau_{\bar{j}}\frac{\partial}{\partial z^i} = Z^i\frac{\partial}{\partial z^i}.$$

It is clear that Z is defined independent of the choice of local coordinates. Let us record the geometric properties of $\mathcal{F}$ in terms of the field Z. For the proofs, *c.f.*, [LS], [PW] (or [B], [S], *mutatis mutandis*), or calculate directly.

Lemma 2.4: *The vector field Z has the following properties:*

1.) Z is tangent to $\mathcal{F}$ on Ω^*.

2.) Z is holomorphic along each leaf of $\mathcal{F}$.

3.) Z has constant length in the τ-metric.

4.) $\nabla'_Z Z = 0$, where ∇' is the (1,0) part of the covariant derivative for the τ-metric on Ω.

Thus, the leaves of $\mathcal{F}$ are totally geodesic submanifolds in Ω, which are flat in the τ-metric.

Define vector fields Ξ, H on $\Omega \setminus \tau^{-1}(0)$ by setting

$$2Z = \Xi - \sqrt{-1}\,\mathrm{H},$$

where Ξ, H are real vector fields. We summarize the properties of Ξ, H in the next lemma. $\phi_t^{\Xi}, \phi_t^{\mathrm{H}}$ will denote the flows of Ξ, H.

Lemma 2.5: *The vector fields Ξ and H have the following properties:*

1.) $\Xi(u) = 0$, $\mathrm{H}(u) = 1$.

2.) Ξ and H are of unit length in the τ-metric.

3.) H is perpendicular to the level sets of u.

4.) The trajectories $\phi_t^{\Xi}(z)$ and $\phi_t^{\mathrm{H}}(z)$ are geodesics in the τ-metric, parametrized by arclength, when defined.

5.) The commutator $[\Xi, \mathrm{H}] = 0$.

We have enough background on the geometry of $\mathcal{F}$ and the τ-metric to begin proving Theorem 1 properly speaking. We have to show that $\tau^{-1}(0)$ is a manifold of real dimension n. Let us first define a map Exp_c from $u^{-1}(c)$ to Ω, for $c \in (0, r)$, as follows: $\mathrm{Exp}_c(z) =$

normal exponential a distance c in the direction of the inward pointing unit normal to $\tau^{-1}(c)$, that is, after Lemma 2.5, parts 1.)-3.), in the direction $-\mathrm{H}$. Also by Lemma 2.5, we have

$$\mathrm{Exp}_c(z) = \lim_{t \nearrow c} \phi^{\mathrm{H}}_{-t}(z).$$

Then, once more by Lemma 2.5, we have

$$u(\mathrm{Exp}_c(z)) = u(\lim_{t \nearrow c} (\phi^{\mathrm{H}}_{-t}(z))) = \lim_{t \nearrow c} (c - t) = 0,$$

that is, $\mathrm{Exp}_c(z) \in \tau^{-1}(0)$.

To prove that $\tau^{-1}(0)$ is a manifold the main point is now to show that the rank of the differential $d\mathrm{Exp}_{c*}$ is n at every point $z \in u^{-1}(c)$. Note that since

$$\mathrm{Exp}_{(c-s)} \circ \phi^{\mathrm{H}}_s = \mathrm{Exp}_c,$$

this is independent of the level $c \in (0, r)$, and can be checked at any point on a trajectory of H. So, let us fix a point $z_0 \in u^{-1}(c)$, and consider the trajectory $\phi^{\mathrm{H}}_{-t}(z_0)$. This trajectory will have a limit, say $0 \in \tau^{-1}(0)$, as $t \nearrow c$, as already noted, and therefore that geodesic will continue smoothly through 0. Let v_0 be its tangent direction at 0. Similarly, it is easy to see that the image $\mathrm{Exp}_c(\phi^{\Xi}_s(z_0))$ is the geodesic $\gamma^{w_0}(s)$ through 0 in the direction w_0, where $w_0 = J \cdot v_0$ by continuity, since $\Xi = -J \cdot \mathrm{H}$ on the leaf of $\mathcal{F}$ through z_0. Since $\tau = 0$ along γ^{w_0}, the vector w_0 is a null vector for the real Hessian of τ at 0. Without loss of generality, we may assume that the vector $w_0 = \frac{\partial}{\partial x^1}$ and $v_0 = -\frac{\partial}{\partial y^1}$, and more generally, the null space of the real Hessian of τ at 0 is the span of the $\frac{\partial}{\partial x^i}, i = 1, ..., n$, while the space orthogonal to the null space is spanned by the $\frac{\partial}{\partial y^i}, i = 1, ..., n$. If U denotes a suffiently small open neighborhood of $0 \in \Omega$, we can define on $U \cap \Omega^*$ vectorfields

$$v_i = \frac{\partial}{\partial x^i} - (\mathrm{H}, \frac{\partial}{\partial x^i}) \cdot \mathrm{H},$$

with the following properties.

Lemma 2.6: *The vectors $v_i, i = 1, ..., n$,*

1.) are tangent to the level sets of u,

2.) are linearly independent along $\phi^{\mathrm{H}}_{-t}(z_0)$ near 0, and

3.) have $\| v_i \|_\tau$ approximately 1 near 0.

Proof: These properties follow directly from the observations just made. Part 1.) is because H is the unit normal to the levels of u. Part 2.) is because $\lim_{t\nearrow c} \mathrm{H} = -\frac{\partial}{\partial y^1}$, and thus $\lim_{t\nearrow c} v_i = \frac{\partial}{\partial x^i}$. This also proves 3.).

Lemma 2.7: *The differential $d\mathrm{Exp}_{c*}$ is of rank n at z_0.*

Proof: We will show that $d\mathrm{Exp}_{c*}$ is injective on the span of the v_i above. For $a := (a^1, ..., a^n) \in \mathbb{R}^n, \sum_{i=1}^n |a^i|^2 = 1$, let $v = a^i v_i$. It is well-known that the differential of Exp_c at z_0 is given by

$$d\mathrm{Exp}_{c*}(v) = V(c),$$

where $V(t)$ is the Jacobi field along $\phi^{\mathrm{H}}_t(z_0)$ with initial conditions $V(0) = v, V'(0) = -\nabla_v \mathrm{H}$. Recall that it suffices to prove this for a sufficiently small $c > 0$. It suffices also to show that the second initial condition $-\nabla_v \mathrm{H}$ remains bounded if we let $c \to 0$. For then it follows from the uniformity of the Jacobi equations on small neighborhoods of the compact set $\tau^{-1}(0)$ and standard continuity results for ordinary differential equations, that a Jacobi field couldn't go from a value at $t = 0$ of length approximately 1, *viz.*, $v(z_0)$, to zero at $t = c$, if c is sufficiently small. So let us examine $\nabla_v \mathrm{H}$.

Write $v = v' + v''$, where v' is of type (1,0), and v'' is of type (0,1). Similarly, H $= \sqrt{-1}(Z - \bar{Z})$. It suffices to bound $\nabla_{v'} Z + \nabla_{v''} Z$. We calculate

$$\begin{aligned}
\nabla_{v'} Z &= \frac{-1}{2\sqrt{-1}} \nabla_{v'} \left(\frac{1}{\sqrt{\tau}} \tau^{s\bar{t}} \tau_{\bar{t}} \frac{\partial}{\partial z^s} \right) \\
&= \left[\frac{-1}{2\tau} a^i \tau_i Z - a^j \tau^{s\bar{l}} \tau_{kj\bar{l}} Z^k + \frac{1}{2\sqrt{-1}\sqrt{\tau}} a^j \tau^{s\bar{t}} \tau_{j\bar{t}} \right] \frac{\partial}{\partial z^s} \\
&= \frac{-1}{2\tau} (a^i \tau_i) Z + \frac{1}{2\sqrt{-1}\sqrt{\tau}} \left(a^i \frac{\partial}{\partial z^i} \right) + \mathrm{O}(|z|).
\end{aligned}$$

In the second equation we use $\nabla_Z Z = 0$. In the third equation we use that the Z^k are bounded and that $\tau_{kj\bar{l}}(0) = 0$, by equation (2.3). Similarly we can calculate

$$\nabla_{v''} Z = \frac{-1}{2\tau} (a^s \tau_{\bar{s}}) Z + \frac{1}{2\sqrt{-1}\sqrt{\tau}} (a^t \tau^{s\bar{l}} \tau_{\bar{t}\bar{l}}) \frac{\partial}{\partial z^s} + \mathrm{O}(|z|).$$

Adding these two equations together gives

$$\nabla_v Z = \frac{-1}{\tau} \left(a^j \frac{\partial \tau}{\partial x^j} \right) Z + \frac{1}{2\sqrt{-1}\sqrt{\tau}} (a^t \tau^{s\bar{l}} (\tau_{t\bar{l}} + \tau_{\bar{t}\bar{l}})) \frac{\partial}{\partial z^s} + \mathrm{O}(|z|).$$

Along the trajectory $\phi_t^{\mathrm{H}}(z_0)$ we have $\tau \asymp (y^1)^2$ and $|\frac{\partial \tau}{\partial x^j}|$, $|\frac{\partial^2 \tau}{\partial x^i \partial x^j}|$, and $|\frac{\partial^2 \tau}{\partial x^i \partial y^j}|$ are all bounded by a constant times $|y^1|$, by the normalizations on τ at $z = 0$. Taken together with the boundedness of Z, this proves the boundedness of the second initial condition for suitably small c, and thus the lemma.

Now it is easy to show that $X = \tau^{-1}(0)$ is a submanifold of Ω of dimension n. Start by noting once again that $d\tau = 0$ along $\tau^{-1}(0)$, and so on a sufficiently small neighborhood U of $0 \in \Omega$, $\tau^{-1}(0) \cap U \subset M := \{ z \in U \mid \tau_{y-j} = 0, j = 1, ..., n \}$. For U appropriate and small enough, M is a connected submanifold of U. By a dimension count, using Lemma 2.7, the image $\mathrm{Exp}_c(u^{-1}(c)) \cap U = M$, and thus, $\mathrm{Exp}_c(u^{-1}(c)) \cap U = \tau^{-1}(0) \cap U$. Since $0 \in \Omega$ was an arbitrary point in $\tau^{-1}(0)$, it follows that $\mathrm{Exp}_c(u^{-1}(c))$ is a connected submanifold of $\tau^{-1}(0)$ which is both open and closed as a subset of $\tau^{-1}(0)$. Since $\tau^{-1}(0)$ is connected, we conclude that $\mathrm{Exp}_c(u^{-1}(c)) = \tau^{-1}(0)$, proving Theorem 1, in light of Theorem 3.1 of [LS].

3 The Kähler-Einstein metric, stability and involutions

We want to ask in this section whether a complex manifold Ω can be represented in more than one way as a Grauert tube, *i.e.*, is there more than one real analytic Riemannian manifold X and real analytic metric g such that Ω is the Grauert tube associated to X, g. Theorems 2 and 2bis answer this negatively in various degrees. To address this question we try to understand the symmetries of Grauert tubes, since the existence of more than one X, g will force extra symmetry of the Grauert tube. We also try to understand what makes the submanifold $X \subset \Omega$ distinguished or even unique. Because of this strategy, in this section we must exploit further the geometric properties of Grauert tubes, not only with respect to the τ-metric, as in §2, but also with respect to the Kähler-Einstein metric on Ω, due to Cheng and Yau, [CY]. In this section we will be working exclusively with Grauert tubes of finite radius r. This is because we will be relying on the result that, in this case, the biholomorphic automorphism group $Aut_C(\Omega)$ is a compact Lie group, as observed by Szöke in [S2]. We also recall that for a Grauert tube, the map $\sigma : T^r(X) \to T^r(X)$ sending v to $-v$ is anti-holomorphic on $\Omega \simeq T^r(X)$ with fixed point set equal to $X \subset \Omega$.

Recall that if Ω is a Stein manifold with compact smooth strictly pseudoconvex boundary, then there exists a complete Kähler-Einstein metric g_{KE} on Ω of constant negative scalar curvature [CY]. This metric is canonical in the sense that any biholomorphic automorphism of such an Ω is an isometry for this metric. The same proof also shows that this is true for any anti-holomorphic automorphism of Ω.

We would like to be able to apply this directly to the case of Grauert tubes. Given X a real analytic manifold, and g a real analytic metric defined on X, there is a maximal radius R for which the Grauert tube structure is defined on $T^R(X)$, and a Kähler-Einstein metric is clearly defined on all $T^r(X)$, for $r < R$. On each such $T^r(X) \simeq \Omega^r$, we have obviously $r < \infty$, and Ω^r is a Stein manifold with a smooth, compact, strictly pseudoconvex boundary. However, we cannot say the same of the original Ω, if $r = R$. This necessitates the following technical indirection.

For any Grauert tube of finite radius r (which may even be the maximal radius R for the underlying Riemannian metric g), the biholomorphic automorphism group $Aut_C(\Omega)$ is compact ([S2], Theorem 6.3). We set $\pm Aut_C(\Omega)$ equal to the group of holomorphic or anti-holomorphic automorphisms of Ω. Note that σ above is in $\pm Aut_C(\Omega) \setminus Aut_C(\Omega)$, and that the composition of any two anti-holomophic transformations of Ω is holomorphic. Thus the index of $Aut_C(\Omega) \subset \pm Aut_C(\Omega)$ is exactly two. In any case, it follows that $\pm Aut_C(\Omega)$ is also a compact Lie group. Now τ is a smooth, strictly plurisubharmonic exhaustion function on Ω. Let $\tilde{\tau}$ be the average of $h^*(\tau)$ over $h \in \pm Aut_C(\Omega)$ with respect to Haar measure on $\pm Aut_C(\Omega)$. Then $\tilde{\tau}$ is also a smooth, strictly plurisubharmonic exhaution of Ω which is $\pm Aut_C(\Omega)$-invariant, which is proper from Ω to $[0, r^2)$. For generic $c \in [0, r^2)$, the set $\tilde{\Omega} := \{z \in \Omega \,|\, \tilde{\tau}(z) \leq c\}$ is a compact Stein manifold with strictly pseudoconvex boundary, and note that if c is chosen sufficiently close to r^2, then X is contained in the interior of $\tilde{\Omega}$. The $\pm Aut_C(\Omega)$-invariance of $\tilde{\tau}$ implies that $\tilde{\Omega}$ is $\pm Aut_C(\Omega)$-invariant. Finally note that there exists a unique, complete Kähler-Einstein metric on $\tilde{\Omega}$ and that $\pm Aut_C(\Omega)$ acts as isometries for this metric.

As noted earlier, the isometry group $Isom(X)$ of the Riemannian manifold X, g acts on Ω, as the differentials of the isometries acting on X itself. Thus we have an embedding $Isom(X) \hookrightarrow Aut_C(\Omega)$, and $Isom(X)$ also acts on the auxiliary subspace $\tilde{\Omega}$ we have just introduced. The following lemma clarifies the relationship between $Isom(X)$ and $Aut_C(\Omega)$, and is contained in [S2].

Lemma 3.1: *The element $\Phi \in Aut_C(\Omega)$ is contained in $Isom(X) \subset Aut_C(\Omega)$ if and only if $\Phi(X) \subset X$, i.e., Φ leaves X stable.*

Recall the standard notation that, if G is a Lie group, then G° denotes the connected component of the identity element of G. We will use Lemma 3.1 in the proof of the following key lemma.

Lemma 3.2: $Isom^\circ(X) \xrightarrow{\sim} Aut_C^\circ(\Omega)$.

Proof: We start with another lemma, which is well-known.

Lemma 3.3: The submanifold $X \subset \hat{\Omega}$ is totally geodesic for the Kähler-Einstein metric, and Lagrangian for the Kähler form $\hat{\omega}_{KE}$ of the Kähler-Einstein metric.

Proof: The first part follows from the fact that X is the fixed point set of the isometry σ of the Kähler-Einstein metric. The second part follows because $\sigma^*\hat{\omega}_{KE} = -\hat{\omega}_{KE}$, since σ is anti-holomorphic.

Returning to Lemma 3.2, we are now in a position to use a result that seems to be due originally to Chen, Leung and Nagano, [C], p. 51, and has been rediscovered a couple of times since, *cf.*, [MW] and [O]. The result states simply that if X is a Lagrangian minimal subvariety in a Kähler-Einstein manifold of (strictly) negative scalar curvature, then X is (strictly) stable. That is, the Jacobi operator, or second variation operator, has (strictly) positive spectrum acting on variation vectors along X. Let us apply that to our $X \subset \Omega$. Let $\mathfrak{aut}_C(\Omega)$ denote the Lie algebra of $Aut_C(\Omega)$. For any $\xi \in \mathfrak{aut}_C(\Omega)$, let ξ also denote the corresponding vector field on $\hat{\Omega}$, the derivative of the one-parameter group generated by $exp(t\xi) \subset Aut_C(\Omega)$ acting on $\hat{\Omega}$. Let V_ξ denote the component of ξ normal to X along X. This is a variation vector along the minimal submanifold X. Since $exp(t\xi)$ acts as isometries of the Kähler-Einstein metric, V_ξ is a null vector for the Jacobi operator on X. Since the spectrum of this operator is strictly positive in our case, we conclude that V_ξ is identically zero along X, that is, the vector field ξ is tangent to X along X. Hence, $exp(t\xi)$ leaves X stable for any $\xi \in \mathfrak{aut}_C(\Omega)$, and hence $Aut^\circ_C(\Omega)$ leaves X stable, acting on $\hat{\Omega}$, and then obviously also as acting on Ω. By Lemma 3.3, $Aut^\circ_C(\Omega) \subset Isom(X)$, which was to be proved.

Now let us consider two possible representations of Ω as a Grauert tube. Thus we will have X and X_0 two compact submanifolds of Ω, the fixed point sets of the anti-holomorphic involutions σ and σ_0, respectively. Consider the composition $\varphi := \sigma \circ \sigma_0 \in Aut_C(\Omega)$. Since $Aut^\circ_C(\Omega) \subset Isom(X)$, the index of $Isom(X) \subset Aut_C(\Omega)$ is finite. Let k be the least integer such that $\varphi^k \in Isom(X)$.

Lemma 3.4: $\varphi^k = id \in Aut_C(\Omega)$.

Proof: Suppose $\varphi^k = d\psi_*$ on Ω, *i.e.*, φ^k acts as the differential of the isometry ψ of X on Ω. Now consider $\sigma_0 \circ \varphi^{k-1} = \sigma \circ d\psi_*$. Regrouping the terms in $\sigma_0 \circ \varphi^{k-1}$ suitably, we get that, if $k = 2\ell + 1$ is odd, then $\sigma \circ d\psi_* = \varphi^{-\ell} \circ \sigma_0 \circ \varphi^\ell$, and when $k = 2\ell$ is even, $\sigma \circ d\psi_* = (\sigma_0 \circ \varphi^{\ell-1}) \circ \sigma \circ (\sigma_0 \circ \varphi^{\ell-1})^{-1}$. In either case, we obtain that $\sigma \circ d\psi_*$ is conjugate in $\pm Aut_C(\Omega)$ to either σ_0 (k odd), or σ (k even). Thus $\sigma \circ d\psi_*$ is an anti-holomorphic involution of Ω which has compact fixed point set $\tilde{X}$ which is equal to $\varphi^{-\ell}(X_0)$ for $k = 2\ell + 1$, and to $(\sigma_0 \circ \varphi^{\ell-1})^{-1}(X)$ for $k = 2\ell$.

Let us now suppose that this anti-holomorphic involution $\sigma \circ d\psi_*$ has a fixed point $z \in \Omega$ which is not contained in X. Let us view the action of $\sigma \circ d\psi_*$ on $\Omega \simeq T^r(X)$. Let $z \in \Omega$ be in $\tilde{X}$, and not in X. Then we may represent z as a pair (x, v), where $x \in X$, and $v \in T^r_x(X)$, $v \neq 0 \in T^r_x(X)$. Then $\sigma \circ d\psi_*(z) = z$ is equivalent to $\psi(x) = x$, and $d\psi_*(v) = -v$. But then the whole interval $\mathbb{R} \cdot v \cap T^r(X) \subset T(X)$ is contained in $\tilde{X}$, and $\tilde{X}$ cannot be compact in $\Omega \simeq T^r(X)$. This contradiction shows that we must have $v = 0 \in T^r_x(X)$, *i.e.*, $z \in X$. Thus $\sigma \circ d\psi_*$ is an anti-holomorphic involution of Ω with fixed point set $\tilde{X} = X$, and so $\sigma \circ d\psi_* = \sigma$. Thus, $d\psi_* = id$ on Ω, and hence $\varphi^k = id \in Aut_C(\Omega)$, as was to be shown.

Thus, for any two anti-holomorphic involutions σ and σ_0 of Ω with compact (non-empty) fixed point sets, the biholomorphism φ is of finite order. Now the submanifold and involution X, σ are unique in Ω if and only if any φ as above is $= id$. We cannot say that in general yet, but we can say something about the order of φ.

Lemma 3.5: *The order k of φ is odd.*

Proof: We show that if $\varphi^k = id$, and $k = 2\ell$, then $\varphi^\ell = id$. If $\varphi^{2\ell} = id$, then we may write

$$\sigma = \varphi^{2\ell} \circ \sigma = \varphi^\ell \circ \sigma \circ \varphi^{-\ell},$$

where we have rearranged factors conveniently, as in the proof of Lemma 3.4. But then

$$\varphi^\ell \circ \sigma = \sigma \circ \varphi^\ell,$$

and φ^ℓ preserves the fixed point set X of σ. By Lemma 3.1, $\varphi^\ell = d\psi_*$, for some $\psi \in Isom(X)$, and hence, $\varphi^\ell = id \in Aut_C(\Omega)$, by Lemma 3.4.

Corollary 3.6: *Given two anti-holomorphic involutions σ, σ_0 of Ω, with compact, non-empty fixed point sets X, X_0, respectively, then there exists a biholomorphic $\Phi \in Aut_C(\Omega)$ such that $\sigma_0 = \Phi \circ \sigma \circ \Phi^{-1}$, and $X_0 = \Phi(X)$.*

Proof: We simply regroup factors again: $\varphi^{2\ell+1} = id$ if and only if

$$\sigma_0 = \varphi^{2\ell+1} \circ \sigma_0 = \varphi^\ell \circ \sigma \circ \varphi^{-\ell}.$$

Taking $\Phi = \varphi^\ell$ proves the lemma.

The proof of Theorems 2 and 2bis are almost complete now. For the proof of Theorem 2, suppose Ω is also represented as a Grauert tube about another $X_0 \subset \Omega$. Let u_0 be the corresponding solution of the homogeneous Monge-Ampère equation on $\Omega \setminus X_0$, and let $\tau_0 = u_0^2$. Then, with $\Phi \in Aut_C(\Omega)$ as in the last lemma, $\Phi^* u_0$ is a solution of the Monge-Ampère equation (1.1) on $\Omega \setminus X$ which tends to 0 as $z \to X$, and r as z leaves any compact set in Ω. Hence, by a uniqueness result of Bedford and Taylor [BT], $\Phi^* u_0 = u$, and thus, $\Phi^* \tau_0 = \tau$, and the mapping Φ is an isometry from the τ-metric to the τ_0-metric, thereby inducing an isometry from X to X_0, as was already remarked in [LS]. This proves Theorem 2.

As to Theorem 2bis, it follows from the arguments above that there are exactly d distinct such σ_i as in the statement of the theorem, where d is equal to the index of $Isom(X)$ in $Aut_C(\Omega)$. The conjugacy statement in Theorem 2bis is contained in Corollary 3.6.

Before concluding this §, let us point out that the assumption in Theorem 2bis that the fixed point set of the σ_i be compact is essential. Consider, for example, X a compact symmetric space. For any point $x \in X$, let ι_x denote the isometry of X fixing x and with $d(\iota_x)_* = -id$ on $T_x(X)$. Then the induced map $d(\iota_x)_*$ commutes with σ on the Grauert tube Ω, and $\sigma_x := d(\iota_x)_* \circ \sigma$ is an anti-holomorphic involution of Ω, with non-compact fixed-point set (it contains the set $T_x^r(X) \subset T^r(X) \simeq \Omega$). This construction can be carried out for any $x \in X$, and each of the σ_x is distinct from $\sigma_{x'}$, for any other $x' \in X$.

4 Homogeneous manifolds

In this § we will prove Theorem 3 by showing that if X, g is a homogeneous, compact, connected, real analytic Riemannian manifold, then there is a unique anti-holomorphic involution σ of the Grauert tube Ω of any finite radius r associated to X, g. Let us simplify notation a bit, and set $G := Isom^{\circ}(X)$. Since we are assuming X connected, G acts transitively on X. Since X is a compact manifold, G is a compact Lie group. We average $h^*\tau$ over $h \in \pm Aut_C(\Omega)$, as in §3 above, and get $\tilde{\tau}$ which will be smooth, strictly plurisubharmonic, $\pm Aut_C(\Omega)$-invariant, and proper from Ω to $[0, r^2)$. We want to examine the critical point behavior of $\tilde{\tau}$. Note that if Theorem 3 is true, then $\tilde{\tau} = \tau$, and the only critical points of $\tilde{\tau}$ are along $X \subset \Omega$. Note also that in case Theorem 3 is true, then restricted to the subset $T_x^r(X) \subset \Omega$, the function $\tilde{\tau} = \tau$ is strictly convex with one critical point at its minimum. The proof below seeks to reproduce these properties for $\tilde{\tau}$ under the assumption that G acts transitively on X.

For any $z \in \Omega$, let $G \cdot z$ denote the orbit of G through z. $G \cdot z$ is a submanifold of Ω. We denote by $p : \Omega \to X$ the vector bundle projection on $T^r(X) \simeq \Omega$, and we set $F_z =$ fiber of p through z. Of course F_z is just $\simeq T^r_{p(z)}(X)$.

Lemma 4.1: *For any $z \in \Omega$, we have*

1.) $dp_* : T_z(G \cdot z) \to T_{p(z)}(X)$ is surjective.

2.) $T_z(G \cdot z) + T_z(F_{p(z)}) = T_z(\Omega)$.

Proof: We first observe that the action of G on Ω is equivariant with respect to the projection p, which is obvious when Ω is viewed as $T^r(X)$, and the action of G as being via the differentials of isometries of X. Since G acts transitively on X, the natural map from the Lie algebra $\mathfrak{g}$ of G to $T_{p(z)}(X)$ is surjective. By the G-equivariance of p, this map factors through $\mathfrak{g} \to T_z(G \cdot z)$, proving part 1.). Part 2.) follows immediately, since $T_z(F_{p(z)}) = \mathrm{Ker}(dp_* : T_z(\Omega) \to T_{p(z)}(X)\,)$.

Suppose there exists another anti-holomorphic involution σ_0 of Ω with compact fixed point set X_0. By the proof of Lemma 3.2, G leaves X_0 invariant. Let us next note that if $z \in X_0$, then $d\tilde{\tau} = 0$, *i.e.*, z is a critical point of $\tilde{\tau}$. Indeed, since $\tilde{\tau}$ is constant on $G \cdot z$, we know $d\tilde{\tau}|_{T_z(G \cdot z)} = 0$. On the other hand, the subspace $W := \{v \in T_z(\Omega) \mid d\sigma_{0*}v = -v\}$ is complementary to $T_z(X_0) = T_z(G \cdot z)$ in $T_z(\Omega)$, and the invariance property $\tilde{\tau} \circ \sigma_0 = \tilde{\tau}$ implies $d\tilde{\tau} = -\tilde{\tau} = 0$ on W.

Now let us fix an $x \in X \subset \Omega$, and let $f := \tilde{\tau}|_{F_x}$. The function $f : F_x \to [0, r^2)$ is proper, since $\tilde{\tau}$ is on Ω. By what we have just observed, f has a critical point at each $z \in F_x$ fixed by an anti-holomorphic involution σ_0. In particular, it has a critical point at x. Note that for any such involution σ_0, the fixed point set X_0 must intersect F_x at some point z, by the G-equivariance of p, and the transitivity of G acting on X. Note finally that if $X_0 \cap X \neq \emptyset$, then $\sigma = \sigma_0$, since X would then be contained in X_0, again by G-invariance. (In fact, by Theorem 2, X_0 must be connected, and $X = X_0$.)

With these preliminaries established, the following lemma will imply Theorem 3 immediately, since the topology of F_x implies that the number of critical points of f must be exactly 1, *viz.*, $z_0 = X \cap F_x$.

Lemma 4.2: *Every critical point of f is a strict local minimum.*

Proof: Start the proof by noting that if $z \in F_x$ is a critical point of f, then $z \in \Omega$ is a critical point for $\tilde{\tau}$ on Ω. This folows from Lemma 4.1, part 2.), since $\tilde{\tau}$ is constant on $G \cdot z$. Again by the G-equivariance of $\tilde{\tau}$, $d\tilde{\tau} = 0$ on all of $G \cdot z$. Thus, if $\mathrm{H}(\tilde{\tau})$ denotes the real Hessian of $\tilde{\tau}$ at z, we have

$$\mathrm{H}(\tilde{\tau})(v, w) = 0, \text{ for all } v \in T_z(\Omega), w \in T_z(G \cdot z).$$

We also know that $\tilde{\tau}$ is strictly plurisubharmonic, and therefore its real Hessian has at least n positive eigenvalues and can be null on a subspace of $T_z(\Omega)$ of dimension at most

n (*c.f.*, *e.g.*, [M]). We conclude that the dimension of $T_z(G \cdot z) \leq n$, and by Lemma 4.1, part 2.), it must be exactly n. It follows then, finally, that $\mathrm{H}(\tilde{\tau})$ is positive definite when restricted to the subspace $T_z(F_{p(z)})$, and this is just the real Hessian of f, proving the lemma.

5 Grauert tubes of infinite radius

In this § we consider the question of the uniqueness of $X \subset \Omega$ in the case of a Grauert tube Ω which has radius $r = \infty$. The real analytic, Riemannian X for which the tube construction is extendible to the whole of $T(X)$ are quite restricted, *c.f.*, [LS], [S1]. One obvious example is X a product of circles, with the product metric. Here Ω for $r = \infty$ is just the product of $\mathbb{C}^*$'s. This simple $X \subset \Omega$ is a counterexample to Theorems 2, 2bis and 3 above (without the assumption of finite radius!). In this section we would like to give a new proof of a theorem of Szöke's on the case of tubes of infinite radius, under further restrictive assumptions. The main point here is to bring out the possible relevance of symplectic geometry for the study of Grauert tubes.

Let X, g be a real analytic, Riemannian manifold whose Grauert tube Ω is of infinite radius. Recall that ω is the Kähler form of the τ-metric on Ω. Ω is identified with the tangent bundle in such a way that ω agrees with the canonical symplectic two-form coming from the cotangent bundle of X, identified with $T(X)$ using the metric g. Thus, Ω, ω is "standard" as a symplectic manifold. Let α be the standard action form on Ω, *i.e.*, viewing Ω as $T(X)$, at $z = (x, v) \in T(X), (x \in X, v \in T_x(X))$, with $p : T(X) \to X$ the projection, and for $w \in T_z(\Omega)$,

$$\alpha_z(w) := g(v, dp_* w).$$

Recall that a Lagrangian submanifold $Y \subset \Omega$ is called exact if the form α is exact when restricted to Y.

Theorem 4 ([S2]): *Let Ω be a Grauert tube of infinite radius, and let $\Phi : \Omega \to \Omega$ be a biholomorphic map which is symplectic, i.e., $\Phi^*\omega = \omega$. Assume further that $\Phi(X)$ is an exact Lagrangian submanifold of Ω. Then $\Phi(X) = X$, and $\Phi = d\psi_*$ for some $\psi \in Isom(X)$.*

Note that Φ symplectic is equivalent to Φ an isometry of the τ-metric. Note also that if $\mathrm{H}^1(X, \mathbb{R}) = 0$, then any Lagrangian $\Phi(X)$ is exact: this was Szöke's original assumption. Finally let us note that there are analogous formulations in terms of involutions σ as earlier which we will leave to the reader.

Proof: The idea of the proof is to use a theorem of Gromov to reduce the question to a simple situation in one complex variable. Under our assumptions, [G], Theorem $2.3\mathrm{B}_4''$, implies that the two exact Lagrangian manifolds X and $\Phi(X)$ have non-empty intersection. Let σ be the standard anti-holomorphic involution of the Grauert tube Ω, and let $\sigma_0 := \Phi \circ \Phi^{-1}$ be the involution about $X_0 := \Phi(X)$. Consider, as in §4, the biholomorphic map $\varphi = \sigma \circ \sigma_0$. (This φ is just the commutator of σ and Φ.) Let us fix a point x in the intersection $X \cap X_0$. Then x is fixed by both σ and σ_0, and hence by φ.

We would like to normalize the action of $d\varphi_*$ on $T_x(\Omega)$. Since Φ and σ are both isometries of the τ-metric on Ω, so also are φ and $\sigma_0 = \sigma \circ \varphi$. We also have that $\sigma \circ \varphi = \varphi^{-1} \circ \sigma$. Putting these two facts together we get that there exist $\theta_j \in [0, 2\pi), j = 1, ..., n$, and two-dimensional J-invariant subspaces $V_j \subset T_x(\Omega)$ such that

$$T_x(\Omega) \;=\; V_1 \oplus ... \oplus V_n,$$

and each V_j is stable under both $d\sigma$ and $d\varphi_*$ and $d\varphi_*$ is rotation by an angle θ_j (in the positive sense) on V_j. The map $d\sigma_{0*} = d\sigma_* \circ d\varphi_*$ will also preserve V_j as well.

Let v_j be a unit vector in $V_j \cap T_x(X)$. Such exists because V_j is σ-stable, and $T_x(X)$ is the set of σ-fixed vectors in $T_x(\Omega)$. There is also a unit vector $v'_j \in V_j$ such that $\sigma_0(v'_j) = v'_j$, that is, $v'_j \in T_x(X_0)$. Elementary geometry shows that the lines through $0 \in V_j$ in the directions v_j, v'_j intersect at an angle $\theta_j/2$.

The unit speed geodesic $\gamma_j : \mathbb{R} \to X \subset \Omega$ with $\gamma_j(0) = x, \gamma'_j(0) = v_j$, extends holomorphically to $\Gamma_j : \mathbb{C} \to \Omega$. The image of $\mathbb{C} \setminus \mathbb{R}$ is the union of two σ-conjugate leaves L and $\sigma(L)$ of the Monge-Ampère foliation $\mathcal{F}$ on Ω^*. Hence, by continuity, $\Gamma_j(\mathbb{C})$ is an immersed submanifold which is totally geodesic in Ω, by Lemma 2.3. It follows then that $\Gamma_j(\mathbb{C})$ is the image of V_j under the geodesic exponential map in the τ-metric from $T_x(\Omega)$ to Ω. Hence the image $\Gamma_j(\mathbb{C})$ is taken to itself by the mappings φ, σ, and σ_0.

We can now pull everything back to $\mathbb{C}$ using Γ_j. The τ-metric is just euclidean on $\mathbb{C}$, as in Lemma 2.2. The function τ pulls back to η^2 where $\zeta = \xi + \sqrt{-1}\eta$ is the usual complex coordinate on $\mathbb{C}$. The τ-geodesics in the directions v_j and v'_j become the real axis and a straight line through 0 making an angle $\theta_j/2$ with the real axis, respectively. Now if $\theta_j \in (0, 2\pi)$, then the function $\tau = \eta^2$ is unbounded on this second line. Back in Ω this represents the τ-geodesic through x in the direction v'_j. Since X_0 is τ-totally geodesic in Ω (it is the fixed point set of the isometry σ_0), this geodesic lies entirely in X_0. Once again, as in §4, we know that X_0 is compact and τ must be bounded on X_0. This proves that each θ_j must be 0. In this case, we have that $T_x(X) = T_x(X_0)$. Since both X and X_0 are totally geodesic, we conclude that $X = X_0$. It follows that $\sigma = \sigma_0$ and Φ commutes with σ. As in §4, we conclude that Φ is induced by an isometry ψ of X, g.

References

[BT] E. Bedford, B.A. Taylor, The Dirichlet problem for a complex Monge-Ampère equation, Inv. Math. **37**, 1976, pp. 1-44.

[B] D. Burns, Curvatures of Monge-Ampère foliations and parabolic manifolds, Ann. Math. **115**, 1982, pp. 349-373.

[C] B. Y. Chen, *Geometry of Submanifolds and Its Applications*, Tokyo, Science Univerity of Tokyo, 1981.

[CY] S-Y. Cheng, S.-T. Yau, On the existence of a complete Kähler-Einstein metric on non-compact complex manifolds and the regularity of Fefferman's equation, Comm. Pure Appl. Math. **33**, 1980, pp. 507-544.

[G] M. Gromov, Pseudo-holomorphic curves in symplectic manifolds, Inv. Math. **82**, 1985, pp. 307-347.

[GS] V. Guillemin, M. Stenzel, Grauert tubes and the homogeneous Monge-Ampère equation, J. Diff. Geom. **34**, 19991, pp. 561-570.

[L] L. Lempert, Complex structures on the tangent bundle of Riemannian manifolds, Chap. 8, pp. 235-251, in *Complex Analysis and Geometry*, ed. V. Ancona, A. Silva, New York, Plenum Press, 1993.

[LS] L. Lempert, R. Szöke, Global solutions of the homogeneous complex Monge-Ampère equation and complex structures on the tangent bundle of Riemannian manifolds, Math. Ann. **290**, 1991, pp. 689-712.

[MW] M. Micallef, J. Wolfson, The second variation of area of minimal surfaces in four manifolds, Math. Ann. **295**, 1993, pp. 245-267.

[M] J. Milnor, Morse Theory, Princeton, N.J., Princeton Univ. Press, 1963.

[O] Y.-G. Oh, Second variation and stabilities of minimal lagrangian submanifolds in Kähler manifolds, Inv. Math. **101**, 1990, pp.501-519.

[PW] G. Patrizio, P.-M. Wong, Stein manifolds with compact symmetric center, Math. Ann. **289**, 1991, pp. 355-382.

[S] W. Stoll, The characterization of strictly parabolic manifolds, Ann. Sc. Nor. Sup. Pisa, Ser. IV, **7**, 1980, pp. 87-154.

[S1] R. Szöke, Complex structures on tangent bundles of Riemannian manifolds, Math. Ann. **291**, 1991, pp. 409-428.

[S2] ———, Automorphisms of Stein tubes of the form $T^r M$ and complexification of the isometry group action on the tangent bundle, preprint MPI/93-23, Bonn, Max Planck Inst.

Hölder Regularity for $\Box_b$ on Hypersurfaces in C^n with Nondiagonalizable Levi Form

Makhlouf Derridj, Mathématiques - Bâtiment 425, Université de Paris-Sud, F-91405 Orsay Cedex, France. mderridj@matups.matups.fr

Abstract. — We establish the Hölder regularity for the boundary Laplacian $\Box_b$, on pseudoconvex hypersurfaces in $\mathbb{C}^n$, $n \geq 3$, whose Levi form can be decomposed into blocs satisfying the so-called "trace condition".

Here we do this in the case $\mathbb{C}^4$. The general case will be done in a next paper. We follow the proof given by Fefferman-Kohn-Machedon in the case of diagonal Levi form.

Résumé. — Nous établissons la régularité Hölderienne pour le Laplacien au bord $\Box_b$, sur des hypersurfaces pseudoconvexes de $\mathbb{C}^n$, $n \geq 3$, dont la forme de Levi peut être décomposée en blocs satisfaisant la "propriété dite de trace". Nous le faisons ici dans le cas de $\mathbb{C}^4$. Le cas général de $\mathbb{C}^n$ sera fait dans un prochain papier.

Nous suivons la méthode de Fefferman-Kohn-Machedon donnée dans le cas d'hypersurfaces à forme de Levi diagonalisable.

I. — Introduction

Our goal is to generalize the result of C. Fefferman, J.J. Kohn, M. Machedon [8], concerning the Hölder regularity for $\Box_b$ (and so for the canonical solution of $\bar{\partial}_b$) in hypersurfaces in $\mathbb{C}^n$, whose Levi form is not necessarily diagonalizable (near a point, say 0).

More precisely, we consider (near 0) hypersurfaces (which we considered first in [3], [4] whose the Levi form can be decomposed in blocks which satisfy the so-called trace condition (see (2.5)). We use the same method as the authors mentionnned; our work consists to show the main differences which appear and to give the lemmas which are needed to make their method work in our case.

We use for that our results given in [3], [4]. We must remark here that our hypersurfaces are not in general with diagonalizable Levi-form, as we showed it in ([3] th. 7.1) .

Note also that this kind of results concerns the Hölder regularity for the boundary Kohn's Laplacian $\bar{\partial}_b$, and so for the Szegö Projector . Many results on Hölder regularity for $\bar{\partial}$– Neumann problem, or on the existence of Hölder solutions for $\bar{\partial}$ with the Kernel method are known, even in some degenerate cases (see for example [6], [12]). But, for $\Box_b$, the degenerate case is mostly open.

As in the case of diagonalizable Levi form, the Hölder regularity is related to some integer m which is determined, naturally, by the blocks of Levi matrix in a suitable basis of tangent holomorphic fields $(L_1, \cdots, L_{n-1})$.(See in next pages the hypothesis (3.1) and the formula (3.2)). In general case, for the study of subelliptic type, see [10] in $\mathbb{C}^2$ and [1], [2] in $\mathbb{C}^n$.

A reading of our work supposes a reading of the work of Fefferman-Kohn-Machedon [8], so we write here just what is really different from that work (parts 4, 5, 6 of our paper).

Also, in this paper, we consider the case of $\mathbb{C}^4$ which shows the new ingredients we need in the proof. We hope, we will write the general case in a next paper, needing some more subtle ways of induction.

Let us, finally, mention that J. Stalker [14] informed us that he obtained a similar result in the particular case where the Levi matrix has one block satisfying the trace condition.

II. — Notations - hypothesis.

In what follows, M is a hypersurface in $\mathbb{C}^n$, near $0 \in M$. We note $(L_1, \cdots, L_{n-1})$ a basis of tangent holomorphic vector fields to M. We denote also by (t, x, y) a coordinate system near 0, where $t \in \mathbb{R}$ is the real direction and (x, y) are the complex directions. Define the functions c_{jk}, a^ℓ_{jk} and b^ℓ_{jk} by

$$[L_j, \bar{L}_k] = -ic_{jk}\partial_t + \sum_{\ell=1}^{n-1} a^\ell_{jk} L_\ell + \sum_{\ell=1}^{n-1} b^\ell_{jk} \bar{L}_\ell. \tag{2.1}$$

Denote also :

$$(2.2)\qquad \begin{cases} T = -i\partial_t, \quad [L_j, T] = d_j T \quad (\text{modulo } L, \bar{L}) \\ [L_\alpha, L_\beta] = \sum_{\ell=1}^{n-1} d^\ell_{\alpha\beta} L_\ell. \end{cases}$$

From $[L_j, \bar{L}_k] = \overline{[L_k, \bar{L}_j]}$ we obtain :

$$(2.3)\qquad a^\ell_{jk} = -\bar{b}^\ell_{kj} \quad , \quad c_{jk} = \bar{c}_{kj}.$$

The class (b) of hypersurfaces we consider is given by (see [4]) :

(2.4) $M \in (b) \Leftrightarrow$ There exist $(L_1, \cdots, L_{n-1})$ such that the Levi matrix (c_{jk}) can be written :

$$(c_{jk}) = \begin{pmatrix} A_1 & & 0 \\ & \ddots & \\ 0 & & A_p \end{pmatrix}$$

where

$$A_j \geq \sigma_j \mathrm{tr}(A_j), \quad \sigma_j > 0.$$

Examples of such hypersurfaces are given in [4].

To simplify, we denote :

$$(2.5)\qquad t_j = \mathrm{tr}(A_j).$$

We recall the writting of $\square^q_b = \bar{\partial}_b \bar{\partial}^*_b + \bar{\partial}^*_b \bar{\partial}_b$, on $(0,q)$ forms. Let $(\bar{\omega}_1, \cdots, \bar{\omega}_{n-1})$ a system of $(0,1)$ forms, dual to the system $(L_1, \cdots, L_{n-1})$. Then, :

(2.6) f is a $(0,q)$ form $\Leftrightarrow f = \sum_{|I|=q} f_I \bar{\omega}_I$, $I = (i_1, \cdots, i_q)$. For f regular we have :

$$(2.7)\qquad \begin{cases} \bar{\partial}_b f = \sum_{(iI)=J} \epsilon^{iI}_J \bar{L}_i f_I \omega^J + O(1) f \\ \bar{\partial}^*_b f = -\sum_{(iJ)=I} \epsilon^{iJ}_I L_i f_I \bar{\omega}^J + O(1) f. \end{cases}$$

Thus we obtain :

$$(2.8)\qquad \begin{cases} \bar{\partial}^*_b \bar{\partial}_b f = -\sum \epsilon^{jJ}_{iI} L_j \bar{L}_i f_I \bar{\omega}_J + O(L, \bar{L}, 1) f \\ \bar{\partial}_b \bar{\partial}^*_b f = -\sum_{i \in I} (\bar{L}_i L_i f_I) \bar{\omega}_I + \sum_{i \neq j} \epsilon^{iI}_{jJ} \bar{L}_i L_j f_I \bar{\omega}_J + O(L, \bar{L}, 1) f \end{cases}$$

$O(L,\bar L,1)f$ means $O(\| f \|, \| Lf \|, \| \bar L f \|)$.

Considering $\Box_b = \bar\partial_b^*\bar\partial_b + \bar\partial_b\bar\partial_b^*$, on $(0,q)$ forms with I varying in the set of q–indices, ordered, we arrive at the $(\Box_b^q)_J^I$ given by :

$$(2.9) \qquad \begin{cases} (\Box_b^q)_J^I = \sum_{j\in J} \bar L_j \bar L_j^* + \sum_{j\notin J} \bar L_j^* \bar L_j + O(L,\bar L,1) \\ (\Box_b^q)_J^I = - \sum_{(iI)=(jJ)} \epsilon_{iI}^{jJ}(L_j\bar L_i - \bar L_i L_j) + O(L,\bar L,1) \ \text{ si } \ I \neq J. \end{cases}$$

We see that $(\Box_b^q)_J^I$ equals 0 if I and J have two different indices. So, it is convenient to consider (I,J) of multi-indices such that :

$$(2.10) \qquad I = K\cup\{j\}, \quad J = K\cup\{i\} \ \text{ with } \ |K| = q-1.$$

In formula (2.9) (jJ) is the multi-index jJ after ordering.

Thus (2.9) can be written :

$$(2.11) \qquad \begin{cases} (\Box_b^q)_J^I = \sum_{j\in J} \bar L_j \bar L_j^* + \sum_{j\notin J} \bar L_j^* \bar L_j + O(L,\bar L,1) \\ (\Box_b^q)_{(K\cup\{j\})}^{(K\cup\{i\})} = -e_{j(K\cup\{j\})}^{i(K\cup\{j\})}[L_i,\bar L_j] + O(L,\bar L,1) \\ (\Box_b^q)_J^I = 0 \ \text{ si } \ |I\cap J| \leq q-2. \end{cases}$$

So, considering (2.11), we see that given any K with $|K| = q-1$, we obtain a matrix with indices (j,k), not in K :

$$(2.12) \qquad \begin{cases} (\Box_b^q)_K = ((\Box_b^q)_K)_{j,k\notin K}, \text{givenby} \\ ((\Box_b^q)_K)_{j,k\notin K} = \begin{cases} \sum_{\ell\in K\cup\{j\}} \bar L_\ell \bar L_\ell^* + \sum_{\ell\notin K\cup\{j\}} \bar L_\ell^* \bar L_\ell + 0(L,\bar L,1) \\ i\epsilon_{k(K\cup\{j\})}^{j(K\cup\{k\})}\partial_t. \end{cases} \end{cases}$$

As announced in the introduction, we consider in this paper the case $\mathbb{C}^4$ our model of hypersurface. The Levi matrix (c_{jk}) has the following form :

$$(2.13) \qquad \begin{pmatrix} c_{11} & 0 & 0 \\ 0 & c_{22} & c_{32} \\ 0 & c_{23} & c_{33} \end{pmatrix} \ \text{ avec } \ \begin{pmatrix} c_{22} & c_{32} \\ c_{23} & c_{33} \end{pmatrix} \geq a(c_{22}+c_{33}), \quad a>0.$$

Now, we write $\Box_b^q$, for $q=3$, $q=2$, $q=1$.

Form of $\square_b^q$ *for* $q = 3$: it is a scalar operator :

$$\square_b^3 = \sum_{j=1}^{3} \bar{L}_j \bar{L}_j^* + 0(L, \bar{L}, 1). \tag{2.14}$$

Form of $\square_b^q$ *pour* q = 2 : As a basis of $(0,2)$ forms, we take $\bar{\omega}_1 \wedge \bar{\omega}_2$, $\bar{\omega}_1 \wedge \bar{\omega}_3$, $\bar{\omega}_2 \wedge \bar{\omega}_3$. Thus modulo $0(L, \bar{L}, 1)$ we have the matrix :

$$\begin{pmatrix} \bar{L}_1\bar{L}_1^* + \bar{L}_2\bar{L}_2^* + \bar{L}_3^*\bar{L}_3 & -ic_{23}\partial_t & 0 \\ -ic_{32}\partial_t & \bar{L}_1\bar{L}_1^* + \bar{L}_3\bar{L}_3^* + \bar{L}_2^*\bar{L}_2 & 0 \\ 0 & 0 & \bar{L}_2\bar{L}_2^* + \bar{L}_3\bar{L}_3^* + \bar{L}_1^*\bar{L}_1 \end{pmatrix}. \tag{2.15}$$

Form of $\square_b^q$ *for* $q = 1$: the matrix $\square_b^q$ (base $\bar{\omega}_1, \bar{\omega}_2, \bar{\omega}_3$) is, modulo $0(L, \bar{L}, 1)$

$$\begin{pmatrix} \bar{L}_1\bar{L}_1^* + \bar{L}_2^*\bar{L}_2 + \bar{L}_3^*\bar{L}_3 & 0 & 0 \\ 0 & \bar{L}_2\bar{L}_2^* + \bar{L}_3^*\bar{L}_3 + \bar{L}_1^*\bar{L}_1 & -ic_{23}\partial_t \\ 0 & -ic_{32}\partial_t & \bar{L}_3\bar{L}_3^* + \bar{L}_2^*\bar{L}_2 + \bar{L}_1^*\bar{L}_1 \end{pmatrix} \tag{2.16}$$

The diagonalizability hypothesis of the Levi form made in [8] means that the preceding matrices are diagonal. Our goal will be to show that as in [8] the proof works, if we establish some corresponding proposition in the case (2.15) and (2.13) are not diagonal.

III. — The statement of the result.

Before the statement of the result, similar to that of [8], we describe the value of the number $\epsilon > 0$ (in the subelliptic estimate). We did that in [5], and we just recall this value :

If I_ℓ(for $1 \leq \ell \leq p$) is the set of indices corresponding to the block A_ℓ, we denote $\mathcal{L}_\ell$ the Lie algebra generated by $\{L_j, \bar{L}_j\}_{j \in I_\ell}$. We define :

(3.1) Let p_ℓ be the smallest integer, (which, we suppose, exists) such that a bracket with length $p_\ell + 2$ in $\mathcal{L}_\ell$ has non-zero component on ∂_t at 0.

$$\epsilon = \inf_{\ell \in \{1, \cdots, p\}} \frac{1}{p_\ell + 2} = \frac{1}{m}. \tag{3.2}$$

Then we have :

$$\| u \|_\epsilon^2 \leq \| \bar{\partial}_b u \|^2 + \| \bar{\partial}_b^* u \|^2 \quad \forall u \in \mathcal{D}^{0,q}(V), \quad q \leq n-2. \tag{3.3}$$

We proved (3.3) in [4] for $q = 1$ $(n \geq 3)$. For our hypersurfaces, m will be called the type at 0 (in [5], we use [9] [13]).

MAIN THEOREM. — *Let Ω be a smooth bounded pseudoconvex domain in* $\mathbb{C}^n$ $(n \geq 3)$. *Assume that the boundary* $\partial\Omega$ *is, near 0, in the class (b). Then :*

a) *The operator* $(\Box_b^q)^{-1}$ *sends* $\mathrm{Lip}(s,0)$ *to* $\mathrm{Lip}(s^{+\frac{2}{m}-\epsilon},0)$, $\forall\epsilon>0$ *if* $1 \leq q \leq n-2$.

b) *The operators* $\bar\partial_b(\Box_b^q)^{-1}$, $\bar\partial_b^*(\Box_b^q)^{-1}$, $(\Box_b^q)^{-1}\bar\partial_b$ *and* $(\Box_b^q)^{-1}\bar\partial_b^*$ *send* $\mathrm{Lip}(s,0)$ *to* $\mathrm{Lip}(s+\frac{1}{m}-\epsilon,0)$, $\forall\epsilon>0$ *to* $1 \leq q \leq n-2$.

c) *The operators* $\bar\partial_b\bar\partial_b^*(\Box_b^q)^{-1}$, $\bar\partial_b^*\bar\partial_b(\Box_b^q)^{-1}$, $(\Box_b^q)^{-1}\bar\partial_b\bar\partial_b^*$, $(\Box_b^q)^{-1}\bar\partial_b^*\bar\partial_b$, $\bar\partial_b(\Box_b^q)^{-1}\bar\partial_b^*$ $\bar\partial_b^*(\Box_b^q)^{-1}\bar\partial_b$ *send* $\mathrm{Lip}(s,0)$ *to* $\mathrm{Lip}(s-\epsilon,0)$, $\forall\epsilon>0$.

The points d) e) of the main theorem in [8] are also true.

Before the proof of this result, with the same method as in [8], we establish some propositions needed in the non-diagonalizable case.

IV. — A comparison theorem beetwen the blocks A_j.

This theorem will be an analogue of the Lemma 13.2 in [8]. The difference is the fact that we have a system of differential equations and so we have to study the interwinings between the vector fields L_j We use the notations (2.1) (2.5).

THEOREM 4.1. — $\forall N \geq 1$, $\exists C = C_N \geq 0$ *t. q.* $\forall\delta>0$, *we have :*

$$\sum_{p,q\in I_i,k\in I_j} | a_{p,q}^k |^2 t_j \leq C(t_i\delta^{-1}+\delta^N). \tag{4.1}$$

Proof : To simplify the notations, let us consider the case of 2 blocks A_1 A_2, and the corresponding I_1 , I_2.

a) *Differential equations satisfied by the coefficients :*
We use the Jacobi identity

$$\left[\bar L_k,[L_p,\bar L_q]\right]+\left[\bar L_q,[\bar L_k,L_p]\right]+\left[L_p,[\bar L_q,\bar L_k]\right]=0.$$

So we got, (noting $T=-i\partial_t$ see (2.2))

$$\left[\bar L_k, c_{pq}T+\sum_\ell a_{pq}^\ell L_\ell+\sum_\ell b_{pq}^\ell \bar L_\ell\right]+\left[\bar L_q, -c_{pk}T-\sum_\ell a_{pk}^\ell L_\ell-\sum_\ell b_{pk}^\ell \bar L_\ell\right] + \left[L_p, \sum_{\ell=1}^{n-1} \bar d_{qk}^\ell \bar d_\ell\right]=0. \tag{4.2}$$

Then, taking the components on T in (4.2), with $p,q\in I_1$, $k\in I_2$, we get :

(4.3) $$\bar{L}_k(c_{pq}) + c_{pq}\bar{d}_k - \sum_{\ell\in I_2} a^\ell_{pq}c_{\ell k} + \sum_{\ell\in I_1} a^\ell_{pk}c_{\ell q} + \sum_{\ell\in I_1} \bar{d}^\ell_{qk}c_{p\ell} = 0.$$

Similarly, using the identity :

(4.4) $$[L_k,[L_p,\bar{L}_q]] + [\bar{L}_q,[L_k,L_p]] + [L_p,[\bar{L}_q,L_k]] = 0$$

and proceeding as above, we obtain :

(4.5) $$L_k(c_{pq}) + \sum_{\ell\in I_1} \bar{a}^\ell_{qk}c_{p\ell} - \sum_{\ell\in I_1} d^\ell_{kp}c_{\ell q} + d_k c_{pq} - \sum_{\ell\in I_2} \bar{a}^\ell_{qp}c_{k\ell}.$$

b) *Proof of(4.1) for $p = q$.*
We take, for example $p = q = 1 \in I_1$. Then

(4.6) $$\bar{L}_k(c_{11}) + \bar{d}_k c_{11} + \sum_{\ell\in I_1} a^\ell_{1k}c_{\ell 1} + \sum_{\ell\in I_1} \bar{d}^\ell_{1k}c_{1\ell} = \sum_{\ell\in I_2} a^\ell_{11}c_{\ell k}.$$

Multiply (4.6) by $\bar{a}^k_{11}$ and add, with $k \in I_2$; then we obtain :

(4.7) $$\sum_{\ell\in I_2} \bar{a}^k_{11}\bar{L}_k(c_{11}) + \left(\sum_{\ell\in I_2} \bar{a}^k_{11}\bar{d}_k\right) c_{11} + \sum_{\ell\in I_1, k\in I_2} \bar{a}^k_{11}a^\ell_{1k}c_{\ell 1} +$$
$$+ \sum_{\ell\in I_1, k\in I_2} \bar{a}^k_{11}\bar{d}^\ell_{1k}c_{1\ell} = \sum_{\ell\in I_2, k\in I_2} c_{\ell k}a^\ell_{11}\bar{a}^k_{11}$$

Now, it follows from the trace property :

(4.8) $$\begin{cases} \displaystyle\sum_{\ell\in I_2, k\in I_2} c_{\ell k}a^\ell_{11}\bar{a}^k_{11} \geq \epsilon t_2 \sum_{\ell\in I_2} | a^\ell_{11} |^2, & \text{with } \epsilon > 0 \\ | c_{11} | \leq \sigma\, c_{11} \quad \text{for} \quad \ell \in I_1 & \text{and } \sigma > 0. \end{cases}$$

Next, we denote by X the following vector field :

(4.9) $$X = \mathcal{R}\mathrm{e} \sum_{\ell\in I_2} \bar{a}^k_{11}\bar{L}_k.$$

The estimates (4.7), (4.8) give :

$$X\, c_{11} + g\, c_{11} \geq ct_2 \sum_{\ell \in I_2} \mid a_{11}^{\ell} \mid^2, \text{near } 0 \tag{4.10}$$

where C is some constant, and g some smooth function which can be deduced from (4.7) and (4.8). We must remark that X can vanish at some points.

Now, let $\varphi(t,x)$ be the flow of X, near (0,0) i.e. :

$$\begin{cases} \varphi'(t,x) = X(\varphi(t,x)) \\ \varphi(0,x) = x. \end{cases} \tag{4.11}$$

Fix x_0. Define

$$F(t) = c_{11}(\varphi(t,x_0)). \tag{4.12}$$

So we obtain the inequality (using (4.10) (4.11))

$$\begin{cases} F'(t) + G(t)F(t) \geq \left(ct_2 \sum_{\ell \in I_2} \mid a_{11}^{\ell} \mid^2 \right) (\varphi(t,x_0)) \\ \text{où} \quad G(t) = g(\varphi(t,x_0)). \end{cases} \tag{4.13}$$

Now, set :

$$H(t) = e^{\int_0^t G(s)ds} F(t). \tag{4.14}$$

Then, one has

$$e^{-\int_0^t G(s)ds} H'(t) \geq \left(ct_2 \sum_{\ell \in I_2} \mid a_{11}^{\ell} \mid^2 \right) (\varphi(t,x_0)) \qquad t \in [0,\epsilon_0] \tag{4.15}$$

An so, with a constant C depending on G :

$$H'(t) \geq C \left(t_2 \sum_{\ell \in I_2} \mid a_{11}^{\ell} \mid^2 \right) (\varphi(t,x_0)) \qquad t \in [0,\epsilon_0]. \tag{4.16}$$

Now, one has $H' \geq 0$ on $[0, \epsilon_0]$, then

$$\text{(4.17)} \qquad \forall N, \ \exists C_N \ \text{ s. t. } \forall \delta > 0 \ : \ H'(0) \leq C_N \left(\frac{H(0)}{\delta} + \delta^N \right).$$

(C_N does not depend on the point 0) (lemm 13.2 in [8]). Using the expression H one has :

$$\text{(4.18)} \qquad \left(t_2 \sum_{\ell \in I_2} \mid a_{11}^{\ell} \mid^2 \right) (x_0) \leq C_N \left(\frac{c_{11}(x_0)}{\delta} + \delta^N \right).$$

The estimate (4.18) is, of course valid for all a_{jj}^{ℓ}, with $j \in I_1$, $\ell \in I_2$.

It remains to show the inequality (4.18) for all the coefficients a_{pq}^{ℓ}, with $p, q \in I_1$, $\ell \in I_2$.

Let us consider $p = 1$, $q = 2$ and try to majorize $t_2 \sum_{\ell \in I_2} \mid a_{12}^{\ell} \mid^2$.

For that, we consider the new basis of holomorphic fields :

$$\text{(4.19)} \qquad \begin{cases} \tilde{L}_1 = L_1 + \alpha L_2 \\ \tilde{L}_j = L_j, \quad j \geq 2 \end{cases}$$

with $\mid \alpha \mid = 1$, α to be chosen.

Let us remark that, in this new basis, the Levi matrix is still decomposed in blocks having the trace property.

For that, we compute coefficients $\tilde{c}_{jk}$, $\tilde{a}_{jk}$: An elementary calculation gives:

$$\text{(4.20)} \qquad \begin{cases} \tilde{c}_{11} = c_{11} + \alpha c_{12} + \bar{\alpha} c_{21} + c_{22} \quad (\text{since } \mid \alpha \mid^2 = 1) \\ \tilde{a}_{11}^{\ell} = a_{11}^{\ell} + a_{22}^{\ell} + \alpha a_{12}^{\ell} + \bar{\alpha} a_{21}^{\ell} \quad \text{for} \quad \ell \in I_2 \\ \tilde{c}_{1\ell} = c_{1\ell} + \alpha c_{2\ell} = 0 \quad \text{for} \quad \ell \in I_2. \end{cases}$$

From (4.18), (in the basis $\tilde{L}_j$, where the Levi matrix has blocks with trace condition) we also have :

$$\text{(4.21)} \qquad t_2 \sum_{\ell \in I_2} \mid \tilde{a}_{jj}^{\ell} \mid^2 \leq \tilde{C}_N \left(\frac{\tilde{t}_1}{\delta} + \delta^N \right) \qquad \forall \delta > 0 \ ; \ j = 1, 2.$$

Using (4.18) and (4.21) we get (since $\tilde{t}_1 \leq 4 t_1$)

$$\text{(4.22)} \qquad t_2 \mid \alpha a_{12}^{\ell} + \bar{\alpha} a_{21}^{\ell} \mid^2 \leq C_N' \left(\frac{t_1}{\delta} + \delta^N \right), \qquad \forall \delta > 0.$$

It suffices now to take 2 values for α, say $\alpha = 1$, and $\alpha = i$. Then, we get the inequalities :

$$(4.23)\qquad \begin{cases} t_2 \mid a_{12}^{\ell} + a_{21}^{\ell} \mid^2 \leq C_{N,1}\left(\frac{t_1}{\delta} + \delta^N\right) & \forall \delta > 0 \\ t_2 \mid a_{12}^{\ell} - a_{21}^{\ell} \mid^2 \leq C_{N,2}\left(\frac{t_1}{\delta} + \delta^N\right) & \forall \delta > 0. \end{cases}$$

From (4.22) , (4.23) one has : $\exists C_N$ s. t.

$$(4.24)\qquad t_2\left(\mid a_{12}^{\ell} \mid^2 + \mid a_{21}^{\ell} \mid^2\right) \leq C_N\left(\frac{t_1}{\delta} + \delta_N\right), \qquad \forall \delta > 0.$$

The proof of theorem 4.1. is thus finished

Our next step is to give a proposition similar to the lemma 13.1 in [8]. Of course, as we are in a new situation,we give a new operator which is an analogue of the one given in lemma 13.1 in [8]

V. — Estimates for some auxilliary operators.

We begin by some notations.

(5.1) Let I a multi-index. Denote by K_I the set of indices j s.t : $I \cap I_j \neq \emptyset$.
Then, with η small, we define :

$$(5.2)\qquad \sum_{I,\delta} = \sum_{i \in K_I}(X_i^2 + Y_i^2) + \sum_{j \notin K_I}\left(\sum_{i \in K_I} t_i + \delta^{\eta}\sum_{\ell \in I_j} \mid a_{ii}^{\ell} \mid^2\right)(X_j^2 + Y_j^2).$$

THEOREM 5.1. — *The operator* $\sum_{I,\delta}$, *statisfies the estimates :*

$$\gamma(x,\delta) \leq C\delta^{\frac{1}{m}}$$
$$t_i(x) \leq C\frac{\delta}{\gamma^2}.$$

Proof : The proof is similar to that of lemma 13.1 in [8]. It needs however some results we established in ([3],[4]) for our class of hypersurfaces. For the sake of completeness in this part, we reproduce in details some parts of [8].

Changing, if necessary, the indices, let $I = \{1, \cdots, p\}$ and note $\sum_p$ the corresponding operator :

$$\sum_p = \sum_{i \in I} (X_i^2 + Y_i^2) + \sum_{j \notin K_p} \left(\sum_{i \in K_p} t_i + \delta^\eta \sum_{\ell \in I_j} | a_{ii}^\ell |^2 \right) (X_j^2 + Y_j^2). \tag{5.3}$$

It suffices to prove the theorem for $\sum_1$: for that, we remark

$$K_{\{1,\cdots,k+1\}} = \{1 \cdots k+1\} \tag{5.4}$$

and then $K_{\{1,\cdots,k\}} \subset K_{\{1,\cdots,k+1\}}$, which gives :

$$\left(\sum_{k+1} u, u \right) \leq C \left((\sum_k u, u) + \| u \|^2 \right). \tag{5.5}$$

From ([8], pp. 74-75), we consider an operator like :

$$L = \frac{\partial^2}{\partial t^2} + \sum_{i,j} c_{ij}(x,t) \frac{\partial^2}{\partial x_i \partial x_j}$$

The ball with center (t_0, x_0) and radius γ associated to L is given by

$$\begin{cases} \{t - t_0 \,|< \gamma\} \times B(x_0, \gamma) \\ B(x_0,\gamma) \text{ associated to} \dfrac{1}{2\gamma} \displaystyle\int_{t_0-\gamma}^{t_0+\gamma} \sum_{i,j} c_{ij}(t,x) \frac{\partial^2}{\partial x_i \partial x_j} dt. \end{cases} \tag{5.6}$$

Now, we have to use our results in ([3][4]). By taking, if necessary linear combinations of X_j, Y_j in the block. $(X_j, Y_j)_{j \in I_1}$, one has

$$X_1^k(c_{11}) \neq 0 \quad \text{for one} \quad k \leq m-2. \tag{5.7}$$

There exists coordinates $(t, x_j, y_j)_{j=1,\cdots,2n-1}$, with

$$\begin{cases} X_1 = \partial_{x_1} \\ X_j(0) = \partial_{x_j}, \quad Y_j(0) = \partial_{y_j} \\ X_j; j \geq 2 \quad , \quad Y_j; j \geq 1 \quad \text{n'have no component on } \partial_{x_1}. \end{cases} \tag{5.8}$$

Now, to prove what we want, it suffices to assume that the coefficients of $\sum_1$ are polynomials.

We can take $(t_0, x_0) = (0,0)$. Here, the variable t of (5.6) is the variable x_1, and x plays the role of the other variables.

Consider the ball

$$(5.9)\quad \begin{cases} \{| x_1 |< \gamma\} \times B(0,\gamma) \\ B(0,\gamma) \text{ relative to : } \dfrac{1}{2\gamma}\displaystyle\int_{-\gamma}^{\gamma}\{Y_1^2 + \sum_{\substack{i\in I_1 \\ i\geq 2}}(X_i^2 + Y_i^2) + \\ \displaystyle\sum_{j<|I_1|}\left[t_1 + \delta^\eta \sum_{\ell\in I_1} | a_{ii}^\ell |^2\right](X_j^2 + Y_j^2)\}dt. \end{cases}$$

Denoting the averaging by M, we obtain that $B(0,\gamma)$ is equivalent to the ball associated to the following operator :

$$(5.10)\quad \begin{cases} (Y_1(x_1 = 0))^2 + \gamma^2 \underset{|x_1|<\gamma}{M}\left\{[X_1, Y_1]^2 + \displaystyle\sum_{2\leq i\in I_2}(X_i^2 + Y_i^2)\right\} \\ + \underset{|x_1|<\gamma}{M}\displaystyle\sum_{j>|I_1|}\left(t_1 + \sum_{\ell\in I_1} | a_{\ell\ell}^j |^2\right)(X_j^2 + Y_j^2). \end{cases}$$

If η is small enough, then $\gamma << \delta^{\eta/2}$; so, we can replace the operator (5.10), by the following one :

$$(5.11)\quad \begin{cases} (Y_1(x_1 = 0))^2 + \displaystyle\sum_{i\in I_1}(X_j(x_1 = 0))^2 + (Y_j(x_1 = 0))^2 \\ +\gamma^2 \underset{|x_1|<\gamma}{M}\{[X_1, Y_1]^2 + \displaystyle\sum_{j\in I_1}([X_1, X_j]^2 + [X_1, Y_j]^2\} \\ + \underset{|x_1|<\gamma}{M}\displaystyle\sum_{j\in I_2}\left(t_1 + \delta^\eta \sum_{\ell} | a_{\ell\ell}^j |^2\right)(X_j^2 + Y_j^2). \end{cases}$$

Now, we have

$$(5.12)\quad \begin{cases} [X_1, Y_\ell] = -\dfrac{1}{2}c_{1\ell}T + \displaystyle\sum_{j>|I_1|} O\left(| a_{1\ell}^j |^2\right)X_j \\ + \displaystyle\sum_{j>|I_1|} O\left(| a_{1\ell}^j |^2\right)Y_j + O(Y_1) + \sum_{j\in I_1}(O(X_j) + O(Y_j)) \end{cases}$$

(5.13) $$Y_j = Y_j(x_1 = 0) + O(\gamma)[X_1, Y_j] + O(\gamma^2) \text{ if } j \in I_2.$$

So, we are reduced to consider an operator of the form :

$$A = (Y_1(x_1 = 0))^2 + \sum_{1<j\in I_1} \{(X_j(x_1 = 0))^2 + (Y_j(x_1 = 0))^2\}$$
$$+\gamma^2 \underset{|x_1|<\gamma}{M} \left(c_{11}^2 \partial_{x_1}^2 + \sum_{\ell\in I_1} c_{1\ell}^2 \partial_{x_1}^2 \right)$$
$$+ \underset{|x_1|<\gamma}{M} \left(\sum_{j>|I_1|} t_1 + \delta^\eta \sum_{\ell\in I_1} | a_{1\ell}^j |^2 \right) \left((X_j(x_1 = 0))^2 + (Y_j(x_1 = 0))^2\right).$$

Now, it suffices, as in [8], to remark that $\partial_{x_1}^k(t_1)(0) \neq 0$, for $k \leq m-2$, to deduce $M(t_1) \geq C\gamma^{m-2}$ and

$$\| u \|^2 - (Au, u) \geq -c\gamma^2(\underset{|x_1|<\gamma}{M} t_1)^2(\Delta u, u) \geq -c\gamma^{2m-2}(\Delta u, u)$$

Δ being the Laplacian. So

(5.14) $$\begin{cases} \delta \geq c_1\gamma^2 \underset{|x_1|<\gamma}{M} (t_1) \geq c_2\gamma^2 t_1(0) \\ \text{et } \delta \geq c\gamma^m. \end{cases}$$

VI. Proof of the main theorem.

As we announced it, our goal, here, is to give a proof for a hypersurface in $\mathbb{C}^4$, with the different forms $\Box_b^q$, for $q = 1, 2, 3$. To give the proof of the analogue of 12.1 in [8], or the theorem (15.1, [8]), we will do it for the parts $\mathcal{P}^+u$, $\tilde{\mathcal{P}}\bar{u}$ (the part $\mathcal{P}^0 u$ corresponding to the good direction $\Box_b$). We will consider the part $\mathcal{P}^+u$, and for the proof, make an induction q (descending induction).

VI.1. Case q = 3.

Rewrite (2.14) in the following form :

(6.1) $$\begin{cases} \Box_b^3 = \frac{1}{2}\{\sum_1^3(\bar{L}_j\bar{L}_j^* + \bar{L}_j^*\bar{L}_j) + \sum_1^3(\bar{L}_j\bar{L}_j^* - \bar{L}_j^*\bar{L}_j) + 0(1, L, \bar{L})\} \\ \quad = \frac{1}{2}\{\sum_1^3(\bar{L}_j\bar{L}_j^* + \bar{L}_j^*\bar{L}_j) + \sum_1^3 c_{jj}T + 0(1, L, \bar{L})\} \\ \quad = \frac{1}{2}\{\sum_1^3(\bar{L}_j\bar{L}_j^* + \bar{L}_j^*\bar{L}_j) + (t_1 + t_2)T + 0(1, L, \bar{L})\}. \end{cases}$$

Now, the hypothesis made on t_1, t_2 gives the subellipticity $\epsilon = \frac{1}{m}$ (see (3.2), (3.3)). $\Box_b^3$ is a scalar operator studied in [8].

VI.2. Case q = 2.

In this case, the Levi matrix in the basis $(\bar\omega_1 \wedge \bar\omega_2, \bar\omega_1 \wedge \bar\omega_3, \bar\omega_2 \wedge \bar\omega_3)$, has two blocks, modulo $0(1, L, \bar L)$ given in (2.15). Consider :

a) The block : $\bar L_1^* \bar L_1 + \bar L_2 \bar L_2^* + \bar L_3 \bar L_3^*$ which is scalar. As in ([8], page 84), we consider this operator in the form :

$$\text{(6.2)} \qquad \begin{cases} \Box_0 = \bar L_2 \bar L_2^* + \bar L_3 \bar L_3^* + \bar L_1^* \bar L_1 = \bar L_2 \bar L_2^* + \bar L_3 \bar L_3^* + \bar\partial_0^* \bar\partial_0 \\ \bar\partial_0^* \bar\partial_0 \text{ work here on functions ; } \bar\partial_0 f = (\bar L_1 f)\bar\omega_1. \end{cases}$$

In other hand, we have, noting $\bar\partial_1$,the operator $\bar\partial$ operating on $(0,1)$ forms $f\bar\omega_1$

$$\text{(6.3)} \qquad (\bar\partial_1^* \bar\partial_1 + \bar\partial_1 \bar\partial_1^*) f\bar\omega_1 = (\bar L_1 \bar L_1^* f)\bar\omega_1.$$

Denote, therefore :

$$\text{(6.4)} \qquad \begin{cases} \tilde\Box_1 = \bar L_1 \bar L_1^* + \bar L_2 \bar L_2^* + \bar L_3 \bar L_3^* \\ \text{operating on } (0,1) \text{ forms } f\bar\omega_1. \end{cases}$$

Now, the operator (6.4) is the operator $\Box_b^3$ seen before, which satisfies the needed estimates.

Our goal, now (see [8], pp. 84-85) is to study the bracket :

$$\text{(6.5)} \qquad [\bar\partial_0, \partial_t - \Box_0].$$

We have :

$$\bar\partial_0(\partial_t - \Box_0) = \bar\partial_0[\partial_t - (\bar L_2 \bar L_2^* + \bar L_3 \bar L_3^* + \bar\partial_0^* \bar\partial_0)]$$

$$= (\partial_t - \tilde\Box_1)\bar\partial_0 - [\bar\partial_0, \bar L_2 \bar L_2^* + \bar L_3 \bar L_3^*].$$

It suffices to consider $[\bar L_1, \bar L_2 \bar L_2^*]$ for example, in the preceeding bracket. We have

$$[\bar L_1, \bar L_2 \bar L_2^*] = [\bar L_1, \bar L_2]\bar L_2^* + \bar L_2[\bar L_1, \bar L_2^*];$$

as $c_{12} = 0$, we obtain :

$$[\bar L_1, \bar L_2 \bar L_2^*] = (\bar L)\bar L_2^* + \bar L_2\ (L \text{ ou } \bar L),$$

$(\bar{L})$ is an expression containing $\bar{L}_j$, at first order.

So, the bracket is an "error" of the form

$$(\bar{L})(\bar{L}_2^*, \bar{L}_3^*) + (\bar{L}_2, \bar{L}_3)(L, \bar{L}). \tag{6.6}$$

So, we get, finally :

$$\bar{\partial}_0(\partial_t - \Box_0) = (\partial_t - \widetilde{\Box}_1)\bar{\partial}_0 + E, \tag{6.7}$$

E of the form (6.6).

The proof given in ([8], pp. 85,86,87) can be rewritten, using the operator in 5.2 with I suitable, and using the relations in (4.1).

The Λ of the page 86 in [8] has to be replaced by $\Lambda = t_2 + \delta^\eta \sum_{j=2}^3 \mid a_{11}^j \mid^2$,which is used in the definition to the analogue operator to A_1 of the page 86 in [8].

Let us consider, now the 2nd block (non diagonal).

b) The non diagonal block $\Box_b^2$: this block is the matrix (2-2) non diagonal given in (2.15).

We will first write the first diagonal term in the form : with $0 < \epsilon < \frac{1}{2}$ modulo $0(1, L, \bar{L})$,

$$\bar{L}_1\bar{L}_1^* + \bar{L}_2\bar{L}_2^*\bar{L}_3 = \frac{1}{2}(\bar{L}_1\bar{L}_1^* + \bar{L}_1^*\bar{L}_1) + \frac{1}{2}(\bar{L}_1\bar{L}_1^* - \bar{L}_1^*\bar{L}_1)$$

$$+(1-\epsilon)(\bar{L}_2\bar{L}_2^* - \bar{L}_2^*\bar{L}_2) + \epsilon(\bar{L}_2\bar{L}_2^* + \bar{L}_2^*\bar{L}_2) + (1-2\epsilon)\bar{L}_2^*\bar{L}_2$$

$$+\epsilon(\bar{L}_3^*\bar{L}_3 - \bar{L}_3\bar{L}_3^*) + \epsilon(\bar{L}_3^*\bar{L}_3 + \bar{L}_3\bar{L}_3^*) + (1-2\epsilon)\bar{L}_3^*\bar{L}_3$$

$$= -[X_1^2 + Y_1^2 + 2\epsilon(X_2^2 + Y_2^2 + X_3^2 + Y_3^2)] + \frac{1}{2}c_{11}T$$

$$+(1-\epsilon)c_{22} - c_{33}T + (1-2\epsilon)(\bar{L}_2^*\bar{L}_2 + \bar{L}_3^*\bar{L}_3) = D_1.$$

The second term (diagonal) is D_2 the analogue of D_1 by reversing indices 2, 3.

$$M_1 = \begin{pmatrix} D_1 & 0 \\ 0 & D_2 \end{pmatrix}$$

$$M_2 = \begin{pmatrix} \frac{1}{2}c_{11} + (1-\epsilon)c_{22} - \epsilon c_{33} & c_{23} \\ c_{32} & \frac{1}{2}c_{11} + (1-\epsilon)c_{33} - \epsilon c_{22} \end{pmatrix}$$

Now, using the hypothesis (2.4), by choosing ϵ small (i.e. $0 < \epsilon \le \epsilon_0$), ϵ_0 depending only on the constants $a_j > 0$ (2.4) we have

$$(6.8)\qquad \begin{cases} \tilde{M_2} \geq \delta_0(c_{11}+c_{22}+c_{33})I,\ \delta_0 > 0\ ;\ \text{(matricial sens)} \\ M_2 = \tilde{M_2}T \end{cases}$$

Now, by chosing ϵ_0 with $1-2\epsilon_0 = \sigma \geq 0$

$$(6.9)\qquad \begin{cases} M = M_1 + M_2,\quad M_1 \text{ diagonal} \\ M_1 \geq -[X_1^2 + Y_1^2 + 2\epsilon_0(X_2^2 + Y_2^2 + X_3^2 + Y_3^2)]I \\ M_2 \text{ satisfies}(6.8). \end{cases}$$

(6.9) is true, by using $\bar{L}_2^*\bar{L}_2 + \bar{L}_3^*\bar{L}_3 \geq 0$. Now, the proof in [8], written for M, gives that M^{-1} satisifies the needed estimates under the conditions (6.8), (6.9). (The only change here is that $\tilde{M_2}$ is a non-diagonal matrix function, but this matrix function satisfies (6.8)).

VI.3. Case q = 1.

The matrix of $\square_b^1$ (in the base $\bar{\omega}_1$, $\bar{\omega}_2$, $\bar{\omega}_3$) is given by (2.16) : So there is a scalar part and a (2–2) block.

a) Scalar part : Denote by S this part :

$$(6.10)\qquad S = \bar{L}_1\bar{L}_1^* + \bar{L}_2^*\bar{L}_2 + \bar{L}_3^*\bar{L}_3.$$

Denote $\bar{\partial}'$ the operator $\bar{\partial}$ related to structure $(\bar{L}_2, \bar{L}_3, \bar{\omega}_2, \bar{\omega}_3)$ i.e.,

$$(6.11)\qquad \begin{cases} \bar{\partial}'f = \bar{L}_2 f\bar{\omega}_2 + \bar{L}_3 f\bar{\omega}_3 \ \text{ if } f \text{ fonction} \\ \bar{\partial}'u = (\bar{L}_2 u_3 - \bar{L}_3 u_2)\bar{\omega}_2 \wedge \bar{\omega}_3 \text{ if } u = u_2\bar{\omega}_2 + u_3\bar{\omega}_3. \end{cases}$$

Then, we have

$$(6.12)\qquad S = \bar{L}_1\bar{L}_1^* + \bar{\partial}'^*\bar{\partial}'.$$

Denote also :

$$(6.13)\qquad \begin{cases} \tilde{S} = \bar{L}_1\bar{L}_1^* + \bar{\partial}'^*\bar{\partial} +' \bar{\partial}'\bar{\partial}'^* \\ \text{working on } (0,1) \text{ forms} : u = (u_2, u_3). \end{cases}$$

Now, looking again at the proof in ([8], pp. 85-88), we have to compare $\bar{\partial}'S$ and $\tilde{S}\bar{\partial}'$, (i.e. we consider there difference). Remark that it is here, in the analogue operator to A_1 to ([8], pp. 86, 87) that we use theorems 4.1 et 5.1.

So, consider $\bar{\partial}' S - \tilde{S}\bar{\partial}'$; we have

$$\begin{aligned} \bar{\partial}' S - \tilde{S}\bar{\partial}' &= \bar{\partial}'(\bar{L}_1 \bar{L}_1^* + \bar{\partial}'^* \bar{\partial}') - (\bar{L}_1 \bar{L}_1^* + \bar{\partial}'^* \bar{\partial}' \bar{\partial}'^*)\bar{\partial}' \\ &= \bar{\partial}'(\bar{L}_1 \bar{L}_1^* + \bar{\partial}'^* \bar{\partial}') - (\bar{L}_1 \bar{L}_1^* + \bar{\partial}' \bar{\partial}'^*)\bar{\partial}' + \bar{\partial}'^* \bar{\partial}' \bar{\partial}' \\ &= [\bar{\partial}', \bar{L}_1 \bar{L}_1^*] + \bar{\partial}'^*(\bar{L}_2, \bar{L}_3). \end{aligned}$$

As $\bar{\partial}'$ is in $(\bar{L}_2, \bar{L}_3)$ and $c_{1k} = 0$ for $k = 2, 3$; we are in the diagonalisable case considered in [8]. So we are reduced to verify that $\tilde{S}$ is in the form just studied.; recall that $\tilde{S}$ works on $(0,1)$ forms : $u = u_2\bar{\omega}_2 + u_3\bar{\omega}_3$, and has a matrix in the basis $(\bar{\omega}_2, \bar{\omega}_3)$:

$$\tilde{S} : \begin{pmatrix} \bar{L}_2 \bar{L}_2^* + \bar{L}_3^* \bar{L}_3 & c_{23} T \\ c_{32} T & \bar{L}_3 \bar{L}_3^* + \bar{L}_2^* \bar{L}_2 \end{pmatrix} \tag{6.14}$$

This matrix is, in fact the matrix $M = M_1 + M_2$ seen before, which is the non-diagonal of $\Box_b^2$, which was studied above.

Here, we take : Λ given by : $\Lambda = c_{11} + \delta^\eta \sum_{j=2}^{3} | a_{11}^j |^2$.

b) Study of the $(2-2)$ block of$\Box_b^1$.

We have to study :

$$T = \begin{pmatrix} \bar{L}_2 \bar{L}_2^* + \bar{L}_3^* \bar{L}_3 + \bar{L}_1^* \bar{L}_1 & c_{23} T \\ c_{32} T & \bar{L}_3 \bar{L}_3^* + \bar{L}_2^* \bar{L}_2 + \bar{L}_1^* \bar{L}_1 \end{pmatrix} \tag{6.15}$$

which can be written, by notation (6.2)

$$T = \begin{pmatrix} \bar{L}_2 \bar{L}_2^* + \bar{L}_3^* \bar{L}_3^* + \bar{\partial}_0^* \bar{\partial}_0 & c_{23} T \\ c_{32} T & \bar{L}_3 \bar{L}_3^* + \bar{L}_2^* \bar{L}_2 + \bar{\partial}_0^* \bar{\partial}_0 \end{pmatrix}.$$

Now, we consider the following $\tilde{T}$ given by :

$$\tilde{T} = \begin{pmatrix} \bar{L}_2 \bar{L}_2^* + \bar{L}_3^* \bar{L}_3 + \bar{\partial}_1 \bar{\partial}_1^* & c_{23} T \\ c_{32} T & \bar{L}_3 \bar{L}_3^* + \bar{L}_2^* \bar{L}_2 + \bar{\partial}_1 \bar{\partial}_1^* \end{pmatrix} \tag{6.16}$$

$\bar{\partial}_1$ is given by (6.2) (i.e. the operator $\bar{\partial}$ working on $(0,1)$ forms $(u\bar{\omega}_1)$, identified to functions u.

Remark $\tilde{T}$ is nothing else than the $(2-2)$ block $\Box_b^2$ (see (2.15)), for which we proved the needed estimates above. So, by using again the proof in ([8], pp. 85-88), it suffices to consider the following operator :

$$\left\{\begin{array}{l}\bar{\partial}_1 T - \tilde{T}\bar{\partial}_1 = A, \text{ working on couples of} \\ (0,1) \text{ forms } (u\bar{\omega}_1), \text{ identified to couples of functions.}\end{array}\right. \tag{6.17}$$

We have in fact :

$$A\begin{pmatrix} f_1 \\ f_2 \end{pmatrix} = \bar{\partial}_1 \begin{pmatrix} (\bar{L}_2\bar{L}_2^* + \bar{L}_3^*\bar{L}_3 + \bar{L}_1^*\bar{L}_1)f_1 + c_{23}Tf_2 \\ (\bar{L}_3\bar{L}_3^* + \bar{L}_2^*\bar{L}_2 + \bar{L}_1^*\bar{L}_1)f_2 + c_{32}Tf_1 \end{pmatrix} - \tilde{T}\begin{pmatrix} \bar{L}_1 f_1 \\ \bar{L}_1 f_2 \end{pmatrix} \tag{6.18}$$

so, after a computation :

$$\left\{\begin{array}{l} A\begin{pmatrix} f_1 \\ f_2 \end{pmatrix} = \begin{pmatrix} g_1 \\ g_2 \end{pmatrix} \quad \text{with} \\ g_1 = \left[\bar{L}_1, \bar{L}_2\bar{L}_2^* + \bar{L}_3^*\bar{L}_3\right] f_1 + [\bar{L}_1, c_{23}T]f_2 \\ g_2 = \left[\bar{L}_1, \bar{L}_3\bar{L}_3^* + \bar{L}_2^*\bar{L}_2\right] f_2 + [\bar{L}_1, c_{32}T]f_1. \end{array}\right. \tag{6.19}$$

In g_j, $j = 1$, 2, the first bracket has the form :

$$g_{j,1} = (L_2, L_3, \bar{L}_2, \bar{L}_3)\cdot(L, \bar{L})f_j + (L, \bar{L}, 1)f_j + (L, \bar{L})(L_2, L_3, \bar{L}_2, \bar{L}_3)f_j. \tag{6.20}$$

This can be seen, by considering, for example

$$[\bar{L}_1, \bar{L}_j\bar{L}_j^*] = [\bar{L}_1, \bar{L}_j]\bar{L}_j^* + \bar{L}_j[\bar{L}_1, \bar{L}_j^*]\cdot \quad j = 2,3$$

and using that $c_{1j} = 0$, $j = 2$, 3. The terms in (6.20) are therefore errors of type $N_\nu \mathcal{N}_\nu$ in ([8], p. 85). It just remains to study the terms

$$\left\{\begin{array}{l} [\bar{L}_1, c_{23}T]f_2 \\ [\bar{L}_1, c_{32}T]f_1 \end{array}\right. \tag{6.21}$$

Or, to consider for example : $[\bar{L}_1, c_{23}T]$ we have :

$$\begin{aligned}[\bar{L}_1, c_{23}T] &= [\bar{L}_1, [\bar{L}_3, \bar{L}_2^*] + 0(L, \bar{L})] \\ &= [\bar{L}_1, \bar{L}_3\bar{L}_2^* - \bar{L}_2^*\bar{L}_3 + 0(L, \bar{L})] \\ &= 0(\bar{L}\bar{L}_2^*) + 0(\bar{L}_3\cdot(L, \bar{L})) + 0((L, \bar{L})\cdot\bar{L}_3) + 0(\bar{L}_2^*\cdot(L, \bar{L})) + 0(L, \bar{L}, 1)\end{aligned}$$

which is, again an error of type entering in the proof of the diagonalizable case (i.e. errors mentionned above)

Our operator A_1 here is not a scalar one. It will be matrix operator of the form :

$$A_1 = B_1 I_{2\times 2} + \tilde{\Gamma}_\delta \tilde{\mathcal{P}} + C_1 \partial_t + c\bar{\partial} \wedge \bar{\partial}^*$$

with

$$B_1 = -2(X_2^2 + Y_2^2 + X_3^2 + Y_3^2) + (1-2\epsilon)(\bar{L}_2^* \bar{L}_2 + \bar{L}_3^* \bar{L}_3)$$

$$C_1 = \begin{pmatrix} (1-\epsilon)c_{22} - \epsilon c_{33} & c_{32} \\ c_{23} & (1-\epsilon)c_{33} - \epsilon c_{22} \end{pmatrix}.$$

The positivity of B_1 and $\tilde{\Gamma}_\delta \tilde{\mathcal{P}}^+ C_1 \partial_t$ for $\epsilon \leq \epsilon_0$ (In fact, we have more $\tilde{\Gamma}_\delta \tilde{\mathcal{P}}^+ C_1 \partial_t \geq \tilde{\Gamma}_\delta \mathcal{P}^+ t_2 \partial_t + 0(1)$) give that the reasoning [8] pp. 86-88 is valid, when we use the theorems. 4.1 and 5.1.

Bibliographie

[1] D. CATLIN : Subelliptic estimates for the $\bar{\partial}$-Neumann problem on pseudoconvex domains. Ann. Math. 126 (1987), 131-191.

[2] J. D'ANGELO : Real hypersurfaces, orders of contact, and applications. Ann. Math. 115 (1982), 615-637.

[3] M. DERRIDJ : Domaines à estimation maximale. Math. Z. 208 (1991), 71-88.

[4] M. DERRIDJ : Estimations par composantes pour le problème $\bar{\partial}$-Neumann pour quelques classes de domaines pseudoconvexes de $\mathbb{C}^n$. Math. Z. 208 (1991), 89-99.

[5] M. DERRIDJ : Microlocalisation et estimation pour $\bar{\partial}_b$ dans quelques hypersurfaces pseudoconvexes. Inv. Math. 104 (1991), 631-642.

[6] K. DIEDERICH, J. E. FORNAESS & J. WIEGERINCK : Sharp hölder estimates for $\bar{\partial}$ on ellipsoïds. Manuscr. Math. 56 (1986), 399-413.

[7] C. FEFFERMAN & J. J. KOHN : Hölder estimates on domains in two complex dimensions and on three dimensional C. R. manifolds. Adv. Math. 69 (1988), 233-303.

[8] C. FEFFERMAN, J. J. KOHN & M. MACHEDON : Hölder estimates on C. R. manifolds with diagonalizable Levi form. Adv. Math. 84 (1990), 1-90.

[9] L. HÖRMANDER : Hypoelliptic second order differential equations. Acta Math. 119 (1967), 147-171.

[10] J. J. KOHN : The range of the tangential Cauchy-Riemann operator. Duke Math. J. 53 (1986), 525-545.

[11] J. J. KOHN : Boundary behavior of $\bar{\partial}$ on weakly pseudoconvex manifolds of dimension two. J. Diff. Geometry 6 (1972), 523-542.

[12] M. RANG : Integral kernels and Hölder estimates for $\bar{\partial}$ on pseudoconvex domains of finite type in $\mathbb{C}^2$. (Preprint).

[13] L. ROTHSCHILD & E. STEIN : Hypoellipticic diff. operators and nilpotent groups. Acta Math. 137 (1976), 247-320.

[14] J. STALKER : Hölder Regularity for $\Box_b$ in the case of comparable eigenvalues. To appear. Princeton University.

Exploding Orbits of Hamiltonian and Contact Structures

John Erik Fornaess, Department of Mathematics, University of Michigan, Ann Arbor, MI 48109-1003, USA. fornaess@math.lsa.umich.edu

Sandrine Grellier, Mathématiques - Bâtiment 425, Université de Paris-Sud, F-91405 Orsay Cedex, France. grellier@anh.matups.fr

1 Introduction

Given a vector field on a space, one natural question is how long its flow is defined.

In this paper we study holomorphic vector fields X on $\mathbb{C}^n$. We say that the orbit of $p \in \mathbb{C}^n$ explodes if the integral curve of p is unbounded on some finite time interval $< 0, t_0 > \subset \mathbb{R}^+$.

Our main question then is whether in a given class of holomorphic vector fields there is a dense subclass of vector fields for each of which a dense set of orbits explode.

We study here the cases of holomorphic Hamiltonian vector fields and holomorphic Reeb vector fields.

Our main results are Theorem 2.1 for the Hamiltonian case in $\mathbb{C}^2$ and Theorem 4.6 for the Reeb vector field case in $\mathbb{C}^3$.

2 Exploding Orbits of Hamiltonian flows in $\mathbb{C}^2$.

Let E denote the space of entire holomorphic functions on $\mathbb{C}^2(z, w)$ with the topology of uniform convergence on compact sets. We consider E as the space of holomorphic Hamiltonians (see [FS]). To each $F \in E$ we associate a holomorphic Hamiltonian vector field $X = (-\partial F/\partial w, \partial F/\partial z)$.

THEOREM 2.1 *There is a dense family $G \subset E$ such that every $F \in G$ has a dense set of points with exploding orbits.*

Proof: First, let us remark that it is sufficient to show that, for any $F \in E$ and any $\beta > 0$, there exists $\tilde{F} \in E$ arbitrarily close to F on $B(0, \beta)$ which has a dense set of points with exploding orbits. To construct $\tilde{F}$, we start with a function $F \in E$ and a dense sequence $\{p_n\} \subset \mathbb{C}^2$. We will make a sequence $\{F_n\}$ of perturbations of F, small on $B(0, \beta)$, so that there exist $\{q_k\}_{k=1}^n$, $\|q_k - p_k\| \leq 1/k$ and the orbit of q_k explodes for F_n and $\{F_n\}$ converges when $n \mapsto \infty$.

Once F_n is chosen, F_{n+1} is chosen so that the orbits of $\{q_k\}_{k=1}^n$ are all unchanged.

We set $F_0 = F$. Suppose F_n has been found. We will define $F_{n+1} = F_n + g_n \prod_{k \leq n}(F_n - F_n(q_k))^2$. The final term will insure that F_{n+1} agrees with F_n to second order on the orbits of q_k, $k \leq n$. We also choose g_n so that $|F_{n+1} - F_n| < 1/2^n$ on $B(0, \beta_n)$ for some sequence $\beta_n \mapsto \infty$.

We will also need to make g_n arbitrarily small on $B(0, \beta)$ to stay in a small neighborhood of F.

If p_{n+1} is not in the Fatou set of F_n then we can find a small perturbation q_{n+1} of p_{n+1} which is not on the level set of any q_k, $k \leq n$, such that the orbit of q_{n+1} is unbounded or a point q_{n+1} in the Fatou set arbitrarily close so we are in the next case.

If p_{n+1} is in the Fatou set, pick at first a small perturbation q_{n+1} on a level set of F_n without critical points. Again assume the point is not on the same level set as any $q_k, k \leq n$.

Let $g : \mathbb{C} \mapsto \mathbb{C}$ be a polynomial such that g vanishes to second order at each point $F_n(q_k)$, $k \leq n$, and $g(z) - z$ vanishes to second order at $F_n(q_{n+1})$. If $h := g \circ F_n$, then h vanishes to second order on all level sets $F_n = F_n(q_k)$, $k \leq n$, and so that $h = F_n + O((F_n - F_n(q_{n+1}))^2)$. Perturbing F_n by $F_n' := F_n + (e^{i\theta} - 1)h$ for very small θ, we can make the orbit of q_{n+1} reach outside of any predescribed compact ball, $\|z\| \leq r$ (see [FS]). We denote F_n' by F_n.

Therefore we can assume in all cases that the orbit of q_{n+1} hits $\|z\| = r$ outward, transversally, $r = \max\{\beta, \beta_n\}$. Let Σ be the level set of q_{n+1} for F_n.

We can find a smooth continuation $\gamma(\tau)$ of the orbit so that $\gamma \subset \Sigma$ and $\|\gamma(\tau)\| = \tau$ everywhere, $\tau \geq r$.

We will modify the orbit to continue in Σ very close to γ and with such a large derivative that it reaches ∞ in finite time.

Choose inductively a sequence of positive numbers $R_m, m \in \mathbb{N}$ satisfying

$$R_0 = r + 1/5, \quad R_{m+1} > R_m \text{ and } R_m \to \infty \text{ as } m \to \infty$$

and so that the projection of $\gamma(\tau)$, $R_{m-1} \leq \tau \leq R_{m+2}$, to the complex line $\mathbb{C}\gamma(R_m)$ is strictly length-increasing. (Set $R_{-1} = r$ say.)

We want to find $g_n = K(F_n - F_n(q_{n+1}))$ where K is a holomorphic function arbitrarily small on $B(0,r)$ and such that the integral curve of some small perturbation of q_{n+1} explodes.
We define K inductively by constructing holomorphic functions k_m, $m \geq 0$, arbitrarily small on $B(0, R_m - 1/5)$ and on a tubular neighborhood of $\gamma[R_m - 1/5, R_{m+2}]$ and such that the integral curve of some small perturbation of q_{n+1} is close to the integral curve of q_{n+1} on $B(0, R_m)$ and goes from $B(0, R_m)$ to $\partial B(0, R_{m+1})$ in arbitrarily small time. Then we just need to continue this inductively to get an exploding orbit.

Let $m \geq 0$. We first outline the steps needed to construct k_m. For simplicity, we will denote k_m by k, R_m by R and assume R_{m+1} and R_{m+2} respectively equal to $R+1$ and $R+2$. We may assume $\gamma(\tau)$ is real analytic for $R-1 < \tau < R+2$ and is tangent to the integral curve of q_{n+1} to at least fourth order at $\tau = R$.

The first step is to find k_1 real-analytic on $\gamma(\tau)$, $R-1 \leq \tau \leq R+2$, so that
k_1 vanishes to at least fourth order at $\gamma(R)$;
the Hamiltonian of $F_n + k_1(F_n - F_n(q_{n+1}))\prod_{k \leq n}(F_n - F_n(q_k))^2$ is tangent to $\gamma(\tau)$, $R-1 \leq \tau \leq R+2$, and
the corresponding integral curve travels from $B(0,R)$ to $\partial B(0, R+1)$ as fast as we wish.

To obtain k from k_1, we construct $\tilde{k}$ on Σ, approximating k_1 on $\gamma \cap \{R \leq ||(z,w)|| \leq R + 3/2\}$, and, then, we globalize it to $\mathbb{C}^2$. This will be done in step 2 and 3. In step 2, we use an approximation lemma and in step 3, Hörmander's L^2-estimates for $\overline{\partial}$ with weights.

Step 1

We want a real-analytic function k_1 defined on a neighborhood of $\gamma(\tau)$ in Σ, $R-1 \leq \tau \leq R+2$, so that

1- k_1 vanishes to at least fourth order at $\gamma(R)$;

2- the Hamiltonian of $F_n + k_1(F_n - F_n(q_{n+1}))\prod_{k \leq n}(F_n - F_n(q_k))^2$ which is equal on Σ to

$$(1 + k_1 \prod_{k \leq n} (F_n - F_n(q_k))^2) \left(-\frac{\partial F_n}{\partial w}, \frac{\partial F_n}{\partial z} \right)$$

must be tangent to $\gamma(\tau)$, $R-1 \leq \tau \leq R+2$, and

3- the corresponding integral curve travels from $B(0,R)$ to $\partial B(0, R+1)$ as fast as we wish.

Denote by $\theta(\tau)$ the $[0, 2\pi[$-valued continuous function so that

$$\exp(i\theta(\tau))\left(-\frac{\partial F_n}{\partial w}, \frac{\partial F_n}{\partial z}\right)$$

is tangent to $\gamma(\tau)$, $R-1 \le \tau \le R+2$ and θ vanishes to fourth order at $\tau = R$. Then θ is real-analytic.

Next, we choose $r_1(\tau)$, $R-1 \le \tau \le R+2$, real-analytic and equal to 1 to fourth order at $\tau = R$ so that r_1 is so large that the integral curve of

$$r_1(\tau)\exp(i\theta(\tau))\left(-\frac{\partial F_n}{\partial w}(\gamma(\tau)), \frac{\partial F_n}{\partial z}(\gamma(\tau))\right)\prod_{k\le n}(F_n - F_n(q_k))^2$$

travels from $B(0,R)$ to $\partial B(0, R+1)$ along γ as fast as we wish. Then, let

$$k_1(\tau) = \frac{r_1(\tau)e^{i\theta(\tau)} - 1}{\prod_{k\le n}(F_n - F_n(q_k))^2}, \quad R-1 \le \tau \le R+2.$$

Remark that by construction, k_1 satisfies all the required properties and extends to a tubular neighborhood of $\gamma[R, R+3/2]$ in Σ denoted by L.
Define

$$k_2 = \begin{cases} 0 & \text{on } B(0,R) \\ k_1 & \text{on } L. \end{cases}$$

Step 2
Now, we want to construct $\tilde{k}$. We prove an approximation lemma.
We can assume that the projection $P(L)$ of L to the complex line $\mathbb{C}\gamma(R)$ is a tubular neighborhood ending by arcs of circles and the arc through $P(\gamma(R))$ is tangent to the circle of radius R around zero with opposite concavity.

LEMMA 2.2 *For any $\varepsilon > 0$ there exists a holomorphic function $\tilde{k}_\varepsilon$ on Σ so that*

$$||k_1 - \tilde{k}_\varepsilon||_{C^1(L)} + ||\tilde{k}_\varepsilon||_{C^1(\overline{B}(0,R)\cap\Sigma} + ||\tilde{k}_\varepsilon||_{C^1(\overline{B}(\gamma(R),\varepsilon)\cap\Sigma} < \varepsilon.$$

In other words, $\tilde{k}_\varepsilon$ is a good approximation of k_2 in C^1-norm.

Proof: We can assume that $\mathbb{C}\gamma(R)$ is the z-axis and that $\gamma(R)$ is on the x-axis.
Let us denote $P(L)$ by L_1 and $P(\overline{B}(0,R)\cap\Sigma)$ by L_2. So P is biholomorphic on a neighborhood of L. Remark that L_2 is included in the circle of radius R around 0. Denote by Γ a circle of radius r around $P(\gamma(R))$ (r will be chosen small enough later). Divide Γ into two pieces Γ_1 and Γ_2 defined by $\Gamma_1 = \Gamma \cap \{x \ge \gamma(R)\}$ and $\Gamma_2 = \Gamma \cap \{x < \gamma(R)\}$.

Define

$$\kappa_1(z) = \int_{\Gamma_1} \frac{k_1 \circ P^{-1}(\zeta)}{\zeta - z} d\zeta \text{ and } \kappa_2(z) = \int_{\Gamma_2} \frac{k_1 \circ P^{-1}(\zeta)}{\zeta - z} d\zeta.$$

Then κ_j, $j = 1$ or 2, defines a holomorphic function outside Γ_j. Note that L_2 (resp. L_1) is contained in the domain of κ_1 (resp. of κ_2).

Let $z \in L_2$, $|\kappa_1(z)| \leq Cr^3$ since, for $\zeta \in \Gamma_1$, $|\zeta - z| \geq cr^2$ and since, by assumption, $k_1 \circ P^{-1}(\zeta) = \mathcal{O}(r^4)$. For the same reasons, $|\kappa_1'(z)| \leq Cr$. The same estimates hold for κ_2 and κ_2'. Furthermore, the $\mathcal{C}^1$-norms of the κ_j's are less than Cr^2 on $B(\gamma(R), r/2) \cap \Sigma$.

Now, define

$$\kappa_r(z) = \begin{cases} \kappa_1(z) & \text{if } z \in L_2 \\ k_1 \circ P^{-1}(z) - \kappa_2(z) & \text{if } z \in L_1. \end{cases}$$

Note that κ_r is holomorphic in a neighborhood of $\overline{\Delta}(0, R) \cup L_1$; since the complement of this set is connected, we can approximate κ_r arbitrarily well on $L_1 \cup L_2 \subset \overline{\Delta}(0, R) \cup L_1$ by a complex polynomial Q_r. Choose r small enough so that $\tilde{k}_\varepsilon = Q_r \circ P$ satisfies the required estimate. ∎

Step 3

We want to extend $\tilde{k} = \tilde{k}_\varepsilon$ (ε small enough) in $\mathbb{C}^2$ by solving a $\overline{\partial}$-problem. Namely, we are going to solve

$$\overline{\partial} h = \frac{\overline{\partial} \chi \, \tilde{k} \circ \Pi}{F_n - F_n(q_{n+1})} = \lambda$$

and the extension will be $\chi \tilde{k} \circ \Pi - h(F_n - F_n(q_{n+1}))$ where χ is a cut-off function equal to 1 around Σ and Π is a holomorphic retraction. More precisely, by Docquier-Grauert (see [DG]), since Σ is a closed submanifold of $\mathbb{C}^2$, there exists a holomorphic map

$$\Pi : U(\Sigma) \to \Sigma, \quad \Pi_{|\Sigma} = Id$$

where $U(\Sigma)$ is a tubular neighborhood of Σ in $B(0, R+2)$. Assume $U(\Sigma)$ small enough so that $\Pi(B(0, R - 1/20) \cap U(\Sigma)) \subset B(0, R) \cap \Sigma$.

We choose two tubular neighborhoods $U_1(\Sigma)$, $U_2(\Sigma)$ of Σ in $B(0, R+2)$ so that $U_1(\Sigma) \subset U_2(\Sigma) \subset U(\Sigma)$.

Pick a cut-off function χ so that

$$\chi = \begin{cases} 1 & \text{on } U_1(\Sigma) \\ 0 & \text{outside } U_2(\Sigma) \end{cases}$$

Let ϱ be a non-negative plurisubharmonic function which is zero on $B(0, R - 1/10)$ and as large as we want outside $B(0, R - 1/20)$. Now, solve $\overline{\partial} h = \lambda$ with weight ϱ on $B(0, R+2)$. This gives ([H])

$$\int_{B(0,R+2)} |h|^2 e^{-\varrho} dV \leq C(R) \int_{B(0,R+2)} |\lambda|^2 e^{-\varrho} dV.$$

But

$$\int_{B(0,R+2)} |\lambda|^2 e^{-\varrho} dV = \int_{B(0,R-1/20)} |\lambda|^2 e^{-\varrho} dV + \int_{(U_2 \setminus U_1)\setminus B(0,R-1/20)} |\lambda|^2 e^{-\varrho} dV$$

is as small as we wish since,

- either $z \in B(0, R - 1/20)$ so, in particular, $\Pi(z) \in B(0, R)$ and $\tilde{k} \circ \Pi$ is as small as we wish and $\varrho(z) \geq 0$;
- or $z \in (U_2 \setminus U_1) \setminus B(0, R - 1/20)$ and $e^{-\varrho}$ is as small as we wish.

So,

$$\int_{B(0,R+2)} |h|^2 e^{-\varrho} dV \text{ is as small as we wish.}$$

Now we look at $k = \chi \tilde{k} \circ \Pi - g(F_n - F_n(q_{n+1}))$ on $B(0, R - 1/20)$. It follows from the preceding estimate and from the fact that $\tilde{k}$ is as small as we want on $B(0, R) \cap \Sigma$ (so that $\tilde{k} \circ \Pi$ is as small as we want on $B(0, R - 1/20)$) that the L^2-norm of k is arbitrarily small on $B(0, R - 1/10)$ from which it follows, by holomorphy, that k is arbitrarily small in $\mathcal{C}^1$-norm on $B(0, R - 1/5)$. Furthermore, by construction $k = \tilde{k}$ on Σ. So, k is arbitrarily close, in $\mathcal{C}^1$-norm, to k_1 on a tubular neighborhood of $\gamma[R, R+1]$ in Σ and the restriction of k to $\Sigma \cap B(0, R)$ is arbitrarily small in $\mathcal{C}^1$- norm. This ends the third step

It remains to check that the orbit through $\gamma(R)$ corresponding to

$$\tilde{F}_{n+1} = F_n + k(F_n - F_n(q_{n+1})) \prod_{k \leq n} (F_n - F_n(q_k))^2$$

goes from a point p_{n+1} close to q_{n+1} to $B(0, R)$ and from $B(0, R)$ to $\partial B(0, R + 1)$ in arbitrarily short time.

Denote by ς the curve in Σ corresponding to the integral curve of F_n starting at q_{n+1}, going to $\gamma(R)$ and continuing as γ from $\gamma(R)$ to $\gamma(R + 2)$.

Denote by G_{n+1} the Hamiltonian defined in a neighborhood of ς by

$$G_{n+1} = F_n + k_2 \circ \Pi \times (F_n - F_n(q_{n+1})) \prod_{k \leq n} (F_n - F_n(q_k))^2.$$

By construction, the integral curve of G_{n+1} starting at q_{n+1} follows ς.

Denote by $\tilde{\varsigma}$ the integral curve of

$$\tilde{F}_{n+1} = F_n + k(F_n - F_n(q_{n+1})) \prod_{k \leq n} (F_n - F_n(q_k))^2$$

passing through $\gamma(R)$ at time $\tau = R$.

The curve $\tilde{\varsigma}$ stays in the level set $\tilde{F}_{n+1} = F_n(\gamma(R))$ which contains Σ. Since by construction the gradient of $\tilde{F}_{n+1} \neq 0$ on

$$W_\varepsilon := (B(0,R) \cap \Sigma) \cup L \cup (B(\gamma(R),\varepsilon) \cap \Sigma),$$

the curve $\tilde{\varsigma}$ stays in Σ (and cannot move to another branch of $\tilde{F}_{n+1} = F_n(\gamma(R))$) as long as it stays in W_ε.

Denote by $\mathcal{C}^j(W_\varepsilon)$ the $\mathcal{C}^j-$ norm inside Σ ignoring derivatives transverse to Σ.

We use the following lemma.

LEMMA 2.3 *Given $\varepsilon > 0$, let $t_{j,\varepsilon} > 0$, $j = 1,2$ be such that $\tilde{\varsigma}(t) \in W_\varepsilon$ for $R - t_{1,\varepsilon} \leq t \leq R + t_{2,\varepsilon}$. Then for any t, $R - t_{1,\varepsilon} \leq t \leq R + t_{2,\varepsilon}$*

$$||\varsigma(t) - \tilde{\varsigma}(t)|| \leq C \frac{||k_2 - \tilde{k}_\varepsilon||_{\mathcal{C}^1(W_\varepsilon)}}{||G_{n+1}||_{\mathcal{C}^2(W_\varepsilon)}} \left[e^{(t-R)||G_{n+1}||_{\mathcal{C}^2(W_\varepsilon)}} - 1 \right].$$

Proof: By definition

$$\begin{aligned} ||\varsigma(t) - \tilde{\varsigma}(t)|| &= \left|\left| \int_R^t \frac{d}{ds}(\varsigma(s) - \tilde{\varsigma}(s)) ds \right|\right| \\ &\leq \int_R^t ||\nabla G_{n+1}(\varsigma(s)) - \nabla \tilde{F}_{n+1}(\tilde{\varsigma}(s))|| ds \\ &\leq \int_R^t ||\nabla G_{n+1}(\varsigma(s)) - \nabla G_{n+1}(\tilde{\varsigma}(s))|| ds \\ &+ \int_R^t ||\nabla G_{n+1}(\tilde{\varsigma}(s)) - \nabla \tilde{F}_{n+1}(\tilde{\varsigma}(s))|| ds \\ &\leq \int_R^t ||G_{n+1}||_{\mathcal{C}^2(W_\varepsilon)} ||\varsigma(s) - \tilde{\varsigma}(s)|| ds \\ &+ \int_R^t ||G_{n+1} - \tilde{F}_{n+1}||_{\mathcal{C}^1(W_\varepsilon)} ds. \end{aligned}$$

But, since $W_\varepsilon \subset \Sigma$,

$$||G_{n+1} - \tilde{F}_{n+1}||_{\mathcal{C}^1(W_\varepsilon)} \leq C ||k_2 - \tilde{k}_\varepsilon||_{\mathcal{C}^1(W_\varepsilon)},$$

we get

$$||\varsigma(t) - \tilde{\varsigma}(t)|| \leq \int_R^t \left(||G_{n+1}||_{\mathcal{C}^2(W_\varepsilon)} ||\varsigma(s) - \tilde{\varsigma}(s)|| ds + C ||k_2 - \tilde{k}_\varepsilon||_{\mathcal{C}^1(W_\varepsilon)} \right) ds.$$

So

$$G(t) = ||\varsigma(t) - \tilde{\varsigma}(t)|| + C \frac{||k_2 - \tilde{k}_\varepsilon||_{\mathcal{C}^1(W_\varepsilon)}}{||G_{n+1}||_{\mathcal{C}^2(W_\varepsilon)}}$$

satisfies

$$G(t) \leq \int_R^t ||G_{n+1}||_{C^2(W_\varepsilon)} G(s)ds + C\frac{||k_2 - \tilde{k}_\varepsilon||_{C^1(W_\varepsilon)}}{||G_{n+1}||_{C^2(W_\varepsilon)}}$$

and Gronwall's Lemma gives the desired estimate ([HS]). ■

We continue with the proof of Theorem 2.1. First, pick an $\varepsilon > 0$ small enough. Then for t close to R, $\tilde{\varsigma}(t) \in B(\gamma(R), \varepsilon) \cap \Sigma$ so the Lemma gives that $\tilde{\varsigma}(t) \in B(0, R) \cap \Sigma$ if $t < R$ and $\tilde{\varsigma}(t) \in L$ if $t > R$. Then, inductively we see that $\tilde{\varsigma}(t) \in W_\varepsilon$ for a sufficiently large interval $a < t < b$. We can even assume that $\tilde{\varsigma}(t)$ crosses $\partial B(0, R+1)$ transversally and that $\tilde{\varsigma}(a)$ is close to q_{n+1}. ■

3 Complex Contact Structures

In this section we recall well known facts. At first we discuss the concept of a real contact structure in $\mathbb{R}^3$ seen as modelled on the restrictions of a Hamiltonian vector field and a symplectic form on $\mathbb{R}^4$ to a level set of the Hamiltonian. After this we show how the same situation in $\mathbb{R}^8$ leads to complex contact structures on $\mathbb{C}^3 \subset \mathbb{C}^4 = \mathbb{R}^8$.

Recall the definition of a contact structure in $\mathbb{R}^3$. If we consider the case of a 4-dimensional phase space $\mathbb{R}^4(p_1, p_2, q_1, q_2)$, one can discuss Hamiltonian vector fields, namely, given a Hamiltonian $H(p_1, p_2, q_1, q_2)$ we consider the Hamiltonian vector field $X = J * \nabla H$ where $J * \partial/\partial p_i = \partial/\partial q_i$ and $J * \partial/\partial q_i = -\partial/\partial p_i$. This gives a vector field X which is tangent to the level sets of H.

Hence we are led to studying vector fields X inside 3 dimensional manifolds S. We can recover the direction up to sign intrinsically from the equation $i(X)(\omega|S) \equiv 0$ where ω is the symplectic form on $\mathbb{R}^4$.

The interior product $i(X)\omega$ of the symplectic form ω and X vanishes on all tangential vector fields to the level set S of H :
If $\omega = \sum dp_i \wedge dq_i$, $X = \sum -H_{q_i}\partial/\partial p_i + H_{p_i}\partial/\partial q_i$ and $Y = \sum a_i\partial/\partial p_i + b_i\partial/\partial q_i$ then the interior product of ω and X is the $1-$ form $\mu = \sum A_i dp_i + B_i dq_i$ given by

$$\mu(Y) = \sum a_i A_i + b_i B_i = \omega(X \wedge Y) =$$

$$\sum -H_{q_i} b_i - H_{p_i} a_i = < -\nabla H, Y > .$$

This gives us the formula $A_i = -H_{p_i}$, $B_i = -H_{q_i}$. Moreover, if Y is tangent to S, $< \nabla H, Y >= 0$, so $\mu(Y) = 0$.

Conversely, one easily shows that if $\nabla H \neq 0$ then the direction of X up to sign is uniquely determined by the condition that $< \omega, X \wedge Y >\equiv 0$ for all Y tangent to the level set of H : If $X = \sum c_i\partial/\partial p_i + d_i\partial/\partial q_i$, $Y = \sum a_i\partial/\partial p_i + b_i\partial/\partial q_i$, then $\omega(X \wedge Y) = \sum c_i b_i - d_i a_i =< (-d_1, -d_2, c_1, c_2), (a_1, a_2, b_1, b_2) >= 0$ for all Y tangent to the level set of H. Hence $(-d_1\partial/\partial p_1 - d_2\partial/\partial p_2 + c_1\partial/\partial q_1 + c_2\partial/\partial q_2)$ is perpendicular to the level set, i.e. $= c(\partial H/\partial p_1 \partial/\partial p_1 + \partial H/\partial p_2 \partial/\partial p_2 + \partial H/\partial q_1 \partial/\partial q_1 + \partial H/\partial q_2 \partial/\partial q_2)$. Therefore X is parallel to $J * \nabla H$.

Note that $d\lambda|S = \omega|S$, where $\lambda = \sum p_i dq_i$, and one way to determine a length of X is to set $i(X)\lambda|S =< \lambda, X >\equiv 1$ (or -1 if we want to reverse the vector field).

For the question whether orbits are singular, that is, unbounded, what is relevant is the direction of X and not its size. (If $S = \{H = 0\}$ and we replace H by gH, we exchange X by gX. Hence intrinsically we need a way to fix a length of X.)

However if we are interested in whether orbits explode in finite time or not, it is more appropriate to use an equation of the form $i(X)\lambda|S \equiv h$ for a (possibly) variable function. This method breaks down if $\lambda|S$ is zero at some point or if X points in the null space of λ.

To be able to work intrinsically in 3 dimensional surfaces S we start with *some* $1-$ form λ on S such that $\lambda \wedge d\lambda \neq 0$. Then λ is said to be a *contact* structure and the associated *Reeb* vector field X is uniquely determined by the equations $i(X)\lambda \equiv 1$, $i(X)d\lambda \equiv 0$.

Example 3.1 *$H = \sum p_i^2 + q_i^2$. Use complex coordinates $p_1 + iq_1$, $p_2 + iq_2$. Consider the level set $H_c := \{H = c\}$, $c > 0$. Then $d^cH = \sum p_i dq_i - q_i dp_i$ (up to constants) and $dd^cH = 2\omega$. Here $d^cH \wedge dd^cH$ is a volume form on the level set.*

We can think of this in the following way: We are given a 1 form λ on $\{H = c\}$, $(\lambda = d^cH)$, and $d\lambda = \omega$, also $\lambda \wedge d\lambda \neq 0$.

This generalizes to strongly plurisubharmonic functions r, where $\lambda = d^cr$, and $d\lambda$ plays the role of the symplectic form on $\{r = c\}$. Here $\lambda \wedge d\lambda \neq 0$. Then using interior products we obtain as above vector fields X similar to Hamiltonian flows.

Example 3.2 *Let $\lambda = dx + ydz$ on $\mathbb{R}^3$. Then $d\lambda = dy \wedge dz$ and $\lambda \wedge d\lambda = dx \wedge dy \wedge dz$. Computing X, we get multiples of $\partial/\partial x$.*

To indicate what comes below, we give a similar *complex* example. But the example has no meaning yet.

Example 3.3 *Let $\lambda = dz + wd\eta$ in $\mathbb{C}^3$. Then $d\lambda = dw \wedge d\eta$ and $\lambda \wedge d\lambda = dz \wedge dw \wedge d\eta$. Looking within the class of holomorphic vector fields,***i.e.** *spanned by $\partial/\partial z$, $\partial/\partial w$, $\partial/\partial \eta$ and with holomorphic functions as coefficients, for solutions to $i(X)d\lambda = 0$ we get multiples of the vector field $\partial/\partial z = (1/2)(\partial/\partial x - i\partial/\partial y)$. To obtain a real vector field, simply take $\Re\partial/\partial z = (1/2)\partial/\partial x$.*

Next we will introduce complex *odd* dimensional contact structures. So these are real *even* dimensional. Contact structures are usually real *odd* dimensional. The

main point is, however, that for holomorphic Hamiltonians the flow is within *complex* hypersurfaces rather than *real* hypersurfaces.
Define

$$z_1 = x_1 + iy_1 = p_1 + ip_2,\ z_2 = x_2 + iy_2 = p_3 + ip_4,$$

$$w_1 = u_1 + iv_1 = q_1 - iq_2,\ w_2 = u_2 + iv_2 = q_3 - iq_4.$$

Let $\Lambda = z_1 dw_1 + z_2 dw_2$ and $\Omega = dz_1 \wedge dw_1 + dz_2 \wedge dw_2$. Then a Hamiltonian H has a holomorphic Hamiltonian vector field if and only if H is pluriharmonic. In this case, $H = \Re F$, F holomorphic and the Hamiltonian vector field is $X = (-\partial F/\partial w_1, -\partial F/\partial w_2, \partial F/\partial z_1, \partial F/\partial z_2)$.

So the flow is within the level sets of F. Next we fix a level set Σ of F, say $F = 0$ and assume $\nabla F \neq 0$. Note that multiplying F by any (invertible) holomorphic function g multiplies the Hamiltonian vector field by g. So we want to recover the direction of X modulo multiplication by complex numbers intrinsically. We use the complex symplectic form and interior products.

LEMMA 3.4 *The vector field X is the unique solution up to complex multiplication, to the equation $< \Omega, X \wedge Y > \equiv 0$ for all holomorphic vector fields $Y = A\ \partial/\partial z_1 + B\ \partial/\partial z_2 + C\ \partial/\partial w_1 + D\ \partial/\partial w_2$ tangent to the level set $F = 0$.*

Here we identify

$$X = (-\partial F/\partial w_1, -\partial F/\partial w_2, \partial F/\partial z_1, \partial F/\partial z_2)$$

with the vector field

$$X' = X - iJX.$$

Here we think of $-F_{w_1}$ as $\Re(-F_{w_1})\partial/\partial x + \Im(-F_{w_1})\partial/\partial y$ etc. Moreover $J\partial/\partial x = \partial/\partial y$ and $J\partial/\partial y = -\partial/\partial x$. So X is a *real* vectorfield and $X = \Re X'$.
More precisely, for a holomorphic function $G(z)$,

$$(\partial G/\partial z) - iJ(\partial G/\partial z) =$$

$$\Re(G_z)\partial/\partial x + \Im(G_z)\partial/\partial y - i(\Im(G_z)(-\partial/\partial x) + \Re(G_z)\partial/\partial y) =$$

$$\Re(G_z)(\partial/\partial x - i\partial/\partial y) + i\Im(G_z)(\partial/\partial x - i\partial/\partial y) = 2G_z\partial/\partial z.$$

Dropping the 2 and the $'$, we get

$$X = -F_{w_1}\partial/\partial z_1 - F_{w_2}\partial/\partial z_2 + F_{z_1}\partial/\partial w_1 + F_{z_2}\partial/\partial w_2.$$

Proof: We show first that if $Y(F) = 0$ then $< \Omega, X \wedge Y >= 0$. We get
$< dz_1 \wedge dw_1 + dz_2 \wedge dw_2, X \wedge Y >= -F_{w_1}C - F_{w_2}D - F_{z_1}A - F_{z_2}B = -Y(F) = 0.$

We need to know that if $< \Omega, Z \wedge Y >= 0$ for all Y for which $Y(F) = 0$ and $Z(F) = 0$, then $Z||_{\mathbb{C}}X$. Since the space of such Y at a point is 3 dimensional, the space of such Z is one dimensional. Since X already is a solution, $Z||_{\mathbb{C}}X$. ■

A *complex* contact form on $\mathbb{C}^3$ is a one form $\Lambda = \sum A_i dz_i$ where the $A_i(z)$ are holomorphic functions and where $\Lambda \wedge d\Lambda \neq 0$.

The associated *Complex Contact Structure* is the zero set of Λ in the tangent bundle of $\mathbb{C}^3$.

LEMMA 3.5 *Given Λ and a holomorphic function f, $f \neq 0$, there is a unique holomorphic vectorfield $X = X_f = \sum a_i \partial/\partial z_i$, a_i holomorphic, so that the interior product $i(X)d\Lambda \equiv 0$ and $< \Lambda, X > \equiv f$.*

Proof: Write $\Lambda = \sum A^i dz_i$ and $X = \sum a_i \partial/\partial z_i$. Then writing down the equations in coordinates we get 4 equations in 3 unknowns. But the last three are dependant and if we eliminate the appropriate one we are left with 3 linear equations in three unknowns $\{a_i\}$ and the condition that the system has a unique solution is precisely that Λ is a contact form.

The equations are:

$$A_1 a_1 + A_2 a_2 + A_3 a_3 = f$$

$$(A_2^1 - A_1^2)a_2 + (A_1^3 - A_3^1)a_3 = 0$$

$$(A_1^2 - A_2^1)a_1 + (A_3^2 - A_2^3)a_3 = 0$$

$$(A_1^3 - A_3^1)a_1 + (A_2^3 - A_3^2)a^2 = 0$$

■

The vector field $\Re X$ is said to be a holomorphic *Reeb* vector field.

We need to see if there is an abundance of complex contact forms Λ. Take any locally injective holomorphic map

$$\Phi : \mathbb{C}^3 \mapsto \mathbb{C}^3, \ \Phi =< Z, W, \Gamma > .$$

Then let $\Lambda = dZ + W d\Gamma$.

LEMMA 3.6 *Given any complex contact form Λ. Then Λ can be written locally as $dZ + W d\Gamma$ for a local injective holomorphic map (Z, W, Γ).*

Proof: Let X be the Reeb vectorfield, near $p = 0 \in \mathbb{C}^3(z', w', \gamma')$ say, $f \equiv 1$. We can assume $X = \partial/\partial z'$ at zero. Choose new coordinates (z, w, γ) by mapping (z, w, γ) to the point (z', w', γ') which is obtained by integrating X to the (complex) time z starting at $(0, w, \gamma)$. In these coordinates X becomes $\partial/\partial z$. Then Λ has the form

$$dz + Adw + Bd\gamma$$

and $d\Lambda = A_z dz \wedge dw + B_z dz \wedge d\gamma + (B_w - A_\gamma) dw \wedge d\gamma$. For $Y = a\partial/\partial z + b\partial/\partial w + c\partial/\partial\gamma$, $< i(X)d\Lambda, Y >= A_z b + B_z c \equiv 0 \ \forall a, b, c$. So $A_z = B_z = 0$.

It follows that both A and B are independant of z. So $A = \sum a_{i,j} w^i \gamma^j$. Let

$$Z' = z + \sum a_{i,j} w^{i+1} \gamma^j/(i+1), \ W" = w, \ \Gamma' = \gamma.$$

Then

$$dz + Adw + Bd\gamma = dZ' - \sum a_{i,j} w^i \gamma^j dw -$$

$$\sum a_{i,j} j w^{i+1} \gamma^{j-1}/(i+1) d\gamma + Adw + Bd\gamma = dZ' + B'd\Gamma'$$

Necessarily $\partial B'/\partial w \neq 0$, so use B' as $W'-$ coordinate. Then $p = (0, W_0', 0)$. Finally, write $dZ' + W'd\Gamma' = d(Z' + W_0'\Gamma') + (W' - W_0')d\Gamma'$ and write $Z = Z' + W_0'\Gamma'$, $W = W' - W_0'$ and $\Gamma = \Gamma'$. Then the contact form becomes $dZ + Wd\Gamma$ near $p = (0, 0, 0)$. ■

4 Exploding Orbits of Complex Contact Structures.

Let Λ be a fixed complex contact form and let C be the space of non-zero holomorphic functions on $\mathbb{C}^3$ with the topology of uniform convergence on compact sets.

Given $f \in C$, we denote by Φ_t^f the flow associated to X_f.

LEMMA 4.1 *Let Λ be a complex contact form and $F \in C$. The volume form*

$$\frac{\Lambda \wedge d\Lambda}{F}$$

is invariant under the flow Φ_t^F.

Proof: Since this property is invariant by change of coordinates, we can work in local coordinates and we may assume that $\Lambda = dZ + Wd\Gamma$. Then $X_F = F\frac{\partial}{\partial Z}$ and

$$\Phi(Z, W, \Gamma) = \Phi_t^F(Z, W, \Gamma) = (Z + tF + \mathcal{O}(t^2), W, \Gamma).$$

So

$$\Phi^* \left(\frac{\Lambda \wedge d\Lambda}{F} \right) = \frac{d(Z + tF + \mathcal{O}(t^2)) \wedge dW \wedge d\Gamma}{F \circ \Phi}$$

$$= \frac{(dZ + tdF + \mathcal{O}(t^2)dZ) \wedge dW \wedge d\Gamma}{F(Z, W, \Gamma) + tF\frac{\partial F}{\partial Z} + \mathcal{O}(t^2)} = \frac{dZ \wedge dW \wedge d\Gamma}{F}.$$

■

Define K_f as the set of points in $\mathbb{C}^3$ with bounded orbits for the flow Φ_t^f. Denote by K the set of $(p, f) \in \mathbb{C}^3 \times C$ so that $p \in K_f$.

Define

$$U_c = \text{int}\left\{(p, f) \in \mathbb{C}^3 \times C; \quad \sup_{t \geq 0} ||\Phi_t^f(p)|| \leq c\right\}.$$

Then, by a category argument, $U := \cup_{c>0} U_c$ is dense in int$K$.

Let $f \in C$ be fixed. Assume Φ_t^f defines an automorphism on a bounded open set $V \subset \mathbb{C}^3$. For any $p \in V$, we can associate an open Riemann surface S_p which is the complex integral curve of X_f.

Then, S_p is foliated by the real integral curves $R_{p'}$, $p' \in S_p$, of X_f.

For any $p' \in S_p$, we denote by $d(p, R_{p'})$ the distance from p to the integral curve of X_f through p'. Then, we have the following lemma.

LEMMA 4.2 *For any compact $K \subset V$, there exists a constant $c \geq 1$ so that, for any $p \in K$ and any $p' \in S_p$ close to p and any $t \in \mathbb{R}$,*

$$\frac{1}{c} d(p, R_{p'}) \leq d(\Phi_t^f(p), R_{\Phi_t^f(p')}) \leq c d(p, R_{p'}).$$

Proof: Since the volume form

$$\frac{\Lambda \wedge d\Lambda}{f}$$

is preserved under the flow Φ_t^f it follows that $\overline{\{\Phi_t^f\}}$ is compact in Aut(V). Hence $\{\Phi_t^f\}$ is uniformly bi-Lipschitz on any compact subset of V. ■

THEOREM 4.3 *For any complex contact form Λ, the interior of K is empty.*

COROLLARY 4.4 *For any complex contact form Λ, there exists $A \subset C$, a dense G_δ, so that, for every $f \in A$, the interior of K_f is empty.*

The theorem follows immediately from the following lemma.

LEMMA 4.5 *Let Λ be a fixed contact form. Given $p \in \mathbb{C}^3$, $f \in C$ and $R > 0$, there exist $\tilde{p} \in \mathbb{C}^3$ arbitrarily close to p and $\tilde{f}$ arbitrarily close to f so that the orbit of $\tilde{p}$ for $\tilde{f}$ goes out of the ball around 0 of radius R.*

Remark: Furthermore, we can always assume that the orbit of $\tilde{p}$ hits $\partial B(0,R)$ transversally.

Proof of the lemma: We may assume that $(p,f) \in \text{int}K$ (otherwise the result is obvious). Therefore, we may assume that $(p,f) \in U_c$ for some $c > 0$ (by wiggling (p,f) a little bit).

Denote by $\gamma(t)$, $t \in \mathbb{R}$, the corresponding integral curve. For $\varepsilon > 0$ small enough, $(p, e^{i\varepsilon}f) \in U_c$. Let $\gamma_\varepsilon(t)$, $t \in \mathbb{R}$, its integral curve. Define

$$\Psi_t : \begin{array}{ll} U_c \rightarrow & U_c \\ (p,f) \rightarrow & (\Phi_t^f(p), f). \end{array}$$

Then, Ψ_t is an automorphism since by Lemma(4.1), Φ_t^f is volume-preserving.

So, $\varepsilon \rightarrow \gamma_\varepsilon(t)$ is holomorphic and uniformly bounded by c, then by Cauchy estimates

$$(*) \qquad ||\gamma_\varepsilon(t) - \gamma(t)|| \leq c\varepsilon \text{ uniformly in } t.$$

Define $d(\gamma_\varepsilon(t), \gamma(t)) = d(R_{\Phi_{-t}^f(\gamma_\varepsilon(t))}, p)$. This means that we measure the distance from p to the integral curve for f which goes through $\gamma_\varepsilon(t)$ at time t.

Since the angle between X_f and $X_{e^{i\varepsilon}f}$ is ε, then $d(\gamma_\varepsilon(t), \gamma(t)) \simeq \varepsilon t$. This contradicts the Cauchy-estimates $(*)$ for t large enough. ■

THEOREM 4.6 *Assume Λ is fixed, then there exists a dense family $G \subset C$ so that for every $F \in G$, the pair (F, Λ) has a dense set of points with exploding orbits.*

Proof: We start with a dense sequence of points $\{p_n\}_{n \in \mathbb{N}^*}$, $p_n \in B(0,n)$, and with a function $F \in C$.

We make an inductive construction. Namely, we construct a sequence $F_n \in C$ and a sequence of points $\{p_{n,k}\}_{1 \leq k \leq n, n \in \mathbb{N}^*}$ so that the following conditions are satisfied:

1- for any $1 \leq k \leq n$, $p_{n,k}$ is close to p_k and $p_{n,k} \in B(0,k)$;

2- F_{n+1} is close to F_n in C^1-norm on $B(0,n)$;

3- the orbits of $\{p_{n,k}\}_{1 \leq k \leq n, n \in \mathbb{N}^*}$ relative to F_n goes out of $B(0,n)$ transversally and the time it takes to go from $B(0,k)$ to $\partial B(0,n)$ is at most $\sum_{l=k+1}^{n} \frac{1}{2^l}$. Moreover, the orbit is strictly norm-increasing between $B(0,k)$ and $\partial B(0,n)$.

This will give the result of the theorem as in the $\mathbb{C}^2$ case. Take $F_0 = F$. The existence of F_1 follows from the preceding lemma.

Now assume $\{p_{n,k}\}_{1\le k\le n}$ and F_n constructed and let us construct $\{p_{n+1,k}\}_{1\le k\le n+1}$ and F_{n+1}.

We are going to construct a sequence of functions $F_{n+1,k}$ for $0 \le k \le n$, close to F_n on $B(0,n)$, so that $F_{n+1,0} = F_n$ and $F_{n+1,n} = F_{n+1}$. Furthermore this sequence will be constructed in such a way that the orbits of $\{p_{n+1,l}\}_{1\le l\le k}$ go from $B(0,n)$ to $\partial B(0,n+1)$ in time strictly less than $\frac{1}{2^{n+1}}$ and are strictly norm-increasing between $B(0,k)$ and $\partial B(0,n+1)$.

Assume $F_{n+1,k}$ chosen for $k \le n-1$. To construct $F_{n+1,k+1}$, we use the following lemma.

LEMMA 4.7 *Let $p \in B(0,R)$ and $f \in C$. Suppose the orbit of p hits $\partial B(0,R)$ transversally at q. Suppose there exists a real-analytic curve γ containing q, included in a complex integral curve of Λ and agreeing to fourth order with the orbit at q. Suppose also that γ is strictly norm-increasing and goes out of $B(0,R+2)$.*
Assume there is a compact set L consisting of finitely many disjoint strictly norm-increasing arcs so that $\gamma \cap L = \emptyset$.
Then, there is a perturbation g of f, arbitrarily small in C^1-norm on a fixed neighborhood of L and on $B(0,R)$, so that the integral curve through q for g goes arbitrarily fast from $B(0,R)$ to $\partial B(0,R+1)$.

Proof of the theorem: Assume this lemma proved. Choose L as the union of the orbits starting at $\{p_{n+1,l}\}_{1\le l\le k}$, $p = p_{n,k+1}$ and $R = n$, then, take $F_{n+1,k+1}$ a small perturbation of $F_{n+1,k}$ given by the Lemma.

It remains to choose $p_{n+1,n+1}$ a small perturbation of p_{n+1} and to make a small perturbation of $F_{n+1,n}$ so that the orbit of $p_{n+1,n+1}$ goes out of $B(0,n+1)$ transversally.

Then the limit of the sequence F_n will satisfy the required properties. ■

Proof of the lemma:

We can always assume that the projection of $\gamma \setminus \{q\}$, denoted by $P(\gamma\setminus\{q\})$, to $\mathbb{C}q$ does not contain q.

First, we choose a real-analytic function g_1 on γ, vanishing to fourth order at q and so that the integral curve corresponding to $e^{g_1}f$ follows γ at arbitrarily high speed. We call also g_1 the holomorphic extension of g_1 to a neighborhood of γ in the complex integral curve of Λ containing γ.

Then, we take a holomorphic function in a neighborhood of γ approximating g_1. We do as follows.

Assume q is on the x-axis and denote by Γ_1 and Γ_2 the two arcs on the circle Γ centered at q of a small radius r, so that

$$\Gamma_1 = \Gamma \cap \{x \ge q\} \text{ and } \Gamma_2 = \Gamma \cap \{x < q\}.$$

Define

$$\kappa_1(z) = \int_{\Gamma_1} \frac{g_1 \circ P^{-1}(\zeta)}{\zeta - z} d\zeta \text{ and } \kappa_2(z) = \int_{\Gamma_2} \frac{g_1 \circ P^{-1}(\zeta)}{\zeta - z} d\zeta.$$

Then, in a neighborhood of q, $\kappa_1(z) = g_1 \circ P^{-1}(z) - \kappa_2(z)$. So, this function extends holomorphically in a neighborhood of γ^* which consists of γ from $\partial B(0,R)$ to $\partial B(0,R+2)$ and of the orbit of f from p to q in the complex integral curve of Λ through q. We denote by g' this extension.

Then g' is defined on a neighborhood of γ^* except for the lines corresponding to $x = q$, $|y| \geq r$ in $\mathbb{C}q$. One next approximates g' in $\mathcal{C}^1$ norm on a fixed neighborhood of γ^* obtained by avoiding $\{|x-q| \leq r^2,\ |y| \geq r - r^2\}$.

This approximation $\tilde{g}$ is arbitrarily close. Note that $\tilde{g}$ is as small as we wish on $B(0,R) \cap Dom(\tilde{g})$ and as close as we wish to g_1 on a neighborhood of $\gamma \cap \{R \leq ||z|| \leq R+3/2\}$ - in $\mathcal{C}^1$−norm. This is possible by choosing r small enough and $\tilde{g}$ close enough to g'. In addition, the $\mathcal{C}^1$−norm of $\tilde{g}$ is smaller than Cr^2 on $B(q, r/2)$. Now, compose $\tilde{g}$ with a holomorphic retraction π to obtain an extension of $\tilde{g}$ in $\mathbb{C}^3$. We assume that π is defined in a neighborhood U in $\mathbb{C}^3$ of γ and of the orbit of f from p to q so that $\pi(B(0, R-1/20) \cap U) \subset B(0,R)$.

We want to solve $\overline{\partial} u = \overline{\partial}\chi\ \tilde{g} \circ \pi$ with Hörmander's L^2-method with weights, where χ is a cut-off function equal to 1 in a neighborhood of γ and in a neighborhood of the orbit from p to q.

Since $B(0, R-1/10) \cup L \cup \gamma^*$ is polynomially convex, we can find a plurisubharmonic non-negative function ϱ which is zero on a neighborhood of $B(0, R-1/10) \cup L \cup \gamma^*$ and which is as large as we wish on $\text{supp}(\overline{\partial}\chi) \backslash B(0, R-1/20)$.

Then, solve $\overline{\partial} u = \overline{\partial}\chi\ \tilde{g} \circ \pi$ with this weight ϱ on $B(0, R+2)$ ([H]).
So

$$\int_{B(0,R+2)} |u|^2 e^{-\varrho} dV \leq C(R) \int_{B(0,R+2)} |\overline{\partial}\chi\ \tilde{g} \circ \pi|^2 e^{-\varrho} dV$$

is arbitrarily small.

So $g = \chi\tilde{g} \circ \pi - u$ is holomorphic in $B(0, R+2)$ and is arbitrarily small on a fixed neighborhood of $L \cup B(0, R-1/5)$ in $\mathcal{C}^1$-norm. Also, inside the complex integral curve of Λ through q, g is arbitrarily small in $\mathcal{C}^1$-norm on $B(0,R)$ and arbitrarily close to g_1 around γ in $\mathcal{C}^1$-norm. This gives the result in the same way as in the proof of Theorem 2.1, using an analogue of Lemma 2.3. ∎

References

[DG] Docquier, F., Grauert, H; *Levisches Problem und Rungescher Satz fü r Teilgebiete Steinscher Mannigfaltigkeiten*, Math. Ann. 140 (1960), 94- 123.

[FS] Fornæss, J. E., Sibony, N; *Holomorphic Symplectomorphisms in* $\mathbb{C}^2$, preprint.

[H] Hörmander, L; *An introduction to Complex Analysis in Several Variables*, (1973), North Holland, Amsterdam.

[HS] Hirsch ,M.W. & Smale, S.*Differential Equations, dynamical Systems and linear Algebra*, (1974) Academic Press.

Holomorphic Automorphisms of C^n: A Survey

Franc Forstneric

Department of Mathematics, University of Wisconsin, Madison, WI 53706, USA

0. Introduction

In this paper we survey some recent results on holomorphic automorphism groups of the complex Euclidean space $\mathbf{C}^n$, with emphasis on question of approximation of biholomorphic maps and approximation of diffeomorphisms on certain classes of submanifolds in $\mathbf{C}^n$ by automorphisms of $\mathbf{C}^n$. We also consider flows generated by holomorphic vector fields, especially by holomorphic Hamiltonian fields, and the global behavior of their orbits. Finally we collect the known classification results for holomorphic flows on the plane $\mathbf{C}^2$.

We do not consider questions of holomorphic dynamics of automorphisms, except for flows of Hamiltonian holomorphic vector fields (sect. 9); for dynamics in several variables we refer the reader to the recent survey [16] by Fornæss and Sibony. For the algebraic aspects of this theory we refer the reader to the survey in preparation by H. Kraft [26].

We denote by $\mathrm{Aut}\mathbf{C}^n$ the group of all holomorphic automorphisms of $\mathbf{C}^n$, and by $\mathrm{Aut}_1\mathbf{C}^n$ its subgroup consisting of automorphisms with Jacobian one ($JF = 1$). While the group $\mathrm{Aut}\mathbf{C}^1$ consists only of affine linear maps $z \mapsto az + b$ ($a \in \mathbf{C}^* = \mathbf{C}\backslash\{0\}$, $b \in \mathbf{C}$), these automorphism groups are very large and complicated when $n > 1$ which we shall assume throughout this paper. In particular, when $n > 1$, both groups are infinite dimensional Lie groups in the topology of uniform convergence on compacts in $\mathbf{C}^n$.

An important study of these groups, and especially of their actions on countable subsets of $\mathbf{C}^n$, was done by Rosay and Rudin [32]. They proved that for any two countable *dense* subsets $X, Y \subset \mathbf{C}^n$ there is an $F \in \mathrm{Aut}\mathbf{C}^n$ such that $F(X) = Y$. On the other hand, they showed that the situation is very different for countable *discrete* subsets $E \subset \mathbf{C}^n$: Every such subset can be mapped onto the standard arithmetic progression in $\mathbf{C}\times\{0\}^{n-1} \subset \mathbf{C}^n$ by an injective holomorphic map $F: \mathbf{C}^n \to \mathbf{C}^n$ with $JF = 1$, but in general not by any automorphism of $\mathbf{C}^n$. (Recall that an injective holomorphic map $F: \mathbf{C}^n \to \mathbf{C}^n$ whose image is a proper subset of $\mathbf{C}^n$ is called a Fatou-Bieberbach map, and its image $F(\mathbf{C}^n) \subset \mathbf{C}^n$ is a Fatou-Bieberbach domain.) If we call two countable discrete sets $E_0, E_1 \subset \mathbf{C}^n$ equivalent when $F(E_0) = E_1$ for some $F \in \mathrm{Aut}\mathbf{C}^n$, then there exist infinitely many equivalence classes, and there exist discrete sets $E \subset \mathbf{C}^n$ such that the only automorphism of $\mathbf{C}^n$ mapping E onto E is the identity. The paper [32] also contains many interesting results on Fatou-Bieberbach maps, but we shall not consider this topic in the present paper.

Supported in part by the NSF grant DMS-9322766 and by a grant from the Ministry of Science of the Republic of Slovenia. This paper is in the final form and no part of it will be submitted elsewhere.

The main tool used by Rosay and Rudin were special automorphisms of $\mathbf{C}^n$, called *shears* (resp. generalized shears); see sect. 1 below. They raised the question whether the subgroup of $\mathrm{Aut}\mathbf{C}^n$ consisting of finite compositions of (generalized) shears is dense in $\mathrm{Aut}\mathbf{C}^n$ or perhaps even equal to it.

Both of these questions were answered by E. Andersén [4] in the volume preserving case and by Andersén and Lempert [5] in the general case: The shear subgroup of $\mathrm{Aut}\mathbf{C}^n$ resp. $\mathrm{Aut}_1\mathbf{C}^n$ is dense in, but not equal to the whole groups (Corollary 2.2 below). The main step in the proof of the density result in [4,5] is a decomposition theorem, Proposition 4.1 below, to the effect that every polynomial holomorphic vector field on $\mathbf{C}^n$ is a finite sum of *complete* polynomial fields whose flows consist of (generalized) shears.

This approach was further developed by J.-P. Rosay and the author in [21]. An automorphism $F \in \mathrm{Aut}\mathbf{C}^n$ is viewed as the time one map in the flow $F_t(z)$ $(0 \le t \le 1)$ of a time-dependent entire holomorphic vector field X_t. Explicitly, $F_t(z)$ is the solution at time $t \in [0,1]$ of an ordinary differential equation

$$\dot{Z} = X_t(Z), \qquad Z(0) = z,$$

where X_t is a holomorphic vector field on $\mathbf{C}^n$ for each fixed t which is of class $\mathcal{C}^1$ in both variables (t, z). We can approximate X_t on short time intervals $[j/N, (j+1)/N] \subset [0,1]$ by time independent polynomial vector fields Y_j. Applying the result of Andersén and Lempert on decomposition of polynomial holomorphic vector fields into finite sums of complete shear fields, one concludes that the flow of each Y_j (and therefore the flow of the original time dependent field X_t) can be approximated by compositions of (generalized) shears. This shows in fact that the flow of any globally defined holomorphic vector field X_t on $\mathbf{C}^n$, wherever it is defined, is approximable by automorphisms of $\mathbf{C}^n$. In [19] we extended these results to symplectic holomorphic automorphisms of $\mathbf{C}^{2n}$ by considering flows of holomorphic Hamiltonian vector fields.

In [21] and in the subsequent papers [17,19,20,22] we used this technique, together with new results on generic polynomial convexity of totally real submanifolds in $\mathbf{C}^n$, to study question of approximation of smooth mappings $F: M \to \mathbf{C}^n$ on compact, totally real, polynomially convex submanifolds $M \subset \mathbf{C}^n$ by holomorphic automorphisms of $\mathbf{C}^n$, as well as the problem of global $\mathcal{G}$-equivalence of such submanifolds in $\mathbf{C}^n$ with respect to various automorphism groups $\mathcal{G} \subset \mathrm{Aut}\mathbf{C}^n$.

In another direction we survey results on dynamics of holomorphic vector fields on $\mathbf{C}^n$, with emphasis on global questions such as completeness in real and complex time, analysis of complex orbits, abundance or nonabundance of bounded resp. exploding orbits, classification of flows induced by complete holomorphic vector fields, etc.

The paper is organized as follows. In section 1 we introduce the shear groups and the holomorphic symplectic groups. In section 2 we collect results on approximation of biholomorphic mappings between domains in $\mathbf{C}^n$ by automorphisms of $\mathbf{C}^n$. In sections 3–4 we recall some properties of flows of holomorphic vector fields and indicate proofs of results of section 2. In sections 5–7 we survey results on approximation by automorphisms on certain classes of real submanifolds in $\mathbf{C}^n$. In section 8 we mention some results on complex orbits of holomorphic vector fields. In section 9 we survey results on dynamics of holomorphic Hamiltonian vector fields on $\mathbf{C}^{2n}$. Section 10 contains results on classification of one-parameter subgroups in the automorphism group $\mathrm{Aut}\mathbf{C}^2$ of the plane.

I wish to thank P. Ahern, G. Buzzard, M. Flores, J.-E. Fornæss, H. Kraft, J-P. Rosay, W. Rudin, and D. Varolin for discussions and communication on this subject.

1. Automorphism groups of $\mathbf{C}^n$

Let $z = (z_1, \ldots, z_n)$ be the complex coordinates on $\mathbf{C}^n$. Recall that $\mathrm{Aut}\mathbf{C}^n$ is the group of all holomorphic automorphisms of $\mathbf{C}^n$ and $\mathrm{Aut}_1\mathbf{C}^n$ is the group of all automorphisms with Jacobian one. These preserve the complex volume form $\Omega = dz_1 \wedge \cdots \wedge dz_n$, in the sense that $F^*\Omega = \Omega$.

On the even dimensional spaces $\mathbf{C}^{2n}$ $(n \geq 1)$ we also have the group of *symplectic* holomorphic automorphisms. Let ω be the complex symplectic form on $\mathbf{C}^{2n}$:

$$\omega = \sum_{j=1}^{n} dz_j \wedge dz_{n+j}. \tag{1}$$

A holomorphic map $F: D \subset \mathbf{C}^{2n} \to \mathbf{C}^{2n}$ is said to be *symplectic holomorphic* if $F^*\omega = \omega$. The symplectic holomorphic automorphism group of $\mathbf{C}^{2n}$ is

$$\mathrm{Aut}_{\mathrm{sp}}\mathbf{C}^{2n} = \{F \in \mathrm{Aut}\mathbf{C}^{2n} : F^*\omega = \omega\}. \tag{2}$$

Since ω^n is a constant multiple of the volume form Ω on $\mathbf{C}^{2n}$, every symplectic holomorphic map has Jacobian one. On $\mathbf{C}^2$ these two classes of maps coincide. Thus

$$\mathrm{Aut}_{\mathrm{sp}}\mathbf{C}^{2n} \subset \mathrm{Aut}_1\mathbf{C}^{2n} \subset \mathrm{Aut}\mathbf{C}^{2n}, \qquad \mathrm{Aut}_{\mathrm{sp}}\mathbf{C}^2 = \mathrm{Aut}_1\mathbf{C}^2.$$

Each of these groups contains the corresponding linear subgroup:

$$GL(n,\mathbf{C}) \subset \mathrm{Aut}\mathbf{C}^n, \quad SL(n,\mathbf{C}) \subset \mathrm{Aut}_1\mathbf{C}^n, \quad Sp(n,\mathbf{C}) \subset \mathrm{Aut}_{\mathrm{sp}}\mathbf{C}^{2n}.$$

The automorphism groups introduced above contain the following complex one parameter subgroups (with complex parameter $t \in \mathbf{C}$):

$$F_t(z) = z + tf(\Lambda z)v, \qquad z \in \mathbf{C}^n, \tag{3}$$

$$G_t(z) = z + \left(e^{tg(\Lambda z)} - 1\right)\langle z, v\rangle v, \qquad z \in \mathbf{C}^n, \tag{4}$$

$$S_t(z) = z + th(\omega(z,v))v, \qquad z \in \mathbf{C}^{2n}. \tag{5}$$

Here $v \in \mathbf{C}^n$ is a fixed vector of length one, $\Lambda: \mathbf{C}^n \to \mathbf{C}^k$ is a $\mathbf{C}$-linear map for some $k < n$ satisfying $\Lambda v = 0$, $\langle z, v\rangle = \sum z_j \bar{v}_j$, f and g are entire functions on $\mathbf{C}^k$, h is an entire function on $\mathbf{C}$, and ω is the symplectic form (1).

Every map of the form (3) or (5) (for a fixed $t \in \mathbf{C}$) is called a *shear*, and maps (4) are *generalized shears*. This terminology was introduced by Rosay and Rudin [32], although automorphisms of this type have been used before. Notice that (5) is a special case of (3).

One can easily verify that in all three cases we have $F_s \circ F_t = F_{s+t}$ for $s, t \in \mathbf{C}$, which means that the above are complex one parameter subgroups of the respective automorphism groups.

Write $z = (z', z_n)$, where $z' = (z_1, \ldots, z_{n-1}) \in \mathbf{C}^{n-1}$.

1.1 Proposition. *(i) The group (3) is $SU(n, \mathbf{C})$-conjugate to a group*

$$\tilde{F}_t(z) = (z', z_n + t\tilde{f}(z')), \qquad \tilde{F}_t \in \mathrm{Aut}_1\mathbf{C}^n.$$

(ii) The group (4) is $U(n)$-conjugate to a group

$$\tilde{G}_t(z) = (z', \exp(t\tilde{g}(z'))z_n), \qquad \tilde{G}_t \in \mathrm{Aut}\mathbf{C}^n.$$

(iii) The group (5) is $Sp(n, \mathbf{C})$-conjugate to a group

$$\tilde{S}_t(z) = (z_1, \ldots, z_n, z_{n+1} + t\tilde{h}(z_1), z_{n+2}, \ldots, z_{2n}), \qquad \tilde{S}_t \in \mathrm{Aut}_{\mathrm{sp}}\mathbf{C}^{2n}.$$

The proof is straightforward; see the appendix in [17] and section 5 in [5]. It follows that the groups (3), (4), and (5) belong to, respectively, $\mathrm{Aut}_1\mathbf{C}^n$, $\mathrm{Aut}\mathbf{C}^n$, and $\mathrm{Aut}_{\mathrm{sp}}\mathbf{C}^{2n}$. The shears of type (5) are called *symplectic*.

1.2 Definition. *(i) $\mathcal{S}_1(n)$ is the subgroup of $\mathrm{Aut}_1\mathbf{C}^n$ consisting of finite compositions of shears (3);*

(ii) $\mathcal{S}(n)$ is the subgroup of $\mathrm{Aut}\mathbf{C}^n$ consisting of finite compositions of (generalized) shears (3) and (4);

(iii) $\mathcal{S}_{sp}(n)$ is the subgroup of $\mathrm{Aut}_{\mathrm{sp}}\mathbf{C}^{2n}$ consisting of finite compositions of symplectic shears (5).

It is easily verified that

$$GL(n, \mathbf{C}) \subset \mathcal{S}(n), \quad SL(n, \mathbf{C}) \subset \mathcal{S}_1(n), \quad Sp(n, \mathbf{C}) \subset \mathcal{S}_{sp}(n).$$

In fact, the group $SL(n, \mathbf{C})$ is generated by the elementary matrices with diagonal entries one and only one other nonzero entry. Clearly such a matrix represents a linear shear (3) in a coordinate direction. To get the group $GL(n, \mathbf{C})$ one has to add to these the dilations in coordinate directions, and these are compositions of linear generalized shears (4). Finally, the linear symplectic group $Sp(n, \mathbf{C})$ is generated by symplectic shears (5) with $h(\zeta) = c\zeta$.

Another group of special interest is the group $\mathcal{P}^n$ of all polynomial holomorphic automorphisms of $\mathbf{C}^n$ and its subgroup $\mathcal{P}^n_1$ of polynomial automorphisms with Jacobian one. Since the Jacobian of every polynomial automorphism is constant, we have $\mathcal{P}^n = \mathcal{P}^n_1 \times \mathbf{C}^*$. On $\mathbf{C}^{2n}$ one also have the symplectic polynomial group $\mathcal{P}^{2n}_{sp} \subset \mathrm{Aut}_{\mathrm{sp}}\mathbf{C}^{2n}$.

Notice that every shear of type (3) and (5) can be approximated by polynomial shears of the same type by approximating the function f resp. h by polynomials. Thus the polynomial shear groups generated by polynomial automorphisms of type (3) resp. (5) are dense in the corresponding shear groups. This is not the case for generalized shears (4) with nonconstant Jacobian.

2. Approximation by shears

Recall that a domain $D \subset \mathbf{C}^n$ (not necessarily pseudoconvex) is Runge if every holomorphic function in D is a limit of polynomials, uniformly on compacts in D. Every convex domain in $\mathbf{C}^n$ is Runge, and so is every starshaped domain [12]. (The last result is already mentioned in Behnke and Thullen [9], p.130, who attribute it to B. Almer.)

We are interested in the problem of approximating a biholomorphic map $F: D \to D'$ between Runge domains in $\mathbf{C}^n$ by automorphisms of $\mathbf{C}^n$. The following result of fundamental importance is due to Andersén [4] and Andersén and Lempert [5] (case (c) was established by the author in [18].) The topology is that of uniform convergence on compact sets in D.

2.1 Theorem. *Let $F: D \to F(D) \subset \mathbf{C}^n$ $(n \geq 2)$ be a biholomorphic mapping from a convex (or starshaped) domain $D \subset \mathbf{C}^n$ onto a Runge domain $F(D)$. Then*

(a) F is the limit of a sequence $\Phi_j|_D$, $\Phi_j \in \mathcal{S}(n)$;

(b) if $F^*\Omega = \Omega$ $(JF = 1)$, then F is the limit of a sequence $\Phi_j \in \mathcal{S}_1(n)$;

(c) if $n = 2m$ and if $F^*\omega = \omega$, then F is the limit of a sequence $\Phi_j \in \mathcal{S}_{sp}(m)$.

2.2 Corollary. *(i) The group $\mathcal{S}_1(n)$ is dense in $\mathrm{Aut}_1\mathbf{C}^n$ for $n \geq 2$;*

(ii) the group $\mathcal{S}(n)$ is dense in $\mathrm{Aut}\mathbf{C}^n$ for $n \geq 2$;

(iii) the group $\mathcal{S}_{sp}(n)$ is dense in $\mathrm{Aut}_{\mathrm{sp}}\mathbf{C}^{2n}$ for $n \geq 1$.

It is known [4,5] that the inclusions in (i) and (ii) are proper, i.e., there exist automorphisms which are not finite compositions of shears. Most likely the same is true in case (iii). It follows that the group of polynomial shears (3) resp. (5) in also dense in $\mathrm{Aut}_1\mathbf{C}^n$ resp. $\mathrm{Aut}_{\mathrm{sp}}\mathbf{C}^{2n}$.

2.3 Corollary. *Every injective holomorphic map (a Fatou-Bieberbach map) $F: \mathbf{C}^n \to \mathbf{C}^n$ onto a Runge domain $F(\mathbf{C}^n) \subset \mathbf{C}^n$ is a limit of automorphisms of $\mathbf{C}^n$.*

It is an open question whether every Fatou-Bieberbach domain in $\mathbf{C}^n$ is Runge.

The following example from [21] shows that, in general, Theorem 2.1 does not hold for an arbitrary Runge domain $D \subset \mathbf{C}^n$. Further examples can be found in [5] and [21].

Example: The curve

$$M = \{(e^{i\theta}, e^{-i\theta}): \theta \in \mathbf{R}\} \subset \mathbf{C}^2$$

is polynomially convex. The mapping

$$F(z, w) = \big(z, w + (1 - zw)(1/z - 1)\big)$$

is biholomorphic near M and fixes M pointwise. There is a basis of strongly pseudoconvex Runge neighborhoods D of M in $\mathbf{C}^2$ such that $F(D)$ is also Runge in $\mathbf{C}^2$. However, the Jacobian of F equals $JF(z, w) = z$, and its winding number along M is one. Hence F can not be approximated by automorphisms of $\mathbf{C}^2$ (or even by diffeomorphisms of $\mathbf{C}^2$!) in any open neighborhood of M.

The following generalization of Theorem 2.1 to 'isotopies' of biholomorphic maps is very useful in applications, and it can be proved by the same methods. Case (i) was proved in [21], case (ii) in [20], and case (iii) in [19].

2.4 Theorem: *Let $D \subset \mathbf{C}^n$ $(n \geq 2)$ be a Runge domain and let $F_t: D \to \mathbf{C}^n$ be a biholomorphic mapping for each $t \in [0,1]$, of class $\mathcal{C}^2$ in $(t,z) \in [0,1] \times D$, such that F_0 is the identity on D and the domain $F_t(D)$ is Runge in $\mathbf{C}^n$ for each $0 \leq t \leq 1$. Then*

(i) F_t is a limit of automorphisms of $\mathbf{C}^n$ for each $t \in [0,1]$.

(ii) If $JF_t = 1$ for each $t \in [0,1]$ and if D is a domain of holomorphy satisfying $H^{n-1}(D;\mathbf{C}) = 0$, then F_t is a limit of volume preserving automorphisms of $\mathbf{C}^n$ for each $t \in [0,1]$.

(iii) If n is even, if $F_t^*\omega = \omega$ for each $t \in [0,1]$, and if D is a domain of holomorphy satisfying $H^1(D;\mathbf{C}) = 0$, then F_t is a limit of symplectic holomorphic automorphisms of $\mathbf{C}^n$ for each t.

Examples in [19] and [20] show that the conclusions in (ii) and (iii) do not hold without the respective cohomology condition on D. See also section 6 below. A sharper version of the last result which includes regularity with respect to the parameter $x \in \mathbf{R}^k$ was proved in [17, Theorem 1.1]. We only consider the group Aut$\mathbf{C}^n$.

2.5 Theorem. *Let $\Omega \subset \mathbf{C}^n$ be a Runge domain, and let B be the closed unit ball in $\mathbf{R}^k$. Assume that $F: B \times \Omega \to \mathbf{C}^n$ is a mapping of class $\mathcal{C}^p$ $(0 \leq p < \infty)$ such that for each $x \in B$, $F_x = F(x,\cdot): \Omega \to \mathbf{C}^n$ is a biholomorphic mapping onto a Runge domain $\Omega_x \subset \mathbf{C}^n$, and the map F_0 is the identity on Ω. Then for each compact set $K \subset \Omega$ and each $\epsilon > 0$ there exists a smooth map $\Phi: B \times \mathbf{C}^n \to \mathbf{C}^n$ such that $\Phi_x = \Phi(x,\cdot)$ is a holomorphic automorphism of $\mathbf{C}^n$ for every $x \in B$ and $||F - \Phi||_{\mathcal{C}^p(B\times K)} < \epsilon$.*

We will indicate the proof of these results at the end of section 4 below.

3. Flows of holomorphic vector fields

In this section we recall some basic notions concerning flows of vector fields.

Suppose that for each $t \in [0,1]$, X_t is a holomorphic vector field on a domain Ω_t in a complex manifold $\mathcal{M}$. The flow $F_{t,s}$ is the solution of the ordinary differential equation

$$\frac{d}{dt}F_{t,s}(z) = X_t\big(F_{t,s}(z)\big), \quad F_{s,s}(z) = z \in \Omega_s.$$

Assume that there is a domain $D_0 \subset \Omega_0$ such that the flow $F_t(z) = F_{t,0}(z)$ exists for all $0 \leq t \leq 1$ and all $z \in D_0$. Then $F_t: D_0 \to F_t(D_0) = D_t \subset \Omega_t$ is a biholomorphic map for each $t \in [0,1]$, F_0 is the identity on D_0, and $F_{s,t} = F_t \circ F_s^{-1}$ on D_s.

If the field $X_t = X$ is time independent, its flow F_t is a local one parameter automorphism group, i.e., $F_{t+s} = F_t \circ F_s$ where both sides are defined. In particular, $F_{t,s} = F_{t-s}$.

3.1 Definition. *A holomorphic vector field X on a complex manifold $\mathcal{M}$ is said to be complete in real (resp. complex) time if for every point $z \in \mathcal{M}$ the flow $F_t(z)$ of X exists for all real (resp. complex) values of t.*

The flow of a complete holomorphic vector field is a real (resp. complex) one parameter subgroup of the automorphism group Aut$\mathcal{M}$.

The Lie derivative of a tensor α with respect to a vector field X_s is defined by

$$L_{X_s}\alpha = \frac{d}{dt}(F_{t,s}^*\alpha)\big|_{t=s}.$$

For any pair t, s we have

$$\frac{d}{dt}F_{t,s}^*\alpha = F_{t,s}^*(L_{X_t}\alpha).$$

If α is a differential form, then

$$L_{X_t}\alpha = d(X_t\rfloor\alpha) + X_t\rfloor(d\alpha),$$

where $\rfloor$ denotes the contraction (see [1, p.121]). Combining the last two identities we get

3.2 Proposition. *Let $F_{t,s}$ be the flow of a time dependent vector field X_t as above, and let α be a closed differential form on $\mathcal{M}$. The following are equivalent:*

(a) *$F_{t,s}^*\alpha = \alpha$ for all $0 \le s, t \le 1$;*

(b) *the form $X_t\rfloor\alpha$ is closed on D_t for all $0 \le t \le 1$.*

From now on let $\mathcal{M} = \mathbf{C}^n$. Let Ω the standard volume form and ω the standard symplectic form (1). Recall that the divergence div$X$ of a holomorphic vector field X with respect to a volume form Ω is the unique holomorphic function satisfying $d(X\rfloor\Omega) = (\operatorname{div} X)\Omega$.

3.3 Definition. *(a) A holomorphic vector field X on a domain $D \subset \mathbf{C}^{2n}$ is Hamiltonian (resp. exact Hamiltonian) if the 1-form $X\rfloor\omega$ is closed (resp. exact) on D.*

(b) A holomorphic vector field X on $D \subset \mathbf{C}^n$ is divergence free (resp. exact divergence free) if the $(n-1)$-form $X\rfloor\Omega$ is closed (resp. exact) on D.

3.4 Corollary. *Let X_t $(0 \le t \le 1)$ be a time dependent holomorphic vector field on $D_t \subset \mathbf{C}^n$, and let $F_{t,s}: D_s \to D_t$ be its flow as above. The following are equivalent:*

(a) *The map $F_{t,s}$ is symplectic (resp. volume preserving) for each $0 \le s, t \le 1$;*

(b) *The vector field X_t is Hamiltonian (resp. divergence free) on D_t for each $0 \le t \le 1$.*

In the first case (X_t Hamiltonian) we assume of course that n is even. For later reference we recall some basic notions of Hamiltonian mechanics, refering the reader to [1] and [6] for details. We denote the coordinates in $\mathbf{C}^{2n}$ by (z, w), with $z, w \in \mathbf{C}^n$. Then $\omega = \sum_{j=1}^n dz_j \wedge dw_j$. A holomorphic vector field X on a domain $D \subset \mathbf{C}^{2n}$ is exact Hamiltonian if

$$X\rfloor\omega = dH \tag{6}$$

for some holomorphic function H on D. If D is a domain of holomorphy satisfying $H^1(D; \mathbf{C}) = 0$, then every Hamiltonian holomorphic vector field on D is exact Hamiltonian. The function H, which is determined up to an additive constant (when D is connected), is called the *energy function* (or simply the Hamiltonian) of X. Conversely, every $H \in \mathcal{O}(D)$ determines an exact Hamiltonian holomorphic vector field $X = X_H$ on D by (6). It is easily verified that

$$X_H = \sum_{j=1}^{n} \frac{\partial H}{\partial w_j}\frac{\partial}{\partial z_j} - \frac{\partial H}{\partial z_j}\frac{\partial}{\partial w_j} = (H_w, -H_z). \tag{7}$$

In particular, every entire functions H on $\mathbf{C}^{2n}$ can be regarded as a holomorphic Hamiltonian, giving rise to a Hamiltonian vector field X_H on $\mathbf{C}^{2n}$.

The (local) flow F_t of the field X_H remains in the level sets of H and it preserves the symplectic form ω ($F_t^*\omega = \omega$) on the set where F_t is defined. This holds both for real as well as complex values of time t.

4. Decomposition of polynomial vector fields into complete shear fields

Simple examples of holomorphic vector fields on $\mathbf{C}^n$ which are complete in complex time are the *shear vector fields* which generate the shear subgroups (3)–(5):

$$X(z) = f(\Lambda z)v, \qquad z \in \mathbf{C}^n, \tag{8}$$
$$Y(z) = g(\Lambda z)\langle z, v\rangle v, \qquad z \in \mathbf{C}^n, \tag{9}$$
$$W(z) = h(\omega(z, v))v, \qquad z \in \mathbf{C}^{2n}. \tag{10}$$

Fields of type (8) have divergence zero, and fields of type (10) are Hamiltonian.

The following decomposition result plays a central role in proof of the approximation theorems in section 2. Part (ii) is due to Andersén [4], part (i) to Andersén and Lempert [5], and part (iii) to the author [18]. (See the appendix in [17] for a short proof of (i) and (ii).)

4.1 Proposition: *Let X be a polynomial holomorphic vector field on $\mathbf{C}^n$.*

(i) X is a finite sum of shear vector fields of type (8) and (9).

(ii) If $\mathrm{div}X = 0$, then X is a finite sum of shear vector fields of type (8).

(iii) If $n = 2m$ and X is Hamiltonian, then X is a finite sum of Hamiltonian shear vector fields of type (10).

Of course there exist other types of complete holomorphic vector fields on $\mathbf{C}^n$. A particular example on $\mathbf{C}^2$ is $X(z, w) = (z^2 w, -zw^2)$, with the flow

$$F_t(z, w) = \left(ze^{tzw}, we^{-tzw}\right), \quad t \in \mathbf{C}.$$

Andersén proved [4] that for $t \neq 0$ the automorphism F_t is not a finite composition of shears. For further examples of complete holomorphic vector fields see [2,6,34] and section 10 below.

The following result is standard; see e.g. [1, p.92]:

4.2 Proposition. *Let X be a vector field of class $\mathcal{C}^1$ on a manifold M which is a finite sum $X = \sum_{j=1}^m X_j$ of $\mathcal{C}^1$ fields X_j. Denote by F_t^j the flow of X_j and by F_t the flow of X. Then*

$$F_t(x) = \lim_{N\to\infty} \left(F_{t/N}^m \circ \cdots \circ F_{t/N}^1\right)^N (x).$$

The convergence is uniform on every compact set $K \subset M$ such that $F_t(x)$ is defined for all $x \in K$.

Combining the above two propositions and observing that the flow of every shear vector field (8)–(10) on $\mathbf{C}^n$ consists of shear automorphisms (3)–(5) we get

4.3 Proposition: *Let X be a holomorphic vector field defined on all of $\mathbf{C}^n$. Let D be an open subset of $\mathbf{C}^n$ and let $t_0 > 0$. Assume that the flow $F_t(z)$ is defined for all $0 \le t \le t_0$ with arbitrary initial condition $F_0(z) = z \in D$. Then the following hold for each t, $0 \le t \le t_0$:*

(i) $F_t: D \to F_t(D) \subset \mathbf{C}^n$ is a biholomorphic map which can be approximated, uniformly on compact sets in D, by automorphisms $\Phi \in \mathcal{S}(n)$.

(ii) If $divX = 0$, then F_t can be approximated by automorphisms $\Phi \in \mathcal{S}_1(n)$.

(iii) If $n = 2m$ and X is Hamiltonian, then F_t can be approximated by symplectic automorphisms $\Phi \in \mathcal{S}_{sp}(n)$.

The same is true for time dependent holomorphic vector fields on $\mathbf{C}^n$.

Sketch of proof of Theorem 2.4: We consider the family of biholomorphic maps $F_t: D \to D_t$ $(0 \le t \le 1)$ as the flow of a time dependent holomorphic vector field X_t, defined on the domain D_t for each $t \in [0,1]$. Since D_t is Runge in $\mathbf{C}^n$ for each t, we can approximate X_t on D_t by a polynomial holomorphic vector field Y_t. Similarly, if X_t is exact divergence zero (resp. exact Hamiltonian) on D_t, we can approximate X_t on D_t by polynomial divergence zero (resp. Hamiltonian) vector fields; see [20] resp. [19]. Such approximation is always possible if D_t is pseudoconvex and $H^{n-1}(D_t; \mathbf{C}) = 0$ (resp. $H^1(D_t; \mathbf{C}) = 0$). Finally, by splitting the time interval into short subintervals and approximating Y_t on every subinterval by a time independent field it suffices to consider the case when Y is time independent. The result now follows from Proposition 4.3.

Theorem 2.1 is proved by first reducing it to the case $F(0) = 0$, $DF(0) = I$, and applying Theorem 2.4 to the isotopy $F_t(z) = F(tz)/t$ for $0 < t \le 1$, $F_0(z) = z$. It is easily verified that, since $F(D)$ is Runge in $\mathbf{C}^n$, the domain $F_t(D)$ is Runge for every t.

5. Global holomorphic equivalence of real-analytic submanifolds

The motivation for results in this section came from a result of J.-P. Rosay [31] to the effect that one can approximately straighten an arbitrary smooth arc in $\mathbf{C}^n$ $(n \ge 2)$ by automorphisms of $\mathbf{C}^n$. He observed that, for real-analytic arcs, this is a consequence of Theorem 2.1 above by Andersén and Lempert. The notion of global holomorphic equivalence was developed further in the papers by Rosay and the author [21] and in [19,20]. In this section we consider the real-analytic manifolds; the case of smooth manifolds is postponed to section 7.

5.1 Definition. *Let $\mathcal{G} \subset \mathrm{Aut}\mathbf{C}^n$ be any group of holomorphic automorphisms of $\mathbf{C}^n$. A real-analytic diffemorphism $F: M_0 \to M_1$ between compact, embedded, real-analytic submanifolds $M_0, M_1 \subset \mathbf{C}^n$ is said to be a $\mathcal{G}$-equivalence if F extends to a biholomorphic mapping in a neighborhood U of M_0 which can be approximated, uniformly on U, by automorphisms $\Phi \in \mathcal{G}$. Two real-analytic embeddings $f_0, f_1: M \hookrightarrow \mathbf{C}^n$ of a compact real-analytic manifold M into $\mathbf{C}^n$ are $\mathcal{G}$-equivalent if the diffeomorphism $F = f_1 \circ f_0^{-1}$ is a $\mathcal{G}$-equivalence.*

Remark 1: Observe that, if a sequence $\Phi_j \in \mathcal{G}$ converges to a biholomorphic map F in a neighborhood of M_0, then by the maximum principle (applied to sequences $\{\Phi_j\}$ and $\{\Phi_j^{-1}\}$) the same sequence converges to a biholomorphic map in a neighborhood of the polynomial hull $\hat{M}_0$. The extended map takes the hull $\hat{M}_0$ biholomorphically onto $\hat{M}_1$. In particular, M_0 is polynomially convex if and only if M_1 is.

Remark 2: If $\mathcal{G}' \subset \mathcal{G}$ are automorphism groups such that $\mathcal{G}'$ is dense in $\mathcal{G}$, then every $\mathcal{G}$-equivalence is at the same time a $\mathcal{G}'$-equivalence.

The following basic result for the case $\mathcal{G} = \mathrm{Aut}\mathbf{C}^n$ is due to Rosay and the author [21, Theorem 3.1]:

5.2 Theorem. *Let $M_0, M_1 \subset \mathbf{C}^n$ $(n \geq 2)$ be compact, embedded real-analytic submanifolds (with or without boundary) which are totally real and polynomially convex. A real-analytic diffeomorphism $F: M_0 \to M_1$ is an $\mathrm{Aut}\mathbf{C}^n$-equivalence if and only if there exists a one parameter family of diffeomorphisms $F_t: M_0 \to M_t \subset \mathbf{C}^n$ $(0 \leq t \leq 1)$, of class $\mathcal{C}^2$ in $(t, z) \in [0,1] \times M_0$, such that F_0 is the identity on M_0, $F_1 = F$, and $M_t = F_t(M_0)$ is totally real and polynomially convex for every $t \in [0,1]$.*

A short proof can be found in [17]. Here is the main idea of the proof. By approximation we may assume that $\{F_t\}$ is real-analytic also in t. Its infinitesimal generator X_t is a vector field of type $(1,0)$ on $\mathbf{C}^n$, defined and real-analytic along M_t. Since M_t is totally real and polynomially convex, X_t can be extended to a holomorphic vector field in a neighborhood of M_t which is a uniform limit of polynomial holomorphic vector fields. The result now follows from Proposition 4.3.

Combining Theorem 5.2 with a result on genericity of polynomial convexity of low dimensional totally real submanifolds [21,17] we obtain ([17], Corollary 1):

5.3 Corollary. *Let M be a compact real-analytic manifold and let $f_0, f_1: M \to \mathbf{C}^n$ be real-analytic, totally real, polynomially convex embeddings. If $\dim M \leq 2n/3$, then f_0 and f_1 are $\mathrm{Aut}\mathbf{C}^n$-equivalent.*

Explicit examples to which Corollary 5.3 applies are curves in $\mathbf{C}^n$ for $n \geq 2$ and surfaces in $\mathbf{C}^n$ for $n \geq 3$. In this case we have the following more precise results (see [21], Theorem 4.2 and Corollary 6.3):

5.4 Theorem. *Let T be the circle. If $f_0, f_1: T \to \mathbf{C}^n$ $(n \geq 2)$ are real-analytic embeddings such that both curves $f_0(T)$ and $f_1(T)$ are polynomially convex, then f_0 and f_1 are $\mathrm{Aut}\mathbf{C}^n$-equivalent. Moreover, given an embedded, real-analytic, polynomially convex curve $\Gamma \subset \mathbf{C}^n$, a biholomorphic map $F: D \to \mathbf{C}^n$ defined in a neighborhood of Γ can be* *approximated by automorphisms of $\mathbf{C}^n$ in some smaller neighborhood of Γ if and only if $F(\Gamma)$ is polynomially convex and the winding number of the Jacobian $J(F)$ along Γ equals zero.*

For a stronger result on equivalence of curves see Theorem 6.1 below.

5.5 Theorem. *Let $M \subset \mathbf{C}^n$ $(n \geq 3)$ be a compact, embedded, real-analytic surface that is totally real and polynomially convex. A biholomorphic mapping $F: U \to \mathbf{C}^n$, defined in a neighborhood of M, can be approximated by automorphisms of $\mathbf{C}^n$ near M if and only if $F(M)$ is polynomially convex and the Jacobian $J(F): M \to \mathbf{C}^*$ is homotopic to a constant on M. In particular, if M is a two dimensional sphere, then F can be approximated by automorphisms of $\mathbf{C}^n$ near M if and only if $F(M)$ is polynomially convex.*

The following result [21, Corollary 4.1] follows directly from Theorem 2.1.

5.6 Corollary. *All manifolds are assumed to be embedded and real-analytic.*

(a) Any two totally real, polynomially convex, k-dimensional discs in $\mathbf{C}^n$ $(k \leq n)$ are Aut$\mathbf{C}^n$*-equivalent.*

(b) Any two arcs in $\mathbf{C}^n$ are Aut$\mathbf{C}^n$*-equivalent.*

(c) Any two embedded analytic discs in $\mathbf{C}^n$ are Aut$\mathbf{C}^n$*-equivalent.*

Moreover, if $M \subset \mathbf{C}^n$ is as in (b) or (c), then every biholomorphic map $F: D \to \mathbf{C}^n$ in a neighborhood D of M in $\mathbf{C}^n$ can be approximated in a smaller neighborhood of M by automorphisms of $\mathbf{C}^n$. If M is a k-disc as in (a), then F can be so approximated near M if and only if the image $F(M)$ is polynomially convex in $\mathbf{C}^n$.

6. Global symplectic and volume preserving equivalence

In this section we recall the results of [19] and [20] on $\mathcal{G}$-equivalence in the cases when $\mathcal{G}$ is either $\mathrm{Aut}_{\mathrm{sp}}\mathbf{C}^{2n}$ or $\mathrm{Aut}_1\mathbf{C}^n$.

Choose a $(1,0)$-form θ on $\mathbf{C}^{2n}$ such that $d\theta = -\omega$; to be specific we will take

$$\theta = \sum_{j=1}^{n} z_{n+j} dz_j.$$

Denote by T the circle. For a smooth map $g: T \to \mathbf{C}^{2n}$ we define the *action integral*

$$\mathcal{A}(g) = -\int_T g^*\theta.$$

The following is the main result of [19].

6.1 Theorem. *Let $g_0, g_1: T \to \mathbf{C}^{2n}$ be two real-analytic embeddings of the circle into $\mathbf{C}^{2n}$. If g_0 and g_1 are* $\mathrm{Aut}_{\mathrm{sp}}\mathbf{C}^{2n}$*-equivalent, then $\mathcal{A}(g_0) = \mathcal{A}(g_1)$. Conversely we have*

(a) if $\mathcal{A}(g_0) = \mathcal{A}(g_1) \neq 0$, then g_0 and g_1 are $\mathrm{Aut}_{\mathrm{sp}}\mathbf{C}^{2n}$*-equivalent;*

(b) if $\mathcal{A}(g_0) = \mathcal{A}(g_1) = 0$, then g_0 and g_1 are $\mathrm{Aut}_{\mathrm{sp}}\mathbf{C}^{2n}$*-equivalent provided that the curves $g_0(T), g_1(T) \subset \mathbf{C}^{2n}$ are both polynomially convex;*

(c) *every two smooth, embedded, real-analytic arcs* $g_0, g_1: [0,1] \to \mathbf{C}^{2n}$ *are* $\mathrm{Aut_{sp}}\mathbf{C}^{2n}$*-equivalent.*

Example 1: The following example shows that in part (b) of Theorem 6.1 the conclusion may fail if the curves are not polynomially convex. Let $g_0, g_1: T \to \mathbf{C}^2$ be the embeddings

$$g_0(\theta) = \left(e^{i\theta}, 0\right),$$
$$g_1(\theta) = \left(e^{2i\theta}, e^{3i\theta}\right).$$

The curve $C_0 = g_0(T)$ bounds the smooth analytic disc $\{(z,0): |z| \leq 1\}$, while $C_1 = g_1(T)$ bounds the analytic curve $\{(z,w): z^3 = w^2,\ |z| \leq 0\}$ with a singularity at the origin. Thus $\mathcal{A}(g_0) = 0 = \mathcal{A}(g_1)$ although the curves C_0 and C_1 are not $\mathrm{Aut}\mathbf{C}^2$-equivalent (see remark 1 following Def. 5.1).

Example 2: The map $F_t(z,w) = (z, w + t/z)$ is a volume preserving automorphism of $\mathbf{C}^* \times \mathbf{C}$ for all $t \in \mathbf{C}$. The circle $T = \{(z,\bar{z}) \in \mathbf{C}^2: |z| = 1\}$ is polynomially convex, hence it has pseudoconvex tubular neighborhoods Ω which are Runge in $\mathbf{C}^2$. The same is true for circles $F_t(T)$ for $t \neq -1$. Theorem 5.4 implies that for such t, F_t is a limit of automorphisms of $\mathbf{C}^2$ in a neighborhood of T. However, F_t for $t \neq 0$ is not the limit of symplectic automorphisms of $\mathbf{C}^2$ in any neighborhood of T. This is because the action integral

$$\mathcal{A}(F_t) = \int_{F_t(T)} w dz = 2\pi i(1+t)$$

depends on t, while Stokes' theorem shows that this integral is preserved by symplectic holomorphic automorphism of $\mathbf{C}^2$ (and therefore by their limits).

Sketch of proof of Theorem 6.1: We first construct a real-analytic isotopy of embeddings $g_t: T \to \mathbf{C}^{2n}$ such that g_0 and g_1 are the given maps, the curve $C_t = g_t(T)$ is polynomially convex for every t, and the action integral $\mathcal{A}(g_t)$ is independent of t. We then show that the flow $g_t \circ g_0^{-1}: C_0 \to C_t$ can be extended to a flow of a (time dependent) exact Hamiltonian vector field X_t, defined in a neighborhood of C_t for each $t \in [0,1]$. This extension is done in complete analogy to the real case. Let H_t be a holomorphic Hamiltonian (the energy function) of X_t, defined in a tube around C_t. Since C_t is polynomially convex, we can approximate H_t near C_t by holomorphic polynomials. The corresponding polynomial Hamiltonian vector field Y_t then approximates X_t near C_t. The flow of Y_t, which by construction approximates the flow of X_t and hence the original flow $g_t \circ g_0^{-1}: C_0 \to C_t$, is itself approximable (at each time t) by compositions of symplectic automorphisms of $\mathbf{C}^{2n}$ according to Proposition 4.3 (iii).

We now consider the problem of $\mathrm{Aut_1}\mathbf{C}^n$-equivalence of real-analytic submanifolds in $\mathbf{C}^n$. Let β be the $(n-1,0)$-form

$$\beta(z) = \frac{1}{n}\sum_{j=1}^{n}(-1)^{j-1} z_j dz_1 \wedge \cdots \widehat{dz_j} \cdots \wedge dz_n,$$

where the hat indicates that the corresponding entry is deleted. Then $d\beta = \Omega$ is the complex volume form on $\mathbf{C}^n$. If M is a smooth, compact, oriented manifold of real dimension

$n-1$ and $f: M \to \mathbf{C}^n$ is a smooth map, we set

$$\mathcal{B}(f) = \int_M f^* \beta.$$

If M is closed, Stokes's theorem implies that *any two embeddings* $f_0, f_1: M \to \mathbf{C}^n$ *which are* $\mathrm{Aut}_1 \mathbf{C}^n$*-equivalent satisfy* $\mathcal{B}(f_0) = \mathcal{B}(f_1)$. The following results were proved in [20].

6.2 Theorem. *Let M be a compact, connected, real-analytic manifold of real dimension m, and let $f_0, f_1: M \to \mathbf{C}^n$ be real-analytic embeddings ($1 \leq m \leq n-1$) such that the submanifolds $f_j(M) = M_j \subset \mathbf{C}^n$ ($j = 0, 1$) are totally real and polynomially convex. Suppose that f_0 and f_1 are $\mathrm{Aut}\mathbf{C}^n$-equivalent. Then f_0 and f_1 are also $\mathrm{Aut}_1\mathbf{C}^n$-equivalent provided that any one of the following conditions holds:*

(i) $m \leq n-2$;

(ii) $m = n-1$ *and* $\mathrm{H}^{n-1}(M; \mathbf{R}) = 0$;

(iii) $m = n-1$, *the manifold M is closed and orientable, and* $\mathcal{B}(f_0) = \mathcal{B}(f_1) \neq 0$.

Remark: We believe that the conclusion in part (iii) holds also when $\mathcal{B}(f_0) = \mathcal{B}(f_1) = 0$.

The proof of Theorem 6.2 is analogous to that of Theorem 6.1, except that we replace Hamiltonian flows by flows of exact divergence zero vector fields. Combining Theorem 6.2 with Corollary 5.3 we get

6.3 Corollary. *Let M be a compact, connected, real-analytic manifold of dimension $m \geq 1$. If $n \geq \max\{m+2, 3m/2\}$, then every two real-analytic embeddings $f_0, f_1: M \to \mathbf{C}^n$ whose images $f_0(M)$ and $f_1(M)$ are totally real and polynomially convex are $\mathrm{Aut}_1\mathbf{C}^n$-equivalent.*

Recall that T is the circle.

6.4 Corollary. *(a) Two real-analytic embeddings $f_0, f_1: T \to \mathbf{C}^2$ with polynomially convex images $f_0(T), f_1(T)$ are $\mathrm{Aut}_1\mathbf{C}^2$-equivalent if and only if $\mathcal{B}(f_0) = \mathcal{B}(f_1)$.*

(b) Let M be a closed orientable surface, and let $f_0, f_1: M \to \mathbf{C}^3$ be real-analytic embeddings whose images $f_0(M), f_1(M) \subset \mathbf{C}^3$ are totally real and polynomially convex. If $\mathcal{B}(f_0) = \mathcal{B}(f_1) \neq 0$, then f_0 and f_1 are $\mathrm{Aut}_1\mathbf{C}^3$-equivalent.

Example [20]: Let S be the $(n-1)$-dimensional sphere, embedded as a hypersurface in $\mathbf{R}^n \subset \mathbf{C}^n$. It is easy to find a closed, but non-exact holomorphic $(n-1)$-form α in a tubular neighborhood U of S such that $\int_S \alpha \neq 0$. Let X be the divergence zero holomorphic vector field in U defined by the formula $\alpha = X \rfloor \Omega$, where Ω is the complex volume form on $\mathbf{C}^n$. Its flow $\{F_t\}$ is a family of volume preserving biholomorphic mappings near S such that F_t for each sufficiently small t can be approximated by automorphisms of $\mathbf{C}^n$ near S (Theorem 5.2), but not by volume preserving automorphisms of $\mathbf{C}^n$ when $t \neq 0$ is small. The reason is that

$$\frac{d}{dt}\mathcal{B}(F_t)|_{t=0} = \int_S \alpha \neq 0$$

(see [20]) and hence the integral $\mathcal{B}(F_t)$ depends on t.

7. Approximation by automorphisms on smooth submanifolds of $\mathbf{C}^n$

In this section we consider the question of approximation of smooth mappings $F: M \to \mathbf{C}^n$ on smooth submanifolds $M \subset \mathbf{C}^n$ by restrictions to M of holomorphic automorphisms of $\mathbf{C}^n$.

Let M be a manifold of class $\mathcal{C}^p$, $p \geq 1$. A $\mathcal{C}^p$ *isotopy of embeddings* of M into $\mathbf{C}^n$ is a map $F: [0,1] \times M \to \mathbf{C}^n$ of class $\mathcal{C}^p$ such that for each fixed $t \in [0,1]$, $F_t = F(t, \cdot): M \to \mathbf{C}^n$ is an embedding. The isotopy $\{F_t\}$ is totally real (resp. polynomially convex) if the submanifold $F_t(M) \subset \mathbf{C}^n$ is totally real (resp. polynomially convex) for every $t \in [0,1]$.

The following result was proved in [17].

7.1 Theorem. *If $M \subset \mathbf{C}^n$ $(n \geq 2)$ is a compact, totally real, polynomially convex submanifold of class $\mathcal{C}^p$ $(2 \leq p < \infty)$ and $F: M \to \mathbf{C}^n$ is a $\mathcal{C}^p$ mapping, then the following are equivalent:*

(i) For each $\epsilon > 0$ there exists a $\Phi \in \mathrm{Aut}\mathbf{C}^n$ such that $||F - \Phi|_M||_{\mathcal{C}^p(M)} < \epsilon$.

(ii) For each $\epsilon > 0$ there exists a totally real, polynomially convex isotopy $F_t: M \to \mathbf{C}^n$ $(t \in [0,1])$ of class $\mathcal{C}^p$ such that F_0 is the identity on M and $||F_1 - F||_{\mathcal{C}^p(M)} < \epsilon$.

Remark: If $M \subset \mathbf{C}^n$ is a compact, totally real, polynomially convex submanifold of class $\mathcal{C}^p$ $(p \geq 1)$, then the set of restrictions to M of holomorphic polynomials on $\mathbf{C}^n$ is dense in the $\mathcal{C}^p(M)$. Thus Theorem 7.1 can be viewed as an analogue of this result for mappings $M \mapsto \mathbf{C}^n$.

The implication (i)$\Rightarrow$(ii) in Theorem 7.1 is a trivial consequence of connectedness of $\mathrm{Aut}\mathbf{C}^n$. The main implication (ii)$\Rightarrow$(i) is proved in a similar way as Theorem 5.2 by constructing an isotopy of biholomorphic mappings in a neighborhood of M which approximates the given isotopy F_t and then applying Theorem 2.4.

Recall from [17] that for compact manifolds M of class $\mathcal{C}^p$ $(p \geq 2)$ of real dimension at most $2n/3$, the set of $\mathcal{C}^p$ embeddings $M \hookrightarrow \mathbf{C}^n$ which are totally real and polynomially convex is open and dense in the space $\mathcal{C}^p(M)^n$ of all $\mathcal{C}^p$ mappings $M \mapsto \mathbf{C}^n$ (in the $\mathcal{C}^p$ topology). This implies

7.2 Corollary. *Let $M \subset \mathbf{C}^n$ be a compact, totally real, polynomially convex submanifold of class $\mathcal{C}^p$, $2 \leq p < \infty$, and of dimension at most $2n/3$. Then the set*

$$\mathrm{Aut}\mathbf{C}^n|_M = \{F|_M : F \in \mathrm{Aut}\mathbf{C}^n\}$$

of restrictions to M of holomorphic automorphisms of $\mathbf{C}^n$ is dense in $\mathcal{C}^p(M)^n$.

Consider now a diffeomorphism $F: M_0 \to M_1$ of class $\mathcal{C}^p$ $(1 \leq p \leq \infty)$ between submanifolds $M_0, M_1 \subset \mathbf{C}^n$. One may ask whether there exists a sequence $\Phi_j \in \mathrm{Aut}\mathbf{C}^n$ such that Φ_j converges to F on M_0 (in the $\mathcal{C}^p$ topology) and, at the same time, the sequence of inverses Φ_j^{-1} converges to the inverse $F^{-1}: M_1 \to M_0$ on M_1. A result of this type was first proved by J.-P. Rosay [31] for $\mathcal{C}^\infty$ arcs in $\mathbf{C}^n$. Unlike in the real-analytic case (section 5), one can not expect the convergence of the approximating sequence Φ_j in any neighborhood of M_0.

The following result was proved by E. Løw and the author [22].

7.3 Theorem: *If $F: M_0 \to M_1$ is a smooth ($\mathcal{C}^\infty$) diffeomorphism between smooth, compact submanifolds $M_0, M_1 \subset \mathbf{C}^n$ which are totally real and polynomially convex, then the following are equivalent:*

(i) There exists a totally real, polynomially convex isotopy of embeddings $F_t: M_0 \to \mathbf{C}^n$ such that F_0 is the identity on M_0 and $F_1 = F$;

(ii) there exists a sequence $\Phi_j \in \mathrm{Aut}\mathbf{C}^n$ ($j \in \mathbf{Z}_+$) such that $\lim_{j\to\infty} \Phi_j|_{M_0} = F$ and $\lim_{j\to\infty} \Phi_j^{-1}|_{M_1} = F^{-1}$ in the $\mathcal{C}^\infty$ topology on the respective manifolds.

We expect to obtain a sharp version of the last result for $\mathcal{C}^p$ diffeomorphisms between $\mathcal{C}^p$ submanifolds for $3 \le p < \infty$. We believe that in this case the approximating sequence Φ_j can be chosen so that the convergence in Theorem 7.3 takes place in the $\mathcal{C}^p$ topology on M_0 resp. M_1.

In the proof of Theorem 7.3 we must solve a certain $\overline{\partial}$-problem in small tubular neighborhoods around totally real submanifolds in $\mathbf{C}^n$. One possible approach is to use the L^2-methods as in Hörmander and Wermer [24] (see also [37]). This method has an inherent loss of derivatives; it can be used to prove Theorem 7.3, but it does not give the sharp result in case of finite smoothness. In that case we intend to use a result of N. Øvrelid (in preparation) on solution of the $\overline{\partial}$-problem in tubes around totally real manifolds by means of an integral kernel.

8. Complex orbits of holomorphic vector fields

Let X be a holomorphic vector field on a complex manifold $\mathcal{M}$. Recall that its local flow $F_t(z)$ ($t \in \mathbf{R}$) is the solution at time t of the differential equation

$$\dot{Z} = X(Z), \qquad Z(0) = z \in \mathcal{M}. \tag{11}$$

There exists a maximal open interval $J(z) = \big(-\beta(z), \alpha(z)\big) \subset \mathbf{R}$, containing the origin, such that $F_t(z)$ is defined for $t \in J(z)$. The functions α and $-\beta$ with values in $(0, \infty]$ are lower semicontinuous on $\mathcal{M}$. If $\alpha(z) < \infty$ then $F_t(z)$ leaves every compact set in $\mathcal{M}$ as $t \in J$ approaches $\alpha(z)$, and similarly for $-\beta(z)$. The set $\{F_t(z): t \in J(z)\}$ is called the *real orbit* of X through z.

We recall the notion of *complex orbits* of a holomorphic vector field; see [18] for details. Locally near $t = 0$ one can solve the equation (11) for complex values of t. This gives a holomorphic mapping $Z(t)$, defined in a neighborhood of 0 in the complex plane, with values in the manifold $\mathcal{M}$. By analytic continuation we can extend the local solution to a maximal global solution $Z: R_z \to \mathcal{M}$, which is defined and holomorphic on a connected Riemann domain R_z spread over $\mathbf{C}$ and which can not be analytically continued to any larger Riemann domain over $\mathbf{C}$. Its image $C_z = Z(R_z) \subset \mathcal{M}$ is called the complex orbit of X through z.

We say that the complex orbit C_z of a point $z \in \mathcal{M}$ is *complete in real time* if $J(z') = (-\infty, +\infty)$ for every point $z' \in C_z$, i.e., the equation (11) can be solved for all real values of time when starting at any point in C_z. The orbit C_z is said to be nontrivial if $C_z \neq \{z\}$; this is true if and only if $X(z) \neq 0$.

A proof of the following simple result can be found in [18] (Proposition 3.2). It depends on the observation that for every complex orbit $C_z \subset \mathcal{M}$ of X which is complete in real time, the Riemann surface R_z is a strip in $\mathbf{C}$ of the form

$$R_z = \{t + is \in \mathbf{C}: -b(z) < s < a(z)\} \subset \mathbf{C},$$

with $a(z), b(z) \in (0, \infty]$.

8.1 Proposition. *Every nontrivial complex orbit of a holomorphic vector field which is complete in real time is isomorphic to one of the following Riemann surfaces:*

(a) *the complex line* $\mathbf{C}$;

(b) *the punctured complex line* $\mathbf{C}^* = \mathbf{C}\backslash\{0\}$;

(c) *a torus;*

(d) *the disc* $U = \{z \in \mathbf{C}: |z| < 1\}$;

(e) *the punctured disc* $U^* = U\backslash\{0\}$;

(f) *an annulus* $A(r) = \{z \in \mathbf{C}: 1 < |z| < r\}$.

Remark: If the manifold $\mathcal{M}$ is Stein then there are no toral orbits. If $\mathcal{M}$ is hyperbolic, then all nontrivial orbits are of types (d)–(e) since the surfaces (a)–(c) are nonhyperbolic. Notice that the fundamental group of C_z has at most one generator unless C_z is a torus.

Suppose now that a holomorphic vector field X on a complex manifold $\mathcal{M}$ is $\mathbf{R}$-complete. Let

$$\tilde{\mathcal{M}} = \{(\zeta, z): z \in \mathcal{M},\ -b(z) < \Im\zeta < a(z)\} \subset \mathbf{C} \times \mathcal{M},$$

where $a(z)$ and $b(z)$ are as above. We call $\tilde{\mathcal{M}}$ the *fundamental domain* of the complex flow of X. The functions a and b are lower semicontinuous on $\mathcal{M}$. The following was proved in [18, Proposition 2.1]:

8.2 Proposition. *If the manifold* $\mathcal{M}$ *is Stein, then the functions* $-a$, $-b$ *are (negative) plurisubharmonic on* $\mathcal{M}$, *and the fundamental domain* $\tilde{\mathcal{M}} \subset \mathbf{C} \times \mathcal{M}$ *is pseudoconvex (hence a Stein manifold).*

8.3 Corollary. *If* $\mathcal{M}$ *is a Stein manifold such that every negative plurisubharmonic function on* $\mathcal{M}$ *is constant, then every* $\mathbf{R}$*-complete holomorphic vector field on* $\mathcal{M}$ *is* $\mathbf{C}$*-complete. This holds in particular if* $\mathcal{M} = \mathbf{C}^n$ *or if* $\mathcal{M} = \mathbf{C}^n\backslash A$ *for some complex hypersurface* $A \subset \mathbf{C}^n$.

The reason is that the plurisubharmonic functions $-a$ and $-b$ defining $\tilde{\mathcal{M}}$ must be constant, and therefore the flow extends to all complex values of time by the group property of flows.

The following result from [18] (Theorem 3.3) shows that holomorphic vector fields on Stein manifolds which are complete in real time have a 'generic type' of complex orbits.

8.4 Theorem. *Let $\mathcal{M}$ be a connected Stein manifold and X a holomorphic vector field on $\mathcal{M}$ which is complete in real time (Def. 3.1). Then there exists a pluripolar set $E \subset \mathcal{M}$, invariant under the flow $\{F_t : t \in \mathbf{R}\}$ of X and containing the zero set of X, such that for every $z \in \mathcal{M} \backslash E$ the complex orbit C_z of X through z is of the same type (a), (b), or (d)–(e). If this generic orbit type is U^* or an annulus, then the flow F_t has a period $\lambda > 0$ and it factors through an action of the circle group $(S, +)$ on $\mathcal{M}$. The action $\{F_t : t \in \mathbf{R}\}$ extends to an action of $\mathbf{C}$ on $\mathcal{M}$ (i.e., X is complete in complex time) if and only if the generic complex orbit is either $\mathbf{C}$ or $\mathbf{C}^*$. This is always the case when $\mathcal{M} = \mathbf{C}^n$.*

For actions of $\mathbf{C}$ on Stein spaces the existence of a generic orbit type ($\mathbf{C}$ or $\mathbf{C}^*$) was proved by Suzuki [34, Proposition 2]. His proof shows that when the generic orbit type is $\mathbf{C}^*$, then C_z is isomorphic to $\mathbf{C}^*$ for every z outside a closed analytic subset of $\mathcal{M}$. For related results see Richardson [30].

Another result which seems important in the study of dynamical properties of complete holomorphic vector fields is that those complex orbits which are not simply connected tend to have at most one limit point, amd this point is a fixed point of the flow. The precise result is as follows ([18, Theorem 4.1], [34]):

8.5 Theorem. *Let X be an $\mathbf{R}$-complete holomorphic vector field on a Stein manifold $\mathcal{M}$, and let C be a nontrivial complex orbit of X. If C is isomorphic to an annulus, then C is closed in $\mathcal{M}$. If C is isomorphic to the punctured disc U^*, or if C is isomorphic to the punctured plane $\mathbf{C}^*$ and X is $\mathbf{C}$-complete, then the limit set of C consists of at most one point which is a critical point of X. In particular, if $\mathcal{M} = \mathbf{C}^n$, then every orbit of type $\mathbf{C}^*$ has at most one limit point.*

The proof in [18] (or in [34]) shows that, if C is a complex orbit of type $\mathbf{C}^*$ and if p is a limit point of C, then $C \cup \{p\}$ is a smooth complex manifold at p. If we had two such limit points p, q, then $C \cup \{p, q\}$ would be a holomorphically embedded Riemann sphere in $\mathcal{M}$ which is impossible since $\mathcal{M}$ is Stein. A similar result holds for orbits of type U^*.

8.6 Corollary. *Let X be a complete holomorphic vector field on $\mathbf{C}^n$ $(n > 1)$ with a hyperbolic critical point p such that the unstable manifold $W^u(p)$ has complex dimension one. Then $W^u(p)$ is closed in $\mathbf{C}^n$. In particular, $W^u(p)$ does not intersect the stable manifold $W^s(q)$ of any other critical point of X.*

Proof: Since $W^u(p)$ is a complex manifold of complex dimension one (an injectively immersed $\mathbf{C}$), the set $C = W^u(p) \backslash \{p\}$ is a complex orbit of X which is necessarily of type $\mathbf{C}^*$. Since p is a limit point of C, C can not have any other limit point according to Theorem 8.5. If $W^u(p) \cap W^s(q) \neq \emptyset$, then q is a limit point of C, a contradiction.

Corollary 8.6 can be used to construct holomorphic vector fields on $\mathbf{C}^n$ which can not be approximated by complete fields. The following example is due to Buzzard and Fornæss (private communication, November 1994).

Example (Buzzard-Fornaess): Let $X(z, w) = \big(z(z-1), -w\big)$. Then $p = (1, 0)$ is a hyperbolic critical point of X whose unstable manifold $W^u(p)$ contains the segment $I = (0, 1) \times \{0\}$, and $q = (0, 0)$ is an attracting critical point which contains I in its basin of attraction. Thus $W^u(p) \cap W^s(q) \neq \emptyset$. Since this behavior is stable under small perturbations of the field, X can not be approximated by complete holomorphic fields. In fact, the closure of the set of complete holomorphic vector fields on $\mathbf{C}^n$ is nowhere dense in the set of all holomorphic vector fields (in the topology of uniform convergence on compacts).

9. Flows of holomorphic Hamiltonian vector fields

Proposition 8.1 is useful in determining whether a given holomorphic vector field is complete or not by examining its complex orbits. We shall illustrate this by looking at Hamiltonian vector fields on the plane $\mathbf{C}^2$.

Suppose that X is a Hamiltonian holomorphic vector field on $\mathbf{C}^2$ with the energy function H (see section 3). Let $\Sigma = \{p \in \mathbf{C}^2 \colon X(p) = 0\}$. It is easily seen that for each point $p = (z_0, w_0) \in \mathbf{C}^2 \backslash \Sigma$ the complex orbit C_p through p is the connected component of the set $\{(z,w) \in \mathbf{C}^2 \backslash \Sigma \colon H(z,w) = H(z_0, w_0)\}$. Thus, to show that the vector field X is not complete, it suffices to find such a component which is not isomorphic to any of the surfaces listed in Proposition 8.1.

We first consider polynomial Hamiltonians H on $\mathbf{C}^2$. In that case every level set $\{H = c\}$ is an affine algebraic curve in $\mathbf{C}^2$ which closes up to a projective algebraic curve in $\mathbf{CP}^2$. Hence every orbit of the Hamiltonian vector field X_H is obtained by removing at most a finite number of points from a compact algebraic curve. The only Riemann surfaces listed in Proposition 8.1 above which have this property are $\mathbf{C}$ and $\mathbf{C}^*$. Together with Theorem 8.2 this proves the following [18, Lemma 7.1].

9.1 Lemma. *If X is a polynomial Hamiltonian vector field on $\mathbf{C}^2$ and if C is a nontrivial complex orbit of X which is $\mathbf{R}$-complete, then the closure $\overline{C}$ in $\mathbf{CP}^2$ is an algebraic curve of genus 0 ($\overline{C}$ is normalized by the Riemann sphere), and C is also $\mathbf{C}$-complete.*

By analyzing the holomorphic type of generic level sets of H, using the Riemann-Hurwitz formula, one gets the following [18, Proposition 7.2].

9.2 Proposition. *Let $H(z,w) = \sum_{j=1}^{d} H^j(z,w)$ be a polynomial of degree $d \geq 3$ on $\mathbf{C}^2$, with H^j its homogeneous part of degree j. Suppose that $(0,0)$ is the only common zero of the following four polynomials: H^d, H^{d-1}, $\partial H^d/\partial z$, $\partial H^d/\partial w$. Then the Hamiltonian vector field $X_H = (H_w, -H_z)$ on $\mathbf{C}^2$ is not complete. In fact, every regular level set of H contains a point p such that the flow $F_t(p)$ of X_H is not defined for all real t.*

To motivate the next result we recall the well known result that, if $Q(x) \geq 0$ is a nonnegative smooth real function on $\mathbf{R}$, the Hamiltonian vector field $X(x,y) = (y, -Q'(x))$ with the energy function $H(x,y) = y^2/2 + Q(x)$ is complete on $\mathbf{R}^2$. In contrast to this we have [18, Proposition 7.3]:

9.3 Proposition. *If f is an entire function on $\mathbf{C}$ which is not affine linear, then the vector field $X(z,w) = \big(w, f(z)\big)$ on $\mathbf{C}^2$ is not complete.*

Notice that X is a Hamiltonian vector field with the energy function $H(z,w) = w^2/2 + Q(z)$, where $Q'(z) = -f(z)$. The proof in [18] uses elementary Morse theory and it shows that every regular level set of H contains a point p such that the flow $F_t(p)$ of X is not defined for all real t.

9.4 Corollary. *If f is an entire function on $\mathbf{C}$ which is not affine linear, there exists a point $(z_0, \dot{z}_0) \in \mathbf{C}^2$ such that the second order ordinary differential equation*

$$\ddot{Z} = f(Z), \qquad Z(0) = z_0, \ \dot{Z}(0) = \dot{z}_0$$

can not be integrated for all real $t \in \mathbf{R}$.

The Corollary follows from Proposition 9.3 by introducing the variable $w = \dot{z}$ and changing this second order equation to the Hamiltonian system $\dot{z} = w$, $\dot{w} = f(z)$.

In theory of Hamiltonian mechanics one basic question is whether most orbits go to infinity. This question was studied in the holomorphic case by Fornæss and Sibony [14,15] who showed that for most Hamiltonian holomorphic vector fields on $\mathbf{C}^{2n}$, most orbits go to infinity.

In order to formulate their results precisely we first recall from [14] the notion of *singular orbit* of a holomorphic vector field X on $\mathbf{C}^n$. At this point X need not be Hamiltonian. Let $\{F_t : t \in \mathbf{R}\}$ be the local flow of X.

9.5 Definition. *The orbit $F_t(z)$ of a point $z \in \mathbf{C}^n$ is said to be singular if there is a sequence $0 < t_1 < t_2 < t_3 < \dots$ such that $\lim_{j\to\infty} |F_{t_j}(z)| = \infty$. The set of points $z \in \mathbf{C}^n$ with nonsingular orbits is denoted by K_X.*

In other words, the orbit is singular if the set $\{F_t(z) : 0 < t < T\}$ is unbounded, where $T \in (0, \infty]$ is the largest number such that the flow $F_t(z)$ is defined on $[0, T)$. If $T < \infty$, then by general properties of flows the orbit of z is necessarily singular since $|F_t(z)| \to \infty$ when $t \to T$. Thus the points $z \in \mathbf{C}^n$ with nonsingular orbits are those for which the flow $F_t(z)$ is defined and bounded for all $t \geq 0$. The set K_X is always of type F_σ, i.e., a countable union of closed sets.

We consider each entire function $H \in \mathcal{O}(\mathbf{C}^{2n})$ as a holomorphic Hamiltonian, giving rise to a holomorphic Hamiltonian vector field X_H (7). Denote the set of nonsingular orbits of X_H by K_H (Def. 9.5). The main result of Fornæss and Sibony [15, Theorem 3.4] is

9.6 Theorem. *There exists a dense G_δ set $\mathcal{E} \subset \mathcal{O}(\mathbf{C}^{2n})$ of holomorphic Hamiltonians such that for each $H \in \mathcal{E}$ the set K_H has empty interior.*

Fornæss and Sibony also obtained a number of interesting results for the *Fatou set* $U_X \subset K_X$ of holomorphic Hamiltonian vector fields on $\mathbf{C}^{2n}$.

9.7 Definition. *Let X be a holomorphic vector field on $\mathbf{C}^n$. For each constant $0 < c < \infty$ we denote by $U_X(c)$ the set of all points $z \in \mathbf{C}^n$ for which there is a neighborhood V of z such that*

$$\sup_{t \geq 0} |F_t(w)| < c \qquad \text{for all } w \in V.$$

The set $U_X = \bigcup_{0<c<\infty} U_X(c)$ is called the Fatou set of X. Any connected component of U_X is called a Fatou component.

Notice that U_X is contained in $\operatorname{int} K_X$, and it is the largest open set on which the flow F_t of X exists for all time $0 \leq t < \infty$ and the family $\{F_t : t \geq 0\}$ is locally bounded. By a category argument U_X is an open dense subset of the interior of K_X. It is shown in [15, Proposition 3.3] that every Fatou component is a Runge domain in $\mathbf{C}^n$.

The following result is proved in [15, Theorem 3.5].

9.8 Theorem. *Let X be a Hamiltonian holomorphic vector field on $\mathbf{C}^{2n}$ with flow F_t and Fatou set U_X. Let $W(c)$ $(0 < c < \infty)$ be a connected component of the set $U_X(c)$ (Def. 9.7). Then F_t is an automorphism of $W(c)$ for each $t \in \mathbf{R}$. The closure G of $\{F_t|_{W(c)}: t \in \mathbf{R}\}$ is a Lie subgroup of $\mathrm{Aut}W(c)$, isomorphic to a torus T^k for some $k \le n$. The vector field X is conjugate on $W(c)$ to a field $Y = (i\theta_1 z_1, \ldots, i\theta_n z_n, 0, \ldots, 0)$.*

A related question is how many orbits of a holomorphic vector field X on $\mathbf{C}^n$ go to infinity in finite time. Following [13] we say that the (real) orbit of a point $z \in \mathbf{C}^n$ *explodes* if the integral curve $\{F_t(z): t \ge 0\}$ through z is unbounded on some finite time interval $[0, t_0) \subset \mathbf{R}_+$. The following result was proved by Fornæss and Grellier [13].

9.10 Theorem. *There is a dense set $\mathcal{E}$ of entire functions on $\mathbf{C}^2$ such that for every $H \in \mathcal{E}$ the Hamiltonian vector field X_H has a dense set of points in $\mathbf{C}^2$ with exploding orbits.*

They proved a similar result for holomorphic Reeb fields on $\mathbf{C}^3$. Observe that Propositions 9.2 and 9.3 above also give exploding orbits of either X or $-X$.

Example: There exist Hamiltonian holomorphic vector fields X on $\mathbf{C}^2$ such that X is a limit of complete Hamiltonian fields, but *every real orbit of X explodes.* One way to construct such fields is as follows (see [18]). Let F be a polynomial automorphism of $\mathbf{C}^2$ with an attracting fixed point p such that the basin of attraction $D(p)$ is not all of $\mathbf{C}^2$. Then $D(p)$ is a Fatou-Bieberbach domain which intersects every complex line in a bounded set [8,32], and there is a biholomorphic map $G: \mathbf{C}^2 \to D(p)$ with Jacobian one which is a limit of automorphisms of $\mathbf{C}^2$ (and hence a limit of automorphisms G_j with Jacobian one). If Y is any constant vector field on $\mathbf{C}^2$, then $X = DG^{-1} \cdot Y$ is a Hamiltonian field which is a limit of complete Hamiltonian fields $X_j = DG_j^{-1} \cdot Y$, but every orbit of X explodes.

10. Flows on the complex plane $\mathbf{C}^2$

In this section we collect some results on classification of actions $F: \mathbf{R} \times \mathbf{C}^2 \to \mathbf{C}^2$ by holomorphic automorphisms $F_t \in \mathrm{Aut}\mathbf{C}^2$. Recall (Corollary 8.3) that every such action extends to an action of $\mathbf{C}$ on $\mathbf{C}^2$. Equivalently, we may consider such actions as flows of the complete holomorphic vector fields on $\mathbf{C}^2$, and also as one parameter subgroups of $\mathrm{Aut}\mathbf{C}^2$.

The following types of holomorphic flows on $\mathbf{C}^2$ have been classified:

1. polynomial flows (Theorem 10.1 below);
2. flows in the shear groups $\mathcal{S}(2)$, $\mathcal{S}_1(2)$, and $\mathcal{S}_1(2) \times \mathbf{C}^*$ (Theorem 10.4 below);
3. proper flows (Theorem 10.5 below);
4. symplectic flows (Theorem 10.6 below);
5. most flows whose time one map is polynomial.

Recall that $\mathcal{P}^2$ is the polynomial automorphism group of $\mathbf{C}^2$. Let (x, y) be the complex coordinates on $\mathbf{C}^2$. The following classification of polynomial flows is due to Suzuki [35] and, independently, to Bass and Meisters [7].

10.1 Theorem. *Every one parameter subgroup $\{F_t : t \in \mathbf{R}\}$ of the polynomial automorphism group $\mathcal{P}^2$ is conjugate in $\mathcal{P}^2$ to one of the following:*

$$\phi_t(x, y) = \big(e^{\mu t}x, e^{\lambda t}y\big), \quad \lambda, \mu \in \mathbf{C}, \tag{12}$$

$$\phi_t(x, y) = \big(x + tf(y), y\big), \quad f \text{ a polynomial}, \tag{13}$$

$$\phi_t(x, y) = \big(e^{n\lambda t}(x + ty^n), e^{\lambda t}y\big), \quad \lambda \in \mathbf{C}^*, \ n \in \mathbf{Z}_+. \tag{14}$$

Previously R. Rentschler has proved [29] that the every algebraic action of $\mathbf{C}$ on $\mathbf{C}^2$ is algebraically conjugate to an action (13).

The methods used in [7] and [35] are quite different from each other. Suzuki first showed that every one parameter subgroup $\{\phi_t : t \in \mathbf{R}\} \subset \mathcal{P}^2$ has bounded degree, i.e., $\deg\phi_t \le N$ for some N independent of $t \in \mathbf{R}$. This implies that there exist a polynomial embedding $H : \mathbf{C}^n \to \mathbf{C}^N$ for some N and an $N \times N$ matrix A such that

$$H \circ F_t(z) = \exp(tA) \cdot H(z), \quad z \in \mathbf{C}^n, \ t \in \mathbf{R}.$$

Suzuki then analysed several cases to obtain the classification.

The proof of Bass and Meisters has two main ingredients. One is the theorem of Van der Kulk [36] and Rentschler [29] (see also Jung [25] and Friedland and Milnor [23]) to the effect that the polynomial automorphism group $\mathcal{P}^2$ is an amalgamated free product $\mathcal{P}^2 = \mathcal{A} * \mathcal{E}$ of the affine automorphism group $\mathcal{A}$ of $\mathbf{C}^2$ and the group $\mathcal{E}$ of all elementary polynomial transformations

$$E(x, y) = \big(\alpha x + p(y), \beta y + \gamma\big),$$

where $\alpha, \beta \in \mathbf{C}^*$, $\gamma \in \mathbf{C}$, and p is a polynomial. This means that every element $g \in \mathcal{P}^2$ which does not belong to $\mathcal{A}$ or $\mathcal{E}$ can be represented by a word $g = \cdots e_r a_r e_{r-1} a_{r-1} \cdots$ of finite length, in which the e_j's belong to $\mathcal{E} \backslash \mathcal{A}$ and the a_j's belong to $\mathcal{A} \backslash \mathcal{E}$. The elements in $\mathcal{A} \cap \mathcal{E}$ are treated as units in this representation.

The second main ingredient is an algebraic theorem of Serre to the effect that every subgroup $\mathcal{H} \subset \mathcal{A} * \mathcal{E}$ in an amalgamated free product group whose elements have uniformly bounded length with respect to the given amalgamated free product structure is conjugate to a subgroup in $\mathcal{A}$ or in $\mathcal{E}$. Since every one parameter subgroup of $\mathcal{P}^2$ has bounded degree, it has bounded length and therefore Serre's theorem aplies. One then identifies and classifies all one parameter subgroups of $\mathcal{A}$ and $\mathcal{E}$.

It is known [26] that the analogous decomposition of the polynomial automorphism group $\mathcal{P}^n$ does not hold when $n \ge 3$. On the other hand, it has been proved recently that the shear groups on $\mathbf{C}^2$ (see section 1) also admit an amalgamated free product decomposition. Let $\mathcal{E} \subset \mathcal{S}(2)$ be the subgroup consisting of all automorphisms of the form

$$E(x,y) = \big(e^{a(y)}x + b(y), \beta y + \gamma\big), \tag{15}$$

where a and b are entire functions on $\mathbf{C}$, $\beta \in \mathbf{C}^*$, and $\gamma \in \mathbf{C}$. Maps of this form will be called *elementary* in analogy to the polynomial case when a is a constant and b is a polynomial (see [23]). Ahern and Rudin proved in [3] that the group $\mathcal{S}(2)$ is an amalgamated free product $\mathcal{S}(2) = \mathcal{A} * \mathcal{E}$, where $\mathcal{A}$ is the affine automorphism group on $\mathbf{C}^2$ and $\mathcal{E}$ is the group (15). Analogous result holds for the groups $\mathcal{S}_1(2)$ (Jacobian one) and $\mathcal{S}_1(2) \times \mathbf{C}^*$ (Jacobian constant but not necessarily one); see de Fabritiis [10,11]. In these cases the group $\mathcal{E}$ consists of automorphisms (15) for which a is constant and, in the volume preserving case, $e^a\beta = 1$.

In order to obtain the analogue of Theorem 10.1 for the shear groups one uses the following result of combinatorial group theory.

10.2 Theorem. *Let $\mathcal{G}$ be a topological group which is a free product $\mathcal{G} = \mathcal{A} * \mathcal{E}$ of subgroups $\mathcal{A}$ and $\mathcal{E}$, amalgamated over their intersection $\mathcal{B} = \mathcal{A} \cap \mathcal{E}$. Suppose that $\mathcal{B}$ is closed in $\mathcal{G}$. Then any topological subgroup of $\mathcal{G}$ which is isomorphic to $\mathbf{R}$ or $\mathbf{C}$ is conjugate in $\mathcal{G}$ to a subgroup in $\mathcal{A}$ or in $\mathcal{E}$.*

Theorem 10.2 follows from a result of Moldavanski ([27] or [38], Theorem 0.3); we refer the reader to the forthcoming paper [2] for the details. The same result holds for connected abelian Lie subgroups in $\mathcal{A} * \mathcal{E}$.

10.3 Corollary. *Every one parameter subgroup of the shear group $\mathcal{S}(2)$ is conjugate in $\mathcal{S}(2)$ to a subgroup of $\mathcal{E}$ (15). The analogous result holds for one parameter subgroups in $\mathcal{S}_1(2)$ and $\mathcal{S}_1(2) \times \mathbf{C}^*$.*

It is immediate that every subgroup in $\mathcal{A}$ is linearly conjugate to a subgroup in $\mathcal{E}$ (by conjugating the relevant matrix to its Jordan form).

One can describe one parameter subgroups $\{F_t : t \in \mathbf{R}\} \subset \mathcal{E}$ in the elementary group $\mathcal{E}$ (15) by using the methods in section 2 of [2]. The infinitesimal generator $V = (V_1, V_2)$ of F_t has the form

$$V_1(x,y) = a(y)x + b(y), \qquad V_2(y) = \lambda y + \gamma.$$

An automorphism Φ of $\mathbf{C}^2$ conjugates V to the field $\tilde{V}$ satisfying $D\Phi \cdot V = \tilde{V} \circ \Phi$. A preliminary linear change of coordinates in the y variable conjugates V to a field in which V_2 is either λy ($\lambda \in \mathbf{C}$) or $V_2 = 1$. Conjugating V with shears $\Phi(x,y) = \big(x + g(y), y\big)$ and generalized shears $\Phi(x,y) = \big(xe^{g(y)}, y\big)$ (and taking into account Corollary 10.3) one obtains the following classification result [2, Theorem 7.1]:

10.4 Theorem. *Every real one parameter subgroup $\{F_t : t \in \mathbf{R}\}$ of the generalized shear group $\mathcal{S}(2)$ is conjugate in $\mathcal{S}(2)$ to one of the following:*

(i) $\phi_t(x,y) = \big(x + tf(y), y\big)$, where f is an entire function on $\mathbf{C}$;

(ii) $\phi_t(x,y) = (e^{a(y)t}(x - b(y)) + b(y), y)$, *where* a *is a nonconstant entire function and* b *is a meromorphic function such that the product* ab *is entire;*

(iii) $\phi_t(x,y) = (e^{\mu t}x, e^{\lambda t}y)$, $\lambda, \mu \in \mathbf{C}$;

(iv) $\phi_t(x,y) = (e^{n\lambda t}(x + ty^n), e^{\lambda t}y)$, $\lambda \in \mathbf{C}^*$, $n \in \mathbf{Z}_+$.

Every subgroup $\{F_t\} \subset \mathcal{S}_1(2) \times \mathbf{C}^*$ $(t \in \mathbf{R})$ *is conjugate in* $\mathcal{S}_1(2) \times \mathbf{C}^*$ *to one of the groups (i), (iii), or (iv). Every subgroup* $\{F_t\} \subset \mathcal{S}_1(2)$ $(t \in \mathbf{R})$ *is conjugate in* $\mathcal{S}_1(2)$ *either to a group (i) or to a linear group (iii) with* $\lambda + \mu = 0$.

Observe that the groups (iii) and (iv) are polynomial, and (i) is polynomial when f is a polynomial. Comparing this with the classification of one parameter polynomial groups (Theorem 10.1) we see that the only group of new type in $\mathcal{S}(2)$ is (ii).

In the paper [2] we attempted to describe all real one parameter subgroups $F_t \in \mathrm{Aut}\mathbf{C}^2$ whose time one map F_1 is a polynomial automorphism $E \in \mathcal{P}^2$. The maps F_t for non-integer values of t need not be polynomial. We succeeded in all cases except when the time one map E is conjugate to an affine aperiodic map. In the last case we identified all flows with polynomial infinitesimal generator. We will not state the results of [2] here since there are too many different cases to consider. In [2] we also identified all generalized shears on $\mathbf{C}^2$ which belong to a flow; it turns out that most do not belong to any flow. This should be compared with results of Palis [28].

In [35] Suzuki classified *proper flows* $\{F_t : t \in \mathbf{C}\}$ on the plane $\mathbf{C}^2$. Recall that a flow is called proper if the complex orbit $C_z = \{F_t(z) : t \in \mathbf{C}\}$ of every point has at most a discrete set of limit points in $\mathbf{C}^2$; hence its closure $\overline{C}_z$ is an analytic curve in $\mathbf{C}^2$. The classification is as follows [35, Theorem 4].

10.5 Theorem: *Every proper holomorphic flow on* $\mathbf{C}^2$ *is conjugate in* $\mathrm{Aut}\mathbf{C}^2$ *to one of the following flows:*

(i) *linear flow* $\phi_t(x,y) = (e^{n\lambda t}x, e^{m\lambda t}y)$, *where* $\lambda \in \mathbf{C}^*$ *and* $m, n \in \mathrm{N} = \{1,2,3,\ldots\}$;

(ii) *shear flow* $\phi_t(x,y) = (x + tf(y), y)$, *where* f *is an entire function on* $\mathbf{C}$;

(iii) $\phi_t(x,y) = (e^{\lambda(y)t}(x - b(y)) + b(y), y)$, *where* b *is a meromorphic function and* λ, λb *are entire functions on* $\mathbf{C}$;

(iv) $\phi_t(x,y) = (e^{n\lambda(z)t}x, e^{-m\lambda(z)t}y)$, *where* $m, n \in \mathbf{N}$, $z = x^m y^n$, *and* λ *is an entire function on* $\mathbf{C}$;

(v) *flows of the form* $\alpha^{-1} \circ \rho_t \circ \alpha$, *where* $\alpha(x,y) = (x, x^l y + P_l(x))$, $l \in \mathbf{N}$, P_l *is a polynomial of degree* $\le l - 1$ *such that* $P_l(0) \neq 0$, *and* ρ_t *is a flow of type (iv) above such that* $\lambda(z)$ *has a zero of order* $\ge l/m$ *at* $z = 0$.

The proof is based on a previous result of Suzuki [33] to the effect that every proper action of $\mathbf{C}$ on $\mathbf{C}^2$ has a meromorphic first integral H. Recall (Theorem 8.2) that every complex orbit of a $\mathbf{C}$-action is isomorphic to either $\mathbf{C}$ or $\mathbf{C}^*$, and most orbits are of the same type. Outside the fixed point set of the action the complex orbits correspond to connected components of the level sets of H; hence most of these are isomorphic to either $\mathbf{C}$ or $\mathbf{C}^*$. Results of Nishino and Saito imply that for such a function H there is an automorphism Φ of $\mathbf{C}^2$ such that $Q = H \circ \Phi$ is a rational function of a special type. Clearly Φ conjugates the flow F_t to a flow whose first integral is Q. Suzuki then identified all flows on $\mathbf{C}^2$ whose first integral is one of these special functions Q.

Every symplectic holomorphic flow on $\mathbf{C}^2$ is proper since the energy function (Hamiltonian) of the infinitesimal generator is a holomorphic first integral. Using this observation we proved in [18] (Theorem 6.1):

10.6 Theorem. *Every symplectic flow $\{F_t : t \in \mathbf{C}\} \subset \mathrm{Aut}_1\mathbf{C}^2$ is conjugate in $\mathrm{Aut}\mathbf{C}^2$ to one of the following:*

$$\phi_t(x,y) = \big(x, y + th(x)\big), \tag{16}$$

$$\psi_t(x,y) = \left(e^{t\lambda(xy)}x, e^{-t\lambda(xy)}y\right), \tag{17}$$

where h resp. λ is an entire function on $\mathbf{C}$. In the first case the flow is conjugated to a flow (16) by a symplectic automorphism of $\mathbf{C}^2$.

We see by inspection that (16) and (17) are the only symplectic flows from the list of equivalence classes of proper flows given by Theorem 10.5. However, since conjugation by non-symplectic automorphisms does not preserve symplectic flows, additional work is needed to prove Theorem 10.6. We don't know whether in the second case the flow is conjugate to (17) by a symplectic automorphism of $\mathbf{C}^2$.

11. Open problems

Problem 1: Classify one parameter subgroups of $\mathrm{Aut}\mathbf{C}^2$. Partial results are given in section 10 above.

Problem 2: Classify one parameter subgroups of the polynomial automorphism group on $\mathbf{C}^3$ (or $\mathbf{C}^n$ for $n \geq 3$).

Problem 3: Which properties of a holomorphic vector field make it nonapproximable by complete fields ? See Corollary 8.6 and the example following it.

Problem 4: Is every Fatou-Bieberbach domain in $\mathbf{C}^n$ Runge ? Equivalently, is every injective holomorphic mapping $F: \mathbf{C}^n \to \mathbf{C}^n$ (Fatou-Bieberbach map) a limit of automorphisms of $\mathbf{C}^n$? This question has already been raised by Rosay and Rudin [32], but it remains unsolved.

Problem 5: Recall that the shear group $\mathcal{S}(n)$ is dense in $\mathrm{Aut}\mathbf{C}^n$. The proof of Theorem 1.1 in [18] implies the following stronger result: *If M is a closed ball in $\mathbf{R}^k$ for some $k \in \mathbf{Z}_+$ and $\Phi: M \times \mathbf{C}^n \to \mathbf{C}^n$ is a map of class $\mathcal{C}^p$ $(0 \leq p \leq \infty)$ satisfying $\Phi_x = \Phi(x, \cdot\,) \in \mathrm{Aut}\mathbf{C}^n$ for every $x \in M$, then Φ can be approximated in the $\mathcal{C}^p$ topology (uniformly on compacts in $M \times \mathbf{C}^n$) by maps $\Phi': M \times \mathbf{C}^n \to \mathbf{C}^n$ such that $\Phi'_x \in \mathcal{S}(n)$ for $x \in M$.* For which other manifolds M is this true ? In particular, does this hold when M is the circle, a sphere, a torus ?

Problem 6: On $\mathbf{C}^n$ one can define $(n-1)$ shear groups $\mathcal{S}^k(n)$ for $1 \leq k \leq n-1$ by letting $\mathcal{S}^k(n)$ be the group generated by all shears of the form (3) and (4), where Λ is a linear projection into $\mathbf{C}^m$ for some $m \leq k$. Are these groups all distinct ?

Problem 7: Let $D \subset \mathbf{C}^n$ be a smoothly bounded, strongly pseudoconvex Runge domain diffeomorphic to the ball, and let $F: D \to \mathbf{C}^n$ be a biholomorphic map onto a Runge domain $F(D) \subset \mathbf{C}^n$. Is F a limit of automorphisms of $\mathbf{C}^n$?

Problem 8: Suppose that Φ is an automorphism of $\mathbf{C}^2$ whose Jacobian $J\Phi$ is a function of the product xy, where (x, y) are complex coordinates on $\mathbf{C}^2$. Does it follow that $J\Phi$ is constant ? If so, then every flow in Theorem 10.5 which is conjugate in Aut$\mathbf{C}^2$ to a flow of the form (16) is already conjugate to such a flow in $\text{Aut}_1\mathbf{C}^2$. More generally, which nonvanishing functions are the Jacobians of automorphisms of $\mathbf{C}^2$ ($\mathbf{C}^n$) ?

Problem 9: Suppose that $\Phi \in \text{Aut}\mathbf{C}^2$ fixes pointwise both coordinate axes and the point $(1, 1)$. Does it follow that $J\Phi(1, 1) = 1$? If there exists a Φ as above such that $(1, 1)$ is an attracting fixed point (this requires $|J\Phi(1, 1)| < 1$), then its basin of attraction is a Fatou-Bieberbach domain in $(\mathbf{C}^*)^2$ [32].

Problem 10: Suppose Φ is an automorphism of $\mathbf{C}^n$ with finite period (i.e., Φ^m is the identity for some $m \in \mathbf{N}$). Is Φ linearizable ? Is every finite subgroup of Aut$\mathbf{C}^n$ linearizable ? For partial results see [3] and [26].

Problem 11: Find characterizations of $\mathbf{C}^n$ in terms of its automorphism group.

Several other problems concerning automorphisms of $\mathbf{C}^n$ and Fatou-Bieberbach maps are mentioned in papers [21], [26], and [32].

References.

1. Abraham, A., Marsden, J.E.: Foundations of Mechanics, 2.ed. Reading: Benjamin 1978
2. Ahern, P., Forstneric, F.: On parameter automorphism groups on $\mathbf{C}^2$. Complex Variables, to appear
3. Ahern, P., Rudin, W.: Periodic automorphisms of $\mathbf{C}^n$. Preprint, 1994
4. Andersén, E.: Volume-preserving automorphisms of $\mathbf{C}^n$. Complex Variables, **14**, 223–235 (1990)
5. Andersén, E., Lempert, L.: On the group of holomorphic automorphisms of $\mathbf{C}^n$. Invent. Math., **110**, 371–388 (1992)
6. Arnold, A.: Mathematical Methods of Classical Mechanics, 2.ed. Berlin-Heidelber-New York: Springer 1989
7. Bass, H., Meisters, G.: Polynomial flows in the plane. Adv. in Math. **55**, 173–208 (1985)
8. Bedford, E., Smillie, J.: Fatou-Bieberbach domains arising from polynomial automorphisms. Indiana Univ. Math. J., **40**, 789–792 (1991)
9. Behnke, H., Thullen, P.: Theorie der Funktionen mehrerer komplexer Verä nderlichen. 2nd ed., Ergebnisse de Mathematik und ihrer Grenzgebiete **51**, Berlin: Springer 1970
10. De Fabritiis, C.: On continuous dynamics of automorphisms of $\mathbf{C}^2$. Manuscripta Math. **77**, 337–359 (1992)

11. De Fabritiis, C.: One-parameter groups of volume-preserving automorphisms of $\mathbf{C}^2$. Rendiconti Inst. Mat. Trieste, to appear.
12. El Kasimi, A: Approximation polynômiale dans les domaines étoilés de $\mathbf{C}^n$. Complex Variables **10**, 179–182 (1988)
13. Fornæss, J.E., Grellier, S.: Exploding orbits of holomorphic hamiltonian and complex contact structures. Preprint, 1994
14. Fornæss, J.E., Sibony, N.: Holomorphic symplectomorphisms in $\mathbf{C}^2$. Preprint, 1994
15. Fornæss, J.E., Sibony, N.: Holomorphic symplectomorphisms in $\mathbf{C}^{2p}$. Preprint, 1994
16. Fornaæss, J.E., Sibony, N.: Complex dynamics in higher dimensions. Preprint, 1994
17. Forstneric, F.: Approximation by automorphisms on smooth submanifolds of $\mathbf{C}^n$. Math. Ann., to appear
18. Forstneric, F.: Actions of $(\mathbf{R}, +)$ and $(\mathbf{C}, +)$ on complex manifolds. Math. Z, to appear
19. Forstneric, F.: A theorem in complex symplectic geometry. J. of Geom. Anal., to appear
20. Forstneric, F.: Equivalence of real submanifolds under volume preserving holomorphic automorphisms of $\mathbf{C}^n$. Duke Math. J., to appear
21. Forstneric, F., Rosay, J.-P.: Approximation of biholomorphic mappings by automorphisms of $\mathbf{C}^n$. Invent. Math., **112**, 323–349 (1993)
22. Forstneric, F., Løw, E.: manuscript in preparation
23. Friedland, S., Milnor, J.: Dynamical properties of plane polynomial automorphisms. Ergod. Th. and Dynam. Sys., **9**, 67–99 (1989)
24. Hörmander, L., Wermer, J.: Uniform approximation on compact sets in $\mathbf{C}^n$. Math. Scand. **23**, 5–21 (1968)
25. Jung, H.W.E.: Über ganze birationale Transformationen der Ebene. J. Reine Angew. Math. **184**, 161–174 (1942)
26. Kraft, H.: Affine n-space, $\mathbf{C}^+$-actions and related problems. Preprint, 1994
27. Moldavanski, D.I.: Certain subgroups of groups with one defining relation. Sib. Mat. Ž. **8**, 1370–1384 (1967)
28. Palis, J.: Vectorfields generate few diffeomorphisms. Bull. Amer. Math. Soc., **80**, 503–505 (1974)
29. Rentschler, R.: Opérations du groupe additif sur le plan affine. C. R. Acad. Sc. Paris, **267**, 384–387 (1968)
30. Richardson, R. W.: Principal orbit type for reductive groups acting on Stein manifolds. Math. Ann., **208**, 323–331 (1974)
31. Rosay, J.-P.: Straightening of arcs. (Colloque d'analyse complexe et géométrie, Marseille, 1992, pp.217–226.) Astérisque **217** (1993)
32. Rosay, J.-P., Rudin, W.: Holomorphic maps from $\mathbf{C}^n$ to $\mathbf{C}^n$. Trans. Amer. Math. Soc. **310**, 47–86 (1988)

33. Suzuki, M.: Sur les intégrales premières de certains feuilletages analytiques complexes. Sém. F. Norguet, 31–57 (1975-76). Berlin-Heidelberg-New York: Springer 1976

34. Suzuki, M.: Sur les opérations holomorphes de **C** et de $\mathbf{C}^*$ sur un espace de Stein. Sém. F. Norguet, 58–66 (1975-76). Berlin-Heidelberg-New York: Springer 1976

35. Suzuki, M.: Sur les opérations holomorphes du groupe additif complexe sur l'espace de deux variables complexes (I). Ann. sc. Éc. Norm. Sup., 4.ser., **10**, 517–546 (1977)

36. Van der Kulk, W.: On polynomial rings in two variables. Nieuw Arch. Wisk. (3) **1**, 33–41 (1953)

37. Wermer, J.: Function Algebras and Several Complex Variables. Berlin-Heidelberg-New York: Springer 1976

38. Wright, D.: Abelian subgroups of $\mathrm{Aut}_k(k[X,Y])$ and applications to actions on the affine plane. Illinois J. Math. **23**, 579–634 (1979)

Complex Structures on Compact Conformal Manifolds of Negative Type

Paul Gauduchon
Centre de Mathématiques
Ecole Polytechnique
91128 Palaiseau Cedex

0. Introduction

A (Riemannian) conformal structure c on a differentiable manifold M of dimension n is an equivalence class of Riemannian metrics on M for the relation : $g' \sim g$ if $g' = \phi^{-2}g$, where ϕ is a positive function. A complex structure on M, viewed as an integrable almost-complex structure J, is *c-compatible* (or *c-orthogonal*) whenever J is c-skew-symmetric, i.e. g-skew-symmetric for any riemannian metric g in the conformal class c. Then, for any g in c, the pair (g,J) determines a Hermitian structure on M.

In this paper, we provide a criterion for the existence of a complex structure J compatible with a given conformal structure c when M is compact. This criterion involves the scalar curvature and the minimal eigenvalue of the Weyl tensor ; in particular, it readily implies the non-existence of a compatible complex structure on a compact, flat conformal manifold, of negative type (=containing a metric with negative total scalar curvature), as soon as n is greater than 2.

In the second part of the paper, we apply this criterion to the case when M is a compact quotient of an irreducible Riemannian symmetric space N of non-compact type (= with negative scalar curvature), of dimension greater than 2. We show that M admits no compatible complex structure except when M is Hermitian-symmetric ; in this case, the complex structure on M induced by the invariant complex structure of N is, up to conjugation, the only compatible complex structure on M.

A similar result has been recently obtained, by using twistorial techniques, for some class of compact Riemannian symmetric spaces, cf. [B-G-M-R].

I warmly thank F. Burstall, F. Labourie, C. Margerin and A. Nannicini for their attention and useful suggestions.

I. A Criterion of existence for a compatible complex structure

In the whole paper, M denotes a connected, oriented, differentiable manifold, of dimension n greater than 2, equipped with a (Riemannian) conformal structure c.

When M is compact, we associate with any Riemannian metric g in the conformal class c the real number $S(g)$ defined by :

$$S(g) = \frac{1}{[V(g)]^{\frac{n-2}{n}}} \int_M \Big[\frac{\mathrm{Scal}^g}{n(n-1)} - \frac{1}{(n-2)}\lambda^g_{\min}(W)\Big] v_g, \tag{1}$$

where : Scal^g denotes the scalar curvature of g (= the trace of the Ricci tensor) ; v_g denotes the volume-form ; $V(g) = \int_M v_g$ is the volume of (M, g) ; $\lambda^g_{\min}(W)$ denotes the smallest eigenvalue of the Weyl tensor W of c, viewed as a symmetric endomorphism of $\Lambda^2 M$ via the metric g (since the trace of W is equal to zero, $\lambda^g_{\min}(W)$ is non-positive and $-\lambda^g_{\min}(W)$ is a norm for W at any point of M).

With the conformal structure c itself, we associate the real number $S(c)$ defined by :

$$S(c) = \inf_{g \in c} S(g). \tag{2}$$

The functional S is closely related to the existence of a compatible complex structure J on M, as shown by the following result.

Theorem 1.— *Let (M, c) be a connected, compact, oriented, conformal manifold, of (even) dimension $n = 2m$ greater than 2.*

If M admits a c-compatible complex structure J, then

a) $S(g)$ is positive or equal to zero for any metric g in the conformal class c.

b) $S(g)$ is equal to zero if and only if the following three conditions are simultaneously satisfied :

(i) $\frac{\mathrm{Scal}}{n(n-1)} = \frac{1}{(n-2)}\lambda^g_{\min}(W)$ at any point ;

(ii) the Kähler form F of the Hermitian structure (g, J) is an eigenform of the Weyl tensor W with respect to $\lambda^g_{\min}(W)$;

(iii) the Hermitian structure (g, J) is semi-Kähler, i.e. the $(n-2)$-form F^{m-1} is closed.

As a immediate consequence of Theorem 1, we get :

Corollary.— *A compact, oriented, conformal manifold (M, c) of (even) dimension n greater than 2, such that $S(c)$ is negative, admits no c-compatible complex structure.*

This holds, in particular, for compact flat conformal manifolds (for which W vanishes identically) of negative type (i.e. containing a metric of negative total scalar curvature), of dimension greater than 2

Remark 1 : the criterion of non-existence given by the above Corollary applies in particular to compact quotients of the real hyperbolic space H^{2m}, for any m greater than 1. As we shall see in the second part, it applies more generally to any compact quotient of an irreducible Riemannian symmetric space G/H of non-compact type, as soon as the centre of the homology group H is discrete and the dimension of G/H is greater than 2, cf. Theorem 2 below.

Proof of Theorem 1

Recall that the Kähler form F of any almost-Hermitian structure (g, J) is the real 2-form on M defined by :

$$(3) \qquad F(X, Y) = g(JX, Y), \quad \forall X, Y \in T_xM, \ \forall x \in M$$

The square norm $|F|^2$ is constant, equal to m, but, in general, F is neither closed, nor co-closed.

Let us consider the real 1-form θ determined by one of the two following equivalent relations :

$$(4) \qquad d(F^{m-1}) = \theta \wedge F^{m-1},$$

$$(4') \qquad \theta = J(\delta F),$$

where δ denotes the codifferential with respect to g (by convention, J acts on the cotangent bundle T^*M by : $(J\alpha)(X) = -\alpha(JX)$.

The 1-form θ is called the torsion 1-form, or Lee form, of the almost-complex structure (g, J) (the denomination Lee form is often used for the normalized 1-form $\theta/(n-2)$ or $-\theta/(n-2)$)

The proof of Theorem 1 lays on the following form of the "Bochner-Weitzenböck formula" for 2-forms, where ψ denotes any (real) 2-form on M :

$$(5) \ \Delta\psi - (D^g)^* D^g \psi = \frac{2(n-2)}{n(n-1)} \mathrm{Scal}^g . \psi - 2W(\psi) + \frac{(n-4)}{(n-2)} [\mathrm{Ric}_0(\Psi.,) - \mathrm{Ric}_0(., \Psi.)],$$

where : Δ denotes the Riemannian Laplace operator, acting on 2-forms ; D^g denotes the Levi-Civita connection of g ; $(D^g)^*$ denotes the adjoint of D^g with respect to g ; Ric_0 the trace-free part of the Ricci tensor Ric of g ; Ψ the c-skewsymmetric endomorphism of TM identified to ψ via the metric g : $g(\Psi\cdot,\cdot) = \psi$; $W(\psi)$ is the image of ψ by the Weyl tensor W, viewed as a symmetric endormorphism of $\Lambda^2 M$ via g. Formula (5) is easily checked by a direct calculation, cf. also [Bou].

When specialized to $\psi = F$, i.e. $\Psi = J$, (5) becomes :

$$\begin{aligned}\Delta F - (D^g)^* D^g F &= \frac{2(n-2)}{n(n-1)} \operatorname{Scal}^g \cdot F - 2W(F) \\ &+ \frac{(n-4)}{(n-2)} [\operatorname{Ric}_0 (J\cdot,\cdot) - \operatorname{Ric}_0 (\cdot, J\cdot)] \ .\end{aligned} \tag{6}$$

By contracting both members of (6) by F, we obtain :

$$(\Delta F, F) - ((D^g)^* D^g F, F) = \frac{(n-2)}{(n-1)} \operatorname{Scal}^g - 2(W(F), F) \ , \tag{7}$$

where $(\cdot,\cdot)$ denotes the scalar product induced on 2-forms by g (Notice that the term involving Ric_0 disappears, as Ric_0 is trace-free).

Since the norm of F is constant, the following holds true (at any point) :

$$((D^g)^* D^g F, F) = |D^g F|^2 \ . \tag{8}$$

On the other hand, we have :

$$\begin{aligned}(\Delta F, F) &= (d\delta F, F) + (\delta d F, F) \\ &= |\delta F|^2 + \delta(J\delta F) + |dF|^2 + \delta((dF, F)) \ ,\end{aligned} \tag{9}$$

where (dF, F) denotes the 1-form which associates with any vector X the scalar product of the 2-forms $dF(X, \cdot\cdot)$ and F.

It follows easily from (4)- (4'), that this 1-form is equal to the torsion 1-form θ :

$$\theta = (dF, F) \ , \tag{10}$$

so that (9) can be written in the following way :

$$(\Delta, F, F) = |dF|^2 + |\theta|^2 + 2\delta\theta \ . \tag{11}$$

We finally get :

$$|dF|^2 - |D^g F|^2 + |\theta|^2 + 2\delta\theta = \frac{(n-2)}{(n-1)} \operatorname{Scal}^g \ - 2(W(F), F) \ . \tag{12}$$

Notice that the identity (12) holds for the Kähler from F of any almost-Hermitian manifold (M, J, g), without assuming M compact or J integrable.

The latter assumption is used in the above computation via the following lemma :

Lemma 1.— *If J is integrable, we have :*

$$|D^g F|^2 = |dF|^2 \ , \tag{13}$$

for any metric g in the conformal class c.

As an immediate consequence of Lemma 1, if J is integrable, the identity (12) is reduced to the following simple form :

$$|\theta|^2 + 2\delta\theta = \frac{(n-2)}{(n-1)} \operatorname{scal}^g \ - 2(W(F), F) \ , \tag{14}$$

which holds true for any Hermitian manifold (M, J, g) .

Note : when $n = 4$, cf. [Boy] Lemma 2.2.

When M is compact, (14) implies the inequality :

$$S(g) \geq \frac{1}{n(n-2)} \int_M |\theta|^2 \cdot v_g \geq 0 \ , \tag{15}$$

which implies the part (a) of Theorem 1.

The equality $S(g) = 0$ holds if and only if the 1-form θ vanishes identically and F is an eigenform of W with respect of the minimal eigenvalue $\lambda^g_{\min}(W)$. It then follows from (14) that we have the equality : $\mathrm{Scal}/n(n-1) - \lambda^g_{\min}(W) = 0$ at any point on M. That proves the part (b) of Theorem 1.

□

Proof of Lemma 1.

Note on norms. In the present paper, the covariant derivative $D^g F$ is viewed as a $\Lambda^2 M$-valued 1-form, and the 3-form dF is equal to 3 times the skew-symmetric part of $D^g F$, i.e. :

$$dF(X, Y, Z) = D^g_X F(Y, Z) + D^g_Y F(Z, X) + D^g_Z F(X, Y) \ . \tag{16}$$

By convention, the square norm $|\psi|^2$ of a (real) p-form ψ is equal to $1/p\,!$ times its tensorial square norm $|\psi|^2_T$. In particular, we have : $|D^g F|^2 = \frac{1}{2}|D^g F|^2_T$ and $|dF|^2 = \frac{1}{6}|dF|^2_T$, so that (13) can be written in the following equivalent way :

$$|D^g F|^2_T = \frac{1}{3}|dF|^2_T = 3|(D^g F)_a|^2_T \ , \tag{13'}$$

where $(D^g F)_a$ denotes the skew-symmetric part of $D^g F$.

Lemma 1 is a consequence of the two following "folkloric" lemmas :

Lemma 2.— *A c-compatible almost-complex structure J is integrable if and only if the identity :*

$$D^g_{JX} J = J_\circ D^g_X J \ , \qquad \forall X \in T_x M, \ \forall x \in M \ , \tag{17}$$

holds for one (hence any) metric g in c.

Lemma 3.— *If J integrable, we have :*

$$D^g_X F(X,Z) = \frac{1}{2}(dF(X,Y,Z) - dF(X,JY,JZ)) \ . \tag{18}$$

Lemma 3 follows easily from Lemma 2, via (16) and (17).

To prove Lemma 2, recall that J is integrable if and only if the Nijenhuis tensor N of J vanishes identically, where N is defined by :

$$4N(X,Y) = [X,Y] + J[JX,Y] + J[X,JY] - [JX,JY] \ , \tag{19}$$

cf [N-N]. Since D^g is torsion-free, (19) can be written :

$$4N(X,Y) = [J(D^g_X J)(Y) - (D^g_{JX} J)(Y)] - [J(D^g_Y J)(X) - (D^g_{JY} J)(X)] \ . \tag{20}$$

From (20), we infer :

$$\begin{aligned} &g((D^g_X J)(Y) + J(D^g_{JX} J)(Y), Z) \\ &= 2[g(N(X,Y),JZ) - g(N(X,Z),JY) - g(N(Y,Z),JX)] \ . \end{aligned} \tag{21}$$

Lemma 2 is a direct consequence of (20) and (21).

When J is integrable, the 3-form dF is of type $(2,1)+(1,2)$, i.e. satisfies the identity :

$$dF(X,Y,Z) = dF(X,JY,JZ) + dF(JX,Y,JZ) + dF(JX,JY,Z) \ . \tag{22}$$

Lemma 1 easily follows from (18) and (22).

□

Remark 2 : with some more work, we obtain the following general formula :

$$|D^g F|^2 = |dF|^2 + 4|N^0|^2 - \frac{2}{3}|(dF)^-|^2 \ , \tag{23}$$

where : $(dF)^-$ denotes the $(3,0)+(0,3)$-component of dF and N^0 denotes the component of the Nijenhuis tensor N satisfying Bianchi identity :

$$(N^0(X,Y),Z) + (N^0(Y,Z),X) + (N^0(Z,X),Y) = 0 \ .$$

In particular, if $n = 4$, we obtain the inequality :

$$|DF| \geq |dF| , \tag{24}$$

with equality if and only if J is integrable.

This fact is specific to dimension 4 ; for higher dimensions, both situations may occur : $|DF| > |dF|$, for example in the so-called (non-Kähler) "almost-Kähler" case, when F is closed (but non-parallel), or : $|DF| < |dF|$, in the (non-Kähler) "nearly-Kähler" case, where N^0, but not $(dF)^-$, vanishes identically (for example, S^6 with its canonical almost-Hermitian structure).

Remark 3 : formula (14) can be interpreted as a property of the Weyl structure D (w.r. to c) attached to the complex structure J, cf. [Gau1] and references enclosed. Recall that D is the linear connection defined by :

$$D_X Y = D^g_X Y - \frac{1}{(n-2)} \left[\theta(X)Y + \theta(Y)X - g(X,Y)\theta^*\right] , \tag{25}$$

where g is any metric in c, θ the corresponding torsion 1-form, θ^* the dual vector field w.r. to g. It is easy to check that the connection D defined by (25) doesn't depend on the choice of g in c, i.e. is uniquely determined by the conformal structure c and any c-compatible almost complex structure J. In addition, D is torsion-free and preserves c : $Dg = \frac{2}{(n-2)} \theta \otimes g$.

The scalar curvature Scal^D of D, defined as the conformal trace of the Ricci tensor of D, appears, when no metric g is chosen in c, as a *scalar of weight* -2, i.e. a section of L^{-2}, where L denotes the (real, rank one) bundle of scalars of weight 1, cf. [Gau2]. Each metric g in c determines a (positive) section of L, through which Scal^D is expressed as an honest scalar function, denoted by Scal^D_g.

From (25), we easily infer the following identity, cf. [Gau1] (38), [Gau2] :

$$\mathrm{Scal}^D_g = \mathrm{Scal}^g - \frac{(n-1)}{(n-2)} \left[|\theta|^2 + 2\delta\theta\right] . \tag{26}$$

Again, the scalar product $(W(F), F)$ appearing in the r.h.s. of (14) is the expression w.r. to g of a section of L^{-2}, so that (14) may be viewed as an equality of sections of L^{-2}. We thus get :

Lemma 4.— *Let (M, c) be an oriented, compact manifold, of (even) dimension n greater than 2 ; J a c-compatible complex structure ; D the corresponding Weyl structure. Then, we have the equality :*

$$\mathrm{Scal}^D = \frac{2(n-1)}{(n-2)} \, (W(F), F) . \tag{27}$$

In particular, we have the following implication :

$$c \text{ is flat } \Longrightarrow \mathrm{Scal}^D \text{ vanishes identically .} \tag{28}$$

For $n = 4$, if J is positive, i.e. induces the chosen orientation of M, the anti-self-dual component W^- of the Weyl tensor W acts trivially on F, so that the implication (28) can be replaced by the following one, cf. [Boy] :

$$c \text{ is anti-self-dual } \Longrightarrow \mathrm{Scal}^D \text{ vanishes identically .} \tag{28'}$$

(Recall that c is called anti-self-dual if the self-dual part W^+ of W vanishes identically).

II. Locally symmetric spaces of non-compact type.

Let $(N = G/H, g)$ be a *irreducible* Riemannian symmetric space, of dimension n greater than 2, where G denotes the isometry group of N and H the isotropy subgroup of the base point $p = [H]$.

Let $\mathfrak{g}, \mathfrak{h}$ denote the Lie algebras of G, H and $\mathfrak{m}$ the orthogonal complement of $\mathfrak{h}$ in $\mathfrak{g}$ with respect of the Killing form $B^{\mathfrak{g}}$ of $\mathfrak{g}$.

Then, $\mathfrak{m}$ is identified to the tangent space T_pN and the adjoint action of $\mathfrak{h}$ on $\mathfrak{m}$ (in $\mathfrak{g}$) coincides with the derivative of the isotropy action of H on N at p. We thus identify the Lie algebra $\mathfrak{h}$ to a Lie subalgebra of the Lie algebra $\mathfrak{o}(\mathfrak{m}) \simeq \mathfrak{o}(n)$ of skew-symmetric endomorphisms of the euclidean vector space $\mathfrak{m} = (T_pN, g)$ by putting :

$$A(X) = [A, X], \qquad \forall A \in \mathfrak{h},\ \forall X \in \mathfrak{m} = T_pN\ , \tag{29}$$

where the bracket in the r.h.s. is the bracket of $\mathfrak{g}$.

The restriction to $\mathfrak{m}$ of the bracket of $\mathfrak{g}$ has its values in $\mathfrak{h}$, hence in $\mathfrak{o}(\mathfrak{m})$, and, as such, coincides with the Riemannian curvature R^N of (N, g) at the base point p :

$$R^N_{X,Y} = [X, Y], \qquad \forall X, Y \in T_pN = \mathfrak{m}\ , \tag{30}$$

where, again, the braket in the r.h.s. is the bracket of $\mathfrak{g}$.

It follows readily that the restriction of $B^{\mathfrak{g}}$ to the vector space $\mathfrak{m}$ is expressed by :

$$B^{\mathfrak{g}}(X, Y) = -2\ \mathrm{Ric}^N\ (X, Y)\ , \qquad \forall X, Y \in \mathfrak{m} = T_pN\ . \tag{31}$$

Since N is irreducible, we infer from (31) that the metric g is Einstein so that (31) is reduced to :

$$B^{\mathfrak{g}}(X,Y) = -\frac{2}{n}\,\mathrm{Scal}^N g(X,Y)\;, \tag{32}$$

where Scal^N denotes the (constant, non-zero) scalar curvature of $\mathfrak{g}$.

Let $(\cdot,\cdot)$ denotes the natural scalar product on $\mathfrak{o}(\mathfrak{m})$ defined by :

$$(A,B) = -\frac{1}{2}\,\mathrm{trace}\,(A \circ B)\;, \qquad \forall A,B \in \mathfrak{o}(\mathfrak{m})\;. \tag{33}$$

Recall that $(\cdot,\cdot)$ is related to the Killing form of $\mathfrak{o}(\mathfrak{m}) = \mathfrak{o}(n)$ by :

$$B^{\mathfrak{o}(\mathfrak{m})}(A,B) = -2(n-2)(A,B)\;. \tag{34}$$

Consider the orthogonal sum :

$$\mathfrak{h} = \mathfrak{c} \oplus \bigoplus_{j=1}^{r} \mathfrak{h}_j\;, \tag{35}$$

where $\mathfrak{c}$ is the center and the $\mathfrak{h}_j$'s the simple components of the Lie algebra $\mathfrak{h}$.

Since N is irreducible, the center $\mathfrak{c}$ of $\mathfrak{h}$ is either reduced to $\{0\}$ or 1-dimensional. The latter case occurs if and only if N is Hermitian-symmetric ; then, the invariant complex structure at p, viewed as an element of $\mathfrak{o}(\mathfrak{m})$, is a generator of $\mathfrak{c}$, cf. [Sal] p. 71.

For each $j = 1, \cdots, r$, define a (positive) real number μ_j by :

$$B_j(A,B) = -2\mu_j\,(A,B)\;, \tag{36}$$

where B_j denotes the Killing form of $\mathfrak{h}_j$.

Then, the curvature R^N at p, viewed as a (symmetric) endomorphism of $\mathfrak{o}(\mathfrak{m})$, is described in the following way.

Lemma 5.—

(i) The restriction of R^N to the orthogonal complement $h^{\perp}$ of $\mathfrak{h}$ in $\mathfrak{o}(\mathfrak{m})$ is trivial.

(ii) The restriction of R^N to $\mathfrak{h}$ is an automorphism of $\mathfrak{h}$, whose inverse R^{-1} is expressed by :

$$R^{-1} = \frac{n}{\mathrm{Scal}}\,(I + \sum_{j=1}^{r} \mu_j I_j)\;, \tag{37}$$

where I denotes the identity of $\mathfrak{h}$ and, for $j = 1, \cdots, r$, I_j denotes the orthogonal projection of $\mathfrak{h}$ to $\mathfrak{h}_j$.

Proof of Lemma 5 : by invariance of $B^{\mathfrak{g}}$, we have, for any X, Y in $\mathfrak{m}$:

$$B^{\mathfrak{g}}(A,[X,Y]) = -B^{\mathfrak{g}}([X,A],Y) . \tag{38}$$

By (29) and (32), the r.h.s. of (38) is equal to $-2\ \frac{\text{Scal}}{n}\ (A(X),Y)$, while, by (30), the l.h.s. is equal to :

$$B^{\mathfrak{g}}(A, R^N_{X,Y}) = B^{\mathfrak{h}}(A, R^N_{X,Y}) + \text{ trace } (A \circ R^N_{X,Y}) .$$

Lemma 5 follows easily.

Nota : the above description of the curvature of an irreducible symmetric space is due to B. Kostant [Kos]. Our exposition closely follows [Sa1] Ch. 5, cf. in particular, [Sa1] Prop.5.1 and Cor.5.2 (where R has to be replaced by R^{-1}).

From (37), we readily obtain the eigenvalues and the eigenforms of the curvature R^N of N at p (hence anywhere), when R^N is viewed as a symmetric endomorphism of $\mathfrak{o}(\mathfrak{m}) = \mathfrak{o}(n)$ (recall that $\mathfrak{h}^\perp$ denotes the orthogonal complement of $\mathfrak{h}$ in $\mathfrak{o}(n)$).

(39)

Eigenvalues of R^N	Eigenspaces of R^N
0	$\mathfrak{h}^\perp$ (if $\mathfrak{h}^\perp \neq 0$) ,
$\dfrac{\text{Scal}}{n(1+\mu_j)}$	$\mathfrak{h}_j, \quad j = 1, \cdots r$,
$\dfrac{\text{Scal}}{n}$	$\mathfrak{c}$ (if $\mathfrak{c} \neq 0$) .

We observe that the eigenvalue $\frac{\text{Scal}}{\text{n}}$ occurs if and only if the center $\mathfrak{c}$ of the isotropy algebra $\mathfrak{h}$ is non-trivial, if and only if N is Hermitian-symmetric, cf. above.

Since the metric g is Einstein, the Weyl tensor W coincides with the trace-free part of the curvature R^N, i.e. is expressed by :

$$W = R^N - \frac{\text{Scal}}{n(n-1)}\ I\ , \tag{40}$$

where I denotes the identity of $\mathfrak{o}(n)$ (notice that Scal is equal to two times the trace of R^N as an endomorphism of $\mathfrak{o}(n)$). The eigenvalues and eigenforms of W are thus given by :

(41)

Eigenvalues of W	Eigenspaces of W
$-\dfrac{\text{Scal}}{n(n-1)}$	$\mathfrak{h}^\perp$ (if $\mathfrak{h}^\perp \neq 0$) ,
$\dfrac{\text{Scal}}{n(n-1)}\dfrac{(n-2-\mu_j)}{(1+\mu_j)}$	$\mathfrak{h}_j, \quad j = 1, \cdots r$,
$\dfrac{\text{Scal}}{n(n-1)}(n-2)$	$\mathfrak{c}$ (if $\mathfrak{c} \neq 0$) .

Again, $\frac{\text{Scal}}{n(n-1)}(n-2)$ occurs if and only if N is Hermitian-symmetric.

We readily infer the following Lemma :

Lemma 6.— *Let (N, g) be a irreducible Riemannian symmetric space of non-compact type, of dimension $n > 2$.*

(a) If (N, g) is Hermitian-symmetric, we have the equality (everywhere) :

$$\lambda^g_{\min}(W) = (n-2)\,\frac{\text{Scal}}{n(n-1)}\,. \tag{42}$$

(b) Si (N, g) is not Hermitian-symmetric, we have :

$$\lambda^g_{\min}(W) = \frac{\text{Scal}}{n(n-1)}\,\frac{(n-2-\mu_1)}{(1+\mu_1)} > (n-2)\,\frac{\text{Scal}}{n(n-1)}\,, \tag{43}$$

where μ_1 denotes the smallest μ_j.

Proof of Lemma 6 : direct consequence of (41) and the fact that N is Hermitian-symmetric if and only if $\mathfrak{c}$ is not reduced to $\{0\}$.

□

Theorem 2.— *Let $(M, g) = (N, g)/\Gamma$ be a compact quotient of an irreducible Riemannian symmetric space of non-compact type, of dimension $n > 2$, where Γ is a discrete, co-compact subgroup of the isometry group G of N.*

(a) If N is not Hermitian-symmetric, M admits no $[g]$-compatible complex structure.

(b) If N is Hermitian-symmetric, the complex structure of M induced by the invariant complex structure of N is, up to conjugation, the only $[g]$-compatible complex structure of M.

Proof :

(a) If N is not Hermitian-symmetric, it follows by Lemma 6 (b) that the integrand of $S(g)$ is (constant) negative. Then, (a) follows from Theorem 1 (a) applied to the compact manifold (M, g).

(b) If N is Hermitian-symmetric, the center $\mathfrak{c}$ of $\mathfrak{h}$ is 1-dimensional, generated by the invariant complex structure, cf. above, and $S(g)$ is equal to zero by Lemma 6 (b). Then, (b) follows from Theorem 1 (b) (ii) applied to (M, g).

Remark 4 : the criterion of non-existence given by Theorem 1 (a) and its Corollary is an open condition. It follows that the conclusion of Theorem 2 (a) still holds for any conformal structure in some neighbourhood of $[g]$ (in the C^2-topology).

NOTE : similar results have been obtained for compact *inner* (not necessarily irreducible) Riemannian symmetric spaces (inner symmetric spaces $N = G/H$ are charactarized by G and H having same rank), cf. [B-G-M-R].

References

[Bou] J.P. Bourguignon - *Formules de Weitzenböck en dimension 4* in : Géometrie riemannienne en dimension 4, Exp. XVI, Cedic/Fernand Nathan, (1981).

[B-G-M-R] F.E. Burstall, G. Grantcharov, O. Muskarov, J.H. Rawnsley - *Hermitian structures on Hermitian symmetric spaces*, J. Geom. Phys. 10, 245-249 (1993).

[Boy] C.P. Boyer - *Conformal duality and Compact Complex Surfaces*, Math. Ann. 274, 517-526 (1986).

[Gau1] P. Gauduchon - *La 1-forme de torsion d'une variété hermitienne compacte*, Math. Ann. 267, 495-518 (1984).

[Gau2] P. Gauduchon - *Structures de Weyl et théorèmes d'annulation sur une variété conforme autoduale*, Ann. Sc. Norm. Sup. Pisa XVIII 4, 563-629 (1991).

[N-N] A. Newlander, L. Nirenberg - *Complex analytic coordinates in almost complex manifolds*, Ann. of Math. 65, 391-404 (1957).

[Kos] B. Kostant - *Holonomy and the Lie algebra of infinitesimal motions of a Riemannian manifold*, Trans. Amer. math. Soc. 80, 528-542 (1955).

[Sal] S. Salamon - *Riemannian Geometry and Holonomy Groups*, Pitman Research Notes in Math. Ser. 201 (1989).

$G_{m,h}$-Bundles over Foliations with Complex Leaves

Giuliana Gigante, Dipartimento di Matematica, Università di Parma, via dell'Università 12, I-43100 Parma, Italy. gigante@prmat.math.unipr.it

Adriano Tomassini, Dipartimento di Matematica, Università di Firenze, Viale Morgagni 67/A, I-50134 Firenze, Italy

In this paper, which must be considered a complement to [6], we are dealing with $G_{m,h}$ – *bundles* over a manifold X which is foliated by complex manifolds.

The fiber of a $G_{m,h}$-bundle $E \to X$ is $\mathbb{C}^m \times \mathbb{R}^h$ and the transition functions are CR and preserve the natural foliation of $\mathbb{C}^m \times \mathbb{R}^h$.

We denote by $\mathcal{E}$ the sheaf of germs of smooth CR-sections of E.

In [6] we have introduced the notion of q-complete foliation. The first result of this paper is that a $G_{m,h}$-bundle over a 1-complete foliation is a 1-complete foliation (Th.1). Using this fact and the vanishing theorem for the structure sheaf of X proved in [6] we then show that $H^j(X, \mathcal{E})$ vanishes for $j \geq 1$ when X and E are real analytic and X is 1-complete.

As an application to the real analytic Levi flat submanifolds of $\mathbb{C}^n$, we obtain a "CR-tubular neighborhood theorem" and an extension theorem for smooth CR-functions (Th.3 and Cor.4).

In the last part of the paper we assume that X is compact and we discuss the notion of positivity for a *CR-line bundle* (i.e for a $G_{1,0}$-bundle). We put in evidence a geometric condition on X (condition (*) of Sect. 5) which combined with a theorem of Hironaka ([12]) guarantees that X embeds in $\mathbb{CP}^N$ by a CR-map. When X is a *foliated torus* the embeddabilty condition is specified algebraically by the Th. 9.

§1. Preliminaries.

1. A *foliation with complex leaves* is a smooth foliation $\mathcal{F}$ on a manifold X of dimension $2n + k$ whose local models are domains $U = V \times B$ of $\mathbb{C}^n \times \mathbb{R}^k$ with $V \subset \mathbb{C}^n$, $B \subset \mathbb{R}^k$

Supported by the Project 40% M.U.R.S.T."Geometria reale e complessa"
This paper is in final form and no part of it will be submitted elsewhere.

and whose local transformations are of the form

$$\begin{cases} z' = f(z,t) \\ t' = h(t) \end{cases}$$

$z, z' \in \mathbb{C}^n$,$t, t' \in \mathbb{R}^k$ where f is holomorphic with respect to z ([15], [3], [13]).
A domain U as above is said to be a *distinguished coordinate domain* and z, t are said to be *distinguished local coordinates.*
We denote by $(X, \mathcal{F})$, or more simply by X , a foliation as above.
A $CR-$ *function* on X is a smooth function which is holomorphic along the leaves of X (i.e. with respect to z).
We denote by $\mathcal{D}$=$\mathcal{D}_X$ the sheaf of germs of CR-functions on X and by $\mathcal{D}_0$ =$\mathcal{D}_{0,X}$ the sheaf of germs of functions which are constant along the leaves.
A *morphism* $\mu : (X, \mathcal{F}) \to (X, \mathcal{F}')$ of foliations with complex leaves (or $CR-map$) is a smooth map $X \to X$ which preserves the leaves is such a way that for each leaf L the restriction of μ to L is holomorphic.

2. In [6] the notion of q-*complete* foliation has been introduced.
X is said to be $q-complete$ if there exists a smooth exhaustion function $\phi : X \to R$ which is strictly q-pseudoconvex along the leaves.
Then the following holds: if X is a real analytic 1-complete foliation, the cohomology groups $H^j(X, \mathcal{D})$ vanish for $j \geq 1$ ([6]).

§2. $\mathbf{G_{m,h}}$ – bundles.

1. Let $G_{m,h}$ be the group of the matrices

$$g = \begin{pmatrix} A & B \\ 0 & C \end{pmatrix}$$

where $A \in GL(m; \mathbb{C})$, $B \in GL(m, h; \mathbb{C})$ and $C \in GL(h; \mathbb{R})$.
We set $G_{m,0} = GL(m; \mathbb{C})$, $G_{0,h} = GL(h; \mathbb{R})$.
To each $g \in G_{m,h}$ we associate the linear transformation $\mathbb{C}^m \times \mathbb{R}^h \to \mathbb{C}^m \times \mathbb{R}^h$ given by $z' = Az + t$, $t' = Ct$, where $z, z' \in \mathbb{C}^m$, $t, t' \in \mathbb{R}^h$.

Let X be a foliation with complex leaves of dimension n and real codimension k and let $\mathcal{D}$ be its structure sheaf. A $G_{m,h}$-*bundle* on X is a vector bundle p: $E \to X$, which has $U \times \mathbb{C}^m \times \mathbb{R}^k$ as local model, and $G_{m,h}$ as structure group (U coordinate neighborhood of X).
The *structure cocycle* of E is a smooth CR-map $g_{ij} : U_i \bigcap U_j \to G_{m,h}$; i.e.,

$$g_{ij}(z,t) = \begin{pmatrix} A_{ij} & B_{ij} \\ 0 & C_{ij} \end{pmatrix}$$

where $A_{ij} = A_{ij}(z,t)$, $B_{ij} = B_{ij}(z,t)$, $C_{ij} = C_{ij}(t)$ are smooth and A_{ij}, B_{ij} are CR.Then E is a foliated manifold with complex leaves of dimension $m + n$ and codimension $h + k$.

If z, t are distinguished coordinates on X, and $(\zeta, \vartheta) \in \mathbb{C}^m \times R^h$, then the coordinate transformations on E are given by $z_i' = f_{ij}(z,t)$, $\zeta_i' = A_{ij}(z,t)\zeta_j + B_{ij}\vartheta_j$, $t_i = h_{ij}(t_j)$, $\vartheta_i = C_{ij}\vartheta_{ij}$. In particular $p : E \to X$ is a morphism of foliated manifolds.
The tangent bundle TX of X is a $G_{n,k}$-bundle with structure cocycle

$$g = \begin{pmatrix} \frac{\partial z'}{\partial z} & \frac{\partial z'}{\partial t} \\ 0 & \frac{\partial t'}{\partial t} \end{pmatrix}$$

where $\frac{\partial z'}{\partial z} = \left(\frac{\partial z'_\alpha}{\partial z_\beta}\right)$, $\frac{\partial z'}{\partial t} = \left(\frac{\partial z'_\alpha}{\partial t_\beta}\right)$, $\frac{\partial t'}{\partial t} = \left(\frac{\partial t'_\alpha}{\partial t_\beta}\right)$.
To see this is sufficient to write a real tangent vector as a linear combination of $\frac{\partial}{\partial z_\alpha}$'s , $\frac{\partial}{\partial \bar{z}_\alpha}$'s $\frac{\partial}{\partial t_\beta}$'s.
The bundle $T_F X$ of the holomorphic tangent vectors to the leaves of X is a $G_{n,o}$-bundle with structure cocycle $\frac{\partial z'}{\partial z}$.
$N_F = TX/T_F X$, the transverse bundle to the leaves of X, is a $G_{n,o}$-bundle with structure cocycle $\frac{\partial t'}{\partial t}$.
If X is embedded in a complex manifold Z, its transverse bundle $N = TZ/TX$ is not a $G_{m,h}$-bundle in general (see Sect. 4).
A *CR − section* of E is given by distinguished open covering $\{U_i\}$ of X and by m+h smooth functions $\lambda_1^j, \ldots, \lambda_m^j, \mu_1^j, \ldots, \mu_h^j$ for each U_j, where $\lambda_1^j, \ldots, \lambda_m^j$ are CR, $\mu_1^j, \ldots, \mu_h^j$ depend only on t and on $U_i \bigcap U_j$

$$\begin{pmatrix} \lambda^i \\ \mu^i \end{pmatrix} = \begin{pmatrix} A_{ij} & B_{ij} \\ 0 & C_{ij} \end{pmatrix} \begin{pmatrix} \lambda^j \\ \mu^j \end{pmatrix}$$

$\lambda^j = (\lambda_1^j, \ldots, \lambda_m^j)$, $\mu^j = (\mu_1^j, \ldots, \mu_h^j)$.
In particular the sheaf $\mathcal{E}$ of germs of CR-sections of E is locally isomorphic to $\mathcal{D}^{\oplus m} \bigoplus \mathcal{D}_{0,X}{}^{\oplus h}$.

Let $p : E \to X$, $p' : E' \to X$ be a $G_{m,h}$-bundle and a $G_{m',h'}$-bundle respectively. A $CR - morphism$ $\varphi : E \to E'$ is a morphism of real vectors bundles locally determined by a matrix $g \in G_{m+m',h+h'}$ where A, B are $m \times m'$ matrices of CR-functions and C is an $h \times h'$ matrix of real smooth functions depending only on t.
The set $Hom(E, E')$ of all CR-morphisms $E \to E'$ is a $G_{m'(m+m',hh'}$-bundle.

Theorem 1. *A $G_{m,h}$-bundle E over a 1-complete foliation X is a 1-complete foliation.*

Proof. For the applications we have in mind it will suffice to prove the theorem in the particular case when E and X are both real analytic.
First of all we observe the following two facts:

(1) if $X = \bigcup X_j$ where $\{X_j\}$ is an exhaustive sequence of domains such that every X_j is a 1-complete foliation, and X_j is Runge in X_{j+1} (i.e. $\mathcal{D}(X_{j+1})$ is dense in $\mathcal{D}(X_j)$) then X is a 1-complete foliation ([6]);

(2) let Z be a complex Stein manifold and $p: W \to Z$ be a holomorphic vector bundle; then if $Y \subset Z$ is a Runge domain, $p^{-1}(Y)$ is a Runge domain ([8]).

Now let $\tilde{X}$ be a complexification of X ([6]). Since $p: E \to X$ is real analytic we can extend it to a neighborhood $\tilde{U}$ of X by a holomorphic vector bundle $p : \tilde{E} \to \tilde{X}$ in such a way that E is a closed subset of $\tilde{E}$.
Let $\phi : X \to R$ be an exhaustion function which is strictly p.sh. along the leaves and let $X_j = \{\phi < j\}$. Then for each j there exists a neighborhood $\tilde{U}_j$ of $\bar{X}_j$ in $\tilde{U}$ with the following property: $\bar{X}_j$ has in $\tilde{U}_{j+1}$ a fundamental system of neighborhoods $\bar{W}_j$ which are relatively compact in $\tilde{U}_{j+1} \bigcap \tilde{U}_j$ and Runge in $\tilde{U}_{j+1}$([6]). In particular X_j is $\mathcal{O}(\tilde{U}_j)$-convex. Now $\tilde{E}_j = \tilde{E}_{|\tilde{U}_j}$ is Stein ([8]) and $\tilde{E}_{|\bar{W}_j}$ is Runge in $\tilde{E}_{j+1}$; it follows that $E_j = E_{|X_j}$ is a 1-complete foliation which is $\mathcal{O}(\tilde{E}_j)$-convex. In view of Freeman's approximation theorem ([4]) every smooth CR-function on E_j can be approximated by global CR-functions.

Thus E_j is Runge in E for every j and we invoke the property (2) above to conclude that E is then a 1-complete foliation.

§3. A vanishing theorerm.

1. With the above notations let X be a real analytic foliation, $p : E \to X$ be a real analytic $G_{m,0}$-bundle and $\mathcal{E}$ be the sheaf of germs of smooth CR-sections of E.

Theorem 2. *If X is 1-complete then*

$$H^j(X, \mathcal{E}) = 0$$

for $j \geq 1$.

Proof. Let us denote by $g_{ij} = A_{ij}$ the structure cocycle of E with respect to a distinguished open covering $\mathcal{U} = \{U_i\}$ of X. Then $g^*_{ij} = (A^t_{ij})^{-1}$ (where A^t_{ij} is the transpose of A_{ij}) is the structure cocycle of a $G_{m,0}$-bundle $p : E^* \to X$. E^* is a real analytic 1-complete foliation.

Let $\mathcal{D}^*$ be the sheaf of smooth CR-functions on E^* and for each domain $U \subset X$ let $\mathcal{D}^*(p^{-1}(U))_r$ denote the subspace of those functions which vanish on U of order r at least. We obtain a filtration of $\mathcal{D}^*(p^{-1}(U))$:

$$\mathcal{D}^*(p^{-1}(U)) = \mathcal{D}^*(p^{-1}(U))_0 \supset \mathcal{D}^*(p^{-1}(U))_1 \supset \cdots.$$

If U is a distinguished domain of X such that $E^*_{|U}$ is trivial and $\xi = (\xi_1, \ldots, \xi_m)$ is the fiber coordinate then a CR-function f belongs to $\mathcal{D}^*(p^{-1}(U))_r$ if and only if the derivatives

$$(D^r_\xi f)_{|U} = \frac{\partial^r f}{\partial \xi_1^{r_1} \cdots \partial \xi_m^{r_m}} |U$$

vanish for $r_1 + \cdots + r_m \leq r$.
Moreover if $U \subset U_i \bigcap U_j$ and $\xi^i = (\xi^i_1, \ldots, \xi^i_m)$, $\xi^j = (\xi^j_1, \ldots, \xi^j_m)$ are fiber coordinates on U_i, U_j respectively, we have

$$(D^r_{\xi^i} f)_{|U} = g^{(r)}_{ij} (D^r_{\xi^j} f)_{|U}$$

where $\{g_{ij}^{(r)}\}$ denotes the structure cocycle of $E^{(r)}$, the rth symmetric power of E. Conversely, if $U = \bigcup U_i$ and $\{S_{i,r_1,\ldots,r_m}\}$, $r_1 + \cdots + r_m = r$, is a smooth CR-section of $E^{(r)}$ then the local functions

$$\sum S_{i,r_1,\ldots,r_m}(\xi_1^i)^{r_1} \cdots (\xi_m^i)^{r_m}$$

define a CR-function on $p^{-1}(U)$.
Thus we obtain an isomorphism

$$\mathcal{E}^{(r)}(U) \simeq \mathcal{D}^*\left(p^{-1}(U)\right)_r / \mathcal{D}^*\left(p^{-1}(U)\right)_{r+1}$$

(where $\mathcal{E}^{(r)}$ is the sheaf of germs of CR-sections of $E^{(r)}$).

Let $\mathcal{U}^*$ be the open covering $\{p^{-1}(U_i)\}$ of E^*. Then from the preceding discussion we derive that each cohomology group $H^j(\mathcal{U}^*, \mathcal{D}^*)$ has a filtration

$$H^j(\mathcal{U}^*, \mathcal{D}^*) = H^j(\mathcal{U}^*, \mathcal{D}^*)_0 \supset H^j(\mathcal{U}^*, \mathcal{D}^*)_1 \supset \cdots$$

whose associated graded group is isomorphic to $\bigoplus H^j(\mathcal{U}, \mathcal{E}^{(r)})$. Consequently, in view of the Leray theorem we have a filtration on each group $H^j(E^*, \mathcal{D}^*)$ whose associated graded group is isomorphic to $\bigoplus H^j(X, \mathcal{E}^{(r)})$.
Since E^* is a 1-complete foliation (Th.1), from the vanishing theorem of [6], we derive $H^j(E^*, \mathcal{D}^*)$=0 for $j \geq 1$ and consequently that $H^j(X, \mathcal{E}^{(r)})$ vanishes for $j \geq 1$ and $r \geq 0$.
In particular for r=1 we obtain $H^j(X, \mathcal{E})$=0 for every $j \geq 1$.

§4. Applications.

1. As an application of the above vanishing theorem we derive the existence of tubular neighborhoods and an extension theorem for smooth CR-functions on a real analytic Levi flat submanifolds of $\mathbb{C}^m$.

We recall that a (locally closed) submanifold X of $\mathbb{C}^m$ is said to be *Levi flat* if each $x \in X$ is contained in some complex submanifold $Z \subset X$ of $\mathbb{C}^m$ with $dim_{\mathbb{R}}X - dim_{\mathbb{R}}Z = m$ and X is "generic" (i.e. if X is locally defined by $f_1 = \cdots = f_{2m-d} = 0$, $d = dim_{\mathbb{R}}X$, then $\partial f_1 \wedge \cdots \wedge \partial f_{2m-d} \neq 0$ on X [13]).
In this situation X carries a foliation $\mathcal{F}$ (the *Levi foliation* of X) whose leaves are complex submanifolds of dimension n and real codimension $k = m - n$.
Moreover if X is real analytic $\mathcal{F}$ is *extendable* : namely there is a neighborhood U of X and a unique holomorphic foliation $\tilde{\mathcal{F}}$ on U such that the leaves of $\mathcal{F}$ are leaves of $\tilde{\mathcal{F}}$ ([13]).
Thus, given a smooth CR-function f on X, we can ask to extend f on a neighborhood $W \subset U$ of X to a smooth function $\tilde{f}$, which is holomorphic along the leaves of $\tilde{\mathcal{F}}_{|W}$.
In the sequel we will answer this question.

2. The key point in the proof is the following "CR-tubular neighborhood theorem"

Theorem 3. *There exists a neighborhood* $W \subset U$ *of* X *and a smooth map* $q : W \to X$ *with the properties:*

(*i*) q *is a morphism* $\tilde{\mathcal{F}}_{|W} \to \tilde{\mathcal{F}}_{|W}$

(*ii*) $q_{|X} = id_{|X}$.

Proof. Let $T = X \times \mathbb{C}^m$, TX be the tangent bundle of X and $p : N \to X$ the transverse bundle of X in $\mathbb{C}^m$. Since $\mathcal{F}$ is extendable on U, there exists on U a distinguished open covering $\{U_i\}$ with holomorphic coordinates (z^i, τ^i) such that if $U_i \bigcap X = V_i \neq \emptyset$ then $\{V_i\}$ is a distinguished open covering of X with coordinates $z_1^i, \ldots, z_n^i, Re\,\tau_1^i, \ldots, Re\,\tau_k^i$. In particular if θ_α^i denotes the imaginary part of τ_α^i, $N_{|V_i}$ is generated by the vector fields $\eta_\alpha^i = \frac{\partial}{\partial \theta_\alpha^i}|X$, $1 \leq \alpha \leq k$. It is immediate to check that N is a $G_{0,k}$-bundle on X, whose structure cocycle is $C_{ij} = \frac{\partial(\theta_1^i, \ldots, \theta_k^i)}{\partial(\theta_1^j, \ldots, \theta_k^j)}|X = \frac{\partial(t_1^i, \ldots, t_k^i)}{\partial(t_1^j, \ldots, t_k^j)}|X$, $t_1^i = \text{Re } \tau_1^i, \ldots, t_k^i = \text{Re } \tau_k^i$.

Let $\tilde{N}$ be the $G_{k,0}$-bundle with the same structure cocycle as N and fiber $\mathbb{C}^k$. Let $\tilde{p} : \tilde{N} \to X$ be the projection; N is a real subbundle of $\tilde{N}$ and $\tilde{p}_{|N} = p$.
We have the following exact sequence of $G_{r,0}$-bundles on X

$$0 \longrightarrow T_F X \longrightarrow T \longrightarrow \tilde{N} \longrightarrow 0$$

and the exact sequence of sheaves

$$0 \longrightarrow \underline{Hom}(\tilde{N}, T_F X) \longrightarrow \underline{Hom}(\tilde{N}, T) \longrightarrow \underline{Hom}(\tilde{N}, \tilde{N}) \longrightarrow 0$$

where $\underline{Hom}$ denotes the sheaf of germs of CR-morphisms of $G_{r,0}$-bundles.
The exact cohomology sequence and Theorem 2 now imply that the homomorphism

$$\Gamma(X, \underline{Hom}(\tilde{N}, T)) \longrightarrow \Gamma(X, \underline{Hom}(\tilde{N}, \tilde{N}))$$

is onto.
Let $\varphi : \tilde{N} \to T$ be the CR-morphism which induces the identity $\tilde{N} \to \tilde{N}$. Then $\varphi(\xi) = (p(\xi), \psi(\xi)) \in X \times \mathbb{C}^m$ for every $\xi \in \tilde{N}$ and $\xi \mapsto \psi(\xi)$ is a smooth CR-map $\tilde{N} \to \mathbb{C}^m$. Moreover ψ has maximum rank along $\tilde{p}^{-1}(x)$, for every $x \in X$. Define $\tilde{\sigma} : \tilde{N} \to \mathbb{C}^m$ by $\tilde{\sigma}(\xi) = \tilde{p}^{-1}(\xi) + \psi(\xi)$. Since $\tilde{\sigma} = id$ on X, $\tilde{\sigma}$ has maximum rank along X. It follows that σ, the restriction of $\tilde{\sigma}$ to N, has maximum rank along X and consequently that it is invertible along X ([7],ch.II,§7).
To conclude our proof it suffices to define $q = p \circ \sigma^{-1}$.

Corollary 4. *Let* $X \subset \mathbb{C}^m$ *be a real analytic Levi flat submanifold,* $\mathcal{F}$ *the Levi foliation of* X *and* $\tilde{\mathcal{F}}$ *the holomorphic extension of* $\mathcal{F}$. *Then every CR-function* f *on* X *extends on a neighborhood* W *of* X *to a smooth function* $\tilde{f}$ *which is holomorphic along the leaves of* $\tilde{\mathcal{F}}$.

It suffices to take $\tilde{f} = f \circ q$.

§5. Positive line bundles.

1. In this section we suppose that X is real analytic and compact.

Consider a distinguished open covering $\{U_i\}$ of X, $U_i = V_i \times B_i$, $V_i \subset \mathbb{C}^n$ $B_i \subset \mathbb{R}^k$ and let

$$\begin{cases} z_i = f_{ij}(z_j, t_j) \\ t_i = h_{ij}(t_j) \end{cases}$$

be the local transformations of coordinates.
Then the matrices

$$\frac{\partial h_{ij}}{\partial t_j} = \frac{\partial(h_{ij}^1, \ldots, h_{ij}^k)}{\partial(t_j^1, \ldots, t_j^k)}$$

determine the structure cocycle of $N_{|F} = TX/T_{|F}$, the transverse bundle to the leaves of X.
Let $\tilde{X}$ be the complexification of X ([6]). We denote by $\{\tilde{U}_i\}$ a distinguished open covering of $\tilde{X}$ such that $\tilde{U}_i = V_i \times \tilde{B}_i$, $\tilde{B}_i \subset \mathbb{C}^k$, $\tilde{U}_i \bigcap X = U_i$ and by

$$\begin{cases} z_i = \tilde{f}_{ij}(z_j, \tau_j) \\ \tau_i = \tilde{h}_{ij}(\tau_j) \end{cases}$$

the local transformations of coordinates where Re $\tau_i^\alpha = t_i^\alpha$, Im $\tau_i^\alpha = \theta_i^\alpha$, $1 \leq \alpha \leq k$ and $\tilde{f}_{ij}$, $\tilde{h}_{ij}$ are obtained from f_{ij},h_{ij}, respectively, by complexification of t_j.
Since $\tilde{h}_{ij}$ is real on X we have $\theta_i = \psi_{ij}\theta_j^*$ where $\theta_i = (\theta_i^1, \ldots, \theta_i^k)$ and $\psi_{ij} = \psi(t_j, \theta_j)$. Moreover, since $\tilde{h}_{ij}$ is holomorphic we also have $\psi_{ij} = \frac{\partial h_{ij}}{\partial t_j}$ on X.
$\{\psi_{ij}\}$ is the structure cocycle for a vector bundle over $\tilde{X}$ of rank k.

2. Let us suppose $k = 1$ for simplicity and let a_i be local smooth functions on $\tilde{X}$ satisfying $a_j = \psi_{ij}^2 a_i$.
Then the local functions $a_i\theta_i^2$ define a smooth non negative function ϕ on $\tilde{X}$ and $\phi \neq 0$ outside X.
Consider the following condition :

$$(*) \qquad \partial\bar{\partial} \log \phi = \partial\bar{\partial} \log a_i + 2\frac{\partial\tau_i \wedge \bar{\partial}\bar{\tau}_i}{(\tau_i - \bar{\tau}_i)^2} \geq 0$$

near X.
Condition $(*)$ implies that ϕ is plurisubharmonic near X and that its Levi form has one positive eigenvalue in the transversal direction τ.
Indeed we have

$$\partial\bar{\partial} \log \phi = \frac{\partial\bar{\partial}\phi}{\phi} - \frac{\partial\phi \wedge \bar{\partial}\bar{\phi}}{\phi^2}$$

and on X, $\partial\bar{\partial}\log\phi = 2\partial\tau \wedge \bar{\partial}\bar{\tau}$.
In particular when $(*)$ holds, X has a fundamental system of weakly 1-complete complex manifolds ([12]).

We observe explicitly that this is certainly the case when X admits a parameter space.

Now let $p : L \to X$ be a real analytic $G_{1,0}$-bundle with structure cocycle $\{g_{ij}\}$. We call L a *CR-line bundle*.
L is said to be *positive* (along the leaves) if there is a smooth metric $h = \{h_i\}$ on (the fibres of) L with *positive curvature*; i.e., such that

$$(**) \qquad \sum_{1\leq\alpha\leq n} \frac{\partial^2 \log h_i}{\partial z^i_\alpha \partial \bar{z}^i_\beta} \xi^\alpha \bar{\xi}^\beta > 0$$

(z^i,t^i are distinguished coordinates).
Since L is real analytic we can extend it on a neighborhood U' of X to a holomorphic line bundle $\tilde{L} \to U'$. The metric h also extends to a smooth metric $\tilde{h} = \{\tilde{h}_i\}$ on $\tilde{L}$ and condition $(**)$ is preserved near X. For every positive λ, consider the new metric $\tilde{h}_{\lambda,i} = e^{\lambda\phi}\tilde{h}_i$ on $\tilde{L}$ and put $\zeta_1 = z^i_1, \ldots, \zeta_n = z^i_n, \zeta_{n+1} = \tau^i$.
We have at a point $x \in X$

$$\sum_{1\leq\alpha\leq n+1} \frac{\partial^2 \log \tilde{h}_{\lambda,i}}{\partial\zeta_\alpha \partial\bar{\zeta}_\beta} \xi^\alpha \bar{\xi}^\beta = \lambda \sum_{1\leq\alpha\leq n+1} \frac{\partial^2 \phi}{\partial\zeta_\alpha \partial\bar{\zeta}_\beta} \xi^\alpha \bar{\xi}^\beta + \sum_{1\leq\alpha\leq n+1} \frac{\partial^2 \log \tilde{h}_i}{\partial\zeta_\alpha \partial\bar{\zeta}_\beta} \xi^\alpha \bar{\xi}^\beta = \mathcal{L}_1 + \mathcal{L}_2.$$

$\mathcal{L}_1$ is positive for $\eta_1 = \cdots = \eta_n = 0$ while $\mathcal{L}_2$ is positive on the linear subspace $\eta_{n+1} = 0$.
It follows that for λ large, $\mathcal{L}_1 + \mathcal{L}_2$ is positive on a neighborhood of X.
Thus $\tilde{L}$ is positive on $\{\phi < c\}$ for c sufficiently small.
A theorem of Hironaka ([12]) now implies that a neighborhood of X embeds in $\mathbb{CP}^N$ by a locally closed holomorphic embedding. In particular X itself embeds in $\mathbb{CP}^N$ by a real analytic embedding which is holomorphic along the leaves.

§6. Applications: foliated tori.

1. A *foliated torus* or a *CR-torus* is defined by $T = \mathbb{C}^n \times \mathbb{R}^k/\Gamma$ where Γ is a lattice in $\mathbb{R}^{2n} \times \mathbb{R}^k$.
T is a real torus, foliated by complex manifolds of dimension n in such a way that the natural projection $\mathbb{C}^n \times \mathbb{R}^k \to T$ and the translations of T are (real analytic and) CR. A CR-torus is in particular a *CR-Lie group*.
A *CR-Lie group* is a real Lie group endowed with a CR-structure for which left and right translations are CR.
This is equivalent to the following: the Lie algebra $\mathcal{G}$ of G splits into $\mathcal{P} \oplus \mathcal{R}$ where $\mathcal{P}$ is an ideal, $\mathcal{R}$ is a vector space of dimension k and there exists a linear map $J : \mathcal{P} \to \mathcal{P}$ such that $J[X,Y] = [JX, JY]$ and $J^2 = -id$.
Equivalently $\mathcal{G}_{\mathbb{C}} = \mathcal{G} \otimes \mathbb{C}$ contains an ideal $\mathcal{Q}$ such that $\mathcal{Q} \bigcap \bar{\mathcal{Q}} = \{0\}$.

Let G be a connected compact CR-Lie group and consider the map $ad_g : G \to G$, $x \mapsto gxg^{-1}$;ad_g is CR and its differential $Ad_g : \mathcal{G} \to \mathcal{G}$ maps $\mathcal{P}$ into $\mathcal{P}$. Moreover $\psi : g \to Ad_g$ is continuous from G to $Hom_R(\mathcal{P}, \mathcal{P}) \simeq \mathbb{C}^{n^2}$.
Let F be the leaf through the identity $e \in G$: then F is a complex Lie group and $\psi_{|F}$ is holomorphic and bounded. It follows that $\psi_{|F}$ is constant ([9]).
Consequently F is (connected and) abelian and $\bar{F}$ is a complex torus T. Now we invoke a theorem of Cartan ([1]) to deduce the following

Theorem 5. *Every $g \in G$ is contained in a maximal CR-torus T_g with the same CR-dimension as G. In particular, if $dim_R G = 2n+1$ then G is solvable.*

2. Let us go back to foliated tori and let $T_1 = \mathbb{C}^n \times \mathbb{R}^k/\Gamma_1$, $T_2 = \mathbb{C}^n \times \mathbb{R}^k/\Gamma_2$.

Proposition 6. *Every (smooth) CR-map $\phi : T_1 \to T_2$ is induced by a map $\Phi : \mathbb{C}^n \times \mathbb{R}^k \to \mathbb{C}^n \times \mathbb{R}^k$, $(z,t) \mapsto \big(Az + Bt + \beta(t), Ct + \gamma(t)\big)$ where A,B,C are constant matrices.*

Before going into the proof we make some remarks.

Let $T = \mathbb{C}^n \times \mathbb{R}^k/\Gamma$ be a foliated torus and denote by $W_1 = (Z_1, r_1), \ldots,$ $W_{2n+k} = (Z_{2n+k}, r_{2n+k})$ a $\mathbb{Z}$-base of Γ, where $Z_j \in \mathbb{C}^n$, $r_j \in \mathbb{R}^k$, $1 \le j \le 2n+k$.
The matrix

$$\Omega = \begin{pmatrix} Z_1 & Z_2 & \ldots & Z_{2n+k} \\ r_1 & r_2 & \ldots & r_{2n+k} \end{pmatrix}$$

is said to be the *period matrix* of T. It is easy to prove that every foliated torus T is CR-isomorphic to a torus T' whose period matrix Ω is such that $Z_1 = \cdots = Z_k = 0$ and $r_1 = e_1 = (1, 0, \ldots, 0), \ldots, r_k = e_k = (0, \ldots, 0, 1)$.
Thus we have

$$r_i = \sum_{j=1}^{k} \lambda_{ij} e_j$$

for $k+1 \le i \le 2n+k$ and we can order the vectors $W'_1 = (0, e_1), \ldots, W'_k = (0, e_k)$, $W'_{k+1} = (Z_{k+1}, r_{k+1}), \ldots, W'_{2n+k} = (Z_{2n+k}, r_{2n+k})$ in such a way that $\lambda_{ij} \in \mathbb{Q}$ for $k+1 \le i \le 2n+k$, $1 \le j \le s$ and if $s+1 \le j \le k$ then $\lambda_{ij} \notin \mathbb{Q}$ for some i.
In this situation $\{W'_1, \ldots, W'_{2n+k}\}$ is said to be an *adapted base* for Γ'.
Write $t = t' + t''$, $t' \in \mathbb{R}^s$, $t'' \in \mathbb{R}^{k-s}$. Then every Γ'-periodic function $f : \mathbb{R}^k \to \mathbb{R}$ depends only on t'.

Now we are ready to prove the proposition 6.

Proof of Proposition 6. ϕ is induced by a smooth map $\Phi(z,t) = \big((U(z,t), V(z,t)\big)$ of $\mathbb{C}^n \times \mathbb{R}^k$ into itself, where $U(z,t) = A(z,t) + B(t)$ and $A(z,t)$ is holomorphic with respect to z.
Let $\{W_1 = (0, e_1), \ldots, W_k = (0, e_k), W_{k+1} = (Z_{k+1}, r_{k+1}), \ldots, W_{2n+k} = (Z_{2n+k}, r_{2n+k})\}$

be an adapted base of Γ_1.
Then arguing as in the complex case we obtain the following:

$$U(z,t) = A(t)z + B(t) \tag{1}$$

where $A(t) = A^{ij}(t)$ $B(t) = \big(B^1(t),\dots,B^n(t)\big)$;

$$V(t) = Ct + \gamma(t) \tag{2}$$

where C is a constant matrix and $\gamma(t)$ is Γ_1-periodic; furthermore for every $(\eta, r) \in \Gamma_1$, $t \in \mathbb{R}^k$, $z \in \mathbb{C}^n$, we have

$$\frac{\partial}{\partial t_l}[B^i(t+r) + \sum_{j=1}^{n} A^{ij}(t+r)\zeta_j - B^i(t))] = 0 \tag{3}$$

$$\frac{\partial}{\partial t_l}[\sum_{j=1}^{n}(A^{ij}(t+r) - (A^{ij}(t))z_j] = 0. \tag{4}$$

It follows that

$$A^{ij}(t) = \sum_{l=1}^{n} A_l^{ij} t_l + \alpha^{ij}(t) \tag{5}$$

where A_l^{ij} is constant and $\alpha^{ij}(t)$ is Γ_1-periodic and that

$$B(t) = Bt + \beta(t) \tag{6}$$

where $B = (B^{ij})$ is constant and $\beta(t) = \big(\beta^1(t),\dots,\beta^k(t)\big)$ is periodic with respect to $\mathbb{Z}^n \simeq \{\sum_{j=1}^{n} m_i e_i, m_i \in \mathbb{Z}\}$.
Let $\Gamma = \{\sum_{j=1}^{2n+k} m_i r_i, m_i \in Z\}$. Then since $\frac{\partial A^{ij}}{\partial t_l}$ is Γ-periodic $\big((3)\big)$, $\frac{\partial}{\partial t_l}\big(\beta^i(t+r) - \beta^i(t)\big)$ also is Γ-periodic $\big((4)\big)$.
Consequently for $r \in \Gamma$, $t \in \mathbb{R}^k$ we have

$$\beta^i(t+r) - \beta^i(t) = \tilde{\beta}^i(t,r) + \sum_{j=i}^{k} b_j^i(r) t_j \tag{7}$$

where $\tilde{\beta}^i(t,r)$ is Γ-periodic (with respect to t).
On the other hand since $\beta^i(t+r) - \beta^i(t) - \tilde{\beta}^i(t,r)$ is $\mathbb{Z}^n$-periodic $b_j^i(r)$ must vanish for every $i,j \in [1,k]$ and $r \in \Gamma$. Moreover, in view of (3) and (6) we have that

$$\tilde{\beta}^i(t,r) + \sum_{j=i}^{n} \alpha^{ij}(t)\zeta_j + \sum_{\substack{1 \le j \le n \\ 1 \le l \le k}} A_l^{ij} t_l \zeta_j$$

is constant with respect to t, $1 \leq i \leq k$. It follows that $A_l^{ij} = 0$ ($\tilde{\beta}^i$, $\alpha^{ij}(t)$ being periodic). Write $t = t' + t''$, $t' \in \mathbb{R}^s$, $t'' \in \mathbb{R}^{k-s}$.Then (see remark above)

$$(8) \qquad \frac{\partial}{\partial t''_l}(\tilde{\beta}^i(t,r) + \sum_{j=1}^{n} \alpha^{ij}(t)\zeta_j) = 0, \frac{\partial}{\partial t_l}\alpha^{ij} = 0$$

and thus

$$(8') \qquad \beta(t) = \beta(t' + t'') = \beta_*(t')t'' + \beta_0(t')$$

where $\beta_*(t')$ is a $n \times (k-s)$-matrix.
Furthermore from $\beta(t + \sum_{j=1}^{k} m_i e_i) = \beta(t)$ we obtain

$$(9) \qquad \beta(t' + \sum_{i=1}^{s} m_i e_i) = \beta_*(t') = 0, \beta_0(t' + \sum_{i=1}^{s} m_i e_i) = \beta_0(t')$$

for every $t' \in \mathbb{R}^s$, $t'' \in \mathbb{R}^{k-s}$, $m_i \in \mathbb{Z}$.
Consider $r_i = \sum_{j=1}^{n} \lambda_{ij} i e_i$ for $k+1 \leq i \leq 2n+k$. Then either $\lambda_{ij} = 0$ for $1 \leq j \leq s$, $k+1 \leq i \leq 2n+k$ or there exist integers $m_i \neq 0$ and m_{ij} such that $m_i r_i = \sum_{j=1}^{s} m_{ij} e_i + \sum_{j=s+1}^{k} c_{ij} e_j$. In both cases there exist integers $m_i \neq 0$ and m_{ij} for which the $\mathbb{R}^s$-component of $r_i^* = m_i r_i - \sum_{j=1}^{s} m_{ij} r_j$ vanishes and $(m_i Z_i, r_i^*) \in \Gamma_1$. Then (9) yelds $\tilde{\beta}(t, r_i^*) = \beta(t + r_i^*) - \beta(t) = \beta_0((t + r_i^*)') - \beta_0(t') = 0$ for every t and consequently from (8) we obtain that $\frac{\partial}{\partial t_l}\alpha(t) Z_h = 0$ for $t \in \mathbb{R}^k$ and $1 \leq l \leq k$, $k+1 \leq h \leq 2n+k$.
Now it is sufficient to observe that $\{Z_{k+1}, \ldots, Z_{2n+k}\}$ is an $\mathbb{R}$-base for $\mathbb{C}^n$ to conclude that $\alpha(t)$ is a constant matrix A.

Thus Φ is the map $(z,t) \longmapsto (Az + Bt + \beta(t'), Ct + \gamma(t'))$ where β and γ are periodic with respect to $\mathbb{Z}^s \simeq \{\sum_{i=1}^{s} m_i e_i\}$.

Let $T = \mathbb{C}^n \times \mathbb{R}^k/\Gamma$ and $T' = \mathbb{C}^n \times \mathbb{R}^k/\Gamma'$ be two foliated tori and let Ω and Ω' be two period matrices of T and T' respectively. Then

Corollary 7. *T and T' are isomorphic CR-Lie groups if and only if*

$$M\Omega = \Omega' M' =$$

where

$$M = \begin{pmatrix} A & B \\ 0 & C \end{pmatrix}$$

is invertible and $M' \in GL(2n+k, \mathbb{Z})$.

Let $s = s(T)$ be the integer defined above. Then

Corollary 8. *T is isomorphic to a product $T_1 \times T_2$ where T_1 is a real torus of dimension s and T_2 is a foliated torus with $s(T_2) = 0$.*

Proof. We change a little our notations and we denote by

$$\Omega = \begin{pmatrix} e_1 & \dots & e_k & r_{k+1} & \dots & r_{2n+k} \\ 0 & \dots & 0 & Z_{k+1} & \dots & Z_{2n+k} \end{pmatrix}$$

the period matrix of T, where

$$r_i = \sum_{j=1}^{k} \lambda_{ij} e_j$$

for $1 \le i \le 2n+k$ and $\lambda_{ij} \in \mathbb{Q}$ for $k+1 \le i \le 2n+k$, $1 \le j \le s$.
We prove that there exists a matrix $P \in GL(2n+k; \mathbb{Z})$ such that the first row of ΩP is $(d, 0, \dots, 0)$, $d \in \mathbb{Q}$.
To do this consider the first row $w_1 = (1, 0, \dots, 0, \lambda_{k+1,1}, \dots, \lambda_{2n+k,1})$ of Ω and set $w_1 = q' v_1$ where $v_1 = (q, 0, \dots, 0, p_{k+1}, \dots, p_{2n+k})$, $q' \in \mathbb{Q}$ and $q, p_{k+1}, \dots, p_{2n+k}$ are integers with no common factors. If $\times$ denotes the scalar product in $\mathbb{R}^{2n+k}$ we can find $v_1' \in \mathbb{Z}^{2n+k}$ such that $v_1 \times v_1' = 1$. Next we find $v_2, w_2' \in \mathbb{Z}^{2n+k}$ such that $v_1 \times v_2 = 0$, $w_2' \times v_2 = 1$ and we set $v_2' = w_2' - (w_2' \times v_1') v_1$. Then $v_1 \times v_1' = 1$, $v_1 \times v_2 = 0$, $v_1' \times v_2' = 0$, $v_2 \times v_2' = 1$. We iterate this process in order to construct vectors $v_1', v_2, \dots, v_{2n+k}$ and $v_1, v_2', \dots, v_{2n+k}'$ with integers components with above properties; then for P we take the matrix whose columns are $v_1', v_2, \dots, v_{2n+k}$ respectively (and P^{-1} is the one whose columns are $v_1, v_2', \dots, v_{2n+k}'$).
In particular the lattice Γ has a base $(d, r_1, Z_1), (0, r_2, Z_2), \dots, (0, r_{2n+k}, Z_{2n+})$ where $d \in \mathbb{Q}, r_j \in \mathbb{R}^{k-1}, Z_j \in \mathbb{C}^n$, $1 \le j \le 2n+k$.
Let $A : \mathbb{R}^k \to \mathbb{R}^k$ be linear, invertible such that $A(d, r_1) = (1, 0)$ and let $B : \mathbb{R}^k \to \mathbb{C}^n$ be linear and such that $B(d, r_1) = -Z_1, B(0, r) = 0$. Then $(t, z) \mapsto (At, z + Bt)$ gives a CR-isomorphism between T and a foliated torus whose period matrix has $(1, 0, \dots, 0)$ as first row and first column. This proves that T is isomorphic to $S^1 \times T'$, where T' is a foliated torus with $s(T') = s(T) - 1$. Our statement follows now by induction on s.

By virtue of Cor.8 we will assume from now on that $s(T) = 0$.

3. Let Γ be a lattice in $\mathbb{C}^n \times \mathbb{R}^k$ generated by $W_1 = (0, e_1), \dots, W_k = (0, e_k)$, $W_{k+1} = (Z_{k+1}, r_{k+1}), \dots, W_{2n+k} = (Z_{2n+k}, r_{2n+k})$. We complexify Γ with respect to $\mathbb{R}^k$ to obtain a new lattice $\tilde{\Gamma}$ in $\mathbb{C}^n \times \mathbb{R}^k$ generated by $\tilde{W}_1 = (0, e_1, 0), \dots, \tilde{W}_k = (0, e_k, 0)$, $\tilde{W}_{k+1} = (0, 0, ie_1), \dots, \tilde{W}_{2k} = (0, 0, ie_k), \tilde{W}_{2k+1} = (Z_{k+1}, r_{k+1}, 0), \dots,$ $\tilde{W}_{2n+k} = (Z_{2n+k}, r_{2n+k}, 0)$.
The complex torus $\tilde{T} = \mathbb{C}^n \times \mathbb{R}^k / \tilde{\Gamma}$ is a complexification of $T = \mathbb{C}^n \times \mathbb{R}^k / \Gamma$.
Due to this particular situation, condition $(*)$ of sec.5 is satisfied in such a way that if T carries a CR-line bundle which is positive along the leaves then T embeds into $\mathbb{C}\mathbb{P}^N$.

For a foliated torus $T = \mathbb{C}^n \times \mathbb{R}^k / \Gamma$ real analytic CR-embeddability in $\mathbb{C}\mathbb{P}^N$ has an algebraic description. In order to give the precise statement we recall that the cohomology of T with values in $\mathcal{D} = \mathcal{D}_T$ is isomorphic to the cohomology of the tangential De Rham-Dolbeault complex of T ([10]). In particular the elements of $H^1(T, \mathcal{D})$ and $H^2(T, \mathcal{D})$ are

represented by differential forms $\sum a_i d\bar{z}_i$, $\sum b_{ij} d\bar{z}_i \wedge d\bar{z}_j$ on $\mathbb{C}^n \times \mathbb{R}^k$ respectively where $a_i = a_i(t)$, $b_{ij} = b_{ij}(t)$ are Γ-periodic and therefore constant since $s(T) = 0$.
Moreover, given a CR-line bundle L on T we denote by $c(L)$ the Chern class of L.

Assume now that T is embedded in $\mathbb{CP}^N$ by a real analytic CR-map $f : T \to \mathbb{CP}^N$ and let $\tilde{T} = \mathbb{C}^n \times \mathbb{R}^k/\tilde{\Gamma}$ be the complexification of T. Then f extends on a neighborhood U of T in $\tilde{T}$ to a holomorphic map $\tilde{f} : U \to \mathbb{CP}^N$. Let $L = \tilde{f}^*\mathcal{O}(1)_{|T}$. Then the space of the real analytic CR-sections of L separates points of X; moreover L is positive along the leaves and its Chern class is represented by a real differential 2-form $c(L)$ on $\mathbb{R}^{2n} \times \mathbb{R}^k$ with constant coefficients. $c(L)$ is associated to an alternating bilinear form E on $\mathbb{R}^{2n} \times \mathbb{R}^k$ such that $E(\Gamma \times \Gamma) \subset \mathbb{Z}$ ([16]).

We want to prove the following

Theorem 9. *The restriction of E to $\mathbb{R}^{2n} \times \{0\}$ is the imaginary part of a non degenerate positive definite hermitian form H on $\mathbb{C}^n \times \{0\}$. Conversely if there exists an alternating bilinear form E on $\mathbb{R}^{2n} \times \mathbb{R}^k$, integral on $\Gamma \times \Gamma$ and such that the restriction of E to $\mathbb{R}^{2n} \times \{0\}$ is the imaginary part of a strictly positive definite hermitian form H on $\mathbb{C}^n \times \{0\}$ then T embeds in $\mathbb{CP}^N$ by a real analytic CR-map.*

We need some preliminary considerations.

Let $\mathcal{A}$ denote the algebra of all invertible CR-functions on $\mathbb{C}^n \times \mathbb{R}^k$. Then a CR-line bundle on T is determined by a map $\mu : \Gamma \to \mathcal{A}$ in the following way: let $f_{(w,r)}$ denote the invertible function $\mu(w, r)$ and suppose that

$$f_{(w,r)}(z + w', t + r')f_{(w,r')}(z, t) = f_{(w+w',r+r')}(z, t);$$

define on $\mathbb{C}^n \times \mathbb{R}^k \times \mathbb{C}$ an equivalence relation $\sim$ by $(z, t, \zeta) \sim (z + w, t + r, f_{(w,r)}(z, t)\zeta)$ for $(z, t, \zeta) \in \mathbb{C}^n \times \mathbb{R}^k \times \mathbb{C}$ and $(w, r) \in \Gamma$; then $\mathbb{C}^n \times \mathbb{R}^k \times \mathbb{C}/\sim$ is a CR-line bundle on T. Moreover two maps μ, μ' as above determine the same CR-line bundle if and only if $f'_{(w,r)}(z, t) = f_{(w,r)}(z, t)g(z + w, t + r)g(z, t)^{-1}$ where $g \in \mathcal{A}$([11]).
The CR-bundles with vanishing Chern class correspond to the forms $\omega = \sum a_i d\bar{z}_i$ on $\mathbb{C}^n \times \mathbb{R}^k$ with constant coefficients and are determined by the complex numbers

$$\rho_{(w,r)} = \exp\left(2\pi i \int_0^w \omega + \bar{\omega}\right)$$

$(w, r) \in \Gamma$, which in fact depend only on w. Furthermore, since $\rho_{(w)} = \rho_{(w,r)} \in S^1$ and $\rho_{(w)}\rho_{(w')} = \rho_{(w+w')}$ we see that every CR-bundle with vanishing Chern class is determined by a homomorphism $\Gamma \to S^1$.

Proof of theorem 9. With the above notations let the Chern class $c(L)$ of L be represented by the form

$$\frac{i}{2}(h_{jl}dz_j \wedge d\bar{z}_l) + 2ReD_{\alpha j}dt_\alpha \wedge dz_j + c_{\alpha\beta}dt_\alpha \wedge dt_\beta$$

and

$$E((z,0),(\zeta,0)) = -Im\, h_{jl} z_j \bar{\zeta}_l.$$

The summation convention that repeated Latin indices (respectively Greek indices) indicate summation from 1 to n (respectively from 1 to k) is followed here as it will be throughout.

We must prove that $h_{jl} z_j \bar{z}_l \leq 0$.
Consider for this the following form

$$K((z,t),(\zeta,s)) = -h_{jl} z_j \bar{\zeta}_l - C_{\alpha\beta} t_\alpha s_\beta + 2i\bar{D}_{\alpha j} t_\alpha \bar{\zeta}_j - 2iD_{\alpha j} s_\alpha z_j$$

where $C_{\alpha\beta} = -ic_{\alpha\beta}$ for $\alpha < \beta$, $C_{\alpha\beta} = ic_{\beta\alpha}$ for $\alpha > \beta$ and $C_{\alpha\alpha} = 0$. We have $K((z,t),(\zeta,s)) = \bar{K}((\zeta,s),(z,t))$ and $E((z,t),(\zeta,s)) = ImK((z,t),(\zeta,s))$.
Given $(w,r) \in \Gamma$ let

$$f_{(w,r)}(z,t) = \exp(\pi K((z,t),(w,r)) + \frac{\pi}{2} K((w,r),(w,r))).$$

Then

$$f_{(w,r)}(z+w', t+r') f_{(w',r')}(z,t) \left[f_{(w+w',r+r')}(z,t)\right]^{-1} = \exp(-iE((w,r),(w',r')));$$

replacing L by $2L$ we may assume that $E(\Gamma \times \Gamma) \subset 2\mathbb{Z}$ in such a way that the set $\{f_{(w,r)}\}$ defines a CR-line bundle L_1 on T. L_1 has the same Chern class as L. It follows that L is determined by $\{f_{(w,r)}\rho_{(w)}\}$ where $\rho : \Gamma \to S^1$ is a character and that the analytic CR-sections of L are determined by the analytic CR-functions θ on $\mathbb{C}^n \times \mathbb{R}^k$ which satisfy

$$\theta(z+w, t+r) = \theta(z,t)\rho_{(w)}(r) \exp(\pi K((z,t),(w,r)) + \frac{\pi}{2} K((w,r),(w,r))).$$

Each function θ extends holomorphically on a neighborhood U of $\mathbb{C}^n \times \mathbb{R}^k$ in $\mathbb{C}^n \times (\mathbb{R}^k \times i\mathbb{R}^k)$, and owing to the above property we may take $U = \mathbb{C}^n \times (\mathbb{R}^k \times iI^k)$ where $I =]-\epsilon, \epsilon[$ for some positive ϵ.
Define on U the function

$$\psi(z,\tau) = \exp(\pi K(z,t))\, |\theta(z,t)|^2$$

where

$$K(z,t) = -h_{jl} z_j \bar{z}_l - C_{\alpha\beta} \tau_\alpha \tau_\beta + 4Im\bar{D}_{\alpha j} \tau_\alpha \bar{z}_j;$$

choose ϵ in such a way that ψ is bounded on U and set $\tilde{\psi}(z,\tau) = \psi(z,\tau) \sum_{\alpha=1}^{k} (\epsilon^2 - t'^2_\alpha)$, $t'_\alpha = Im\tau_\alpha, 1 \leq \alpha \leq k$. $\tilde{\psi}$ is periodic along the subspaces $t'_\alpha = const, 1 \leq \alpha \leq k$ and it vanishes on ∂U. Let $(z^\circ, \tau^\circ) \in U$ be a maximum point for $\tilde{\psi}$; then $Hess(\log \tilde{\psi})$, the complex Hessian of $\log \tilde{\psi}$ at (z°, τ°) must be negative definite. We have

$$Hess(\log \psi) = -\pi K(\xi, \eta) + a_{\alpha\beta} \eta_\alpha \eta_\beta \leq 0$$

and from this we deduce that $h_{j l}\xi_j\bar{\xi}_l = -K(\xi, 0) \leq 0$.
Suppose that $K(\xi, 0)$ is degenerate; then $D_{\alpha j} = 0$ (otherwise $Hess(\log\tilde{\psi})$ would be strictly positive for some ξ and η); consequently every function θ satisfies

$$\theta(z+w, t+r) = \theta(z,t)\rho_{(w)}(r)\exp(-\pi i C_{\alpha\beta}t_\alpha r_\beta - \frac{\pi}{2} i C_{\alpha\beta}r_\alpha r_\beta)$$

for every $(z,t) \in \mathbb{C}^n \times \mathbb{R}^k, (w,r) \in \Gamma$, where $C_{\alpha\beta} \in \mathbb{R}$. In particular we have $|\theta(z+w, t+r)| = |\theta(z,t)|$ and then from the maximum principle we deduce that $\theta = \theta(t)$. This is in contradiction to our assumption that the global sections of L separate points on T.

In particular a foliated torus T whose leaves are complex curves is always embeddable in $\mathbb{CP}^N$. We observe explicitly that this could not be the case for $\tilde{T}$, the complexification of T.
As an example we consider in $\mathbb{C} \times \mathbb{R}$ the lattice Γ generated by $\omega_1 = (0,0,1), \omega_3 = (1,0,0), \omega_4 = (0,1,\sqrt{2})$. All leaves of $T = \mathbb{C} \times \mathbb{R}/\Gamma$ are everywhere dense. Let $\tilde{\Gamma}$ be the lattice generated by $\tilde{\omega}_1 = (0,0,1,0), \tilde{\omega}_2 = (0,0,0,1), \tilde{\omega}_3 = (1,0,0,0), \tilde{\omega}_4 = (0,1,\sqrt{2},0)$ and $\tilde{T} = \mathbb{C}^2/\tilde{\Gamma}$; T is a subfoliation of $\tilde{T}$ and $\tilde{T}$ is not algebraic ([14]). Consider the matrix $E = \big(E(\tilde{\omega}_i, \tilde{\omega}_j)\big)$ where $E(\tilde{\omega}_1, \tilde{\omega}_2) = 1$, $E(\tilde{\omega}_2, \tilde{\omega}_1) = -1$ and the remainders are zero and define on $\mathbb{C}^2$ the hermitian form H by $H(\zeta, \zeta') = E(i\zeta, \zeta') + iE(\zeta, \zeta')$, $\zeta = (z, \tau)$, $\zeta' = (z', \tau')$. To H corresponds a line bundle $\tilde{L} \to \tilde{T}$ whose restriction to T is positive.

REFERENCES

[1] J.F. Adams: *Lectures on Lie groups,* W. A. Benjamin, New York, 1969;
[2] A. Andreotti-H.Grauert: *Théorèmes de finitude pour la cohomologie des espaces com plexes,* Bull. Soc. Math. France 90 (1962), 193-259;
[3] M. Freeman: *Local complex foliations of real submanifolds*, Math. Ann. 209 (1970), 1-30
[4] M. Freeman: *Tangential Cauchy-Riemann equations and uniform approximation,* Pac. J. Math. 33, N.1 (1970), 101-108;
[5] G. Gigante-G. Tomassini: *CR-structures on a Real Lie Algebra*, Adv. in Math. vol. 94, N. 1, July 1992;
[6] G. Gigante-G. Tomassini: *Foliations with complex leaves*, to appear in Diff. Geom. and its applic.;
[7] M. Golubitsky-V. Guillemin: *Stable Mappings and Their Singularities*, Springer-Verlag New York Heidelberg Berlin 1973;
[8] R. Gunnung-H. Rossi: *An Introduction to Complex Analysis in Several variables*, North-Holland 1973;

[9] S. Kobayashi: *Hyperbolic Manifolds and Holomorphic Mappings,* Dekker, New York, 1970;
[10] C. C. Moore-C. Schochet: *Global Analysis on Foliated Spaces*, Math. Sc. Res. Inst. Pub. Springer-Verlag 1988;
[11] D. Mumford: *Abelian Varieties*, Oxford University Press 1970;
[12] S. Nakano: *Vanishing theorems for weakly 1-complete manifolds,*Number Theory, Algebraic Geometry and Comm. Algebra in honour of Y. Akizuki, Kinokuniya Tokio 1973, 169-179;
[13] C. Rea: *Levi flat submanifolds and biholomorphic extension of foliations*, Ann. Sc. Nor. Sup. Pisa, Serie III-Vol. XXVI (1972), 664-681;
[14] I.R.Shafarevich: *Basic Algebraic Geometry,* Die Grund. der math. Wiss .213, Springer-Verlag 1974;

[15] F. Sommer: *Komplexe analytische Blätterung reeler Maningfaltigkeiten in* $\mathbb{C}^n$; Math. Ann. 136 (1958), 111-133;
[16] A. Weil: *Variétés abéliennes et courbes algébriques*, Hermann, Paris, 1948.

A Real Analytic Version of Abels' Theorem and Complexifications of Proper Lie Group Actions

Peter Heinzner, Mathematisches Institut, Ruhr-Universität Bochum, Universitätsstrasse 150, D-44780 Bochum, Germany. heinzner@ruba.rz.ruhr-uni-bochum.de

Alan T. Huckleberry, Mathematisches Institut, Ruhr-Universität Bochum, Universitätsstrasse 150, D-44780 Bochum, Germany. huckleberry@ruba.rz.ruhr-uni-bochum.de

Frank Kutzschebauch, Mathematisches Institut, Ruhr-Universität Bochum, Universitätsstrasse 150, D-44780 Bochum, Germany. kutzschebauch@ruba.rz.ruhr-uni-bochum.de

I. Introduction

By Whitney's Theorem (see [Hi] and sec. VI) a differentiable manifold M possesses a real analytic structure. If it has countable topology, which we shall assume from now on, M can be realized as a closed real analytic totally real submanifold of a complex manifold X with $\dim_{\mathbb{C}} X = \dim_{\mathbb{R}} M$ ([WB],[Sh]). If furthermore every component of X has non-empty intersection with M, we refer to X as a complexification of M. Let (M_1, X_1) and (M_2, X_2) be complexifications and $\phi : M_1 \to M_2$ a real analytic map. Then there exists a neighborhood U_1 of M_1 in X_1 and a unique holomorphic map $\tilde{\phi} : U_1 \to X_2$ with $\tilde{\phi}|M_1 = \phi$. If ϕ is in addition a real analytic isomorphism, then U_1 can be chosen so that $\phi : U_1 \to U_2$ is a biholomorphic map onto its image U_2.

Utilizing his solution of the Levi-Problem, Grauert showed that X *can be chosen to be a Stein manifold* ([G]). From Remmert's embedding theorem for Stein manifolds it follows that M can be realized as a real analytic submanifold of some $\mathbb{R}^N$. If M_1 and M_2 are real analytic manifolds which are diffeomorphic via $\varphi : M_1 \to M_2$, then, regarding M_2 as a closed submanifold of $\mathbb{R}^N$, we may approximate φ in the Whitney topology (see sec. II) by a real analytic map $\tilde{\varphi} : M_1 \to \mathbb{R}^N$ so that the composition $\pi \circ \tilde{\varphi}$ with the (real analytic) normal bundle projection is still a diffeomorphism. In particular, the real analytic structure is unique.

Let G be a Lie group which is acting as a group of real analytic diffeomorphisms of M, i.e. the action map $G \times M \to M$ is real analytic. If this action extends to a real analytic action $G \times X \to X$ so that the individual automorphisms are holomorphic, then we refer to X as a *G-complexification of M*.

Examples. (1) Let M be compact and V be a smooth vector field on M which vanishes on an open set. It follows from the compactness of M that V can be globally integrated, i.e., it comes from an action of $G \cong \mathbb{R}$. However, unless $V \equiv 0$, the identity principle implies that there is no G-complexification. Of course this is only an example in the differentiable category.

(2) For a real analytic example which can not be G-complexified, we let $M := S^1$ and let V be a real analytic vector field with at least three zeros. This yields a $G \cong \mathbb{R}$-action on S^1.

This project was partially supported by Sonderforschungsbereich 237 and the Schwerpunkt Komplexe Mannigfaltigkeiten of the Deutscheforschungsgemeinschaft and the Graduiertenkollegs Geometrie und mathematische Physik.

However, it follows from the uniformization theorem that there is no Riemann surface with a non-trivial $\mathbb{R}$-action of holomorphic transformations having at least three fixed points.

We consider here *proper G-actions*, i.e., the map $G \times M \to M \times M$, $(g, x) \mapsto (g(x), x)$, is proper. This implies that the geometric quotient M/G, i.e., the space of G-orbits, is Hausdorff in the quotient topology. In particular, the *orbits* Gx are closed and furthermore the *isotropy groups* G_x are compact for all $x \in M$. See sec. VI for a general discussion of proper actions.

In the present paper we restrict to the case where G is a Lie group having only finitely many components. The main theorems are valid for proper actions of arbitrary Lie groups ([H1],[K1]), e.g., infinite discrete groups, but their proofs require a more technical approach which does not fit into our concept here. In virtually all proofs the difference between *finitely many components* and *connected* is only notational and the *generalization* to the former case is left to the reader. If care is needed, we supply additional remarks (see e.g. Proposition III.1)

One of the main goals of this exposition is to underline the importance of the following result of Abels ([A]):

Let G be a Lie group with finitely many components which is acting properly as a group of diffeomorphisms on a differentiable manifold M. Let K be a maximal compact subgroup. Then there is a smooth equivariant map $\pi : M \to G/K$.

In sec. VI we prove a real analytic version of this theorem.

By complexifying the fiber and base of a *real analytic Abels fibration* we construct a complexification of a proper action (sec. VII):

Let G be a Lie group with finitely many components which is acting properly on M as a group of real analytic diffeomorphisms. Then there exists a Stein G-complexification Z where G is acting properly.

Remarks. (1) The proof of Whitney's theorem as given in [Hi] on the existence of a real analytic structure can be adapted to the case of proper actions: *If a Lie group G is acting properly as a group of diffeomorphisms of a smooth manifold M, then there is a G-stable real analytic structure on M* (see [I] and sec. III). In fact, using a differential geometric averaging technique, the third author has recently shown that this structure is *unique* ([K2]). Of course, due to the existence of G-stable real analytic structures, our complexification result is valid for smooth actions as well.

(2) In the case where $M = G/K$ is a symmetric homogeneous space of a semi-simple Lie group G, the existence of a Stein G-complexification where G acts properly was proved by Akhiezer and Gindikin ([AG]).

(3) In ([H1]) the first author constructs a type of universal Stein complexification X^* where the universal complexification $G^{\mathbb{C}}$ of the group G acts. In this case, in particular due to the fact that $G \to G^{\mathbb{C}}$ is in general not injective, the universal $G^{\mathbb{C}}$ complexification

$M \to X^*$ is also not necessarily injective. Furthermore, in [H1] there is no discussion of the *properness* of the action of G on X^*.

(4) Our complexification has the following universality property: *Any G-equivariant real analytic map of M into a complex manifold where G acts can be extended holomorphically and G-equivariantly to an open neighborhood of M in its complexification.* This yields the desired unicity (*as a G-germ along M*) of the complexification in the above Theorem (see sec. VII).

We conclude this paper by proving the following *retraction* result (see sec. VIII).

Theorem. *Let G act properly on M and let X be a G-complexification where G is acting properly. Then there exists an open G-invariant Stein neighborhood of M in X which contains M as a strong G-equivariant deformation rectract.*

We wish to underline the fact that, once the principle of the real analytic slice is understood, the complexification of proper actions is a relatively simple consequence of known results. Those results which are needed from the theory of actions of compact groups are scattered through the literature in various contexts (e.g., [H1],[H2],[HHL]). In sec. II-V we have attempted to give a systematic account of the results which are used here. We hope that this also serves as a good introduction to certain recent developments in the area of compact Lie groups of holomorphic transformations.

II. Actions of compact groups.

After introductory remarks on approximation by real analytic functions and by special functions related to actions of compact groups, we summarize the basic techniques for constructing equivariant maps. For actions of a compact group K of holomorphic transformations on a Stein space X, the algebraic nature of the local $K^{\mathbf{C}}$-action is described. The case where X possesses only the constant invariant holomorphic functions is analyzed in some detail.

1. Spaces of functions. Suppose that M is a real analytic manifold which is countable at infinity and K is a compact Lie group acting as a group of real analytic diffeomorphisms on M. Let $C^\infty(M)$ denote the space of smooth complex valued functions and $A(M)$ the subspace of real analytic functions. A function f is said to be *K-finite* if its orbit Kf is contained in a finite-dimensional subspace of the function space at hand. The C^∞-Fourier theorem shows that the K-Fourier series of $f \in C^\infty(M)$, $f = \sum f_\rho$, is convergent in the C^∞-topology (see e.g. [W]). The projections f_ρ are K-finite. So in particular the K-finite functions are dense. Analogously, if X is a complex space equipped with a K-action of holomorphic automorphisms, then functions in the Fréchet space $\mathcal{O}(X)$ have a convergent Fourier development and the K-finite functions are likewise dense.

Proposition 1. *Let X be a Stein space which is equipped with an action of a compact group K of holomorphic transformations and let Ω be a K-stable relatively compact open subset. Then there exists a holomorphic K-equivariant map $F : X \to V$ to a finite-dimensional complex representation space such that $F|\Omega$ is an injective immersion.*

Proof. Since $\bar{\Omega}$ is compact, there exists a globally defined holomorphic map $G : X \to W$ to some complex vector space which is an injective immersion on Ω. By the approximation result described above, we may assume that the component functions of G with respect to some basis of W are K-finite. The union of the orbits of these functions generate a finite-dimensional vector space V^* in $\mathcal{O}(X)$. Point evaluation $f \mapsto f(x)$ yields the desired map. □

In order to prove embedding results or results of the above type for the real analytic manifold M it is useful to apply Stein theory. For example, for a given Stein complexification X of M we have the distinguished subspace $\mathcal{O}(X) \hookrightarrow A(M)$. Recall that, by the embedding theorem for Stein manifolds, X can be realized as a closed submanifold of some complex vector space. Of course the restriction embeds M. Hence approximation problems for real analytic functions on M can be reduced to the analogous problems for domains in a vector space where classical approximation results can be applied. We now formulate the essential result in this context.

For K compact in a domain $\Omega \subset \mathbf{R}^n$ and $0 \leq m \leq k$, define

$$\|f\|_m^K := \sum_{|\alpha| \leq m} sup_K |D^\alpha f| \quad for\ all\ f \in C^k(\Omega).$$

We refer to an increasing sequence $\{K_n\}$ of compact subsets of Ω as an exhaustion if $K_n \uparrow \Omega$, $K_0 = \emptyset$, and $int(K_{n+1}) \supset K_n$.

Whitney's Approximation Theorem. *Let m_n be a sequence of positive integers with $0 \leq m_n \leq k$ and ϵ_n an arbitrary sequence of positive numbers. Then for all $f \in C^k(\Omega)$ there exists $g \in A(\Omega)$ with*

$$\|f - g\|_{m_n}^{K_{n+1} \setminus K_n} \leq \epsilon_n \quad for\ all\ n.$$

□

For a detailed account of this important result see ([N]). One defines the Whitney- or strong-topology on $C^k(\Omega)$ by regarding the set of $g \in C^k(\Omega)$ which satisfy the above inequalities to be a neighborhood of f, i.e., for every exhaustion, choice of derivatives to be approximated, and the closeness of approximation $\{\epsilon_n\}$ we have an open neighborhood of f. Here we usually consider C^∞-functions, in which case $0 \leq m < \infty$ is required in the definitions. With respect to the Whitney topology *the real analytic functions are a dense subset* of the set of C^∞-functions.

Of course, using coordinate charts, one can define the Whitney topology on an abstract manifold. However, because of the embedding theorem it is enough to consider closed submanifolds of a vector space. It follows that *$A(M)$ is dense in $C^\infty(M)$ with respect to the Whitney-topology.*

We now wish to give a real analytic version of Proposition 1 above. For this we need the ***averaging technique.***

Let K act continuously on a topological space X. If $d\mu$ is the left- and right-invariant Haar measure on K, $\rho : K \to V$ a representation, and $F : X \to V$ a continuous map, then the ***averaged map*** $\tilde{F} : X \to V$, defined by

$$\tilde{F}(x) := \int_{k \in K} \rho(k) F(k^{-1}x) d\mu(k),$$

is an equivariant map. If X is a complex space and F is holomorphic, then $\tilde{F}$ is likewise holomorphic. The analogous result for real analytic functions is also true, but is perhaps not as easy to prove. Of course, it follows from the local version of the equivarant complexification theorem (see sec. IV).

Using the averaging technique it is possible to deduce equivariant mapping theorems in the real analytic case from the corresponding results in the C^∞ category. For this recall that the *orbit type of the K-action at a point $x \in M$ is the conjugacy class of the isotropy group K_x.*

Proposition 2. *Let K be a compact group acting as a group of real analytic diffeomorphisms of M and let $\Omega \subset M$ be a K-invariant open relatively compact subset. Then there is a complex representation space V and a real analytic equivariant map $F : M \to V$ such that $F|\Omega$ is an injective immersion. If K has only finite many orbit types in M, then there is an equivariant closed embedding of M in a representation space.*

Proof. We take for granted the result for C^∞-maps ([Wa]) and let $G : M \to V$ be a C^∞-map which has either of the required properties. Then there is a neighborhood N of G in the Whitney topology so that every mapping $F \in N$ has this property (see [Hi], p.38). Since the *Whitney size* of the difference $\tilde{\epsilon} = G - \tilde{F}$ can be estimated from the difference $\epsilon = G - F$ and the derivatives of the action map, there is a perhaps smaller neighborhood N' of G so that $F \in N'$ implies $\tilde{F} \in N$. Since every Whitney neighborhood of G contains real analytic elements, the result follows.□

In reality the main point which was used here is the following

Zusatz. *Every Whitney neighborhood of a K-equivariant C^∞-map to a representation space contains a K-equivariant real analytic map.*

2. Applications of the averaging technique to Stein spaces. Let X be a Stein space with a K-action of holomorphic transformations and let Y be a K-stable closed complex subspace. Suppose that $F : Y \to V$ is a K-equivariant holomorphic map to a complex K-representation space.

Proposition 3. *There exists an equivariant holomorphic map* $\tilde{F} : X \to V$ *with* $\tilde{F}|Y = F$.

Proof. By Stein theory there exists a holomorphic map $G : X \to V$ with $G|Y = F$. Define $\tilde{F}$ to be the averaged map $\tilde{G}$. □

Remark. If the complexification $K^{\mathbf{C}}$ is acting on X, then it follows from the identity principle that F and $\tilde{F}$ are $K^{\mathbf{C}}$-equivariant.

A typical application of Prop.3 is the construction of invariant functions. Notation: $\mathcal{O}(X)^K := \{f \in \mathcal{O}(X) : f(kx) = f(x)\ \forall k \in K\}$, i.e., $\mathcal{O}(X)^K$ is the set $Fix_K(\mathcal{O}(X))$ of functions fixed under the natural K-action.

Corollary 4. *Let* Y_1 *and* Y_2 *be disjoint closed* K*-invariant analytic subsets of* X*. Then there exists* $h \in \mathcal{O}(X)^K$ *with* $h|Y_i = i$, $i = 1, 2$.

Proof. Let V be the trivial 1-dimensional representation, $Y = Y_1 \cup Y_2$, and $f : Y \to V$ be defined by $f|Y_i := i$. Then $h := \tilde{f}$ does the job. □

We now apply the above type of result to a particularly simple quotient construction. For this let $G = L^{\mathbf{C}}$ be the complexification of a compact Lie group, i.e., a *linear reductive group*, and suppose that G acts holomorphically and properly on a complex space X. Consider the geometric quotient X/G defined by the equivalence relation $x \sim y :\iff$ *there exists* $g \in G$ *with* $g(x) = y$. In this case we have a *holomorphic equivalence relation* and with Hausdorff quotient X/G. Thus, when equipped with the sheaf of germs of G-invariant functions, X/G is a complex space (see e.g. ([Ho]) or the *geometric slice theorem* in VI).

Proposition 5. *Let* G *be a reductive complex Lie group which is acting holomorphically and properly on a Stein space* X*. Then the complex space* $Z := X/G$ *is Stein.*

Proof. Let $\pi : X \to X/G := Z$ be the (holomorphic) quotient map and $\{z_j\}$, with $z_i \neq z_j$ if $i \neq j$, be a divergent sequence in Z. Let $Y_j := \pi^{-1}(z_j)$ and $Y := \bigcup Y_j \hookrightarrow X$. Then Y is a closed analytic subset of X and $f : Y \to \mathbf{C}$ defined by $f|Y_j = j$ is holomorphic. The extended invariant function $\tilde{f}$ pushes down to a holomorphic function on Z. Thus there exists $h \in \mathcal{O}(Z)$ with $h(z_j) = j$ for all j. This shows that Z is holomorphically convex. The same argument applied to two points shows that $\mathcal{O}(Z)$ separates points. □

A special case of this quotient construction is of particular interest. For this let L be a closed subgroup of K and $L^{\mathbf{C}} < K^{\mathbf{C}}$ the inclusion of reductive groups. Let V be a complex representation space for $L^{\mathbf{C}}$ via $\rho : L^{\mathbf{C}} \to Gl(V)$. Since the $L^{\mathbf{C}}$-action on $K^{\mathbf{C}}$ is proper, we can apply Proposition 5 to the free proper $L^{\mathbf{C}}$-action on $K^{\mathbf{C}} \times V$ defined by $g(k, v) = (kg^{-1}, \rho(g)v)$.

Corollary 6. *The quotient* $Z = (K^{\mathbf{C}} \times V)/L^{\mathbf{C}} =: K^{\mathbf{C}} \times_{L^{\mathbf{C}}} V$ *is a Stein manifold.* □

Remarks. 1.) If V is taken to be 0-dimensional, this shows that the quotient $K^{\mathbf{C}}/L^{\mathbf{C}}$ is Stein. In fact it is even affine-algebraic ([M], [O]).

2.) If $E \to K^{\mathbf{C}}/L^{\mathbf{C}}$ is a homogeneous vector bundle with a given lifting of the $K^{\mathbf{C}}$-action, then, with respect to the representation of $L^{\mathbf{C}}$ on the fiber V over the neutral point $L^{\mathbf{C}}$, it is of the form $E = K^{\mathbf{C}} \times_{L^{\mathbf{C}}} V \to K^{\mathbf{C}}/L^{\mathbf{C}}$.

3.) If K is acting real analytically on M, then every orbit Kx has a neighborhood which is real analytically K-equivariantly isomorphic to $K \times_L S$, where S is an L-stable open neighborhood of the origin in a representation space V (see VI). In other words, every real analytic K-manifold is locally of the form $M = K \times_L S$. The $K^{\mathbf{C}}$-Stein manifold $X = K^{\mathbf{C}} \times_{L^{\mathbf{C}}} V^{\mathbf{C}}$ is Stein complexification of $K \times_L V^{\mathbf{C}}$ where even $K^{\mathbf{C}}$ acts holomorphically as a complex Lie group.

3. The local $K^{\mathbf{C}}$-action. The simplest example of a K-action on a Stein manifold is an action on a complex vector space V given by a representation $\rho : K \to Gl(V)$; the representation ρ extends uniquely to an *algebraic* representation of the algebraic group $K^{\mathbf{C}}$. In particular $K^{\mathbf{C}}$ acts algebraically on V and consequently every $K^{\mathbf{C}}$-orbit is Zariski open in its closure (see [Ch]).

Now let X be a Stein space equipped with a K-action of holomorphic transformations. Since the space of K-finite functions in $\mathcal{O}(X)$ is dense, given $x \in X$, there is an equivariant holomorphic map $F : X \to V$ to a K-representation space which maps a neighborhood of x biholomorphically onto its image which is analytic in an open subset in V (see §2 above).

Let $A(x)$ be the smallest K-invariant closed analytic subset of X which contains the orbit Kx. Since the $K^{\mathbf{C}}$-action on V is algebraic, $\Omega(x) := A(x) \cap F^{-1}(K^{\mathbf{C}}F(x))$ is Zariski open in $A(x)$. The complement $E(x) := A(x) \setminus \Omega(x)$ is exactly the inverse image of lower-dimensional $K^{\mathbf{C}}$-orbits in the closure of $K^{\mathbf{C}}F(x)$ via $F|A(x)$. Since $\Omega(x)$ does not depend on the map F, we refer to $\Omega(x)$ as the *local $K^{\mathbf{C}}$-orbit of x in X* and summarize these remarks in the following

Proposition 7. *Let X be a Stein space equipped with a K-action of holomorphic transformations. Then every local $K^{\mathbf{C}}$-orbit is Zariski open in its complex analytic Zariski closure.*
□

Remark. The local orbits of minimal dimension are closed and any two K-invariant closed analytic sets can be separated by an invariant holomorphic function (Corollary 4). Thus if $\mathcal{O}(X)^K \cong \mathbf{C}$, then there exists a unique *closed* local $K^{\mathbf{C}}$-orbit in X. Moreover, every local $K^{\mathbf{C}}$-orbit in X has the closed local $K^{\mathbf{C}}$-orbit in its closure.

Proposition 8. *For every $x \in X$ there exists a globally defined holomorphic K-equivariant map $F : X \to V$ to a representation space such that $\Omega(x)$ is biholomorphically mapped onto an open subset of $K^{\mathbf{C}}(F(x))$.*

Proof. We begin with $F_1 : X \to V_1$ which is biholomorphic at x. If F_1 is not injective on $\Omega(x)$ we *add* to it a map F_2 which improves the situation, e.g., if $F_1(p) = F_1(q)$ for some $p, q \in \Omega(x)$, then some K-finite function f separates p and q and we define F_2 by a basis of the vector space generated by the orbit Kf. Our *new* map is $F_1 \oplus F_2 : X \to V_1 \oplus V_2$.

Let H_j denote the $K^{\mathbf{C}}$-isotropy at $F_j(x)$, j=1,2. Then the projection on the first factor $V_1 \oplus V_2 \to V_1$ induces an equivariant map of the $K^{\mathbf{C}}$-orbits containing the images of $\Omega(x)$:

$$(*) \qquad K^{\mathbf{C}}(F_1(x) \oplus F_2(x)) : K^{\mathbf{C}}/H_1 \cap H_2 \to K^{\mathbf{C}}/H_1 = K^{\mathbf{C}}(F(x)).$$

Now, since F_1 was chosen to be biholomorphic at x and since both orbits contain the image of $\Omega(x)$ as an open set, they have the same dimension. Furthermore, the $K^{\mathbf{C}}$-actions on the representation spaces is algebraic. Thus the isotropy groups are algebraic and the map in $(*)$ is finite. Hence, after finitely many steps our map can not be *improved* and thus is biholomorphic from $\Omega(x)$ to an open set of a $K^{\mathbf{C}}$-orbit. □

Remark. Note that it is possible that $\Omega(x) := \Omega$ is contained as a K-invariant domain in an algebraic homogeneous space $K^{\mathbf{C}}/H$ where the embedding can still be *improved* as above, i.e., there might be K-finite holomorphic functions on Ω which can not be extended to $K^{\mathbf{C}}/H$. For example, for a finite group H, in the non-abelian situation it is often the case that $K_x = (e)$ for x generic. If Ω is a small enough K-invariant neighborhood of such an orbit, then, via the finite covering map $K^{\mathbf{C}} \to K^{\mathbf{C}}/H$, it can be regarded as a domain in the group $K^{\mathbf{C}}$ itself. Hence, if H is non-trivial, then there are indeed K-finite holomorphic functions on Ω which can not be extended to the homogeneous space $K^{\mathbf{C}}/H$.

Repeating the above process $(*)$ we obtain a quasi-affine homogeneous space $K^{\mathbf{C}}/H := \Omega^{\mathbf{C}}$ and a holomorphic K-equivariant embedding $\Omega \hookrightarrow \Omega^{\mathbf{C}}$ so that every K-finite holomorphic function on Ω extends to $\Omega^{\mathbf{C}}$. Let $\Omega \hookrightarrow \Omega_j^{\mathbf{C}}$, $j = 1, 2$, be two such *universal* $K^{\mathbf{C}}$-*complexifications*. Since the identity map $\iota : \Omega \to \Omega$ is defined by K-finite functions, it and its inverse extend to a $K^{\mathbf{C}}$-equivariant isomorphism, i.e., the *complexification* $\Omega^{\mathbf{C}}$ is uniquely defined.

Remark. In general, given a K-action by holomorphic transformations on a Stein space X, there exists a Stein space $X^{\mathbf{C}}$ with a holomorphic $K^{\mathbf{C}}$-action and an open K-equivariant holomorphic embedding $X \hookrightarrow X^{\mathbf{C}}$ which is universal with respect to this property. In this way X is realized as a Runge subset of $X^{\mathbf{C}}$. The above construction is one of the first steps in the proof of a more general result in [H2]. In §4 we present a *complexification* which will suffice for our purposes here.

We conclude this section with the result in the simplest case, i.e., where X is a closed local $K^{\mathbf{C}}$-orbit.

Proposition 9. *Let Ω be a closed local $K^{\mathbf{C}}$-orbit and $\Omega^{\mathbf{C}} = K^{\mathbf{C}}/H$ its universal $K^{\mathbf{C}}$-complexification. Then H is reductive and $\Omega^{\mathbf{C}}$ is affine.*

Proof. Let $\varphi : \Omega \to \mathbf{R}^{\geq 0}$ be a strictly pluri-subharmonic exhaustion. By averaging we may assume that φ is K-invariant. Note that $dim_{\mathbf{R}}\ Kx \geq dim_{\mathbf{C}}\ K^{\mathbf{C}}x = dim_{\mathbf{C}}\ \Omega$ for all $x \in \Omega$. If $x \in \Omega$ is a point where φ takes on its minimum, and therefore its minimum is taken on along the orbit Kx, the strictly pluri-subharmonicity of φ implies that $dim_{\mathbf{R}} Kx \leq dim_{\mathbf{C}} \Omega$ (see e.g. Lemma VIII.1). Hence, $Kx = K/K_x$ is exactly half-dimensional and consequently $(K_x)^{\mathbf{C}}$ is open in H. Thus H is conjugate to $L^{\mathbf{C}}$, where L is a finite extension of K_x. In particular H is reductive. □

4. A construction of $X^{\mathbf{C}}$ in a special case. The purpose of this section is to prove the following special case of the main result of ([H2]).

Proposition 10. *Let X be a Stein K-space with $\mathcal{O}(X)^K \cong \mathbf{C}$. Then there exists an affine algebraic $K^{\mathbf{C}}$-variety $X^{\mathbf{C}}$ and an open holomorphic K-equivariant embedding $\iota : X \hookrightarrow X^{\mathbf{C}}$ so that the K-finite holomorphic functions on X extend uniquely and holomorphically to $X^{\mathbf{C}}$.*

Remark. Since K-equivariant maps from X to Stein spaces are defined by K-finite functions, it follows that $X^{\mathbf{C}}$ is the unique Stein space with this property. In fact $\iota : X \to X^{\mathbf{C}}$ is universal with respect to equivariant maps to any complex space where $K^{\mathbf{C}}$ acts (see [H2]).

The principle of the proof is similar to that of Proposition 8 in that, using approximation by K-functions, we construct a K-equivariant locally biholomorphic map $\iota_1 : X \to Z_1$ into a $K^{\mathbf{C}}$-stable affine algebraic variety in a representation space which is a good first candidate for $X^{\mathbf{C}}$. By adding K-finite functions to ι_1, we obtain an *improvement* $\iota_2 : X \to Z_2$. By construction we have a projection $\rho : Z_2 \to Z_1$ with $\rho \circ \iota_2 = \iota_1$. This procedure is continued until ρ is an isomorphism, in which case $Z_1 \equiv Z_2$ is the desired complexification.

Proof of Proposition 10

To begin with we consider the equivariant map of the closed local $K^{\mathbf{C}}$-orbit $\Omega \hookrightarrow \Omega^{\mathbf{C}}$ as constructed above. Realizing $\Omega^{\mathbf{C}}$ as a closed $K^{\mathbf{C}}$-orbit in a representation space, we extend the resulting map $\Omega \hookrightarrow V$ holomorphically and equivariantly to X by Proposition 2. We fix $x \in \Omega$ and, via Proposition 1, choose an equivariant holomorphic map of X to another representation space which gives coordinates in a neighborhood of x. Add this to the one above and denote the resulting map by $\iota_1 : X \to V$.

Lemma. *The map ι_1 is locally biholomorphic at every point of X, its restriction to every local $K^{\mathbf{C}}$-orbit has finite fibers, and $Z := K^{\mathbf{C}}.\iota_1(X)$ is a closed algebraic subvariety of V. In fact $p_1 = p_2$.*

Proof. Since every local $K^{\mathbf{C}}$-orbit in X has $x \in \Omega$ in its closure, ι_1 is biholmorphic in a neighborhood of x and the set where ι_1 is not locally biholomorphic is invariant with respect to the local $K^{\mathbf{C}}$-action, ι_1 is biholomorphic in a neighborhood of every point in X. Thus, just as in Proposition 8, we see that its restriction to every local $K^{\mathbf{C}}$-orbit in X has finite fibers.

It remains to prove that Z is a closed algebraic subvariety. For this first note that $Z := K^{\mathbf{C}}.\iota_1(X) = K^{\mathbf{C}}.\iota_1(W)$, where W is an arbitrarily small neighborhood of x. Thus we may assume that $\iota_1 : W \to \iota_1(W)$ is an injective immersion and that $\iota_1(W)$ is a locally closed analytic subset of V. Furthermore, since W may be taken arbitrarily small, we may assume that it has only finitely many irreducible components. Arguing one component at a time, we may therefore assume that W is irreducible.

We let $\bar{Z}^{alg}$ be the Zariski closure of Z which is then also irreducible. Recall the $K^{\mathbf{C}}$-quotient $q : \bar{Z}^{alg} \to C$ of Rosenlicht ([R]). This is a meromorphic map into the Chow scheme which is regular on a Zariski open dense subset $U \subset \bar{Z}^{alg}$ consisting of certain $K^{\mathbf{C}}$-orbits of maximal dimension and $q|U$ is just the geometric quotient defined by $K^{\mathbf{C}}$.

Since $\iota_1(W)$ is not contained in a proper analytic subset of $\bar{Z}^{alg}$, and therefore the generic $K^{\mathbf{C}}$-orbit dimensions in W and in $\bar{Z}^{alg}$ agree, it follows that $q_W := q|\iota_1(W)$ is a well-defined meromorphic map.

Of course $q_W(\iota_1(x))$ is a compact *algebraic* set in C. Therefore $q^{-1}(q_W(\iota_1(x))) := A$ is a closed subvariety of V. But every local $K^{\mathbf{C}}$-orbit in $W \cap U$ has $\iota_1(x)$ in its closure. Hence $W \cap U \subset q_W(\iota_1(x))$ and therefore $W \subset A$.

Now *dim* A can be computed as the dimension of a generic $K^{\mathbf{C}}$-orbit plus the codimension of such in W. It follows that W is open in A. Hence $K^{\mathbf{C}}\iota_1(X) = K^{\mathbf{C}}\iota_1(W)$ is open in $A = \bar{Z}^{alg}$. Therefore $K^{\mathbf{C}}\iota_1(X) = K^{\mathbf{C}}\iota_1(W) = A = \bar{Z}^{alg}$, as otherwise the complement of $K^{\mathbf{C}}\iota_1(X)$ in $\bar{Z}^{alg}$ would contain an additional closed $K^{\mathbf{C}}$-orbit which is ruled out by the fact that $\mathcal{O}(\bar{Z}^{alg})^K \cong \mathbf{C}$. □

We are now in a position to *improve* the map $\iota_1 : X \to Z_1$, where $Z_1 := Z$ in the above Lemma. Each time we improve by a map $\iota_2 : X \to Z_2$, we check if $\rho : Z_2 \to Z_1$ is injective. If so, then, since it is automatically locally biholomorphic, we are finished. If not, we replace $\iota_1 : X \to Z_1$ by $\iota_2 : X \to Z_2$ and continue improving.

The following observation reduces the question of injectivity of ρ to the level of orbits.

Lemma. *Let $\iota_2 : X \to Z_2$ be an improvement with projection $\rho : Z_2 \to Z_1$. For $q \in Z_1$ suppose that $p_1, p_2 \in \rho^{-1}(q)$ are such that the induced maps $\rho|K^{\mathbf{C}}p_1$ and $\rho|K^{\mathbf{C}}p_2$ are isomorphisms. Then $K^{\mathbf{C}}p_1 = K^{\mathbf{C}}p_2$.*

Proof. Let $U = U(\iota_2(x))$ be a neighborhood of $x \in \Omega$ where ρ is biholomorphic. Since $\iota_2(x)$ is in the closure of both of the orbits, $\overline{K^{\mathbf{C}}p_i} \cap U \neq \emptyset$ for $i = 1, 2$. The result follows from the injectivity of the maps $\rho : \overline{K^{C}p_j} \to \overline{K^{C}q}$. □

If $p \in \rho^{-1}(q)$ and the induced map of $K^{\mathbf{C}}$-orbits is not injective, then the number of components of $(K^{\mathbf{C}})_p$ is less than that of $(K^{\mathbf{C}})_q$. But, since the action of $K^{\mathbf{C}}$ on Z is algebraic, the number of $K^{\mathbf{C}}$-isotropy components is bounded.

Starting with $Z = Z_1$, let $n(q)$ be the maximum number of points in a fiber of $K^{\mathbf{C}}p \to K^{\mathbf{C}}q$ of any improvement. Since $\rho : Z_2 \to Z_1$ is locally biholomorphic and therefore for every n the set $\{p : n(p) \geq n\}$ is open.

Let m be the maximum of the values of $n(p)$ over all possible improvements. In order to prove Proposition 10 we apply the following result to the irreducible components of X and the Z_i's.

Lemma. *Assume that X is irreducible. Let $\iota_2 : X \to Z_2$ be such that for some $z_1 \in Z_1$ there exists $z_2 \in \rho^{-1}(z_1)$ so that the induced map $\rho|K^{\mathbf{C}}z_2$ is an m to 1 cover. If $\iota_3 : X \to Z_3$ is a further improvement, then the induced projection $\tau : Z_3 \to Z_2$ is an isomorphism.*

Proof. Let W_1 be the open $K^{\mathbf{C}}$-invariant subset of Z_1 so that if $W_2 := \rho^{-1}(W_1)$, then $\rho|W_2$ is m *to* 1 at the $K^{\mathbf{C}}$-orbit level. It follows from the maximality of m and the previous Lemma that τ is injective over W_2. Thus the generic fiber of τ consists of one point. Since τ is locally biholomorphic, the result follows. □

III. The Kempf-Ness set

The basic properties of the categorical quotient $\pi : X \to X//K$ of a Stein K-space are summarized. In particular the Kempf-Ness set $R(\varphi) \subset X$ associated to a strictly pluri-subharmonic function which is proper along the π-fibers is introduced. This is shown to be the 0-level set of the associated moment map. It follows that π induces a homeomorphism from the geometric quotient $R(\varphi)/K$ to $X//K$.

1. The categorical quotient. Let X be a reduced Stein space and K a compact Lie group acting via holomorphic transformations on X. Let $\mathcal{O}(X)^K$ denote the space of invariant holomorphic functions and define an *analytic equivalence relation* by $x_1 \sim x_2$ whenever $f(x_1) = f(x_2)$ for all $f \in \mathcal{O}(X)^K$.

The quotient $X/ \sim =: X//K$, called the *categorical quotient*, is equipped with the structure of a Stein space and the projection $\pi : X \to X//K$ is holomorphic. It has the following universal property: for every *invariant* holomorphic map $F : X \to Y$ to a complex space, there exists a unique holomorphic map $\tau_F : X//K \to Y$ so that $F = \tau_F \circ \pi$.

If $F_x = \pi^{-1}(\pi(x))$ is a π-fiber, then the local $K^{\mathbf{C}}$-orbits of minimal dimension in F_x are closed in X. Since any two closed K-invariant analytic subsets can be separated by some $f \in \mathcal{O}(X)^K$ (see II Corolarry 4), it follows that there is a unique closed local $K^{\mathbf{C}}$-orbit in F_x. Hence $X//K$ can be identified with the set of closed local $K^{\mathbf{C}}$-orbits in X.

Considerations related to the categorical quotient first appeared in representation theoretic and algebraic geometric contexts where X is affine and $\mathcal{O}(X)^K$ is the ring of invariant regular functions. In that case the universal complexification $K^{\mathbf{C}}$, which is itself an affine algebraic group, automatically acts algebraically on X. In particular, since $\mathcal{O}(X)^K = \mathcal{O}(X)^{K^{\mathbf{C}}} =: I$, one remains in the category of algebraic geometry over $\mathbf{C}$. The quotient $X//K$, which in this case is denoted by $X//K^{\mathbf{C}}$, can be constructed by purely algebraic means, because *the ring of invariants is finitely generated.* Hence $X//K^{\mathbf{C}}$ is defined as $Spec(I)$ (see ([K]) for proofs and a historical discussion).

In the complex analytic setting one can not a priori assume that $K^{\mathbf{C}}$ is acting, e.g., for X a bounded domain. Furthermore, even if it is acting, there is no known direct method for analyzing the algebra structure of the ring of holomorphic invariants. However, using ideas which are of a more geometric nature, D. Snow ([Sn]) showed that if $K^{\mathbf{C}}$ is acting on a Stein manifold X, then $X//K^{\mathbf{C}}$ carries a canonical structure of a normal Stein space such that every invariant holomorphic map from X to a complex space factors through the *categorical quotient* $\pi : X \to X//K^{\mathbf{C}}$. An elementary, but very important, by-product of Snow's construction is that the fibers of this quotient are *affine algebraic varieties* on which the $K^{\mathbf{C}}$-action is algebraic (see also II Proposition 10).

In ([H2]) the first author constructed the categorical quotient $\pi : X \to X//K$ of a Stein space equipped with a compact group of holomorphic transformations. This opens up a wide range of examples, e.g., complexifications as we consider them here, where the algebraic group $K^{\mathbf{C}}$ does not act.

2. Invariant plurisubharmonic functions. Critical points of invariant strictly plurisubharmonic functions play a major role in the study of the categorical quotient. Here we begin with some standard information in the 1-dimensional case and proceed to the study of critical sets in fibers of categorical quotients.

Let S^1 act on $\mathbf{C}^*$ by rotations. An S^1-invariant plurisubharmonic function φ is of the form

$\varphi(z) = \chi(log|z|)$, where $\chi : \mathbf{R} \to \mathbf{R}$ is convex. Strict pluri-subharmonicity of φ corresponds to strict convexity of χ.

The following facts are immediate consequences of the definitions.

Basic Fact 1. *Suppose that φ is an S^1-invariant strictly plurisubharmonic function on $\mathbf{C}^*$ with a critical point. Then the critical set consists of exactly one orbit, φ has an absolute minimum there, and*

$$\lim_{z \to \infty} \varphi(z) = \infty.$$

Here $z \to \infty$ is meant in the set-theoretic sense, i.e., z goes to either end of $\mathbf{C}^*$.

Basic Fact 2. *Let φ be an S^1-invariant plurisubharmonic function on a disk Δ containing the origin $0 \in \mathbf{C}$. Then φ takes its minimum at 0. Unless φ is locally constant at 0, this is an absolute minimum.*

The following is an important tool in the invariant theory of reductive groups.

Hilbert Lemma.(see e.g. [K]) *Let V be a complex $K^{\mathbf{C}}$-representation space, $v \in V$, and Z the closure of $\Omega := K^{\mathbf{C}}v$. If $Z \setminus \Omega$ is non-empty, then there exists a 1-dimensional multiplicative subgroup $G_m \cong \mathbf{C}^*$ in $K^{\mathbf{C}}$ which is the complexification of some $S^1 \subset K$ such that* $\lim_{t \in G_m} t(v)$ *exists and is in the closed $K^{\mathbf{C}}$-orbit in $Z \setminus \Omega$.*

Remark. The group G_m is acting algebraically on V and therefore the closure of its orbits in the projective space containing V are rational curves. Hence an orbit in V is either closed or its closure is formed by adding exactly one point, i.e., a fixed point, to it. The existence of the limit in the Hilbert Lemma means that as t goes to the appropriate end of G_m the flow $t(v)$ converges to the unique fixed point in the closure.

Elementary properties of plurisubharmonic functions combined with the Hilbert Lemma yield remarkably strong statements. The first of these is in the context of a unitary representation V of K where the strictly plurisubharmonic function $\varphi : V \to \mathbf{R}^{\geq 0}$ is just the invariant norm.

Kempf-Ness Lemma.([KN]) *Let $\Omega = K^{\mathbf{C}}v$ be an orbit in a unitary representation and suppose that the restriction of the norm function $\varphi|\Omega$ is critical at v. Then Ω is closed.*

Proof. If the orbit is not closed, then we take G_m as in the Hilbert Lemma. The closure $\overline{G_m v}$ is a perhaps singular affine curve which we normalize at z to obtain a copy of $\mathbf{C}$ where the S^1 coming from G_m acts and z corresponds to the origin. The pull-back of φ to this curve is a smooth S^1-invariant plurisubharmonic function which is strictly plurisubharmonic outside the origin. However, contrary to Basic Fact 1, it has a critical point at the point corresponding to v. □

In fact we may replace the norm by any strictly plurisubharmonic function.

Zusatz. *Let X be a Stein $K^{\mathbf{C}}$-space and $\varphi : X \to \mathbf{R}$ a strictly plurisubharmonic function. For $x \in X$ and $\Omega := K^{\mathbf{C}}x$, if $\varphi|\Omega$ is critical at x, then Ω is closed.* □

Since in general we do not assume that the group $K^{\mathbf{C}}$ acts on X, we can only consider local orbits, i.e., K-invariant open sets in quasi-affine $K^{\mathbf{C}}$-homogeneous spaces. Let ψ be a K-invariant strictly plurisubharmonic function on a Stein K-space X whose restriction φ to a local $K^{\mathbf{C}}$-orbit Ω is critical at x_0. Of course φ is critical along the entire orbit Kx_0. As we have seen in the proof of II.Proposition 9, it follows that $\Omega^{\mathbf{C}}$ is an affine homogeneous space $K^{\mathbf{C}}/L^{\mathbf{C}}$ and, at least in a neighborhood of the critical orbit Kx_0, the critical set agrees with Kx. Since the real hessian in the directions normal to Kx_0 agrees with the Levi-form which is positive-definite, φ is strictly increasing in these directions. Then, for $r > 0$ sufficiently small, the *tube* $T(r) := \{\varphi < \varphi(x_0) + r\}$ is a relatively compact neighborhood of Kx_0 in Ω with smooth boundary.

Proposition 1. *Let Ω be a K-connected K-invariant open subset of $\Omega^{\mathbf{C}} = K^{\mathbf{C}}/L^{\mathbf{C}}$ and suppose that $\varphi : \Omega \to [0, m)$ is a proper K-invariant strictly pluri-subharmonic exhaustion. Suppose that $x_0 \in \Omega$ is a point at which φ takes its minimum. Then $\{d\varphi = 0\} = Kx_0$ and $T(r)$ is K-connected for all $r > \varphi(x_0)$.*

Proof. Let $r > \varphi(x_0)$. Since for every $x \in \partial T(r)$ there are points y arbitrarily near to x with $\varphi(y) < \varphi(x)$, the above argument with the Levi-form shows that $d\varphi(x) \neq 0$. Let $T_1(r)$ be the connected component of $T(r)$ which contains x_0. Then $T_1(r)$ is relatively compact with smooth boundary for $\varphi(x_0) < r < m$. Note that, since $\varphi : \Omega \to [0, m)$ is a proper exhaustion, $lim_{r \uparrow m} T_1(r) = \Omega$. Thus for every $T(r)$ there exists $s \geq r$ with $T(r) \subset T_1(s)$.

Let s^* be the infimum of such values. It follows that $T(r) \subset T_1(s^*)$. But $T_1(s^*)$ has smooth compact boundary. Hence, using the standard gradient flow argument, if $r < s^*$, one sees that $T(r) \subset T_1(s^* - \epsilon)$ for ϵ sufficiently small. This is contrary to the definition of s^* and therefore $T(r) = T_1(r)$ as desired. □

Remarks. 1.) A set is K-connected if K act transitively on its components.

2.) If $\Omega = \Omega^{\mathbf{C}}$ and φ is a K-invariant strictly pluri-subharmonic function with non-empty critical set, then φ is already an exhaustion ([AL1], see also [AL2] and [H2]).

Let X be a Stein K-space equipped with a strictly plurisubharmonic K-invariant function φ. In the next section we shall show that if $\mathcal{O}(X)^K \cong \mathbf{C}$ and $\varphi : X \to \mathbf{R}^{\geq 0}$ is proper, then, of the restrictions $\varphi|\Omega$ to the local $K^{\mathbf{C}}$-orbits, there is exactly one with a non-empty critical set, i.e., the closed local $K^{\mathbf{C}}$-orbit. For this we use the space $X^{\mathbf{C}}$ constructed in II Proposition 10 along with following

Extension Lemma. *Let Z be a Stein space and φ be a plurisubharmonic function defined in a neighborhood of a relatively compact open subset $W = \{\varphi < t\} \subset Z$. Assume that W is Runge in Z. Then there exists a plurisubharmonic exhaustion $\psi : Z \to \mathbf{R}\geq 0$ with $\psi \equiv \varphi$ on $\bar{W}$. If φ is strictly pluri-subharmonic, then ψ can also be chosen to be strictly pluri-subharmonic. If Z and φ are smooth then ψ can also be chosen to be smooth.*

Proof. Let U be a relatively compact neighborhood of $\bar{W}$ in Z so that φ is defined and pluri-subharmonic on $\bar{U}$. Since W is Runge in Z, there exist finitely many functions $f_1, \ldots, f_N \in \mathcal{O}(Z)$ so that

$$U \supset\supset P := \{|f_j|^2 < 1 \ \forall j\} \supset\supset W.$$

By taking appropriate powers of these functions we may assume in addition that $\rho := \sum |f_j|^2 < 1$ on $\bar{W}$. Let $M := max\varphi|\partial U$ and $1 < m := min\rho|\partial U$. Define $\chi : \mathbf{R}^{\geq 0} \to \mathbf{R}^{\geq 0}$ to be a convex function with $\chi(t) \equiv 0$ for $0 \leq t \leq 1$ and $\chi(m) > M$. Finally, define $\psi := max(\varphi, \chi \circ \rho)$ on U and $\psi := \chi \circ \rho$ on $Z \setminus U$.

As the maximum of two pluri-subharmonic functions, the function ψ is pluri-subharmonic on U. Furthermore, in a neighborhood of ∂U, $\psi \equiv \chi \circ \rho$. Thus the extension to Z is pluri-subharmonic.

In order to insure that ψ is an exhaustion, we let $\eta : Z \to \mathbf{R}^{\geq 0}$ be a strictly pluri-subharmonic exhaustion and $\alpha : \mathbf{R}^{\geq 0} \to \mathbf{R}^{\geq 0}$ a convex exhaustion which is identically zero on $[0, a]$, where $a := max(\eta|\bar{W})$, and which is strictly convex thereafter. We then replace ψ by $\psi + \alpha \circ \eta$.

Note that by moving the functions f_j in the definition of ρ above by appropriate small automorphisms and taking futher linear combinations of their absolute values squared, we may assume in addition that ρ is strictly plurisubharmonic on any given compact set, e.g. $\{\eta \leq a\}$. Thus, if φ is strictly pluri-subharmonic, then so is ψ.

Finally, suppose that Z and φ are smooth. Here we choose $U := \{\varphi < t + \epsilon\}$ and $V := \{\varphi < t + \delta\}$, where $0 < \epsilon < \delta$ are small enough so that U and V are still relatively compact. Let χ_1 be a cut-off function with $\chi_1 \equiv 1$ on $[0, \epsilon]$ and which has compact support in $[0, \delta)$. Consider $\psi = \chi_1 \circ \varphi + \chi \circ \rho$, where χ and ρ are as above.

Arguing as before, we can choose χ and ρ such that the Levi-form of $\chi \circ \rho$ is *large enough* to cancel out the negativity caused by the cut-off function χ_1. Furthermore, if so desired, we can again add $\alpha \circ \eta$ to insure that the resulting function is a strictly pluri-subharmonic function in the case when φ itself is strictly pluri-subharmonic. □

The following fact is the basis of many results which involve the categorical quotient (see [H2]). For this it is convenient to let $C_\varphi(\Omega)$ denote the set of critical points on $\varphi|\Omega$ in the local $K^{\mathbf{C}}$-orbit Ω.

Proposition 2. *Let X be a Stein K-space with $\mathcal{O}(X)^K = \mathbf{C}$ and suppose that $\varphi : X \to \mathbf{R}^{\geq 0}$ is a proper strictly plurisubharmonic K-invariant exhaustion. Then $C_\varphi(\Omega) \neq \emptyset$ if and only if Ω is the closed local $K^{\mathbf{C}}$-orbit and in that case $C_\varphi(\Omega)$consists of a single K-orbit.*

Proof. Let Z be the complexification of X constructed in II Proposition 10. Since φ is proper strictly plurisubharmonic exhaustion of X, the sets $\{\varphi < c\}$ are Runge in X. Consequently they are Runge in Z and every critical level $C_\varphi(\Omega)$ is contained in an open set W of the type where the Extension Lemma can be applied. Let $\psi : Z \to \mathbb{R}^{\geq 0}$ be the resulting extension of $\varphi|\bar{W}$ to Z. By ***averaging*** (see II.1-2), we may assume that ψ is K-invariant. From the Kempf-Ness Lemma (Zusatz) it follows that, for $z \in C_\varphi(\Omega)$, the orbit $K^{\mathbb{C}}z$ is closed in Z. In particular $\Omega(z)$ is the closed local $K^{\mathbb{C}}$-orbit. □

3. The Kempf-Ness set. Let X be a Stein manifold equipped with a compact group K of holomorphic transformations and let $\pi : X \to X//K$ denote the categorical quotient. We consider a K-invariant non-negative strictly plurisubharmonic function φ on X which is *proper on the π-fibers*. This means that φ is regarded as a map from X to the half-open interval $I := [0, m)$, where $m = \infty$ is allowed, and the map $\pi \times \phi$ is proper.

Remarks. 1.) If φ is an unbounded exhaustion, then it is clearly proper along the π-fibers. Thus there exist such functions. However, in the context of complexifications, where one further demands that $\varphi^{-1}(0) = M$, if M is not compact, such functions are never exhaustions of the full space X.

2.) It is possible to carry out all of our work here in the context of complex spaces (see [HHL],[HL]).

Let φ be a non-negative strictly plurisubharmonic function which is proper along the π-fibers. For $y \in Y := X//K$ let $F_y := \pi^{-1}(y)$ and

$$R(\varphi)_y := \{x \in F_y : \varphi(x) = min_{z \in F_y} \varphi(z)\}.$$

The above Proposition 2 shows that $R(\varphi)_y$ consists of exactly one K-orbit which is totally real and half the dimension in the unique closed local $K^{\mathbb{C}}$-orbit in F_y.

We refer to the union $R(\varphi) := \bigcup_{y \in Y} R(\varphi)_y$ as the *Kempf-Ness set* associated to φ. Note that the categorical quotient $X//K$ is at least set theoretically identified with the geometric quotient $R(\varphi)/K$. It is of course important to understand the structure involved in this identification.

An additional tool for understanding the Kempf-Ness set is the *moment map* associated to the Kähler form $\omega := i\partial\bar{\partial}\varphi$. Of course, since φ is K-invariant, the symplectic form ω is likewise invariant. If there is a Lie algebra map $\lambda : Lie(K) \to C^\infty(X)$ so that $\lambda(v)$ is an associated Hamiltonian function to the vector field $\tilde{v}$ on X, then one defines a moment map $\mu : X \to Lie(X)^*$ by $\mu(x)(v) = \lambda(v)(x)$ (see ([GS])). In general there may not exist such *liftings*, but here the potential function φ gives a lifting free of charge: $\mu_v^\varphi(x) := \mu^\varphi(x)(v) := \frac{1}{2}d\varphi(J\tilde{v}(x))$. Here J denotes the complex structure on X.

We call μ^φ *the moment map with respect to* φ. It has the following properties:

i) *With respect to the given K-action on X and the co-adjoint action on $Lie(K)^*$ it is K-equivariant.*

ii) *On each K-invariant complex submanifold Y of X we have $d\mu_v^\varphi(w) = \omega_Y(\tilde{v}(x), w)$ for all $x \in Y$, $w \in T_x Y$ and $v \in Lie(K)$, where $\omega_Y = i\partial\bar{\partial}\varphi|Y$ is the restricted Kähler form.*

Note that ii) means that $J\tilde{v}$ is the gradient field of the function $\mu_v^\varphi|Y$ with respect to the Riemannian metric g_Y which is given by the Kähler form ω_Y.

A point $x \in X$ is in the zero fiber $\mu^{-1}(0)$ if and only if it is a critical point of $\varphi|\Omega(x)$. In particular, assuming that φ is proper along the π-fibers so that $R(\varphi)$ makes sense, it follows that $\mu^{-1}(0) \supset R(\varphi)$. But again we may apply Proposition 2:

$$\mu^{-1}(0) = R(\varphi).$$

Corollary 3. *The Kempf-Ness set $R(\varphi)$ is closed.*

We equip $R(\varphi)$ with the induced topology and consider the geometric quotient $R(\varphi) \to R(\varphi)/K$, equipping the base with the quotient topology. Using the identification with the zero-fiber of the moment map, this is the *reduced phase space* or *Marsden-Weinstein reduction* (see e.g. ([GS])).

Notice that the construction of $R(\varphi)$ yields a continuous bijective map $\iota : R(\varphi)/K \to X//K$.

Proposition 4. *The map ι is a homeomorphism. Equivalently, the induced map $\pi : R(\varphi) \to X//K$ is proper.*

Proof. Without loss of generality we may assume that $\pi \times \varphi : X \to X//K \times \mathbb{R}^{\geq 0}$ is proper. Consider a sequence $\{x_n\} \subset R(\varphi)$ with $\pi(x_n) := y_n \to y \in X//K$. We wish to show that $\{x_n\}$ is bounded.

Take $z_n \in \pi^{-1}(y_n)$ with $z_n \to z \in \pi^{-1}(y)$. It follows that $\varphi(z_n) \to \varphi(z)$, and, since $\varphi(z_n) \geq \varphi(x_n)$, it follows that $\varphi(x_n)$ is bounded. Hence $(y_n, \varphi(x_n))$ is bounded and, since $\pi \times \varphi$ is proper, it follows that $\{x_n\}$ is bounded. □

IV. Complexifications for actions of compact groups

For a compact group K of real analytic diffeomorphisms of M the existence of a Stein K-complexification X is proved. A K-invariant strictly pluri-subharmonic function $\varphi : X \to \mathbb{R}^{\geq 0}$ with $\varphi^{-1}(0) = M$ and which is proper along the fibers of the categorical quotient is constructed.

1. K-complexifications. Let M be a real analytic manifold and K a compact group acting real analytically on M. Let X be a Stein K-complexification of M and $\varphi : X \to \mathbb{R}^{\geq 0}$ a K-invariant strictly plurisubharmonic function with $\varphi^{-1}(0) = M$. We refer to the triple

(M, X, φ) as a *gauged Stein K-complexification.* Our first goal here is, given a K-action on M, to construct such a triple.

We begin by extending the K-action to some neighborhood in a given complexification X.

Proposition 1. *Let X be a complexification of M. Then there is a neighborhood W of M in X which is a K-complexification of M.*

Proof. We cover M locally finitely with K-stable relatively compact open sets U_j: neighborhoods which are K-equivariantly injectively real analytically immersed in some real representation space $(V_{\mathbf{R}})_j$ (II Proposition 2). Of course such an immersion can be continued to a biholomorphic map of a neighborhood W_j of $\bar{U}_j$ in X. Regarding W_j to be in the complexification $(V_{\mathbf{C}})_j$, we may choose it to be K-invariant. Thus we have the desired K-action on each set of the open cover $\mathcal{W} = \{W_j\}$. Using a Riemannian metric on X, let V_j be an embedded neighborhood of the normal bundle of U_j which is contained in W_j. For each $k \in K$ we have a holomorphic map $k_j : V_j \to X$ which is defined by the extension of the K-action to W_j and then restriction to V_j. Now, for all i, j every component of $V_i \cap V_j$ is connected and has non-empty intersection with M. Thus, since $k : M \to M$ is globally defined, the identity principle shows that it extends to a well-defined holomorphic map $k : V \to X$, where $V := \bigcup V_j$, for all $k \in K$.

Of course V may not be K-invariant, but, we may choose smaller neighborhoods, $\tilde{W}_j \subset W_j$, where the K-action is defined and such that $\tilde{W}_j \subset V$ for all j. The set $\tilde{W} := \bigcup \tilde{W}_j$ does the job. □

Now take a K-invariant Riemannian metric on W and let $V_j \subset W$ be a K-invariant tubular neighborhood in the normal bundle which we regard as being in X. In particular we have the K-equivariant normal bundle projections $\tau_j : V_j \to U_j$.

We now follow Grauert's construction([G]). In every $(V_{\mathbf{C}})_j$ as above, let $N_j := (y_1)^2 + \ldots + (y_{n_j})^2$ be the ***norm-squared of the imaginary parts*** in some unitary basis, where $(V_{\mathbf{R}})_j$ is embedded as the real points and K is acting as a group of orthogonal transformations. Let ψ_j be the pull-back of N_j to V_j, $\{\chi_j\}$ be a K-invariant partition of unity subbordinate to $\{U_j\}$, and take $\varphi_j := (\chi_j \circ \tau_j)\psi_j$. Finally, let

$$\varphi := \sum c_j \varphi_j,$$

where c_j are positive constants to be chosen later.

Since N_j vanishes to the second order along $\mathbf{R}^{n_j}$,

$$i\partial\bar{\partial}\varphi|M = i \sum c_j \chi_j \partial\bar{\partial}\psi_j > 0.$$

Tube Lemma. *Let X be a complexification of M. Then there is a K-invariant neighborhood W of M in X which is a K-complexification and a non-negative K-invariant strictly plurisubharmonic function φ on W so that $\varphi^{-1}(0) = M$. Furthermore, the closure of the tube $T := \{w \in W : \varphi(w) < 1\}$ in X is contained in W.*

Proof. The function φ constructed above is K-invariant, strictly plurisubharmonic in some neighborhood W of M, and $\varphi^{-1}(0) = M$. The constants c_j can be chosen large enough so that the tube T has the required properties. □

Remark. If we take X to be Stein, then T is likewise Stein.

In order to construct a guaging φ which is proper along the π-fibers we need a technical result which is implicit in the above considerations.

Zusatz. *Let $\{U_j\}$ be a locally finite covering of M by K-invariant open sets. For each j suppose that W_j is a K-invariant open neighborhood of U_j in X with a K-invariant strictly plurisubharmonic function $\varphi_j : \overline{W_j} \to \mathbb{R}^{\geq 0}$ with $\varphi_j^{-1}(0) = U_j = M \cap W_j$. Let $\{m_j\}$ be an arbitrary sequence of positive numbers. Then, for $W = \bigcup W_j$, the function φ can be chosen so that $\varphi_j < m_j$ on $W_j \cap \overline{T}$.*

Proof. Since φ_j is defined on a neighborhood of the closure $\overline{W_j}$, the normal bundle neighborhood V_j over U_j may be chosen in the domain of definition of φ_j. The result follows by choosing the constants $\{c_j\}$ appropriately. □

The strictly pluri-subharmonic function φ constructed above is locally the *sum of norm-functions in the directions transversal to M*. For example if we equip X with a Riemannian metric, then, for every relatively compact open subset $U \subset M$ and any tubular neighborhood N_U of its normal bundle which is embedded in X with projection $r : N_U \to U$, there exists $\epsilon > 0$ so that

$$r \times \varphi : N_U \cap \{\varphi < \epsilon\} \to U \times I(\epsilon)$$

is proper. Here $I(\epsilon)$ denotes the interval $[0, \epsilon)$. We summarize this property by saying that *φ is locally proper in the directions transversal to M*. In fact much more is true (see VIII Lemma 1).

2. Anti-holomorphic involutions. Here we construct a Stein K-complexification X which possesses a K-equivariant anti-holomorphic involution $\sigma : X \to X$ with $Fix(\sigma) = M$. We will apply this in V. to obtain information on the image $\pi(M)$ in $X//K$. Of more immediate importance is the construction of a guaged Stein K-complexification (M, X, φ), where φ is proper along the fibers of $\pi : X \to X//K$. For this the existence of such an involution is also of value (see §3 below).

For the construction of σ note that, if we take coordinates $z^\alpha = (x^\alpha, y^\alpha)$ along M so that M is locally defined by the equations $y^\alpha = 0$, then at least locally $\sigma^\alpha(z^\alpha) := \bar{z}^\alpha$ does the job. Now $\sigma^\alpha \circ \sigma^\beta$ is holomorphic and fixes the points of M whereever both involutions are defined. Choosing the domains of definition so that every component of their intersections has non-empty intersection with M, the identity principle yields a well-defined anti-holomorphic map $\sigma : W \to X$ on some neighborhood of M. We may replace W by $W \cap \sigma(W)$ so that $\sigma : W \to W$. Of course $Fix(\sigma) = M$.

Now there exists an M-connected neighborhood $\tilde{W} \subset W$ such that for all $k \in K$ the holomorphic maps $k : \tilde{W} \to X$ and $\sigma k \sigma : \tilde{W} \to X$ are defined. Since $\sigma k \sigma = k$ on M, it follows that they are equal on $\tilde{W}$.

It follows from the Tube Lemma in §1 above that there exists a K-invariant strictly pluri-subharmonic function $\varphi : X \to \mathbb{R}^{\geq 0}$ so that the the Stein tube $T = \{\varphi < 1\}$ is contained in $\tilde{W}$. In fact we can choose φ so that, for $\tilde{\varphi} := \varphi + \varphi \circ \sigma$, $\tilde{T} = \{\tilde{\varphi} < 1\}$ is also a Stein neighborhood of M in $\tilde{W}$ so that $\sigma \circ k(\tilde{T})$ is in the domain of definition of φ for all $k \in K$.

Thus $\tilde{\varphi}$ is K- and σ-invariant and, making the obvious redefinitions, we have the following

Proposition 2. *For a real analytic K-manifold M there exists a guaged Stein K-complexification (M, X, φ) and an anti-holomorphic involution $\sigma : X \to X$ which commutes with the K-action, leaves φ invariant and has the property that $Fix(\sigma) = M$.* □

Let (M, K, φ) and σ be as above. For $f \in \mathcal{O}(X)$, note that $x \mapsto \overline{f(\sigma(x))}$ is likewise holomorphic. Therefore the equivalence relation which defines $\pi : X \to X//K$ is stabilized by σ, i.e., we have an induced map $\tilde{\sigma} : X//K \to X//K$.

Proposition 3. *There exists a unique anti-holomorphic involution $\tilde{\sigma} : X//K \to X//K$ with $\pi \circ \tilde{\sigma} = \tilde{\sigma} \circ \pi$. Furthermore, $\pi(M) \subset Fix(\tilde{\sigma})$.*

Proof. Let ψ be any K-invariant strictly plurisubharmonic function which is proper along the fibers of the categorical quotient. Of course $\tilde{\sigma}$ can also be defined by restriction of σ to the Kempf-Ness set $R(\psi)$ where it is K-invariant. Using the fact that $\pi | R(\psi) : R(\psi) \to X//K$ is proper and surjective (see III Proposition 4), it follows that $\tilde{\sigma}$ is continuous. Now a function on $X//K$ is anti-holomorphic whenever its lift to X is anti-holomorphic. For $\tilde{f} \in \mathcal{O}(X//K)$, the lift $\tilde{f} \circ \tilde{\sigma} \circ \pi = \tilde{f} \circ \pi \circ \sigma$ is obviously anti-holomorphic. Hence $\tilde{\sigma}$ is anti-holomorphic, and, since $\pi \circ \sigma = \tilde{\sigma} \circ \pi$, it follows that $Fix(\tilde{\sigma}) \supset \pi(Fix(\sigma)) = \pi(M)$. □

The following observation is of use in our construction of a *proper guaging.*

Proposition 4. *Let X be a Stein K-complexification of M equipped with a K-equivariant anti-holomorphic involution σ such that $Fix(\sigma) = M$. Let $\psi : X \to \mathbb{R}^{\geq 0}$ be a K- and σ-invariant strictly pluri-subharmonic function which is proper along the fibers of the categorical quotient $\pi : X \to X//K$. Then M is contained in the Kempf-Ness set $R(\psi)$. In particular, $\pi(M) = M/K$.*

Proof. Since ψ is K-invariant, $d\psi | T_x Kx = 0$ for all $x \in X$. For $x \in M$ let N_x be a σ-stable complement to $T_x Kx$ in the tangent space $T_x \Omega(x)$ of the local $K^{\mathbb{C}}$-orbit. Of course $\sigma_*(v) = -v$ for all $v \in N_x$. Since ψ is σ-invariant, it follows that $d\psi | N_x = 0$ as well, i.e., $\psi | \Omega(x)$ is critical on Kx and consequently $Kx \subset R(\psi)$.□

3. Existence of a proper guaging. We begin by introducing some notation. Let (M, X, φ) be a guaged Stein K-complexification. Suppose that there exist families $\{W_j\}$

of K-invariant relatively compact open sets in X, $\{Z_j\}$ of open neighborhoods in $X//K$, and $\{I_j\}$ of intervals $I_j = [0, r_j)$ such that

i) $W_j \cap M =: U_j \neq \emptyset$, $W := \bigcup W_j \supset M$ and $\{W_j\}$ is locally finite

ii) $(\pi \times \varphi)|W_j : W_j \to Z_j \times I_j$ is proper, where $\pi : X \to X//K$ is the categorical quotient. Then we say that φ ***is proper on the fibers of*** π ***in*** $\{W_j\}$. Our first goal is to construct such a covering.

Let X and σ be as in Proposition 2 above. Take $\psi : X \to \mathbb{R}^{\geq 0}$ to be a strictly pluri-subharmonic K-invariant exhaustion. Replacing ψ by $\psi + \sigma \circ \psi$, we may assume that it is also σ-invariant. Thus $M \subset R(\psi)$.

We shall make use of the following

General Fact. *Let $x \in R(\psi)$ and $V = V(x)$ be a K-invariant relative compact neighborhood of x in X. Then there exists a K-invariant neighborhood $U = U(R(\psi))$ so that, for $F_x := \pi^{-1}(\pi(x))$,*

$F_x \cap \partial V$ is compact in $\partial V \setminus U$.

Proof. If this were not the case, then $F_x \cap \partial V \cap R(\psi) \neq \emptyset$, contrary to $F_x \cap R(\psi) = Kx$.□

Let X,σ, and ψ be as above, in particular ψ is σ-invariant so that $M \subset R(\psi)$ and, as before, let $\varphi : X \to \mathbb{R}^{\geq 0}$ to be a K-invariant strictly pluri-subharmonic function with $\varphi^{-1}(0) = M$.

Let $x \in M$, $V = V(x)$ a relatively compact neighborhood and let U be as in the above General Fact. For W_0 a neighborhood of $\pi(x)$ in $X//K$, define

$$W := V \cap \pi^{-1}(W_0) \cap \{\varphi < \epsilon\}.$$

Lemma A. *For W_0 and ϵ small enough, it follows that*

$\pi \times \varphi : W \to W_0 \times I(\epsilon)$ is proper.

Proof. Recall that φ is locally proper transversal to M and that $M \subset R(\varphi)$. Thus, for ϵ small enough, $\{\varphi < \epsilon\} \subset U$. Shrink W_0 to $\pi(x)$ and simultaneously let ϵ tend to 0. If each of the corresponding sets W were not relatively compact, then $F_x \cap \partial V \cap U \neq \emptyset$, contrary to the General Fact, i.e., if W_0 and ϵ are small enough, then W is relatively compact.

If W is relatively compact as above, then $\partial W = B_1 \cup B_2$, where

$B_1 := V \cap \pi^{-1}(\partial W_0) \cap \{\varphi < \epsilon\}$

and

$B_2 := V \cap \pi^{-1}(W_0) \cap \{\varphi = \epsilon\}$.

Let $\{w_n\}$ be an infinite sequence in W with $w_n \to w \in \partial W$. It follows that $w \in \pi^{-1}(\partial W_0)$ or $\varphi(w) = \epsilon$. This shows that $\pi \times \varphi : W \to W_0 \times I(\epsilon)$ is proper. □

A guaged Stein K-complexification (M, X, φ) is called ***proper*** if φ is proper along the fibers of the categorical quotient $\pi : X \to X//K$. As an immediate consequece of Lemma A, there exists a covering $\{W_j\}$ so that φ is proper along the fibers of π in $\{W_j\}$.

Lemma B. *Let (M, X, φ) be a guaged Stein K-complexification so that, with the data $\{W_j\}$, $\{Z_j\}$, $\{I_j\}$, the function φ is proper along the fibers of π on $\{W_j\}$. Let $(M, \tilde{X}, \tilde{\varphi})$ be a guaged Stein K-complexification with $\tilde{X} \subset W = \bigcup W_j$ and $sup(\varphi|(W_j \cap \tilde{X})) < r_j$ for all j. Then $(M, \tilde{X}, \tilde{\varphi})$ is a proper guaged Stein K-complexification.*

Proof. The condition $sup(\varphi|(W_j \cap \tilde{X}) < r_j$ for all j, along with the fact that $(\pi \times \varphi)|W_j$ is proper, implies that $F_j(x) := (\pi^{-1}(\pi(x)) \cap W_j) \cap \tilde{X}$ is closed in $\tilde{X}$ for all $x \in W_j$ and for all j.

Now let $\{x_n\} \subset \tilde{X}$ be an infinite discrete subset with $\varphi(x_n) \leq s < 1$ for all n. Since $\{x_n\} \cap W_j$ is finite for all j, if $x_n \in W_{j_n}$, it follows that $Y_n := F_{j_n}(x_n)$ is a discrete sequence of closed K-invariant analytic sets in $\tilde{X}$, e.g., $Y := \bigcup Y_n$ is closed and there exists $f \in \mathcal{O}(\tilde{X})^K$ with $f|Y_n \equiv n$. Thus $\pi(x_n)$ is divergent in $\tilde{X}//K$. □

The main result of this section is now immediate.

Proposition 5. *Let K be a compact group of real analytic diffeomorphisms of a real analytic manifold M. Then there exists a proper guaged Stein K-complexification (M, X, φ).*

Proof. Let (M, X, φ) be as in Lemma A, i.e., the assumptions of Lemma B are fullfilled for $(M, \tilde{X}, \tilde{\varphi})$ which is given by applying the Zusatz to the Tube Lemma in §1 for, e.g., $m_j := \frac{1}{2} r_j$. The desired proper complexification is $(M, \tilde{X}, \tilde{\varphi})$. □.

Remark. If (M, K, φ) is a proper guaged Stein K-complexification, then, by III Proposition 4, $R(\varphi)/K$ and $X//K$ are naturally homeomorphic. It follows immediately that *the categorical quotient map $\pi : X \to X//K$ realizes the geometric quotient M/K as a closed subset of $X//K$.* We will study this embedding more carefully in the next chapter.

V. On the structure of $M/K \hookrightarrow X//K$.

We discuss some aspects of the singularities of M/K. A dense set of smooth points is characterized. In particular, if K acts freely, then it is shown that $X//K$ is a complexification of M/K.

1. The anti-holomorphic involution on $X//K$. Throughout this section we assume that X is equipped with an anti-holomorhphic involution σ as in IV Proposition 2. Let $\pi : X \to X//K$ be the categorical quotient.

Summarizing the information in IV., we have the following

Proposition 1. *The map $\pi|M$ induces a topological isomorphism $M/K \to \pi(M)$ and $\pi(M)$ is a closed subset of $X//K$ which is contained in $Fix(\tilde{\sigma})$.* □

One of the goals of this chapter is to show that M/K and $Fix(\tilde{\sigma})$ are locally equal at generic points of M/K.

Example. Let $K := SO_2(\mathbf{R})$ act on $M := \mathbf{R}^2$ via its standard representation. Of course $K^{\mathbf{C}} = SO_2(\mathbf{C})$ and we may take $X := \mathbf{C}^2$ as a K-complexification. The ring of invariants of the (linear) $K^{\mathbf{C}}$-action on X is generated by the complex norm $f(z,w) = z^2 + w^2$. Hence the categorical quotient $\pi : X \to X//K$ is given by $\pi : \mathbf{C}^2 \to \mathbf{C}$, $(z,w) \mapsto z^2 + w^2$.

The anti-holomorphic involution $\sigma : X \to X$ which defines $M = Fix(\sigma)$ is the standard one, $\sigma(z,w) = (\bar{z}, \bar{w})$. This induces the standard involution $\tilde{\sigma} : X//K \to X//K$, i.e., $\zeta \mapsto \bar{\zeta}$. Now $M/K = \pi(M) = \mathbf{R}^{\geq 0}$. However, $Fix(\tilde{\sigma}) = \mathbf{R}$.

This example is typical in that M/K is defined by inequalities in $Fix(\tilde{\sigma})$.

2. Generic orbits. Recall that for $x \in X$ we have the local $K^{\mathbf{C}}$-orbit $\Omega(x)$. For k a non-negative integer, let $X_k := \{x : dim_{\mathbf{C}}\ \Omega(x) \leq k\}$. Noting that the usual proof for globally defined actions is really of a local nature, we recall that X_k is a closed analytic subset of X for all k. Analogous considerations for the K-action on M yield closed real analytic subvarieties $M_k := \{x \in M : dim_{\mathbf{R}}\ Kx \leq k\}$.

Note, for example, that in every connected component of X it follows that there is a Zariski open set, i.e., the complement on an appropriate X_k, where the local $K^{\mathbf{C}}$-orbits are of maximal dimension in that component. It would be convenient to only consider this *connected* case. However, we do not wish to assume that K is connected. Consequently the components of X or M may not be K-stable

Nevertheless we may assume that X and M are K-connected, i.e., that K acts transitively on the components, and handle the K-components one at a time. Thus, without reference, we always make this assumption in our arguments.

For example, after reducing to the K-connected case, it follows that there is a unique m so that $\Omega_X := X_m \setminus X_{m-1}$ is open, dense, and K-connected in X. Similarly, the space Ω_M of maximal dimensional K-orbits in M is open and dense. We refer to these as the sets of *dimension-theoretically* generic orbits.

Remark. In order to prove that Ω_X is K-connected, it is useful to recall that the complement of an analytic set in a complex manifold is connected.

Warning. For any smooth G-action on a manifold M we have the decomposition of M by *G-orbit types*, i.e., the equivalence relation defined by $x \sim y := \iff$ *G_x and G_y are conjugate in G*. The equivalence relation defined by ***orbit-dimension*** can be rougher.

Example. Let $K := S^1$ act on $\tilde{M} := \mathbf{R} \times S^1$ by multiplication on the second factor and let $\Gamma := \mathbf{Z}_2$ act by $\rho(x,z) = (-x, z^{-1})$. Then K acts on $M := \tilde{M}/\Gamma$ which we regard as a fiber-space over $\mathbf{R}$ by projection on the first factor. Of course these fibers are stabilized by K and for those over points in $\mathbf{R} \setminus \{0\}$ the action is free. However, over the origin the isotropy is Γ.

We now return to complexifications.

Proposition 2. *If X is a K-complexification of M, then $\Omega_X \cap M = \Omega_M$ and*

$$dim_{\mathbf{R}}\ Kx = dim_{\mathbf{C}}\ \Omega(x)$$

for all $x \in \Omega_M$.

Proof. First note that $\Omega_X \cap M$ is the complement of a proper real analytic subset in M, because M is nowhere locally contained in a proper complex analytic subset of X. Furthermore, since the statements are local near M in X, we may choose (M, X, φ) to be a guaged Stein K-complexification.

For $x \in \Omega_X \cap M$ the strictly plurisubharmonic function $\varphi|\Omega(x)$ has its minimum along Kx and $\Omega(x)$ is K-equivariantly holomorphically embeddable in an affine homogenous space $K^{\mathbf{C}}/L^{\mathbf{C}}$, where $L^\circ = K_x^\circ$ (II Proposition 9). In particular, for $x \in M$ it follows that $dim_{\mathbf{R}}\ Kx = dim_{\mathbf{C}}\ \Omega(x)$.

Now $\Omega_X \cap M$ is dense in M and therefore $\Omega_X \cap \Omega_M \neq \emptyset$. Consequently $dim_{\mathbf{R}}\ Kx = dim_{\mathbf{C}}\ \Omega(y) =: m$ for all $x \in \Omega_M$ and $y \in \Omega_X$. Therefore $\Omega_M \subset \Omega_X \cap M$. Conversely, if $x \in \Omega_X \cap M$, then $dim_{\mathbf{R}}\ Kx = m$, i.e., $x \in \Omega_M$. □

We wish to localize our considerations to neighborhoods of generic K-orbits in M and the corresponding neighborhoods in the categorical quotient. For this we need some general information on the categorical quotient $\pi : X \to X//K$.

General Fact 1. *Let X be a Stein space equipped with a K-action and let S be a K-invariant closed analytic subset of X. Then $\pi|S$ realizes $S//K$ as a closed analytic subset of $X//K$.*

Proof. Using the *extension* and *averaging* procedures as discussed in II., one proves the following statements.

i) The injection $S \hookrightarrow X$ induces a surjection $\mathcal{O}(X)^K \to \mathcal{O}(S)^K$.

ii) $\pi(S)$ is defined as the zero set of the ideal in $\mathcal{O}(X//K) = \mathcal{O}(X)^K$ of invariant holomorphic functions which vanish on S.

By the universality of $S \to S//K$ we have a surjective holomorphic map $\iota : S//K \to \pi(S)$. By *i*) it is injective. By *ii*) $\pi(S)$ is a closed Stein subspace of $X//K$. But, again by *i*), ι induces an isomorphism $\iota^* : \mathcal{O}(\pi(S)) \to \mathcal{O}(S//K)$ of algebras of global functions. Since global functions give local coordinates, it follows that ι is an isomorphism of Stein spaces. □

General Fact 2. *Let D be a Stein domain in $X//K$. Then $\pi : \pi^{-1}(D) \to D$ realizes the categorical quotient $\pi^{-1}(D) \to \pi^{-1}(D)//K$.*

Proof. Using the same argument as above, it is enough to show that the natural map $\iota : \pi^{-1}(D)//K \to D$ is bijective. But surjectivity is obvious and injectivity follows from the fact that the π-fibers have no non-constant K-invariant functions. □

The desired localization is now a simple matter.

Proposition 3. *Let X be a Stein K-complexification of M with categorical quotient $\pi : X \to X//K$. Then, for every $x \in \Omega_M$ there exists a Stein neighborhood D of $\pi(x)$ in $X//K$ so that the following hold.*

1. $\pi : \pi^{-1}(D) \to D$ realizes the categorical quotient $\pi^{-1}(D) \to \pi^{-1}(D)//K$.
2. $\pi^{-1}(D) \subset \Omega_X$.
3. Every local $K^{\mathbf{C}}$-orbit in $\pi^{-1}(D)$ is closed.

Proof. Let S be the set of non-generic orbits in X, i.e., $S = X \setminus \Omega_X$. Since S is K-stable, it follows that $\pi(S)$ is a closed analytic subset of $X//K$ (General Fact 1). Let $D = D(\pi(x))$ be a Stein domain in its complement. Property 1. is a consequence of General Fact 2. By construction $\pi^{-1}(D) \subset \Omega_X$, and of course 3. follows from 2., because if a $K^{\mathbf{C}}$- orbit would not be closed, then it would have a smaller orbit in its closure. □

Remark. If we are in the setting of Proposition 1, we can choose D so that it is in addition $\tilde{\sigma}$-stable.

As soon as we have localized as in Proposition 3, we can further localize without conditions. This follows from

General Fact 3. *Let X be a Stein K-manifold such that all local $K^{\mathbf{C}}$-orbits are closed. Then $\pi : X \to X//K$ has constant fiber dimension, i.e., it is an open map. If in addition $\varphi : X \to \mathbf{R}^{\geq 0}$ is a K-invariant strictly pluri-subharmonic function which is proper on the π-fibers and if $W := \{\varphi < r\}$ for some r, then $\pi_W = \pi|W$.*

Proof. Since all local $K^{\mathbf{C}}$-orbits are closed, π-fibers coincide with local $K^{\mathbf{C}}$-orbits. Thus the dimension k of a generic fiber agrees with the dimension of a generic local $K^{\mathbf{C}}$-orbit. Now let F be a π-fiber. Thus $dim(F) \geq k$. If we regard $F = \Omega(x)$ as an orbit, then $dim(F) \leq k$, i.e. π has constant fiber-dimension.

Now consider $W = \{\varphi < r\}$. Of course the fibers of $\pi_W : W \to W//K$ are just the connected components of the intersections of the π-fibers with W. By universality we have a surjective map $\iota : W//K \to \pi(W)$. Since $\pi(W)$ is open in $X//K$ which is a normal complex space, in order to prove $W//K \equiv \pi(W)$ it is enough to prove that ι is injective.

If ι were not injective, then some π-fiber F intersects W in more than one component. But φ must have critical points on each of these components, contrary to the critical points of $\varphi|F$ being exactly one K-orbit (III Proposition 4). □

3. The Kempf-Ness set at a generic orbit. Let (M, K, φ) be a proper guaged Stein K-complexification and define $R(\varphi)_{gen} := \Omega_X \cap R(\varphi)$ to be the part of the Kempf-Ness set which corresponds to the closed local $K^{\mathbf{C}}$-orbits which are dimension theoretically generic.

Remark. *$R(\varphi)_{gen}$ is an open dense subset of $R(\varphi)$.*

Proof. Let X_{m-1} be the set of lower-dimensional local $K^{\mathbf{C}}$-orbits as above and define $S := \pi^{-1}(\pi(X_{m-1}))$. Now $\Omega_M \cap S = \emptyset$. In particular $\pi(S)$ is a proper analytic set in $\pi(X)$. If $x \in X \setminus S$, then $\Omega(x)$ is closed and of generic dimension, i.e. $R(\varphi)_{gen} = R(\varphi) \setminus S$. Since this is the inverse image under $\pi|R(\varphi)$ of a Zariski open set in $X//K$, the result follows. □

The following well known result is the main goal of this section.

Proposition 4. *$R(\varphi)_{gen}$ is smooth.*

For this we first observe that if we localize as in §2. above by taking an open Stein set $D \subset X//K$, then $(\pi|\pi^{-1}(D)) \times \varphi$ is still proper. So, for our purposes here, it is enough to consider the case where $\Omega_X = X$ and $\Omega_M = M$, in particular all local $K^{\mathbf{C}}$-orbits are closed and equi-dimensional.

Before proceeding, we need some general information on *fixed point sets*. First note that, since $\Omega_M = M$, there exists a connected subgroup $L \subset K$ which is open in some K_x, $x \in M$, so that $M = K \cdot F_M$, where $F_M := \{x \in M : Lx = x\}$. Let $F = F_X$ be the L-fixed points in X. It follows that $X = \bigcup_{x \in F} \Omega(x)$, because the right hand side is an analytic set which contains M. Of course we use here the assumption that M intersects every component of X. In other words, L has a fixed point in every local $K^{\mathbf{C}}$-orbit and F is a sort of *thick section* for the categorical quotient.

Now consider $\Omega(x)$ embedded in its maximal complexification $Z = K^{\mathbf{C}}/H$, where H corresponds to x (see II.). Let N be the normalizer $N_K(L)$ and recall that $N_{K^{\mathbf{C}}}(L^{\mathbf{C}}) = N^{\mathbf{C}}$. Furthermore note that the L-fixed point set F_Z in Z is just the orbit $N^{\mathbf{C}}x$.

General Fact 4. *Let $\Omega^{\mathbf{C}} = K^{\mathbf{C}}/H$ be the $K^{\mathbf{C}}$-complexification of a closed local $K^{\mathbf{C}}$-orbit Ω. Let $N \subset K$ be a connected subgroup, $x \in \Omega$, and suppose that the complex orbit $N^{\mathbf{C}}x$ is closed in $\Omega^{\mathbf{C}}$. Then $N^{\mathbf{C}}x \cap \Omega$ is connected.*

Proof. Let $\varphi : \Omega \to \mathbf{R}^{\geq 0}$ be a K-invariant strictly pluri-subharmonic exhaustion function and $T(r) := \{\varphi < r\}$. It is sufficient to prove that $N^{\mathbf{C}}x \cap T(r)$ is connected for $r > R$ for some R. Of course we have the $K^{\mathbf{C}}$-equivariant finite maps $T(r)^{\mathbf{C}} \to \Omega^{\mathbf{C}}$ for all r. For $r > R$ these must all be the same map $\rho : K^{\mathbf{C}}/H_1 \to K^{\mathbf{C}}/H$. This gives us a K-equivariant embedding of Ω in $K^{\mathbf{C}}/H_1$, i.e., ρ is an isomorphism. Therefore the K-finite holomorphic functions on $T(r)$ extend to $\Omega^{\mathbf{C}}$ and consequently $T(r)$ is Runge in $\Omega^{\mathbf{C}}$ for $r > R$.

We may therefore apply the Extension Lemma of III. to obtain a globally defined strictly plurisubharmonic exhaustion $\psi_r : \Omega^{\mathbf{C}} \to \mathbf{R}^{\geq 0}$ which agrees with φ in a neighborhood of $T(r)$. By applying III Proposition 1 to the restriction $\psi_r|N^{\mathbf{C}}x$, the desired connectedness result follows. □

Let us summarize our situation. The group $N := N_K(L)$ acts on the fixed point set F which is a complex submanifold of X. It is a symplectic manifold with respect to the restriction of $\omega = i\partial\bar{\partial}\varphi$. Thus we have the associated moment map μ_N. Now L acts trivially on F. Hence we regard this as a moment map for the action of N/L which is acting almost freely. Since the rank of a moment map at a point is just the codimension of the orbit through that point, μ has constant rank and in particular the Kempf-Ness set $\mu_N^{-1}(0)$ of this action is smooth.

Observe that, since $N^{\mathbf{C}}x$ is closed in $K^{\mathbf{C}}/H$, the local $N^{\mathbf{C}}$-orbit $\Omega_{N^{\mathbf{C}}}(x) \subset F$ is likewise closed. In order to avoid confusion, let $\Omega_{K^{\mathbf{C}}}(x)$ denote the local $K^{\mathbf{C}}$-orbit.

Lemma 5. *If $x \in \mu_K^{-1}(0) \cap F$, then*

$$(*) \qquad \mu_N^{-1}(0) \cap \Omega_{K^{\mathbf{C}}}(x) = Nx.$$

Proof. In every component of $\Omega_{N^{\mathbf{C}}}(x)$ there is a unique N°-orbit in $\mu_N^{-1}(0)$. By General Fact 4 $(N^{\mathbf{C}})^\circ x \cap \Omega_{K^{\mathbf{C}}}(x)$ is connected. It follows that N acts transitively on these components, i.e.,

$$(**) \qquad F \cap \Omega_{K^{\mathbf{C}}}(x) = \Omega_{N^{\mathbf{C}}}(x).$$

Thus

$$\mu_N^{-1}(0) \cap \Omega_{K^{\mathbf{C}}}(x) = \mu_N^{-1}(0) \cap \Omega_{N^{\mathbf{C}}}(x) = Nx.$$

□

Corollary 6. $K \cdot \mu_N^{-1}(0) = \mu_K^{-1}(0)$.
Proof. Since $X = \bigcup \Omega_{K^{\mathbf{C}}}(x)$, where this is taken over $x \in \mu_K^{-1}(0) \cap F$, the proof follows by applying K to both sides of $(*)$. □

Using this description of $\mu_K^{-1}(0)$ we show that it is smooth. In the following, the orthogonal complement $\perp$ is computed with respect to the Riemannian metric associated to ω. The symbol $\angle$ refers to the *skew-complement* with respect to ω itself.

Lemma 7. *For $x \in \mu_N^{-1}(0)$,*

$$T_x \mu_N^{-1}(0) = (T_x Nx) \oplus (T_x \Omega_{K^{\mathbf{C}}}(x))^{\perp}.$$

Proof. Since $Ker(\mu_N)_*(x) = (T_x Nx)^{\angle}$ and Nx is isotropic, it follows that $T_x\mu_N^{-1}(0) \supset (T_x Nx) \oplus (T_x \Omega_{K^{\mathbf{C}}}(x))^{\perp}$. Equality follows from $(**)$ by a dimension count. □

From Lemma 7 it follows that locally the set $K \cdot \mu_N^{-1}(0)$ is smooth with tangent space $(T_x Kx) \oplus (T_x \Omega_{K^{\mathbf{C}}}(x))^{\perp}$. Thus, for the global smoothness, it is enough to prove

Lemma 8. *Let $x \in \mu_N^{-1}(0)$, $k \in K$, and suppose that $kx \in \mu_N^{-1}(0)$. Then $k \in N$.*
Proof. Regard the discussion as taking place in $\Omega_{K^{\mathbf{C}}}(x) \subset K^{\mathbf{C}}/H$ as above. Since $kx \in F$, it follows that $k \in N^{\mathbf{C}}$. But $N^{\mathbf{C}} \cap K = N$. □

The smoothness of $R(\varphi)_{gen}$, i.e. Proposition 4, has now been proved. In fact we have given a precise discription of this set.

Proposition 9. *Under the assumption that the local $K^{\mathbf{C}}$-orbits in X are equidimensional, it follows that*

$$R(\varphi) = \mu_K^{-1}(0) = K \times_N \mu_N^{-1}(0).$$

□

Before closing this section we would like to comment on the Cauchy-Riemann structure of $R(\varphi)_{gen}$. Recall that a *CR-submanifold* of a complex manifold is said to be *generic* if the rank of its complex tangent bundle is minimal.

Proposition 10. *The manifold $R(\varphi)_{gen}$ is a generic Cauchy-Riemann submanifold of X. For $x \in R(\varphi)_{gen}$ the maximal complex tangent space in $T_x R(\varphi)$ is*

$$H_x = (T_x \Omega(x))^{\perp},$$

where the closed local $K^{\mathbf{C}}$-orbit $\Omega(x)$ is the fiber of the categorical quotient $\pi : X \to X//K$ at x.

Proof. We showed above that

$$T_x R(\varphi) = (T_x Kx) \oplus (T_x \Omega(x))^{\perp}.$$

Since $T_x Kx$ is totally real and $T_x \Omega(x)$ is complex, the computation of H_x is immediate. Genericity follows by the obvious dimension count. □

4. The quotient at generic points. Here we are particularly interested in the points of $R(\varphi)$ which are generic from the point of view of orbit type. These are the points $x \in R(\varphi)_{gen}$ where the local slice representation of the isotropy is trivial, i.e., a slice neighborhood $K \times_{K_x} S$ of the orbit Kx is equivariantly a trivial product $(K/K_x) \times S$. Since inclusion of isotropy induces a partial ordering of orbit type where the points of generic orbit type are maximal, we denote the set of such points by $R(\varphi)_{max}$.

Proposition 11. *For $x \in R(\varphi)_{max}$ it follows that $X//K$ is smooth at $\pi(x)$ and $\pi_*(x)$ has maximal rank.*

Proof. Let S be a slice as above. In particular $T_x S$ is Riemannian orthogonal to $T_x Kx$ in $T_x R(\varphi)_{gen}$ and therefore $T_x S = H_x$, the maximal complex subspace (see Proposition 10). Take Z to be a (local) complex submanifold X with $T_x Z = H_x$.

Now $\pi|S$ maps S bijectively onto an open neighborhood of $\pi(x) \in X//K$. Hence, since $T_x S = T_x Z$, it follows that $\pi|Z$ is injective near x. But $X//K$ is normal and consequently $\pi|Z$ is biholomorphic near x.□

Remarks. 1.) If $x \in R(\varphi)_{gen}$, then one can reverse the above argument, i.e., $\pi_*(x)$ *has maximal rank if and only if $x \in R(\varphi)_{max}$.*

2.) The set of smooth points $(X//K)_{reg}$ may be bigger than $\pi(R(\varphi)_{max})$, e.g., any 1-dimensional categorical quotient of a normal space is smooth.

Let $x \in M_{max} = R(\varphi)_{max} \cap M$ and choose $D \subset X//K$ to be a $\tilde{\sigma}$-invariant Stein neighborhood of $\pi(x)$ so that $\pi|\pi^{-1}(D) : \pi^{-1}(D) \to D$ realizes the categorical quotient and $\pi^{-1}(D) \cap R(\varphi) \subset R(\varphi)_{max}$ (see General Fact 2 in 3. above).

Corollary 12. *The geometric quotient $(M \cap \pi^{-1}(D))/K = D \cap \pi(M)$ is embedded as a smooth totally real submanifold of half the dimension in D and is defined by the anti-holomorphic involution: $Fix(\tilde{\sigma}) = D \cap \pi(M)$.*

Proof. For $x \in M_{max}$, the complementary space $H_x \subset T_x R(\varphi)$ is just the complexification of the Riemannian orthogonal complement $H_x(\mathbf{R})$ to $T_x Kx$ in $T_x M$. Since π is holomorphic and $\pi_* : H_x \to T_{\pi(x)}(X//K)$ is an isomorphism, it follows that $T_{\pi(x)}M/K = \pi_*(H_x(\mathbf{R})$ is totally real and half-dimensional in $T_{\pi(x)}(X//K)$. Since $Fix(\tilde{\sigma})$ is itself half-dimensional, the result follows. □

The following special case is of particular use in applications.

Corollary 13. *If $M = M_{max}$, then $X//K$ is a Stein complexification which is smooth in a neighborhood of the smooth geometric quotient M/K.* □

If the group is acting freely, then of course $M = M_{max}$. For example, let H be a closed subgroup of K which acts real analytically on a real analytic manifold S. Let $M := K \times_H S$ be the associated K-manifold. Now take $S^{\mathbf{C}}$ to be a Stein H-complexifiction of S and $K^{\mathbf{C}}$ to be the universal complexification of the group K. Of course $K^{\mathbf{C}}$ is a $K \times K$-complexification of K as a manifold and in particular $K \times H$ acts on $K^{\mathbf{C}}$. Finally, consider $K^{\mathbf{C}} \times S^{\mathbf{C}}$ as an H-complexification of $K \times S$, i.e., with respect to the right H-action on $K^{\mathbf{C}}$.

Corollary 14. *The complex manifold $X = (K^{\mathbf{C}} \times S^{\mathbf{C}})//H$ is a Stein K-complexification of the geometric quotient $K \times_H S = M$.*

Proof. By Corollary 13 the manifold X is a Stein complexification of M. Furthermore, we may choose φ to be $K \times H$-invariant. In particular $R_H(\varphi)$ is K-invariant and therefore K acts real analytically on the geometric quotient $R_H(\varphi)/H = X$. Since K stabilizes the H-invariant holomorphic functions on $K^{\mathbf{C}} \times S^{\mathbf{C}}$, this action is holomorphic. □

Remark. In general, if Z is a Stein H-manifold, then we have the associated Stein K-manifold $X = (K^{\mathbf{C}} \times Z)//H$. A K-invariant function on X lifts to a K-invariant function on the product $K^{\mathbf{C}} \times Z$. In this way we have the natural isomorphism $\mathcal{O}(X)^K \cong \mathcal{O}(Z)^H$ and it follows that $X//K = Z//H$. □

We now briefly consider the categorical quotient in a neighborhood of $x \in R(\varphi)_{gen}$. For this we may assume that we have already localized so that $M = K \times_H S$, where $H = K_x$ and S is a closed real analytic H-stable submanifold. We take $S^{\mathbf{C}}$ to be a Stein H-complexification which is embedded as a locally closed complex submanifold of X.

Now $K^{\mathbf{C}} \times S^{\mathbf{C}}//H$ is a Stein K-complexification of $M = K \times_H S$. Note that, since $x \in R(\varphi)_{gen}$, it follows that H° acts trivially on S and therefore also on $S^{\mathbf{C}}$. Thus the $K^{\mathbf{C}}$-orbits in $K^{\mathbf{C}} \times S^{\mathbf{C}}//H$ are equi-dimensional, i.e., up to finite coverings they are of the form $K^{\mathbf{C}}/H^{\mathbf{C}}$. Hence, by General Fact 3 above, if we are interested in the embedding

$M/K \hookrightarrow X//K$ at $x \in R(\varphi)_{gen}$, we may replace the given complexification by such a *slice complexification.*

Proposition 15. *Let X be a Stein K-complexification of M. Then, for every $x \in \Omega_M$, there exists a finite group Γ represented on a real vector space $V_{\mathbf{R}}$ so that the embedding $M/K \hookrightarrow X//K$ in a neighborhood of $\pi(x)$ is locally given by $V_{\mathbf{R}}/\Gamma \hookrightarrow V_{\mathbf{C}}/\Gamma$, where $V_{\mathbf{C}} = V_{\mathbf{R}} \otimes \mathbf{C}$.*

Proof. The slice S is an open subset of the origin in a $H/H^\circ =: \Gamma$ representation space $V_{\mathbf{R}}$. Thus

$$M/K = K \times_H S = S/H = S/\Gamma \hookrightarrow V_{\mathbf{R}}/\Gamma.$$

For the analogous statement for $X//K$, note that the categorical quotient with respect to a finite group is just the geometric quotient. Thus

$$X//K = S^{\mathbf{C}}//H = S^{\mathbf{C}}//\Gamma = S^{\mathbf{C}}/\Gamma \hookrightarrow V_{\mathbf{C}}/\Gamma. \square$$

5. Pushing down functions. Let us return to the general setting of a proper gauged Stein K-complexification (M, X, φ). Recall that the restriction $\pi|R(\varphi)$ induces a homeomorphism $R(\varphi)/K \to X//K$. This gives a surjective morphism at the level of continuous functions $C(X)^K \to C(X//K)$, $f \mapsto \tilde{f}$, where $\tilde{f}$ is the function on $R(\varphi)/K$ induced from $f|R(\varphi)$. For holomorphic functions this is of course an isomorphism.

Even in the special case where all local $K^{\mathbf{C}}$-orbits are closed and where $R(\varphi)$ is a smooth submanifold of X, quotients via finite groups play a role. In fact, even when the quotient is smooth, the transform of a smooth function may not be smooth. For example, let $K = \mathbf{Z}_2$ act on $X = \mathbf{C}$ by $z \mapsto -z$. If $\varphi(z) = |z|^2$, then $\tilde{\varphi}(\zeta) = |\zeta|$.

If we restrict to $R(\varphi)_{max}$, then, in the notation of Proposition 11 above, $\pi|S : S \to \pi(S)$ is a diffeomorphism. Hence K-invariant smooth functions defined in a neighborhood of $R(\varphi)_{max}$ push down to smooth functions on the complex manifold $\pi(R(\varphi)_{max})$. This of course applies for the function φ itself.

Proposition 16. *The function $\tilde{\varphi}$ is a smooth strictly plurisubharmonic function on $\pi(R(\varphi)_{max})$.*

Proof. It is enough to compute the Levi-form of $\tilde{\varphi}$ at $\pi(x)$ for $x \in R(\varphi)$ (see also [HHL]). Using the local holomorphic section Z in the proof of Proposition 11, X is locally the product $X = \Delta_1 \times \Delta_2$ of polydisks where π is given by the projection on the first factor, $(z, w) \mapsto z$. Here Z corresponds to $\Delta_1 \times \{0\}$ and our point of interest x to the origin $(0, 0)$.

Recall that Z is tangent to the slice $S \subset R(\varphi)$. Thus, localizing further if necessary, S is parameterized by $\sigma : \Delta_1 \to S$, $z \mapsto (z, \eta(z))$, where $\eta(0) = 0$ and $\eta_*(0)$ is the 0-map. Of course $\tilde{\varphi}(z) = \varphi(z, \eta(z))$. An explicit calculation shows that

$$\frac{\partial^2 \tilde{\varphi}}{\partial z_i \partial \bar{z}_j}(0) = \frac{\partial^2 \varphi}{\partial z_i \partial \bar{z}_j}(0, 0).$$

$\square$

Remarks. 1.) It should be underlined that the calculation indicated above strongly relies on the fact that $\varphi|\{0\} \times \Delta_2$ is critical at the origin, i.e., we really do use that $x \in R(\varphi)$.

2.) If X is a complex space, it is possible to define ***strictly plurisubharmonic*** in an appropriate way. Using a similar argument to that above, it can be shown that if X is a Stein K-space and φ is a strictly plurisubharmonic function on X which is proper along the π-fibers, then the transform $\tilde{\varphi}$ is likewise strictly plurisubharmonic ([HHL]). □

As a last observation in this section we note a result which is particularly useful in slice considerations.

Corollary 17. *Let (M, X, φ) be a proper gauged Stein K-complexification where K is acting freely. Then the Stein manifold $\tilde{X} := X//K$ is a complexification of the smooth geometric quotient $\tilde{M} := M/K$ and $\tilde{\varphi} : \tilde{X} \to \mathbb{R}^{\geq 0}$ is a strictly plurisubharmonic function with $\tilde{\varphi}^{-1}(0) = \tilde{M}$.* □

VI. Proper actions

After some introductory discussion of proper actions, we prove the following real analytic version of a result which was proved by H. Abels in the case of smooth actions([A]): Let M be a real analytic manifold and $G \times M \to M$ be a proper real analytic action of a Lie group G which has finitely many components. Then there exists a G-equivariant real analytic map $\pi : M \to G/K$. Here K denotes a maximal compact subgroup of G. It should be noted that compatible G-invariant real analytic structures on a C^∞-manifold exist and are unique.

1. Slice neighborhoods. We begin this section by recalling some basic facts on proper actions. Throughout G denotes a real Lie group with finitely many components and K a maximal compact subgroup. At first we only assume that M is a differentiable manifold and that G acts properly as a group of diffeomorphisms.

Example. Let H be a compact subgroup of G and $\rho : H \to Gl(V)$ a finite-dimensional representation. Consider the G-vector bundle $E := G \times_H V \to G/H$ which is defined as the geometric quotient of the trivial bundle $G \times V \to G$ by the diagonal H-action $h(g, v) := (gh^{-1}, \rho(h)v)$. The left action of G on itself induces a *proper* G-action on the vector bundle E. Let S be an H-stable open neighborhood of $0 \in V$ and $U := G \cdot S$. Then U is open in E and of course the restricted G-action on U is proper.

Tubular neighborhoods of zero sections of the type U above are the local models for general proper G-actions. To see this, let $x \in M$ and consider the orbit $Gx = G/H$, where

$H := \{g : g(x) = x\}$ is the G-isotropy at x. Since the action is proper, Gx is closed and H is compact.

Let g be a Riemannian metric on M and consider its *average* over H:

$$\tilde{g}(v, w) := \int_{h \in H} g(h_* v, h_* w) dh,$$

where dh is the bi-invariant probability measure on H. By replacing g with $\tilde{g}$, we may choose the metric to be H-invariant.

In the tangent space $T_x M$ we consider the subspace $V := T_x Gx$ and, with respect to the H-invariant metric g, its orthogonal complement N, $T_x M = V \oplus N$. Let $B = \{v : \|v\|_g^2 < r\}$ be a ball in $T_x M$ so that the differential geometric exponential map is a diffeomorphism onto its image $W = W(x)$. Of course we have the natural representation $\tau : H \to Gl(T_x M)$, $\tau(h) = h_*(x)$, and, from the uniqueness of geodesics, it follows that $exp : B \to U$ is H-equivariant: $exp(\tau(h)v) = h(exp(v))$.

The *normal bundle* $E := G \times_H N \to G/H \cong Gx$ to the orbit contains the tubular neighborhood $G \times_H S_N$, where $S_N := N \cap B$. Of course $exp : S_N \to S \subset W$ is a diffeomorphism onto its image S which we regard as a *slice* on the G-orbit Gx. After shrinking S appropriately, the map $G \times S_N \to M$, $(g, v) \mapsto g(exp(v))$, establishes a G-equivariant diffeomorphism $G \times_H S_N \to U := G \cdot S$.

As a local converse, observe that, if the compact group H acts real analytically on S, e.g., where S is an open H-stable neighborhood of the origin in a representation space, then G acts real analytically on $M := G \times_H S$.

The fact that a slice neighborhood possesses a real analytic structure where G is acting as a group of real analytic diffeomorphism allows an adaptation of Whitney's theorem on the existence of a compatible real analytic structure to the case of proper actions (compare [I]).

Proposition 1. *Let M be a differentiable manifold and G a Lie group of diffeomorphisms acting properly on M. Then there exists a real analytic structure on M so that G is acting real analytically as a group of real analytic diffeomorphisms.* □

2. Abels' Theorem. Now the above *local slice description* depends on the isotropy $H = G_x$ which varies from point to point. The only common denominator is that, after conjugation, these isotropy groups lie in a fixed maximal compact subgroup K. Thus one could hope to find some sort of *thick global slice* which is K-stable and contains the local slices. This has been realized in the following basic result of Abels([A]).

C^∞-global slice theorem. *Let G be a Lie group with finitely many components and K a maximal compact subgroup. Suppose that G is acting smoothly and properly on a differentiable manifold M. Then there exists a G-equivariant differentiable map $\pi : M \to G/K$.*

Our goal here is to prove a real analytic version of Abels' Theorem. Before doing so, we would like to ellaborate a bit on its statement.

Let $y_0 \in G/K$ be a neutral point corresponding to the isotropy group K and, abusing the notation of the local slice theorem, $S := \pi^{-1}(y_0)$. The K-action on S yields the G-fiber bundle $G \times_K S \to G/K$ and the map $[(g,s)] \mapsto g(s)$ defines a bundle equivalence to $\pi : M \to G/K$. For $s \in S$, the projection π induces an equivariant fibration of the orbit $Gs = G/G_s \to G/K$ with compact fiber $Ks = K/G_s$ contained in S. Since the orbit Gs intersects S in this K-orbit, which may very well be positive dimensional, we refer to S as a *thick slice.*

On the other hand, S is a perfect slice for a smaller subgroup. For this let $G = KAN$ be an Iwasawa decompostion. The solvable group $R := AN$ acts freely and transitively on the cell G/K. Thus the R-orbits $R(s)$, $s \in S$, are sections of $\pi : M \to G/K$. Looking at things from the other direction, the proper, free R-action defines a principal bundle $M \to M/R$. Of course S is a section and therefore M has the differentiable product structure $M \cong S \times R$.

The real analytic version of Abels' theorem follows from an elementary application of K-equivariant Whitney approximation.

Real analytic global slice theorem. *Let G be a Lie Group with finitely many components and K a maximal compact subgroup. Suppose that G is acting real analytically and properly on a real analytic manifold M. Then there is a G-equivariant real analytic map $\pi : M \to G/K$.*

Before proceeding with the proof, we mention the following Lemma which is a direct consequence of Rolle's Theorem.

Lemma. *Let $\phi_0 : U \to V$ be a diffeomorphism of domains in $\mathbb{R}^n$ and let $\tilde{U} \subset U$ be a relatively compact convex open subset of U. Then there exists $\epsilon > 0$ so that, if ϕ is a C^∞-map on U with $\phi(\tilde{U}) \subset V$ and $\|D_x\phi - D_x\phi_0\| \leq \epsilon\|D_x\phi_0\|$ for all $x \in \tilde{U}$, then $\phi : \tilde{U} \to \tilde{V} \subset V$ is likewise a diffeomorphism.* □

Proof of the real analytic slice theorem. Consider a smooth G-equivariant map $\pi_a : M \to G/K$, the existence of which is guaranteed by Abels' Theorem. Let y_0 be a point in the base with isotropy K and $S := \pi_a^{-1}(y_0)$. Take $\tilde{D}$ and D to be relatively compact neighborhoods of the identity $e \in R$ with $\tilde{D} \subset\subset D \subset\subset R$. We may regard R as being identified with G/K as the orbit of y_0 and may assume that $\tilde{V} := \tilde{D}(y_0)$ and $V := D(y_0)$ are K-stable.

Take $s_0 \in S$ and define $\tilde{U} := \tilde{D}(s_0) \subset\subset U := D(s_0)$. Finally, let $\tilde{W} := \pi_a^{-1}(\tilde{V}) \subset\subset W := \pi_a^{-1}(V)$.

Now let π_w be a K-equivariant real analytic approximation to π_a in the Whitney topology so that

i) The conditions of the Lemma are fulfilled for $\phi_0 := \pi_a|U$ and $\phi := \pi_w|U$ with $y_0 \in \pi_w(\tilde{U})$.
ii) At every $w \in W$ the map π_w has maximal rank and is transversal to the orbit $R(w)$.

iii) $y_0 \notin \pi_{\mathfrak{w}}(W \setminus \tilde{W})$.

Note that, since $G/K = \mathbf{R}^n$, this is just a matter of approximating the map $\pi_{\mathfrak{a}}$ which has values in a K-representation space. Thus the Zusatz to II Proposition 2 can be applied.

Define $S_{\mathfrak{w}} := \pi_{\mathfrak{w}}^{-1}(y_0)$. It follows from *ii)* that $S_{\mathfrak{w}}$ is a real analytic submanifold of M. By *iii)*, $S_{\mathfrak{w}} \subset \tilde{W}$. Thus, $S_{\mathfrak{w}} \cap D(s) = S_{\mathfrak{w}} \cap \tilde{D}(s)$ for all $s \in S_{\mathfrak{a}}$. From *ii)* it follows that the intersection $S_{\mathfrak{w}} \cap D(s)$ is transversal. Therefore the cardinality $n > 0$ of $S_{\mathfrak{w}} \cap D(s)$ is constant *independent* of s. Since *i)* is fulfilled, we may apply the Lemma to show that $\pi_{\mathfrak{w}}|\tilde{U}$ is a diffeomorphism with y_0 in its image. Thus $n = 1$ and consequently $S_{\mathfrak{w}} \times R \to M$, $(r, s) \mapsto r(s)$, is a real analytic homeomorphism. Since the R-orbits are transversal to $S_{\mathfrak{w}}$, it is a real analytic isomorphism. In particular M is thereby real analytically the product $S_{\mathfrak{w}} \times R$.

The action map $G \times S_{\mathfrak{w}} \to M$ factors through the real analytic geometric quotient $G \times_K S_{\mathfrak{w}}$ and the above considerations show that the induced map $G \times_K S_{\mathfrak{w}} \to M$ is a real analytic isomorphism. Thus the standard bundle map $G \times_K S_{\mathfrak{w}} \to G/K$ yields the desired fibration of M. □

Remarks. 1.) Note that in the above proof, in order to construct the equivariant diffeomorphism with the bundle space $G \times_K S_{\mathfrak{w}}$, one only uses that K is acting real analytically. Hence, given a K-stable structure, we obtain a G-stable structure. As we noted in I., the G-stable real analytic structure is in fact unique ([K2]).

2.) It should be noted that the real analytic slice allows us to construct a G-invariant real analytic metric in a simple way: Take any K-invariant real analytic metric and restrict it as a tensor to $S_{\mathfrak{w}}$. If τ is the resulting tensor, for $s \in S_{\mathfrak{w}}$ and $h \in G$, define the metric g at $h^{-1}(s)$ by $g(v, w) := \tau(h_*(v), h_*(w))$. □

VII. Complexifications for proper actions.

We construct a complexification X of a proper G-action. It is the K-categorical quotient of the product of the complexifications of the fiber S of a real analytic global slice fibration and of the group G itself. The group G acts properly on X as a group of holomorphic transformations. While the fiber complexification is a K-complexification of the type discussed in IV., our work here is spent in constructing a $G \times K$ complexification of G, i.e. a proper guaged Stein K-complexification where the pluri-subharmonic function is in addition G-invariant. It is noted that the G-categorical quotient of X is just the K-categorical quotient of the K-complexification of S. In conclusion we prove that X has the desired universality property.

1. Complexifications of G and proper guagings. Throughout this section G denotes a real Lie group with finitely many components. It should first be underlined that the universal complexification $G \to G^{\mathbf{C}}$ from Lie group theory may not be a G-complexification of G in our sense.

Example. Let $\widetilde{Sl_2(\mathbf{R})} \to \widetilde{Sl_2(\mathbf{R})}/\Gamma = Sl_2(\mathbf{R})$ be the universal cover of $Sl_2(\mathbf{R})$. The fundamental group Γ is isomorphic to $\mathbf{Z}$ and can therefore be realized as a dense subgroup of S^1. Define the Lie group $G = (S^1 \times \widetilde{Sl_2(\mathbf{R})})/\Gamma$ as the quotient by the diagonal action.

Since SU_2 and $Sl_2(\mathbf{R})$ are the only real forms of $Sl_2(\mathbf{C})$, it follows that the projection of Γ into $\widetilde{Sl_2(\mathbf{R})}$ is in the kernel of $G \to G^{\mathbf{C}}$. The density of the projection of Γ into S^1 then implies that the complexification factors through the obvious map to $Sl_2(\mathbf{R})$:

$$G \to Sl_2(\mathbf{R}) \hookrightarrow Sl_2(\mathbf{C}) = G^{\mathbf{C}}.$$

Despite these difficulties, Winkelmann has shown that G-complexifications in our sense do exist.([Wi]):

Proposition 1. *Let G be a Lie group acting on itself, i.e., $M := G$, by left multiplication. Then there exists a Stein, complete hyperbolic G-complexification X of M where G acts properly and freely.* □

Let K be a maximal compact subgoup of G and note that $G \times K$ acts properly on $M := G$, i.e., G from the left and K from the right. For our purposes it is necessary to prove the above result for the action of this larger group. It is possible to do so by modifying Winkelmann's proof, but we find it more instructive to carry this out via the *adapted complex structure.*

Let (M, g) be a complete Riemannian manifold. A geodesic in M yields a map of tangent bundles $\gamma_* : T\mathbf{R} \to TM$. Now, $T\mathbf{R}$ possesses a *standard* complex structure, i.e., $z := x + iy \in \mathbf{C} = T\mathbf{R}$ denotes the *tangent vector* y at the *point* $x \in \mathbf{R}$. A complex structure on TM is said to be *adapted* if all maps induced by geodesics are holomorphic. It is too much to demand this globally and so one asks only for an *adapted complex structure* on a neighborhood Ω of the zero-section $M \hookrightarrow TM$, i.e., a map induced by a geodesic is holomorphic at least on some neighborhood of $\mathbf{R}$ in $\mathbf{C}$.

If g is a real analytic metric, then there exists a unique adapted complex structure on a neighborhood Ω of the zero-section of TM ([S],[GSt]). In our *group setting* we let $M := G$ and g to be $G \times K$-invariant. Note that any such metric is real analytic and complete.

Now $G \times K$ acts on TM in the natural way and of course this action is proper. From the uniqueness of the adapted structure, since it is acting as a group of isometies on M, $G \times K$ acts holomorphically on Ω. In particular, if $B := B_x(r) \subset T_xM$ is a *ball of radius* r which is contained in Ω, then the *tubular neighborhood of radius r of the zero-section*

$$T(r) = \{v \in TM : \|v\|_g \le r\} = G \cdot B$$

is likewise contained in Ω.

The function $\phi(v) := \|v\|_g^2$ is strictly plurisubharmonic on Ω (see e.g.[LS]). Thus $\partial T(r)$ is strongly pseudoconvex. However, in the case where M is non-compact, i.e., $T(r)$ is not relatively compact, it is not clear whether or not $T(r)$ is Stein. In our particular case, if r is sufficiently small, we can apply Proposition 1 to prove that $T(r)$ is indeed Stein.

Proposition 1'. *There exists $r_0 > 0$ so that for $0 < r < r_0$ the tube $T(r)$ is a Stein hyperbolic manifold.*

Proof. Let X be the complexification of M guaranteed by Proposition 1. For r sufficiently small and $D = D(e)$ a sufficiently small neighborhood of the identity $e \in G$, the real analytic identity map $\iota : D \to D$ extends to a holomorphic map from the open set $U := D \cdot B_e(r)$ in $T(r)$ into X, $\iota^{\mathbf{C}} : U \to X$. By taking r smaller if necessary, we may assume that $\iota^{\mathbf{C}}$ is biholomorphic onto its image, $\iota^{\mathbf{C}}(B_e(r))$ is a geometric slice for the G-action on X, and

$$(*) \qquad \iota^{\mathbf{C}}(g(x)) = g(\iota^{\mathbf{C}}(x))$$

for all $x \in U$. This *equivariance* follows from the equivariance of ι and the identity principle.

Now G acts freely on both $T(r)$ and X. Hence, $\iota^{\mathbf{C}}$ has a well-defined extension given by $(*)$, i.e. we may assume that $\iota^{\mathbf{C}} : T(r) \to X$ is an equivariant diffeomorphism onto its image. Now $\iota^{\mathbf{C}}$ is holomorphic on U and G acts on both X and $T(r)$ as a group of biholomorphic transformations. Thus this extension is biholomorphic.

Finally, since $T(r)$ has a strongly pseudoconvex boundary and is embedded in the Stein manifold X, it follows that $T(r)$ itself is Stein. Since holomorphic maps are distance decreasing for the Kobayashi metric, $T(r)$ is likewise hyperbolic. □

We now come to the main point of this section. For this we let $I(r)$ denote the interval $[0, r)$, $\varphi : T(r) \to I(r)$ the norm-function, $v \mapsto \|v\|_g^2$, and $\pi_r : T(r) \to T(r)//K$ the categorical quotient.

Proposition 2. *There exists $r_0 > 0$ so that for $r < r_0$ $(M, T(r), \varphi)$ is a proper guaged Stein K-complexification of $M = G$. The group G acts properly and freely as a group of holomorphic transformations on $T(r)$, the K-action on $T(r)$ is free, and the G- and K-actions commute.*

Proof. Let $x \in M$ be given and apply Lemma A in IV.3 to obtain K-invariant neighborhoods $W = W(x)$, $Z(\pi_r(x))$ and $I(s)$, for some $s \leq r$, so that $(\pi_r \times \varphi)|W : W \to Z \times I(s)$ is proper. For an appropriate choice of $\{g_j\} \subset G$, the sets $W_j := g_j W$, $Z_j := g_j Z$, and $I_j = I(s)$ provide the data so that φ is proper along the fibers of π_r in $\{W_j\}$.

Take $r_0 < s$ small enough so that $T(r_0) \subset G \cdot W$ and apply Lemma B of *IV*.2. It follows that $(M, T(r), \varphi)$ is a proper Stein K-complexification for all $r < r_0$. □

We note that this gives explicit constructions of G-complexifications of G/K.

Corollary 3. *Take $(M, T(r), \varphi)$ with $M = G$ and $r < r_0$ as in Proposition 2. Let $\tilde{M} = M/K$, $\tilde{T} = T(r)//K$, and take $\tilde{\varphi}$ to be the "pushed down" strictly pluri-subharmonic* *function (see V.5). Then $(\tilde{M}, \tilde{T}, \tilde{\varphi})$ is a guaged smooth Stein G-complexification of $\tilde{M} = G/K$ where G acts properly as a group of holomorphic transformations.*

Proof. From Corollary V.17 it follows that $(\tilde{M}, \tilde{T}, \tilde{\varphi})$ is a guaged smooth Stein complexification. Since the G- and K-actions on T commute, the individual elements of G act as holomorphic transformations on $\tilde{T} = T//K$. Furthermore, since φ is G-invariant, G stabilizes $R(\varphi)$. But $\pi : T \to T//K$ induces a real analytic isomorphism $R(\varphi) \cong \tilde{T}$. Hence, the G-action pushes down via the geometric quotient $R(\varphi) \to R(\varphi)/K$. □

2. Complexification of a proper G-action. We now are in a position to to construct a complexification of a proper G-action of real analytic diffeomorphisms.

Proposition 4. *Let G be a Lie group with finitely many components which is acting properly as a group of real analytic diffeomorphisms on a real analytic manifold M. Then there exists a guaged Stein G-complexification (M, X, φ) where G acts properly as a group of holomorphic transformations.*

We would also like to state a technical more constructive version of this result.

Proposition 4'. *Let S be a real analytic slice, i.e., the fiber over the neutral point in a real analytic G-equivariant map $M \to G/K$, and let (S, X_S, φ_S) be a proper guaged Stein K-complexification as in IV Propposition 2. Let (G, X_G, φ_G) be a proper guaged G-complexification as in Proposition 2 above. Define $X := (X_G \times X_S)//K$ and let φ be the push-down of $\varphi_G \times \varphi_S$ as in Corollary V.17. Then $(G \times_K S, X, \varphi)$ is the G-complexification as required in Proposition 4.*

Proof. Since $\varphi_G \times \varphi_S$ is proper along the fibers of the $(K \times K)$-categorical quotient $(X_G \times X_S) \to X_G//K \times X_S//K$, it is also proper along the fibers of the categorical quotient $X_G \times X_S \to (X_G \times X_S)//K$ defined by the diagonal action. By construction M can be identified with the geometric quotient $(G \times S)/K = G \times_K S$. Thus, by Corollary V.17, it only remains to show that G acts properly. The proof of this is exactly the same as in Cor.3 above. □

The description in Proposition 4' of the G-complexification allows us to construct its categorical quotient.

Proposition 5. *Let $\pi : X_G \times X_S \to (X_G \times X_S)//K = X$ be the K-categorical quotient which defines the G-complexification X of M. Then the G-categorical quotient $X \to X//G$ exists and $X//G$ and is naturally identifiable with $X_S//K$.*

Proof. Let $f \in \mathcal{O}(X)^G$ and consider its lift as a $(G \times K)$-invariant function on $X_G \times X_S$, i.e., a K-invariant function on X_S. Hence f can be regarded as a function on the base of the map $q : X \to X_S//K$ which is induced by projection on the second factor. The universality for G-invariant maps $F : X \to Y$ follows from the universality of the K-quotient $X_S \to X_S//K$. □

Remark. The geometric quotient M/G is, via a given global slice S, identified with S/K. Hence, any statement which can be proved in the context of compact groups for $S/K \hookrightarrow X_S//K$ (see V.) carries over to the same statement for $M/G \hookrightarrow X//G$. For example, applying the Tarski-Seidenberg theorem to a *Luna slice* along points of M, it is possible to show that S/K is a *semi-analytic* subset of $X_S//K$. It then follows that M/G is a semi-analytic subset of $X//G$.

3. Universality. We close this chapter by proving that the G-complexifications of the type above have the desired universality property. In particular, *they are unique along* M. For this it is convenient to first prove a technical result.

Lemma. *Let X be a G-complexification of M where G is acting properly and U be an arbitrary neighborhood of M in X. Then there exists a connected neighborhood W of M which is contained in U so that, for every $u \in W$, there exists an open connected neighborhood $W(u) \subset W$ and a point $m \in W(u) \cap M$ with*
i.) $G_u \subset G_m$
ii.) $W(u)$ is G_m-stable.

Proof. Let $x \in M$ and $\Sigma(x)$ be a slice on the orbit $G(x)$ in X. Take $S(x)$ and $B(x)$ to be relatively compact G_x stable connected neighborhoods of $x \in \Sigma(x)$ and $e \in G$ respectively so that $N(x) := B(x) \cdot S(x)$ is relatively compact in U. For $u \in N(x)$ there exists $g \in B(x)$ so that $u \in S(g(x)) := gS(x)$.

Let $m := g(x)$ and take B to be a G_m-stable neighborhood of $e \in G$ so that $W(u) := B \cdot S(m)$ is contained in $N(x)$. Of course $W(u)$ is G_m-stable. The proof is completed by letting $W := \bigcup_{x \in M} N(x)$. □

We now prove the desired "universality" property.

Proposition 6. *Let X be a G-complexification of real analytic manifold M where G is acting properly as a group of real analytic diffeomorphisms. Then, for every real analytic G-equivariant map $F : M \to Y$ to a complex space Y where G acts as a group of holomorphic transformations, there exists a G-stable neighborhood Z of M in X and a holomorphic G-equivariant map $F^{\mathbf{C}} : Z \to Y$ so that $F^{\mathbf{C}}|M = F$.*

Proof. Since F is real analytic, it extends as a holomorphic map $F^{\mathbf{C}} : U \to Y$, where U is some neighborhood of M in X. We take $W \subset U$ as in the Lemma and let $Z := G \cdot W$. Thus we must prove that

$$\iota^{\mathbf{C}}(g(u)) := g(\iota^{\mathbf{C}}(u)) \tag{1}$$

is a well-defined map. For this it is enough to show that (1) holds for $u \in W(u)$ and $g \in G_u$. Now $W(u)$ is connected, has non-empty intersection with M and is G_m stable for some $m \in M$ with $G_u \subset G_m$. The identity principle implies that $\iota^{\mathbf{C}}|W(u)$ is G_m-equivariant. Thus it is in particular G_u-equivariant and (1) holds as desired.□

As a consequence, as is shown below, *every G-complexification contains a neighborhood of M where G is acting properly.*

Corollary 7. *Let X be a G-complexification of M where G is acting properly and $\tilde{X}$ any G-complexification. Then there exists a G-stable neighborhood Z of M in X so that the natural injection $\iota : M \to \tilde{X}$ extends to a biholomorphic G-equivariant map $\iota^{\mathbf{C}} : Z \to \tilde{Z}$, where $\tilde{Z}$ is a neighborhood of M in $\tilde{X}$.*

Proof. Take U to be a neighborhood to which ι extends to an map $\iota^{\mathbf{C}} : U \to X$ which is biholomorphic onto its image $\tilde{U}$. Let W be as in the Lemma and $\tilde{W} := \iota^{\mathbf{C}}(W)$. It follows that the image sets $\tilde{W}(u) := \iota^{\mathbf{C}}(W(u))$ have the properties of the Lemma. Consequently, $(\iota^{\mathbf{C}})^{-1}$ is well-defined on $\tilde{Z} := G \cdot \tilde{W}$ and $\iota^{\mathbf{C}} : Z \to \tilde{Z}$ is the desired equivariant extension. □

VIII. The retraction of a G-stable Stein neighborhood

Let G be a connected Lie group acting properly as a group of real analytic diffeomorphisms on a real analytic manifold M and let X be a G-complexification of M. Here we show that there is a G-stable Stein neighborhood W of M in X on which the G-action is proper and which is strongly G-equivariantly retractible to M. The retraction statement is proved here by adapting the proof in ([HW]) to our setting.

A *strong G-equivariant deformation retraction* of a G-space W to a G-subset A is a continous map $R : [0,1] \times W \to W$ with the following properties:
(1) $R(0,z) = z \ \ \forall z \in W$
(2) $R(1,z) \in M \ \ \forall z \in W$
(3) $R(t,z) = z \ \ \forall z \in A, t \in [0,1]$
(4) $R(t,gz) = gR(t,z) \ \ \forall z \in W, t \in [0,1]$

Our first aim is to construct a deformation retraction of *some* neighborhood of M (Prop 2.). As mentioned above, by shrinking X if necessary, we may assume that it is equivariantly an open neighborhood of M in the complexification constructed in VII. (Prop.VII.4). In particular we may assume that the action is proper and that there exists a strictly plurisubharmonic (s.p.s.h.) G-invariant smooth function $\varphi : X \to \mathbf{R}^{\geq 0}$ with $M = \{\varphi = 0\}$

The following Lemma is a well known fact which we need for the proof of Prop.2 below.

Lemma 1. *Let W be an open neighborhood of 0 in $\mathbf{C}^n$ and $\varphi : W \to \mathbf{R}^{\geq 0}$ be a smooth s.p.s.h. function with $\{\varphi = 0\} = W \cap \mathbf{R}^n$. Then there are local real C^∞ coordinates $x^1, \dots, x^n, y^1, \dots, y^n$ in a neighborhood U of 0 such that the function φ in these coordinates has the form $\varphi(x^1, \dots, x^n, y^1, \dots, y^n) = \sum_{i=1}^n (y^i)^2$.*

Proof. Since $d\varphi(x,0) = 0 \ \forall (x,0) \in W$, it follows that

$$\frac{\partial^2}{\partial z_i \partial \overline{z_j}} \varphi(x,0) = \frac{1}{4} \frac{\partial^2}{\partial y_i \partial y_j} \varphi(x,0) \ \forall i,j.$$

Since φ is s.p.s.h., the matrix $(\frac{\partial^2}{\partial y_i \partial y_j}\varphi)(x,0)$ is positive definite for all $(x,0) \in W$. Now the proof is the same as the proof of the classical lemma of Morse (c.f. [Mi]). □

Let $\langle \cdot, \cdot \rangle$ denote the Riemanian metric induced from the s.p.s.h. function φ on X, i.e., the metric associated to the Kähler form $i\partial\bar{\partial}\varphi$. All gradients are computed with respect to this metric.

Proposition 2. *There is an open G-stable neighborhood U of M in X such that the gradient flow of φ defines a strong G-equivariant deformation retraction of U to M.*

Proof. Since φ is invariant, the induced Riemannian metric is invariant. Using the uniqueness theorem for the associated ordinary differential equations, it follows immediately that the gradient flow defined by

$$\frac{\mathrm{d}\gamma_z}{\mathrm{dt}}(t) = -\mathrm{grad}\varphi(\gamma_z(\mathrm{t})) \quad \gamma_z(0) = \mathrm{z}$$

satisfies $\gamma_{g\cdot z}(t) = g \cdot \gamma_z(t)$.

From Lemma 1 it follows immediately that every point $a \in M$ has a neighborhood U_a such that for all $z \in U_a$ the flow $\gamma_z(t)$ exists for all $t \geq 0$ and preserves U_a. The equivariance of the flow implies that for all $g \in G$ the set $g \cdot U_a$ has the same property. So we set $U := \bigcup_{a\in M\ g\in G} g \cdot U_a$ and define the deformation retraction $R : [0,1] \times U \to U$ by $R(t,z) = \gamma_z(\frac{1}{1-t} - 1)\ \forall t < 1$ and $R(1,z) := \lim_{t\to\infty} \gamma_z(t) = \lim_{t\to 1-} R(t,z)$. □

Remark. *The G-stable neighborhood U of M is such that* $\{z \in U : \mathrm{d}\varphi(\mathrm{z}) = 0\} = \{\mathrm{z} \in \mathrm{U} : \varphi(\mathrm{z}) = 0\}$.

To construct a G-invariant *Stein* neighborhood of M in X which is preserved by the deformation retraction we need several technical facts. These are contained in the following two lemmas.

Lemma 3. *Let S be a compact subset of X and V be an open subset of X with $S \subset V$. Then there exists a C^∞-function $\epsilon : X \to \mathbf{R}^{\geq 0}$ with*

(1) $\mathrm{supp}(\epsilon) \Subset \mathrm{V}$

(2) $\epsilon(z) > 0\ \forall z \in M \cap S$

(3) There is a G-neighborhood W of M in X such that ϵ is constant on the curves $\gamma_z(t)$, $t \geq 0$, of the flow through points $z \in W$.

Proof. Choose an open subset S_1 of X with $S \subset S_1 \Subset V$. Let $\chi : M \to \mathbf{R}^{\geq 0}$ be a smooth function with $\chi(z) > 0\ \ \forall z \in M \cap S$ and $\mathrm{supp}(\chi) \Subset \mathrm{M} \cap \mathrm{V} \cap \mathrm{G} \cdot \mathrm{S}_1$. Define $f_1 : X \to \mathbf{R}^{\geq 0}$ by

$$f_1(z) = \chi(R(1,z)),$$

where $R : [0,1] \times X \to X$ denotes the deformation retraction. From the proof of Prop. 2 it follows that the function $f_2 : X \to \mathbf{R}^{\geq 0}$ defined by

$$f_2(z) = \int_0^\infty ||\mathrm{grad}\varphi(\gamma_z(\mathrm{t}))||\ \mathrm{dt}$$

is a G-invariant smooth function which vanishes pricisely on M. Its easy to see that the number $b := \inf_{z\in(X\setminus(V\cap G\cdot S_1))\cap R_1^{-1}(\mathrm{supp}(\chi))} f_2(z)$ is positive.

Let $\alpha : \mathbb{R} \to \mathbb{R}^{\geq 0}$ be a smooth function with

$$\alpha(x) = 1 \ \forall x \in [0, \frac{b}{2}] \text{ and } \alpha(x) = 0 \ \ \forall x \geq \frac{3}{4}b.$$

Then the function $f_3 : X \to \mathbb{R}^{\geq 0}$, $f_3(z) = \alpha(f_2(z))$, is constant 1 in the open G-invariant neighborhood $W := \{z \in X : f_2(z) < \frac{b}{2}\}$ of M in X. We define the desired function ϵ by $\epsilon(z) = f_1(z) \cdot f_3(z)$. In W we have $\epsilon(z) = f_1(z) = \chi(R(1, z))$, so ϵ fulfills condition (3).

Next we show $\mathrm{supp}(\epsilon) \subset (\mathrm{V} \cap \mathrm{G} \cdot \mathrm{S}_1)$. Let $z \in \mathrm{supp}(\epsilon) \subset \mathrm{supp}(\mathrm{f}_1) \cap \mathrm{supp}(\mathrm{f}_3)$. Then $f_2(z) \leq \frac{3b}{4} < b$ and $z \in R_1^{-1}(\mathrm{supp}(\chi))$. By the definition of b this implies $z \in (V \cap G \cdot S_1)$.

The G-equivariance of R_1 implies

$$H := \{g \in G : g \cdot S_1 \cap \{\epsilon > 0\} \neq \emptyset\} \subset \{g \in G : g \cdot (R_1(S_1)) \cap \mathrm{supp}(\chi) \neq \emptyset\} =: \mathrm{H}_1$$

Because the G-action is proper and $R_1(S_1)$ and $\mathrm{supp}(\chi)$ are relatively compact in X, H_1 and hence H are relatively compact in G. It follows that the set $\{\epsilon > 0\} \subset H \cdot S_1$ (use the fact that $\{\epsilon > 0\} \subset G \cdot S_1$) is relatively compact in X. This implies condition (1).

On M we have $\epsilon = \chi$. So condition (2) is fulfilled. □

From now on we fix a right-invariant Haar measure μ on the Lie group G. For two subsets A_1, A_2 of X, define

$$[A_1, A_2] := \{g \in G : g \cdot A_1 \cap A_2 \neq \emptyset\} \subset G$$

Recall that the properness of the G-action means $[C_1, C_2] \Subset G$ for any two relatively compact subsets C_1, C_2 of X. In particular $\mu([C_1, C_2]) < \infty$. The proof of the following fact is straightforward.

Fact. *For every compact subset C of X there exists a real number $A > 0$ such that every point $z \in X$ has a relatively compact open neighborhood U_z with the property*

$$\mu([\overline{U_z}, C]) \leq A.$$

□

Using this fact we prove

Lemma 4. *Let $f : X \to \mathbb{R}^{\geq 0}$ be a C^∞-function with compact support. Then, for every closed subset L of G,*

$$f_L(z) := \int_L f(g \cdot z) d\mu(g)$$

defines a C^∞-function $f_L : X \to \mathbb{R}^{\geq 0}$. Moreover there exists a number $b > 0$ depending on f but not depending on L such that the function $\varphi - bf_L$ is s.p.s.h. and the function $b \cdot f_L$ is not greater than 1.

Proof. Let $A > 0$ and U_z be as in the above *fact* for $C := supp(f)$. First note

$$f_L(z) = \int_{[\{z\},\mathrm{supp(f)}]\cap \mathrm{L}} f(g \cdot z) \mathrm{d}\mu(\mathrm{g}) \tag{*}$$

and therefore for all $y \in \overline{U_z}$

$$f_L(y) = \int_{[\overline{U_z},\mathrm{supp(f)}]\cap \mathrm{L}} f(g \cdot z) \mathrm{d}\mu(\mathrm{g}). \tag{**}$$

Interchanging the order of integration and differentiation shows that f_L is a C^∞-function.

Since f has compact support, for all sufficiently small $c > 0$ the function $\varphi - cf$ is s.p.s.h. and the function $c \cdot f$ is not greater than 1. Set $b = \frac{c}{A}$. Fix a point $z \in X$. The definition of U_z together with $(**)$ immediately yields $bf_L(z) \leq 1$. To prove that $\varphi - bf_L$ is s.p.s.h. in z, we may assume $D := \mu([\overline{U_z}, \mathrm{supp(f)}] \cap \mathrm{L}) > 0$. The G-invariance of φ together with $(**)$ implies

$$(\varphi - bf_L)(y) = \int_{[\overline{U_z},\mathrm{supp(f)}]\cap \mathrm{L}} H(g \cdot z) \mathrm{d}\mu(\mathrm{g}) \ \ \forall \mathrm{y} \in \mathrm{U_z}, \tag{***}$$

where the function $H(z) := \frac{1}{D}\varphi(z) - bf(z)$ is s.p.s.h. in X. The last statement follows from $D \leq A$ and the fact that the functions φ and $\varphi - cf$ are s.p.s.h. Because G acts on X by holomorphic automorphisms for all $g \in G$ the function $H_g(z) := H(g \cdot z)$ is s.p.s.h. Thus differentiation of $(***)$ ($\frac{\partial^2}{\partial w \partial \bar{w}}$, where $w \in \mathbf{C}$ is the parameter of an analytic disc through z in X) and interchanging the order of integration and differentiation shows that $\varphi - bf_L$ is s.p.s.h. in z. □

Remarks. (1) If L is compact, then the function f_L has compact support. (2) The function f_G is G-invariant. □

Let $\pi : X \to X/G$ denote geometric quotient and recall that $\langle \cdot, \cdot \rangle$ is the Riemannian metric induced by the s.p.s.h. function φ on X. All gradients are computed with respect to this metric.

Proposition 5. *For any two open G-invariant subsets V', V'' of X with $\pi(V'') \Subset \pi(V')$ there exist a G-invariant C^∞-function $\epsilon_0 : X \to [0,1]$ and a sequence of C^∞-functions with compact support $\epsilon_n : X \to [0,1], n = 1, \ldots,$ satisfying the following properties:*
(1) The function $\varphi - \epsilon_n$ is s.p.s.h.
(2) For all $z \in X$ it follows that $\epsilon_n(z) \leq \epsilon_{n+1}(z)$ and $\lim_{n\to\infty} \epsilon_n(z) = \epsilon_0(z)$.
(3) $\langle \mathrm{grad}\varphi, \mathrm{grad}\varphi \rangle \geq \langle \mathrm{grad}\varphi, \mathrm{grad}\epsilon_0 \rangle$.
(4) $\epsilon_0(z) > 0 \ \ \forall z \in M \cap V''$ and $\mathrm{supp}(\epsilon_n) \subset \mathrm{V}'$..

Proof. Using the slice-theorem for proper actions (see §3), we find a compact subset S of V' with $G \cdot S = \overline{V''}$. Let $\epsilon : X \to \mathbf{R}^{\geq 0}$ be a smooth function with $\mathrm{supp}(\epsilon) \Subset \mathrm{V}'$ and

$\epsilon(z) > 0\ \forall z \in M \cap S$ as constructed in Lemma 3. Denote by W the G-stable neighborhood of M in X where ϵ is constant on the curves $\gamma_z(t),\ t \geq 0$, through points $z \in W$. Let $b > 0$ be as in Lemma 4. Define

$$\epsilon_0(z) := b \cdot \int_G \epsilon(g \cdot z) d\mu(g).$$

To define the ϵ_n we choose a sequence K_n of compact subsets of the group G with $K_n \subset K_{n+1}$ and $\bigcup_{n \in \mathbb{N}} K_n = G, = 1, 2, \ldots$. Now set

$$\epsilon_n(z) := b \cdot \int_{K_n} \epsilon(g \cdot z) d\mu(g).$$

Except for (3) it is clear that the ϵ_n's have all the desired properties. For (3) note that, since the function ϵ is (in W) constant on the curves $\gamma_z(t)$ of the gradient flow of φ, it follows from the definition of ϵ_0 (and the equivariance of the flow), that the function ϵ_0 has the same property. This means

$$\langle \mathrm{grad}\varphi(z), \mathrm{grad}\epsilon_0(z) \rangle_z = \frac{d\epsilon_0(\gamma_z(t))}{dt}|_{t=0} = 0 \quad \forall z \in W.$$

From the G-invariance of φ we have

$$A_1 := \inf_{z \in \mathrm{supp}(\epsilon_0) \setminus W} \langle \mathrm{grad}\varphi(z), \mathrm{grad}\varphi(z) \rangle_z = \min_{z \in \mathrm{supp}(\epsilon) \setminus W} \langle \mathrm{grad}\varphi(z), \mathrm{grad}\varphi(z) \rangle_z$$

Note that $A_1 > 0$ (Corollary). Set

$$A_2 := \sup_{z \in \mathrm{supp}(\epsilon_0) \setminus W} \langle \mathrm{grad}\varphi(z), \mathrm{grad}\epsilon(z) \rangle_z = \max_{z \in \mathrm{supp}(\epsilon) \setminus W} \langle \mathrm{grad}\varphi(z), \mathrm{grad}\epsilon(z) \rangle_z.$$

Note that $A_2 < \infty$. In case $A_2 \leq A_1$ the function ϵ_0 fulfilles condition (3). Otherwise replace the functions $\epsilon_i, i = 0, 1, 2, \ldots$ by $\frac{A_1}{A_2}\epsilon_i$. □

Now we are now in a position to prove the main result of this section.

Proposition 6. *Let G be a connected Lie group acting properly as a group of real analytic diffeomorphisms of a real analytic manifold M and let X be a G-complexification of M. Then there exists a Stein G-stable neighborhood W of M in X on which the G-action is proper and such that M is a strong G-equivariant deformation retract of W.*

Proof. By Prop.VII.2 we may assume that the G-action on X is proper and that there is a s.p.s.h. function $\varphi : X \to \mathbb{R}^{\geq}0$, whose gradient flow $\gamma_z(t)$ defines a deformation retraction R from X to M. As above let $\pi : X \to X/G$ denote the geometric quotient. Using the topological properties of X and X/G, we can construct two locally finite countable coverings $X = \bigcup_{i \in \mathbb{N}} V_i' = \bigcup_{i \in \mathbb{N}} V_i''$ of X consisting of G-invariant open subsets V_i' resp.

V_i'' with $\pi(V_i'') \Subset \pi(V_i)'$. For all $i \in \mathbb{N}$ let $\epsilon_k^i : X \to [0,1]$ denote the functions constructed in Prop.5 for V_i' and V_i'' and define $E_k : X \to \mathbb{R}^{\geq 0}$ by

$$E_k(z) = \sum_{i=1}^{\infty} \frac{1}{2^i} \epsilon_k^i(z), k = 0, 1 \ldots .$$

Since the covering $X = \bigcup_{i \in \mathbb{N}} V_i'$ is locally finite, property (4) in Prop.5 implies that these functions are well-defined and smooth. Further, we define open subsets W_k of X by

$$W_k := \{z \in X : \varphi(z) < E_k(z)\}.$$

Property (2) in Prop.5 implies that $W_k \subset W_{k+1}$, $k = 1, 2, \ldots$, and $\bigcup_{k=1}^{\infty} W_k = W_0$. The invariance of the functions ϵ_0^i and φ, together with property (4) of Prop. 5 and the fact that $M = \{\varphi = 0\}$, imply that W_0 is an open G-stable neighborhood of M in X. If we show that the subsets W_k $(k \geq 1)$ which exhaust W_0 are Stein and $W_k \subset W_{k+1}$ is a Runge pair for each $k \geq 1$, a classical result of K. Stein ([St]) implies that W_0 is likewise Stein. It therefore remains to establish the following three claims:

Claim 1. *W_k is Stein.*

Claim 2. *$W_k \subset W_{k+1}$ is a Runge pair.*

Claim 3. *The deformation retraction R preserves W_0.*

Proof of Claim 1. (c.f. [HW] Theorem 2.2. (a)) Property (1) in Prop.5 implies together with $\sum_{i=1}^{\infty} \frac{1}{2^i} = 1$ that the functions $\varphi - E_k$ are s.p.s.h. We are finished if we show that the function $f_k : W_k \to \mathbb{R}^{\geq 0}$ defined by $f_k(z) := \frac{1}{E_k(z) - \varphi(z)}$ is a s.p.s.h. exhaustion function on W_k. Let z be the coordinate on a complex line in local coordinates around a point in W_k. Then we have:

$$\frac{\partial^2}{\partial z \partial \bar{z}} f_k = \frac{\partial^2}{\partial z \partial \bar{z}} \left(\frac{1}{E_k - \varphi}\right) = \frac{1}{(E_k - \varphi)^2} \frac{\partial^2}{\partial z \partial \bar{z}} (\varphi - E_k) + \frac{2}{(E_k - \varphi)^3} \left|\frac{\partial}{\partial z} (\varphi - E_k)\right|^2$$

The right side is positive on W_k, so f_k is s.p.s.h. A straightforward computation using $\mathrm{supp}(\epsilon_k^i) \Subset \mathrm{X}$ and $\sum_{i=1}^{\infty} \frac{1}{2^i} = 1$ shows that f_k is an exhaustion function on W_k. □

Proof of Claim 2. see [HW] Theorem 2.2. (b) □

Proof of Claim 3. Property (3) in Prop.5 implies

$$\langle \mathrm{grad}\varphi(\mathrm{z}), \mathrm{grad}\varphi(\mathrm{z}) \rangle_{\mathrm{z}} \geq \langle \mathrm{grad}\varphi(\mathrm{z}), \mathrm{grad}\mathrm{E}_0(\mathrm{z}) \rangle_{\mathrm{z}} \quad \forall \mathrm{z} \in \mathrm{X}.$$

By definition of the gradient flow $\gamma_z(t)$ this implies

$$\frac{\mathrm{d}(\varphi - \mathrm{E}_0)(\gamma_{\mathrm{z}}(\mathrm{t}))}{\mathrm{dt}} \leq 0.$$

□

References

[A] Abels, H.: *Parallelizability of proper actions, global K-slices and maximal compact subgroups.* Math. Ann. 212, 1-19 (1974)

[AG] Akhiezer, D.N., Gindikin S.G.: *On Stein extensions of real symmetric spaces.* Math. Ann. 286, 1-12 (1990)

[AL1] Azad, H.; Loeb, J.J.: *Pluri-subharmonic functions and the Kempf-Ness theorem.* Bull. London Math. Soc. 25, 162–168 (1993)

[AL2] Azad, H.; Loeb, J.J.: *On a theorem of Kempf and Ness.* Indiana Univ. Math. Journal 39, 61-65 (1990)

[Ch] Chevalley, C.: Theory of Lie groups I. Princeton: Princeton University Press 1946

[G] Grauert, H.: *On Levi's problem and the imbedding of real-analytic manifolds.* Ann. of Math. 68,2, 460–472 (1958)

[GSt] Guillemin, V.; Stenzel, M.: *Grauert tubes and the homogenous Monge–Ampére equation.* J. Differential Geometry 34, 561–570 (1991)

[GS] Guillemin, V.; Sternberg, S.: Symplectic techniques in physics. Cambridge: Cambridge University Press 1984

[HW] Harvey., R.F., Wells, R.O.: *Holomorphic approximation and hyperfunctions theory on a C^1 totally real submanifold of a complex manifold.* Math. Ann. 197, 287–318 (1972)

[H1] Heinzner, P.: *Equivariant holomorphic extensions of real analytic manifolds.* Bull. Soc. Math. France 121, 101–119 (1993)

[H2] Heinzner, P.: *Geometric invariant theory on Stein spaces.* Math. Ann. 289, 631–662 (1991)

[HL] Heinzner, P.; Loose, F.: *Reduction of complex Hamiltonian G-spaces.* Geometric and Functional Analysis 4, 288–297 (1994)

[HHL] Heinzner, P.; Huckleberry, A. T.; Loose, F.: *Kählerian extensions of the symplectic reduction.* Preprint Bochum 1992, 1–18. To appear in J. reine und angew. Math.

[Hi] Hirsch, M.: Differential topology. Berlin Heidelberg New York: Springer 1976

[Ho] Holmann, H.: *Quotienten komplexer Räume.* Math. Ann. 142, 407–440 (1961)

[I] Illman, S.: *Every proper smooth action of a Lie group is equivalent to a real analytic action: A contribution to Hilberts fifth problem.* Preprint MPI/93-3 Bonn (1993)

[Kr] Kraft, H.: Geometrische Methoden in der Invariantentheorie. Vieweg–Verlag, Braunschweig, 1984

[K1] Kutzschebauch, F.: *Eigentliche Wirkungen von Liegruppen auf reell-analytischen Mannigfaltigkeiten.* Schriftenreihe des Graduiertenkollegs Geometrische und Mathematische Physik 5 (1994) Ruhr-Universität Bochum

[K2] Kutzschebauch, F.: *On the uniqueness of the analyticity of a proper G-action.* Preprint Bochum 1994

[LS] Lempert, L.; Szöke, R.: *Global solutions of the homogeneous complex Monge-Ampére equation and complex structures on the tangent bundle of Riemannian manifolds.* Math. Ann. 290, 689–712 (1991)

[M] Matsushima, Y.: *Espaces homoènes de Stein des groupes de Lie complexes I.* Nagoya Nathematical Journal 16, 205–218 (1960)

[Mi] Milnor, J.: Morse Theory. Ann. of Math. Studies 51, Princeton, 1963

[N] Narasimhan, R.: Analysis on real and complex manifolds. Adv. Studies in pure math., North Holland, 1968

[O] Onishchik, A.L.: *Complex envelopes of compact homogeneous spaces.* Dokl. Acad. Nauk SSSR 130. 726–729 (1960) (russian)

[R] Rosenlicht, M.: *Some basic theorems on algebraic groups.* Amer. J. of Math. 78 , 401–443 (1956)

[Sh] Shutrick, H.B.: *Complex extensions.* Quart. J. of Math. Series 2, 189–201 (1958)

[Sn] Snow, D. M.: *Reductive group actions on Stein Spaces.* Math. Ann. **259**, 79–97 (1982)

[S] Szöke, R.: *Complex structures on tangent bundles of Riemannian manifolds.* Math. Ann. 291, 409–428 (1991)

[W] Wallach N.R.: Harmonic analysis on homogeneous spaces. New York: Marcel Dekker 1973

[Wa] Wassermann, A. G.: *Equivariant differential topology.* Topology 8, 127–150 (1969)

[WB] Whitney, H., Bruhat, F.: *Quelques propriétés fondamentales des ensembles analytiques réels.* Comment. Helv. 33, 132–160 (1959)

[W] Winkelmann, J.: *Invariant hyperbolic Stein domains.* Manu. Math. 79, 329-334 (1993)

Pseudoconcave *CR* Manifolds

C. Denson Hill, Department of Mathematics, SUNY at Stony Brook, Stony Brook, NY 11794, USA. Dhill@math.sunysb.edu

Mauro Nacinovich, Dipartimento di Matematica, Università di Pisa, via F. Buonarroti 2, I-56127 Pisa, Italy. nacinovi@dm.unipi.it

Introduction

In this paper we initiate the global study of pseudoconcave CR manifolds of arbitrary CR dimension and CR codimension. By a pseudoconcave CR manifold we mean one which is smooth, is locally embeddable, and is at least 1-pseudoconcave. In making this definition, we are guided by the analogy with the theory of analytic spaces, in which one has a natural local model. In some sense these manifolds bear the same relation to complex manifolds as plurisubharmonic functions bear to holomorphic functions. The authors feel that the investigation of this subject will eventually prove to be fruitful in complex geometry.

After fixing notation and definitions in §1, we discuss in §2 Riemann domains for pseudoconcave CR manifolds. In §3 we construct tubular neighborhoods which we later utilize in the proof of finiteness and vanishing theorems for the natural global cohomology groups in these manifolds. We treat the compact case in §4, and for the noncompact case, we discuss pseudoconcavity at infinity in §5, and pseudoconvexity at infinity in §6. Some of the global results are new even for the case of hypersurfaces. The last §7 is devoted to a study of abstract pseudoconcave CR manifolds in the compact case. We believe the results we obtained are optimal in the compact case.

§1 Pseudoconcave CR manifolds

(a) *Abstract CR manifolds*

Let M be a smooth real manifold of dimension m, countable at infinity. A *partial complex structure* of type (n, k) on M is the pair consisting of a vector subbundle HM of the tangent bundle TM and a smooth vector bundle isomorphism $J: HM \to HM$ such that

$$(1.1)\qquad \begin{cases} J^2 = -Id : HM \longrightarrow HM \\ JH_PM = H_PM \quad \forall P \in M \\ [X,Y] - [JX,JY] \in \Gamma(M,HM) \quad \forall X,Y \in \Gamma(M,HM) \\ \dim_{\mathbf{R}} H_pM = 2n \\ m = \dim_{\mathbf{R}} M = 2n + k. \end{cases}$$

We say that the partial complex structure (HM, J) is *formally integrable* if

$$(1.2)\qquad [X,JY] + [JX,Y] = J([X,Y] - [JX,JY]) \quad \forall X,Y \in \Gamma(M,HM).$$

We denote respectively by $T^{1,0}M = \{X - iJX \mid X \in HM\}$ and by $T^{0,1}M = \{X + iJX \mid X \in HM\}$ the complex vector subbundles of the complexification $\mathbb{C}HM$ of HM corresponding to the eigenvalues i and $-i$ of J.

This paper is in final form and no part of it will be submitted elsewhere.

Then the conditions (1.1) and (1.2) can also be expressed by

$$[\Gamma(M,T^{1,0}M),\Gamma(M,T^{1,0}M)]\subset\Gamma(M,T^{1,0}M)$$

or, equivalently, by

$$[\Gamma(M,T^{0,1}M),\Gamma(M,T^{0,1}M)]\subset\Gamma(M,T^{0,1}M).$$

We note that the following relations hold:

$$\begin{aligned}&T^{0,1}M=\overline{T^{1,0}M}\\&T^{1,0}M\cap T^{0,1}M=0\\&T^{1,0}M\oplus T^{0,1}M=\mathbb{C}HM.\end{aligned}$$

Definition An abstract CR manifold $\mathbf{M}=(M,HM,J)$ of type (n,k) is the triple consisting of a smooth paracompact manifold M of dimension $m=2n+k$ and a formally integrable partial complex structure (HM,J) of type (n,k) on it. We call n the *CR dimension* and k the *CR codimension* of $\mathbf{M}$.

(b) *The characteristic bundle and the Levi form.*

Let $\mathbf{M}=(M,HM,J)$ be a CR manifold of type (n,k). The *characteristic bundle* H^0M is the annihilator of HM in T^*M: it is the rank k subbundle of T^*M

$$H^0M=\{\xi\in T^*M\mid\xi(X)=0\quad\forall X\in\Gamma(M,HM)\}.$$

Given $\xi\in H^0_PM$, $X,Y\in H_PM$, let us choose $\tilde{\xi}\in\Gamma(M,H^0M)$, $\tilde{X}$, $\tilde{Y}\in\Gamma(M,HM)$ such that $\tilde{\xi}(P)=\xi$, $\tilde{X}(P)=X$, $\tilde{Y}(P)=Y$.

Then we have

$$d\tilde{\xi}(X,Y)=-\xi([\tilde{X},\tilde{Y}])\tag{1.3}$$

and hence the two sides of (1.3) only depend on ξ, X, Y. In this way we associate to $\xi\in H^0_PM$ a quadratic form

$$L(\xi,X)=\xi([J\tilde{X},\tilde{X}])=d\tilde{\xi}(X,JX)$$

on H_PM. This form is hermitian for the complex structure of H_PM defined by J. Indeed, by (1.1) we have

$$L(\xi,JX)=L(\xi,X)\quad\forall X\in H_PM\,.$$

We denote by $\sigma(\xi)=(\sigma^+(\xi),\sigma^-(\xi))$ the signature of L_ξ as a hermitian form for the complex structure of H_PM defined by J: the numbers $\sigma^+(\xi)$ and $\sigma^-(\xi)$ are respectively the number of positive and negative eigenvalues of $L(\xi,\cdot)$.

We say that $\mathbf{M}$ is *q-pseudoconcave at* $P\in M$ if for every $\xi\in H^0_PM\setminus\{0\}$ we have $\sigma^-(\xi)\geq q$ and that $\mathbf{M}$ is *q-pseudoconcave* if it is *q-pseudoconcave* at all points.

The hermitian form $L(\xi,\cdot)$ is called the *Levi form* of $\mathbf{M}$ at $\xi\in H^0M$.

(c) *CR complexes and $\bar\partial_M$-cohomology groups.*

Let $A(M) = \oplus_{h=0}^{m} A^{(h)}(M)$ denote the Grassman algebra of differential forms on M, with smooth complex valued coefficients.

We denote by $\mathcal{J}$ the ideal of $A(M)$ that annihilates $T^{0,1}M$:

$$\mathcal{J} = \{\alpha \in \oplus_{h>0} A^{(h)}(M) \,|\, \alpha|_{T^{0,1}M} = 0\} .$$

From the integrability condition (1.2) we have

$$\mathrm{d}\mathcal{J} \subset \mathcal{J} . \tag{1.4}$$

We consider also the powers $\mathcal{J}^p$ of $\mathcal{J}$: we obtain a decreasing sequence of ideals of $A(M)$:

$$A(M) = \mathcal{J}^0 \supset \mathcal{J}^1 \supset \ldots \supset \mathcal{J}^N \supset \mathcal{J}^{N+1} = \{0\}$$

where $N = n + k$. By (1.4) we also have:

$$\mathrm{d}\mathcal{J}^p \subset \mathcal{J}^p \quad \forall p = 0, \ldots, N .$$

Passing to the quotient, the exterior differential defines linear maps

$$\bar\partial_M^{(p)} = \bar\partial_M : \mathcal{J}^p/\mathcal{J}^{p+1} \to \mathcal{J}^p/\mathcal{J}^{p+1} .$$

We note that the grading of $A(M)$ induces a grading on $\mathcal{J}^p/\mathcal{J}^{p+1}$:

$$\mathcal{J}^p/\mathcal{J}^{p+1} = \oplus_{q=0}^{n} Q^{p,q}(M) .$$

As

$$\bar\partial_M \circ \bar\partial_M = 0$$

and

$$\bar\partial_M Q^{p,q}(M) \subset Q^{p,q+1}(M)$$

we obtain *the Cauchy-Riemann complexes:*

$$(Q^{p,*}(M), \bar\partial_M) = \left\{\{0 \to Q^{p,0}(M) \xrightarrow{\bar\partial_M} Q^{p,1}(M) \to \ldots \to Q^{p,n}(M)\right\} \tag{1.5}$$

for $0 \le p \le N$. We also note that $Q^{p,q}(M) = \Gamma(M, Q^{p,q})$ are smooth sections of a smooth complex vector bundle $Q^{p,q}$ of rank $\binom{N}{p} + \binom{n}{q}$ and that (1.5) is a complex of first order partial differential operators on smooth vector bundles on M.

The cohomology groups $H^q(Q^{p,*}(M), \bar\partial_M)$ will be denoted in the following by $H^{p,q}(M)$ and called the *CR cohomology groups* of **M**. These are the analogues of the Dolbeault cohomology groups of a complex manifold, for smooth forms.

The vector spaces $H^{p,0}(M)$ will also be denoted by $\mathbf{\Omega}_M{}^p(M)$ and called the spaces of *CR p-forms* on **M**. For $p = 0$ we set also $\mathcal{O}_M(M) = \mathbf{\Omega}^0_M(M)$ and call the elements of $\mathcal{O}_M(M)$ the *CR functions* on M.

We also notice that the exterior product on $A(M)$ induces a bilinear map

$$Q^{p,q}(M) \times Q^{h,k}(M) \to Q^{p+h,q+k}(M) .$$

Indeed, if $\alpha, \alpha_1 \in \mathcal{J}^p \cap A^{(p+q)}(M)$, $\beta, \beta_1 \in \mathcal{J}^h \cap A^{(h+k)}(M)$ and $\alpha - \alpha_1 \in \mathcal{J}^{p+1}$, $\beta - \beta_1 \in \mathcal{J}^{h+1}$, then $\alpha \wedge \beta - \alpha_1 \wedge \beta_1 \in \mathcal{J}^{p+h+1}$.

Denoting by $\alpha \wedge \beta$ this map corresponding to the exterior product, we have then the formula:

$$\bar{\partial}_M(\alpha \wedge \beta) = \bar{\partial}_M \alpha \wedge \beta + (-1)^{p+q} \alpha \wedge \bar{\partial}_M \beta \quad \forall \alpha \in Q^{p,q}(M), \beta \in Q^{h,k}(M) .$$

Therefore the exterior product of forms defines, by passing to the quotient, the structure of a cohomology ring on $\mathbf{H}(M) = \oplus H^{p,q}(M)$.

(d) *CR sheaf cohomology groups.*

It is clear that an open submanifold of an abstract CR manifold is again an abstract CR manifold of the same type. So we use $\mathcal{O}_M(U)$ and $\Omega^p_M(U)$ to denote smooth CR functions and forms defined in an open subset U of M. The assignment $U \to \mathcal{O}_M(U)$ ($U \to \Omega^p_M(U)$) is a presheaf, and we use $\mathcal{O}_M$ (Ω^p_M) for the corresponding sheaf of germ of CR functions (CR p-forms).

The Čech cohomology groups on M with coefficients in these sheaves are denoted by $H^q(M, \mathcal{O}_M)$ and $H^q(M, \Omega^p_M)$. In general, we have homomorphisms

$$H^q(M, \Omega^p_M) \to H^{p,q}(M) .$$

As we have no information about the validity of the Poincaré lemma for the Cauchy Riemann complexes in this abstract setting, we cannot claim these maps to be isomorphisms.

(e) *Local embeddability of abstract CR manifolds.*

A CR map of a CR manifold $\mathbf{M}_1 = (M_1, H_1, J_1)$ into a CR manifold $\mathbf{M}_2 = (M_2, H_2, J_2)$ is a differentiable map $\phi: M_1 \to M_2$ such that $\phi_*(H_1) \subset H_2$ and $\phi_*(J_1 X) = J_2 \phi_*(X)$ for every $X \in H_1$.

In view of the Newlander-Nirenberg theorem, a complex manifold is exactly the same thing as an abstract CR manifold of CR codimension zero. For instance CR maps from $\mathbf{M}$ to $\mathbb{C}$ are CR functions.

A *CR embedding* ϕ of an abstract CR manifold $\mathbf{M}$ into a complex manifold X is a CR map which is an embedding. We say that a CR embedding $\phi: M \to X$ is generic if the complex dimension of X is $n + k$, where (n, k) is the type of $\mathbf{M}$.

For a point p of an abstract CR manifold $\mathbf{M}$ of type (n, k), the following three statements are equivalent:

(i) We can find an open neighborhood U of p in M and a CR embedding $\phi: U \to \mathbb{C}^N$ for some N;

(ii) We can find an open neighborhood U of p in M and smooth $\mathbb{C}$-valued functions $f_1, \ldots, f_{n+k} \in \mathcal{O}_M(U)$ with $\mathrm{d} f_1, \ldots, \mathrm{d} f_{n+k}$ linearly independent over $\mathbb{C}$ at each point of U;

(iii) We can find an open neighborhood U of p in M and a generic CR embedding $\phi: U \to \mathbb{C}^{n+k}$.

If any one of these equivalent conditions is satisfied, we say that $\mathbf{M}$ is *locally embeddable at p*. $\mathbf{M}$ is called *locally embeddable* if this holds for every $p \in M$.

(f) *Pseudoconcave CR manifolds.*

Definition 1.1 An abstract CR manifold $\mathbf{M}$ of type (n, k) is called *pseudoconcave* provided that it is *(i)* locally embeddable, and *(ii)* 1-pseudoconcave.

If *(ii)* is replaced by q-pseudoconcavity, then $\mathbf{M}$ will be called a q-pseudoconcave CR manifold.

(g) *A remark about orientability.*

A complex manifold is always orientable, but a pseudoconcave CR manifold may not be orientable.

A simple example, which is q-pseudoconcave, can be obtained as follows. We denote by (u, v), with $u, v \in \mathbf{C}^{q+1}$, homogeneous coordinates in $\mathbf{CP}^{2q+1}$. Let X be the complex manifold

$$X = \{(u,v) \in \mathbf{CP}^{2q+1} \,|\, u \neq \pm v\} \,.$$

Then the involution

$$\lambda(u,v) = (v,u)$$

operates on X without fixed points and transforms the CR manifold

$$\hat{M} = \{(u,v) \in X \,|\, {}^t\bar{u}u = {}^t\bar{v}v\}$$

of type $(2q, 1)$ into itself.

Let M be obtained from $\hat{M}$ by identifying the pairs of points that are transformed one into the other by λ.

Then M has a natural structure of a nonorientable q-pseudoconcave CR manifold of type $(2q, 1)$.

§2. Riemann domains for pseudoconcave CR manifolds

Let $\mathbf{M} = (M, H, J)$ be a pseudoconcave CR manifold of type (n, k). We recall that this means that $\mathbf{M}$ is locally embeddable and 1-concave.

Let us consider a pseudoconcave CR manifold $\mathbf{M}$ which is globally and generically embedded in a complex manifold X of complex dimension $n + k$.

PROPOSITION 2.1 (REGULARITY). *If T is an $\mathcal{J}^p$-current $(0 \leq p \leq n+k)$ on $\mathbf{M}$ satisfying the tangential CR equations $\bar{\partial}_M T = 0$ on M, then T is equivalent to a smooth section g of $\mathcal{J}^p$ over M, and $g \in \Omega_M{}^p(\mathbf{M})$. Moreover there is a connected neighborhood U of M in X with the property that every such g extends as an element $\tilde{g} \in \Omega^p(U)$, such that $i^*\tilde{g} = g$ ($i\colon M \to X$ being the embedding map).*

PROOF: The regularity of T and the fact that g uniquely holomorphically extends to a fixed neighborhood of each point of M is a consequence of [AH,BP,NV]. By the existence of a tubular neighborhood of M in X, we can find an open covering $\{U_\alpha\}$ of M in X such that: *(i)* every $g \in \Omega_M{}^p(\mathbf{M})$ uniquely holomorphically extends to a $\tilde{g}_\alpha \in \Omega^p(U_\alpha)$ and *(ii)* $U_\alpha \cap U_\beta$ is either empty or else each component meets M. By unique continuation the $\tilde{g}_\alpha$'s agree on intersections, and therefore we find a unique holomorphic extension $\tilde{g} \in \Omega^p(U)$, where $U = \cup_\alpha U_\alpha$.

As a simple consequence we obtain

PROPOSITION 2.2 (COHERENCE OF THE STRUCTURE SHEAF). *Let* $\mathbf{M}$ *be a pseudoconcave* CR *manifold. Then, for every* $0 \le p \le n+k$, *the sheaf* $\mathbf{\Omega}^p_M$ *of germs of smooth* CR *p-forms on* $\mathbf{M}$ *has a Hausdorff topology and is a coherent sheaf of* $\mathcal{O}_M$*-modules.*

PROOF: As the properties in the proposition are local, we can reduce to the previous proposition, and use the corresponding properties for holomorphic p-forms.

PROPOSITION 2.3 (UNIQUE CONTINUATION). *Let* $\mathbf{M}$ *be a pseudoconcave* CR *manifold, and* $\omega \subset M$ *be a connected open set. Then if* $g \in \mathbf{\Omega}_M{}^p(\omega)$ *and* g *vanishes to infinite order at a point* $p \in \omega$, *then* $g \equiv 0$ *in* ω.

PROOF: We take a local holomorphic extension $\tilde{g}$ of g to a neighborhood, with respect to a local generic embedding, of a point where g vanishes to infinite order. Then $\tilde{g}$ also vanishes to infinite order at the point, and therefore on its connected component. Hence the set of points where g vanishes to infinite order is open and closed in ω.

We next explain the general concept of a Riemann domain. If M is a differentiable manifold (countable at infinity), a *Riemann domain spread over* M consists of the following data: *(i)* a connected Hausdorff topological manifold R, and *(ii)* a continuous map $\pi: R \to M$ which is a local homeomorphism. Then R and M necessarily have the same dimension, and R inherits a unique differentiable structure, which makes π a local diffeomorphism. By the theorem of Poincaré-Volterra [AN1,p.73] it follows that R has a countable topology, and hence is also countable at infinity. In particular, the fibers $\pi^{-1}(p)$ are countable.

If M is the underlying differentiable manifold of a CR manifold $\mathbf{M}$, there is a unique CR structure $\mathbf{R}$ on R with respect to which π is a local CR diffeomorphism. By the invariance of the Levi form under local CR diffeomorphisms, $\mathbf{R}$ is a q-pseudoconcave CR manifold if $\mathbf{M}$ is such.

Given two Riemann domains $\pi: R \to M$ and $\sigma: S \to M$ spread over the same CR manifold $\mathbf{M}$, a *morphism* of the first into the second is a local homeomorphism $\lambda: R \to S$ making the diagram

$$\begin{array}{ccc} R & \xrightarrow{\lambda} & S \\ & {\scriptstyle\pi}\searrow \quad \swarrow{\scriptstyle\sigma} & \\ & M & \end{array}$$

commute. Then λ is a local CR diffeomorphism of $\mathbf{R}$ into $\mathbf{S}$.

We now follow closely the arguments of [AN1]. Let $f \in \mathbf{\Omega}_M{}^p(\omega)$ be a smooth CR p-form defined an open connected $\omega \subset M$, with $\mathbf{M}$ a pseudoconcave CR manifold. It defines a section $F_f: \omega \to \mathbf{\Omega}_M{}^p_M$ of the sheaf of germs of CR p-forms. The image $F_f(\omega)$ is a connected open subset of $\mathbf{\Omega}^p_M$. Let $\tilde{\omega}_f$ be the connected component which contains $F_f(\omega)$, and $\pi_f: \tilde{\omega}_f \to M$ be the natural projection. Since $\mathbf{\Omega}^p_M$ has a Hausdorff topology, $\tilde{\omega}_f$ is a connected Hausdorff topological space, and π_f is a local homeomorphism. Therefore $\pi_f: \tilde{\omega}_f \to M$ is a Riemann domain spread over M, and hence a pseudoconcave CR manifold. Moreover, for each $\tilde{p} \in \tilde{\omega}_f$, the pullback of the evaluation of the germ $\tilde{p}$ defines a smooth CR p-form $\tilde{f}$ on $\tilde{\omega}_f$. It is clear that $\tilde{f}$ extends the pullback of f to $F_f(\omega)$.

Moreover let $\sigma: S \to M$ be a Riemann domain spread over M provided with a CR p-form $g \in \mathbf{\Omega}_M{}^p(S)$ such that there is a section $G: \omega \to S$ satisfying $g|_{G(\omega)} = \sigma^* f$. By associating to each point of S the germ of $\sigma^*(g)$ we obtain a local homeomorphism $\lambda: S \to \mathbf{\Omega}^p_M$. Then $\lambda(S)$ is a connected open subset of $\mathbf{\Omega}^p_M$. Since $\lambda(S)$ contains $F_f(\omega)$, it is contained in $\tilde{\omega}_f$.

The map λ therefore specializes to a morphism of Riemann domains spread over M, and we obtain a commutative diagram

$$\begin{array}{ccc} S & \xrightarrow{\lambda} & \tilde{\omega}_f \\ & \sigma\searrow \quad \swarrow\pi_f & \\ & M & \end{array} \tag{2.1}$$

of local CR diffeomorphisms. We summarize with the

THEOREM 2.4. *Let* $\mathbf{M}$ *be a pseudoconcave* CR *manifold of type* (n, k). *Let* F *be a smooth* CR *p-form defined on an open connected subset* ω *of* M. *Then there is a pseudoconcave* CR *Riemann domain* $\pi_f: \tilde{\omega}_f \to M$ *spread over* M, *unique up to isomorphisms, with the following properties:*

(i) There is a section $F_f: \omega \to \tilde{\omega}_f$ *and* $\tilde{f} \in \mathbf{\Omega}_M{}^p(\tilde{\omega}_f)$ *such that* $\tilde{f}|_{F_f(\omega)} = \pi_f^* f$.

(ii) If $\sigma: S \to M$ *is another Riemann domain spread over* M *satisfying the analogous property (i), then there is a unique* CR *embedding* $\lambda: S \to \tilde{\omega}_f$ *making the diagram (2.1) commute.*

Remark. If $\mathbf{M}$ is q-pseudoconcave, then every Riemann domain spread over M is a q-pseudoconcave CR manifold.

§3. Tubular neighborhoods of pseudoconcave CR manifolds

We shall need the following:

PROPOSITION 3.1. *Let* $\mathbf{M}$ *be a pseudoconcave* CR *manifold of type* (n, k). *Then we can find a* $(n + k)$*-dimensional complex manifold* X, *and a* CR *generic embedding* $\phi: M \to X$ *such that the restriction maps* $\phi^*: \mathbf{\Omega}_M{}^p(X) \to \mathbf{\Omega}_M{}^p(M)$ *are isomorphisms for* $0 \leq p \leq n + k$.

PROOF: The proof of the existence of the tubular neighborhood X follows exactly the argument of Andreotti-Fredricks (cf. [AF], pp. 296-300). It suffices to observe that an assumption of local embeddability and 1-pseudoconcavity can just be used to replace their assumption of real analyticity, in view of the ability to extend CR functions to a full neighborhood of any point on M (cf. [AH], [BP], [NV]). For the same reason, we can choose X so as to obtain the desired isomorphism ϕ^*.

Next we investigate the shape of a fundamental system of tubular neighborhoods of $\mathbf{M}$.

PROPOSITION 3.2. *Let* $\mathbf{M}$ *be a* q*-pseudoconcave* CR *manifold of type* (n, k), *and* X *be an* $(n + k)$*-dimensional tubular neighborhood of* M. *Then there exists a fundamental system* $\{U\}$ *of tubular neighborhoods of* M, *with* $U \subset X$, *having smooth boundary* ∂U, *and with the Levi form of* ∂U *having at least* q *negative and at least* $q + k - 1$ *positive eigenvalues.*

PROOF: We can find a smooth nonnegative function ρ on X which vanishes to second order on M, and is positive on $X \setminus M$, and such that: corresponding to each point $p \in M$ there is a neighborhood U_p of p in X, and an $r_p > 0$, so that the Levi form of the smooth hypersurface $\{\rho = \epsilon\} \cap U_p$, for $0 < \epsilon < r_p$, has at least q negative and at least $q + k - 1$ positive eigenvalues. Indeed such a function ρ can be constructed locally by taking the sum of squares of k local defining functions for M, and then gluing by a partition of unity. Our fundamental system of tubular neighborhoods $\{U\}$ will be defined by open sets in X

of the form $\{h\rho < 1\}$, for suitable smooth positive functions h defined on X. We choose a Hermitian metric on X, whose restriction to M is a complete Riemannian metric, and with respect to it we shall utilize functions h which have the property that they grow rapidly at infinity, and for all sufficiently small $\epsilon > 0$,

$$|h'| + |h''| \leq \epsilon h^{1+\epsilon}, \tag{2.2}$$

outside of some compact set $K = K(h, \epsilon)$. The fact that in this way one obtains a fundamental system of tubular neighborhoods of M in X is a consequence of Lemma 1 in [N2]. Let us consider the hypersurface $\{h\rho = \delta\}$. It is smooth if h is very large. The Levi form of this hypersurface is proportional to

$$-2\delta h^{-3}|\langle \partial h, w\rangle|^2 + \delta h^{-2}\langle \partial\bar{\partial} h, w \wedge \bar{w}\rangle + \langle \partial\bar{\partial}\rho, w \wedge \bar{w}\rangle$$

on $\langle \partial(hg), w\rangle = 0$. By (2.2) we can take care of points outside a compact set, and by taking $\delta > 0$ small, we can take care of the points in a compact set.

A complex manifold X of complex dimension N is called *r-pseudoconvex* (*r-pseudoconcave*) if there is a real valued smooth function ϕ on X, a compact set K and a constant $c_0 \in \mathbf{R} \cup \{\infty\}$, such that

(i) $\phi < c_0$ on $X \setminus K$,

(ii) for every $c < c_0$, $\{\phi \leq c\}$ is compact in X,

(iii) the complex Hessian of ϕ has at least $N - r$ positive ($r + 1$ negative) eigenvalues at each point of $X \setminus K$.

If, in the definition of r-pseudoconvexity, we can choose $K = \emptyset$ the manifold X is called *r-complete*.

If X is r-pseudoconvex then for any coherent sheaf $\mathcal{F}$ on X we have

$$\dim_{\mathbf{C}} H^j(X, \mathcal{F}) < \infty \quad if \quad j > r.$$

Moreover if X is r-complete then $H^j(X, \mathcal{F}) = 0$ for $j > r$. If X is r-pseudoconcave and $\mathcal{F}$ is a locally free coherent sheaf, then

$$\dim_{\mathbf{C}} H^j(X, \mathcal{F}) < \infty \quad for \quad j < r\,.$$

For an r-pseudoconcave X and any coherent sheaf $\mathcal{F}$ on X, we have

$$\dim_{\mathbf{C}} H^j_k(X, \mathcal{F}) < \infty \quad for \quad j > n + k - r,$$

where H^j_k denotes cohomology with compact supports. For these results see [AG].

As a consequence of Proposition 3.2 we obtain

COROLLARY 3.3. *Let* M *be a compact q-pseudoconcave CR manifold of type (n, k). Then the fundamental system $\{U\}$ of tubular neighborhoods in Proposition 3.2 can be taken to be $(n - q)$-pseudoconvex and q-pseudoconcave.*

PROOF: Let ρ be the function constructed in the proof of Proposition 3.2. In fact we can take the tubular neighborhood U of the form $\{\rho < \epsilon\}$ for $\epsilon > 0$ sufficiently small. In the definition of $(n-q)$-pseudoconvexity we can take $K = M$ and $\phi = \rho$. For q-pseudoconcavity we consider the function $\phi = 1 - e^{-\lambda\rho}$, for $\lambda > 0$ large, and $K = \{\rho \leq \epsilon/2\}$.

§4. Cohomology of compact pseudoconcave CR manifolds

THEOREM 4.1. *Let M be a compact q-pseudoconcave CR manifold of type (n,k). Then for all $0 \leq p \leq n+k$:*

(i) for $j < q$ and $j > n-q$, $\dim_{\mathbf{C}} H^j(M, \Omega^p_M) < \infty$ and indeed the natural maps

$$H^j(U, \Omega^p) \to H^j(M, \Omega^p_M)$$

are isomorphisms, where $\{U\}$ is as in Corollary 3.3.

(ii) for $j < q$ there are isomorphisms $H^j(M, \Omega^p_M) \simeq H^{p,j}(M)$.

(iii) for $j > n-q$, $\dim_{\mathbf{C}} H^{p,j}(M) < \infty$, and $H^{p,q}(M)$ is Hausdorff.

PROOF: We have $H^j(M, \Omega^p_M) = \varinjlim_{U \supset M} H^j(U, \Omega^p)$. By [AG],

$$H^j(U', \Omega^p) \xrightarrow{\simeq} H^j(U, \Omega^p) \quad for \quad j < q \quad and \quad j > n-q$$

if $U' \supset U$ are tubular neighborhoods as in Corollary 3.3, and hence we obtain *(i)*. The isomorphisms in *(ii)* follow by the proof of the abstract deRham theorem, since we have the Poincaré lemma for $\bar{\partial}_M$ in the correct range [AjHe], [N1]. For *(iii)* see [NV, p.155].

Examples

1°. We set $z^* = {}^t\bar{z}$ where $z = (z_0, z_1, \ldots, z_N)$ denotes homogeneous coordinates in $\mathbf{CP}^N$. Let H be a nondegenerate $(N+1) \times (N+1)$ Hermitian matrix having ν^+ positive and ν^- negative eigenvalues, and $q = \min(\nu^+, \nu^-) - 1 > 0$.

Then $M = \{z \mid z^*Hz = 0\} \subset \mathbf{CP}^N$ is a compact q-pseudoconcave CR manifold of type $(N-1, 1)$.

Actually for this example of a real nondegenerate codimension 1 quadric in $\mathbf{CP}^N$, we may obtain more precise results than those given in Theorem 4.1 by using the results of §18 of [AH] and of [AG], [ANg1], [ANg2] as follows:

(a) For $j \neq q$ and $j \neq N-q-1$, the complex vector spaces $H^j(M, \Omega^p_M)$ and $H^{p,j}(M)$ are finite dimensional for all $0 \leq p \leq N$.

(b) For $j = q$ and $j = N-q-1$, they are infinite dimensional for all $0 \leq p \leq N$.

2°. Now let K be another $(N+1) \times (N+1)$ Hermitian matrix, and consider $M = \{z \mid z^*Hz = 0,\ z^*Kz = 0\} \subset \mathbf{CP}^N$. If all the eigenvalues of $H^{-1}K$ are distinct, then M is a smooth real codimension 2 submanifold of $\mathbf{CP}^N$. If in addition they are all real, then M is a CR manifold of type $(N-2, 2)$. It is possible to arrange that M be q-pseudoconcave, for various values of q, with an appropriate choice of H and K. To be specific, we will give an example having maximal pseudoconcavity in codimension 2. Let N be odd with $N+1 = 2m$, and

$$H = \begin{bmatrix} I & 0 \\ 0 & -I \end{bmatrix}, \quad I \quad m \times m \textit{ identity matrix}$$

$$K = \begin{bmatrix} \lambda_1 & & & & & \\ & \ddots & & & & \\ & & \lambda_m & & & \\ & & & -\mu_1 & & \\ & & & & \ddots & \\ & & & & & -\mu_m \end{bmatrix},$$

with $0 \leq \lambda_1 < \mu_1 < \lambda_2 < \mu_2 < \ldots < \lambda_m < \mu_m$. It is then easy to check that every real linear combination $sH + tK$, with $s^2 + t^2 \neq 0$, has at least $m - 1$ negative eigenvalues. It turns out that M is $(m-2)$-pseudoconcave.

3°. If the eigenvalues of $H^{-1}K$ are not all real, then M fails to be a CR manifold at a finite number of points, corresponding to eigenvectors of $H^{-1}K$ that are null vectors for both H and K. With this finite number of points removed, M becomes a noncompact CR manifold of type $(N-2, 2)$.

4°. Finally we give an example in codimension 3 having maximal pseudoconcavity: we split the homogeneous coordinates $z = (u, v)$ in $\mathbf{CP}^N$ by $u, v \in \mathbf{C}^m$, with $N + 1 = 2m$, and choose any $A \in \mathrm{GL}(m, \mathbf{C})$. Let

$$M = \{(u,v) \,|\, u^*u = v^*v \,,\, \Re u^*Av = 0 \,,\, \Im u^*Av = 0\} \subset \mathbf{CP}^N \,.$$

This M is a compact CR manifold of type $(N-3, 3)$, as the vectors $(u, -v)$, $(Av, 0)$, $(0, A^*u)$ are linearly independent over $\mathbf{C}$ for $(u, v) \in M$. We take arbitrary real linear combinations of the Hermitian forms given by the matrices

$$\begin{bmatrix} I & 0 \\ 0 & -I \end{bmatrix}, \begin{bmatrix} 0 & A \\ A^* & 0 \end{bmatrix}, \begin{bmatrix} 0 & iA \\ -iA^* & 0 \end{bmatrix}$$

and for each Hermitian form in the pencil, we compute the restriction to the 3-dimensional complex subspace generated by the vectors $(u, -v)$, $(Av, 0)$, $(0, A^*u)$. It turns out that this form has rank 2 with signature $(1,1)$. This implies that the corresponding Levi form has $(m-2)$ positive and $(m-2)$ negative eigenvalues on the holomorphic tangent space to M. Thus M is a compact $(\frac{N-3}{2})$-pseudoconcave CR manifold of type $(N-3, 3)$. By Theorem 4.1 we have that:

(a) The complex vector spaces $H^j(M, \Omega^p_M)$ and $H^{p,j}(M)$ are finite dimensional for $0 \leq p \leq N$ and $j \neq \frac{N-3}{2}$.

By Theorem 4.2 below we also obtain

(b) $H^{p,j}(M)$ is infinite dimensional for $0 \leq p \leq N$ and $j = \frac{N-3}{2}$.

THEOREM 4.2. *Let* M *be a compact q-pseudoconcave CR manifold of type (n, k), which is generically embedded in its tubular neighborhood U. Moreover assume that there is a smooth function $\phi \colon U \to \mathbf{R}$, with $\phi = 0$, $\mathrm{d}\phi \neq 0$ on M, such that the complex Hessian $\sqrt{-1}\partial\bar{\partial}\phi$ has q negative and $n - q$ positive eigenvalues on $T^{0,1}_p(M)$, for every $p \in M$. Then $H^{pq}(M)$ is infinite dimensional, for $0 \leq p \leq n + k$.*

Remark. A more intrinsic formulation of the theorem can be obtained using the notion of a transversal 1-jet to M, as in the next section.

PROOF: Let ρ be the function from the proof of Proposition 3.2. Then for large $c > 0$ the hypersurface $S = \{p \in U \,|\, \phi(p) + c\rho(p) = 0\}$ is compact, and is the boundary of a complex manifold which is q-pseudoconvex and q-pseudoconcave. Indeed the Levi form of S has signature $(q, n+k-q-1)$ and $(n+k-q-1, q)$ according to the two choices of normals to S. It was proved in [AH] that the restriction homomorphism $H^{pq}(S) \to H^{pq}_{(P)}(S)$ has infinite dimensional image, for every $P \in S$. Here $H^{pq}_{(P)}(S) = \varinjlim H^{pq}(V)$, where the direct limit is taken over the net of open neighborhoods V of P in S. We know that the map $H^{pq}_{(P)}(S) \to H^{pq}_{(P)}(M)$ is injective, for every $P \in M$, by a theorem of Ajrapetyan and Henkin (see [AjHe], [N3]). This proves the theorem.

§5. Pseudoconcavity at infinity

We now turn our attention to the study of pseudoconcave CR manifolds $\mathbf{M} = (M, H, J)$ of type (n, k) which are not compact, but which are pseudoconcave at infinity.

We shall need the notion of the complex Hessian $\sqrt{-1}(\partial\bar{\partial})_M$ on M, which acts on transversal 1-jets to $\mathbf{M}$. Since $\mathbf{M}$ is locally embeddable, we may consider a sufficiently small neighborhood of any point $p \in M$ to be a generic submanifold of an open neighborhood of the origin in $\mathbf{C}^{n+k}$, where we use coordinates $(z, w) \in \mathbf{C}^n \times \mathbf{C}^k$. The local defining equations of M can be taken of the form

$$\rho_j \equiv s_j - g_j(z_1, \dots, z_n, t_1, \dots, t_k), \quad j = 1, \dots, k$$

with $w_j = t_j + is_j$ and g_j vanishing to the second order at $(0,0) \leftrightarrow p$. We may take $d z_1, \dots, d z_n, \partial\rho_1, \dots, \partial\rho_k$ and $d \bar{z}_1, \dots, d \bar{z}_n, \bar{\partial}\rho_1, \dots, \bar{\partial}\rho_k$ as bases for forms of type $(1,0)$ and $(0,1)$, respectively. In particular

$$\begin{aligned} d w_j &= a_j^r \partial\rho_r + \alpha_j^m d z_m \\ d \bar{w}_j &= \bar{a}_j^r \bar{\partial}\rho_r + \bar{\alpha}_j^m d \bar{z}_m , \end{aligned}$$

where we use summation convention with $1 \le r \le k$ and $1 \le m \le n$. Here

$$[a_j^r] = -2\left[i\delta_j^r + \frac{\partial g_j}{\partial t_r}\right]^{-1} \quad \textit{and} \quad \alpha_j^m = a_j^r \frac{\partial g_r}{\partial z_m} .$$

With respect to these coordinates a transversal 1-jet ψ to M has a unique expression as

$$\psi = \psi_0 + \sum_{j=1}^{k} \rho_j \psi_j ,$$

with ψ_0, ψ_j smooth functions of $z_1, \dots, z_n, t_1, \dots, t_k$. Note that ψ_0 is well defined on M, independent of the choice of coordinates. Set

$$\ell_{pq}^j = \frac{\partial^2 \rho_j}{\partial z_p \partial \bar{z}_q}, \quad 1 \le j \le k, \quad 1 \le p, q \le n$$

$$\mathrm{L}_{pq} = \frac{\partial^2}{\partial z_p \partial \bar{z}_q} + \frac{1}{2} \sum_{j=1}^{k} \left\{ \bar{\alpha}_j^q \frac{\partial^2}{\partial z_p \partial t_j} + \alpha_j^p \frac{\partial^2}{\partial \bar{z}_p \partial t_j} \right\} + \frac{1}{4} \sum_{i,j=1}^{k} \alpha_i^p \bar{\alpha}_j^q \frac{\partial^2}{\partial t_i \partial t_j} ,$$

$1 \le p, q \le n$. Here $\ell_{pq}^j(0)$ are the coefficients of the Levi form of $\mathbf{M}$ at $d^c\rho_j(0)$. The complex Hessian $\sqrt{-1}(\partial\bar{\partial})_M \psi$ is defined by

$$(\partial\bar{\partial})_M \psi = \sum_{p,q=1}^{n} \left\{ \{\mathrm{L}_{pq}\psi_0 + \sum_{j=1}^{k} \ell_{pq}^j \psi_j\} \right\} d z_p \wedge d \bar{z}_q .$$

If ψ is real valued this defines a Hermitian form on the holomorphic tangent space to $\mathbf{M}$. Note that both the notion of a transversal 1-jet to $\mathbf{M}$ and that of the complex Hessian $\sqrt{-1}(\partial\bar{\partial})_M\psi$ on $\mathbf{M}$ are invariant concepts, and can even be defined on a CR manifold without the assumption of local embeddability. For a more detailed discussion in the hypersurface case, see [AN2].

Definition 5.1. $\mathbf{M}$ is said to be *q-pseudoconcave at infinity* if there is a real valued transversal 1-jet ψ on M, a compact subset K of M, and a constant $c_0 \in \mathbf{R} \cup \{\infty\}$, such that:

(i) $\psi_0 < c_0$ on $M \setminus K$,
(ii) $\{p \in M \mid \psi_0(p) \leq c\}$ is compact for every $c < c_0$,
(iii) the Hermitian form $\sqrt{-1}(\partial\bar{\partial})_M\psi + L(\xi, \cdot)$ has at least q negative eigenvalues on $\{Z \in T^{0,1}M \mid \bar{\partial}_M\psi_0(Z) = 0\}$, at each point $p \in M \setminus K$ and for each $\xi \in H_p^0(M)$.

Recall that $L(\xi, \cdot)$ is the Levi form of $\mathbf{M}$. In the case of a complex manifold M, $\psi = \psi_0$ and H^0M reduces to the zero bundle; hence this definition coincides with the classical definition of a q-pseudoconcave complex manifold. So the following theorem is a generalization of the corresponding classical one.

THEOREM 5.1. *Let* $\mathbf{M}$ *be a* q*-pseudoconcave* CR *manifold of type* (n, k) *which is also* q*-pseudoconcave at infinity. Then*

$$H^{p,j}(M) \simeq H^j(M, \Omega_M^p)$$

are finite dimensional for $j < q$ *and* $0 \leq p \leq n + k$.

PROOF: First we remark that the isomorphism follows by the proof of the abstract deRham theorem, since the Poincaré lemma for $\bar{\partial}_M$ is valid for $j < q$; see [AjHe], [N1]. By Proposition 3.2 we may assume that M is generically embedded in a complex tubular neighborhood U. By the theorem on extension of CR forms (see [BP]), the sheaf Ω_M^p is isomorphic to the restriction to M of the sheaf Ω^p on U, and therefore we obtain the isomorphisms

$$H^j(M, \Omega_M^p) \simeq \varinjlim H^j(V, \Omega^p),$$

where the direct limit is taken over the net of open neighborhoods V of M contained in U. It will therefore be sufficient to prove that M has in U a fundamental system of neighborhoods V that are q-pseudoconcave.

These V can be obtained in the following way:
Let ρ be the function defined on U from the proof of Proposition 3.2. By hypothesis we have a real valued transversal 1-jet ψ on M, as in Definition 5.1. This can be extended to a smooth real valued function on U, which we still denote by ψ. By composing ψ with an increasing function of a real variable, we may assume that $c_0 < \infty$. Moreover we can arrange that $\psi < c_0$ on M. The set $\{p \in U \mid \psi(p) < c_0\}$ is an open neighborhood of M in U. Therefore we can assume that $\psi < c_0$ on U. We shall consider smooth functions χ, defined on $\{-\infty < t < c_0\}$ and taking positive real values, with $\chi' > 0$ and $\chi'' > 0$. Consider $\phi = \psi + \chi(\psi)\rho$ and $V = \{p \in U \mid \phi(p) < c_0\}$. These V we have constructed can be made to be q-pseudoconcave complex manifolds, of complex dimension $n + k$, as in the definition in §3:

Indeed

$$\begin{aligned}\partial\bar{\partial}\phi = &[1+\rho\chi'(\psi)]\partial\bar{\partial}\psi + \chi(\psi)\partial\bar{\partial}\rho \\ &+\chi'(\psi)[\partial\psi\wedge\bar{\partial}\rho + \partial\rho\wedge\bar{\partial}\psi] + \rho\chi''(\psi)\partial\psi\wedge\bar{\partial}\psi\ .\end{aligned}$$

On $T^{0,1}M$, $\partial\bar{\partial}\rho$ is a linear combination of Levi forms $L(\xi,\,\cdot)$ of M. The complex Hessian $\sqrt{-1}\partial\bar{\partial}\psi$ of the various extensions ψ of the transversal 1-jet to M differ from $\sqrt{-1}(\partial\bar{\partial})_M\psi$ by linear combinations of Levi forms of M. By Definition 5.1 the first two terms in $\partial\bar{\partial}\phi$ bring q-pseudoconcavity for ϕ on a neighborhood in U of the complement of K in M. With a fixed choice of χ, we can obtain the q-pseudoconcavity of V simply by adding a sufficiently large constant to χ. Notice that the last two terms in $\partial\bar{\partial}\phi$ cause no trouble since the set of tangent vectors to V of type $(0,1)$, which are annihilated by $\bar{\partial}\psi$ and $\bar{\partial}\rho$, contains a distribution that is a smooth extension of $T^{0,1}M$. This completes the proof of the theorem.

Examples.
1°. We consider in $\mathbf{C}^3$ the hypersurface

$$M = \{w\in\mathbf{C}^3\,|\,w_1\bar{w}_1 + w_2\bar{w}_2 = 1 + w_3\bar{w}_3\}\,,$$

its compactification

$$\hat{M} = \{z\in\mathbf{CP}^3\,|\,z_1\bar{z}_1 + z_2\bar{z}_2 = z_0\bar{z}_0 + z_3\bar{z}_3\}$$

in $\mathbf{CP}^3$, and we denote

$$\Sigma = \{z\in\hat{M}\,|\,z_3 = 0\}\subset M\ .$$

Note that Σ is the standard S^3 in $\mathbf{C}^2$, and that $\hat{M}$ is a compact pseudoconcave (1-pseudoconcave) CR manifold of type $(2,1)$. As it is a special case of example 1° of §4, we know that for $0\le p\le 3$, $H^0(M,\Omega^p_M)\simeq H^{p,0}(M)$, $H^2(M,\Omega^p_M)$, $H^{p,2}(M)$ are finite dimensional; whereas $H^1(M,\Omega^p_M)$ and $H^{p,1}(M)$ are infinite dimensional.

First we observe that $\hat{M}\setminus\Sigma$ is a noncompact pseudoconcave CR manifold, but it is not 1-pseudoconcave at infinity. In fact

$$\{\left(\frac{z_0}{z_3}\right)^m\,|\,m=1,2,\dots\}$$

are linearly independent CR functions on $\hat{M}\setminus\Sigma$; hence

$$H^0\left(\hat{M}\setminus\Sigma,\mathcal{O}_M|_{\hat{M}\setminus\Sigma}\right)\simeq H^{0,0}(\hat{M}\setminus\Sigma)$$

are infinite dimensional.

Second we point out that one of the key points in the theory of several complex variables fails to hold for 1-pseudoconcave CR manifolds; namely the Hartogs phenomenon: If K is a compact subset of a Stein manifold X of complex dimension $n\ge 2$, with $X\setminus K$ connected, then every holomorphic function on $X\setminus K$ has a unique holomorphic extension to X. Here we may consider the CR analogue with either $X = \hat{M}$ or $X = M$, and $K = \Sigma$. Even though $\hat{M}$ (M) is a CR submanifold of $\mathbf{CP}^3$ $(\mathbf{C}^3)$, as we have seen, there is an infinite dimensional space of CR functions on $\hat{M} \setminus \Sigma$ $(M \setminus \Sigma)$ which do not have CR extensions across the compact set Σ.

2°. Let $\hat{M}$ be a compact q-pseudoconcave CR manifold of type (n, k), with $q \geq 2$. Let M be obtained from $\hat{M}$ by removing a finite number of either points, or else sufficiently small disjoint balls in the tubular neighborhood U of M, centered about points of $\hat{M}$. Then M is $(q-1)$-pseudoconcave at infinity. Since the essential requirements on the transversal 1-jet ψ are only needed outside a compact subset K, and therefore are local in this example, it will suffice to consider the case where we have removed either one point or else one small ball. Indeed it then suffices to take ψ to be minus the square of the distance from the origin, in a coordinate patch in $\mathbf{C}^{n+k}$, where the point (or the center of the ball) corresponds to the origin.

3°. Let $\hat{M}$ be a compact smooth real codimension k submanifold of a complex manifold X. Assume that, outside of a finite number of points, $\hat{M}$ is a q-pseudoconcave CR manifold of type (n, k), with $q \geq 2$ (cf. example 3° in §4). The manifold M obtained from $\hat{M}$ by removing these finite number of points is then $(q-1)$-pseudoconcave at infinity. This follows by using the same construction of ψ as in the example above.

§6. Pseudoconvexity at infinity

In this section we consider noncompact pseudoconcave CR manifolds $\mathbf{M} = (M, H, J)$ of type (n, k) which have some pseudoconvexity at infinity.

Definition 6.1. $\mathbf{M}$ is said to be *r-pseudoconvex* at infinity if there is a real valued transversal 1-jet ψ on M, a compact subset K of M, and a constant $c_0 \in \mathbf{R} \cup \{\infty\}$, such that:

(i) $\psi_0 < c_0$ on $M \setminus K$,

(ii) $\{p \in M \mid \psi_0(p) \leq c\}$ is compact for every $c < c_0$,

(iii) the Hermitian form $\sqrt{-1}(\partial\bar{\partial})_M\psi + L(\xi, \cdot)$ has at least $n - r$ positive eigenvalues on $T^{0,1}M$, at each point $p \in M \setminus K$ and for each $\xi \in H_p^0(M)$.

If we can take $K = \emptyset$ then we say that $\mathbf{M}$ is *r-complete*.

THEOREM 6.1. *Let $\mathbf{M}$ be an $(n-r)$-pseudoconcave CR manifold of type (n, k) which is also r-pseudoconvex at infinity. Then $H^j(M, \Omega_M^p)$ are finite dimensional for $j > r$ and $0 \leq p \leq n + k$. These cohomology groups vanish if M is also r-complete.*

PROOF: Again using Proposition 3.2, we may assume that M is generically embedded in a complete tubular neighborhood U. As in the proof of Theorem 5.1 , it will suffice to show that M has in U a fundamental system of neighborhoods V which are r-pseudoconvex (r-complete). To this aim we consider the same function $\phi = \psi + \chi(\psi)\rho$ as before, after arranging that $c_0 < \infty$ and $\psi < c_0$ on M. We set $V = \{p \in U \mid \phi(p) < c_0\}$. By checking the formula for $\partial\bar{\partial}\phi$, as written in the proof of Theorem 5.1 , it is easy to see that we can obtain the r-pseudoconvexity (r-completeness) of V. Indeed it suffices to choose a χ with $\chi' > 0$, $\chi'' > 0$ and satisfiyng $\chi' + \chi'' < const(\epsilon)\chi^{1+\epsilon}$, $\chi\chi'' > (\chi')^2$. If $K \neq \emptyset$ we must also add a sufficiently large constant to χ. This completes the proof of the theorem.

Remark. Let V be an r-pseudoconvex (r-complete) neighborhood of M in U. From the formal Cauchy-Kovalevsky theorem (cf. [AH], [AFN]) we have an exact sequence

$$\cdots \to H^j(V, \Omega^p) \to H^{p,j}(M) \to H^{j+1}\left(\mathcal{F}^{p,*}(M,V), \bar\partial\right) \to \cdots,$$

where $\mathcal{F}^{p,j}(M,V)$ denotes the space of smooth (p,j)-forms in V which are flat on M, and where the last term on the right denotes the $(j+1)$-cohomology group of the complex

$$0 \to \mathcal{F}^{p,0}(M,V) \xrightarrow{\bar\partial} \mathcal{F}^{p,1}(M,V) \xrightarrow{\bar\partial} \cdots \to \mathcal{F}^{p,n+k}(M,V) \to 0\,.$$

Then the maps $H^{p,j}(M) \to H^{j+1}\left(\mathcal{F}^{p,*}(M,V), \bar\partial\right)$ have a finite dimensional kernel and cokernel (are isomorphisms) for $j > r$ and $0 \le p \le n+k$. Thus to prove the analogue of Theorem 6.1 for $\bar\partial_M$-cohomology with smooth forms, it is equivalent to prove that the natural maps

$$H^s(V, \Omega^p_{V\setminus M}) \to H^s\left(\mathcal{F}^{p,*}(M,V), \bar\partial\right)$$

have finite dimensional kernels and cokernels (are isomorphisms) for $s > r+1$. Therefore the comparison of cohomology with coefficients in the sheaf Ω^p_M and $\bar\partial_M$-cohomology is reduced in this case to the study of a Cauchy problem for cohomology classes.

We conjecture that under the same assumptions as in Theorem 6.1 , the same conclusions hold for $\bar\partial_M$-cohomology. This could perhaps be proved by the technique used in [N3].

Examples.

1°. Let M be a noncompact q-pseudoconcave closed smooth CR submanifold of a Stein manifold X, which is of type (n,k). By the embedding theorem for Stein manifolds, we may regard M as being a closed CR submanifold of $\mathbb{C}^N$, for some N $(n+k \le N)$. Then M is $(n-q)$-complete, as we may use $\psi|_M M$, when ψ is the square of the distance from the origin in $\mathbb{C}^N$. Therefore by Theorem 6.1 we have that $H^j(M, \Omega^p_M) = 0$ for $j > n-q+1$ and $0 \le p \le n+k$.

2°. Let $\hat M$ be the quadric hypersurface in $\mathbb{CP}^N$ defined in homogeneous coordinate by $\sum_0^q z_j\bar z_j - \sum_{q+1}^N z_j \bar z_j = 0$, where $2q \le N-1$. Then $\hat M$ is a compact q-pseudoconcave CR manifold of type $(N-1,1)$.

(a) Let $M = \{z \in \hat M \mid z_0\bar z_0 + z_N \bar z_N \ne 0\}$. By taking

$$\psi = \log\left(\sum_0^N z_j\bar z_j\right) - \log(z_0\bar z_0 + z_N\bar z_N)$$

it is easily checked that M is $(N-q-1)$-complete.

(b) Let $M = \{z \in \hat M \mid z_0\bar z_0 + z_1\bar z_1 \ne 0\}$. In the same way we see that M is $(N-q)$-complete.

§7. Abstract pseudoconcave CR manifolds

We now turn our attention to the study of abstract CR manifold of type (n,k); that is, we drop the assumption of local embeddability. In this context we shall consider only the case where $\mathbf{M} = (M, H, J)$ is compact.

THEOREM 7.1. *Let M be a compact q-pseudoconcave abstract CR manifold of type (n,k). Then for all $0 \leq p \leq n+k$ and for $j < q$ and $j > n-q$, the cohomology groups $H^{p,j}(M)$ are finite dimensional. Moreover $H^{pq}(M)$ is Hausdorff. If $q \geq 1$ then any locally defined $\mathcal{J}^p$current T satisfying $\bar{\partial}_M T = 0$ is smooth.*

PROPOF: The proof will be divided into several steps:

(a) The Laplace-Beltrami operator for a CR manifold.

We choose a Riemannian metric on M and a Hermitian metric on $\mathbb{C}TM$, which then gives compatible Hermitian metrics on each bundle $Q^{p,j}$. Thus we have L^2-norms and inner products on the fibers of $Q^{p,j}$, and can define the formal adjoints $\bar{\partial}^*_M: Q^{p,j}(M) \to Q^{p,j-1}(M)$, for $j \geq 1$, setting $\bar{\partial}^*_M = 0$ for $j = 0$. Then the Laplace-Beltrami operator $\Box_M: Q^{p,j}(M) \to Q^{p,j}(M)$ is defined by $\Box_M = \bar{\partial}_M \bar{\partial}^*_M + \bar{\partial}^*_M \bar{\partial}_M$. We denote by $L^2 Q^{p,j}(M)$ the space of L^2-forms of type (p,j) on M, and by $WQ^{p,j}(M)$ the subspace of all $u \in L^2 Q^{p,j}(M)$ such that $\bar{\partial}_M u$ and $\bar{\partial}^*_M u$ also belong to L^2, where derivatives are taken in the sense of currents. On $WQ^{p,j}(M)$ we consider the sesquilinear form

$$a(u,v) = \left(\bar{\partial}_M u, \bar{\partial}_M u\right) + \left(\bar{\partial}^*_M v, \bar{\partial}^*_M v\right) .$$

Here $(\cdot,\cdot)$ denotes the various L^2-inner products on the spaces $L^2 Q^{p,j}(M)$, with $\|\cdot\|$ being the associated norm. Since M is compact it follows that the subspaces $L^2 Q^{p,j}(M)$ and $WQ^{p,j}(M)$ are actually independent of the choice of the Riemannian metric on M and the Hermitian metric on $\mathbb{C}TM$. Likewise the norm

$$\|u\|_W = \left(\|u\|^2 + a(u,u)\right)^{\frac{1}{2}}$$

on $WQ^{p,j}(M)$ is uniquely defined, up to equivalence.

(b) Reduction to an a priori estimate.

On each space $Q^{p,j}(M)$ we fix once and for all a choice of the 1/2-norm $|\cdot|_{1/2}$, and recall that any other choice gives an equivalent norm. Our aim will be to prove an estimate of the form

$$|u|_{1/2} \leq c\|u\|_W, \quad \forall u \in Q^{p,j}(M) . \tag{7.1}$$

Given such an estimate for fixed (p,j), it follows from the general theory of subelliptic estimates (cf. [Hö1]) that:

(i) The space of (p,j)-harmonic forms

$$\mathcal{H}^{p,j}(M) = \{u \in WQ^{p,j}(M) \,|\, \bar{\partial}_M u = 0\,,\, \bar{\partial}^*_M u = 0\}$$

is contained in $Q^{p,j}(M)$ and is finite dimensional.

(ii) For every $f \in L^2 Q^{p,j}(M)$ which is L^2-orthogonal to $\mathcal{H}^{p,j}(M)$, the equation

$$a(u,v) = (f,v), \quad \forall v \in WQ^{p,j}(M) \tag{7.2}$$

has a unique solution which is L^2-orthogonal to $\mathcal{H}^{p,j}(M)$.

(iii) Any solution u of (7.2) for $f \in Q^{p,j}(M)$ is also in $Q^{p,j}(M)$.

(iv) If moreover $\bar{\partial}_M f = 0$, then $v = \bar{\partial}_M^* u$ is a solution of $\bar{\partial}_M v = f$.
(v) Every cohomology class in $H^{p,j}(M)$ has a unique harmonic representative.
(vi) $\bar{\partial}_M$ and $\bar{\partial}_M^*$ have closed range in $Q^{p,j+1}(M)$ and $Q^{p,j-1}(M)$, respectively.

As the last statement *(vi)* is not standard, we give an explicit argument for the range of $\bar{\partial}_M$, the argument being analogous for $\bar{\partial}_M^*$. Let $u \in WQ^{p,j}(M)$. First we decompose $u = h+v$ where $h \in \mathcal{H}^{p,j}(M)$ and $v \in \mathcal{H}^{p,j}(M)^\perp$. By *(ii)* and *(iii)* we can find $w \in Q^{p,j}(M)$ such that $\Box_M w = v$. We set $v_1 = \bar{\partial}_M^* \bar{\partial}_M w$ and $v_2 = \bar{\partial}_M \bar{\partial}_M^* w$ and obtain the decomposition (Hodge decomposition)

$$u = h + v_1 + v_2\,, \tag{7.3}$$

where $h \in \mathcal{H}^{p,j}(M)$, v_1 and $v_2 \in \mathcal{H}^{p,j}(M)^\perp$, $\bar{\partial}_M^* v_1 = 0$ and $\bar{\partial}_M v_2 = 0$. Take a sequence $\{f_n\} \subset \bar{\partial}_M Q^{p,j}(M)$ which converges in the C^∞ topology of $Q^{p,j}(M)$ to a form f. For each n there is a unique $u_n \in Q^{p,j}(M) \cap \mathcal{H}^{p,j}(M)^\perp$ with $\bar{\partial}_M^* u_n = 0$ and $\bar{\partial}_M u_n = f_n$. By (7.1) we have $|u_n|_{1/2} \le c\|f_n\|$ and therefore $\{u_n\}$ converges weakly in $WQ^{p,j}(M)$ to an element u with $\bar{\partial}_M u = f$ and $\bar{\partial}_M^* u = 0$. Therefore $\Box_M u = \bar{\partial}_M^* f$, so u is smooth and hence the range of $\bar{\partial}_M$ is closed.

It follows that $H^{p,j+1}(M)$ is Hausdorff because it is the quotient of a Fréchet-Schwartz space by a closed subspace.

(c) Localization of the estimate.

In order to prove the desired estimate (7.1) it will suffice to prove a localized version; namely

$$|u|_{1/2} \le c_p \|u\|_W, \quad \forall u \in Q_c^{p,j}(U)\,. \tag{7.4}$$

This means that for each point $p \in M$ we can find an open neighborhood U of p in M, and a constant $c_p > 0$, such that (7.4) holds, where $Q_c^{p,j}(U)$ denotes (p,j)-forms with compact support in U. Suppose we have the local estimate (7.4) . We take a finite covering $\{U_i\}$ of M by open sets U_i where (7.4) holds, and a smooth partition of unity $\{\phi_i^2\}$ subordinate to the covering $\{U_i\}$. By finiteness we have the equivalence of norms

$$c|u|_{1/2} \le \sum_i |\phi_i u|_{1/2} \le C|u|_{1/2}\,,$$

$$c\|u\|_W \le \sum_i \|\phi_i u\|_W \le C\|u\|_W\,,$$

for some constants $c, C > 0$, and for every $u \in Q^{p,j}(M)$.

(d) A local computation.

Fix a point $p \in M$ and choose a local coordinate patch V, with local coordinates $x_1, \ldots, x_n, y_1, \ldots, y_n, t_1, \ldots, t_k$, vanishing at p, in such a way that the complex vector fields of type $(0,1)$ on V are spanned by

$$\bar{L}_j = \frac{1}{2}\left(\frac{\partial}{\partial x_j} + i\frac{\partial}{\partial y_j}\right) + \bar{R}_j, \quad j = 1, \ldots, n\,.$$

Here the $\bar{R}_j$ are complex vector fields vanishing at p, and

$$\mathrm{L}_1, \dots, \mathrm{L}_n, \bar{\mathrm{L}}_1, \dots, \bar{\mathrm{L}}_n, \frac{\partial}{\partial t_1}, \dots, \frac{\partial}{\partial t_k}$$

provide a basis for $\mathbf{C}T(V)$. As we may assume that all bundles are trivial over V, it will suffice to prove the local estimate (7.4) for forms of type $(0, j)$, and we will use the notation $Q^j = Q^{0,j}$, etc. Then the elements of $Q^j(V)$ can be uniquely represented by $u = \sum_{|I|=j}{}' u_I d\bar{z}^I$, where $I = (i_1, \dots, i_j)$ with $1 \le i_1, \dots, i_j \le n$ and the prime indicates summation over increasing multiindices. Here we have set $d\bar{z}_i = dx_i - i\,dy_i$ and $d\bar{z}^I = d\bar{z}_{i_1} \wedge \dots \wedge d\bar{z}_{i_j}$. With this notation we have that

$$\bar{\partial}_M u = \sum_I{}' \sum_{h=1}^{n} \bar{\mathrm{L}}_h u_I d\bar{z}_h \wedge d\bar{z}^I + \textit{terms of order zero in } u\,.$$

In view of *(a)* we may use the standard Euclidean metric with respect to the coordinates $x_1, \dots, x_n, y_1, \dots, y_n, t_1, \dots, t_k$ in V, and the standard Hermitian metric on the fibers of $Q^j(V)$. Then the formal adjoint of $\bar{\partial}_M$ is given by

$$\bar{\partial}_M^* u = \sum_{(i_1, \dots, i_j)}{}' \sum_{h=1}^{j} (-1)^h \mathrm{L}_h u_{i_1 \dots \hat{i}_h \dots i_j} d\bar{z}_{i_1} \wedge \widehat{d\bar{z}_{i_h}} \wedge \dots \wedge d\bar{z}_{i_j}$$
$$+ \textit{terms of order zero in } u\,.$$

For $1 \le j,\, k \le n$ we have

$$[\mathrm{L}_j, \bar{\mathrm{L}}_k] = i \sum_h c_{jk}^h \frac{\partial}{\partial t_h} + \sum_s \left(A_{jk}^s \bar{\mathrm{L}}_s + B_{jk}^s \mathrm{L}_s \right)$$

in V. For each fixed h, $1 \le h \le k$, the $[c_{jk}^h]$ is a Hermitian matrix, which at the origin is the Levi form at dt_h.

Recall that M is q-pseudoconcave at the point p. If $0 \le j < q$ or $n - q < j \le n$, then for each ξ^0 with $|\xi^0| = 1$, we can find vectors $(w_I) \in \mathbf{C}^n \otimes \bigwedge^j (\mathbf{C}^n)^*$ with $|w_I| = 1$ for all I, and a constant $\nu(\xi^0) > 0$ such that

$$\sum_K{}' \sum_{j,k} \sum_h \xi_h^0 c_{jk}^h(0) \eta_{jK} \bar{\eta}_{kK} - \sum_I{}' \sum_{j,k} \sum_h \xi_h^0 c_{jk}^h(0) w_I^j \bar{w}_I^k |\eta_I|^2$$
$$\ge \nu(\xi^0) \sum_I{}' |\eta_I|^2\,,$$

for all $\eta \in \bigwedge^j (\mathbf{C}^n)^*$.

With ξ^0 fixed and the (w_I) as above, we can find $\epsilon > 0$ sufficiently small that: if $|\xi| = 1$ and $|\xi - \xi^0| < \epsilon$ then

$$
\begin{aligned}
&\sum_K{}' \sum_{j,k} \sum_h \xi_h c_{jk}^h(0) \eta_{jK} \bar{\eta}_{kK} - (1-\epsilon') \sum_I{}' \sum_{j,k} \sum_h \xi_h c_{jk}^h(0) w_I^j \bar{w}_I^k |\eta_I|^2 \\
&\quad - \epsilon'' \sum_I{}' \sum_j \sum_h \xi_h c_{jj}^h(0) |\eta_I|^2 \geq \frac{1}{2}\nu(\xi^0) \sum_I{}' |\eta_I|^2 ,
\end{aligned}
\tag{7.5}
$$

for all $\eta \in \bigwedge^j (\mathbf{C}^n)^*$ and $0 \leq \epsilon', \epsilon'' < \epsilon$.

(e) Proof of the local estimate.

For $u \in Q_c^j(V)$ we introduce the norm $|\cdot|$ defined by

$$
|u|^2 = \sum_I{}' \left\{ \sum_j \|\bar{\mathrm{L}}_j u_I\|^2 + \|\mathrm{L}_j u_I\|^2 + \|u_I\|^2 + \int |\xi| |\tilde{u}_I(x,y,\xi)|^2 \,\mathrm{d}\,x \,\mathrm{d}\,y \,\mathrm{d}\,\xi \right\}
$$

where

$$
\tilde{u}_I(x,y,\xi) = \int_{\mathbf{R}^k} u_I(x,y,t) e^{-i\langle t,\xi\rangle} \mathrm{d}\,t
$$

is the partial Fourier transform of u_I with respect to t. Since

$$
\mathrm{L}_1, \ldots, \mathrm{L}_n, \bar{\mathrm{L}}_1, \ldots, \bar{\mathrm{L}}_n, \frac{\partial}{\partial t_1}, \ldots, \frac{\partial}{\partial t_k}
$$

form a basis for $\mathbf{C}T(V)$, we have that

$$
|u|_{1/2} \leq C|u|
$$

for some constant $C > 0$. By repeating the formula obtained from integrating by parts, as in [Hö2, p.123] we obtain, for some constant $c > 0$, the estimate

$$
\begin{aligned}
c \cdot a(u,u) \geq &\sum_I{}' \sum_j \int |\bar{\mathrm{L}}_j u_I|^2 \mathrm{d}\,\lambda + i \sum_K{}' \sum_{j,k} \int \sum_h c_{jk}^h(0) \frac{\partial u_{jK}}{\partial t_h} \bar{u}_{kK} \mathrm{d}\,\lambda \\
&- \mathrm{c}\|u\|^2 - c\delta|u|^2 - \mathrm{c}\|u\| \cdot |u| ,
\end{aligned}
\tag{7.6}
$$

provided that $u \in Q_c^j(V)$ and u has its support contained in a coordinate ball of radius δ centered about 0. Here $\mathrm{d}\,\lambda$ denotes Lebesgue measure in $\mathbf{R}^{2n+k}$.

To complete the proof of the local estimate, we shall need to microlocalize with respect to the characteristic direction. To this aim we construct a partition of unity on $\mathbf{R}^k$ as follows. First we choose a C^∞ funtion ϕ_0, defined for $\xi \in \mathbf{R}^k$, which satisfies :

$$
\phi_0(\xi) = 1 \quad \textit{for} \quad |\xi| \leq \frac{1}{2},
$$

$$0 < \phi_0(\xi) < 1 \quad for \quad \frac{1}{2} < |\xi| < 1,$$
$$\phi_0(\xi) = 0 \quad for \quad |\xi| \geq 1.$$

Then $\sqrt{1 - \phi_0(\xi)^2}$ is a smooth function on $\mathbf{R}^k$ that equals 1 for $|\xi| \geq 1$ and is zero for $|\xi| \leq 1/2$.

Next we take real valued smooth functions $\psi_1, \dots, \psi_M$ defined on the unit sphere $\Sigma = \{|\xi| = 1\}$ which satisfy

(i) $\psi_1^2, \dots, \psi_M^2$ is a partition of unity on Σ,

(ii) there are points $\xi_1, \dots, \xi_M$ on Σ such that

$$\mathrm{supp}\psi_s \subset \{\xi \in \Sigma \,|\, |\xi - \xi_s| < \epsilon(\xi_s)\},$$

where $\epsilon(\xi_s)$ is the ϵ found at the end of *(d)*.

For $j = 1, \dots, M$ we define

$$\phi_j(\xi) = \sqrt{1 - \phi_0(\xi)^2}\psi_j\left(\frac{\xi}{|\xi|}\right) \quad for \quad \xi \in \mathbf{R}^k \setminus \{0\},$$
$$\phi_j(0) = 0.$$

Note that the functions $\phi_0^2, \phi_1^2, \dots, \phi_M^2$ give a partition of unity on $\mathbf{R}^k$.

Next we fix a C^∞ function χ having compact support in V, and equal to 1 in a neighborhood W of 0 in V. Then for $0 \leq j \leq M$,

$$\Phi_j v = (2\pi)^{-k}\chi \int e^{i\langle t, \xi\rangle}\phi_j(\xi)\tilde{v}\, d\,\xi$$

defines a properly supported pseudodifferential operator on V, of order 0 for $1 \leq j \leq M$ and of order $-\infty$ for $j = 0$. For fixed I and for $u \in Q_c^j(W)$,

$$\begin{aligned} c_1\|u_I\|^2 + \sum_j \|\bar{\mathrm{L}}_j u_I\|^2 &\geq \sum_j \sum_{s=0}^{M} \|\Phi_s \bar{\mathrm{L}}_j u_I\|^2 \\ &\geq \sum_j \sum_{s=0}^{M} \|\bar{\mathrm{L}}_j \Phi_s u_I\|^2 - c_2\left(\|u_I\|^2 + \|u_I\| \cdot |u|\right) \end{aligned}$$

for constants $c_1, c_2 \geq 0$. For $s = 1, \dots, M$ and $w \in \mathbf{C}^n$ with $|w| \leq 1$,

$$\begin{aligned} \sum_j \|\bar{\mathrm{L}}_j \Phi_s u_I\|^2 &\geq \|\sum_j w^j \bar{\mathrm{L}}_j \Phi_s u_I\|^2 \\ &\geq \|\sum_j \bar{w}^j \mathrm{L}_j \Phi_s u_I\|^2 \\ &\quad - i\sum_{j,k}\int \sum_h c_{jk}^h(0) w^j \bar{w}^k \frac{\partial}{\partial t_h}(\Phi_s u_I) \cdot \overline{\Phi_s u_I}\, d\,\lambda \\ &\quad - c_3\left\{\{\|u_I\|^2 + \|u_I\| \cdot |u_I| + \delta |u_I|^2\}\right\}, \end{aligned} \tag{7.7}$$

provided that the support of u_I is contained in a coordinate ball of radius δ centered about 0, and contained in W. It will be convenient, in what follows, to use the notation $A \gtrsim B$ as a shorthand for an estimate of the type $A \geq B - c(\|u\|^2 + \delta|u|^2)$, where $c > 0$ is some constant, and where u is understood to have its support contained in a coordinate ball of some radius $\delta > 0$ centered about the origin.

Integration by parts leads to

$$\int |\bar{\mathrm{L}}_j u_I|^2 d\lambda \gtrsim \int |\mathrm{L}_j u_I|^2 d\lambda - i \int \sum_h c_{jj}^h(0) \frac{\partial}{\partial t_h} u_I \cdot \bar{u}_I d\lambda . \tag{7.8}$$

For small $\sigma > 0$ we rewrite (7.6) as

$$\begin{aligned} c \cdot a(u,u) &\gtrsim \sigma \sum_I{}' \sum_j \int |\bar{\mathrm{L}}_j u_I|^2 d\lambda + (1-\sigma) \sum_I{}' \sum_j \int |\bar{\mathrm{L}}_j u_I|^2 d\lambda \\ &\quad + i \sum_K{}' \sum_{j,k} \int \sum_h c_{jk}^h(0) \frac{\partial u_{jK}}{\partial t_h} \cdot \bar{u}_{kK} d\lambda \\ &\gtrsim \frac{\sigma}{2} \sum_I{}' \sum_j (\|\bar{\mathrm{L}}_j u_I\|^2 + \|\mathrm{L}_j u_I\|^2) \\ &\quad - i \frac{\sigma}{2} \sum_I{}' \sum_j \int \sum_h c_{jj}^h(0) \frac{\partial}{\partial t_h} u_I \cdot \bar{u}_I d\lambda \\ &\quad + (1-\sigma) \sum_I{}' \sum_j \int |\bar{\mathrm{L}}_j u_I|^2 d\lambda \\ &\quad + i \sum_K{}' \sum_{j,k} \int \sum_h c_{jk}^h(0) \frac{\partial u_{jK}}{\partial t_h} \cdot \bar{u}_{kK} d\lambda . \end{aligned} \tag{7.9}$$

To handle the next to the last term above, we note that

$$\begin{aligned} \sum_j \|\bar{\mathrm{L}}_j u_I\|^2 &\gtrsim \sum_s \sum_j \|\bar{\mathrm{L}}_j \Phi_s u_I\|^2 \\ &\gtrsim \sum_s \| \sum_j w_I^j(s) \bar{\mathrm{L}}_j \Phi_s u_I \|^2 \\ &\gtrsim -i \sum_s \int \sum_{j,k} \sum_h w_I^j(s) \bar{w}_I^k(s) c_{jk}^h(0) \frac{\partial}{\partial t_h} (\Phi_s u_I) \cdot \overline{\Phi_s u_I} d\lambda , \end{aligned}$$

for $|w_I(s)| \leq 1$, where the last term was obtained using integration by parts. Hence we obtain

$$\begin{aligned}
&c{\cdot}a(u,u)\\
&\gtrsim \frac{\sigma}{2}\sum_I{}' \sum_j (\|\bar{\mathrm{L}}_j u_I\|^2 \mathrm{d}\,\lambda + \|\mathrm{L}_j u_I\|^2)\\
&+ i\sum_K{}' \sum_{j,k} \int \sum_h c^h_{jk}(0)\frac{\partial u_{jK}}{\partial t_h}\cdot \bar{u}_{kK}\mathrm{d}\,\lambda\\
&- i(1-\sigma)\sum_I{}' \sum_s \sum_{j,k} \int \sum_h w^j_I(s)\bar{w}^k_I(s)c^h_{jk}(0)\frac{\partial}{\partial t_h}(\Phi_s u_I)\cdot \overline{\Phi_s u_I}\mathrm{d}\,\lambda\\
&- i\frac{\sigma}{2}\sum_I{}' \sum_s \sum_j \int \sum_h c^h_{jj}(0)\frac{\partial}{\partial t_h}(\Phi_s u_I)\cdot \overline{\Phi_s u_I}\mathrm{d}\,\lambda\,.
\end{aligned}$$

We estimate the last three terms above making a partial Fourier transform in the t variables (Plancherel formula). Finally we choose the vectors $(w_I(s))$ as at the end of *(d)*, and employ (7.5) with σ chosen sufficiently small. We then obtain the local estimate (7.1) by choosing δ sufficiently small with respect to σ. The proof is complete.

With the same proof we obtain in the hypersurface case the following (cf.[FK])

THEOREM 7.2. *Let* M *be a compact abstract CR manifold of type $(n,1)$. Assume that for every $\xi \in H^0M$, $\xi \neq 0$, the Levi form $L(\xi,\cdot)$ has either at least $j+1$ negative or at least $n-j+1$ positive eigenvalues. Then $H^{p,j}(M)$ is finite dimensional for* $0 \leq p \leq n+1$.

Bibliography

[AjHe] R.A. Ajrapetyan, G.M. Henkin, Integral Representation of differential forms on Cauchy-Riemann manifolds and the theory of *CR* functions I,II. Usp. Mat. Nauk 39, n.3, (1984) 39-106; English transl.: Russ. Math. Surv. 39, n.3, (1984) 41-118; Mat. Sb. Nov. Ser. 127, n.1, (1985) 92-112; English transl.: Math. USSR, Sb. 55. n.1, (1986) 91-111.

[AF] A. Andreotti, G.A. Fredricks, Embeddability of real analytic Cauchy-Riemann manifolds, Ann. Scuola Norm. Sup. Pisa, 6 (1979), 285-304.

[AFN] A. Andreotti, G.A. Fredricks, M. Nacinovich, On the absence of Poincaré lemma in tangential Cauchy-Riemann complexes, Ann.Scuola Norm. Sup. Pisa, 8 (1981), 365-404.

[AG] A. Andreotti, H. Grauert, Théorèms de finitude pour la cohomologie des espaces complexes, Bull. Soc. Math. France, 90 (1962), 193-259.

[AH] A. Andreotti, C.D. Hill, E.E. Levi Convexity and the Hans Lewy problem I, II, Ann. Scuola Norm. Sup. Pisa, 26 (1972), 325-363, 747-806.

[AN1] A. Andreotti, M. Nacinovich, On the envelope of regularity for solution of homogeneous systems of linear partial differential operators,Ann. Scuola Norm. Sup. Pisa, 6 (1979), 69-141.

[AN2] A. Andreotti, M. Nacinovich, Noncharacteristic hypersurface for complexes of differential operators, Ann. Math. Pura Appl., 125 (1980), 18-83.

[ANg1] A. Andreotti, F. Norguet, Problème de Levi et convexité holomorphe pour les classes de cohomologie, Ann. Scuola Norm. Sup. Pisa, 20 (1966), 197-241.

[ANg2] A. Andreotti, F. Norguet, La convexité holomorphe dans l'espace analytique des cycles d'une variété algébrique, Ann. Scuola Norm. Sup. Pisa, 21 (1967), 31-82.

[BP] A. Boggess, J. Polking, Holomorphic extensions of *CR* functions, Duke Math. J., 49 (1982), 757-784.

[FK] G.B. Folland,, J. Kohn, The Neumann problem for the Cauchy Riemann complex, Princeton, 1972.

[Hö1] L. Hörmander, Pseudo-differential operators and non-elliptic boundary problems, Ann. of Math., 83 (1966), 129-209.

[Hö2] L. Hörmander, L^2 estimates and existence theorems for the $\bar{\partial}$ operator, Acta Math., 113 (1965), 89-152.

[N1] M. Nacinovich, Poincaré lemma for tangential Cauchy-Riemann complexes, Math. Ann., 268 (1984), 449-471.

[N2] M. Nacinovich, On weighted estimates for some systems of partial differential equations, Rend. Sem. Mat. Padova, 69 (1983), 221-232.

[N3] M. Nacinovich, On a theorem of Ajrapetyan and Henkin, in Seminari di Geometria 1988-1991 Univ. Bologna, pp 99-135, Bologna, (1991).

[NV] M. Nacinovich, G. Valli, Tangential Cauchy-Riemann complexes on distributions, Ann. Math. Pura Appl., 146 (1987), 123-160.

Exotic Structures on C^n and C^*-Action on C^3

SHULIM KALIMAN Department of Mathematics and Computer Science, University of Miami, Coral Gables, Florida 33124, USA

1 INTRODUCTION.

The object of this survey is to discuss recent papers on exotic structures on $\mathbf{C}^n$. A smooth complex affine algebraic variety is called an exotic algebraic (resp. analytic) structure on $\mathbf{C}^n$ if it is diffeomorphic to $\mathbf{R}^{2n}$ as a real manifold but not isomorphic (resp. biholomorphic) to $\mathbf{C}^n$. Of course, every exotic analytic structure on $\mathbf{C}^n$ is an exotic algebraic structure on $\mathbf{C}^n$, but it is unknown whether vice versa is true. It was C.P. Ramanujam (1971) who first tried to study these structures. His theorem implies that there is no exotic analytic or algebraic structure on $\mathbf{C}^2$. In the case of $\mathbf{C}^3$ or higher dimensions the situation is different. First examples of exotic analytic structures on $\mathbf{C}^n (n \geq 3)$ were constructed by Zaidenberg in 1990. In the same year Petrie and tom Dieck presented some interesting examples of exotic algebraic structures. Further development leads to a wide list of smooth contractible hypersurfaces in $\mathbf{C}^n (n \geq 4)$ (e.g., see the papers of A. Dimca (1990), S. Kaliman (1993), and P. Russell (1992)). These hypersurfaces are diffeomorphic to real Euclidean spaces by the h-cobordism theorem. It is proved that some of them are exotic algebraic structures on $\mathbf{C}^{n-1}$, but for the rest this fact is still unknown. Meanwhile the problem of recognizing exotic structures may be crucial for several conjectures in affine algebraic geometry. In particular, P. Russell (1992) found recently a remarkable connection between this problem and the problem of linearizing of a $\mathbf{C}^*$-action on $\mathbf{C}^3$. A. Dimca (1990) found some connection with the Abhyankar-Sathaye conjecture. The connection with the Zariski cancellation conjecture is obvious. We shall discuss below these problems and constructions of exotic structures and contractible hypersurfaces.

2. In this section we give a sketch of a proof of the following well-known fact (Ramanujam, 1971).

THEOREM. *Let X be a contractible affine algebraic manifold of dimension $n \geq 3$. Then X is diffeomorphic to $\mathbf{R}^{2n}$.*

We follow the argument of A. Dimca (1990). Consider X as a closed affine algebraic submanifold of $\mathbf{C}^m$. One may suppose that the distance-function has finitely many critical points on X (Milnor, 1963). Therefore, the intersection $X \cap B$ of X with a ball B of large radius is diffeomorphic to X, by the Morse theory. The Lefschetz type theorem proved by H. Hamm (1983) implies that the boundary $X \cap \partial B$ has the same fundamental group as $X \cap B$ has, i.e. it is trivial. Hence the h-cobordism theorem (Smale, 1962) implies that $X \cap B$ and, therefore, X are diffeomorphic to $\mathbf{R}^{2n}$.

COROLLARY. *Every contractible affine algebraic manifold X of dimension $n \geq 3$ is either $\mathbf{C}^n$ or an exotic structure on $\mathbf{C}^n$.*

3. In 1971 Ramanujam constructed the first example of smooth complex contractible algebraic surfaces which is not homeomorphic to $\mathbf{R}^4$. In turns out that every surface of this type (we call them Ramanujam surfaces) is affine rational and has Kodaira logarithmic dimension either 1 or 2 (see the papers of R.V. Gurjar and M. Miyanishi, 1987, R. V. Gurjar and A.R. Shastri, 1989, T. Fujita, 1982, and T. Petrie and T. tom Dieck, 1990). The direct product $X = X_1 \times \cdots \times X_l \times \mathbf{C}^m$ of Ramanujam surfaces X_i and $\mathbf{C}^m$ is already diffeomorphic to a Euclidean space, by Theorem 2, as soon as $\dim X \geq 3$. The question whether X is an exotic structure or not leads to the following cancellation theorems.

3.1. M. Zaidenberg proved (1990a, 1990b, 1991) that X was an exotic analytic structure on $\mathbf{C}^n$ when every X_i had Kodaira logarithmic dimension 2. It turns out that a stronger fact holds. If one of the factors X_i has Kodaira dimension 2, then X is an exotic analytic structure (Kaliman, to appear). The last fact is due to intrinsic measures of Eisenman (1970) (see also the paper of Kobayashi, 1976) since these analytic invariants survive the operation of taking direct products. We do not know if X is an exotic analytic structure when all factors X_i have Kodaira logarithmic dimension 1. The answer would be positive, if every Ramanujam surface of Kodaira logarithmic dimension 1 has either a nontrivial Kobayashi-Eisenman pseudovolume or a nontrivial Kobayashi-Royden pseudometric.

3.2. Nevertheless for every Ramanujam surface X_1 (regardless of its Kodaira logarithmic dimension) the product $X_1 \times \mathbf{C}^m$ ($m \geq 1$) is an exotic algebraic structure on $\mathbf{C}^{m+2}$ by the Iitaka-Fujita cancellation theorem (1977). All Ramanujam surfaces of Kodaira logarithmic dimension 1 are known and some of them can be represented as hypersurfaces in $\mathbf{C}^3$ (Petrie and tom Dieck, 1990). Here is the construction of these hypersurfaces. Let l, k be coprime natural and let $l > k > 1$. Consider the polynomial $h_{l,k}(x,y) = (x+1)^l - (y+1)^k$. Put $f_{l,k}(x,y,z) = h_{l,k}(xz, yz)/z$. Then for each pair (l,k) the fiber $V(l,k) = \{(x,y,z) \in \mathbf{C}^3 \mid f_{l,k}(x,y,z) = 1\}$ is a desired hypersurface. Note that the product $V(l,k) \times \mathbf{C}^m$ can be viewed as a contractible hypersurfaces in $\mathbf{C}^{m+3}$. It is diffeomorphic to $\mathbf{R}^{2m+4}$, by Theorem 2, and it is an exotic algebraic structure on $\mathbf{C}^{m+2}$, by the Iitaka-Fujita theorem.

4. The other known examples of contractible hypersurfaces in $\mathbf{C}^4$ admit a $\mathbf{C}^*$-action with one fixed point. Consider a complex algebraic variety X and a $\mathbf{C}^*$-action on it, $(\lambda, p) \to \lambda \cdot p$, where $\lambda \in \mathbf{C}^*$ and $p \in X$. Recall that this action is algebraic if the map $\mathbf{C}^* \times X \to X, (\lambda, p) \to \lambda \cdot p$ is a morphism of complex algebraic varieties. The simplest example of an algebraic $\mathbf{C}^*$-action on $\mathbf{C}^3$ is a linear action on $\mathbf{C}^3$ given by

$$(x, y, z) \to (\lambda^a x, \lambda^b y, \lambda^c z) \tag{4.1}$$

where (x, y, z) is a coordinate system on $\mathbf{C}^3$, $\lambda \in \mathbf{C}^*$, a, b, c are integers. We call a, b, c the weights of the linear representation.

THE LINEARIZING CONJECTURE. *Every algebraic $\mathbf{C}^*$-action on $\mathbf{C}^3$ is equivalent to a linear one up to a polynomial coordinate substitution.*

A function f on an algebraic manifold X is semi-invariant (or quasi-invariant) of weight $l \in \mathbf{Z}$ relative to a $\mathbf{C}^*$-action G if for every $\lambda \in \mathbf{C}^*$ we have $f \circ G(\lambda) = \lambda^l f$. Note that the linearizing conjecture just claims the existence of a semi-invariant coordinate system. This conjecture is true in all cases except for one, the answer to which is unknown yet (see the papers of H. Bass and W. Haboush (1985), A. Bialynicki-Birula (1973), M. Koras and P. Russell (1979), and P. Russell (1992) for details). Following Russell (1992) we call this case the "hard-case" $\mathbf{C}^*$-action. This "hard-case" $\mathbf{C}^*$-action on a threefold X can be described by the two following conditions.

(a) The $\mathbf{C}^*$-action has only one fixed point o. Therefore, one may consider the tangent representation of $\mathbf{C}^*$ on T_oX which is of form (4.1).

(b) The weights a, b, c of the tangent representation are nonzero and the sign of a is different from the signs of b and c.

5. First examples of contractible hypersurfaces which admit a "hard-case" $\mathbf{C}^*$-action were obtained by A. Dimca (1990), and can be described as follows. Let k, d be natural such that $1 \leq k < d-1$, $(k, d) = (k, d-1) = 1$. Let X be a hypersurface in $\mathbf{C}^{2n}$ defined by $f(x) = x_0^k\, x_1^{d-k} + x_1 x_2^{d-1} + \cdots + x_{2n-3}x_{2n-2}^{d-1} + x_{2n-2} + x_{2n-1}^d = 0$. Then X is a smooth contractible hypersurface. Moreover, if $k = 1$ and $n = 2$ then X is isomorphic to $\mathbf{C}^3$ (this case was also known to A. Libgober, 1977), and if $k > 1$ the nonzero fibers of the polynomial f are noncontractible. Dimca noted that if X is isomorphic to $\mathbf{C}^{2n-1}$ for $k > 1$ then this is a counterexample to the Abhyankar-Sathaye conjecture. Recall that according to this conjecture every irreducible polynomial on $\mathbf{C}^m$ whose zero fiber is isomorphic to $\mathbf{C}^{m-1}$ becomes a linear polynomial after a suitable polynomial coordinate substitution. This is true for $m = 2$ (S.S. Abhyankar and T.T. Moh, 1975, and M. Suzuki ,1974). Moreover if X is isomorphic to $\mathbf{C}^3$ for $n = 2$ and $k > 1$ we have a counterexample to the linearizing conjecture. It suffices to consider the simplest nontrivial Dimca's example in order to check this. Then $X = \{(x_1, x_2, x_3, x_4) \in \mathbf{C}^n \mid x_1 + x_1^4 x_2 + x_2^2 x_3^3 + x_4^5 = 0\}$. The hypersurface X admits the $\mathbf{C}^*$-action that is the restriction of the linear action $(x_1, x_2, x_3, x_4) \to (\lambda^{15}x_1, \lambda^{-45}x_2, \lambda^{35}x_3, \lambda^3 x_4)$ on $\mathbf{C}^4$. The only fixed point of this action is the origin o. Since (x_2, x_3, x_4) is a semi-invariant local coordinate system in a neighborhood of o, one can see that we deal with a "hard-case" $\mathbf{C}^*$-action. Consider a semi-invariant regular function p of the greatest negative weight l on X whose zero fiber is smooth irreducible and contains o. It is a simple exercise to check that $l = -45$ and p coincides with x_2 up to a constant factor. Therefore, if X is $\mathbf{C}^3$ and the action is linearizable x_2 can be viewed as a semi-invariant coordinate. But it can be shown that the Euler characteristic of nonzero fiber of x_2 on X is different from 1, which leads to a contradiction.

6. The following construction of P. Russell (1992) generalizes Dimca's examples. Let a', b', c' be pairwise prime natural such that $b' \leq c'$, and let ω be the group of a'-roots of unity. Consider an ω-action on $\mathbf{C}^2$ given by $(u,v) \to (\lambda^{b'}u, \lambda^{c'}v)$, where $\lambda \in \omega$, and (u,v) is a coordinate system. Consider a polynomial f which is semi-invariant relative to ω and is satisfying two assumptions.

ASSUMPTION (a). The fiber $L = \{f = 0\}$ is isomorphic to a line and L meets the axis $u = 0$ normally at the origin and $r-1$ other points ($r \geq 2$). Hence, without loss of generality one may suppose that $f(u,v) = v+$ high order terms. In particular the weight of f is c'. (Here we consider c' as an element of $\mathbf{Z}_{a'}$).

ASSUMPTION (b). Consider the Laurent polynomial $\tilde{F}(s, s^{-1}, u, v) = s^{-c'} f(s^{b'}u, s^{c'}v)$. The first assumption on f implies that $\tilde{F} = F(w, w^{-1}, u, v)$ with $w = s^{a'}$. Assume that the function F does not depend on w^{-1}, i.e. F is a polynomial $F(w,u,v)$.

Note that F is semi-invariant of weight c' under the $\mathbf{C}^*$-action $(w,u,v) \to (\lambda^{-a'}w, \lambda^{b'}u, \lambda^{c'}v)$. Choose pairwise prime natural $\alpha_1, \alpha_2, \alpha_3$ such that $(\alpha_1, a') = (\alpha_2, b') = (\alpha_3, c') = 1$.

6.1. THEOREM (Russell, 1992). *The hypersurface* $X = \{(x,y,z,t) \in \mathbf{C}^4 \mid t^{\alpha_3} = F(y^{\alpha_1}, z^{\alpha_2}, x)\}$ *is smooth contractible.*

Put $a = a'\alpha_2\alpha_3$, $b = b'\alpha_1\alpha_3$, $c = c'\alpha_1\alpha_2$. Then we have the $\mathbf{C}^*$-action on X given by $(x,y,z,t) \to (\lambda^{c\alpha_3}x, \lambda^{-a}y, \lambda^{b}z, \lambda^{c}t)$. The origin is the only fixed point. Since (y,z,t) is a semi-invariant local coordinate system in a neighborhood of the origin, it is again a "hard-case" $\mathbf{C}^*$-action.

EXAMPLE: $a' = b' = c' = 1$, $f(u,v) = v + v^2 + u$. Then $F(w,u,v) = v + yv^2 + u$ and $X = \{(x,y,z,t) \in \mathbf{C}^4 \mid t^{\alpha_3} = x + x^2y^{\alpha_1} + z^{\alpha_2}\}$.

Again as in section 5 one can check that if one of these hypersurfaces is isomorphic to $\mathbf{C}^3$ then we have a counterexample to both the Abhyankar-Sathaye conjecture and the linearizing conjecture.

6.2. We say that the weak linearizing conjecture holds if every "hard-case" $\mathbf{C}^*$-action on $\mathbf{C}^3$ is linearizable under the following additional condition: the weights of the tangent representation at the fixed point are pairwise prime. In many case the "weak" linearizing conjecture was investigated by P. Russell and M. Koras (1989).

THEOREM (Russell, 1992). *If all hypersurfaces in Russell's construction are exotic algebraic structures, then the linearizing conjecture can be reduced to the "weak" linearizing conjecture.*

7. All Dimca's and Russell's hypersurfaces can be described by the following theorem.

THEOREM (Kaliman, 1993). *Let G be a regular $\mathbf{C}^*$-action on $\mathbf{C}^{n-1} (n \geq 2)$ and let $q(x')$ be a polynomial on $\mathbf{C}^{n-1}$ which is semi-invariant of positive weight l under this action (where $x' = (x_1, \ldots, x_{n-1})$ is a coordinate system on $\mathbf{C}^{n-1}$). Suppose that the fiber $F_0 = \{q(x') = 0\}$ is smooth non-multiple, and $\pi_1(\mathbf{C}^{n-1} - F_0) = \mathbf{Z}$. Suppose that for some natural k the manifold F_0 is $\mathbf{Z}_k$-acyclic and $(k, l) = 1$. Then the hypersurface $X = \{q(x') + x_n^k = 0\}$ in $\mathbf{C}^n$ is smooth contractible. Moreover if the nonzero fibers of the polynomial q are noncontractible, then the nonzero fibers of the polynomial $x_n^k + q(x')$ are noncontractible.*

8. It is unknown yet whether all hypersurfaces constructed in 6.1 are exotic algebraic structures on $\mathbf{C}^3$ or not. In this and the next sections we present all known cases in which they are exotic algebraic structures. Suppose that the notation from section 6.1 holds and let $k(M)$ denote the Kodaira logarithmic dimension of an algebraic manifold M. Recall that $k(M)$ is invariant under isomorphisms and $k(\mathbf{C}^n) = -\infty$ (see the paper of Iitaka, 1977).

THEOREM. *Let X be a hypersurface in $\mathbf{C}^4$ described in Theorem 6.1.*

(a) If $a \leq b + c$ and F is the fiber $y = 1$ of the function y on X, then $k(X) \geq k(F)$;

(b) (Russell, 1992) If $\alpha_2 >> 0$, $\alpha_3 >> 0$, then $k(F) = 2$. If $\alpha_1 \geq \alpha_2\alpha_3$ then $k(X) \geq k(F)$.

EXAMPLES. It follows from (a) that the hypersurface $x + x^4y^{46} + y^{92}z^3 + t^5 = 0$ has Kodaira logarithmic dimension 2 and thus it is an exotic algebraic structure on $\mathbf{C}^3$. There exist hypersurfaces from section 6.1 whose Kodaira logarithmic dimension are $-\infty$. One of them is $x + x^2y + z^3 + t^2 = 0$, and we do not know if this hypersurface is isomorphic to $\mathbf{C}^3$ or not.

In some other cases this dimension is not found yet. For instance, we do not know the Kodaira logarithmic dimension of Dimca's hypersurfaces in $\mathbf{C}^4$. Nevertheless none of them is isomorphic to $\mathbf{C}^3$ as follows from

9. THEOREM (Kaliman and Makar-Limanov, to appear). *Let X be a hypersurface described in 6.1, and let $\alpha_1 \geq 2, \alpha_2 > 3, \alpha_3 > 3$. Then X is not isomorphic to $\mathbf{C}^3$.*

10. The next procedure is essential for the rest of the paper. Let M be a smooth algebraic variety of dimension $n \geq 2$, let E be a smooth hypersurface in M, and let C be a smooth closed subvariety in E such that $\mathrm{codim}_M C \geq 2$. Consider the blow-up $\tilde{\tau} : \tilde{M} \to M$ with locus at C and the proper transform $\tilde{E}$ of E. Put $N = \tilde{M} - \tilde{E}$ and $\tau = \tilde{\tau}\mid_N$. Then we say that $\tau : N \to M$ is a "half locus attachment" over the divisor E with locus C. When $\mathit{dim} M = 2$, E is isomorphic to $\mathbf{C}$, and C is a point, then this procedure is similar to a half point attachment described by Fujita (1982).

THEOREM (Kaliman, to appear). *Let M be a smooth affine contractible algebraic variety of dimension ≥ 2 and let E be a smooth contractible hypersurface in M. Suppose that C is a smooth closed proper contractible subvariety in E and $\tau : N \to M$ is a half locus attachment over E with locus C. Then N is affine contractible.*

Now we can produce an example of a family of exotic analytic structures on $\mathbf{C}^3$ with given number of moduli. Consider $M = \mathbf{C} \times X$ where X is a Ramanujam surface of Kodaira dimension 2. Let $a_1, \ldots, a_n$ be different points in $\mathbf{C}$ and let $E_k = a_k \times X$. Choose points

$x_1, ..., x_n$ in X and put the point $C_k = (a_k, x_k) \in E_k$. Make a half locus attachment over E_k with locus C_k for each $k = 1, ..., n$. Then we obtain a new threefold $N(\overline{a}, \overline{x})$ which depends on parameters $(\overline{a}, \overline{x}) = (a_1, ..., a_n, x_1, ..., x_n) \in \mathbf{C}^n \times X^n$. The theorem above implies that $N(\overline{a}, \overline{x})$ is affine contractible, i.e. it is diffeomorphic to $\mathbf{R}^6$, by Theorem 2. It can be proved (Kaliman, to appear) that for general points $(\overline{a}, \overline{x})$, $(\overline{a}', \overline{x}') \in \mathbf{C}^n \times X^n$ the threefolds $N(\overline{a}, \overline{x})$ and $N(\overline{a}', \overline{x}')$ are not biholomorphic. Let $\overline{\tau}: N(\overline{a}, \overline{x}) \to M$ be the natural projection and let $H_k = \overline{\tau}^{-1}(C_k)$. It can be also shown that, except for $H_1, \ldots, H_n$, there is no closed surfaces in $N(\overline{a}, \overline{x})$ which are biholomorphic to $\mathbf{C}^2$.

11. In this section we describe one more construction which is essential for the proof of Theorem 8(a). Suppose that X is a contractible affine algebraic threefold with a "hard-case" $\mathbf{C}^*$-action G. Let $-a, b, c$ be the weights of G at the fixed point o (a, b, c are natural and $b \leq c$). Recall some facts (e.g., see the paper of Russell and Koras, 1989). There exists a $\mathbf{C}^*$-invariant neighborhood U of o and analytic coordinates (w, u, v) on U such that $G(\lambda)(w, u, v) = (\lambda^{-a}w, \lambda^b u, \lambda^c v)$. Define $X^+ = \{x \in X \mid \lim_{\lambda \to 0} G(\lambda)(x) = o\}$ and $X^- = \{x \in X \mid \lim_{\lambda \to \infty} G(\lambda)(x) = o\}$. Then $X^+ \simeq \mathbf{C}^2$ and $X^- \simeq \mathbf{C}$. Moreover X^+ coincides with zeros of a semi-invariant regular function f on X of weight $-a$ and the restriction of this function to U can be viewed as the coordinate w. The line X^- is also contained in U and it is given by $v = u = 0$.

Put $F = \{f = 1\}$, $F' = U \cap F$, $q = F' \cap X^-$. Consider the direct products $Y_0 = \mathbf{C} \times F$ and $U_0 = \mathbf{C} \times F' \subset Y_0$. Note that U_0 admits a coordinate system (w_0, u_0, v_0), where w_0 is a coordinate on $\mathbf{C}$ and (u_0, v_0) coincides with the coordinate system (u, v) on F'. Let a $\mathbf{C}^*$-action on $\mathbf{C}$ be given by $w_0 \to \lambda^{-1}w_0$. It generates a $\mathbf{C}^*$-action G_0 on Y_0, and the restriction of G_0 to U_0 is given by $(w_0, u_0, v_0) \to (\lambda^{-1}w_0, u_0, v_0)$. Put $E_0 = 0 \times F$, $Y_0^- = \mathbf{C} \times q$, $q_0 = 0 \times q$. All these sets and U_0 are invariant under the action of G_0. Let $\tau_1 : Y_1 \to Y_0$ be a half locus attachment over E_0 with locus q_0, let the proper transform of Y_0^- be Y_1^-, let $\tau_1^{-1}(U_0) = U_1$, $\tau_1^{-1}(q_0) = E_1$, and let $q_1 = E_1 \cap Y_1^-$. Note that $E_1 \simeq \mathbf{C}^2$ and U_1 admits a coordinate system (w_1, u_1, v_1) such that $w_1 = w_0 \circ \tau_1$, $u_1 = \dfrac{u_0}{w_0} \circ \tau_1$, $v_1 = \dfrac{v_0}{w_0} \circ \tau_1$. In particular, q_1 is the origin of this system and $E_1 = \{w_1 = 0\}$. The action G_0 on Y_0 generates a $\mathbf{C}^*$-action G_1 on Y_1 whose restriction to U_1 has form $(w_1, u_1, v_1) \to (\lambda^{-1}w_1, \lambda u_1, \lambda v_1)$. Recall that $b \leq c$. Consider a half locus attachment $\tau_2 : Y_2 \to Y_1$ over E_1 with locus at q_1 and keep doing this procedure till we obtain $\tau_b : Y_b \to Y_{b-1}$. Put $\tau = \tau_b \circ \cdots \tau_1 : Y_b \to Y_0$, $\tau^{-1}(U_0) = U_b$, $\tau^{-1}(q_0) = E_b$, and let G_b be the $\mathbf{C}^*$-action on Y_b generated by G_0. One can see that U_b admits a coordinate system (w_b, u_b, v_b) such that $w_b = w_0 \circ \tau$, $u_b = \dfrac{u_0}{w_0^b} \circ \tau$, $v_b = \dfrac{v_0}{w_0^b} \circ \tau$, $E_b = \{w_b = 0\}$ and the restriction G_b to U_b has form $(w_b, u_b, v_b) \to (\lambda^{-1}w_b, \lambda^b u_b, \lambda^b v_b)$. Again $E_b \simeq \mathbf{C}^2$. Consider the line $C_b = \{v_b = 0\}$ in E_b. Make the half locus attachment $\rho_1 : Y_{b+1} \to Y_b$ over E_b with locus C_b and keep doing similar half locus attachments with loci along suitable lines till we get $\rho_{c-b} : Y_c \to Y_{c-1}$. Put $\rho = \rho_{c-b} \circ \cdots \circ \rho_1 \circ \tau : Y_c \to Y_0$ and $U_c = \rho^{-1}(U_0)$. The action G_0 generates the action G_c on Y_c such that U_c is $\mathbf{C}^*$-invariant. There is a coordinate system (w_c, u_c, v_c) on U_c such that $w_c = w_0 \circ \rho$, $u_c = \dfrac{u_0}{w_0^b} \circ \rho$, $v_c = \dfrac{v_0}{w_0^c} \circ \rho$. In these coordinates the action G_c has form $(w_c, u_c, v_c) \to (\lambda^{-1}w_c, \lambda^b u_c, \lambda^c v_c)$.

Now consider again the $\mathbf{C}^*$-action G on X. Put $\nu = \exp \frac{2\pi i}{a}$. Then the restriction of $G(\nu)$ to the fiber $F = \{f = 1\}$ is an isomorphism φ. The restriction of φ to U' has form $(u, v) \to (\nu^b u, \nu^c v)$. Consider a $\mathbf{Z}_a$-action g on $Y_0 = \mathbf{C} \times F$ given by $(w_0, p) \to (\nu^m w_0, \varphi^m(p))$

where $m \in \mathbf{Z}_a$, $w_0 \in \mathbf{C}$, $p \in F$. It generates a $\mathbf{Z}_a$-action g_c on Y_c whose restriction to U_c can be written as $(w_c, u_c, v_c) \to (\nu^m w_c, u_c, v_c)$.

THEOREM. *The quotient space Y_c/g_c is isomorphic to X. Moreover, under this isomorphism the $\mathbf{C}^*$-action G_c generates a $\mathbf{C}^*$-action on Y_c/g_c which coincides with G.*

PROOF. Consider $\tilde{X} = \{(\tilde{w}, x) \mid \tilde{w} \in \mathbf{C},\ x \in X,\ \tilde{w}^a = f(x)\}$. Let $\pi : \tilde{X} \to X$ be the natural projection, let $\tilde{U} = \pi^{-1}(U)$, and let $\tilde{G}$ be the $\mathbf{C}^*$-action on $\tilde{X}$ generated by G. Note $\tilde{U}$ admits a coordinate system $(\tilde{w}, \tilde{u}, \tilde{v})$ such that $\tilde{u} = u \circ \pi$, $\tilde{v} = v \circ \pi$ and $\tilde{G}$ has form $(\tilde{w}, \tilde{u}, \tilde{v}) \to (\lambda^{-1}\tilde{w}, \lambda^b \tilde{u}, \lambda^c \tilde{v})$. There is also a $\mathbf{Z}_a$-action $\tilde{g}$ on $\tilde{X}$, for which its restriction to $\tilde{U}$ is given by $(\tilde{w}, \tilde{u}, \tilde{v}) \to (\nu^m \tilde{w}, \tilde{u}, \tilde{v})$. Clearly, $\tilde{X}/\tilde{g} = X$. Thus it suffices to find an isomorphism $\Phi : \tilde{X} \to Y_c$ mapping the actions $\tilde{G}$ and $\tilde{g}$ to G_c and g_c respectively. Consider an isomorphism $\Phi' : \tilde{U} \to U_c$ given by $w_c = \tilde{w}$, $u_c = \tilde{u}$, $v_c = \tilde{v}$. Since both $\tilde{X} - \{\tilde{w} = 0\}$ and $Y_c - \{w_c = 0\}$ are isomorphic to the direct product $\mathbf{C}^* \times F$, there is also the natural isomorphism $\Phi'' : \tilde{X} - \{\tilde{w} = 0\} \to Y_c - \{w_c = 0\}$. It is easy to see that the restriction Φ'' to $\tilde{U} - \{\tilde{w} = 0\}$ coincides with Φ'. Therefore we obtain an isomorphism $\Phi : \tilde{X} \to Y_c$ with the desired properties.

12. We shall sketch the proof of Theorem 8(a). Denote by $K_{\overline{M}}$ a canonical divisor on a compact complex manifold $\overline{M}$, by $k(D, \overline{M})$ the Kodaira dimension of a divisor D on $\overline{M}$, by $H^0(D, \overline{M})$ the vector space of holomorphic sections of the linear bundle $[D]$ associated with D (see the paper of S. Iitaka, 1977, and the book of P. Griffiths and J. Harris (1978) for definitions). Let $P_m(D, \overline{M}) = dim H^0(mD, \overline{M})$. Recall that $k(D, \overline{M}) = \overline{\lim}_{m\to\infty} \log P_m(D)/\log m$. If M is a complex algebraic manifold, $\overline{M}$ is its completion so that $D = \overline{M} - M$ is of normal simple crossing type, then the Kodaira logarithmic dimension of M is defined by $k(M) = k(K_{\overline{M}} + D, \overline{M})$. Iitaka shows (1977) that this definition does not depend on the choice of a completion $\overline{M}$.

Let the assumption of Theorem 8(a) hold. Suppose that $k(F) \geq 0$, otherwise there is nothing to prove. Put as in section 11 $Y_0 = \mathbf{C} \times F$, $E_0 = 0 \times F$. Let $\overline{Y}_0$ be a smooth completion of $Y_0 - E_0$ for which the divisor $D = \overline{Y}_0 - (Y_0 - E_0)$ is of normal simple crossing type, and D contains the closure $\overline{E}_0$ of E_0 as a component, i.e. $D = \overline{E}_0 + D_0$ where $D_0 = \overline{Y}_0 - Y_0$. Since $Y_0 - E_0 \simeq \mathbf{C}^* \times F$, we have $k(Y_0 - E_0) = k(F)$. Moreover it is easy to see that $P_m(K_{\overline{Y}_0} + D_0 + \overline{E}_0, \overline{Y}_0) = P_m(L(\overline{F}), \overline{F})$ where $\overline{F}$ is a completion of F with the divisor $\overline{F} - F$ of simple normal crossing type and $L(\overline{F}) = K_{\overline{F}} + (\overline{F} - F)$.

Let $q_0 \in Y_0$, E_k, Y_k $(k = 1, ..., c)$ have the same meaning as in section 10. Then a completion $\overline{Y}_c$ of Y_c may be treated as a blow-up $\overline{\rho} : \overline{Y}_c \to \overline{Y}_0$ at q_0 and infinitely near points and lines. Recall that $E_1, ..., E_b$ are the exceptional divisors of the blow-ups $\tau_1, ..., \tau_b$ with loci of codimension 3, and $E_{b+1}, ..., E_c$ are the exceptional divisors of the blow-ups $\rho_1, ..., \rho_{b-c}$ with loci of codimension 2. Put $D_c = \overline{Y}_c - Y_c$. By construction, D_c contains the closures $\overline{E}_0', \ldots, \overline{E}_{c-1}'$ of the proper transforms $E_0', ..., E_{c-1}'$ in $\overline{Y}_c$ of the divisors $E_0, ..., E_{c-1}$, $\overline{\rho}^*(\overline{E}_0) = \overline{E}_0' + \cdots + \overline{E}_{c-1}' + \overline{E}_c$, and $D_c = \overline{\rho}^*(\overline{E}_0 + D_0) - \overline{E}_c$ (where $\overline{E}_c$ is the closure of E_c in $\overline{Y}_c$). Recall also that if $\sigma : \tilde{M} \to \overline{M}$ is a blow-up with locus C, then $K_{\tilde{M}} = \sigma^*(K_{\overline{M}}) + (l-1)E$, where E is the exceptional divisor and $l = codim_{\overline{M}} C$. Hence $P_m(L(\overline{F}), \overline{F}) = P_m(K_{\overline{Y}_0} + D_0 + \overline{E}_0, \overline{Y}_0) = P_m(\overline{\rho}^*(K_{\overline{Y}_0} + D_0 + \overline{E}_0), \overline{Y}_c) = P_m(\overline{\rho}^*(K_{\overline{Y}_0}) + D_c + \overline{E}_c, \overline{Y}_c) = P_m(K_{\overline{Y}_c} + D_c - 2\overline{E}_1{}' - \ldots - 2b\overline{E}_b{}' - (2b+1)\overline{E}_{b+1}{}' - \ldots - (b+c-1)\overline{E}_c{}' - (b+c-1)\overline{E}_c, \overline{Y}_c) \leq P_m(K_{\overline{Y}_c} + D_c - (b+c-1)\overline{E}_c, \overline{Y}_c)$. Note that elements of $H^0(m(K_{\overline{Y}_c} + D_c - (b+c-1)\overline{E}_c), \overline{Y}_c)$ can be treated as holomorphic sections of $[m(K_{\overline{Y}_c} + D_c)]$ that have zeros of order at least $m(b+c-1)$ along E_c. Let $\hat{Y}$ be another compactification of Y_c such that the divisor $\hat{D} = \hat{Y} - Y_c$ is of simple normal crossing type. The identical mapping of Y_c generates a strictly rational map $\psi : \hat{Y} \to \overline{Y}_c$. By Iitaka's result (1977), ψ generates an isomorphism $\psi^* : H^0(m(K_{\overline{Y}_c} + D_c), \overline{Y}_c) \to H^0(m(K_{\hat{Y}} + \hat{D}), \hat{Y})$, i.e. it transforms holomorphic sections of $[m(K_{\overline{Y}_c} + D_c)]$ into holomorphic sections of $[m(K_{\hat{Y}} + \hat{D})]$. Since this isomorphism preserves the multiplicity of zeros of holomorphic sections along E_c we have $P_m(L(\overline{F}), \overline{F}) \leq P_m(K_{\hat{Y}} + \hat{D} - (b+c-1)\hat{E}_c, \hat{Y})$ (where $\hat{E}_c$ is the closure of E_c in $\hat{Y}$), and thus $k(K_{\hat{Y}} + \hat{D} - (b+c-1)\hat{E}_c, \hat{Y}) \geq k(F)$. By Theorem 11, Y_c can be viewed as a ramified covering h of X with ramification divisor E_c and the order of ramification a. Let $\overline{X}$ be completion of X, for which $B = \overline{X} - X$ and $B \cup h(E_c)$ are of normal simple crossing type. Then, by Hironaka's theorem, one may suppose that h generates a regular morphism $\hat{h} : \hat{Y} \to \overline{X}$. Then $K_{\hat{Y}} + \hat{D} \sim \hat{h}^*(K_{\overline{X}} + B) + R_h$, where R_h is an effective divisor which is called the logarithmic ramification divisor of h (see the paper of Iitaka, 1977). Let h_0 be the restriction of h to $Y_c - E_c$. Then h_0 is étale over $X - h(E_c)$ and, therefore, $K_{\hat{Y}} + \hat{D} + \hat{E}_c = \hat{h}^*(K_{\overline{X}} + B + \hat{h}(\hat{E}_c)) = \hat{h}^*(K_{\overline{X}} + B) + \hat{h}^*(\hat{h}(\hat{E}_c))$. Hence $R_h = \hat{h}^*(\hat{h}(\hat{E}_c)) - \hat{E}_c = (a-1)\hat{E}_c + Q$ where Q is exceptional for $\hat{h}$, i.e. $codim h(Q) \geq 2$. Lemma 1 from Iitaka's paper (1977) implies $k(X) = k(K_{\overline{X}} + B, \overline{X}) = k(h^*(K_{\overline{X}} + B) + Q, \hat{Y}) = k(K_{\hat{Y}} + \hat{D} - R_h + Q, \hat{Y}) = k(K_{\hat{Y}} + \hat{D} - (a-1)\hat{E}_c, \hat{Y})$. It remains to note that, since $a \leq b+c$, $k(X) = k(K_{\hat{Y}} + \hat{D} - (a-1)\hat{E}_c, \hat{Y}) \geq k(K_{\hat{Y}} + \hat{D} - (b+c-1)\hat{E}_c, \hat{Y}) \geq k(F)$, which is the desired conclusion.

REFERENCES

S.S. Abhyankar, T.T. Moh (1975). *Embedding of the line in the plane*, J. Reine Angew. Math., **276**.

H. Bass, W. Haboush (1985). *Linearizing certain reductive group actions*, Trans. AMS, **292**.

A. Bialynicki-Birula (1973). *Some theorems on action of algebraic groups*, Ann. Math. **98**.

A. Dimca (1990). *Hypersurfaces in* $\mathbf{C}^{2n}$ *diffeomorphic to* $\mathbf{R}^{4n-2} (n \geq 2)$, Max-Plank Institute, preprint.

D.A. Eisenman (1970). *Intrinsic measures on complex manifolds and holomorphic mappings*, Mem. AMS, No. 96, AMS, Providence, R.I.

P. Griffiths, J. Harris (1978). *Principles of algebraic geometry*, John Wiley, New York, p. 813.

R.V. Gurjar, M. Miyanishi (1987) *Affine surfaces with* $\bar{k} \leq 1$, Algebraic geometry and commutative algebra in honor of Masayoshi Nagata, pp. 99–124.

R. V. Gurjar, A.R. Shastri (1989). *On rationality of complex homology 2-cells: I and II*, J. Math. Soc. Japan, **41**.

H. Flenner, M. Zaidenberg (1992). $\mathbf{Q}$*-acyclic surfaces and their deformations* , Max-Plank Institute, preprint.

T. Fujita (1982). *On the topology of non-complete algebraic surfaces*, J. Fac. Sci. Univ. Tokyo (ser. 1A) **29**.

H. Hamm (1983). *Lefschetz theorems for singular varieties*, Proc. Symp. Pure Math. **40** , part 1, AMS, pp. 547–557.

S. Iitaka (1977). *On logarithmic Kodaira dimension of algebraic varieties*, Complex Analysis and Algebraic Geometry, Tokyo, Iwanami, pp. 175–189.

S. Iitaka, T. Fujita (1977). *Cancellation theorem for algebraic varieties*, J. Fac. Sci. Univ., Tokyo (Sec. 1A), **24**.
S. Kaliman (1993). *Smooth contractible hypersurfaces in* $\mathbf{C}^n$ *and exotic algebraic structures on* $\mathbf{C}^3$, Math. Zeitschrift, **214**.
S. Kaliman (to appear). *Exotic analytic structures and Eisenman intrinsic measures*, Israel Math. J..
S. Kaliman, L. Makar-Limanov (to appear). *On some family of contractible hypersurfaces in* $\mathbf{C}^4$, Seminair d'algèbre. Journées Singulières et Jacobiénnes 26–28 mai 1993, Prépublication de l'Institut Fourier, Grenoble.
S. Kobayashi (1976). *Intrinsic distances, measures and geometric function theory*, Bull. AMS, **82** .
M. Koras, P. Russell (1979) *On linearizing "good"* $\mathbf{C}^*$*-action on* $\mathbf{C}^3$, Canadian Math. Society Conference Proceedings, **10**.
A. Libgober (1977). *A geometrical procedure for killing the middle dimensional homology groups of algebraic hypersurfaces*, Proc. AMS, **63**.
J. Milnor (1963). *Singular Points of Complex hypersurfaces*, Princeton University Press, Princeton, p.122
T. Petrie, T. tom Dieck (1990). *Contractible affine surfaces of Kodaira dimension one*, Japan. J. Math., **16**.
C.P. Ramanujam (1971). *A topological characterization of the affine plane as an algebraic variety*, Ann. Math. **94**.
P. Russell (1992). *On a class of* $\mathbf{C}^3$*-like threefolds*, Preliminary Report.
S. Smale (1962). *On the structure of manifolds*, Amer. J. Math., **84**.
M. Suzuki (1974). *Propiétes topologiques des polynômes de deux variables complexes, et automorphismes algébrigue de l'espace* $\mathbf{C}^2$, J. Math. Soc. Jpn., **26**.
M. Zaidenberg (1990a). *Affine acyclic curves and surfaces. Exotic algebraic structures on* $\mathbf{C}^n$, **29**, Mathematisches Arbeitstagung, Max–Planck–Institut fur Mathematik, Bonn, preprint MPI/90-52.
M. Zaidenberg (1990b) *Ramanujam surfaces and exotic algebraic structures on* $\mathbf{C}^n$, Dokl. AN SSSR, **314**, Enlglish transl. in Soviet Math. Doklady, **42**(1991), no. 2.
M. Zaidenberg (1991). *An analytic cancellation theorem and exotic algebraic structures on* $\mathbf{C}^n$, $n \geq 3$, preprint, Max–Planck–Institut fur Mathematik, Bonn, preprint MPI/91-26.

Anti-involutions, Symmetric Complex Manifolds, and Quantum Spaces

JULIAN ŁAWRYNOWICZ Institute of Mathematics, Polish Academy of Sciences and the University of Łódź, Łódź, Poland

ENRIQUE RAMIREZ DE ARELLANO Centro de Investigación y de Estudios Avanzados, México, D.F.

Summary. A duality between one-dimensional symmetric complex manifolds and phase spaces related to the quantum plane is pointed out. In both cases the division of an associative algebra by an anti-involution is involved. In analogy to the only classical Klein projective line $\mathbb{C}P^1$ we find the only differential calculus, related to the anti-involution $x^* = p$, $p^* = x$, such that the q-anticommutator of dx, dp vanishes. In analogy to the non-classical Klein projective line $\Delta = \{z \in \mathbb{C} : |z| \leq 1\}$ we find the only differential calculus, related to the same anti-involution, such that the q/r-anticommutator of dx, dp vanishes, where r has to be chosen so that the r-commutator of p, dp vanishes. The remaining non-classical Klein projective line $\mathbb{R}P^2$ has its counterpart in the only still possible differential calculus related to the same anti-involution: it satisfies the condition that the q-commutator of dx, dp vanishes.

Introduction

A *symmetric* complex manifold is a complex manifold $\mathbb{M}$ together with an *anti-involution* c on $\mathbb{M}$, i.e. an antiholomorphic mapping c on $\mathbb{M}$ such that $c^2 = \mathrm{id}$. A *morphism* of symmetric complex manifolds from $(\mathbb{M}, c)$ to $(\mathbb{M}', c')$ is a holomorphic mapping $\phi : \mathbb{M} \to \mathbb{M}'$ such that $c' \circ \phi = \phi \circ c$. The symmetric complex manifolds and their morphisms form a category. If $\mathcal{O}_{\mathbb{M}}$ is the structural sheaf of the symmetric complex manifold $\mathbb{M}$, the set $\Gamma(\mathbb{M}, \mathcal{O}_{\mathbb{M}})$ of the global sections f of $\mathcal{O}_{\mathbb{M}}$ is a commutative $\mathbb{C}$-algebra equipped with an involution $f \mapsto \overline{f \circ c}$.

Jurchescu and Succi [4] called symmetric complex manifolds *involutive* manifolds and had shown the equivalence of their category to a category of $\mathbb{R}$-ringed spaces which is "local" in the precise sense that an $\mathbb{R}$-ringed space is in this category if and only if it locally is. The objects of the new category are called *Klein manifolds* and they are obtained, loosely speaking, as quotients of involutive manifolds with respect to their anti-involutions. The underlying $\mathbb{R}$-ringed spaces of complex manifolds, the bordered Riemann surfaces, and the Klein surfaces of Alling and Greenleaf [1] are specific categories of Klein manifolds.

For every complex manifold $\mathbb{M}$ of dimension $n \geq 1$, the underlying $\mathbb{R}$-ringed space is a Klein manifold: the *underlying* Klein manifold of $\mathbb{M}$. Every Klein manifold which is underlying to a connected complex manifold is called *classical*. All other Klein manifolds are called *non-classical*. Thereafter we confine ourselves to the case $n = 1$. In particular, taking $\mathbb{M} = \mathbb{C}$ provided with the canonical anti-involution c (the usual conjugation map), we may consider a sheaf $\mathcal{F}$ of non-commutative rings over $\mathbb{C}$, such that the set of its global sections $\Gamma(\mathbb{C}, \mathcal{F})$ be an associative algebra with unit element generated over $\mathbb{C}$ by two indeterminates x and p related by the "regulated non-commutativity" rule $xp = qpx$ for a non-zero complex number q. This means that $\Gamma(\mathbb{C}, \mathcal{F})$ is the quotient of an associative algebra with unit element, freely generated over $\mathbb{C}$ by x and p, with respect to the two-sided ideal $I[x,p]_q$ spanned in $\mathcal{A}$ by monomials containing the expressions $([x,p]_q)^k$, $k > 0$, where $[x,p]_q$ denotes the q-commutator of x, p :

$$[x,p]_q := xp - qpx, \qquad [x,p]_1 \equiv [x,p],$$

and q is a fixed complex number. The object $\mathbb{M}_q^2 := \mathcal{A}/I[x,p]_q$ is the *quantum plane* of Manin [8]. A non-commutative *phase space* is obtained by introducing in $\mathbb{M}_q^2$ the antilinear anti-involution c^* by the relations

$$c^*(x) = p, \quad c^*(p) = x; \tag{1}$$

in the case when x and p are complex numbers and $q \in \mathbb{R}$ we may consider the antilinear involution

$$\overline{c}(x) = \overline{x}, \qquad \overline{c}(p) = \overline{p} \quad \text{(complex conjugation)}, \quad q \in \mathbb{R}. \tag{2}$$

The basic reordering rule reads $[x,p]_q = 0$ or, equivalently, for $q \neq 0$,

(3)
$$\mathbf{x}\otimes\mathbf{x} = B\mathbf{x}\otimes\mathbf{x}, \quad \text{where } B = \begin{bmatrix} 1 & 0 & 0 & 0 \\ 0 & 1-1/r & q/r & 0 \\ 0 & 1/q & 0 & 0 \\ 0 & 0 & 0 & 1 \end{bmatrix}, \quad \mathbf{x} = \begin{bmatrix} x \\ p \end{bmatrix}, \quad \mathbf{x}\otimes\mathbf{x} = \begin{bmatrix} x^2 \\ xp \\ px \\ p^2 \end{bmatrix},$$
$r \in \mathbb{C}\setminus\{0\}$.

If $q \notin \mathbb{R}$, the hermicity in $[x,p]_q = 0$ requires that $|q| = 1$. The q-anticommutator of x, p is defined as

$$\{x,p\}_q := xp + qpx, \qquad \{x,p\}_1 \equiv \{x,p\},$$

while the first-order differential forms dx, dp have to satisfy the conditions

(4)
$$(B\otimes I_2)C_{23}C_{12} = C_{23}C_{12}(I_2\otimes B),$$
$$\mathbf{x}\otimes d\mathbf{x} = c d\mathbf{x}\otimes\mathbf{x}, \quad \text{where} \quad d\mathbf{x} = \begin{bmatrix} dx \\ dp \end{bmatrix}, \quad C_{12}C_{23}C_{12} = C_{23}C_{12}C_{21}, \qquad (dx)^2 = 0,$$
$$C_{12} = C\otimes I_2, \quad C_{23} = I_2\otimes C, \quad (dp)^2 = 0.$$
$$C = [C_{jk}]_{j,k=1,\ldots,4}, \quad (B\otimes I_2 - I_2)(C_{12}+I_2) = 0.$$

We are going to consider, subsequently, three cases, corresponding to the only Klein projective lines $\mathbb{C}P^1$, $\Delta = \{z \in \mathbb{C} : |z| \leq 1\}$, and $\mathbb{R}P^2$ [2, 5, 10]. The first-named author is greatly indebted to Professor Francesco Succi for fruitful discussions. They plan to investigate the correspondence or duality mentioned above in an explicit form in a subsequent, joint paper [7].

1. THE CASE WHEN THE q-ANTICOMUTATOR OF dx, dp VANISHES

Consider first the case when

(5)
$$\{dx, dp\}_q = 0.$$

Then we have

THEOREM 1. *Suppose* (1), (3) *and* (4), *and consider the case when the relation* (5) *holds. Then*

$$C = \begin{bmatrix} s & 0 & 0 & 0 \\ 0 & 0 & q & 0 \\ 0 & 1/q & 0 & 0 \\ 0 & 0 & 0 & t \end{bmatrix}, \quad \begin{array}{l} [x, dx]_s = 0, \quad |s| = 1, \quad s \in \mathbb{C}, \\ [p, dx]_{1/q} = 0, \\ [x, dp]_q = 0, \\ [p, dp]_t = 0, \quad |t| = 1, \quad t \in \mathbb{C}. \end{array} \tag{6}$$

If the derivatives f_x, f_p *of a differentiable function* f *of* x, p *are defined by*

$$f_x dx + f_p dp = df, \tag{7}$$

then

$$f_x(x,p) = \lim_{x' \to x} \frac{f(sx',p) - f(x',p)}{(s-1)x'}, \quad f_p(x,p) = \lim_{p' \to p} \frac{f(x,tp') - f(x,p)}{(t-1)p'} \quad \text{for } s,t \neq 1. \tag{8}$$

In the case of (2) *and* q, $r \in \mathbb{R} \setminus \{0,1\}$, *for* z, $\overline{z}$ *and* dz, $d\overline{z}$ *defined by*

$$z = x + ip, \quad \overline{z} = x - ip \quad \text{and} \quad dz = dx + idy, \quad d\overline{z} = dx - idy, \tag{9}$$

we get

$$(dz)^2 = \frac{1-q}{1+q} dz d\overline{z} = \frac{q-1}{q+1} d\overline{z} dz = -(d\overline{z})^2 \tag{10}$$

and the Cauchy-Riemann equations $f_{\overline{z}}(z,\overline{z}) = 0$ *for a mapping* f, *holomorphic in a neighbourhood in* $\mathbb{C}$, *in the form* $qf_x = -if_p$.

Remark 1. Symbols of the type

$$\lim_{x' \to x} \frac{F(x',p)}{x'} \quad \text{resp.} \quad \lim_{p' \to p} \frac{G(x,p')}{p'}$$

mean decreasing by one the powers of x resp. p in the series defining the function $F(x,p)$ resp. $G(x,p)$ without assuming the existence of $1/x$ resp. $1/p$.

Remark 2. As usuall $f_z := \frac{1}{2}(f_x - if_y)$ and $f_{\overline{z}} := \frac{1}{2}(f_x + if_y)$.

Proof. The system of equations (4) for C involves four independent real parameters which can be chosen as

$$\arg q, \quad \arg r, \quad \arg s, \quad \arg t \quad \text{if} \quad |q| = |r| = |s| = |t| = 1,$$

or just q, r, s, t if they are real, thus yielding (6), (8), (10), and the desired form of the Cauchy-Riemann equations.

2. THE CASE WHEN q/r-ANTICOMUTATOR OF dx, dp VANISHES WITH $[p, dp]_r = 0$

Consider next the case when

$$\{dx, dp\}_{q/r} = 0 \quad \text{with} \quad [p, dp]_r = 0. \tag{11}$$

Then we have

THEOREM 2. *Suppose* (1), (3) *and* (4), *and consider the case when the relation* (11) *holds. Then*

$$C = \begin{bmatrix} r & 0 & 0 & 0 \\ 0 & 1/r & q & 0 \\ 0 & r/q & 0 & 0 \\ 0 & 0 & 0 & r \end{bmatrix}, \quad \begin{array}{l} [x, dx]_r = 0, \\ [p, dx]_{r/q} = 0, \\ [x, dp]_q = (t-1)dx\ p, \\ |r| = 1,\ r \in \mathbb{C}. \end{array} \tag{12}$$

If the derivatives f_x, f_p *of a differentiable function* f *of* x, p *are defined by* (7), *then*

$$f_x(x,p) = \lim_{x' \to x} \frac{f(rx', rp) - f(x, rp)}{(r-1)x'}, \quad f_p(x,p) = \lim_{p' \to p} \frac{f(x, rp') - f(x,p)}{(r-1)p'} \quad \text{for } r \neq 1. \tag{13}$$

In the case of (2) *and* $q \in \mathbb{R} \setminus \{0\}$, $r \in \mathbb{R} \setminus \{0, 1\}$, *for* z, $\overline{z}$ *and* dz, $d\overline{z}$ *defined by* (9), *we get*

$$(dz)^2 = \frac{r-q}{r+q} dz d\overline{z} = \frac{q-r}{q+r} d\overline{z} dz = -(d\overline{z})^2 \tag{14}$$

and the Cauchy-Riemann equations $f_{\overline{z}}(z,\overline{z}) = 0$ for a mapping f holomorphic in a neighbourhood in $\mathbb{C}$, in the form $qf_x = -if_p$.

Proof. The system of a equations (4) for C involves now two real parameters which can be chosen as

$$\arg q, \quad \arg r \quad \text{if} \quad |q| = |r| = 1;$$

or just q, r if they are real, thus yielding (12), (13), (14), and the desired form of Cauchy-Riemann equations.

3. THE CASE WHEN THE q-COMMUTATOR OF dx, dp VANISHES

Finally, consider the only still possible case when

$$[dx, dp]_q = 0. \tag{15}$$

Then we have

THEOREM 3. *Suppose (1), (3) and (4), and consider the case when the relation (15) holds. Then*

$$C = \begin{bmatrix} 1/r & 0 & 0 & 0 \\ 0 & 0 & q/r & 0 \\ 0 & 1/q & 1/r-1 & 0 \\ 0 & 0 & 0 & 1/r \end{bmatrix}, \quad \begin{aligned} &[x, dx]_r = 0 \quad |r| = 1, \ r \in \mathbb{C}, \\ &\{p, dx\}_{1/q} = 0, \\ &[x, dp]_q = -2dx\ p, \\ &\{p, dp\} = 0. \end{aligned} \tag{16}$$

If the derivatives f_x, f_p of a differentiable function f of x, p are defined by (7), then

$$f_x(x,p) = \lim_{x' \to x} \frac{f(x'/r, p) - f(x,p)}{(1/r-1)x'}, \quad f_p(x,p) = \lim_{p' \to p} \frac{f(x/r, p'/r) - f(x/r, p')}{(1/r-1)p'}. \tag{17}$$

In the case of (2) and $q \in \mathbb{R} \setminus \{0\}$, $r \in \mathbb{R} \setminus \{0,1\}$, for z, $\overline{z}$ and dz, $d\overline{z}$ defined by (9), we get

$$(dz)^2 = \frac{q-r}{q+r} dz d\overline{z} = \frac{r-q}{r+q} d\overline{z} dz = -(d\overline{z})^2 \tag{18}$$

and the Cauchy-Riemann equations $f_{\overline{z}}(z,\overline{z}) = 0$ *for a mapping* f *holomorphic in a neighbourhood in* $\mathbb{C}$, *in the form* $qf_x = if_p$.

Proof. Theorem 3 is a corollary to Theorem 2, if we replace there everywhere x, p and q, r by p, x and $1/q$, $1/r$, respectively.

4. PERSPECTIVES FOR FURTHER GENERALIZATIONS

The n-dimensional case can be introduced by replacing a pair of two generators x, p satisfying $[x,p]_q = 0$, $|q| = 1$, $q \in \mathbb{C}$ or $q \in \mathbb{R}$, by a sequence of $2n$ generators $x_1, p_1, \ldots, x_n, p_n$, satisfying the conditions [10]:

$$[\xi^j, \xi^k]_{q_{jk}} = 0, \quad \text{where} \quad q_{kj} = q_{jk}^{-1}, \quad q_{jk} \in \mathbb{C} \text{ (or } \mathbb{R}), \quad q_{jj} = 1, \quad j,k = 1,\ldots,2n;$$
$$\xi^j = x_j, \qquad \xi^{j+n} = p_j, \quad j = 1,\ldots,n.$$

The even number of generators is essential only in the case if we like to have a counterpart of the quantum differential calculi in the theory of symmetric complex manifolds.

Now, if γ^α, $\alpha = 1,\ldots,p'-1$, $I_{2\nu}$ are generators of a Clifford algebra $C^{(r',s')}$, $r'+1+s' = 2n$, then the deformation [6]:

$$\{\gamma^j, \gamma^k\}_{q_{jk}} = 2g^{jk} I_{2\nu},$$

where the metric g is involved in the real "line element"

$$ds^2 = \sum_{j,k=1}^{2n} d\xi^j g_{jk} d\xi^k,$$

leads to a generalization to the curved manifolds, in particular to replacing in Section 1 the algebra $\Gamma(\mathbb{C}, \mathcal{O}_{\mathbb{C}})$ by $\Gamma(\mathbb{M}, \mathcal{O}_{\mathbb{M}})$, where $\mathbb{M}$ is a symmetric complex manifold.

Now, on one side symmetric complex manifolds may be replaced by more general double complex manifolds [3]. On the other side, the relations (3) and (4) may be put in the frames of supergeometry [9] and quantum spaces [11, 12]. Namely, we can introduce a *quantum space* V as the quotient algebra $\mathcal{F}/J$, where $\mathcal{F}$ is an associative algebra with a unit element, freely generated by the elements $\xi^1, \ldots, \xi^{2n}$ or ξ^{2n+1}, while J is a two-sided ideal in $\mathcal{F}$.

Usually one restricts to the so-called *Yang-Baxter category* of quantum spaces. In that case the ideal J is defined by a collection of bilinear reordering rules for the generators $\xi^1, \ldots, \xi^{2n}$ or ξ^{2n+1}. Denoting the column $(\xi^1, \ldots, \xi^{2n}$ or $\xi^{2n+1})^T$ by $\mathbf{x}$, we can write the reordering rules as follows:

$$\mathbf{x} \otimes \mathbf{x} = B\mathbf{x} \otimes \mathbf{x}, \tag{19}$$

where $\otimes$ denotes the usual direct product and $B \subset \mathrm{End}(\mathbb{C}^{2n} \times \mathbb{C}^{2n})$. The matrix B is assumed to satisfy the *Yang-Baxter equation* guaranteeing the associativity of V:

$$\begin{aligned} &B_{12}B_{23}B_{12} = B_{23}B_{12}B_{23}, \quad \text{where} \quad B_{12} = B \otimes I_{2n}, \quad B_{23} = I_{2n} \otimes B, \\ &B = [B_{jk}]_{j,k=1,\ldots,2n}. \end{aligned} \tag{20}$$

Besides, we need the *twisted algebra associated with* V; it is generated by the totality of ξ^j and $d\xi^k$, where d is the exterior differential operator obeying the standard conditions of linearity, nilpotency, and the Leibniz rule with gradation. It is assumed that the generators ξ^j and differentials $d\xi^k$ satisfy the reordering rules

$$\mathbf{x} \otimes d\mathbf{x} = C d\mathbf{x} \otimes \mathbf{x}, \tag{21}$$

where C belongs to $\mathrm{End}(\mathbb{C}^{2n} \times \mathbb{C}^{2n})$ and satisfies the consistency conditions

$$\begin{aligned} (B_{12} - I_{2n})(C_{12} + I_{2n}) &= 0, & C &= [C_{j,k}]_{j,k=1,\ldots,2n}, \\ B_{12}C_{23}C_{12} &= C_{23}C_{12}B_{23}, & C_{12} &= C \otimes I_{2n}, \\ C_{12}C_{23}C_{12} &= C_{23}C_{12}C_{23}, & C_{23} &= I_{2n} \otimes C. \end{aligned} \tag{22}$$

It is clear that the relations (19)-(22) are direct generalizations of (3)-(4).

References

[1] N. L. Alling and N. Greenleaf, *Foundations of the Theory of Klein Surfaces* (Lecture Notes in Math. 219), Springer, Berlin-Heidelberg-New York 1971.

[2] T. Brzeziński, H. Dąbrowski, and J. Rembieliński, *On the quantum differential calculus and the quantum holomorphicity*, J. Math. Phys. **33** (1992), 19–24.

[3] M. Jurchescu, K. Spallek, and F. Succi, *DC-spaces*, Ms., to be published.

[4] M. Jurchescu and F. Succi, *Involutive complex manifolds and localization*, Rev. Roumaine Math. Pures Appl. **31** (1986), 479-488.

[5] J Ławrynowicz, *Quantized complex and Clifford structures*, Ber. Univ. Jyväskylä Math. Inst. **55** (1993), 113-120.

[6] J. Ławrynowicz, L. Papaloucas, and J. Rembieliński, *Quantum braided Clifford algebras*, in *Clifford Algebras and Spinor Structures* (Crumeyrolle memorial volume) (R. Ablamowicz and P. Lounesto, eds.), Kluwer Academic, Dordrecht-Boston-London, to appear.

[7] J. Ławrynowicz, J. Rembieliński, and F. Succi, *Generalized Hurwitz mappings of the type* $S \times V \rightarrow W$ *and quantum braided Clifford algebras*, in *Generalizations of Complex Analysis and Their Applications in Physics* (J. Ławrynowicz, ed.), Banach Center Publications, Warszawa.

[8] Yu. I. Manin, *Quantum Groups and Non-Commutative Geometry*, Technical Report, Centre de Recherches Math., Montréal 1988.

[9] Yu. I. Manin, *Gauge Field Theory and Complex Geometry*, (Grundlehren der mathematischen Wissenschaften 289), Springer, Berlin-Heidelberg-New York-Paris-Tokyo 1989.

[10] F. Succi, *Symmetric complex manifolds, dianalytic structures and Klein manifolds*, in *Ergebnisse der Tagung Komplexe Analysis und komplexe Differentialgeometrie* (A. Duma, ed.), Hagen 1986, pp. 9-16.

[11] A. Sudbery, *Consistent multiparameter quantization of* $GL(n)$, Preprint ANL-HEP -CP-90-32.

[12] J. Wess and B. Zumino, *Covariant differential calculus on the quantum hyperplane*, Recent Advances in Field Theory, Proceedings, Annecy-le-Vieux 1990, Nuclear Phys. B, Proc. Suppl. **18B** (1990), 302–312..

Newlander–Nirenberg Type Theorem for Analytic Algebras

Tomasz Maszczyk

Institute of Mathematics, Warsaw University
ul. Banacha 2, 02–097 Warszawa, Poland
e-mail:maszczyk@mimuw.edu.pl

Abstract

The criterion of solvability and the solution of the following problem is presented:

Given n commuting endomorphisms of the tangent bundle, satisfying relations generated by n generic polynomial identities with functional coefficients, find at every point a germ of a holonomic map diagonalizing them.

Introduction

Our point of departure for this paper is a classic theory of real-analytic integrable almost complex (or almost product) structures. It had been intensively investigated in fifties [10][9][12], until Newlander and Nirenberg gave the proof in differentiable case [29] (in [23] proof was reduced to the real-analytic case). The next period in this topic consisted of several examples of endomorphisms satisfying polynomial identities with constant coefficients coming from geometry, for which the problem of integrability was considered in terms of the Nijenhuis tensor [36][17][37][41][21][42][22][18][43][26][27][28] [13][19][25] [20] [3][39][31][38][15][14][16][5][7][2][6][1][30]. (The only paper we have found, considering functional coefficients of a polynomial identity is

This paper is in final form and no part of it will be submitted elsewhere.

[33].) Next, the remarkable application of the Nijenhuis tensor in mechanics was found [32][4][35].

Our problem is the following one: how, given commuting endomorphisms $e_1, \ldots, e_n$ of the tangent bundle of a complex manifold X satisfying algebraic (over the sheaf $\mathcal{O}_X$ of holomorphic functions) relations , generated by n generic polynomial identities, to find if they are $\mathcal{O}_X$-linear combinations of endomorphisms represented in some coordinate system by standart diagonal idempotent matrices.

There are the two new points in this problem where previous methods fail

- an arbitrary number of endomorphisms instead a single one
- general functional coefficients of polynomial identities instead constants

We show that there is an invariant which is a generalization of the Nijenhuis tensor (but tensorial with respect to an endomorphism), and which is an obstruction for that integrability.

If this invariant vanishes we construct some multivalued regular (locally of constant rank) Pfaff system, which integrals give such coordinate systems. Even in the case of arbitrary complex space (singular, with nilpotents etc.) this Pfaff system is well defined and satisfies the Frobenius condition if the above invariant vanishes.

As an illustration of the general method we integrate some tautological endomorphism of a tangent bundle of the space of separable monic polynomials of degree n. The topology of such spaces has been intensively studied in connection with the topology of configuration spaces, the cohomology of braid groups and estimates of the complexity of algorithms of finding roots of polynomials in one variable ([40]). However, the geometry of such spaces has not been explored as deeply as their topology, so our example of the tautological endomorphism of the tangent bundle, with very special properties, should be of its own interest. Another application of the theory considered here is given in [24]. (We introduce there the notion of *hyperradicals* - extensions of special type, which solve algebraic equations of arbitrary degree over complex numbers.)

In the case of the real-analytic almost complex structure, after complexification, the problem can be reformulated in terms of complex almost product structures and solved immediately using only the Frobenius theorem. In our problem the way to the Frobenius theorem is much longer and main concepts and difficulties are rather of algebraic or geometric nature - since in our problem we allow arbitrary functional coefficients, then we must use analytic geometry instead linear algebra. It is also convenient not to consider a couple of commuting endomorphisms itself (even if there is given only one endomorphism) but rather the whole algebra generated by them. To enable natural geometric constructions with such data we care about functoriality of all our geometro-analytic constructions. We restrict our considerations to the analytic case to avoid a mixture of C^∞ and analytic arguments, however many points could be easily generalized onto C^∞-case.

Analytic Newlander-Nirenberg Type Theorem for Finite Étale Algebras

$\mathcal{E}$-structures Let X be a complex space and $\mathcal{E}$ be a finite étale $\mathcal{O}_X$-algebra (i.e. finite, flat and unramified $\mathcal{O}_X$-algebra). We will call an $\mathcal{E}$ – *structure* a structure of a right $\mathcal{E}$-module on the sheaf of differential 1-forms Ω^1_X

$$\Omega^1_X \otimes \mathcal{E} \to \Omega^1_X$$

Example 1.Constant $\mathcal{E}$-structure.

Let $X = Y_1 \times \ldots \times Y_n$ with projections $X \xrightarrow{f_i} Y_i$ and $\mathcal{E} = \prod_{i=1}^n f_i^* \mathcal{O}_{Y_i}$ (it is the product of n copies of $\mathcal{O}_X$) Then $\Omega^1_X = \prod_{i=1}^n f_i^* \Omega^1_{Y_i}$ and we define an $\mathcal{E}$-structure by the diagonal action

$$f_i^* \Omega^1_{Y_i} \otimes f_i^* \mathcal{O}_{Y_i} \to f_i^* \Omega^1_{Y_i}$$

Example 2.

Let X be a manifold and $e_1, \ldots, e_n$ be commuting endomorphisms of the tangent bundle of X, satisfying polynomial relations generated by the polynomial identities

$$f_1(e_1, \ldots, e_n) = 0, \ldots, f_n(e_1, \ldots, e_n) = 0$$

(we consider polynomials over $\mathcal{O}_X$), such that the Jacobian $det[\frac{\partial f_i}{\partial e_j}]$ is invertible in

$$\mathcal{E} := \mathcal{O}_X[e_1, \ldots, e_n]$$

The latter condition is obviously a generic property of the couple of polynomials $f_1(e_1, \ldots, e_n), \ldots, f_n(e_1, \ldots, e_n)$.

Since the tangent bundle is locally free of finite rank then $\mathcal{E}$ is finite, by the jacobian criterion ([34], Chap. V, §2, Th. 5) $\mathcal{E}$ is an étale $\mathcal{O}_Y$-algebra and the structure of a right $\mathcal{E}$-module on the $\mathcal{O}_X$-module Ω^1_X arises in a canonical way.

Example 3. Tautological $\mathcal{E}$-structure on the space of separable monic polynomials.

This example is an interesting particular case of the example 2).

Let $\sigma_1, \ldots, \sigma_n$ be coordinates in the complex affine space $\mathbb{A}^n$ and $D_n = D_n(\sigma_1, \ldots, \sigma_n)$ be the discriminant of the polynomial

$$P(T) = T^n - \sigma_1 T^{n-1} + \ldots + (-1)^n \sigma_n$$

We will regard $\mathbb{A}^n$ as a space of monic polynomials of degree n parametrized by their coefficients. Let X be a space of separable monic polynomials of degree n, that is the complement of a zero-set of the discriminant function. We define an endomorphism e of the tangent bundle by means of coordinates σ_i which differentials form at every point coordinates on the tangent space: for $i = 1, \ldots, n-1$

$$d\sigma_i \circ e = \sigma_i d\sigma_1 - d\sigma_{i+1}$$

and

$$d\sigma_n \circ e = \sigma_n d\sigma_1.$$

Here the little circle denotes the composition of a differential form with an endomorphism of a tangent bundle. We will keep this notation in the abstract case.

It is easy to see that these formulas imply that the minimal polynomial of e is equal to $P(T)$ so they define an $\mathcal{E}$-structure,where $\mathcal{E} = \mathcal{O}_X[e] = \mathcal{O}_X[T]/(P(T))$. (Moreover, since $P(T)$ is a characteristic polynomial of e, we get the following trivial but, as it seems, not popular observation about location of roots of a polynomial

Proposition: *Let us choose an arbitrary hermitian metric on X and $c_1, \ldots, c_N \in \mathbb{C}$. Then*

$$| f - c_1 | \cdot \ldots \cdot | f - c_N | \leq \| (e - c_1) \ldots (e - c_N) \|$$

Let us note that the right hand side expression is an explicit function of coefficients of a polynomial.

Let us note the difference comparing with the classical case: coefficients of the minimal polynomial of e are not constant: they are even functionally independent.

After introducing natural notions of integrability of an $\mathcal{E}$-structure and $\mathcal{E}$-holomorphicity, the following facts will be established for this $\mathcal{E}$-structure

Theorem. *There exists the unique integrable $\mathcal{E}$-structure on X such that the root of polynomial $P(T)$ is an $\mathcal{E}$-holomorphic function.*

Theorem *f is a solution of the algebraic equation $P(T) = 0$ if and only if $df \neq 0$ and f is a solution of a partial differential equation*

$$df \circ e - f df = 0$$

Analysis of symmetries of this partial differential equation provides the mentioned above notion of hyperradicals.

Torsion of an $\mathcal{E}$-structure. Next we introduce an invariant generalizing the Nijenhuis tensor on "the case of non-constant coefficients". We wil call the *torsion* of an $\mathcal{E}$-structure some morphism $\tau_{\mathcal{E}/X}$ of $\mathcal{O}_X$-modules

$$\Omega_X^1 \otimes \bigodot^2 \mathcal{E} \overset{\tau_{\mathcal{E}/X}}{\rightarrow} \Omega_X^2$$

i.e. a quadratic form on an $\mathcal{O}_X$-module $\mathcal{E}$ with values in $\mathcal{H}om(\Omega_X^1, \Omega_X^2)$.

Let us emphasize that the Nijenhuis tensor is a quadratic form on $\mathcal{E}$ but only over constants.

We will say that an $\mathcal{E}$-structure is *integrable* if its torsion vanishes.

Newlander-Nirenberg Type Theorem for $\mathcal{E}$-structures. After precise definition of the torsion it will be easy to show that the torsion of a constant $\mathcal{E}$-structure vanishes. Our main theorem states that locally the inverse is also true.

Theorem. *An $\mathcal{E}$-structure is integrable if and only if it is locally constant.*

This theorem is effective - there is a *multivalued analog of the Dolbeault complex*, which is a resolution of a sheaf of local multivalued functions lifted from factors of local decompositions into cartesian products of local constant $\mathcal{E}$-structures.

Let us illustrate the above theorem by the example 3): It turns out that this $\mathcal{E}$-structure is integrable, locally $X = \mathbb{A}^n = \mathbb{A}^1 \times \ldots \times \mathbb{A}^1$ and local projections on $\mathbb{A}^1$ glue together to the multivalued choice of the root of the polynomial $P(T)$.

Geometry of $\mathcal{E}$-structure Let us define $X' := \mathbf{specan}\mathcal{E}$ (i.e. *analytical spectrum*, see.[11], p.59). Then we have the finite étale cover $X' \xrightarrow{p} X$ and the natural isomorphism of $\mathcal{O}_X$-algebras $\mathcal{E} \to p_*\mathcal{O}_{X'}$. For instance:

- in the example 1) $X' = \coprod_{i=1}^n X_i$, where X_i are copies of X. If we put $Y := \coprod_{i=1}^n Y_i$, then we obtain an n-valued submersion from X to Y

$$\begin{array}{ccc} X' & \xrightarrow{f} & Y \\ \downarrow & & \\ X & & \end{array}$$

where $f\,|_{X_i}$ is equal to the composition

$$X_i = X \xrightarrow{f_i} Y_i \hookrightarrow Y$$

- in the example 3) X' is a zero-set in $X \times \mathbb{A}^1$ of the polynomial $P(T)$ with the canonical projection on X.

In fact the proof of the theorem will pass through the Frobenius theorem for some differential Pfaff ideal on the covering space X', so it is convenient to reformulate our definitions in geometric terms.

So, in a new manner we consider the structure consisting of a finite étale cover $X' \xrightarrow{p} X$ and a structure of a right $p_*\mathcal{O}_{X'}$-module on Ω^1_X

$$\Omega^1_X \otimes p_*\mathcal{O}_{X'} \to \Omega^1_X$$

Then we will call such a structure X'-structure.

The main stress we lay on the question how to find practically whether an $\mathcal{E}$-structure is integrable given finite sets of generators and relations for $\mathcal{O}_X$-algebra $\mathcal{E}$. This is the main reason for introducing the invariant tensorial with respect to $\mathcal{E}$, which we shall call the *torsion.* Since the torsion is $\mathcal{O}_X$-linear and $\mathcal{E}$ is a finitely generated $\mathcal{O}_X$-module then it is sufficient to check the vanishing of the torsion on a finite set of monomials of multiplicative generators spanning the $\mathcal{O}_X$-module $\mathcal{E}$.

However, from the theoretical point of view, our torsion is very closely related to the Nijenhuis tensor, at least on manifolds:

From the isomorphism $p^{-1}\mathcal{O}_X \to \mathcal{O}_{X'}$ we obtain (because p is proper) the isomorphism

$$p_*\mathbb{C}_{X'} \otimes_{\mathbb{C}_X} \mathcal{O}_X \to p_*\mathcal{O}_{X'} = \mathcal{E}$$

and $p_*\mathbb{C}_{X'}$ is a $\mathbb{C}_X$-algebra locally isomorphic to a product of copies of $\mathbb{C}_X$ (number of them is equal to the degree of the cover p). Using the property of unique extension of derivations for p (because p is étale) we obtain the connection $\mathcal{E} \to \mathcal{E} \otimes \Omega^1_X$, which extends to a differential of a complex $\mathcal{E} \otimes \Omega^\bullet_X$ canonically isomorphic to the complex $p_*\Omega^\bullet_{X'}$. But on the manifold X' the complex $\Omega^\bullet_{X'}$ is a resolution of a constant sheaf $\mathbb{C}_{X'}$, so the complex $\mathcal{E} \otimes \Omega^\bullet_X$ is a resolution of $p_*\mathbb{C}_{X'}$ because p_* is exact. After computing in this way the sheaf $p_*\mathbb{C}_{X'}$ we can use the fact that the torsion is determined by the $\mathbb{C}_X$-linear morphism

$$\bigodot^2_{\mathbb{C}_X} p_*\mathbb{C}_{X'} \to \mathcal{H}om(\Omega^1_X, \Omega^2_X)$$

which is nothing but the Nijenhuis tensor.

But the solution of the problem from the example 3) shows that in fact this way of constructing of $p_*\mathbb{C}_{X'}$ requires one to split the monic polynomial with general coefficients, hence for sufficiently high degree is useless.

This is the main difficulty comparing with the case of the almost complex structure, which can be regarded as a special case "of constant coefficients" and of degree two.

Functorial Properties of Torsion The torsion of such a structure is functorial in the following sense. We define a morphism of the two structures as a morphism of finite étale covers $p \to q$

$$\begin{array}{ccc} X' & \xrightarrow{f'} & Y' \\ \downarrow & & \downarrow \\ X & \xrightarrow{f} & Y \end{array}$$

such that the morphism

$$f^*\Omega^1_Y \to \Omega^1_X$$

is $f^*q_*\mathcal{O}_{Y'}$-linear. Then we obtain the commutative diagram

$$\begin{array}{ccc} f^*(\Omega^1_Y \otimes \bigodot^2 q_*\mathcal{O}_{Y'}) & \overset{f^*\tau_{Y'/Y}}{\rightarrow} & f^*\Omega^2_Y \\ \downarrow & & \downarrow \\ \Omega^1_X \otimes \bigodot^2 p_*\mathcal{O}_{X'} & \overset{\tau_{X'/X}}{\rightarrow} & \Omega^2_X \end{array}$$

Corollary *1)f - immersion of manifolds and $X' \times_X X' \rightarrow X' \times_Y Y'$ - closed immersion. Then*

$$\tau_{X'/X} = 0 \Leftarrow \tau_{Y'/Y} = 0$$

2)f - surjective submersion of manifolds. Then

$$\tau_{X'/X} = 0 \Rightarrow \tau_{Y'/Y} = 0$$

3)f - finite étale cover. Then

$$\tau_{X'/X} = 0 \Leftrightarrow \tau_{Y'/Y} = 0$$

Functorial Properties of Analytic $\mathcal{E}$-Dolbeault Complex Not every morphism from an $\mathcal{E}$-structure to an $\mathcal{F}$-structure is compatible with the construction of analytic $\mathcal{E}$-Dolbeault complex. For that one has to complete the definition of such a morphism with a condition, that the induced morphism of $\mathcal{O}_X$-algebras

$$f^*(\mathcal{F} \otimes \mathcal{F}) \rightarrow \mathcal{E} \otimes \mathcal{E}$$

is an epimorphism, or equivalently, that the induced morphism of complex spaces over X'

$$X' \times_X X' \rightarrow X' \times_Y Y'$$

is a closed immersion.

This kind of functoriality works well with finite 'etale Galois covers. Namely, if X' over X and Y' over Y are finite étale Galois covers, $G = Aut(X'/X)$, $H = Aut(Y'/Y)$, $G \rightarrow H$ is a homomorphism of finite groups and a morphism between covers is G-equivariant, then $X' \times_X X' \rightarrow X' \times_Y Y'$ is a closed immersion if and only if $G \rightarrow H$ is injective. Hence, in particular, the analytic $\mathcal{E}$-Dolbeault complex is a functor on the category of

X'-structures for finite étale Galois covers $X' \to X$ with fixed complex space X' and morphisms compatible with injective homomorphisms of their Galois groups.

1 Local Theory of $\mathcal{E}$-Structures

We will work with $\mathcal{E}$-structures locally. We choose an open Stein subspace $U \subset X$ embedded as a closed subspace into an affine complex space and denote $A = \mathcal{O}_X(U)$. Then the A-module of Pfaff differentials induced by this embedding depends actually only on A and we denote it by Ω^1_A. Then $U' = p^{-1}U = X' \times_X U$ is also Stein, $E := \mathcal{E}(U) = \mathcal{O}_{X'}(U')$ is a finite étale A-algebra, hence in particular a projective A-module, so $\Omega^1_E = \Omega^1_{X'}(U') = p^*\Omega^1_X(U') = p_*p^*\Omega^1_X(U) = (p_*\mathcal{O}_{X'} \otimes \Omega^1_X)(U) = (\mathcal{E} \otimes \Omega^1_X)(U) = E \otimes_A \Omega^1_A$ because p is étale and proper and E is a locally free $\mathcal{O}_X$-module. Then we have on Ω^1_A the structure of a *right* E-module (we will call it later simply an E-structure)

$$\Omega^1_A \otimes_A E \to \Omega^1_A$$

$$\omega \otimes e \mapsto \omega \circ e$$

Tensoring from the left by E over A the above morphism we obtain on $\Omega^1_E = E \otimes_A \Omega^1_A$ the structure of E-bimodule. We emphasize that it does not need to be a symmetric E-bimodule independently of the commutativity of E.

We will also consider the stalk of the sheaf $\mathcal{O}_X$ (resp. of other $\mathcal{O}_X$-modules) at a given point. We will also denote it by A (resp. E, Ω^1_A, $\Omega^1_E = E \otimes_A \Omega^1_A$), because all our constructions commute with localization.

Let M be any A-module. Then we denote by $\bigodot^p_A M$ the submodule of symmetric elements in $\bigotimes^p_A M$.

Proposition 1 *The structure of a right E-module on Ω^1_A induces the following product*

$$\Omega^n_A \otimes_A \bigodot^n_A E \to \Omega^n_A$$

*Proof.*The statement follows from the fact that the map

$$\bigotimes^n_A \Omega^1_A \otimes_A \bigotimes^n_A E \to \bigotimes^n_A \Omega^1_A$$

$$(\omega_1 \otimes \ldots \otimes \omega_n) \otimes (e_1 \otimes \ldots \otimes e_n) \mapsto \omega_1 \circ e_1) \otimes \ldots \otimes (\omega_n \circ e_n)$$

induces the map

$$Ker^n \otimes_A \bigodot\nolimits_A^n E \to \Omega_A^n$$

where Ker^n denotes the kernel of the canonical projection

$$0 \to Ker^n \to \bigotimes\nolimits_A^n \Omega_A^1 \to \Omega_A^n \to 0$$

This fact follows from the following: Let

$$\tau_{ij} : \bigotimes\nolimits_A^n \Omega_A^1 \to \bigotimes\nolimits_A^n \Omega_A^1$$

$$\sigma_{ij} : \bigotimes\nolimits_A^n E \to \bigotimes\nolimits_A^n E$$

denote transpositions of i-th and j-th places in the tensor product. If $e \in \bigodot_A^n E$, $\omega \in \bigotimes_A^n \Omega_A^1$ and for some i, j $\tau_{ij}(\omega) = \omega$ then

$$\omega \circ e = \tau_{ij}(\omega) \circ e = \tau_{ij}(\omega) \circ \sigma_{ij}(e) = \tau_{ij}(\omega \circ e)$$

This equality ends the proof of the proposition, since Ker^n is generated by elements $\omega \in \bigotimes_A^n \Omega_A^1$ such that $\tau_{ij}(\omega) = \omega$ for some i, j □

We consistently denote all above products by the same symbol for all n. Using the isomorphism of E-modules $E \otimes_A \Omega_A^1 \to \Omega_E^1$ we can define the natural product

$$\Omega_E^n \otimes_A \bigodot\nolimits_A^n E \to \Omega_E^n$$

and the product

$$\Omega_E^1 \otimes_A \Omega_E^1 \to \Omega_E^2$$

$$(e \otimes \omega) \otimes (e' \otimes \eta)) \mapsto e \otimes (\omega \circ e') \wedge \eta$$

both also denoted by a small circle as above.

Definition. We define *universal elementary homogeneous polynomials* of degree $(n-k)$:

$$\sigma_k^n : E \to \bigodot\nolimits_A^n E$$

such that $\sigma_0^0 = 1$ and

$$(e + t \cdot 1) \underbrace{\otimes \ldots \otimes}_{n-times} (e + t \cdot 1) = \sum_k t^k \cdot \sigma_k^n(e)$$

for $n \geq 1$, $e \in E$ and a formal variable t.

In particular

$$\sigma_n^n = 1 \otimes \ldots \otimes 1, \sigma_{n-1}^n = e \otimes 1 \otimes \ldots \otimes 1 + \ldots + 1 \otimes \ldots \otimes 1 \otimes e$$

Lemma 1 *If $\omega \in \Omega_E^p$, $\eta \in \Omega_E^q$ then for all $e \in E$*

$$(\omega \wedge \eta) \circ \sigma_k^{p+q}(e) = \sum_{i+j=k} (\omega \circ \sigma_i^p(e)) \wedge (\eta \circ \sigma_j^q(e))$$

In particular, for $p = 1, k = q$ we get

$$(\omega \wedge \eta) \circ \sigma_k^{k+1}(e) = (\omega \circ e) \wedge \eta + \omega \wedge (\eta \circ \sigma_{k-1}^k(e)) \tag{1}$$

Proof. From the definition of universal elementary homogeneous polynomials σ_k^p we obtain

$$\sum_{k=0}^{p+q} t^k \cdot (\omega \wedge \eta) \circ \sigma_k^{p+q}(e) = (\omega \wedge \eta) \circ ((e + t \cdot 1) \underbrace{\otimes \ldots \otimes}_{(p+q)-times} (e + t \cdot 1)) =$$

$$= (\omega \circ ((e + t \cdot 1) \underbrace{\otimes \ldots \otimes}_{p-times} (e + t \cdot 1))) \wedge (\eta \circ ((e + t \cdot 1) \underbrace{\otimes \ldots \otimes}_{q-times} (e + t \cdot 1))) =$$

$$= \sum_{i=0}^{p} t^i \cdot (\omega \circ \sigma_i^p(e)) \wedge \sum_{j=0}^{q} t^j \cdot (\eta \circ \sigma_j^q(e)) =$$

$$= \sum_{k=0}^{p+q} t^k \cdot \sum_{i+j=k} (\omega \circ \sigma_i^p(e) \wedge (\eta \circ \sigma_j^q(e))$$

Comparing expressions at the same power of t we obtain the lemma □

2 Torsion

Next we introduce the invariant called torsion, which modifies the Nijenhuis tensor but is tensorial with respect to an endomorphism.

Proposition 2 *The following map*

$$\Omega_A^1 \otimes_{\mathbb{C}} \bigodot_{\mathbb{C}}^2 E \to \Omega_A^2$$

$$(\omega \otimes (e \otimes e)) \mapsto d\omega \circ \sigma_0^2(e) - d(\omega \circ e) \circ \sigma_1^2(e)$$
$$+ d(\omega \circ e^2) - \omega \circ (de \circ e - e \cdot de)$$

factorises uniquely through the morphism of A-modules

$$\Omega_A^1 \otimes_A \bigodot_A^2 E \xrightarrow{\tau_{E/A}} \Omega_A^2$$

Definition We call $\tau_{E/A}$ the *torsion* of a given E-structure.

Proof. Since the last summand in the above expression is obviously A-linear with respect to ω we will check the remaining part (the Nijenhuis tensor expressed in terms of differential forms) After the multiplication ω by $a \in A$ we have

$$d(a \cdot \omega) \circ \sigma_0^2(e) - d(a \cdot \omega \circ e) \circ \sigma_1^2(e) + d(a \cdot \omega \circ e^2)$$
$$= (da \wedge \omega + a \cdot d\omega) \circ \sigma_0^2(e) - (da \wedge (\omega \circ e) + a \cdot d(\omega \circ e)) \circ \sigma_1^2(e)$$
$$+ da \wedge (\omega \circ e^2)$$
$$= (da \circ e) \wedge (\omega \circ e) + a \cdot d\omega \circ \sigma_0^2(e) - (da \circ e) \wedge (\omega \circ e)$$
$$- da \wedge (\omega \circ e^2) - a \cdot d(\omega \circ e) \circ \sigma_1^2(e)$$
$$= a \cdot (d\omega \circ \sigma_0^2(e) - d(\omega \circ e) \circ \sigma_1^2(e) + d(\omega \circ e^2))$$

Thus our map is A-linear with respect to ω. What we have now to prove is that the map

$$\bigodot_{\mathbb{C}}^2 E \to Hom_A(\Omega_A^1, \Omega_A^2)$$
$$e \mapsto (\omega \mapsto d\omega \circ \sigma_0^2(e) - d(\omega \circ e) \circ \sigma_1^2(e) + d(\omega \circ e^2)$$
$$- \omega \circ (de \circ e - e \cdot de))$$

is a quadratic form over A with respect to e. Of course it is a quadratic form over $\mathbb{C}$ with the corresponding bilinear symmetric form

$$e \otimes e' + e' \otimes e \mapsto (\omega \mapsto d\omega \circ (e \otimes e' + e' \otimes e) - d(\omega \circ e) \circ \sigma_1^2(e')$$
$$- d(\omega \circ e') \circ \sigma_1^2(e) + d(\omega \circ (e \cdot e' + e' \cdot e))$$
$$- \omega \circ (de \circ e' + de' \circ e - e \cdot de' - e' \cdot de))$$

But if we take the difference between the above expression with e replaced by $a \cdot e$ and the same multiplied by a we obtain

$$
\begin{aligned}
&d\omega \circ (a \cdot e \otimes e' + e' \otimes a \cdot e) - d(\omega \circ a \cdot e) \circ \sigma_1^2(e') \\
&\quad -d(\omega \circ e') \circ \sigma_1^2(a \cdot e) + d[\omega \circ a \cdot (e \cdot e' + e' \cdot e)] \\
&\quad -\omega \circ [d(a \cdot e) \circ e' + de' \circ a \cdot e - a \cdot e \cdot de' - e' \cdot d(a \cdot e)] \\
&\quad -a \cdot \{d\omega \circ (e \otimes e' + e' \otimes e) - d(\omega \circ e) \circ \sigma_1^2(e') \\
&\quad -d(\omega \circ e') \circ \sigma_1^2(e) + d[\omega \circ (e \cdot e' + e' \cdot e)] \\
&\quad -\omega \circ [de \circ e' + de' \circ e - e \cdot de' - e' \cdot de]\} \\
&= -[da \wedge (\omega \circ e)] \circ \sigma_1^2(e') + da \wedge [\omega \circ (e \cdot e' + e' \cdot e)] \\
&\quad -\omega \circ [e \otimes (da \circ e') - e' \cdot e \otimes da] \\
&= -(da \circ e') \wedge (\omega \circ e) - da \wedge (\omega \circ e \cdot e') + da \wedge (\omega \circ (e \cdot e' + e' \cdot e)) \\
&\quad +(da \circ e') \wedge (\omega \circ e) - da \wedge (\omega \circ (e'e) = 0
\end{aligned}
$$

So the mentioned $\mathbb{C}$-bilinear form is A-bilinear □

The torsion is functorial in the following sense. Let us assume that we have a commutative square of $\mathbb{C}$-algebras

$$
\begin{array}{ccc}
E & \leftarrow & F \\
\uparrow & & \uparrow \\
A & \leftarrow & B
\end{array}
$$

where E, F are étale over A, B respectively and an appropriate commutative square of B-modules

$$
\begin{array}{ccc}
\Omega_B^1 \otimes_B F & \rightarrow & \Omega_B^1 \\
\downarrow & & \downarrow \\
\Omega_A^1 \otimes_A E & \rightarrow & \Omega_A^1
\end{array}
$$

Then we obtain the following commutative square of B-modules

$$
\begin{array}{ccc}
\Omega_B^1 \otimes_B \bigodot_B^2 F & \stackrel{\tau_{F/B}}{\rightarrow} & \Omega_B^2 \\
\downarrow & & \downarrow \\
\Omega_A^1 \otimes_A \bigodot_A^2 E & \stackrel{\tau_{E/A}}{\rightarrow} & \Omega_A^2
\end{array}
$$

Let us now assume that A and F are étale over B and put $E = A \otimes_B F$. Then $\Omega_A^p = A \otimes_B \Omega_B^p$ and $\bigodot_A^p E = A \otimes_B \bigodot_B^p F$. Thus naturally arises the E-structure such that $\tau_{E/A} = id_A \otimes_B \tau_{F/B}$.

That construction can be regarded as a localization in the étale sense of an E-structure.

3 Integration of Integrable $\mathcal{E}$-structure by Means of Analytic $\mathcal{E}$-Dolbeault Complex

In this section we construct a generalization of the complex-analytic extension of the classical real-analytic Dolbeault complex on the complexification of the underlying real-analytic manifold. In particular, this complex provide a resolution of the sheaf which sections give rise a coordinate system diagonalizing the representation of $\mathcal{E}$ on Ω^1_X.

Namely, we show that if the torsion vanishes, then some graded ideal in the algebra of differential forms on the covering space (which will turn out to be a Pfaff ideal by étality of our algebra $\mathcal{E}$) satisfies the Frobenius condition, or equivalently, it is a differential graded ideal. Our $\mathcal{E}$-analytic Dolbeault complex will be a graded object obtained from the filtration on the De Rham complex by powers of this ideal.

Let us consider the short exact sequence of $E \otimes_A E$-modules

$$0 \to J_{E/A} \to E \otimes_A E \to E \to 0$$

where the surjective arrow is the multiplication in the A-algebra E.

Since $J_{E/A}$ is an E-submodule in $E \otimes_A E$ then $\bigodot^n_E J_{E/A}$ is an E-submodule in $\bigodot^n_E (E \otimes_A E) = E \otimes_A \bigodot^n_A E$.

$J_{E/A}$ is generated as a left E-module by terms of the form $1 \otimes e - e \otimes 1$. Indeed, if $\sum e_i e'_i = 0$ then $\sum e_i \otimes e'_i = \sum (e_i \otimes 1) \cdot (1 \otimes e'_i - e'_i \otimes 1)$. Therefore $\bigodot^n_E J_{E/A}$ as a left E-submodule in $E \otimes_A \bigodot^n_A E$ is generated by terms of the form

$$\sum_k (-e)^k \otimes \sigma^n_k(e)$$

Then we have the product

$$\Omega^n_E \otimes_E \bigodot^n_E J_{E/A} \to \Omega^n_E$$

Definition We define $I^{\bullet}_{E/A}$ as a graded E-submodule in $\Omega^{\bullet}_E$ such that for grading n the following natural sequence is exact

$$0 \to I^n_{E/A} \to \Omega^n_E \to Hom_E\left(\bigodot^n_E J_{E/A}, \Omega^n_E\right)$$

In particular

$I^0_{E/A} = 0$

$I^1_{E/A} = \{\omega \in \Omega^1_E \mid \forall e \in E \;\; \omega \circ e - e\omega = 0\}$
$I^2_{E/A} = \{\omega \in \Omega^2_E \mid \forall e \in E \;\; \omega \circ (e \otimes e) - e\omega \circ (1 \otimes e + e \otimes 1) + e^2\omega = 0\}$
By E-linearity the above sequence defines the appropriate graded $\mathcal{O}_{X'}$-submodule $I^\bullet_{X'/X}$ of $\Omega^\bullet_{X'}$.

Theorem 1 *$I^\bullet_{X'/X}$ is a graded ideal and if $\tau_{X'/X} = 0$ then it is a graded differential ideal in the De Rham complex $\Omega^\bullet_{X'}$.*

Proof. What we need are the following properties
(a) if $\omega \in I^\bullet_{E/A}$ and ϑ is any form then $\omega \wedge \vartheta \in I^\bullet_{E/A}$
(b) if $\omega \in I^\bullet_{E/A}$ then $d\omega \in I^\bullet_{E/A}$ Let us fix $e \in E$ and introduce the

operator on $\Omega^\bullet_E$

$$\Phi_e(\omega) = \sum_{i=0}^{k}(-e)^i \cdot \omega \circ \sigma^k_i(e) \tag{2}$$

where ω is a k-form. From the above remarks one can easily see that

$$I^\bullet_{E/A} = \bigcap_{e \in E} ker(\Phi_e)$$

So, the property (a) follows from the following lemma

Lemma 2 *For all $\omega, \eta \in \Omega^\bullet_E$*

$$\Phi_e(\omega \wedge \eta) = \Phi_e(\omega) \wedge \Phi_e(\eta)$$

Proof. If $\omega \in \Omega^p_E$, $\eta \in \Omega^q_E$ then by Lemma 1

$$\Phi_e(\omega \wedge \eta) = \sum_k (-e)^k (\omega \wedge \eta) \circ \sigma^{p+q}_k(e)$$

$$= \sum_k \sum_{i+j=k} ((-e)^i \omega \circ \sigma^p_i(e)) \wedge ((-e)^j \eta \circ \sigma^q_j(e)) = \Phi_e(\omega) \wedge \Phi_e(\eta)$$

□

Lemma 3 *If E is generated over A by a single element then $I^\bullet_{E/A}$ is generated by $I^1_{E/A}$.*

Proof. By the assumption $E = A[x]/(P(x))$. Let $e = xmod(P(x))$. Then $J_{E/A}$ is generated as an $(E \otimes_A E)$-module by the element $1 \otimes e - e \otimes 1$. Indeed, as we have already known, $J_{E/A}$ is generated by terms of form $(1 \otimes e' - e' \otimes 1)$ where $e' \in E$. But every such an e' is a combination of powers of e so it is enough to decompose these terms as follows

$$1 \otimes e^k - e^k \otimes 1 = \sum_{i+j=k-1} (e^i \otimes e^j) \cdot (1 \otimes e - e \otimes 1)$$

Since E is étale over A then $J_{E/A}/J^2_{E/A} = \Omega^1_{E/A} = 0$ hence $J_{E/A} = J^2_{E/A}$. Since $J^2_{E/A}$ is generated by $(1 \otimes e - e \otimes 1)^2$ then exists $\varepsilon \in E \otimes_A E$ such that

$$1 \otimes e - e \otimes 1 = \varepsilon \cdot (1 \otimes e - e \otimes 1)^2$$

Then the element $p = \varepsilon \cdot (1 \otimes e - e \otimes 1)$ is an idempotent in $E \otimes_A E$ generating $J_{E/A}$. Thus $\Omega^1_E = I^1_{E/A} \oplus D^{0,1}_{E/A}$ and $D^{0,1}_E = p \cdot \Omega^1_E$ where we identify E-bimodules symmetric with respect to A with left $E \otimes_A E$-modules.

From that we obtain the following bigradation on the graded E-module $\Omega^\bullet_E$

$$\Omega^n_E = \bigoplus_{p+q=n} \left(\bigwedge\nolimits^p_E I^1_{E/A} \otimes_E \bigwedge\nolimits^q_E D^{0,1}_{E/A}\right)$$

Our proof will be done if we show that the operator Φ_e is invertible on the prime summand $D^{0,1}_{E/A}$ in Ω^1_E. Indeed, then by Lemma 2 Φ_e will be invertible on the prime summand $\bigwedge^n_E D^{0,1}_{E/A}$ in Ω^n_E so its kernel equal to $I^n_{E/A}$ will be included in

$$\bigoplus_{k>0} \left(\bigwedge\nolimits^k_E I^1_{E/A} \otimes_E \bigwedge\nolimits^{n-k}_E D^{0,1}_{E/A}\right) = I^1_{E/A} \wedge \Omega^{n-1}_E$$

But $D^{0,1}_{E/A} = p \cdot \Omega^1_E$ so the multiplication by $\varepsilon \cdot (1 \otimes e - e \otimes 1)$ is an identity on $D^{0,1}_{E/A}$. Since Φ_e on Ω^1_E is simply a multiplication by $(1 \otimes e - e \otimes 1)$ then we see that it is invertible on $D^{0,1}_{E/A}$ □

Since over a henselian ring a finite étale algebra is generated by a single element (a lift of a primitive element of the extension of the residual field, [34], Chap. I, §, Prop. 4) then we obtain

Corollary 1 *If A is a henselian ring then $I^\bullet_{E/A}$ is generated by $I^1_{E/A}$.*

Next, applying this fact to the local ring of a complex space ([34], Chap. Vii, §4, Example 4) we obtain

Corollary 2 *The sheaf-ideal $I^{\bullet}_{X'/X}$ is generated by $I^{1}_{X'/X}$*

Lemma 4 *If the torsion $\tau_{X'/X}$ vanishes then the sheaf $I^{1}_{X'/X}$ is a Pfaff system satisfying the Frobenius condition.*

Proof. It is enough to show that if $\omega \in I^{1}_{E/A}$ then for all $e \in E$ $\Phi_e(d\omega) = \tau_{E/A}(\omega \otimes (e \otimes e))$ Since $\omega \in I^{1}_{E/A}$ if and only if for all $e \in E$ $\omega \circ e = e\omega$ then

$$\begin{aligned}
\tau_{E/A}(\omega \otimes (e \otimes e)) &= d\omega \circ \sigma_0^2(e) - d(\omega \circ e) \circ \sigma_1^2(e) \\
&\quad + d(\omega \circ e^2) - \omega \circ (de \circ e - e \cdot de) \\
&= d\omega \circ \sigma_0^2(e) - d(e\omega) \circ \sigma_1^2(e) \\
&\quad + d(e^2\omega) + (de \circ e - e \cdot de) \wedge \omega \\
&= d\omega \circ \sigma_0^2(e) - (de \wedge \omega + ed\omega) \circ \sigma_1^2(e) \\
&\quad + 2ede \wedge \omega + e^2 d\omega + (de \circ e - e \cdot de) \wedge \omega \\
&= d\omega \circ \sigma_0^2(e) - (de \circ e) \wedge \omega - de \wedge (\omega \circ e) + ed\omega \circ \sigma_1^2(e) \\
&\quad + 2ede \wedge \omega + e^2 d\omega + (de \circ e - e \cdot de) \wedge \omega \\
&= d\omega \circ \sigma_0^2(e) - (de \circ e) \wedge \omega - de \wedge (e \cdot \omega) + ed\omega \circ \sigma_1^2(e) \\
&\quad + 2ede \wedge \omega + e^2 d\omega + (de \circ e - e \cdot de) \wedge \omega \\
&= \Phi_e(d\omega)
\end{aligned}$$

Let us recall that then $I^{1}_{E/A}$ is locally an image of a projector, hence a prime summand in Ω^1_E. Thus we obtain

Corollary 3 *If X is a manifold then $I^{1}_{X'/X}$ is locally free.*

From these latter two corollaries and the Frobenius theorem we obtain our main theorem

Theorem 2 *An $\mathcal{E}$-structure on a manifold is integrable if and only if it is locally constant.*

Proof.

Let $U \subset X$ be an open subset trivializing the cover $X' \xrightarrow{p} X$,

$$U' = X' \times_X U = \coprod_{i=1}^{n} U_i$$

where U_i are copies of U and such that the foliation defined by the Pfaff system $I^1_{X'/X}$ restricted to U' arises from a submersion $U' \xrightarrow{f} Y$. Let $U \xrightarrow{s_i} U'$ be sections of $p\,|_{U'}$ which are isomorphisms onto U_i. Then the compositions $f_i = f \circ s_i$ are submersions. Since for every $x \in X$ $(p_*\mathcal{O}_{X'})_x$ is generated over $\mathcal{O}_{X,x}$ by orthogonal idempotents $e_1, \ldots, e_n$ which act as projectors decomposing the tangent bundle into prime sum of subbundles which are tangent to the fibers of submersions f_i. Hence we obtain locally a desired decomposition of X into a cartesian product □

Definition. We define the *analytic $\mathcal{E}$-Hodge filtration* on the De Rham complex on X' by powers of the graded differential ideal $I^\bullet_{X'/X}$

$$F^p\Omega^\bullet_{X'} = {I^\bullet_{X'/X}}^p$$

Let us note that since $I^\bullet_{X'/X}$ is generated by $I^1_{X'/X}$ then the $\mathcal{E}$-Hodge filtration has the classic form

$$F^p\Omega^n_{X'} = \bigwedge^p I^1_{X'/X} \wedge \Omega^{n-p}_{X'}$$

Next we define the p-th *analytic $\mathcal{E}$-Dolbeault complex*

$$D^{p,\bullet}_{X'/X} = Gr^p\Omega^\bullet_{X'}[p]$$

Cohomology sheaf of the p-th analytic $\mathcal{E}$-Dolbeault complex of an $\mathcal{E}$-structure we will call the *sheaf of $\mathcal{E}$-holomorphic forms* (or functions if p=0) and denote

$$\Omega^{n,hol}_{X'} := \bigoplus_{p+q=n} H^q(D^{p,\bullet}_{X'/X})$$

Using the integral complex convention we can also write

$$\Omega^{\bullet,hol}_{X'} := H(D^\bullet_{X'/X})$$

Let us note that in the case of a constant $\mathcal{E}$-structure the sheaf-complex of $\mathcal{E}$-holomorphic forms is isomorphic to the inverse image of the sheaf-complex of forms on Y:

$$\Omega^{\bullet,hol}_{X'} = f^{-1}\Omega^\bullet_Y$$

because then

$$I^1_{X'/X} = f^*\Omega^1_Y$$

hence the analytic $\mathcal{E}$-Hodge filtration on the De Rham complex turns to be the Leray filtration with respect to the submersion $f : X' \to Y$.

4 Functorial Properties of Analytic $\mathcal{E}$-Dolbeault Complex

Now we will consider the functoriality of that analytic E-Dolbeault complex. Let us take the commutative squares as above

$$\begin{array}{ccc} E & \leftarrow & F \\ \uparrow & & \uparrow \\ A & \leftarrow & B \end{array}$$

and

$$\begin{array}{ccc} \Omega^1_B \otimes_B F & \to & \Omega^1_B \\ \downarrow & & \downarrow \\ \Omega^1_A \otimes_A E & \to & \Omega^1_A \end{array}$$

and let us assume additionally that the appropriate morphism of E-algebras

$$E \otimes_A E \leftarrow E \otimes_B F$$

is surjective.

The latter condition is a generalization of the structure of a morphism of finite Galois extensions.

Namely, let us assume that for an A-algebra E there is given a co-group object $G_{E/A}$ in the category of A-algebras and a *principal* co-action

$$E \otimes_A G_{E/A} \leftarrow E$$

which means that the appropriate morphism of A-algebras (co-graph of co-action)

$$E \otimes_A G_{E/A} \leftarrow E \otimes_A E$$

is an isomorphism. If for a B-algebra F with a principal co-action of a co-group object $G_{F/B}$ one considers the commutative diagrams

$$\begin{array}{ccc} E & \leftarrow & F \\ \uparrow & & \uparrow \\ A & \leftarrow & B \end{array}$$

$$\begin{array}{ccc} G_{E/A} & \leftarrow & A \\ \uparrow & & \uparrow \\ G_{F/B} & \leftarrow & B \end{array}$$

and

$$\begin{array}{ccc} E \otimes_A G_{E/A} & \leftarrow & E \\ \uparrow & & \uparrow \\ F \otimes_B G_{F/B} & \leftarrow & F \end{array}$$

such that the appropriate morphism of co-group objects over A

$$G_{E/A} \leftarrow A \otimes_B G_{F/B}$$

is surjective, then one obtains the surjective homomorphism of E-algebras

$$E \otimes_A E \leftarrow E \otimes_B F$$

Indeed, ofter tensorising the surjective morphism of co-group objects over A by E one obtains the surjective morphism of E-algebras

$$E \otimes_A G_{E/A} \leftarrow E \otimes_B G_{F/B}$$

But since mentioned co-actions are principal then $E \otimes_A E = E \otimes_A G_{E/A}$ and $E \otimes_B G_{F/B} = E \otimes_F (F \otimes_B G_{F/B}) = E \otimes_F (F \otimes_B F) = E \otimes_B F$ what proves the desired surjectivity.

If now we assume that $A \to E$ and $B \to F$ are finite Galois extensions with a given injective homomorphism of Galois groups $Gal(E/A) \to Gal(F/B)$ and the diagram

$$\begin{array}{ccc} E & \leftarrow & F \\ \uparrow & & \uparrow \\ A & \leftarrow & B \end{array}$$

is $Gal(E/A)$-equivariant then we can define the following co-group objects

$$G_{E/A} = \prod_{g \in Gal(E/A)} A_g$$
$$G_{F/B} = \prod_{g \in Gal(F/B)} B_g$$

where

$$A_g = A$$
$$B_g = B$$

with co-group structures which are products of identities

$$A_{g'} \otimes_A A_{g''} \leftarrow A_{g'g''}$$
$$B_{g'} \otimes_B B_{g''} \leftarrow B_{g'g''}$$

and with right principal co-actions

$$E \otimes_A G_{E/A} = \prod_{g \in G} E_g \leftarrow E$$
$$F \otimes_B G_{F/B} = \prod_{g \in G} F_g \leftarrow F$$

which are induced by group actions.

Then the induced morphism of algebras $G_{E/A} \leftarrow A \otimes_B G_{F/B}$ is of course surjective.

We are coming back to considerations about functorial properties of the analytic $\mathcal{E}$-Dolbeault complexes. Let us tensorise the short exact sequence

$$0 \to J_{F/B} \to F \otimes_B F \to F \to 0$$

of left F-modules by E and put it into the exact commutative diagram

$$\begin{array}{ccccccccc} & & & & & & 0 & & \\ & & & & & & \downarrow & & \\ & & E \otimes_F J_{F/B} & \to & E \otimes_B F & \to & E & \to & 0 \\ & & \downarrow & & \downarrow & & \downarrow & & \\ 0 & \to & J_{E/A} & \to & E \otimes_A E & \to & E & \to & 0 \\ & & \downarrow & & \downarrow & & \downarrow & & \\ & & 0 & & 0 & & 0 & & \end{array}$$

From that we obtain the surjectivity of the appropriate morphism of E-modules

$$J_{E/A} \leftarrow E \otimes_F J_{F/B}$$

Then the morphism

$$\bigodot_E^n J_{E/A} \leftarrow \bigodot_E^n (E \otimes_F J_{F/B}) = E \otimes_F \bigodot_F^n J_{F/B}$$

is also surjective. So, by the definition of $I^n_{E/A}$ as a $\bigodot^n_E(E \otimes_A E)$-submodule in Ω^n_E annihilated by the ideal $\bigodot^n_E J_{E/A}$ the morphism of F-modules $\Omega^n_E \leftarrow \Omega^n_F$ determines the morphism of graded ideals $I^n_{E/A} \leftarrow I^n_{F/B}$, hence the morphism of E-Hodge filtrations

$$F^p\Omega^\bullet_E \leftarrow F^p\Omega^\bullet_F$$

and consequently the morphisms of analytic E-Dolbeault complexes and complexes of E-holomorphic p-forms.

5 Calculations for Tautological $\mathcal{E}$-Structure on Space of Monic Separable Polynomials

Now let us show how the whole machinery works in that particular case. So, let us recall that now X is a complement in $\mathbb{A}^n$ (with coordinates $\sigma_1, \ldots, \sigma_n$) to the zero-set of the discriminant $D_n = D_n(\sigma_1, \ldots, \sigma_n)$ of the polynomial

$$P(T) = T^n - \sigma_1 T^{n-1} + \ldots + (-1)^n \sigma_n$$

Then X is a Stein manifold. If we denote

$$A := \mathcal{O}_X(X)$$

and define the tautological finite étale A-algebra

$$E := A[T]/(P(T))$$

then we can realize $X' = \mathbf{specan}\mathcal{E}$ as a zero-set of the polynomial $P(T)$ in $X \times \mathbb{A}^1$, where we regard the variable T as a coordinate on $\mathbb{A}^1$. Hence we have canonical projections

$$X' \xrightarrow{p} X$$

$$(\sigma_1, \ldots, \sigma_n, T) \mapsto (\sigma_1, \ldots, \sigma_n)$$

and

$$X' \xrightarrow{f} \mathbb{A}^1$$

$$(\sigma_1, \ldots, \sigma_n, T) \mapsto T$$

where p (resp. f)is a finite étale cover (resp. a submersion) of Stein manifolds. Let us denote $F := \mathcal{O}_{\mathbb{A}^1}(\mathbb{A}^1)$.

Proposition 3 *There exists the unique integrable $\mathcal{E}$-structure on X such that the root of polynomial $P(T)$ is an $\mathcal{E}$-holomorphic function.*

Proof. Let us assume that such an E-structure exists. Then the structure of a right E-module on Ω^1_A

$$\Omega^1_A \otimes_A E \to \Omega^1_A$$

provides a morphism of graded differential $\mathbb{C}$-algebras

$$\mathbb{C}[T] \to D^{0,\bullet}_{E/A}$$

extending the canonical homomorphism of $\mathbb{C}$-algebras

$$\mathbb{C}[T] \to E = D^{0,0}_{E/A}$$

$$T \mapsto e$$

Let us denote the image of T in E by e. Since $\mathbb{C}[T]$ is a graded differential algebra with trivial derivation, then $dT = 0$ in $\mathbb{C}[T]$. (Here dT does not denote the element of $\Omega^1_{\mathbb{A}^1}(\mathbb{A}^1)$, but an element of the graded differential algebra $\mathbb{C}[T]$ without nonzero elements of grading different from zero.) Therefore in $D^{0,1}_{E/A}$ must be $de = 0$ or, equivalently, in Ω^1_E must be fulfilled the following relation $de \circ e = e \cdot de$. But in Ω^1_E

$$de = P'(e)^{-1} \cdot (T^{n-1} d\sigma_1 + \ldots + (-1)^{n-1} d\sigma_n)$$

from where, since $1, T, \ldots, T^{n-1}$ are linearly independent over A and Ω^1_A is a free A-module with a basis $d\sigma_1, \ldots, d\sigma_n$, we obtain the following relations:

for $i = 1, \ldots, n-1$

$$d\sigma_i \circ T = \sigma_i d\sigma_1 - d\sigma_{i+1}$$

and

$$d\sigma_n \circ T = \sigma_n d\sigma_1.$$

Hence such a structure, if exists, is unique. But these formulas imply that $d\sigma_i \circ P(T) = 0$ so really define such a structure because the minimal polynomial of e is equal to $P(T)$. Let us observe that then for all $e_1 \in E$ $de \circ e_1 - e_1 de = 0$ because $E = A[e]$. Let us denote by $1 \otimes \tau_{E/A}$ the morphism

$$\Omega^1_E \otimes_A \bigodot^2_A E \to \Omega^2_E$$

induced by $\tau_{E/A}$. Then we obtain

$$(1 \otimes \tau_{E/A})(de \otimes (e_1 \otimes e_1))$$

$$= -d(de \circ e_1) \circ \sigma_1^2(e_1) + d(de \circ e_1^2) - de \circ (de_1 \circ e_1 - e_1 \cdot de_1)$$

$$= -d(e_1 \cdot de) \circ \sigma_1^2(e_1) + d(e_1^2 \cdot de) + (de_1 \circ e_1 - e_1 \cdot de_1) \wedge de$$

$$= -(de_1 \wedge de) \circ \sigma_1^2(e_1) + 2e_1 \cdot de_1 \wedge de + (de_1 \circ e_1 - e_1 \cdot de_1) \wedge de$$

$$= -(de_1 \circ e_1) \wedge de - de_1 \wedge (e_1 \cdot de) + 2e_1 d \cdot de_1 \wedge de + (de_1 \circ e_1 - e_1 \cdot de_1) \wedge de = 0$$

Since $1 \otimes \tau_{E/A}$ is E-linear we obtain

$$\sum_{i=1}^{n} (-1)^i e^{n-i} \tau_{E/A}(d\sigma_i \otimes (e_1 \otimes e_1)) = 0$$

Since Ω^2_A - free A-module and $1, \ldots, e^{n-1}$ - linearly independent over A then $\tau_{E/A}(d\sigma_i \otimes (e_1 \otimes e_1)) = 0$. Since elements of the form $(e_1 \otimes e_1)$ generate $\bigodot^2_A E$ then consequently $\tau_{E/A} = 0$ □

Proposition 4

$$D^{p,\bullet}_{X'/X} = f^* \Omega^p_{\mathbb{A}^1} \otimes \Omega^\bullet_{X'/\mathbb{A}^1}$$

$$\Omega^{\bullet,hol}_{X'} = f^{-1} \Omega^\bullet_{\mathbb{A}^1}$$

Here $\Omega^\bullet_{X'/\mathbb{A}^1}$ denotes the relative De Rham complex of the morphism

$$X' \xrightarrow{f} \mathbb{A}^1$$

Since the differential in this complex is $f^{-1}\mathcal{O}_{\mathbb{A}^1}$-linear and

$$f^*\Omega^p_{\mathbb{A}^1} \otimes \Omega^\bullet_{X'/\mathbb{A}^1} = f^{-1}\Omega^p_{\mathbb{A}^1} \otimes_{f^{-1}\mathcal{O}_{\mathbb{A}^1}} \Omega^\bullet_{X'/\mathbb{A}^1}$$

then one can extend this differential on the right-hand side tensor product acting trivially on the first factor, which defines the structure of a complex on the left-side tensor product. In the first equality in the above proposition we mean equality of complexes.

Proof. Since $\mathcal{E}$ is generated over $\mathcal{O}_{X'/X}$ by a single element e (an image of T), then it is enough to prove that

$$I^1_{E/A} = E \otimes_F \Omega^1_F$$

which is equivalent to

$$I^1_{X'/X} = f^*\Omega^1_{\mathbb{A}^1}$$

because then the analytic $\mathcal{E}$-Hodge filtration on the De Rham complex turns to be the Leray filtration with respect to the submersion $f : X' \to \mathbb{A}^1$.

We will show that if $\omega \circ e - e\omega = 0$ then ω is proportional to $de = f^*dT$. Let

$$\omega = \sum_{i=1}^{n}(-1)^i e_{n-i} d\sigma_i$$

Then

$$\omega \circ e - e\omega =$$

$$= \sum_{i=1}^{n}(\sum_{j=1}^{n}(-1)^j e_{n-j}\sigma_j)\delta_{i1} + (-1)^i(e_{n-i+1} - ee_{n-i}))d\sigma_i$$

Since $d\sigma_i$ form a basis of Ω^1_E over E, then

$$\sum_{i=1}^{n}(\sum_{j=1}^{n}(-1)^j e_{n-j}\sigma_j)\delta_{i1} + (-1)^i(e_{n-i+1} - ee_{n-i}) = 0$$

Given e_0 this equation has the unique solution $e_k = e^k e_0$. Thus

$$\omega = e_0 \sum_{i=1}^{n}(-1)^i e^{n-i} d\sigma_i$$

But

$$\sum_{i=1}^{n}(-1)^i e^{n-i} d\sigma_i = -P'(e)de$$

because $P(e) = 0$, hence

$$\omega = -e_0 P'(e) de$$

□

Corollary 4 *The local submersions f_i glue together (as a multivalued function) to the root of the tautological polynomial $P(T)$.*

This means that in this way arises on X an atlas with local charts $(f_1, \ldots, f_n)$ with permutations of coordinates as transition maps, in which

$$P(T) = (T - f_1) \ldots (T - f_n)$$

where $(f_1, \ldots, f_n)$ are these ones from the proof of the main theorem. In such coordinates the endomorphism e is a diagonal matrix with the diagonal entries $(f_1, \ldots, f_n)$.

Corollary 5 *A section $f \in H^0(X', \mathcal{O}_{X'}) = H^0(X, p_* \mathcal{O}_{X'})$ is a solution of the algebraic equation $P(T) = 0$ if and only if $df \neq 0$ and f is a solution of a partial differential equation*

$$df \circ e - f df = 0$$

Proof. We have to prove only that if f satisfies the above partial differential equation and $df \neq 0$ then it is a solution of the equation $P(T) = 0$. But by induction we then get

$$df \circ e^k = f^k df$$

which implies

$$P(f) df = df \circ P(e) = 0$$

hence $P(f) = 0$, because $df \neq 0$ □

References

[1] Al-Aqeel, A.: *On Riemannian 4-symmetric manifold.* Tensor, N.S. 48, No. 2, 120–127 (1989)

[2] Atanasiu, Gh.; Klepp, F.C.: *p-structures of constant rank.* Lucr. Semin. Mat. Fiz. 1987, 41–46

[3] Bureŝ, J.; Vanẑura, J.: *Metric Polynomial structures.* Kôdai Math. Sem. Rep. 27 (1976), 345–352

[4] Cabau, P.; Grifone, J.; Mehdi, M.: *Existence de lois de conservation dans le cas cyclique.* Ann. Inst. Henri Poincaré, Phys. Théor. 55, No.3, 789–803, (1991)

[5] Demetropoulou-Psomopoulou, D. *On integrability conditions of a structure* f *satisfying* $f^{2\nu+3}+f=0$. Tensor, N.S. 42, 252–257 (1985)

[6] Demetropoulou-Psomopoulou, D. *Linear connections on manifolds admitting an* $f(2\nu+3,-1)$*-structure or* $f(2\nu+3,1)$*-structure.* Tensor, N.S. 47, No.3, 235–239 (1988)

[7] Deszyński, K. *A theorem on polynomial Lorentz structures.* Glasg. Math. J. 28, 229–235 (1986)

[8] Deszyński, K. *Notes on polynomial structures equipped with a Lorentzian metric.* Univ. Iagell. Acta Math., Fasc. XXVIII (1991), 11–18

[9] Eckmann B., Frölicher A. *Sur l'intégrabilité de structures presques complexes.* C. r. Acad. Sci. Paris, 1951, 232, p.2284–2286

[10] Ehresmann, C.: *Sur les varietés presque complexes.* Proc. Intern. Congr. Math., Cambridge (Mass.), 1950, p.412–419

[11] Fischer, G.: *Complex Analytic Geometry.* Lecture Notes in Math. 538, Springer - Verlag, 1976

[12] Frölicher A. *Zur Differentialgeometrie der komplexen strukturen.* Math. Ann., 1955, 129, 50–95

[13] Goldberg, S.I.; Yano, K. *On normal globally framed* f*-manifolds.* Tôhoku Math. J. 22 (1970), 362–370

[14] Gouli-Andreou, F.: *On a structure defined by a tensor field* f *satisfying* $f^5+f=0$ Tensor, N.S. 36 (1982), 79–84

[15] Gouli-Andreou, F.; Demetropoulou-Psomopoulou, D.: *On connections on manifolds admitting* $F(5,1)$*-structure.* Demonstr. Math. 21, No. 1, 127–140 (1980)

[16] Gouli-Andreou, F.: *On integrability conditions of a structure f satisfying $f^5 + f = 0$.* Tensor, N.S.,40 (1983), 27–31

[17] Ishihara, S.: *On a tensor field Φ_i^h satisfying $\Phi^n = \pm 1$.* Tôhoku Math. J., 13 (1961), 443–454

[18] Ishihara, S.; Yano, K. *On integrability conditions of a structure f satisfying $f^3 + f = 0$.* Q. J. Math., Oxf.II., Ser.15, 217–222 (1964)

[19] Kanemaki, S.: *Twisted f-structures. I, II* Tensor, N.S. 27 (1973), 63–69, 70–72

[20] Kanemaki, S.:*Product of f-manifolds.* TRU Math. 10 (1974), 11–17

[21] Kobayashi, E.: *A remark of the Nijenhuis Tensor.* Pac. J. Math., 12, No. 3 (1962), 963-977

[22] Kotô, S.: *Infinitesimal transformations of a manifold with f-structure.* Kôdai Math. Sem. Rep. 16 (1964), 117–126

[23] Malgrange, B.: *Sur l'intégrabilité des structures presque-complexes.* Symp. Math., vol. II, (INDAM, Rome, 1968), Academic Press, London, 1969, 289–296

[24] Maszczyk, T.:*Solving algebraic equations in hyperradicals.* (to appear)

[25] Mishra, R.S.: *A differentiable manifold with f-structure of rank r.* Tensor, N.S., 27 (1973), 369–378

[26] Nakagawa, H.:*f-structures induced on submanifolds in spaces, almost Hermitian or Kaehlerian.* Kôdai Math. Sem. Rep. 18 (1966), 161–183

[27] Nakagawa, H.:*On the automorphism groups of f-manifolds.* Kôdai Math. Sem. Rep. 18 (1966), 251–257

[28] Nakagawa, H.:*On framed f-manifolds.* Kôdai Math. Sem. Rep. 18 (1966), 293–305

[29] Newlander A., Nirenberg L. *Complex analytic coordinates in almost complex manifolds.* Ann. Math., 65, 391–404 (1954)

[30] Nikić, J.: *On integrability conditions of an $f(2k+1,-1)$-structure.* Diff. Geom.. and it's applications, Proc. Conf. Dubrovnik/ Yugosl. 1988, 253–265 (1989)

[31] Nivas,R., Prasad, C.S. *Integrability conditions of a structure satisfying* $f^5 - a^2 f = 0$. Demonstr. Math. 18, 823–831

[32] Okubo, S. *The Nijenhuis tensor, BRST cohomology and related algebras.* Algebras Groups Geom. 6, No.1, 65–112 (1989)

[33] Opozda, B.: *Almost product and almost complex structures generated by polynomial structures.* Univ. Iagell. Acta Math., Fasc. XXIV (1984), 23–32

[34] Raynaud, M.: *Anneaux Locaux Henséliens.*Lect. Notes in Math. 169, Berlin 1970, Springer - Verlag

[35] Roger, C.; Donato, P.; Duval, C.; Elhalad, J.; Tuynman, G.M.; (Souriau, J.-M.) *Symplectic geometry and mathematical physics.* Actes du colloque de géométrie symplectique et physique mathématique en l'honneur de Jean-Marie Souriau, Aix-en-Provence, France, June 11-15, 1990, (Graded Lie algebras and quantization.) Proc.Colloq., Aix-en-Provence/Fr.1990, Prop. Math. 99, 374–421 (1991)

[36] Sasaki, S.: *On differentiable manifolds with certain structures which are closely related to almost contact structure, I.* Tôhoku Math. J. 12 (1960), 459–476

[37] Sasaki, S.: Hatekayama, Y.: *On differentiable manifolds with certain structures which are closely related to almost contact structure, II.* Tôhoku Math. J. 13 (1961), 281–294

[38] Upadhay, M.D., Gupta, V.C.: *Integrability conditions of a structure f satisfying* $f^3 - \lambda^2 f = 0$. Math. Debrecen, 24 (1977), 249–255

[39] Vanẑura, J.: *Integrability Conditions for Polynomial Structures.* Kôdai Math. Sem. Rep. 27 (1976), 42–50

[40] Vassiliev, V. A.: *Complements of Discriminant of Smooth Map.* Topology and Applications. (Rev. ed.) Transl. of Math. Monographs, vol. 98, AMS, Providence RI, 1994

[41] Yano, K.: *On a structure f satisfying $f^3 + f = 0$*. Technical Report, No. 2, June 20 (1961), Univ. of Washington

[42] Yano, K.: *On a structure defined by a tensor field of type (1,1) satisfying $f^3 + f = 0$*. Tensor, N.S. 14 (1963), 99–109

[43] Yano, K., Ishihara, S.: *The f-structure on submanifolds of complex and almost complex spaces*. Kôdai Math. Sem. Rep. 18 (1966), 120–160

Semi-positivity and Cotangent Bundles

Keiji Oguiso, Department of Mathematics, Ochanomizu University, Otsuka Bunkyo, 112 Tokyo, Japan. oguiso@math.ocha.ac.jp

Thomas Peternell, Mathematisches Institut der Universität Bayreuth, Postfach 101251, D-95440 Bayreuth, Germany. thomas.peternell@uni-bayreuth.d400.de

Introduction

Let (X, L) be a polarised projective manifold, i.e. a projective manifold with an ample line bundle L. Suppose that the cotangent bundle Ω^1_X of X is not nef, i.e. the natural line bundle $\mathcal{O}(1)$ on the projectivised bundle $\mathbf{P}(\Omega^1_X)$ is not nef. Then there exists a unique positive real number α such that the "virtual bundle" $\Omega^1_X(\alpha)$ is nef but not ample. Since ampleness is a numerical property it does not matter whether this twist of the cotangent bundle has a geometric meaning or not, i.e. is rational or not . We call the number α the *cotangent value* of (X, L) or the *nef value* of the cotangent bundle with respect to the polarisation L. The main theme of this paper is the question whether the nef value is of importance for the geometry of X. We try to compute nef values mainly on surfaces and on Calabi-Yau manifolds as we explain in a moment.

The paper is organised as follows. The first section is devoted to the introduction of nef values for arbitrary vector bundles. Then we prove some Nakai-Moishezon type results on surfaces which are used later to investigate nef values. In sect.2 we explicitly compute the cotangent value of ruled surfaces with respect to any polarisation. It turns out that the cotangent value is always rational. So one might ask whether this is always the case, in particular having in mind the rationality theorem of Kawamata-Shokurov saying that the nef value of the canonical bundle (if it is not nef) is rational for every polarisation, which i s one the cornerstones of the birational classification of projective varieties. However we show in sect.3 that every quartic surface S in $\mathbf{P}_3$ with Picard number 1 has cotangent value $\sqrt{2}$ with respect to the embedding line bundle. The number $\sqrt{2}$ comes from the vanishing of the Segre class $s_2(\Omega^1_S(\alpha))$, which is the same as the highest (= third) self-intersection number of the line bundle $L = \mathcal{O}(1)$ on the threefold $X = \mathbf{P}(\Omega^1_S(\alpha))$. The vanishing of s_2 turns out to be the only obstacle for $\Omega^1_S(\sqrt{2})$ to be ample: $\Omega^1_S(\sqrt{2})|C$ is ample for every curve $C \subset S$ and $L^2.Y > 0$ for every surface $Y \subset X$. We now transfer this information to the projectivised bundle X. Here we look at the ample cone, the closed cone K spanned by the numerical equivalence classes of ample divisors. This cone has dimension 2, so has 2 extremal rays at the boundary. One of them is of course given by the pull-back of the ample generator on the quartic, while the other is generated by L. The non-rationality of the cotangent value says that there is no $\mathbf{Q}$−divisor on this ray. By dualising we get an analogous statement on the dual cone $\overline{NE}(X)$ of effective curves . Namely, one boundary is given by lines in the fibers of the projective bundle while there is no curve lying on the other boundary R. Consequently there mus t be a sequence C_ν of irreducibles curves whose classes converge (after dividing suitably by real numbers, i.e. normalising) to R. It would be interesting to say something on the geometry of those curves which are sufficiently near to R. What actually can be said is that given a curve $C \subset X$ sufficiently near to R, then C is "instable", i.e. $\Omega^1_S|C$ is instable after possibly normalising (which is the same as to say that $\Omega^1_S|C$ is not nef) , moreover C gets more and more instable the nearer C comes to R.

Note that for K3 surfaces which appear as 2:1 coverings of the projective plane the situation is quite different; here the cotangent value is 3 and the corresponding twist $\Omega^1_S(3)$ is even generated by global sections. This should however not be typical, we tend to believe that for all the other families of K3 surfaces (with Picard number 1) the geometry of the ample and effective cone s should be as in the quartic case (although sometimes α will be rational as solution of $s_2(\Omega^1_S(\alpha)) = 0$).

Turning to surfaces of general type with ample canonical class (sect.4) we first observe that for hypersurfaces in $\mathbf{P_3}$ the situation is completely the same as for quartics. If however $c_1(\Omega^1_S(\alpha)) > 2c_2(\Omega^1_S(\alpha))$, where α is the cotangent value with respect to the canonical class, then α is rational and moreover there exists a curve $C \subset X$ such that $\Omega^1_S(\alpha)|C$ is nef but not ample. Much more can be said about the structure of the cones on the projectivised cotangent bundle, see sect.4.

Now we consider higher dimensions. Hypersurfaces S in $\mathbf{P_n}$ of degree $n+1$ have cotangent value 2; the reason is that there are lines on S, preventing

$\Omega^1_S(2)$ from being ample. These lines are the only curves in S on which $\Omega^1_S(2)$ is not ample (5.4). We let again $X = \mathbf{P}(\Omega^1_S(2))$ and $L = \mathcal{O}_{\mathbf{P}(\Omega^1_S(2))}(1)$. Since $\Omega^1_S(2)$ is even generated by global sections, so is L and we can consider the map Φ given by the global sections of L. In case $n = 4$, i.e. for quintic threefolds, we explicitly describe the structure of Φ. Indeed, Φ is birational, it contracts only smooth rational curves whose images in S are lines. Except in the case where the normal bundle of the line in S is of the form $\mathcal{O}(2) \oplus \mathcal{O}(-4)$, the fibers of Φ are isolated. The situation for Calabi-Yau hypersurfaces of any dimension is analogous.

This description of $\Omega^1_S(\alpha)$ for hypersurfaces of degree $n+1$ of course raises the question whether something similar holds for every Calabi-Yau manifold with Picard number 1 (of dimension ≥ 3.) This seems to be a very hard but also very interesting problem, in particular since it is connected with the existence problem for rational curves in Calabi-Yau manifolds.

Notations

A polarised manifold (X, L) is a projective manifold X with an ample line bundle L. We denote by Ω^1_X the cotangent bundle on X.

A line bundle L is nef iff $L.C \geq 0$ for every curve $C \subset X$. A vector bundle E is called nef if $\mathcal{O}(1)$ on the projectivised bundle $\mathbf{P}(E)$ is nef. The projective bundle is always understood in Grothendieck's sense, i.e. taking hyperplanes.

We let $N^1(X)$ be the vector space generated by the numerical equivalence classes of irreducible divisors on X and $N_1(X)$ that one generated by the classes of irreducible curves. The Picard number is defined by $\rho(X) = \dim \ N^1(X)$. Inside $N^1(X)$ we have the ample cone K, i.e. the closed cone generated by the classes of ample divisors, and analogously the effective cone $\overline{K}_{eff}$. Inside $N_1(X)$ we have the closed cone $\overline{NE}(X)$ generated by the classes of irreducible curves. Let E be a vector bundle of rank r and L a line bundle on X; furthermore let α be a real irrational number. Then by $\mathcal{O}_{\mathbf{P}(E \otimes L^\alpha)}(1)$ we understand the class $c_1(\mathcal{O}_{\mathbf{P}(E)}(1)) + \alpha\pi^*(L)$ in $\mathrm{N}^1(\mathbf{P}(E))$. Concerning Chern classes we define

$$c_1(E \otimes L^\alpha) = c_1(E) + r\alpha c_1(L),$$

similarly for $c_2(E \otimes L^\alpha)$. Then numerical Chern class inequalities which hold for integer or rational α remain still true (by taking limits).

This paper is in final form and no part of it will be submitted elsewhere.

1.Nef Values

In this section we introduce the notion of the nef value of a vector bundle E on a polarised projective manifold (X, L). The most important special case is $E = \Omega^1_X$ when we obtain the "cotangent value " of (X, L). In order to be able to compute nef values explicitly we then discuss Nakai - Moishezon type theorems on surfaces.

1.1 Definition

Let (X, L) be a polarised projective manifold, i.e. L is an ample line bundle on X. Let E be a vector bundle on X. The nef value of E on (X, L) or nef value of E with respect to L is the uniquely determined real number α such that $E \otimes L^\alpha$ is nef but not ample. We write $\alpha = \alpha(X, L)$. In case $E = \Omega^1_X$ we say that α is the cotangent value of (X, L) or of X with respect to L.

1.2 Remark Of course, L^α is only a formal expression. If we write $L = \mathcal{O}_X(D)$ with a divisor D, then L^α should be understood as the real divisor αD or as element of $N^1(X)$. By definition, $E \otimes L^\alpha$ is nef if the real divisor

$$\mathcal{O}_{\mathbf{P}(E \otimes L^\alpha)}(1) = \mathcal{O}_{\mathbf{P}(E)}(1) \otimes \pi^*(L^\alpha)$$

is nef on $\mathbf{P}(E)$. Again this equation should be read in $N^1(\mathbf{P}(E))$.
Alternatively, $E \otimes L^\alpha$ is nef iff for all rational β, $\beta > \alpha$, the $\mathbf{Q}$-vector bundle $E \otimes L^\beta$ is ample. If $\beta = \frac{p}{q}$ with $p \in \mathbf{Z}$, $q \in \mathbf{N}$, this is to say that $S^q E \otimes L^p$ is ample. Usually we will write $\mathcal{O}_X(1)$ instead of L; obviously we consider only non-divisible line bundles. Note also that the nef value of E depends only on the numerical equivalence class of L.

1.3 Remark If $E = \Omega^n_X, n = \dim X$, then the rationality theorem of Shokurov says that the nef value of E w.r.t. any L is always rational. For $E = \Omega^1_X$ this is not always the case as we shall see in sect.3, however the cotangent value still has some geometric significance. Our first aim is to try to compute nef values of rank 2-bundles on surfaces. For this we will need the real version of the Nakai-Moishezon criterion as proved in [CP90]:

1.4 Proposition

Let X be a projective manifold, $D \in N^1(X)$. D is ample, i.e. in the interior of the ample cone of X iff $(D^s.Y) > 0$ for every $0 \leq s \leq \dim X$ and every irrreducible s-dimensional algebraic subset $Y \subset X$.

We start with the following real analog of a result of Schneider - Tancredi [ST85].

1.5 Proposition

Let X be an algebraic surface, $\mathcal{O}_X(1)$ an ample line bundle on X and E a vector bundle

of rank 2 on X. Let $\alpha \in \mathbf{R}$. Assume :
(1) $c_1^2(E(\alpha)) > 2c_2(E(\alpha))$, $c_2(E(\alpha)) > 0$
(2) $E(\alpha)|C$ is ample for every irreducible curve $C \subset X$
(3) E is semistable with respect to $\det E(\alpha)$.
Then $E(\alpha)$ is ample.

Proof. Consider the real divisor $L = \mathcal{O}_{\mathbf{P}(E(\alpha))}(1)$ on $\mathbf{P}(E)$. We are going to verify that L is ample, following the lines of [ST85] in the "rational case". By (1.4) we have to prove
(a) $c_1(L)^3 > 0$
(b) $c_1(L)^2.Y > 0$ for every surface $Y \subset \mathbf{P}(E)$
(c) $c_1(L).C > 0$ for every curve $C \subset \mathbf{P}(E)$.

(a) and (c) are straightforward and completely analogous to [ST85]. Note that (a) is nothing than the inequality $c_1^2(E(\alpha) > c_2(E(\alpha))$ which holds by (1). (c) is immediate from (2). The only not obvious point is (b) and here semi-stability comes into the game. Pick an irreducible surface $Y \subset \mathbf{P}(E)$. Write

$$\mathcal{O}_{\mathbf{P}(E)}(Y) = \mathcal{O}_{\mathbf{P}(E(\alpha))}(m) \otimes \pi^*(F) = \mathcal{O}_{\mathbf{P}(E)}(m) \otimes \pi^*(F \otimes \mathcal{O}_X(m\alpha))$$

with some $m \in \mathbf{Z}$ and $F \in \mathrm{Pic}(X) \otimes \mathbf{R}$. If $m = 0$,then (b) is easily verified, so we concentrate on $m > 0$. The surface Y defines a section in $S^m E \otimes F \otimes \mathcal{O}_X(m\alpha)$, hence we have an inclusion

$$F^* \otimes \mathcal{O}_X(-m\alpha) \longrightarrow S^m E.$$

In order to make computations easier, i.e. similar to [ST85], we write formally

$$F^* \longrightarrow S^m(E(\alpha)).$$

Alternatively, one may consider the Harder-Narasimhan filtration for real polarizations. By semi-stability (which is invariant under twists) we obtain as in [ST85]

$$\frac{m}{2} c_1^2(E(\alpha))) + c_1(F).c_1(E(\alpha)) \geq 0,$$

and the conclusion is as in [ST85]; here the strong inequality $c_1^2 > 2c_2$ comes into the game:

$$\begin{aligned} c_1(L)^2.Y &= m(c_1(E(\alpha))^2 - c_2(E(\alpha))) + c_1(F).c_1(E(\alpha)) \\ &\geq m(c_1(E(\alpha))^2 - c_2(E(\alpha))) - \frac{m}{2} c_1(E(\alpha))^2 \\ &= \frac{m}{2}(c_1(E(\alpha))^2 - 2c_2(E(\alpha))) > 0. \end{aligned}$$

Of course, semi-stability is not easily verified. In order to avoid this assumption notice that condition (2) in (1.5) already implies for the line bundle $L = \mathcal{O}_{\mathbf{P}(E(\alpha))}(1)$:
(α) $c_1(L)^3 \geq 0$

(β) $c_1(L)^2.Y \geq 0$ for every surface $Y \subset \mathbf{P}(E)$.

Hence we have :

1.6 Proposition

Let E and α be as in

(1.5). Assume:

(1) $E(\alpha))|C$ is ample for every irreducible curve $C \subset X$

(2) $c_1^2(E(\alpha)) \neq c_2(E(\alpha))$

(3) there is no irreducible surface $Y \subset \mathbf{P}(E)$ with $c_1^2(L).Y = 0$ where $L = \mathcal{O}_{\mathbf{P}(E(\alpha))}(1)$.

Then $E(\alpha))$ is ample.

1.7 Corollary

If $E(\alpha)$ is nef but not ample, then one of the following assertions holds .

(1) $c_1^2(E(\alpha)) = c_2(E(\alpha))$

(2) there is a curve $C \subset X$ such that $E(\alpha)|C$ is nef but not ample

(3) there is a surface $Y \subset \mathbf{P}(E)$ such that $c_1(\mathcal{O}_{\mathbf{P}(E(\alpha))}(1))^2.Y = 0$.

1.8 Remark 1.7(1) and 1.7(3) give quadratic equations for α, however 1.7(2) is a linear condition. In fact, let us consider $\mathbf{P}(E|C)$. If C is not smooth, normalise and do the following computation on the normalisation. We shall use the notations of sect. 2. We may assume E to be normalised in the sense of sect.2 and let $e = -c_1(E)$. Let C_0 be a section with $C_0^2 = -e$ and F a fiber of the ruling. Then $E(\alpha)$ is nef but not ample iff $C_0 + F$ is. Then [Ha77,chap.5.2] implies

$$\alpha = e \text{ if } \; e \geq 0$$

$$\alpha = \frac{e}{2} \text{ if } \; e < 0.$$

We discuss finally the semi-continuity of the nef value. So let (X_t) be a family of projective manifolds over the unit disc Δ in $\mathbf{C}$, le t (L_t) be a family of ample line bundles, and (E_t) be a family of vector bundles on the X_t. We consider the function α associating to every $t \in \Delta$ its nef value $\alpha_t(E_t, L_t)$. The openness of ampleness immediately implies that the function α is upper semi-continous. Actually we have more :

1.9 Proposition

Given $t_0 \in \Delta$ there is a neighborhood U of t_0 in Δ such that $\alpha(t) \leq \alpha(t_0)$ for all $t \in U$.

Proof. Assuming the contrary there is a sequence (t_ν) converging to t_0 such that $\alpha(t_\nu) > \alpha(t_0)$ for all ν. Now pass to the family $(\mathbf{P}(E_\nu))$. Let H_ν be the corresponding natural line bundle. Then by Nakai-Moishezon applied to the real divisor $D_{\alpha,\nu} = H_\nu \otimes \pi_\nu^*(L_\nu^{\alpha(t_\nu)})$ we obtain an irreducible subvariety $Y_\nu \subset \mathbf{P}(E_\nu)$ of dimension n_ν such that

$$D_{\alpha,\nu}.Y_\nu = 0.$$

Passing to a subsequence we may assume that all Y_ν have the same dimension n_0. Moreover we may assume that the Y_ν appear in an analytic family (the equations give a boundedness property), so by closedness of the components of the relative cycle space we find Y_0 as limit of the Y_ν. Hence $D_{\alpha,0}.Y_0 = 0$. But since all Y_{t_ν} are homologically the same, it is clear that also the divisors $D_{\alpha,\nu}$ have the same Chern classes which contradicts $\alpha(t_\nu) > \alpha(t)$.

Finally let us remark that by the real Nakai-Moishezon criterion the nef value of a vector bundle is always an algebraic number (actually a zero of a polynomial with integral coefficients of degree at most $2\dim X - 1$).

2. Ruled Surfaces

In this section we begin computing cotangent values of surfaces. Of course the cotangent value of $(\mathbf{P_n}, \mathcal{O}_{\mathbf{P}_n}(1))$ is 2 for all $n \in \mathbf{N}$. Furthermore the cotangent value of $(\mathbf{P}_1 \times \mathbf{P}_1, \mathcal{O}(1,1))$ is 2.

We now turn to ruled surfaces. So let C be a compact Riemann surface, let E be a vector bundle of rank 2 on C and let $S = \mathbf{P}(E)$ with projection map π to C. In the following we make freel use of the theory of ruled surfaces as presented in [Ha77, chap.5,sect.2]. Let E be normalised, i.e. $H^0(E) \neq 0$, but $H^0(E \otimes L) = 0$ for all line bundles L on C of negative degree. Put $e = -c_1(E)$. We fix an ample line bundle L on S and write for numerical equivalence

$$L \equiv \mathcal{O}_{\mathbf{P}(E)}(\mu) \otimes \pi^*(\mathcal{O}_C(\nu)).$$

We consider the exact sequence of cotangent sheaves

$$0 \longrightarrow \pi^*(\Omega^1_C) \longrightarrow \Omega^1_S \longrightarrow \Omega^1_{S|C} \longrightarrow 0 \qquad (*).$$

The relative cotangent sheaf $\Omega^1_{S|C}$ is given by

$$\Omega^1_{S|C} = \mathcal{O}_{\mathbf{P}(E)}(-2) \otimes \pi^*(detE).$$

Let $\beta \in \mathbf{R}_+$ be the unique number such that

$$\Omega^1_{S|C} \otimes L^\beta = \mathcal{O}_{\mathbf{P}(E)}(\beta\mu - 2) \otimes \pi^*(\mathcal{O}_C(\beta\nu - e))$$

is nef but not ample, i.e. the nef value of $\Omega^1_{S|C}$ with respect to L. Then we have

$$\beta = \frac{2}{\mu} \qquad (**).$$

In fact, if $\beta = \frac{2}{\mu}$, then $\beta\nu - e = \frac{2\nu}{\mu} - e > 0$ since L is ample ([Ha77,5.2]). So $\Omega^1_{S|C} \otimes L^\beta$ is nef but not ample (use again [Ha77,5.2]). First we assume that the genus of C satisfies

$g(C) \geq 1$. Since Ω^1_C is nef, we conclude from (*) by (2.2) below that $\Omega^1_S \otimes L^\beta$ is nef but not ample, so the cotangent value $\alpha = \alpha(X, L) = \frac{2}{\mu}$. This settles the problem for $C \neq \mathbf{P_1}$. Suppose now $C = \mathbf{P_1}$. The difficulty is that Ω^1_C is not nef, hence we do not know whether $\pi^*\Omega^1_C \otimes L^\beta$ is nef. If $\pi^*\Omega^1_C \otimes L^\beta$ is nef, the same arguments as above apply and we obtain $\alpha = \frac{2}{\mu}$. So assume that $\pi^*\Omega^1_C \otimes L^\beta$ is not nef. This is equivalent to $\nu > \mu e$ (by ampleness of L we also have $\nu > \mu e$.) We claim

$$\alpha = \frac{max(e, 2)}{\nu - \mu e} \qquad (***).$$

First observe that for $e = 0$, i.e. $S = \mathbf{P_1} \times \mathbf{P_1}$, we obtain obviously $\alpha = \frac{2}{min(\nu,\mu)}$ verifying (***) in the range $\nu < \mu$. So let $e < 0$. Let C_0 be the unique section with $C_0^2 = -e$. We determine $\gamma > 0$ so that $\Omega^1_S \otimes L^\gamma | C_0$ is nef but not ample : since $\Omega^1_S | C_0 = \mathcal{O}(-2) \oplus \mathcal{O}(-e)$ we have

$$\gamma = \frac{e}{\nu - \mu e} \quad \text{if } \ e \geq 2$$

$$\gamma = \frac{2}{\nu - \mu e} \quad \text{if } \ e = 1.$$

We immediately verify that $\pi^*\Omega^1_C \otimes L^\gamma$ is nef. Tensorising (*) by L_γ proves that $\Omega^1_S \otimes L^\gamma$ is nef. Since $\Omega^1_S \otimes L^\gamma | C_0$ is not ample , we conclude $\alpha = \gamma$.

In summary we have proved

2.1 Proposition

Let $S = \mathbf{P}(E)$ be a ruled surface over the curve Cwith E a normalised rank 2-bundle. Let $L = \mathcal{O}_{\mathbf{P}(E)}(\mu) \otimes \pi^*(\mathcal{O}_C(\nu))$ be an ample line bundle where $c_1(\mathcal{O}(1))$ generates $H^2(C, \mathbf{Z})$.
Then the cotangent value $\alpha = \alpha(S, L)$ is given by

$$\alpha = \frac{2}{\mu} \quad \text{if } \ g(C) \geq 1 \ \text{ or } \ \nu \geq \mu(e+1)$$

and

$$\alpha = \frac{max(e, 2)}{\nu - \mu e} \quad \text{if } \ g = 0 \ \text{ and } \ \nu < \mu(e+1)$$

In the proof of (2.1) we used

2.2 Lemma

Let X be a projective manifold and

$$0 \longrightarrow V_1 \longrightarrow V_2 \longrightarrow V_3 \longrightarrow 0$$

an exact sequence of vector bundles. Let D be a divisor on X and $\mu \in \mathbf{Q}$.
(1) If $V_i \otimes \mathcal{O}(\mu D)$ is nef for $i = 1, 3$, then $V_2 \otimes \mathcal{O}(\mu D)$ is nef, too.
(2) If $V_2 \otimes \mathcal{O}(\mu D)$ is ample, then $V_3 \otimes \mathcal{O}(\mu D)$ is ample.

Proof : (2) is easy by passing to $\mathbf{P}(V_2)$ containing $\mathbf{P}(V_3)$. For (1) write $\mu = \frac{p}{m}$ with $p \in \mathbf{Z}$ and $m \in \mathbf{N}$. Choose a finite covering $f : X^* \longrightarrow X$ such that $f^*(\mathcal{O}_X(D))$ is divisible by m. Hence $f^*(\mathcal{O}_X(\mu D)) := \mu f^*(\mathcal{O}_X(D))$ is a well defined line bundle on X^*. Since $f^*(V_i) \otimes f^*(\mathcal{O}_X(\mu D))$ is nef for $i = 1, 3$, the exact sequence

$$0 \longrightarrow f^*(V_1) \otimes f^*(\mathcal{O}_X(\mu D)) \longrightarrow f^*(V_2) \otimes f^*(\mathcal{O}_X(\mu D)) \longrightarrow f^*(V_3) \otimes f^*(\mathcal{O}_X(\mu D)) \longrightarrow 0$$

gives the nefness of $f^*(V_2) \otimes f^*(\mathcal{O}_X(\mu D))$, i.e. of $S^m(f^*(V_2) \otimes f^*(\mathcal{O}_X(\mu D)) = f^*(S^m V_2 \otimes \mathcal{O}(pD)$, so $S^m V_2 \otimes \mathcal{O}(pD)$ is nef which gives our claim.

3. K3 surfaces

We now turn to K3 surfaces. For simplicity we will assume $\rho(S) = 1$, i.e. $\text{Pic}(S) = \mathbf{Z}$, an assumption which is satisfied by the general algebraic K3 surface. Let $\mathcal{O}_S(1)$ be the ample generator of $\text{Pic}(S)$. Let α_0 be the unique positive number such that

$$c_1^2(\Omega_S^1(\alpha_0)) = c_2(\Omega_S^1(\alpha_0)),$$

i.e. the Segre class $s_2(\Omega_S^1(\alpha_0))$ vanishes, which can also be expressed as

$$\alpha_0 = \sqrt{\frac{8}{c_1(\mathcal{O}_S(1))^2}}\ .$$

Formally, we let for a rank 2-bundle E :

$$c_1(E \otimes L^\alpha) = c_1(E) + 2\alpha c_1(L)$$

and

$$c_2(E \otimes L^\alpha) = c_2(E) + \alpha c_1(E) c_1(L) + \alpha^2 c_1(L)^2.$$

We let $d = c_1(\mathcal{O}_S(1)^2$ be the "degree" of S. The number α_0 is a candidate for the cotangent value $\alpha = \alpha(S, \mathcal{O}_S(1))$. We will however prove $\alpha = 3 > \alpha_0 = 2$ if $d = 2$, i.e. if S is a 2:1-covering over $\mathbf{P_2}$ branched along a sextic, but on the other hand $\alpha = \alpha_0$ if $d = 4$, i.e. S is a quartic in $\mathbf{P_3}$ (in general the K3 surfaces fall into infinitely many irreducible families $\mathcal{F}_n, n \in \mathbf{N}$, where $S \in \mathcal{F}_1$ has degree $d = 2$, $S \in \mathcal{F}_2$ has degree 4, and in general for $n \geq 2$, $S \in \mathcal{F}_n$ is embedded in $\mathbf{P_{n+1}}$ of degree $2n$, so $d = 2n$).

Of course we always have $\alpha \geq \alpha_0$. Let us first fix some notations. We set $X = \mathbf{P}(\Omega_S^1)$ and $L = \mathcal{O}_{\mathbf{P}(\Omega_S^1(\alpha))}(1) \in \text{Pic}(X) \otimes \mathbf{R}$. By definition, L is nef but not ample. The boundary of the ample cone $K \subset N^1(X)$ is given by $\mathbf{R}_+[\pi^*(\mathcal{O}_S(1))]$, and $\mathbf{R}_+[L]$ where $\pi : X \longrightarrow S$ is the projection.

3.1 Proposition

Let S be a K3 surface with $\rho(S) = 1$ and assume that there is a 2:1 covering $f : S \longrightarrow \mathbf{P_2}$ branched along a sextic $\Delta \subset \mathbf{P_2}$. Then the cotangent value is computed by $\alpha = \alpha(S, \mathcal{O}_S(1)) = 3$, while $\alpha_0 = 2$.

Proof. Observe $\mathcal{O}_S(1) = f^*(\mathcal{O}_{\mathbf{P_2}}(1))$. Let $\tilde{\Delta} = f^{-1}(\Delta)$; then $\mathcal{O}_S(\tilde{\Delta}) = \mathcal{O}_S(3)$. Consider the exact sequence

$$0 \longrightarrow f^*\Omega^1_{\mathbf{P_2}} \longrightarrow \Omega^1_S \longrightarrow \Omega^1_{S|\mathbf{P_2}} \longrightarrow 0 \qquad (*).$$

The sheaf $\Omega^1_{S|\mathbf{P_2}}$ is supported on $\tilde{\Delta}$ and in fact a line bundle on $\tilde{\Delta}$. An easy computation shows

$$\Omega^1_{S|\mathbf{P_2}} = N^*_{\tilde{\Delta}|S} = \mathcal{O}_S(-3)|\tilde{\Delta}.$$

Since $f^*(\Omega^1_{\mathbf{P_2}}(2))$ is nef, the sheaf $\Omega^1_S(2)$ is nef on every curve except $\tilde{\Delta}$ due to the tensorised sequence (*) and the following easy fact :

If C is an irreducible curve, $\mathcal{F}, \mathcal{G}$ locally free sheaves on C of the same rank with an injective map $\mathcal{F}$Ł$r\mathcal{G}$, then $\mathcal{G}$ is nef if $\mathcal{F}$ is. Restricting to $\tilde{\Delta}$ we finally see that $\alpha = 3$.

Now we turn to quartics $S \subset \mathbf{P_3}$. Here $\alpha_0 = \sqrt{2}$, hence we are aiming for

3.2 Theorem

The general quartic $S \subset \mathbf{P_3}$ (i.e. with $\rho(S) = 1$) has cotangent value $\alpha = \sqrt{2}$.

The proof needs some preparations.

3.3 Lemma

There is no effective divisor D on the boundary of the cone of effective divisors of X. In other words : there is no $\lambda \in \mathbf{Q}_+$ with

$$H^0(S^m(\Omega^1_S(\lambda))) \neq 0 \qquad (a)$$

for some $m \in \mathbf{N}$ sufficiently divisible (such that the symmetric power makes sense), and

$$H^0(S^m(\Omega^1_S(\mu))) = 0 \qquad (b)$$

for all rational $\mu < \lambda$ and all $m \in \mathbf{N}$ sufficiently divisible.

We introduce some terminology .

3.4 Definition

Let (X, L) be a polarised manifold, E a vector bundle on X. The effective value of E with respect to L is the largest real number λ such that $H^0(S^m(E \otimes L^\mu))) = 0$ for all rational $\mu < \lambda$ and all $m \in \mathbf{N}$ sufficiently divisible.

Using this terminology Lemma 3.3 is an immediate consequence of

3.5 Lemma

(1) $H^0(S^m(\Omega^1_S(1))) = 0$ for all $m \in \mathbf{N}$,
(2) The effective value of Ω^1_S (with respect to $\mathcal{O}_S(1)$) is 1.

Proof. (1) We use the exact sequence

$$0 \longrightarrow S^{m-1}\Omega^1_{\mathbf{P_3}}|S \otimes \mathcal{O}_S(-4) \longrightarrow S^m\Omega^1_{\mathbf{P_3}}|S \longrightarrow S^m\Omega^1_S \longrightarrow 0$$

and tensorise it by $\mathcal{O}_S(m)$ to obtain

$$0 \longrightarrow S^{m-1}(\Omega^1_{\mathbf{P_3}}(1))|S \otimes \mathcal{O}(-3) \longrightarrow S^m(\Omega^1_{\mathbf{P_3}}(1))|S \longrightarrow S^m(\Omega^1_S(1)) \longrightarrow 0 \qquad (*).$$

From the Euler sequence we have easily :

$$H^0(S^m(\Omega^1_{\mathbf{P_3}}(1))) = H^1(S^m(\Omega^1_{\mathbf{P_3}}(1)) \otimes \mathcal{O}(-4)) = 0,$$

hence $H^0(S^m(\Omega^1_{\mathbf{P_3}}(1))|S) = 0$. In the same way we observe that

$$H^1(S^{m-1}(\Omega^1_{\mathbf{P_3}}(1)|S \otimes \mathcal{O}(-3))) = 0,$$

hence the sequence (*) yields the claim.

(2) By(1) it is now sufficient to show the following : for all $\epsilon > 0$ rational there is $m \in \mathbf{N}$ with

$$H^0(S^m(\Omega^1_S(1+\epsilon))) \neq 0.$$

From the sequence (*) tensorised by $\mathcal{O}(m\epsilon)$ we see that it is sufficient to show that, given $\epsilon > 0$, we have

$$H^0(S^m(\Omega^1_{\mathbf{P_3}}(1+\epsilon)|S)) \neq 0$$

for some m (choose then m minimal with this property !). To this extend it is now sufficient to know

$$H^0(S^m(\Omega^1_{\mathbf{P_3}}(1+\epsilon))) \neq 0$$

for some m. This however is obtained easily using representation theory (Bott).

3.6 Corollary

Let $S \subset \mathbf{P_3}$ be a quartic with $\rho(S) = 1$. There is no surface $Y \subset X = \mathbf{P}(\Omega^1_S)$ such that $L^2.Y = 0$.

Proof. Assume $L^2.Y = 0$ for some surface Y. Since $L^2.W \geq 0$ for every surface $W \subset X$, L being nef, Y has to be on the boundary of the effective cone of X, contradicting (3.3).

Now assume that our quartic S has cotangent value $\alpha > \sqrt{2}$.

Since $L^2.Y > 0$ for every surface $Y \subset X$, we conclude by (1.7) that there must be a curve $C_0 \subset X$ with $L.C_0 = 0$. Hence for finishing the proof of Theorem 3.2 it is sufficient to prove

3.7 Lemma

Let $S \subset \mathbf{P_3}$ be a quartic with $\rho(S) = 1$. Then there is no curve on the boundary of $\overline{NE}(X)$ except for fibers of the projection map $\pi : X \longrightarrow S$.

Proof. Assume that the class $[C] \in \partial\overline{NE}(X)$, with a curve C not being a fiber of π. Since $\dim N_1(X) = 2$, there exists a real number d such that

$$[C] = d\, c_1(\mathcal{O}_{\mathbf{P}(\Omega^1_S(\beta))}(1))^2$$

for some β. Write for short $L_\beta = \mathcal{O}_{\mathbf{P}(\Omega^1_S(\beta))}(1)$, so $L = L_\alpha$. Since $L.C = 0$ ($\overline{NE}(X)$ being the dual cone of the cone of ample divisors), we obtain

$$L.L_\beta^2 = 0.$$

Since $L = L_\beta \otimes \pi^*(\mathcal{O}_S(\alpha - \beta))$, it follows

$$0 = L_\beta^3 + (\alpha - \beta)L_\beta^2.\pi^*(\mathcal{O}_S(1)) = 12\beta^2 - 24 + 8(\alpha - \beta)\beta,$$

hence

$$\beta^2 + 2\alpha\beta - 6 = 0.$$

Now observe $\alpha \leq 2$, since $\Omega^1_S(2)$ is nef (Euler sequence), hence we conclude

$$\beta > 1.$$

Since the effective value of S is 1 with no surface on the boundary, there exists a sequence of irreducible divisors (Y_ν) in X such that

$$\frac{[Y_\nu]}{\|[Y_\nu]\|} \longrightarrow \frac{c_1(L_1)}{\|c_1(L_1)\|}$$

as ν approaches ∞ where $\| \ \|$ is some norm on $N^1(X)$ and $L_1 = \mathcal{O}_{\mathbf{P}(\Omega^1_S(1))}(1)$. Thus

$$\frac{(Y_\nu.Y_{\nu+1})}{\|Y_\nu\| \ \|Y_{\nu+1}\|} \longrightarrow \frac{c_1(L_1)^2}{\|c_1(L_1)\|^2}.$$

Since $(Y_\nu.Y_{\nu+1}) \in \overline{NE}(X)$, we conclude that $c_1(L_1)^2 \in \overline{NE}(X)$. But this implies $\beta \leq 1$, since $[C] \in \partial\overline{NE}(X)$, contradiction.

The proof of Theorem 3.2 is now complete.

(3.8) We can now describe the ample cone K and the cone of curves $\overline{NE}(X)$ of $X = \mathbf{P}(\Omega^1_S), S \subset \mathbf{P}_3$ a quartic with $\rho = 1$. First, K has boundary rays spanned by $\pi^*(\mathcal{O}_S(1))$ and by $\mathcal{O}_{\mathbf{P}(\Omega^1_S(\sqrt{2}))}(1)$ - whereas the closed cone $\overline{K}_{eff}$ of effective divisors is spanned by $\pi^*(\mathcal{O}_S(1))$ and by $\mathcal{O}_{\mathbf{P}(\Omega^1_S)}(1)$. Second, $\overline{NE}(X)$ has boundary rays $\mathbf{R}_+[F], F$ a fiber of the projection π, and $R = \{x \in N_1(X) | c_1(\mathcal{O}_{\mathbf{P}(\Omega^1_S(\sqrt{2}))}(1).x) = 0\}$.

(3.9) We shall describe the relation to stability. Let things be as in (3.8). Take an irreducible curve $C \subset S$ with normalisation $F : \tilde{C} \longrightarrow C$. By [Mi87] the pull-back $f^*(\Omega^1_S|C)$ is semi-stable if and only if it is nef, which is the same as to say that $\Omega^1_S|C$ is nef. Assume now that $\Omega^1_S|C$ is not nef. Let e be the invariant of the ruled surface $\mathbf{P}(f^*\Omega^1_S|C)$ in the sense of sect.2. The instability is then translated into $e > 0$. Now write for numerical equivalence in X

$$C_0 \equiv aZ + bF$$

where C_0 is the image of any multi-section of the ruled surface in X, F a fiber of the projection to X and Z a fixed 1-cycle in the "non-trivial" boundary ray of $\overline{NE}(X)$. Letting d be defined by $\mathcal{O}_S(C) = \mathcal{O}_S(1)$, we see easily

$$a = d \deg(\pi|C_0), b = 4\sqrt{2}d - \frac{e}{2}, \qquad (*)$$

where Z is assumed to be normalised in such way that $\pi_*(Z) = \mathcal{O}_S(1)$. If λ is the nef value of $\Omega^1_S|C$ with respect to $\mathcal{O}_S(1)|C$, then the formula

$$e = 8\lambda d$$

holds. If (C_ν) is a sequence of irreducible curves with

$$\lim \frac{[C_\nu]}{\|[C_\nu]\|} \in R,$$

then $\lim \frac{b_\nu}{a_\nu} = 0$ and $\lim \frac{e_\nu}{d_\nu} = 8\sqrt{2}$. Since $\lim d_\nu = \infty$, we conclude $\lim e_\nu = \infty$, i.e. $\Omega^1_S|\pi(C_\nu)$ gets "more and more instable". Since $\Omega^1_S|C$ is nef if and only if $e \leq 0$, the equation (*) implies the following

$\Omega^1_S|C$ is nef if and only if $\frac{b}{a} \leq 4\sqrt{2}$ for some section $C_0 \subset \mathbf{P}(\Omega^1_S|C)$.

(Here "section" means : image of a section under the normalisation map

$$\mathbf{P}(f^*(\Omega^1_S|C)) \longrightarrow \mathbf{P}(\Omega^1_S|C)).$$

If we introduce the subcone

$$I = \{aZ + bF \in \overline{NE}(X) | \frac{b}{a} > 4\sqrt{2}\},$$

then we can state :

$\Omega^1_S|C$ is not nef if and only if $f^*(\Omega^1_S|C)$ is instable if and only if there is a section $C_0 \subset \mathbf{P}(\Omega^1_S|C)$ (necessarily the exceptional section) with $[C_0] \in I$.

(3.10) We want to say some words on the structure of X in case S has degree 2, i.e. is a $2:1-$ covering of the projective plane. Then the nef value is 3, as seen before, in fact $\Omega^1_S(3)$ is even generated by global sections. From our discussion in (3.1) it follows easily that the map defined by the global sections of $L = \mathcal{O}_X(1) \otimes \pi^*(\mathcal{O}_S(3))$ contracts exactly the exceptional section of $\mathbf{P}(\Omega^1_S|\tilde{\Delta})$, which is a curve of genus 10 and nothing else. Hence $\mathrm{Proj}(\bigoplus H^0(X, mL)$ is smooth except for only one non-rational singularity.

So the behaviour in degree 2 and degree 4 are quite different, but we think that degree 4 should be typical in view of higher degree.

4. Surfaces of general type

We begin treating hypersurfaces in $\mathbf{P_3}$. Here the situation is completely the same as for quartics.

4.1 Theorem

Let $S \subset \mathbf{P_3}$ be a hypersurface in $\mathbf{P_3}$ of degree $d \geq 5$ and with $\rho(X) = 1$.

(1) The cotangent value $\alpha = \alpha(S, \mathcal{O}_S(1))$ is given by the equation

$$c_1^2(\Omega_S^1(\alpha)) = c_2(\Omega_S^1(\alpha)),$$

i.e. as the positive solution of

$$\alpha^2 + (d-4)\alpha - \frac{4}{3}d + \frac{10}{3} = 0.$$

In particular $1 < \alpha \leq 2$.

(2) There is no curve $C \subset S$ with $\Omega_S^1(\alpha)|C$ not ample.

(3) For all surfaces $Y \subset \mathbf{P}(\Omega_S^1)$ we have :

$$c_1(\mathcal{O}_{\mathbf{P}(\Omega_S^1(\alpha))}(1)^2.Y > 0.$$

Proof. Exactly the same as for the analogous statements in case $d = 4$ (sect.3). Of course some numerical calculations get more complicated. We omit details.

4.2 Corollary

There are infinitely many curves $C \subset S$ such that $\Omega_S^1|C$ is not ample, even more : $\Omega_S^1(\beta)|C$ is not ample for any fixed number $\beta < \alpha$.

In contrast to (4.1) we have

4.3 Proposition

Let S be a surface with ample canonical bundle K_S. Let $\alpha = \alpha(S, K_S)$ be the cotangent value of S with respect to K_S. Assume $c_1^2(\Omega_S^1(\alpha)) > 2c_2(\Omega_S^1(\alpha))$. Then there exists a curve $C \subset S$ with $\Omega_S^1(\alpha)|C$ not ample . In particular $\alpha \in \mathbf{Q}$.

Proof. Since Ω_S^1 is semi-stable with respect to K_S, this follows from (1.5).

4.4 Corollary

In the situation of (4.3) there exists a curve $C \subset X = \mathbf{P}(\Omega_S^1)$ such that $[C] \in \partial\overline{NE}(X)$.

4.5 Remarks

(1) The condition $c_1^2(\Omega_S^1(\alpha)) > 2c_2(\Omega_S^1(\alpha))$ can be rewritten as

$$(2\alpha^2 + 2\alpha + 1)c_1^2(S) > 2c_2(S) \qquad (*).$$

If Ω^1_S is not ample then we have $\alpha \geq 0$, so (*) is weaker than $c_1^2(S) > 2c_2(S)$. Observe also that in (4.3) the canonical bundle can be replaced by any ample line bundle L such that Ω^1_S is $L-$ semi-stable.

(2) Let S be a surface of general type with $c_1^2 > 2c_2$. Then Miyaoka proved in [Mi82] that Ω^1_S is "ample almost everywhere" (to be defined in (4.6)), in particular there are only finitely many curves C such that $\Omega^1_S|C$ is not ample. Given any ample line bundle L and letting $\alpha = \alpha(S, L)$ it is then obvious that there is a curve $C \subset S$ with $\Omega^1_S(\alpha)|C$ is not ample and thus $\alpha \in \mathbf{Q}$.

(3) The curves C occuring in (4.3) should again be thought of the "most" instable curves concerning Ω^1_S. We do not know whether one should expect a special structure of these curves.

The question arises in the situation of (4.4) to determine the structure of $\overline{NE}(X)$ near to the boundary : are there curves $C' \subset X$ such that the half ray $\mathbf{R}_+[C]$ is arbitrary near to $\mathbf{R}_+[C]$? The answer is given in terms of the notion of "almost everywhere ampleness" (see [Mi82]).

4.6 Definition

Let X be a projective manifold, E a vector bundle on X. Then E is said to be ample almost everywhere (ample a.e.) if the following holds. Fix an ample line bundle H on $\mathbf{P}(E)$. Then there exists a proper algebraic subset $T \subset X$ and $\epsilon > 0$ such that

$$c_1(\mathcal{O}_{\mathbf{P}(E)}(1)).C \geq \epsilon\ H.C$$

for all curves $C \subset \mathbf{P}(E)$ with $\pi(C) \subset T$.

4.7 Theorem

Let S be a surface with ample canonical bundle K_S. Assume that Ω^1_S is not ample and let α be the cotangent value with respect to K. Assume $c_1^2(\Omega^1_S(\alpha)) > 2c_2(\Omega^1_S(\alpha))$. Then $\Omega^1_S(\alpha)$ is ample a.e., moreover there exists $\epsilon > 0$ such that $\Omega^1_S(\beta)$ is ample a.e. for all β with $\alpha - \epsilon < \beta < \alpha$.

Proof. (cp. [Mi82]). Of course the second statement follows from the first. By (4.3) α is rational. Let $X = \mathbf{P}(\Omega^1_S)$ and let $L_\gamma = \mathcal{O}_{\mathbf{P}(\Omega^1_S(\gamma))}(1)$ for $\gamma \in \mathbf{R}$. Since L_α is big and nef and since

$$H^2(\mathcal{O}_S(mL_{\alpha-\epsilon})) = H^2(S^m(\Omega^1_S) \otimes \mathcal{O}_S(m(\alpha - \epsilon)K_X)) = 0$$

for $m \gg 0, \epsilon \ll 1$, we have $\kappa(L_{\alpha-\epsilon}) = 3$ by Riemann-Roch for small rational ϵ (and m such that $m(\alpha - \epsilon) \in \mathbf{N}$). Let $\beta = \alpha - \epsilon$. Choose $s \in H^0(mL_\beta)$ and set $D_\beta = \{s = 0\}$. We may assume that

$$D_\beta = \sum n_i A_i\ +\ \sum m_j B_j$$

where A_i are multi-sections , $\dim\pi(B_j) = 1$ and $B_1 = \pi^*(B'_1)$, B'_1 a multicanonical divisor, $n_i, m_j > 0$. Now take a curve $C \subset X$ with $L_\beta.C < 0$. Then $C \subset A_i$ for some i or $C \subset B_j$

for some j. Since $L_\alpha^2.A_i > 0$ by (1.5) we have $L_{\alpha-\epsilon}^2.A_i > 0$ if ϵ is sufficiently small. So if $L_\beta^2.A_i \leq 0$ we replac e ϵ by a smaller ϵ' and we replace D_β by

$$D_{\alpha-\epsilon'} = \sum n_i A_i + (m_1 + \frac{\epsilon - \epsilon'}{k})B_1 + \sum_{j\geq 2} m_j B_j$$

(where $B_1 \in |kK_S|$.) So we may assume $L_\beta^2.A_i > 0$ from the beginning (for all i). So either $\kappa(L_\beta^{-1}|A_i) = 2$ or $\kappa(L_\beta^{-1}|A_i) = 2$. But clearly $L_\beta.H > 0$ for a general hyperplane section with respect to a very ample divisor which is far from $L_\alpha^\perp$ in $\overline{NE}(X)$. So we must have $\kappa(L_\beta|A_i) = 2$. Hence $L_\beta.C < 0$ can occur only for finitely many curves $C \subset A_i$. Now let

$$T_\beta = \pi(\cup B_j) \cup \bigcup_{C\subset A_i, L_\beta.C<0} \pi(C).$$

Then dim$T_\beta \leq 1$ and $L_\beta.C \geq 0$ for all $C \subset X$ with $\pi(C) \not\subset T_\beta$. Then we claim that L_τ is ample a.e. for all $\tau > \beta$ (with $T = T_\beta$). In fact, if $\tau > \beta$, then there exists $\epsilon > 0$ such that

$$L_\tau \geq \epsilon\pi^*(K_X).C$$

for all C with $\pi(C) \not\subset T_\beta$. Hence L_τ is ample a.e. by [Mi82,lemma1.1].

4.8 Remarks

(1) In (4.7) K_S can be replaced by any ample line bundle or ample $\mathbf{Q}$-divisor L on S such that Ω_S^1 is semi-stable with respect to L. If S is merely minimal but with (-2)-curves so that K_S is only big and nef then $\Omega_S^1(\lambda K_S)$ is never ample, so a cotangent value does not exist. However, there is a substitute. Let $f : S \longrightarrow S'$ be the blow down of the (-2)-curves. Then S' has only ADE-singularities, $K_{S'}$ is ample and $K_S = f^*(K_{S'})$. Now consider the sheaf of Kähler differentials $\Omega_{S'}^1$. Still it makes perfect sense to speak of nef and ample coherent sheaves, hence we can define the cotangent value α of $\Omega_{S'}^1$ with respect to $K_{S'}$. Then we define α to be the "almost nef" or cotangent value of Ω_S^1 with respect to K_S. Observe that $f^*(\Omega_{S'}^1)$/torsion is a nef (but not ample) subsheaf of Ω_S^1, the inclusion map being an isomorphism outside the (-2)-curves. One might hope that (4.7) stills holds with this definition of α.

(2) (4.7) (partly) generalises Theorems 2 and 3 in [Mi82] in case K_S is ample.

4.9 Corollary

Assume the situation of (4.7). Let α be the cotangent value with respect to K_S. Let $C \subset \mathbf{P}(\Omega_S^1) = X$ be a curve whose class is in $\partial\overline{NE}(X)$ (such a curve exists by (4.3)).

(1) Then there exists an open convex neighborhood U of $\mathbf{R}_+[C] = R$ in $N_1(X)$ such that no $R' \subset U$ different from R is represented by an irreducible curve In other words: the classes of irreducible curves not in R are far away from R; the closed convex cone generated by the classes of curves not in R is a prope r subcone of $\overline{NE}(X)$.

(2) There are only finitely many curves $C_1, ..., C_s \subset S$ such that $\Omega_S^1(\alpha)|C_i$ is not ample. In particular there are only finitely many curves in X whose class is in R, namely the "exceptional sections" of the "ruled" surfaces $\mathbf{P}(\Omega_S^1|C_i)$.

We now turn to the following problem :

assume that $\Omega^1_S|C$ is ample for every curve $C \subset S$. Is then Ω^1_S already ample?

In [ST85] it is shown that the answer is yes if additionally $c_1^2(S) > 2c_2(S)$ (a consequence of (1.3)). We give the following partial answer.

4.10 Proposition

Let S be a surface such that $\Omega^1_S|C$ is ample for every curve $C \subset S$. Assume $\rho(S) = 1$ and $c_1^2(S) \neq c_2(S)$. Assume furthermore that the effective value of Ω^1_S with respect to the ample generator $\mathcal{O}_S(1)$ of $N^1(S)$ is not $k(\frac{c_2(S)}{c_1^2(S)} - 1)$ where $K_S \equiv \mathcal{O}_S(k)$. Then Ω^1_S is ample.

Proof. Assume Ω^1_S not to be ample. From our conditions follows the existence of a surface $Y \subset X = \mathbf{P}(\Omega^1_S)$ such that $L^2.Y = 0$, where as usual $L = \mathcal{O}_{\mathbf{P}(\Omega^1_S)}(1)$. Write for numerical equivalence

$$Y \equiv L^\rho \otimes \pi^*(\mathcal{O}_S(\tau)).$$

Letting $\xi = \frac{\tau}{\rho}$ we deduce from $L^2.Y = 0$:

$$c_1^2(S) - c_2(S) + \xi c_1(S).c_1(\mathcal{O}_S(1)) = 0.$$

With $K_S \equiv \mathcal{O}_S(k)$ we obtain

$$\xi = k(\frac{c_2(S)}{c_1^2(S)} - 1).$$

Since Y is on the boundary of the cone of effective divisors, ξ is the effective value of Ω^1_S with respect to $\mathcal{O}_S(1)$. Hence we conclude by contradiction.

The assumption $c_1^2(S) > c_2(S)$ in (4.10) can be removed:

4.11 Proposition

Let S be a surface such that $\Omega^1_S|C$ is ample for every curve $C \subset S$. Then $c_1^2(S) > c_2(S)$.

Proof. Since Ω^1_S is nef by assumption, we have $c_1^2(S) \geq c_2(S)$. Assume that equality holds. Let X and L be as in (4.10). By Riemann-Roch $\chi(L^m)$ grows at least linearly. Let $L^m = \mathcal{O}(D)$. Then $L^3 = 0$ translates into $L^2.D = 0$. Hence we find a lot of surfaces $Y \subset X$ with $L^2.Y = 0$. The same argument shows that $\kappa(L \otimes \pi^*(\mathcal{O}_S(-\epsilon H))) = -\infty$ for all rational positive ϵ and all effective divisors H on S. We conclude that the effective value of Ω^1_S with respect to every effective divisor H is 0 (observe $L^2.\pi^*(C) > 0$). Now let B be the stable base locus of the linear system $|mL|, m \in \mathbf{N}$. It follows that $\dim B \leq 1$. In fact, assume that B has a divisor component B_0. Write

$$B_0 = L^\mu \otimes \pi^*(\mathcal{O}_S(\lambda)).$$

Then $\mu \leq \lambda$ since the cotangent value of (X, L) is 1. Now

$$mL - B_0 = L^{m-\mu} \otimes \pi^*(\mathcal{O}(m - \lambda)).$$

Since $mL - B_0$ is effective, we have $m - \mu \leq m - \lambda$, contradicting $\mu \leq \lambda$. Thus $\dim B \leq 1$.

Assume now that $\dim B = 1$. Pick $D_1, D_2, D_3 \in |mL|$, all three different and general. Then $D_1.D_2$ is a curve, hence $D_1.D_2.D_3 > 0$ by our assumption that Ω^1_S is ample on every curve (which implies the same for L), contradicting $L^3 = 0$. Hence B is at most finite. But then B is actually empty and by $L.C > 0$ for every curve we conclude that L, hence Ω^1_S, is ample, contradicting $c_1^2(S) = c_2(S)$. Alternatively, the fact $\dim B \leq 0$ implies that $L^2.Y > 0$ for every surface $Y \subset X$, yielding again a contradiction.

4.12 Remark Let S be a smooth ample divisor in a projective smooth 3-fold X with $c_1(X) = 0$, which is not covered by a torus. Then we have $c_1^2(S) = S^3$ and $c_2(S) = c_2(X).S + S^3$. Hence $c_1^2(S) \leq c_2(S)$ and equality holds if $c_2(X).S = 0$. Thus 4.11 implies that $\Omega^1_S|C$ is not ample for some curve $C \subset S$. Using a result of Miyaoka [Mi87] that $c_2(X)$ is in the closure of the cone of effective curves for threefolds with K_X nef, we conclude even without using 4.11 that Ω^1_S cannot be nef, because otherwise we would have $c_1^2(S) = c_2(S)$, hence $c_2(X).S = 0$, hence $c_2(X) = 0$ and X would be covered by a torus (Yau) contradiction.

5. Higher dimensions: Calabi-Yau manifolds

We are now going to investigate the cotangent value α of certain Calabi-Yau manifolds (with $\rho = 1$).

5.1 Proposition

Let $S \subset \mathbf{P_n}$ be a smooth hypersurface of degree $n+1, n \geq 4$. Then $\alpha = \alpha(S, \mathcal{O}_S(1)) = 2$.

Proof. Since $\Omega^1_{\mathbf{P_n}}(2)$ is nef and since $\Omega^1_S(2)$ is a quotient of $\Omega^1_{\mathbf{P_n}}(2)|S$, we have $\alpha \leq 2$. By [BV78] there exists a line $l \subset S$. The conormal sequence of $l \subset S$ leads to a sequence

$$0 \longrightarrow N^*_{l|S}(2) \longrightarrow \Omega^1_S(2)|l \longrightarrow \mathcal{O}_l \longrightarrow 0,$$

so $\Omega^1_S(2)|l$ is not ample, hence $\alpha = 2$.

5.2 Remark

The reasoning of (5.1) shows more generally the following. Let (S, L) be a polarised manifold with L very ample. If there is a smooth rational curve $l \subset S$ with $L.l = 1$, then $\alpha(S, L) \geq 2$. Compare 5.5(2). It should be possible to apply this to complete intersections in $\mathbf{P_n}$ or to hypersurfaces in other Fano manifolds. Note also that the inequality $\alpha \leq 2$ in (5.1) holds for hypersurfaces of any degree.

(5.3) We now investigate quintics $S \subset \mathbf{P_4}$. As usual let $X = \mathbf{P}(\Omega^1_S)$ and let $L = \mathcal{O}_{\mathbf{P}(\Omega^1_S(2))}(1)$. By (5.1), L is nef but not ample. An easy calculation shows

$$c_1(L)^5 = s_3(\Omega^1_S(2)) = 100 > 0,$$

so L is big. Here s_i is the i–th Segre class. Moreover $\Omega^1_{\mathbf{P_n}}(2)$ is globally generated, so is $\Omega^1_S(2)$. We let $\phi : X \longrightarrow Y$ be the Stein factorisation of the map associated to the linear system $|L|$, so Y is a normal projective 5–fold. In order to investigate the fibers of ϕ, we note that $\dim\phi(C) = 0$ for a curve $C \subset X$, if and only if $L.C = 0$, in particular $\Omega^1_S(2)|\pi(C)$ is not ample. In this situation we will use the following

5.4 Lemma

Let $C \subset S$ be an irreducible curve such that $\Omega^1_S(2)|C$ is not ample. Then C is a line (with respect to the embedding $S \subset \mathbf{P_4}$).

Proof. It is sufficient to show that $\Omega^1_{\mathbf{P_4}}(2)|C$ is ample for every curve $C \subset \mathbf{P_4}$ except for lines. Since $\Omega^1_{\mathbf{P_4}}(2) = \Lambda^3(T_{\mathbf{P_4}}(-1))$, we first note that $T_{\mathbf{P_4}}(-1)|C$ is ample if and only if C is non-degenerate. This is well-known and follows immediately from the geometry of $\mathbf{P}(T_{\mathbf{P_4}})$ which can be identified with the flag manifold

$$\{(x,H)|x \in \mathbf{P_4}, H \subset \mathbf{P_4} \text{ a hyperplane}, x \in H\}.$$

$\mathbf{P}(T_{\mathbf{P_4}})$ has two projections, the obvious one and that one to the dual projective space which is given by the global sections of $\mathcal{O}_{\mathbf{P}(T_{\mathbf{P_4}}(-1))}(1)$.

So we are reduced to degenerate curves C. Note that for every hyperplane $H \subset \mathbf{P_4}$ there is a splitting

$$T_{\mathbf{P_4}}|H = \mathcal{O}(1) \oplus T_H.$$

Since

$$\Lambda^3(T_{\mathbf{P_4}}(-1))|H = \Lambda^2(T_H(-1)) \oplus \Lambda^3(T_H(-1))$$

it is sufficient to show that $\Lambda^2(T_H(-1)|C$ is ample except C is a line. By the same reasoning for $\mathbf{P_3}$ instead of $\mathbf{P_4}$ we may assume $C \subset H' = \mathbf{P_2}$, a hyperplane in $\mathbf{P_2}$. Now we have

$$\Lambda^2(T_H(-1))|H' = T_{H'}(-1) \oplus \Lambda^2(T_{H'}(-1)),$$

and $T_{H'}(-1)|C$ is ample if and only if C is non-degenerate, i.e. not a line.

5.5 Remarks

(1) The arguments of (5.4) work of course in all dimensions. Hence: $\Lambda^{n-1}(T_{\mathbf{P_n}}(-1)) = \Omega^1_{\mathbf{P_n}}(2)$ is ample on all curves except lines.

(2) Hence (5.4) holds in all dimensions ≥ 4 and for hypersurfaces S of any degree. It follows that $\alpha(S, \mathcal{O}_S(1)) = 2$ if and only if S contains a line.

(5.6) Coming back to our quintic $S \subset \mathbf{P_4}$ and a curve $C \subset X$ with $L.C = 0$, we conclude that $\pi(C)$ is a line in S. Writing

$$\Omega^1_S(2)|\pi(C) = \mathcal{O}(a) \oplus \mathcal{O}(b) \oplus \mathcal{O}(c)$$

with $0 = a \leq b \leq c$ (and $b + c = 6$), we moreover see that if

(a) $b > 0$, then $C = \mathbf{P}(\mathcal{O}) \subset X$.

(b) $b = 0$, then $C \subset \mathbf{P}(\mathcal{O} \oplus \mathcal{O}) = \mathbf{P_1} \times \mathbf{P_1} \subset X$.

In both cases the normal bundle $N = N_{C|X}$ is given by the exact sequence of normal bundles

$$0 \longrightarrow \mathcal{O}(-b+1) \oplus \mathcal{O}(-c+1) \longrightarrow N \longrightarrow \mathcal{O}(-b+2) \oplus \mathcal{O}(-c+2) \longrightarrow 0.$$

Thus we obtain the following picture : all non-trivial fibers of ϕ are smooth rational curves lying over lines in S; they are isolated except in situation (b) when the fibers have normal bundle $\mathcal{O} \oplus \mathcal{O}(-1)$ lying over lines $l \subset S$ with normal bundle $\mathcal{O}(2) \oplus \mathcal{O}(-4)$. In summary the picture is completely different from the case of K3-surfaces.

(5.7) Turning to hypersurfaces $S \subset \mathbf{P_n}$ of degree $n+1$ in general, we can imitate most of (5.3) and (5.6). Since we cannot prove that $c_1(L)^{2n-1} > 0$ a priori, the contraction map ϕ might at first be a fiber space. By (5.5) however we know that $\dim \phi(C) = 0$ implies that $\pi(C)$ is a line and this proves that ϕ must be birational (otherwise there would be a line through every point of S). Hence $c_1(L)^{2n-1} > 0$ a posteriori. The description of ϕ is now analogous to (5.6).

Of course our discussion leads to a number of very interesting but hard problems.

5.8 Problems

Let S be a Calabi-Yau manifold (which has by definition trivial canonical class and finite fundamental group) with $\rho(S) = 1$.

(1) Is the cotangent value $\alpha = \alpha(S, \mathcal{O}_S(1))$ a rational number ?

In case $\alpha \in \mathbf{Q}$, we consider $X = \mathbf{P}(\Omega^1_S)$ and $L = \mathcal{O}_{\mathbf{P}(\Omega^1_S(\alpha))}(1)$.

(2) Is L^m generated by global sections for some m?

(3) Is L big ?

(4) If (2) has a positive answer : what is the structure of the non-trivial fibers of the corresponding map ?

If one could prove that every (or some) non-trivial fiber contains a rational curve, then we would obtain a rational curve in S itself. One might even ask whether one can find an effective $\mathbf{Q}$-divisor D on X such that (X, D) is (weakly) log-terminal ([KMM87]) and such that $(K_X + D.C) < 0$ for a curve C with $\dim \phi_{mL}(C) = 0$. In that case we would find a rational curve in every non-trivial fiber of ϕ. However we do not know whether D exists even in the simplest case, when S is a quintic in $\mathbf{P_4}$. On the other hand , if the K3-surface occurs as a $2:1-$ covering of the projective plane, the answers to (1),(2) and (3) are all positive but the only non-trivial fiber of the map in question is a curve of genus 10.

References

[BV78] Barth,W.;van de Ven,A.: Fano varieties of lines on hypersurfaces . Arch.d.Math.31,96-104(1978)

[CP90] Campana,F.;Peternell,T.: Algebraicity of the ample cone of algebraic varieties. Crelles J.407,160-166(1990)

[Ha77] Hartshorne,R.: Algebraic Geometry. Springer1977

[KMM87] Kawamata,Y.;Matsuda,K.;Matsuki,K.: Introduction to the minimal model problem. Adv.Stud.Pure Math.10,283-360(1987)

[Mi82] Miyaoka,Y.: Algebraic surfaces of positive index. Classification of algebraic and analytic manifolds. Katata Symposium Proc.1982. Progress in Math. vol.39,281-301.Birkhäuser 1983

[Mi87] Miyaoka,Y.: The Chern classes and Kodaira dimension of a minimal variety. Adv. Stud. Pure Math. 10,449-476 (1987)

[ST85] Schneider,M.;Tancredi,A.: Positive vector bundles on complex surfaces. manuscr. math. 50, 133-144(1985).

On a Class of (—3,1)-Exceptional P^1

TAKEO OHSAWA Department of Mathematics, School of Science, Nagoya University, Chikusa-ku Nagoya 464-01, Japan

INTRODUCTION

A compact complex submanifold A of a complex manifold M is said to be exceptional if there exist a complex space (allowing singular points) $\widehat{M}$ and a proper holomorphic map $\pi : M \to \widehat{M}$ such that $\pi|M \setminus A$ is a biholomorphism. H. Grauert [G] showed that A is exceptional whenever the normal bundle $N_{A/M}$ is weakly negative in the sense that its zero section admits a strongly pseudoconvex neighbourhood system. He also gave an example of (A, M) for which A is exceptional but $N_{A/M}$ is not weakly negative. In 1981, H. Laufer [L] started to classify the normal bundles of exceptionally embedded $\mathbf{P}^1$. On one hand he showed that the Chern class of the determinant of $N_{A/M}$ does not exceed $-\dim M + 1$ if A $(= \mathbf{P}^1)$ is exceptional. On the other hand he showed that $\mathcal{O}(-3)\oplus\mathcal{O}(1)$ and $\mathcal{O}(-30)\oplus\mathcal{O}(2)$ can occur as the normal bundles of exceptional embeddings of $\mathbf{P}^1$. More recently, T. Ando gave systematic constructions of exceptional embeddings of $\mathbf{P}^1$ whose normal bundles are $\mathcal{O}(-2k-1) \oplus \mathcal{O}(k)$, $k > 0$ (cf. [A-1] and [A-2, Example 3.1]). Quite recently, Vo Van Tan found an intriguing link between the exceptional embeddings of $\mathbf{P}^1$ and the existence of Kähler metrics on M (cf. [V]). Therefore it seems worthwhile to study the exceptional embeddings of $\mathbf{P}^1$ more in detail, but not so sporadically as before. The purpose of the present note is to give a class of exceptional embeddings of $\mathbf{P}^1$ whose normal bundle is $\mathcal{O}(-3) \oplus \mathcal{O}(1)$. It will turn out that examples of Laufer and Ando are members of an analytic family whose parameter space is a Zariski open subset of a weighted projective space.

§1 ELEMENTARY DEFORMATION OF RANK TWO BUNDLES

Let E be any holomorphic vector bundle of rank two over $\mathbf{P}^1$. By Birkhoff-Grothendieck's theorem, $E \cong \mathcal{O}(a) \oplus \mathcal{O}(b)$ for some $a, b \in \mathbf{Z}$. Here $\mathcal{O}(a)$ denotes the line bundle of degree a over $\mathbf{P}^1$. Let z denote an inhomogeneous coordinate of $\mathbf{P}^1$ and let $w = z^{-1}$. We put $U_0 = \{z \mid |z| < 2\}$ and $U_1 = \{w \mid |w| < 2\}$. Then we may identify E with a complex manifold obtained from $U_0 \times \mathbf{C}^2$ and $U_1 \times \mathbf{C}^2$ by patching them along $(U_0 \cap U_1) \times \mathbf{C}^2$ by

$$\begin{cases} u_0 = w^{-a} u_1 \\ v_0 = w^{-b} v_1. \end{cases}$$

Here (z, u_0, v_0) and (w, u_1, v_1) denote the coordinates of $U_0 \times \mathbf{C}^2$ and $U_1 \times \mathbf{C}^2$, respectively. By an elementary deformation of E we mean a complex manifold obtained from $U_0 \times \mathbf{C}^2$ and $U_1 \times \mathbf{C}^2$ by patching them together through

$$\begin{cases} u_0 = w^{-a} u_1 + \sum_{j=2}^{k} F_j(w) v_1^j, & k \in [2, \infty) \cap \mathbf{Z} \\ v_0 = w^{-b} v_1. \end{cases} \tag{1}$$

Here $F_j(w)$ are holomorphic functions on $U_0 \cap U_1$. The term "elementary" is taken from the fact that any polynomial automorphism of $\mathbf{C}^2$ is a composite of linear transformations and automorphisms of the form $(x, y) \to (x + cy^k, y)$, $c \in \mathbf{C}$, $k \geq 2$ (cf. [J]). Letting $\mathbf{F}(w) = (F_2(w), \ldots, F_k(w))$, we shall denote the $\mathbf{C}^2$-bundle defined by (1) as $E(a, b : \mathbf{F}(w))$.

Let

$$F_j(w) = \sum_{i \in \mathbf{Z}} c_{i,j} w^i$$

be the Laurent expansion of F_j, and put

$$u_0' = u_0 - \sum_{j=2}^{k} \sum_{i \geq 0} c_{-i+jb,j} z^i v_0^j$$

and

$$u_1' = u_1 + \sum_{j=2}^{k} \sum_{i \geq \mu_j} c_{i,j} w^{i+a} v_1^j,$$

where $\mu_j = \max(-a, -jb + 1)$. Then, by the coordinate changes $(u_0, v_0) \to (u_0', v_0')$ and $(u_1, v_1) \to (u_1', v_1)$ we see that

$$E(a, b : \mathbf{F}(w)) \cong E\left(a, b : \left(\sum_{i=\mu_j}^{\nu_j} c_{i,j} w^i\right)_{j=2}^{k}\right).$$

Here $\nu_j = -jb+1$. It is clear that

$$E\left(a,b:\left(\sum_{i=\mu_j}^{\nu_j} c_{i,j}w^i\right)_{j=2}^{k}\right) \cong E\left(a,b:\left(\sum_{i=\mu_j}^{\nu_j} c'_{i,j}w^i\right)_{j=2}^{k}\right)$$

if and only if there exists a $\lambda \in \mathbf{C}\setminus\{0\}$ such that $\lambda^j c_{i,j} = c'_{i,j}$ for $2 \le j \le k$ and $\mu_j \le i \le \nu_j$. Thus, nontrivial elementary deformations of E correspond to the points of weighted projective spaces.

§2 QUADRATIC ELEMENTARY DEFORMATIONS OF $\mathcal{O}(-3)\oplus\mathcal{O}(1)$

From now on we shall focus our attention to $E(-3,1:\mathbf{F}(w))$. First we would like to investigate the case $k=2$ to understand the circumstance. As we have seen above, we may assume $F_2(w)=\sum_{i=-1}^{2} c_i w^i$. For brevity we set

$$E_{\mathbf{c}} = E\left(-3,1:\sum_{i=-1}^{2} c_i w^i\right)$$

for any $\mathbf{c}=(c_{-1}:c_0:c_1:c_2)\in\mathbf{P}^3$.

PROPOSITION 1. *$E_{\mathbf{c}}$ is holomorphically convex if $c_{-1}\neq 0$ and $c_0^2-c_1c_{-1}\neq 0$.*

Proof. We put

$$(2)\qquad \begin{cases} f_1(z,u_0,v_0) = \left(z-\dfrac{c_0}{c_{-1}}\right)u_0 - c_{-1}v_0^2 \\ f_2(z,u_0,v_0) = \left(z^2-\dfrac{c_1}{c_{-1}}\right)u_0 - (c_{-1}z+c_0)v_0^2 \\ f_3(z,u_0,v_0) = \left(z^3-\dfrac{c_2}{c_{-1}}\right)u_0 - (c_{-1}z^2+c_0z+c_1)v_0^2. \end{cases}$$

Since $u_0 = w^3u_1+\sum_{i=-1}^{2} c_iw^iv_1^2$ and $v_0=w^{-1}v_1$ we have

$$\begin{cases} f_1(z,u_0,v_0) = w^2u_1+\displaystyle\sum_{i=1}^{2} c_iw^{i-1}v_1^2 - \dfrac{c_0}{c_{-1}}\left(w^3u_1+\sum_{i=0}^{2} c_iw^iv_1^2\right) \\ f_2(z,u_0,v_0) = wu_1+c_2v_1^2-\dfrac{c_1}{c_{-1}}\left(w^3u_1+\displaystyle\sum_{i=0}^{2} c_iw^iv_1^2\right) \\ f_3(z,u_0,v_0) = u_1-\dfrac{c_2}{c_{-1}}\left(w^3u_1+\displaystyle\sum_{i=0}^{2} c_iw^iv_1^2\right). \end{cases}$$

Hence f_i extend to holomorphic functions on $E_{\mathbf{c}}$. In order to prove that $E_{\mathbf{c}}$ is holomorphically convex, it suffices to show that $\phi := (f_1,f_2,f_3)$ is a proper holomorphic map from $E_{\mathbf{c}}$ to $\mathbf{C}^3$.

For that, we note that the restriction of ϕ to the fibers of $E_{\mathbf{c}} \to \mathbf{P}^1$ are linear with respect to (u_i, v_i^2). Therefore the properness of ϕ follows from the following two identities:

$$\begin{vmatrix} \dfrac{\partial f_1}{\partial u_0} & \dfrac{\partial f_1}{\partial v_0^2} \\ \dfrac{\partial f_2}{\partial u_0} & \dfrac{\partial f_2}{\partial v_0^2} \end{vmatrix} = \begin{vmatrix} z - \dfrac{c_0}{c_{-1}} & -c_{-1} \\ z^2 - \dfrac{c_0}{c_{-1}} & -c_{-1}z - c_0 \end{vmatrix} = \frac{c_0^2}{c_{-1}} - c_1$$

$$\begin{vmatrix} \dfrac{\partial f_1}{\partial u_1} & \dfrac{\partial f_1}{\partial v_1^2} \\ \dfrac{\partial f_2}{\partial u_1} & \dfrac{\partial f_2}{\partial v_1^2} \end{vmatrix} = \begin{vmatrix} 0 & c_1 - \dfrac{c_0^2}{c_{-1}} \\ 1 & -\dfrac{c_2 c_0}{c_{-1}} \end{vmatrix} = c_1 - \frac{c_0^2}{c_{-1}}.$$

PROPOSITION 2. *The zero section of $E_{\mathbf{c}} \to \mathbf{P}^1$ is not exceptional for any $\mathbf{c}$.*

Proof. First we assume that $c_{-1} \neq 0$ and $c_0^2 - c_1 c_{-1} \neq 0$. Then we eliminate u_0 and v_0 from (2). Namely we have

$$\begin{aligned} &\left(z^2 - \frac{c_1}{c_{-1}}\right) f_1 - \left(z - \frac{c_0}{c_{-1}}\right) f_2 \\ = &- c_{-1}\left(z^2 - \frac{c_1}{c_{-1}}\right) v_0^2 + (c_{-1}z + c_0)\left(z - \frac{c_0}{c_{-1}}\right) v_0^2 = \left(c_1 - \frac{c_0^2}{c_{-1}}\right) v_0^2 \end{aligned}$$

and

$$\begin{aligned} &\left(z^3 - \frac{c_2}{c_{-1}}\right) f_1 - \left(z - \frac{c_0}{c_{-1}}\right) f_3 \\ = &- c_{-1}\left(z^3 - \frac{c_2}{c_{-1}}\right) v_0^2 + (c_{-1}z^2 + c_0 z + c_1)\left(z - \frac{c_0}{c_{-1}}\right) v_0^2 \\ = &\left\{\left(c_1 - \frac{c_0^2}{c_{-1}}\right) z + c_2 - \frac{c_1 c_0}{c_{-1}}\right\} v_0^2, \end{aligned}$$

so that

$$\begin{aligned} &\left\{\left(z^2 - \frac{c_1}{c_{-1}}\right) f_1 - \left(z - \frac{c_0}{c_{-1}}\right) f_2\right\}\left\{z + \left(c_1 - \frac{c_0^2}{c_{-1}}\right)^{-1}\left(c_2 - \frac{c_1 c_0}{c_{-1}}\right)\right\} \\ &-\left(z^3 - \frac{c_2}{c_{-1}}\right) f_1 + \left(z - \frac{c_0}{c_{-1}}\right) f_3 = 0, \end{aligned}$$

or equivalently

$$\begin{aligned} c_{-1}(Af_1 - f_2)z^2 + \{ - c_1 f_1 + (c_0 - c_{-1}A) f_2 + f_3\} z \\ + (-c_1 A + c_2) f_1 + c_0 A f_2 - c_0 f_3 = 0. \end{aligned}$$

Here $A = (c_{-1}c_2 - c_1 c_0)(c_{-1}c_1 - c_0^2)^{-1}$. This quadratic polynomial of z vanishes for the values of ϕ lying in a complex line because

$$\begin{vmatrix} A & -1 & 0 \\ -c_1 & c_0c_{-1}A & c_{-1} \\ -c_1A+c_2 & c_0A & -c_0 \end{vmatrix}$$

$$=A-(c_0-c_{-1}A)c_0-c_0c_{-1}A+c_1c_0+c_1c_{-1}A-c_{-1}c_2$$

$$=(c_1c_{-1}-c_0^2)A+c_1c_0-c_{-1}c_2=0.$$

Therefore the zero section Z of $E_{\mathbf{c}}$ is not exceptional.

Z is not exceptional for all $\mathbf{c}$ since the exceptionality of Z is an open condition by Grauert's characterization of exceptional sets (cf. [G]).

Next we shall study the case $c_{-1}=0$.

PROPOSITION 3. *$E_{\mathbf{c}}$ is holomorphically convex if $c_{-1}=0$ and $c_0 \neq 0$.*

Proof. We put

$$\begin{cases} g_1(z,u_0,v_0) = u_0 = w^3u_1 + \sum_{i=0}^{2} c_i w^i v_1^2 \\ g_2(z,u_0,v_0) = c_1zu_0 - c_0z^2u_0 + c_0^2v_0^2 \\ \qquad = c_1w^2u_1 - c_0wu_1 + c_1^2v_1^2 + c_1c_2wv_1^2 - c_0c_2v_1^2 \\ g_3(z,u_0,v_0) = c_2zu_0 - c_0z^3u_0 + (c_0c_1 + c_0^2z)v_0^2 \\ \qquad = c_2w^2u_1 - c_0u_1 + c_1c_2v_1^2 + c_2^2wv_1^2. \end{cases}$$

Then

$$\begin{vmatrix} \dfrac{\partial g_1}{\partial u_0} & \dfrac{\partial g_1}{\partial v_0^2} \\ \dfrac{\partial g_2}{\partial u_0} & \dfrac{\partial g_2}{\partial v_0^2} \end{vmatrix} = \begin{vmatrix} 1 & 0 \\ * & c_0^2 \end{vmatrix} = c_0^2$$

and

$$\begin{vmatrix} \dfrac{\partial g_1}{\partial u_1} & \dfrac{\partial g_1}{\partial v_1^2} \\ \dfrac{\partial g_3}{\partial u_1} & \dfrac{\partial g_3}{\partial v_1^2} \end{vmatrix}_{w=0} = \begin{vmatrix} 0 & c_0 \\ -c_0 & c^* \end{vmatrix} = c_0^2.$$

Therefore, (g_1,g_2,g_3) is a proper holomorphic map from $E_{\mathbf{c}}$ to $\mathbf{C}^3$. Hence $E_{\mathbf{c}}$ is holomorphically convex.

Combining Propositions 1 and 3 we obtain

PROPOSITION 4. *$E_{\mathbf{c}}$ is holomorphically convex if $c_{-1}c_1 - c_0^2 \neq 0$.*

We also have the following.

PROPOSITION 5. *$E_{\mathbf{c}}$ is not holomorphically convex if $c_{-1}c_1 - c_0^2 = 0$ and $c_2 = c_0^3$.*

Proof. Under the above hypothesis, the transition relation (1) is equivalently written as

$$\begin{cases} u_0 = w^3 u_1 + (w^{-1} + c + c^2 w + c^3 w^2) v_1^2 \\ v_0 = w^{-1} v_1. \end{cases}$$

By a straightforward computation it is verified that $E_{\mathbf{c}}$ contains then a family of rational curves $\{A_t\}_{t \in \mathbf{C}}$ given by

$$\begin{cases} u_1 = tc^4(1 - cw) \\ v_1 = t(1 - cw). \end{cases}$$

Since every A_t intersects with the zero section of $E_{\mathbf{c}}$ and the union of $\{A_t\}$ is not compact, it follows that $E_{\mathbf{c}}$ is not holomorphically convex.

Let us leave the following as an open question.

CONJECTURE. *$E_{\mathbf{c}}$ is not holomorphically convex if $c_{-1}c_1 - c_0^2 = 0$.*

Note. By Proposition 2, it suffices to show that if $c_2 \neq c_0^3$ $E_{\mathbf{c}}$ does not contain any compact curve other than the zero section.

§3 A CLASS OF CUBIC ELEMENTARY DEFORMATIONS OF $\mathcal{O}(-3) \oplus \mathcal{O}(1)$

Our goal of this section is to prove that certain deformations of $E_{\mathbf{c}}$ contain the zero section as an exceptional submanifold. Explicitly, we shall establish the following.

THEOREM. *For any $(c_1, c_2) \in \mathbf{C}^2$ and $(a, b) \in \mathbf{C}^2 \setminus \{(0,0)\}$, the zero section of $E(-3, 1 : (1 + c_1 w + c_2 w^2, aw + bw^2))$ is exceptional.*

Proof. As before we set

$$\begin{cases} u_0 = w^3 u_1 + (1 + c_1 w + c_2 w^2) v_1^2 + (aw + bw^2) v_1^3 \\ v_0 = w^{-1} v_1. \end{cases}$$

For brevity we set $M = E(-3, 1 : (1 + c_1 w + c_2 w^2, aw + bw^2))$ and let $Z \subset M$ be the zero section.

First we note that

$$H_1 := c_1 z u_0 - z^2 u_0 - v_0^2 - aw^{-1} v_1^3$$

and

$$H_2 := c_2 z u_0 - z^3 u_0 - z v_0^2 - c_1 v_0^2 - a v_0^3 - bw^{-1} v_1^3$$

are regular on $U_1 \times \mathbf{C}^2$, which is directly verified from (3). Utilizing the identity

$$u_0 v_0 = w^2 u_1 v_1 + (w^{-1} + c_1 + c_2 w) v_1^3 + (a + bw) v_1^4,$$

we see that $H_1 + au_0v_0$ and $H_2 + bu_0v_0$ are regular on M.

We put

$$\begin{cases} h_1 = u_0 \\ h_2 = H_1 + au_0v_0 \\ h_3 = H_2 + bu_0v_0 \end{cases}$$

and consider the holomorphic map $\psi = (h_1, h_2, h_3)$ from M to $\mathbf{C}^3$. Since $h_1 = u_0$ and the coefficient of v_0^2 in h_2 is -1, the restrictions of ψ to the fibers of $u_0 \times \mathbf{C}^2 \to U_0$ are proper. On the other hand,

$$h_1|_{w=0} = v_1^2$$

and

$$h_3|_{w=0} = -u_1 + c_1c_2v_1^2 + ac_2v_1^3.$$

From this it is obvious that the restriction of ψ to $\{w = 0\}$ is proper. Furthermore the above computation shows that $\psi^{-1}((0,0,0)) = Z$. Therefore, in order to show that Z is exceptional, it suffices to show that the fibers of ψ are finite sets outside Z. First we want to show that

$$\#\psi^{-1}((\alpha, \beta, \gamma)) < \infty$$

if $\alpha \neq 0$. For that it suffices to show that

$$F := -\beta + (c_1z - z^2)\alpha + a\alpha X - X^2 - az^{-2}X^3$$

and

$$G := -\gamma + (c_2z - z^3)\alpha + b\alpha X - (z + c_1)X^2 - (a + bz^{-2})X^3$$

are relatively prime over $\mathbf{C}(z)$. But obviously F and G are irreducible polynomials in $\mathbf{C}(z)[X]$. Since they are non-proportional, F and G are relatively prime. To show that

$$\#\psi^{-1}((0, \beta, \gamma)) < \infty \qquad for \quad (\beta, \gamma) \neq (0,0),$$

we need to consider the solutions to

$$\begin{cases} -\beta - X^2 - az^{-2}X^3 = 0 \\ -\gamma - (z + c_1)X^2 - (a + bz^{-2})X^3 = 0. \end{cases}$$

But the argument goes similarly as above, since $(\beta, \gamma) \neq (0,0)$.

Remark. From the above result, we recover the examples of Laufer and Ando by letting

$$(c_0, c_1, c_2, a, b) = (1,0,0,1,0), (1,0,0,0,1),$$

respectively.

[A-1] Ando, T., An example of an exceptional $(-2n-1, n)$-curve in an algebraic 3-fold, Proc. Japan Acad., **66** (1990), 269–271.

[A-2] ________, On the normal bundle of $\mathbf{P}^1$ in a higher dimensional projective variety, Amer. J., **113** (1991), 949–961.

[G] Grauert, H., Über modifikationen und exceptionelle analytische Mengen, Math. Ann., **146** (1962), 331–368.

[L] Laufer, H., On $\mathbf{CP}^1$ as an exceptional set, Ann. of Math. Stud., Princeton Univ., 1981, 261–275.

[J] Jung, H.W.E., Über ganze birationale Transformationen der Ebene, J. Reine Angew. Math., **184** (1942), 161–174.

[V] Vo Van Tan, On the non-Kählerian structure of strongly pseudoconvex 3-folds, C.R. Acad. Sci., Paris, t.**317** (1993), 599–604.

$\partial\bar{\partial}$-Closed Positive Currents and Special Metrics on Compact Complex Manifolds

ALESSANDRO SILVA

Dipartimento di Matematica G.Castelnuovo,
Università di Roma La Sapienza, 00185 Roma (Italy)

Introduction

A very remarkable result of Miyaoka, [21], asserts that if a (compact) complex manifold M possesses a Kähler metric in the complement of a point, it is itself Kählerian. It is remarkable also that the positive d-closed $(1,1)$-form giving the Kählerianity of M has no a priori relationship with the given form having the same properties in the complement of a point. It is not possible to expect, as simple examples readily show, that a compact complex manifold which is Kählerian outside a closed analytic subset of positive dimension is itself Kähler. Another remarkable result due to Barlet, [6], says that a compact analytic subset of dimension q of a complex manifold has an open neighbourhood which is $q+1$-complete in the sense of Andreotti-Grauert, [5]. It is not difficult to construct a d-closed positive $(q+1, q+1)$- form on the same neighbourhood. This fact, together with a positive answer obtained by inspecting several examples, leads to ask whether a compact complex manifold which is Kählerian outside a closed analytic subset of dimension q possesses a d-closed postive $(q+1, q+1)$-form and whether this form can be put in relationship with a *special* , in some sense, metric, in general non Kählerian, on M. It should be noticed, incidentally, that positive forms always exist on a manifold (Sullivan, [24]), but in general they are not d-closed.

A frequent, important geometrical situation in which the manifold has a Kähler metric outside a closed analytic subset is the case of smooth modifications. This is the main motivation for studying the answer to the question asked above.

A way of producing smooth proper modifications is by blowing up a complex manifold M along an analytic subset Y. In the case of a blow up points are always replaced by projective varieties, while in complex geometry it is necessary to consider modifications with arbitrary fibres. The case in which M or the modified manifold M' has a Kähler metric will show this necessity. It is known since [Blanchard, 7] that if $f : M' \to M$ is the blow up of M with center the submanifold Y, and M is compact Kähler, then M' is also Kähler. But if $f : M' \to M$ is just a smooth modification between compact complex manifolds with M Kähler M' is in general not Kähler as many examples show. The basic one was given by [Hironaka, 16], enlightening the phenomenon particularly. In this case M is Kähler and M' is Kählerian outside the exceptional divisor, but not Kähler. It is natural to ask therefore for metric properties of the modified manifold, assuming metric properties on M.

Conversely, let $f : M'' \to M$ be a smooth modification of compact complex manifolds. We suppose now that M'' has a Kähler metric.

In this case even when f is the blow up of M with smooth center Y, M is not necessarily Kähler : that is there is no converse, in the sense of putting a Kähler metric on the domain of a modification, of Blanchard's theorem. This is the most interesting situation since important classes of (non-Kähler) compact complex manifolds M are obtained in this way, namely the class $\mathcal{C}$ of Fujiki, or, more particularly, the class of Moišezon manifolds. We have in this case that M is Kählerian outside a codimension two analytic subset and we ask for the existence of a $(n-1, n-1)$-form which is positive and d-closed on M. This is the case in which it is known that the existence of such forms is equivalent to the existence of a special (non-Kähler) metric on M.

Let us describe such metrics. For a given hermitian metric $\mathbf{h}$ on M we can consider two natural connections preserving $\mathbf{h}$, the Riemannian connection and the $(1,0)$-connection or hermitian connection which in addition preserves the complex structure tensor J. In terms of the torsion tensor, the first is characterized by its vanishing and the second by the vanishing of its $(1,1)$ part. When the Riemannian and the hermitian connections coincide, the metric is Kähler and this property is then equivalent with the condition that its torsion tensor is identically zero.

Let $(z_1, \ldots, z_n)$ be a local coordinate system and in these coordinates let

$$\mathbf{h} = \sum h_{ij}\, dz_i \otimes d\bar{z}_j$$

and write the hermitian connection $\mathcal{D}$ as

$$\mathcal{D}(\partial_{z_j}) = \sum \omega_{jk} \otimes \partial_{z_k}, \text{ where } \partial_{z_k} = (\frac{\partial}{\partial z_k}).$$

Let $H = (\mathbf{h}_{kl})$. As a $\mathbf{T}M$-valued 2-form, the torsion tensor $T^{\mathcal{D}}$ has no (1,1) component and its (2,0) and (0,2) components are J conjugates. Hence

$$T^{\mathcal{D}} = \sum_{j,k,l} T^l{}_{jk} dz_j \wedge d\bar{z}_k \otimes \partial_{z_l},$$

with

$$T^l{}_{jk} = \sum_a (\partial_{z_j} h_{\bar{a}k} - \partial_{z_k} h_{\bar{a}j}) h^{l\bar{a}},$$

where (h^{kl}) denotes the inverse of H. We define now the *torsion 1-form* $\tau_{\mathbf{h}}$ or the *trace of the torsion tensor* as

$$\tau_{\mathbf{h}} = \sum \tau_k dz_k,$$

where $\tau_k = \sum_j T^j{}_{kj}$.

We say that the hermitian metric $\mathbf{h}$ is *balanced* if $\tau_{\mathbf{h}} = 0$. It turns out that if ω is the Kählerform associated with a balanced metric and n is the dimension of M, ω^{n-1} is a positive d-closed $(n-1, n-1)$-form, and if M admits a form of that kind, M has a balanced metric. The following results hold:

THEOREM, [Alessandrini-Bassanelli, 3]. *Let $f : M' \to M$ be a smooth modification between compact complex manifolds of dimension n with M (n-1)-Kähler. Then M' is (n-1)-Kähler.*

THEOREM, [Alessandrini-Bassanelli, 4]. *Let M and M' be compact complex manifolds of dimension n.*

Let $f : M' \to M$ be a modification. Then if M' is (n-1)-Kähler M is (n-1)-Kähler.

Therefore a balanced metric is a bimeromorphic invariant, to the contrary of a Kähler metric. Their proofs are based on a generalization of Harvey-Lawson intrinsic characterization of Kähler manifolds,[15]:

THEOREM,[Alessandrini-Andreatta, 1]. *Let M be a compact complex manifold. There exists on M a d-closed positive (p,p)-form if and only if there is no non zero strongly positive (p,p)-current which is the (p,p)-component of a boundary.*

In order to show the existence of a balanced metric, it is then enough to show that if T is a positive $(n-1, n-1)$-current which is the $(n-1, n-1)$ component of a boundary, then T must be identically zero. This class of currents coincides locally with the class of $\partial\bar{\partial}$-closed positive currents and its properties are the ones that are needed in the proofs. In particular it is important to understand how a $\partial\bar{\partial}$-closed current is determined by its support. In the classical d–closed positive case this is well known and based on the property that such a current is locally flat. A *locally normal* current T is a current such that T and dT have measure coefficients. A *locally flat* current on the compact set K is the limit in the flat norm of locally normal currents supported in K. A d-closed positive current is locally flat.

Locally flat currents have a very important property given by the Support Theorem [Federer, 9, 4.1.15, 4.1.20] :

a locally flat current T of total dimension r whose support has Hausdorff measure zero in dimension r has to be zero; moreover if the support of T is an r-dimensional oriented submanifold, T is the current of integration on it.

Unfortunately a $\partial\bar{\partial}$-closed positive current is not locally flat; furthermore they are not d-closed in general, so that the classical theory of closed positive currents cannot be applied to them. Support Theorems especially made for the class of $\partial\bar{\partial}$-closed positive currents are given in section I.2, and, since currents do not have in general nice pullbacks, it has to be shown that this is not the case for proper modifications, section III.3, and other Support Theorems relative to modifications are given in section I.4.

This survey reports on the work of Alessandrini-Bassanelli, [2,3,4]. Their work follows a program set up, together with the basic conjectures, in a Seminar run by the author held at the University of Trento in the year 87/88.

I

SUPPORT THEOREMS FOR $\partial\bar\partial$ - CLOSED POSITIVE CURRENTS

I.1. Positive Currents

Let V be a complex vector space of dimension n and let p, $0 \le p \le n$, be a fixed integer.

$\eta \in \bigwedge^{p,0} V$ is called *simple or decomposable* if there exists $a_1, \ldots, a_n \in \bigwedge^{1,0} V$ such that

$$\eta = a_1 \wedge \ldots \wedge a_n.$$

Let $e_1, \ldots, e_n$ be a basis of V and $\mathbf{v} \in \bigwedge^{n,n}$.

$\mathbf{v}$ is called a *positive (strictly positive) volume* if

$$\mathbf{v} = k\sqrt{-1}\, e_1 \wedge \bar{e}_1 \wedge \ldots \wedge e_n \wedge \bar{e}_n$$

with $0 \le k$, $(0 < k)$.

$\zeta \in \bigwedge^{p,p} V$ is *real* if $\zeta = \bar{\zeta}$ (where the bar denotes the usual conjugation in V extended to its exterior algebra).

Let $\bigwedge_{\mathbb{R}}^{p,p} V$ denote the subspace of all real (p,p)-vectors and let

$$\sigma_p = \sqrt{-1^{p^2}}\, 2^{-p}$$

and $k = n - p$. Then:

- $\zeta \in \bigwedge_{\mathbb{R}}^{p,p} V$ is called *(strictly) positive* if $\zeta \wedge \sigma_k \eta \wedge \bar{\eta}$ is a (strictly) positive volume for all non zero $\eta \in \bigwedge^{k,0} V$,

- $\zeta \in \bigwedge_{\mathbb{R}}^{p,p} V$ is called *(strictly) strongly positive* if $\zeta \wedge \eta$ is a (strictly) positive volume for all non zero weakly positive $\eta \in \bigwedge_{\mathbb{R}}^{k,k} V$,

- $\zeta \in \bigwedge_{\mathbb{R}}^{p,p} V$ is called *(strictly) weakly positive* if $\zeta \wedge \sigma_k \eta \wedge \bar{\eta}$ is a (strictly) positive volume for all non zero simple $\eta \in \bigwedge^{k,0} V$.

We easily have:

I.1.1 PROPOSITION. *$\zeta \in \bigwedge_{\mathbb{R}}^{p,p}$ is strongly positive if and only if*

$$\zeta = \sigma_p \sum_j \eta_j \wedge \bar{\eta}_j$$

with simple non zero $\eta_j \in \bigwedge^{k,0} V$.

Let $WP^p(V)$, $P^p(V)$, $SP^p(V)$ denote the cones of weakly positive, positive and strongly positive (p,p) - vectors, respectively.
Their interiors are the sets of strictly weakly positive, strictly positive and strictly strongly positive (p,p)- vectors, respectively.

It is clear that

$$WP^p(V) \supseteq P^p(V) \supseteq SP^p(V),$$

the equality being valid only for $p = 1$ or $p = n-1$, because in that case every element in $\bigwedge^{p,0} V$ is simple.

Let now M be a complex manifold of dimension n. A (p,p)-form α is *real* if $\alpha = \bar{\alpha}$ and we shall write $\alpha \in \mathcal{E}^{p,p}{}_{\mathbb{R}}(M)$.

$\alpha \in \mathcal{E}^{p,p}{}_{\mathbb{R}}(M)$ is called a *weakly positive (p,p)-form* (resp. *positive, strongly positive*) if

$$\alpha_x \in WP^p(\mathbf{T}_x^*(M)), (\text{resp } \alpha_x \in P^p(\mathbf{T}_x^*(M)), \alpha_x \in SP^p(\mathbf{T}_x^*(M))),$$

for every $x \in M$.

There is no uniformity in the literature as to which real (p,p)-forms are to be called just *positive.* The definitions given above, following [Harvey-Knapp, 14], contain them all. One particular subset, nonetheless, is worthy of a special name:

I.1.2 DEFINITION, [Sullivan, 24]. If

$$\alpha_x \in interior(WP^p(\mathbf{T}_x^*(M))),$$

that is α is a strictly weakly positive (p,p)-form, it will be called *transverse* .

I.1.3 PROPOSITION. *A transverse (p,p)-form α is strictly positive in the sense of Lelong and Siu,* [18, 22], *that is*

$$\alpha_x(\sigma_p^{-1} \eta \wedge \bar{\eta}) \geq 0,$$

for every simple non zero $\eta \in \bigwedge^{p,0}(\mathbf{T}_x(M))$, *for every* $x \in M$.

PROOF. Indeed, if we fix $x \in M$ and a non zero $\mathbf{v} \in \bigwedge^{n,n}(\mathbf{T}_x(M))$, the non degenerate bilinear map

$$\Phi_{\mathbf{v}} : \bigwedge^{p,p}(\mathbf{T}_x^*(M) \times \bigwedge^{k,k}(\mathbf{T}_x^*(M) \to \mathbb{C}$$

given by $\Phi_{\mathbf{v}}(\eta, \xi) = \mathbf{v}(\eta \wedge \xi)$, , gives an isomorphism

$$G : \bigwedge^{p,p}(\mathbf{T}_x(M) \to \bigwedge^{k,k}(\mathbf{T}_x^*(M).$$

G maps simple (p,p)-vectors of the form $\sigma_p^{-1}\eta \wedge \bar{\eta}$ to simple (k,k)-covectors of the form $\sigma_k \zeta \wedge \bar{\zeta}$, hence

$$\alpha_x(\sigma_p^{-1}\eta \wedge \bar{\eta}) = \alpha_x \wedge g(\sigma_p^{-1}\eta \wedge \bar{\eta}) = \alpha_x \wedge \sigma_k \zeta \wedge \bar{\zeta}. \quad \square$$

A (p,p)-current[1] $T \in \mathcal{E}'_{p,p}(M)$ is *real*, and we shall write $T \in \mathcal{E}'_{p,p}(M)_{\mathbb{R}}$, if $\bar{T}(\omega) = T(\bar{\omega})$, for every (p,p) -form ω with compact support.

$T \in \mathcal{E}'_{p,p}(M)$ is called a *weakly positive (resp. a positive, strongly positive) current*
if $T(\omega) \geq 0$ for every strongly positive (resp. positive, weakly positive) (p,p)-form ω with compact support. (Note that a positive, in any sense, (p,p)-current has to be real).

We have:

I.1.4 PROPOSITION [Harvey, 13,1.20]. *A weakly positive* (p,p)*-current* T *on* M *has measure coefficients.*

Let T be a current with measure coefficients on the open suset W of M; if A is a subset of W, let χ_A denote its characteristic function. For each compact subset K of W define the mass $M_K(T)$ of T on K as:

$$M_K(T) = \sup_{||\phi||^* \leq 1} |\chi_K T(\phi)|,$$

[1] We are using the following terminology: a (p,p)-current is a continuous linear functional on the compactly supported smooth (p,p)-forms (test forms). Locally it is a (k,k)-form with coefficients distributions. (p,p) is called the *bidimension* of the current, while (k,k) is called the *bidegree* of the current. We recall also that the support of a current T is the smallest closed set F such that $T(\phi) = 0$, for every test form ϕ with support in $M \setminus F$.

for each test form ϕ,
(where

$$||\phi||^* = \sup_{x \in K} ||\phi||_x^*,$$

and

$$||\phi||_x^* = \sup_{\xi} < \phi_x, \xi_x >, \ \xi \text{ simple unitary } (p,p)\text{-vector}$$

is the co-mass of ξ).

The measure that assignes $M_K(T)$ to each compact subset K is called the *mass measure* and it is denoted by $||T||$.

We have:

I.1.5 PROPOSITION. *Let M be a complex manifold, T a (p,p)-current with measure coefficients.*

There exists a $||T||$ -measurable function

$$\overrightarrow{T} : M \to \bigwedge_{\mathbb{R}}^{p,p}(\mathbf{T}^*(M))$$

such that

$$||\overrightarrow{T}(x)|| = 1, for \ ||T|| - \ a.e. \ x \in M$$

and

$$T(\phi) = \int_{suppT} < \overrightarrow{T}(x), \phi(x) > ||T||.$$

And:

I.1.6 PROPOSITION. *Let M be a complex manifold. T is a weakly positive (positive, strongly positive), (p,p)-current if and only if $\overrightarrow{T}(x)$ is a weakly positive (positive, strongly positive) (p,p)-vector for $||T||$-a.e. $x \in M$.*

Let $P_{p,p}$ be the cone of strongly positive (p,p)-currents in the subspace of all (p,p)-currents whose coefficients are positive measures . Then:

I.1.7 PROPOSITION. *Let M be a complex manifold. Then $P_{p,p}$ has compact base in the weak topology on $\mathcal{E}'_{p,p,}(M)$.*

I.2 Support Theorems for $\partial\bar\partial$-closed positive currents.

A *locally normal* current T is a current such that T and dT have measure coefficients. A *flat* current on the compact set K is the limit in the flat norm (defined as the sup of T taken on all test forms with co-mass ≤ 1 on K, whose differential have co-mass ≤ 1 on K) of locally normal currents supported in K.

T is *locally flat* if each cut-off ϕT is flat on some compact set containing the support of ϕ. A d-closed positive current is locally flat.

Locally flat currents have a very important property given by the Support Theorem [Federer, 9, 4.1.15, 4.1.20] :

a locally flat current T of total dimension r whose support has Hausdorff measure zero in dimension r has to be zero; moreover if the support of T is an r-dimensional oriented submanifold, T is the current of integration on it.

This theorem has very important applications to holomorphic chains. For instance every holomorphic p-chain is d-closed and if T is d-closed and locally flat of total dimension $2p$ with support in a subvariety V of complex dimension p, then T is a linear combination with real coefficients of the currents of integration on the irreducible components of V.

It happens, unfortunately, that a $\partial\bar\partial$-closed positive current is not locally flat; furthermore they are not d-closed in general, so we cannot apply to them the classical theory of closed positive currents.

We have, nonetheless, the following results, beginning with a support theorem:

I.2.1 THEOREM. *Let M be a connected complex manifold of dimension n, p a fixed integer with $0 \leq p \leq n$, T a weakly positive (p,p)-current on M such that $\partial\bar\partial T = 0$ and the 2p-Hausdorff measure, $\mathcal{H}^{2p}(\text{supp } T)$, is zero.*

Then $T = 0$.

PROOF. Let $x^0 \in \text{supp}\, T$. Choose a coordinate neighbourhood $(U, z_1, \dots, z_n)$ with center x^0 such that for every increasing multi-index $I = (i_1, \dots, i_p)$ it holds on U :

(1) if Π_I is the projection

$$\Pi_I(z_1, \dots, , z_n) = (z_{i_1}, \dots, z_{i_p}),$$

then

$$supp\, T \cap ker\, \Pi_I = \{x^0\},$$

(2) there are polidisks $\Delta'_I \subset \mathbb{C}^p$ and $\Delta''_I \subset \mathbb{C}^{n-p}$ such that

$$(\Delta'_I \times \partial\Delta''_I) \cap supp\, T = \emptyset.$$

(Hence

$$\Pi_I \restriction_{supp\, T}: (\Delta'_I \times \Delta''_I) \cap (supp\, T) \to \Delta'_I$$

is a proper map.) Fix I of length p and let $T_I = (\Pi_I \restriction_{supp\, T})_*(T)$. T_I is a $\partial\bar{\partial}$ -closed (p,p)-current on Δ'_I hence it can be identified with a $\partial\bar{\partial}$ -closed distribution f_I on Δ'_I i.e. a smooth pluriharmonic function.

But Π_I is a Lipschitzian map and therefore $\mathcal{H}^{2p}(\text{supp } f_I) = 0$, so that $f_I = 0$.

Let:

$$\omega = \sigma_1 \sum_{\alpha=1}^{n} dz_\alpha \wedge d\bar{z}_\alpha$$

in $B = \bigcap_I(\Delta'_I \times \Delta''_I)$.

Since T is weakly positive,

$$T(\chi_B \frac{\omega^p}{p!}) \geq 0.$$

But in B,

$$\frac{\omega^p}{p!} = \sum_I \sigma_p dz_I \wedge d\bar{z}_I = \sum_I \Pi_I^*(\sigma_p dz_I \wedge d\bar{z}_I),$$

therefore:

$$T(\chi_B \frac{\omega^p}{p!}) \leq \sum_I T(\chi_{\Delta'_I \times \Delta''_I} \Pi_I^*(\sigma_p dz_I \wedge d\bar{z}_I) = \sum_I \Pi_{I_*} T(\chi_{\Delta'_I \times \Delta''_I} \sigma_p dz_I \wedge d\bar{z}_I)$$

$$= \sum_I T_I(\sigma_p dz_I \wedge d\bar{z}_I) = \sum_I \int_{\Delta'} f_I \sigma_p dz_I \wedge d\bar{z}_I = 0. \quad \square$$

For the next lemma, cfr. [Siu, 22] :

I.2.2 LEMMA. *Let p, q be integers with $1 \le p, q \le n$. Let*

$$\eta = \sigma_{n-p} \sum_{A,B} \eta_{A\bar{B}} dz_A \wedge d\bar{z}_B$$

(with A and B increasing multiindices of length k) in $\bigwedge^{k,k} \mathbb{C}^n$ be weakly positive and satisfying:

(P) A does not contain I and B does not contain I imply $\eta_{A\bar{B}} = 0$ (where $I = \{q+1, \dots, n\}$).

Then:

(Q) A does not contain I or B does not contain I imply $\eta_{A\bar{B}} = 0$.

PROOF. If $p > q$, (P) holds for each A and B, thus $\eta_{A\bar{B}} = 0$.

Suppose then $p \le q$; it is enough to show (Q) for

$$A = \{\alpha_1, \dots, \alpha_s, \gamma_{q+1}, \dots, \gamma_{q+t+1}, q+t+2, \dots, n\},$$

with $s = q - p$, $t \ge 0$, $1 \le \alpha_1 < \dots < \alpha_s < \gamma_{q+1} < \dots < \gamma_{q+t+1} \le q$, and

$$B = \{\beta_1, \dots, \beta_s, q+1, \dots, n\},$$

with $1 \le \beta_1 < \dots < \beta_s \le q$.

Set

$$\partial_k = \frac{\partial}{\partial z_k}$$

and

$$X = (\partial_{\alpha_1} + u_1 \partial_{\beta_1}) \wedge \dots \wedge (\partial_{\alpha_s} + u_s \partial_{\beta_s}) \wedge \dots \wedge (w_{q+1} \partial_{\gamma_{q+1}} + \partial_{q+1}) \wedge \dots \\ \wedge (w_{q+t+1} \partial_{\gamma_{q+t+1}} + \partial_{q+t+1}) \wedge \partial_{q+t+2} \wedge \dots \wedge \partial_n,$$

with $u_j, w_k \in \mathbb{C}$.

Then

$$\eta(\sigma_{n-p}^{-1} X \wedge \overline{X}) = p(u,w) + \overline{p(u,w)},$$

where p is a polynomial in $u_1, \dots, u_s, \bar{u}_1, \dots, \bar{u}_s, w_{q+1}, \dots, w_{q+t+1}$.

Property (P) actually implies that no term containing $w_k \bar{w}_j$ appears in $\eta(\sigma_{n-p}^{-1} X \wedge \bar{X})$. It follows that for fixed $u \in \mathbb{C}^s$, $p(u,w)$ is a constant, otherwise there would exist w with $p(u,w) < 0$.

The lemma then follows from the fact that $\eta_{A\bar{B}}$ is the coefficient of $\bar{u}_1, \dots, \bar{u}_s, w_{q+1}, \dots, w_{q+t+1}$ in $p(u,w)$. □

I.2.3 THEOREM. *Let $U \subset \mathbb{C}^n$ be a domain and T a $\partial\bar{\partial}$-closed weakly positive (p,p)-current on U.*

Suppose that supp $T \subset V = \{(z_1, \dots, z_n) \in U : z_k = 0,\ k = q+1, \dots, n\}$. Then:

(1) *T induces a weakly positive current on V,*

(2) *if $p > q$, $T = 0$,*

(3) *if $p = q$, there exists a pluriharmonic function $h : V \to \mathbb{R}$ such that $h \geq 0$ and*

$$T(\psi) = \int_V h\psi,$$

for each test form ψ on U.

PROOF. Let $U = U' \times U''$, and $z \in U$, z=(z', z'') with $z' = (z_1, \dots, z_q)$, $z'' = (z_{q+1}, \dots, z_n)$. Let T be given in U by

$$T = \sigma_{n-p} \sum_{A,B} \tau_{A\bar{B}} dz_A \wedge d\bar{z}_B$$

with A and B increasing multindices of length $n - p$, and $\tau_{A\bar{B}}$ measure coefficients.

Since supp $T \subset V$,

$$\tau_{A\bar{B}} = \tau_{A\bar{B}}(z') \otimes \delta(z'').$$

To prove (1) it is enough to show that

$$\overrightarrow{T}(x) \in \bigwedge_{\mathbb{R}}^{p,p} \mathbf{T}_x^*(V) \forall x \in V,$$

since $T = \int_V \overrightarrow{T}\, ||T||$, by I.1.5 .

This means :

A does not contain I or B does not contain I imply $\tau_{A\bar{B}} = 0$ (where $I = \{q+1, \dots, n\}$.)

Let us choose $\alpha \in I \setminus A, \beta \in I \setminus B$ and let $A' = A \cup \{\alpha\}$, $B' = B \cup \{\beta\}$, with indices arranged in increasing order.

Compute $\partial\bar{\partial}T$ and notice that in the coefficient of $dz'_A \wedge d\bar{z}''_B$, the only addendum containing

$$\partial^2_{\alpha\bar{\beta}} \delta(z'')$$

is

$$\tau_{A\bar{B}}(z') \otimes \partial^2_{\alpha\bar{\beta}}\delta(z'').$$

As $\partial\bar{\partial}T = 0$, $\tau_{A\bar{B}} = 0$, hence :

A does not contain I and B does not contain I imply $\tau_{A\bar{B}} = 0$, so that by lemma I.2.2, (1) is proved.

If $p = q$, $\tau_{A\bar{B}} \neq 0$ implies A = B = I. Let

$$h = \tau_{I\bar{I}} \neq 0;$$

since $\partial\bar{\partial}T = 0$, h is pluriharmonic and satisfies (3).

(2) comes at once from theorem I.2.1. □

It follows directly from I.2.3:

I.2.4 COROLLARY. *Let M be a complex manifold of dimension n, N a submanifold of dimension q. Let T be a $\partial\bar{\partial}$-closed weakly positive (p,p)-current on M with supp $T \subset N$.*

Then T induces a weakly positive $\partial\bar{\partial}$-closed current on N and, if $p > q$, $T = 0$.

We have already said, cfr [Harvey, 13,1.9], that a locally flat d-closed current of the appropriate dimension, supported on a subvariety, is a linear combination with real coefficients of the currents of integration on the irreducible components of the subvariety. In the $\partial\bar{\partial}$-closed and weakly positive case, we obtain that such a current differs from a linear combination of the currents of integration on the irreducible components of a subvariety by a weakly positive $\partial\bar{\partial}$-closed current.

Let us denote by $[E]$ the current of integration on the analytic subset E of the complex manifold M. We have:

I.2.5 THEOREM. *Let M be a complex manifold of dimension n, E a compact analytic subset of M. Let $\{E_j\}$, $j = 1, \dots, s$, be the irreducible components of E of dimension p. Let T be a weakly positive $\partial\bar{\partial}$-closed (p,p)-current on M with supp $T \subset E$.*

There exist then real constants $c_j \geq 0$ such that

$$S = T - \sum_{j=1}^{s} c_j[E_j]$$

is a weakly positive $\partial\bar{\partial}$-closed current on M, which support is contained in the union of irreducible components of E of dimension bigger than p.

PROOF. Let $E = A \cup B \cup C$, where A is the union of the irreducible components of E of dimension bigger than p, $B = \cup_j E_j$, C is the union of the irreducible components of E of dimension less than p.

Let $x \in E_j \cap Reg(E)$; if we choose coordinates near x as in theorem 1.2.3 there are then pluriharmonic functions $c_j : E_j \cap Reg(E) \to \mathbb{R}$ such that c_j are non negative and

$$T(\psi) = \sum_{j=1}^{s} c_j [E_j](\psi),$$

for all test forms ψ with support in $B \cap Reg(E)$.

Since c_j is pluriharmonic and non negative on $E_j \cap Reg(E)$ it extends to E_j by [Grauert-Remmert, 10]. But E_j is compact, hence c_j is a constant.

Let ω be the Kähler form of an hermitian metric on M. Let $x \in Sing\,(E)$, K a compact neighbourhood of x and $\{C_\epsilon\}$, $\epsilon > 0$, a fundamental neighbourhood system of $K \cap Sing\,(E)$ in K.

Now,

$$S = T - \sum_{j=1}^{s} c_j [E_j]$$

is a well defined $\partial\bar{\partial}$-closed weakly positive current on M, in fact:

$$0 \leq \int_{(K \setminus C_\epsilon) \cap E_j} c_j \omega^p \leq \sum_i \int_{(K \setminus C_\epsilon) \cap E_i} c_i \omega^p \leq T(\chi_{(K \setminus Sing(E))} \omega^p)$$
$$\leq T(\omega^p) \leq \infty.$$

$E_{(p)} = E \setminus (B \cap Reg(E))$ is an analytic subset without any irreducible component of dimension p and we have just shown that supp $S \subset E_{(p)}$. Let $E_{(p)} = A \cup B_{(p)} \cup C_{(p)}$ where $B_{(p)}$ is the union of the irreducible components of E of dimension $p-1$ and $C_{(p)}$ is the union of the irreducible components of E of dimension smaller than $p-1$.

If $x \in B_{(p)} \cap Reg E_{(p)}$, by theorem I.2.3 x is not in supp S hence

$$\text{supp } S \subset E_{(p-1)} = E_{(p)} \setminus (B_{(p)} \cap Reg E_{(p)}).$$

In a finite number of steps we get that supp $S \subset A$. $\square$

I.3 (p,p)-components of boundaries and $\partial\bar{\partial}$-cohomology.

A particular and for us important class of $\partial\bar{\partial}$-closed currents is given by (p,p)-components of boundaries.

I.3.1. DEFINITION. Let M be a complex manifold. $T \in \mathcal{E}'_{p,p}(M)_{\mathbb{R}}$ is called the *(p,p)-component of a boundary* if there exists a real $(p,p+1)$-current S such that

$$T = \partial\bar{S} + \bar{\partial}S.$$

We shall write in this case $T = d_{p,p}S$.

$d_{p,p}$ can be thought of as

$$d_{p,p} : \mathcal{E}'_{p,p+1}(M)_{\mathbb{R}} \oplus \mathcal{E}'_{p+1,p}(M)_{\mathbb{R}} \to \mathcal{E}'_{p,p}(M)_{\mathbb{R}}$$

given by the restriction of

$$d = \partial + \bar{\partial} : \mathcal{E}'_{2p+1}(M)_{\mathbb{R}} \to \mathcal{E}'_{2p}(M)_{\mathbb{R}}$$

to $\mathcal{E}'_{p,p+1}(M)_{\mathbb{R}} \oplus \mathcal{E}'_{p+1,p}(M)_{\mathbb{R}}$ composed with the projection

$$p : \mathcal{E}'_{2p}(M)_{\mathbb{R}} \to \mathcal{E}'_{p,p}(M)_{\mathbb{R}}.$$

I.3.2. PROPOSITION. *$d_{p,p}$ is the adjoint operator to*

$$d : \mathcal{E}'_{p,p}(M)_{\mathbb{R}} \to \mathcal{E}'_{p,p+1}(M)_{\mathbb{R}} \oplus \mathcal{E}'_{p+1,p}(M)_{\mathbb{R}}.$$

PROOF. For any test form ϕ and $S \in \mathcal{E}'_{p,p+1}(M)_{\mathbb{R}} \oplus \mathcal{E}'_{p+1,p}(M)_{\mathbb{R}}$, one has in fact:

$$< S, d\phi >=< dS, \phi >=< (p \circ d)S, \phi > . \quad \square$$

I.3.3. If T is the (p,p)-component of a boundary, obviously T is $\partial\bar{\partial}$- closed. Interestingly enough, locally the two classes coincide. Indeed, let T be a $\partial\bar{\partial}$-closed (p,p)-current; locally,

$$\ker \partial\bar{\partial} = Im\,\partial + Im\,\bar{\partial},$$

hence write $T = \partial S' + \bar{\partial}S''$; T real implies that $S' = \bar{S}''$, taking $S = S''$, S is a $(p,p+1)$-current such that

$$T = \partial\bar{S} + \bar{\partial}S.$$

Moreover if $\phi \in \mathcal{E}'_{p,p}(M)_{\mathbb{R}}$ is d-closed, $T(\phi) = 0$. Hence the orthogonal subspace to $\ker d$ is the closure of $Im\,\partial \oplus Im\,\bar{\partial}$ in the weak topology.

When M is compact the following theorem tells that every current which is the limit of currents which are components of boundaries is the component of a boundary itself. We have indeed:

I.3.4 THEOREM. *Let M be a compact complex manifold.*

$$d_{p,p} : \mathcal{E}'_{p,p+1}(M)_{\mathbb{R}} \oplus \mathcal{E}'_{p+1,p}(M)_{\mathbb{R}} \to \mathcal{E}'_{p,p}(M)_{\mathbb{R}}$$

has closed rank in the weak topology on $\mathcal{E}'_{p,p}(M)_{\mathbb{R}}$.

PROOF. Let $\mathcal{H}$ be the sheaf of germs of pluriharmonic functions on M. Let us consider the following sequence of sheaf homomorphisms on M, for $p \geq 1$:

$$0 \to \mathcal{H} \xrightarrow{f} \mathcal{L}^0 \xrightarrow{f_0} \mathcal{L}^1 \to \ldots \to \mathcal{L}^k \xrightarrow{f_k} \mathcal{L}^{k+1} \to \ \ldots \mathcal{L}^{p-1} \xrightarrow{g} \mathcal{B}^p$$
$$\to \ldots \to \mathcal{B}^k \xrightarrow{g_k} \mathcal{B}^{k+1} \to \ldots \to \mathcal{B}^{2p-1} \xrightarrow{h} \mathcal{E}^{p,p}(M)_{\mathbb{R}}$$
$$\xrightarrow{i(\partial-\bar\partial)} \mathcal{E}^{p+1,p+1}(M)_{\mathbb{R}} \xrightarrow{d} [\mathcal{E}^{p+1,p+2}(M) \oplus \mathcal{E}^{p+2,p+1}(M)]_{\mathbb{R}} \to \ldots$$

where

(i) $$\mathcal{L}^k = \overline{\Omega^{k+1}} \oplus \mathcal{E}^{0,k-\alpha}(M)_{\mathbb{R}} \oplus \ldots \oplus \mathcal{E}^{k,0}(M)_{\mathbb{R}} \oplus \Omega^{k+1}$$

with $0 \leq k \leq p-1$ and Ω^k is the sheaf of germs of holomorphic k-forms,

(ii) $$\mathcal{B}^k = \mathcal{E}^{k-p,p}(M)_{\mathbb{R}} \oplus \ldots \oplus \mathcal{E}^{p,k-p}(M)_{\mathbb{R}},$$

for $p \leq k \leq 2p-1$,

And

(I) $$f : \mathcal{H} \to \mathcal{L}^0$$

is given by

$$f(\phi) = (-\bar\partial\phi, \phi, -\partial\phi),$$

(II) $$f_k : \mathcal{L}^k \to \mathcal{L}^{k+1}$$

is given by

$$f_k(\bar\phi, \oplus a^{0,k} \oplus \ldots \oplus a^{k,0}, \phi) = (-\bar\partial\bar\phi, \phi + \bar\partial a^{0,k}, \partial a^{0,k} + +\bar\partial a^{1,k-1} \oplus \ldots$$
$$\oplus \partial a^{k-1,1} + +\bar\partial a^{k,0}, \phi + \partial a^{k,0}, -\partial\phi),$$

for $0 \leq k \leq p-2, p > 1$,

(III)
$$g : \mathcal{L}^{p-1} \to \mathcal{B}^p$$

is given by

$$g(\phi, a^{0,p-1}, \phi) = (\phi + \bar{\partial} a^{0,p-1}, \partial a^{0,p-1} + +\bar{\partial} a^{1,p-2} \oplus \ldots \oplus \partial a^{p-2,1} + +\bar{\partial} a^{p-1,0}, \phi + \partial a^{p-1,0}),$$

(IV)
$$g_k : \mathcal{B}^k \to \mathcal{B}^{k+1}$$

is given by

$$g_k(a^{k-p,p} \oplus \ldots \oplus a^{p,k-p}) = \partial a^{k-p,p} + \bar{\partial} a^{k-p+1,p-1} \oplus \ldots \oplus \partial a^{k-2,2} + \bar{\partial} a^{k-1,1},$$

for $p \leq k \leq 2p-1$.

It is not simple but direct to show that the above sequence is exact, [Alessandrini-Andreatta, 1].

The sheaves $\mathcal{L}^k$ are not fine but if we call $\mathcal{Z} = \ker g_p$, the following is a fine resolution of $\mathcal{Z}$:

$$0 \to \mathcal{Z} \to \mathcal{B}^p \to \ldots \to \mathcal{B}^k \to \mathcal{B}^{k+1} \to \ldots \to \mathcal{B}^{2p-1} \to \mathcal{E}^{p,p}(M)_{\mathbb{R}} \to \mathcal{E}^{p+1,p+1}(M)_{\mathbb{R}} \to [\mathcal{E}^{p+1,p+2}(M) \oplus \mathcal{E}^{p+2,p+1}(M)]_{\mathbb{R}} \to \ldots$$

so that:

$$H^{p+2}(M, \mathcal{Z}) = \frac{\{\mu \in [\mathcal{E}^{p+1,p+2} \oplus \mathcal{E}^{p+2,p+1}]_{\mathbb{R}} : d\mu = 0\}}{\{\mu = d\pi, \pi \in \mathcal{E}^{p+1,p+1}_{\mathbb{R}}\}}.$$

We are going to prove that $H^{p+2}(M, \mathcal{Z})$ is a finite dimensional real vector space.

We split the original exact sequence into short exact sequences:

$$0 \to \mathcal{H} \to \mathcal{L}^0 \to \ker f_0 \to 0,$$

$$0 \to ker f_0 \to \mathcal{L}^1 \to \ker f_1 \to 0$$

....

$$0 \to \ker g \to \mathcal{L}^{p-1} \to \mathcal{Z} \to 0;$$

if we take the long exact cohomology sequence in the last one we obtain that

$$H^k(M, \mathcal{Z}) \to H^{k+1}(M, ker\, g)$$

has finite dimensional kernel and cokernel since

$$H^s(M, \mathcal{L}^r) = H^s(M, \overline{\Omega^{r+1}} \oplus \Omega^{r+1}),$$

which has finite dimension for each $r, 0 \leq r \leq p-1, s \geq 0$, , since M is compact.
Applying the same argument to the other short exact sequences, we obtain that

$$H^{k+r}(M, \ker f_{p-r}) \to H^{k+r+1}(M, \ker f_{p-r-1}),$$

for $1 \leq r \leq p-2$, and

$$H^{k+p-1}(M, \ker f_r) \to H^{k+p}(M, \mathcal{H})$$

have finite dimensional kernel and cokernel too, so that

$$H^k(M, \mathcal{Z}) \to H^{k+p}(M, \mathcal{H})$$

has finite dimensional kernel and cokernel , but $H^s(M, \mathcal{H})$ is a finite dimensional real vector space for $s \geq 1$, by the compactness of M and the exact sequence

$$0 \to \mathbb{R} \to \mathcal{O}_M \to \mathcal{H} \to 0,$$

hence $H^k(M, \mathcal{Z})$ is finite dimensional too, for $k \geq 1$.

Then

$$d : \mathcal{E}_{\mathbb{R}}^{p,p} \to [\mathcal{E}^{p,p+1} \oplus \mathcal{E}^{p+1,p}]_{\mathbb{R}}$$

has closed rank in the weak topology by the open mapping theorem.

Let $B_{p,p}$ be the rank of $d_{p,p}$. Then $B_{p,p}$ is closed too, since $d_{p,p}$ is the adjoint of d, by I.3.2. □

I.3.5. Let us define now the *real (p,q)-Aeppli groups of M* as:

$$\mathcal{V}^{p,q}(M)_{\mathbb{R}} = \frac{ker(\sqrt{-1}\partial\bar{\partial} : \mathcal{E}^{p,q}(M)_{\mathbb{R}} \to \mathcal{E}^{p+1,q+1}(M)_{\mathbb{R}}}{\partial\mathcal{E}^{p-1,q}(M)_{\mathbb{R}} \oplus \bar{\partial}\mathcal{E}^{p,q-1}(M)_{\mathbb{R}}}$$

and

$$\Lambda^{p,q}(M)_{\mathbb{R}} = \frac{ker(d : \mathcal{E}^{p,q}(M)_{\mathbb{R}} \to \mathcal{E}^{p+1,q}(M)_{\mathbb{R}} \oplus \mathcal{E}^{p,q+1}(M)_{\mathbb{R}}}{im(\partial\bar{\partial} : \mathcal{E}^{p-1,q-1}(M)_{\mathbb{R}} \to \mathcal{E}^{p,q}(M)_{\mathbb{R}})}.$$

Thus a $\partial\bar{\partial}$-closed (p,p)-current is the component of a boundary if and only if its class in $\mathcal{V}^{p,p}(M)_{\mathbb{R}}$ is zero.

A class in $\mathcal{V}^{p,p}(M)_{\mathbb{R}}$ is *positive* if it can be represented by a positive current. Let $H^{p,p}(M,\mathbb{R})$ denote as usual the set of classes in $H^{2p}(M,\mathbb{R})$ which have a (p,p) representative. There are natural maps:

$$\alpha : \Lambda^{p,p}(M)_{\mathbb{R}} \to H^{p,p}(M,\mathbb{R})$$

and

$$\beta : H^{p,p}(M,\mathbb{R}) \to \mathcal{V}^{p,p}(M)_{\mathbb{R}}$$

Since α is always surjective, we immediately have that:

I.3.6 PROPOSITION. *If $\beta \circ \alpha$ is an isomorphism, then α and β are isomorphisms too.*

I.4 Relative Support Theorems for $\partial\bar{\partial}$-closed currents.

We study in this section the behaviour of $\partial\bar{\partial}$-closed currents of *bidegree* (1,1) and of order zero (in particular positive $\partial\bar{\partial}$-closed currents), whose support is contained in the exceptional set of a proper modification.

Let $f : M' \to M$ be a holomorphic map between complex spaces.
f is called a *proper modification* if it is proper and if there is a thin subset Y of M with the following properties:

(1) $E := f^{-1}(Y)$ is thin in M'

(2) the restriction of f between $M' \setminus E \to M \setminus Y$ is biholomorphic.

It is easy to see that Y can be chosen an analytic subset of M of codimension ≥ 2, otherwise f is a biholomorphism, so that if M' is reduced and M is a manifold, E is an analytic set of pure dimension $n-1$, called the *exceptional set* of the modification. This is going to be always our case since we will consider *proper smooth modifications,* i.e. modifications with M' and M both complex manifolds.

A way of producing smooth proper modifications is by blowing up M along Y. If Y is a closed analytic subspace of the complex space M, there exist a complex space M' and a proper modification $f : M' \to M$ with the property that f is universal with respect to E, i.e. if there is any holomorphic map $g : M'' \to M$ such that $g^{-1}(Y)$ is of pure codimension 1, then there is a unique holomorphic map $h : M'' \to M'$, such that $g = h \circ f$ and if M is a manifold and Y a closed submanifold, M' is a manifold. Such f is called the *blow up of M with center Y, or along Y.*

In the case of a blow up points are always replaced by projective varieties, but in complex geometry it is necessary to consider modifications with arbitrary fibres. The case in which M has a Kähler metric will show this necessity.

Firstly we are collecting a few facts about the local structure of modifications that will be useful in what follows.

Let M' and M be complex manifolds of dimension d and $f : M' \to M$ be the blow up of M with smooth center Y. Let $E = f^{-1}(Y)$.

Let B_k be the euclidean ball in $\mathbb{C}^k$ with center the origin and radius 1, and set $B_0 = \{0\}$. For every $y \in Y$, choose an open neighbourhood $U = B_m \times B_n$ with $m = \dim Y$ and $n = d - m$, such that $Y \cap U = B_m \times \{0\}$. Let us call $U' = f^{-1}(U)$ and $B_n{}'$ the blow up of B_n at $\{0\}$, i.e.

$$B_n{}' = \{(z, \zeta) \in B_n \times \mathbb{P}_{n-1} : z_j \zeta_k = z_k \zeta_j 1 \leq j, k \leq n\}.$$

We can identify $f \restriction'_U : U' \to U$ with the projection $\pi : B_m \times B'_n \to B_m \times B_n$.

The natural Kähler form on U' is given by

$$\omega' = \frac{\sqrt{-1}}{2}(\partial\overline{\partial}||t||^2 + \partial\overline{\partial}||z||^2 + \partial\overline{\partial}log||\zeta||^2)$$

with $t \in \mathbb{C}^m$, therefore:

$$\pi_* \omega' = \frac{\sqrt{-1}}{2}(\partial\overline{\partial}||t||^2 + \partial\overline{\partial}||z||^2 + \partial\overline{\partial}log||z||^2).$$

Let us take $y \in Y \cap U = B_m$ and identify the singular fibre $\pi_1(y)$ with $\mathbb{P}_{n-1}$.

Set $E_U = E \cap U$; the *conormal bundle* of the singular fibre $\mathcal{N}^*{}_{\mathbb{P}_{n-1}} \restriction_{E_U}$ is defined by means of the exact sequence of vector spaces:

$$0 \to \mathcal{N}^*{}_{\mathbb{P}_{n-1}} \restriction_{E_U,x} \to \mathbf{T}'^*_{E_U,x} \to \mathbf{T}'^*_{\mathbb{P}_{n-1},x} \to 0,$$

for $x \in \mathbb{P}_{n-1}$, and it is of course trivial on $E_U = B_m \times \mathbb{P}_{n-1}$.

We have:

I.4.1 LEMMA. *Let M' and M be complex manifolds od dimension n and $f : M' \to M$ be the blow up of M with smooth center Y. Let $E = f^{-1}(Y)$. Then*

$$H^0(E, \Omega^1_E(\mathcal{N}_E \restriction_{M'})) = 0$$

PROOF. From the exact sequence defining the normal bundle by tensoring with $\mathcal{N}_{\mathbb{P}_{n-1}} \restriction_{B'm}$ we obtain the exact sequence of sheaves on $\mathbb{P}_{n-1}$:

$$0 \to \mathcal{O}(\mathcal{N}^*{}_{\mathbb{P}_{n-1}} \restriction_{E_U}) \otimes \mathcal{O}(\mathcal{N}_{\mathbb{P}_{n-1}} \restriction_{B'm}) \to \Omega^1{}_{E_U} \restriction_{\mathbb{P}_{n-1}} \otimes \mathcal{O}(\mathcal{N}_{\mathbb{P}_{n-1}} \restriction_{B'm})$$
$$\to \Omega^1{}_{\mathbb{P}_{n-1}} \otimes \mathcal{O}(\mathcal{N}_{\mathbb{P}_{n-1}} \restriction_{B'm}) \to 0$$

By the triviality of $\mathcal{N}^*{}_{\mathbb{P}_{n-1}} \restriction_{E_U}$ and the standard identification

$$\mathcal{N}_{\mathbb{P}_{n-1}} \restriction_{B'm} = [\mathbb{P}_{n-1}] \restriction_{\mathbb{P}_{n-1}} = [-H],$$

we obtain that

$$\mathcal{O}(\mathcal{N}^*{}_{\mathbb{P}_{n-1}} \restriction_{E_U}) \otimes O(\mathcal{N}^*{}_{\mathbb{P}_{n-1}} \restriction_{E_U}) = \mathcal{O}(-1).$$

The long exact cohomology sequence starts with:

$$0 \to H^0(\mathbb{P}_{n-1}, \mathcal{O}(-1)) \to H^0(\mathbb{P}_{n-1}, \Omega^1_{E_U} \restriction_{\mathbb{P}_{n-1}}) \otimes \mathcal{O}(-1))$$
$$\to H^0(\mathbb{P}_{n-1}, \Omega^1(-1)) \to \ldots$$

hence by the Kodaira-Nakano vanishing theorem plus some easy facts about Riemann surfaces if $n = 2$, we obtain that:

$$H^0(\mathbb{P}_{n-1}, \mathcal{O}(-1)) = H^0(\mathbb{P}_{n-1}, \Omega^1(-1)) = 0,$$

so that

$$H^0(\mathbb{P}_{n-1}, \Omega^1_{E_U} \restriction_{\mathbb{P}_{n-1}}) \otimes \mathcal{O}(-1)) = 0.$$

Now let

$$h \in H^0(E, \Omega^1_E(\mathcal{N}_E \restriction_{M'}));$$

$h \restriction_{\mathbb{P}_{n-1}}$ is a section of $\Omega^1_{E_U} \restriction_{\mathbb{P}_{n-1}} \otimes \mathcal{O}(-1)$ since $\mathcal{N}_{E \restriction U'} \restriction_{\mathbb{P}_{n-1}} = \mathcal{N}_{\mathbb{P}_{n-1}} \restriction_{B'm}$ hence it is zero on each fibre over y, i.e. $h = 0$. $\square$

The following result about the structure of modifications is folklore, but apparently cannot be found in the literature:

I.4.2 LEMMA. *Let $f : M' \to M$ be a proper modification of complex spaces; for every $x \in M$ there exist an open neighbourhood V of x in M, a complex manifold Z and holomorphic maps $g : Z \to M'$, $h : Z \to V$ such that $h = f \circ g$; $g : Z \to f^{-1}(V)$ is a blow up and h is obtained as a finite sequence of blow-up's with smooth centers.*

PROOF. By [17, lemma 8 pag. 321], for every $x \in M$ there exist an open neighbourhood V and a complex subspace $(D, \mathcal{O}_D)$ of V such that if $h' : V' \to V$ is the blow up with center $(D, \mathcal{O}_D)$ there exists a holomorphic map $g' : V' \to M'$ with $h' = f \circ g'$.

Let $\mathcal{I}$ be the coherent ideal sheaf of $\mathcal{O}_V$ which defines $(D, \mathcal{O}_D)$; by applying [17, lemma 7 p. 320 of] to V and $\mathcal{I}$, we obtain a finite sequence $h_j : V_{j+1} \to V_j, 0 \leq j < r$, of blow-up's with smooth centers such that $Z = V_r$ is smooth and if $h = h_0 \circ \ldots \circ h_{j-1}, h^{-1}(\mathcal{I})$ is invertible [cfr. remark after lemma 7 in 17]. We may also suppose $V = V_0$ by shrinking V if necessary.

The universal property of modifications will give a holomorphic map $g'' : Z \to V'$ such that $h = h' \circ g''$; if $g = g' \circ g'' : Z \to M'$, we get $h = f \circ g$ and $g : Z \to f^{-1}(V)$ is a blow up, since $h : Z \to V$ is obtained by a finite sequence of blow-up's and $f : f^{-1}(V) \to V$ is a proper modification [cor. 1 p. 320 and lemma 4, p. 318 of 17]. □

We prove now support theorems relative to modifications. We start with the following relative version of I.2.3:

I.4.3 THEOREM. *Let M' and M be complex manifolds of dimension n and $f : M' \to M$ be the blow up of M with smooth center Y. Let T be a real $\partial\bar{\partial}$-closed current on M' of order zero and of bidegree $(1,1)$ whose support is contained in the exceptional set E.*

Then there exists a pluriharmonic function $h : Y \to \mathbb{R}$ such that

$$T = (h \circ f)[E].$$

Moreover, if T is the limit in the weak topology of currents that are components of boundaries, then

$$T = 0.$$

PROOF. Let us fix a coordinate neighbourhood $\{V, (v_1, \ldots, v_n)\}$ in M' such that

$$V \cap E = \{v_n = 0\}.$$

In V, T has then the following expression:

$$T = \frac{\sqrt{-1}}{2} \sum_{\alpha,\beta=1}^{n} \tau_{\alpha\bar\beta}(v') \otimes \delta(v_n)\, dv_\alpha \wedge d\bar v_\beta,$$

where $v' = (v_1, \ldots, v_{n_1})$ and $\tau_{\alpha\bar\beta}$ is a measure with $\tau_{\bar\alpha\beta} = \overline{\tau_{\alpha\bar\beta}}$.

Fix $\alpha, \beta < n$. The coefficient of

$$d\,v_\alpha \wedge d\,v_n \wedge d\bar v_\beta \wedge d\bar v_n$$

in $\sqrt{-1}\,\partial\bar\partial\, T$, which has to vanish, is given by

$$(*) \quad -\tau_{\alpha\bar\beta}(v') \otimes \partial^2{}_{n\bar n}\,\delta(v_n) + \partial_\alpha \tau_{n\bar\beta}(v') \otimes \partial_{\bar n}\,\delta(v_n) + \partial_{\bar\beta}\tau_{\alpha\bar n}(v') \otimes \partial_n\,\delta(v_n)$$
$$- \partial^2{}_{\alpha\bar\beta}\,\tau_{n\bar n}(v') \otimes \delta(v_n)$$

Let us choose test functions as follows:

$$\psi = \psi(v_n) \in \mathcal{C}_0{}^\infty(V) \text{ with } \psi = 1 \text{ near the origin}$$

and

$$\phi = \phi(v') \in \mathcal{C}_0{}^\infty(V).$$

Using $\psi\phi|v_n|^2$, $\psi\phi v_n$, $\psi\phi$ in (*), we obtain:

$$0 = -<\tau_{\alpha\bar\beta}\,,\, \phi> = <\partial_{\bar\beta}\tau_{\alpha\bar n}\,, \phi> = -<\partial^2{}_{\alpha\bar\beta}\,\tau_{n\bar n}\,, \phi>.$$

So that:

$$(**) \qquad \begin{cases} \tau_{\alpha\bar\beta} = 0 & \text{for } 1 \le \alpha, \beta < n \\ \tau_{\alpha\bar n} & \text{is holomorphic for } 1 \le \alpha < n \\ \tau_{n\bar n} & \text{is pluriharmonic.} \end{cases}$$

In another coordinate neighbourhood $\{W, (w_1, \ldots, w_n)\}$ with

$$W \cap V \neq \emptyset \text{ and } W \cap E = \{w_n = 0\},$$

if we set $w' = (w_1, \ldots, w_{n-1})$, we assume:

$$T = \frac{\sqrt{-1}}{2} \sum_{\lambda,\mu=1}^{n} \sigma_{\lambda\bar\mu}(w') \otimes \delta(w_n)\, dw_\lambda \wedge d\bar w_\mu.$$

By (**) and the similar results for $\sigma_{\lambda\bar{\mu}}$, and using the fact that

$$\frac{\partial v_n}{\partial w_\lambda} = \frac{\partial w_n}{\partial v_\alpha} = 0 \, \text{on} \, E, \text{ for } \alpha, \lambda < n,$$

we obtain the following relations:

$$\tau_{\alpha\bar{n}} = \sum_{\lambda=1}^{n-1} \sigma_{\lambda\bar{n}} \left(w'\left(v'\right)\right) \frac{\partial v_n}{\partial w_n} \frac{\partial w_\lambda}{\partial v_\alpha}$$

if $\alpha < n$, and

$$\begin{aligned} \tau_{n\bar{n}}\left(v'\right) &= \sum_{\lambda=1}^{n-1} \sigma_{\lambda\bar{n}}\left(w'\left(v'\right)\right) \frac{\partial v_n}{\partial w_n} \frac{\partial w_\lambda}{\partial v_n} \\ &+ \sum_{\lambda=1}^{n-1} \overline{\sigma_{\mu\bar{n}}}\left(w'\left(v'\right)\right) \overline{\frac{\partial v_n}{\partial w_n} \frac{\partial w_\mu}{\partial v_n}} + \sigma_{n\bar{n}}\left(w'\left(v'\right)\right) \end{aligned}$$

If we cover M' by coordinate charts of the form $(V, (v', v))$, then $\tau_{1\bar{n}}, \ldots, \tau_{n-1\bar{n}}$ is a global section of $\Omega^1_E(\mathcal{N}_E \restriction_{M'})$ on $E \cap U$, since the cocycles of $\mathcal{N}_E \restriction_{M'}$ are given by

$$\frac{\partial v_n}{\partial w_n}.$$

By Lemma I.4.1, we obtain:

$$\tau_{\alpha\bar{n}} = \sigma_{\lambda\bar{n}} = 0,$$

and

$$h = \tau_{n\bar{n}} = \sigma_{n\bar{n}}$$

is a pluriharmonic function on M'. Since the fibres of f are compact, h depends only on the coordinates of Y.

As for the second part of the statement, we need only to show that if T is not identically zero, hence if $h : Y \to \mathbb{R}$ is not identically zero, then T is not in the orthogonal subspace to

$$\ker\left(d : \mathcal{E}'_{n-1,n-1}(M')_{\mathbb{R}} \to \mathcal{E}'_{n,n}(M')_{\mathbb{R}}\right),$$

by I.3.3, that is there exists a d-closed $(n-1, n-1)$-form on M', Θ such that

$$T(\Theta) = h \circ f[E](\Theta) = \int_E (h \circ f)\Theta > 0.$$

Let $y \in Y$ such that $h(y) > 0$, and choose an open neighbourhood U of y, biholomorphic to $B^l \times B^m$ and such that $U \cap Y = B^l \times \{0\}$ and $h > 0$ in $U \cap Y$; then identify $f \restriction_{f^{-1}(U)}$ with the blow up π.
Let us choose real functions u and v such that:

$$u \in \mathcal{C}_0{}^{\infty}(B^l),\ u \neq 0,\ u(t) \geq 0,$$

$$v \in \mathcal{C}_0{}^{\infty}(B^m),\ v(z) = 1 \text{ near the origin.}$$

Then set:

$$\psi(z) = \frac{\sqrt{-1}}{2\pi}\partial\bar{\partial}((1 - v(z))\log ||z||^2)$$

and

$$\Theta = \left(\frac{\sqrt{-1}}{2\pi}\partial\bar{\partial}\log ||\zeta||^2 - \pi^*\psi\right)^{m-1} \wedge u(t)\left(\frac{\sqrt{-1}}{2\pi}\partial\bar{\partial}\,||t||^2\right)^l.$$

Θ then satisfies the requirements. In particular, if $i : E \cap f^{-1}(U) \to f^{-1}(U)$ is the inclusion,

$$i^*\Theta = \left(\frac{\sqrt{-1}}{2\pi}\partial\bar{\partial}\log ||\zeta||^2\right)^{m-1} \wedge u(t)\left(\frac{\sqrt{-1}}{2\pi}\partial\bar{\partial}\,||t||^2\right)^l,$$

thus

$$h \circ f[E](\Theta) = \int_E (h \circ f \circ i)(i^*\Theta) = C\int_{B^l} h(t)u(t)\left(\frac{\sqrt{-1}}{2\pi}\partial\bar{\partial}\,||t||^2\right)^l > 0. \quad \square$$

I.4.4. Apparently, an example of a current of order zero which is $\partial\bar{\partial}$ - closed but not locally flat was not available in the literature. The proof of theorem I.4.3 provides an explicit one. Let

$$T = \frac{\sqrt{-1}}{2\pi}C\,\delta(v_n)(dv_1 \wedge d\bar{v}_n + dv_n \wedge d\bar{v}_1),\ C \in \mathbb{R}.$$

T is a real $\partial\bar{\partial}$ - closed current of bidegree $(1,1)$ with supp $T \subset E \cap V$.
If T were locally flat, a pluriharmonic function would exist $h : E \to \mathbb{R}$ such that $T = h[E]$, but this is not the case if $C \neq 0$.

We give also the relative versions of I.2.5:

I.4.5 PROPOSITION. *Let $f : M' \to M$ be a proper smooth modification between manifolds of dimension n obtained as a finite sequence of blow-ups with smooth centers. Let $\{C_\alpha\}$ be the set of the irreducible components of the exceptional set E.*

Then:

(i) *Every real $\partial\bar{\partial}$-closed current T of order zero and bidegree (1,1) on M', such that supp $T \subset E$ is of the form*

$$\sum_\alpha u_\alpha[C_\alpha],$$

where u_α is a real pluriharmonic function defined on C_α,

(ii) *T is the limit in the weak topology of currents which are components of boundaries if and only if every u_α vanishes.*

PROOF. We have by assumption a finite sequence

$$f_j : V_{j+1} \to V_j \, , \; 0 \leq j < r$$

of blow-ups with smooth centers $Y_j \subseteq V_j$ and exceptional sets $E_{j+1} \subseteq V_{j+1}$ such that

$$V_0 = M \, , \; V_r = M' \, , \; f = f_0 \circ \ldots \circ f_{r-1}.$$

By theorem I.4.3 then we obtain

$$(f_1 \circ \ldots \circ f_{r-1})_* T = (u_1 \circ f_0)[E_1]$$

with

$$u_1 : Y_0 \to \mathbb{R}$$

pluriharmonic.

If $E^*{}_1$ is the strict transform of E_1 under f_1, then

$$(u_1 \circ f_0 \circ f_1)[E^*{}_1]$$

is $\partial\bar{\partial}$ - closed. Therefore we can apply Theorem I.4.3 again and obtain

$$(*) \qquad (f_2 \circ \ldots \circ f_{r-1})_* T - (u_1 \circ f_0 \circ f_1)[E^*{}_1] = (u_2 \circ f_1)[E_2]$$

with

$$u_2 : Y_1 \to \mathbb{R}$$

pluriharmonic. We obtain eventually:

$$T = \sum_{j=1}^{r} (u_j \circ f_{j-1} \circ \ldots \circ f_{r-1})[E_j^*]$$

where $E_j^*$is the strict transform of E_j under $f_{j-1} \circ \ldots \circ f_{r-1}$, $1 \leq j < r$, $E_r^* = E_r$ and

$$u_j : Y_{j-1} \to \mathbb{R}$$

is pluriharmonic.

As for (ii), if we suppose that T is the limit of components of boundaries, then

$$(f_1 \circ \ldots \circ f_{r-1})_* T = (u_1 \circ f_0)[E_1]$$

is limit of components of boundaries too, so that by theorem I.4.3

$$u_1 = 0.$$

From (*) and theorem I.4.3 we also have

$$u_2 = 0,$$

and so on. □

For the case of a general smooth modification we have:

I.4.6 THEOREM. *Let $f : M' \to M$ be a proper modification between complex manifolds of dimension n, and let $\{C_\alpha\}$ be the set of the irreducible components of the exceptional set E.*

If

$$T = \sum_\alpha c_\alpha [C_\alpha],\ c_\alpha \in \mathbb{C} \quad \forall\, \alpha,$$

and for every $x \in M$ there exists an open neighbourhood V of x such that $T \restriction_{f^{-1}(V)}$ is the limit in the weak topology of currents which are components of boundaries, then

$$c_\alpha = 0 \quad \forall\, \alpha.$$

PROOF. Fix a component C_γ and choose $x \in f(C_\gamma)$.
For a suitable neighbourhood V of x in M, by lemma I.4.2 we have a complex manifold Z and holomorphic maps

$$g : Z \to f^{-1}(V)\,,\; h : Z \to V.$$

We have that:

$$T \restriction_{f^{-1}(V)} = \sum_{\gamma_j} c_{\gamma_j}[E_{\gamma_j}]$$

where $\{E_{\gamma_j}\}$ is the set of connected components of $C_\gamma \cap f^{-1}(V)$.
Let $\{F_\beta\}$ be the set of irreducible components of the exceptional set of $g : Z \to f^{-1}(V)$ and denote by $E^*_{\gamma_j}$ the strict transform of E_{γ_j} under g.

Thus $\{F_\beta\} \cup \{E^*_{\gamma_j}\}$ is the set of irreducible components of the exceptional set of $h : Z \to V$.

Therefore the total transform $\hat{T}$ of $\sum_{\gamma_j} c_{\gamma_j}[E_{\gamma_j}]$ under g is of the form

$$\hat{T} = \sum_{\gamma_j} c_{\gamma_j}[E^*_{\gamma_j}] + \sum_{\beta} c'_\beta[F_\beta].$$

By proposition I.4.5 we need only to show that $\hat{T} \restriction_{h^{-1}(V)}$ satisfies our assumptions.

Let

$$\phi \in \mathcal{E}_{1,1}(f^{-1}(V))_{\mathbb{R}}$$

be a representative of the class of

$$\sum_{\gamma_j} c_{\gamma_j}[E_{\gamma_j}]$$

in $\Lambda^{1,1}(f^{-1}(V))_{\mathbb{R}}$; hence ϕ is also a representative of the fundamental class of $\sum_{\gamma_j} c_{\gamma_j}[E_{\gamma_j}]$ in $H^2(f^{-1}(V), \mathbb{R})$, i.e.

$$\phi = \sum_{\gamma_j} c_{\gamma_j}[E_{\gamma_j}] + dQ$$

for a suitable current Q in $f^{-1}(V)$.

Then $g^*\phi$ represents the fundamental class of the total transform $\hat{T}$, i.e.

$$g^*\phi = \hat{T} + dQ',$$

for a suitable current Q' on Z.

By assumption we have a sequence $\{R_\mu\}$ of currents of bidegree $(1,0)$ on $f^{-1}(V)$ such that

$$\sum_{\gamma_j} c_{\gamma_j}[E_{\gamma_j}] = \lim_\mu \bar{\partial} R_\mu + \partial \overline{R}_\mu$$

weakly. By smoothing R_μ and Q we have also

$$\phi = \lim_\mu \bar{\partial}\rho_\mu + \partial\overline{\rho}_\mu$$

where ρ_μ are smooth $(1,0)$-forms on $f^{-1}(V)$.

Let S be a d - closed current of bidegree $(n-1, n-1)$ with compact support in $f^{-1}(V)$, and let

$$\psi \in \mathcal{E}_{n-1,n-1}(f^{-1}(V))_{\mathbb{R}}$$

be a d - closed compactly supported representative of the class of S in $\Lambda^{n-1,n-1}(f^{-1}(V))_{\mathbb{R}}$, i.e.

$$S = \psi + \sqrt{-1}\,\partial\bar{\partial}u$$

for a suitable current u with compact support in $f^{-1}(V)$. Now,

$$\begin{aligned} S(\phi) &= \int_{f^{-1}(V)} \phi \wedge \psi + \sqrt{-1}\,\partial\bar{\partial}u(\phi) \\ &= \lim_\mu \int_{f^{-1}(V)} (\bar{\partial}\rho_\mu + \partial\overline{\rho}_\mu) \wedge \psi + \sqrt{-1}\,\bar{\partial}u(\partial\phi) \\ &= 0. \end{aligned}$$

Hence ϕ belongs to the orthogonal subspace to

$$\ker\left(d : \mathcal{E}'_{1,1}(f^{-1}(V))_{\mathbb{R}} \to \mathcal{E}'_{0,1}(f^{-1}(V))_{\mathbb{R}} \oplus \mathcal{E}'_{1,0}(f^{-1}(V))_{\mathbb{R}}\right)$$

then by I.3.3 ϕ is limit of components of boundaries and the same holds for $g^*\phi$ and also for $\hat{T}$. $\square$

II

CRITERIA FOR THE EXISTENCE OF POSITIVE d-CLOSED FORMS ON COMPACT COMPLEX MANIFOLDS

II.1 A theorem of Harvey-Lawson and its generalizations.

If M is a compact complex manifold, the Kähler form ω of any hermitian metric on M has the property that restricted to any complex submanifold N of M is a (positive) volume form on N. If in addition M is Kahler , no submanifold can be a boundary.

Indeed, if $N = \partial N'$, then, up to a constant factor,

$$Vol\,(N) = \int_{\partial N'} \omega \restriction_N = \int_{N'} d\omega = 0.$$

The same argument shows that a positive analytic cycle T must represent a non zero homology class of M. Whether this kind of condition could be also sufficient for the existence of a Kähler metric on M is not known, possibly it is not even a well posed question. Harvey-Lawson in [15], strengthen the condition on submanifolds in order to find a sufficient condition for the existence of a Kähler metric in two ways: firstly T is allowed to be any d-closed positive $(1,1)$-current, secondly, the conclusion that T represents a non zero class in homology is replaced by: T is not the $(1,1)$-component of a boundary.

This condition is necessary for the complex manifold M to be Kähler. Indeed,

II.1.1 PROPOSITION. *If M is Kähler there are no positive (1,1)-currents which are (1,1)-components of boundaries.*

PROOF. Suppose that T is the $(1,1)$-component of a boundary. In I.3.2 we have seen that d and $d_{1,1}$ are adjoint operators to each other hence $d\omega = 0$ is equivalent to the fact that ω annihilates the range of $d_{1,1}$, but the positivity (in any sense, since ω is $(1,1)$!) of T implies that $T(\omega) > 0$, since by I.1.5 $T(\omega) = M(T)$. $\square$

The theorem of Harvey-Lawson consists in showing that this condition is also sufficient for the existence of a Kähler metric on M :

II.1.2 THEOREM. *If there is no non zero positive (1,1)-current T on M which is the (1,1)-component of a boundary then M admits a Kähler metric.*

Theorem II.1.2 has been generalized by [Alessandrini-Andreatta, 1] to the following :

II.1.3 THEOREM. *Let M be a compact complex manifold. There exists on M a d-closed transverse (p,p)-form if and only if there is no non zero strongly positive (p,p)-current which is the (p,p)-component of a boundary.*

PROOF. Let ω be a transverse (p,p)-form on M. If T is a strongly positive (p,p)-current, then

$$T(\omega) = \int_M \omega_x(\overrightarrow{T}(x))||T||$$

by I.1.3. and I.1.5. Hence, if $T \neq 0$,

$$\omega_x(\overrightarrow{T}(x)) > 0, \text{ and } T(\omega) > 0.$$

If we suppose that $T = d_{p,p}S$, we have :

$$0 =< d\omega, S >=< \omega, d_{p,p}S >=< \omega, T >= T(\omega),$$

a contradiction.

Conversely, it follows from I.3.4 that if $B_{p,p}$ is the rank of $d_{p,p}$, then $B_{p,p}$ is closed in the weak topology . Since $P_{p,p}$ is compact, I.1.7, and convex and $B_{p,p}$ is closed, both in the space $\mathcal{E}'_{p,p}(M)_{\mathbb{R}}$, the Hahn-Banach separation theorem gives a real (p,p)-form ω which annihilates $B_{p,p}$ and it is positive and non zero on each element in $P_{p,p}$.
Now, $T(\omega) = 0$ for each $T \in B_{p,p}$ implies $d\omega = 0$.

It remains to be shown that ω is transverse. Choose $\overrightarrow{T}(x) \in \bigwedge_{\mathbb{R}}^{p,p} \mathbf{T}_x(M)$ for some x in M. Set

$$T = \overrightarrow{T}(x)\,\delta_x$$

(where δ_x is the Dirac measure in x). Thus $T \in P_{p,p}$ so that $\omega(T) > 0$.

By I.1.6 $\overrightarrow{T}(x)$ is a weakly positive (p,p)-vector and

$$\omega_x(\overrightarrow{T}(x)) = \int_M \omega(\overrightarrow{T}(x))\delta_x$$

which is equal to $T(\omega)$ by I.1.5 hence it is positive nowhere zero.

Then the theorem will follow from the arbitrariness of x and $\overrightarrow{T}(x)$. □

REMARK. An earlier generalization of the theorem of Harvey-Lawson is due to [Michelson, 20] for the case $p = n - 1$. Richer and very interesting properties of these manifolds shall be discussed in the next sections.
In terms of I.3.5, Theorem II.1.3. becomes:

II.1.4. THEOREM. *Let M be a compact complex manifold. There exists on M a d-closed transverse (p,p)-form if and only if every non zero $\partial\bar{\partial}$ - closed strongly positive (p,p)-current represents a non zero class in* $\mathcal{V}^{p,p}(M)_{\mathbb{R}}$.

II.2 p-Kähler manifolds

Let M be a compact complex manifold of dimension n, and p an integer, $1 \le p \le n$.

II.2.1 DEFINITION. M is said to be a *p-Kähler manifold,* or that M has *Kähler degree p,* if it carries a d - closed transverse (i.e. a strictly weakly positive) (p,p)-form, called a *p-Kahler form.*

Transverse forms always exist on a manifold,[Sullivan, 24]; it is required here that one should be closed.
Theorem II.1.3 then characterizes p-Kähler manifolds by means of positive currents.

A compact manifold M of dimension n is obviously n-Kähler and M is a Kähler manifold if and only if it is 1-Kähler.
A Kähler manifold M with Kähler form ω is p-Kähler for each p, $1 \le p \le n$, since ω^p is a d - closed transverse (p,p)-form, therefore the Kähler degree is a meaningful notion only in non Kähler geometry.

There is an important difference between the cases $p = n - 1$ and $1 < p < n - 1$: let Ω be a p-Kahler form on M, then, if $p = n - 1$, $\Omega = \omega^{n-1}$ for a suitable strictly positive $(1,1)$-form ω as we shall see in the next section, while for $1 < p < n - 1$ and M not Kähler Ω cannot be the power of a $(1,1)$-form, indeed $\Omega = \omega^p$ implies that ω is strictly positive and d - closed

[12]. We shall prove that important classes of compact manifolds are $(n-1)$-Kähler in chapter III. For general p the known examples suggest that it can be conjectured that M p-Kähler implies M q-Kähler for every $q \geq p$.
If the dimension of M is 2, the only intermediate Kähler degree is Kählerianity.

For higher dimensions non trivial p-Kähler manifolds exist:

II.2.2 THEOREM. *For every m, $m > 2$, there exists a manifold M which is p-Kähler for $m \leq p < 2m-1$ and not p-Kähler for $1 \leq p \leq m-1$.*

PROOF. We divide the proof into several steps:

(i) If M is a compact p-Kähler manifold of dimension r, there exists no non-zero d-exact simple holomorphic $(r$-$p)$-form on M.

Let Ξ be a p-Kähler form on M, $d\alpha$ a simple holomorphic $(r-p)$-form. Transversality of Ξ implies that

$$\Xi \wedge \sigma_{r-p} d\alpha \wedge d\bar{\alpha}$$

is a strongly positive volume form, hence

$$0 < \int_M \Xi \wedge \sigma_{r-p} d\alpha \wedge d\overline{\alpha}.$$

The integral is equal to

$$\pm \int_M d(Xi \wedge \sigma_{r-p} \alpha \wedge d\overline{\alpha})$$

which is zero by Stokes' theorem. Conversely:

(ii) Suppose that M is a compact complex manifold of dimension r with trivial holomorphic tangent bundle (i.e. M is a holomorphically parallelizable manifold). Then if there exists no non-zero d-exact simple holomorphic $(r$-$p)$-form on M for some p, $1 \leq p \leq r-1$, M is p-Kähler.

Let $\{\phi_\alpha\}_{1 \leq \alpha \leq r}$ be a set of holomorphic 1-forms that give a basis for $\mathbf{T}_x{}^*(M)$ for every $x \in M$. Let $\Theta = \phi_1 \wedge \ldots \wedge \phi_r$.
Define a non degenerate bilinear form

$$F : \Omega^p(M) \times \Omega^{r-p}(M) \to \mathbb{C}$$

by

$$F(\psi,\zeta)\Theta = \psi \wedge \zeta$$

and an isomorphism

$$G : \bigwedge^{p,0}(\mathbf{T}_x{}^*(M)) \rightarrow \Omega^{r-p}(M)$$

by

$$\psi \wedge G(V) = \psi_x(V)\Theta.$$

Let now $\{\rho_\alpha\}_{1\leq\alpha\leq s}$ be a basis for

$$\ker\,(d : \Omega^{r-p-1}(M) \rightarrow \Omega^{r-p}(M))$$

and complete it to a basis $\{\rho_\alpha\}_{1\leq\alpha\leq t}$ of $\Omega^{r-p-1}(M)$.
Let W be the span of $\{d\rho_\alpha\}_{s+1\leq\alpha\leq t}$ and let $W^\perp$ its orthogonal with respect to F. If $\{\psi_\alpha\}_{1\leq\alpha\leq k}$ is a basis of $W^\perp$ the ψ_α's are closed forms. Indeed,

$$\forall\alpha,\,\beta,\,1\leq\alpha\leq k,\,1\leq\beta\leq t,$$

$\psi_\alpha \wedge \rho_\beta$ is a holomorphic $(r-1)$-form on a compact manifold so that

$$0 = d(\psi_\alpha \wedge \rho_\beta).$$

But

$$d(\psi_\alpha \wedge \rho_\beta) = d\psi_\alpha \wedge \rho_\beta,$$

since $d\rho_\beta = 0$ for $1 \leq \beta \leq s$, and $\psi_\alpha \wedge d\rho_\beta = 0$ for $s+1 \leq \beta \leq t$. Therefore:

$$\Xi = \sigma_p \sum \psi_\alpha \wedge \overline{\psi_\alpha}$$

is a d-closed (p,p)-form.

Ξ is transverse. Indeed, fix x in M and a non zero simple vector $V \in \bigwedge^{p,0}(\mathbf{T}_x(M))$.

If

$$\Xi_x(\sigma_p{}^{-1}V \wedge \overline{V}) \leq 0,$$

then by definition of Ξ,

$$\psi_{\alpha x}(V) = 0$$

i.e.

$$\psi_\alpha \wedge G(V) = 0$$

for $1 \leq \alpha \leq k$. This means that $G(V)$ being a non zero simple holomorphic $(r-p)$-form belongs to W hence it is d - exact. Then it must be

$$\Xi_x({\sigma_p}^{-1} V \wedge \overline{V}) \geq 0.$$

(iii) For every m, $m > 2$, let G be the subgroup of $\mathbf{GL}(m+1, \mathbb{C})$ *of the matrices*

$$A = \begin{pmatrix} 1 & X & z \\ 0 & I_{m-1} & Y \\ 0 & 0 & 1 \end{pmatrix}$$

where X and Y are in $\mathbb{C}^{m-1}$, $z \in \mathbb{C}$ and I_{m-1} is the $(m-1)\times(m-1)$ identity matrix. Let Γ be subgroup of G of all A's with entries in $\mathbb{Z}[\sqrt{-1}]$. Let M_m be the space $G\backslash\Gamma$. Then M_m is a compact holomorphically parallelisable non Kähler manifold.

M_m is of course compact and holomorphically parallelisable.

To show the non Kählerianity of M_m identify G with $(C^{2m-1}, \odot)$, where

$$\begin{aligned}(x_1, \ldots, x_{m-1}, y_1, \ldots, y_{m-1}, z) \odot (u_1, \ldots, u_{m-1}, v_1, \ldots, v_{m-1}, w) = \\ = (x_1 + u_1, \ldots, y_{m-1} + v_{m-1}, z + w + x_1 v_1 + \ldots + x_{m-1} v_{m-1}).\end{aligned}$$

A straightforward calculation shows that we can choose a basis $\{\phi_\alpha\}_{1\leq\alpha\leq 2m-1}$ for $\Omega^1(M)$ such that

$$d\phi_1 = \ldots = d\phi_{2m-2} = 0,$$

and

$$d\phi_{2m-1} = \phi_1 \wedge \phi_2 + \ldots + \phi_{2m-3} \wedge \phi_{2m-2}.$$

By (i) M_m cannot be a Kähler manifold.

(iv) M_m *is p-Kähler for $p \geq m$, and not p-Kähler for* $1 \leq p \leq m-1$.

Let $p < m-1$ and

$$\begin{aligned}\zeta &= d(\phi_{2m-1} \wedge \phi_1 \wedge \phi_3 \wedge \ldots \wedge \phi - 2m - 5 \wedge \phi_2 \wedge \ldots \wedge \phi_{2(m-1-p)}) \\ &= \pm\phi_{2m-3} \wedge \wedge\phi 2m - 2 \wedge \phi_1 \wedge \phi_3 \wedge \ldots \wedge \phi_{2m-5} \wedge \phi_2 \wedge \ldots \wedge \phi_{2(m-1-p)},\end{aligned}$$

hence ζ is a non zero d-exact simple holomorphic $(2m-1-p)$-form, so that M_m cannot be p-Kähler.

If $p = m-1$ take

$$\zeta = d(\phi_{2m-1} \wedge \phi_1 \wedge \phi_3 \wedge \ldots \wedge \phi_{2m-5}),$$

hence M_m cannot be $(m-1)$-Kähler.

Suppose now $p \geq m$.
Let $k = 2m-1-p$ and ζ be a simple holomorphic d-exact k-form. If $k=1$, $\zeta = 0$ because M is compact. Otherwise, let $\chi \in \Omega^{k-2}(M)$ such that

$$\zeta = \chi \wedge d\phi_{2m-1} = \zeta_1 \wedge \ldots \zeta_k.$$

Then

$$\zeta_\lambda \wedge \chi \wedge d\phi_{2m-1} = 0$$

for every λ. Since $k-1 \leq m-2$, we obtain

$$\zeta_\lambda \wedge \chi = 0$$

for every λ, see for instance [14, p.47], so that $\chi = 0$ and therefore $\zeta = 0$.

It remains to be shown that M_m is q-Kähler for $q \geq p$. By induction we have only to show that M_m is $(p+1)$-Kähler.

Let us suppose that β is a non trivial d-exact simple holomorphic $2m-2-p$-form. Then

$$\dim \{v \in \mathbf{T}^*_x(M_m) : \beta_x \wedge v = 0\} = 2m-2-p,$$

$\forall x \in M_m$ [Federer, 9, p.23]. Since M_m is holomorphically parallelizable, there exists $\alpha_1, \ldots, \alpha_{2m-1-p}$ d-closed holomorphic 1-forms, which are linearly independent at each $x \in M_m$, by the proof of (ii). Then

$$\beta \wedge \alpha_j \quad \forall j,\ 1 \leq j \leq 2m-1-p$$

is a d - exact simple holomorphic $(2m-1-p)$-form, hence it is zero by (i). A contradiction. $\square$

As it was pointed out in I.1, the three notions of positivity of a current are dual to those for a form, thus Theorem II.2.2 can be easily restated for strictly

positive forms and positive currents or also for strictly strongly positive forms and weakly positive currents.
A natural issue of this remark is that, from an abstract point of view, it would be possible to introduce three different definitions of a p-Kähler manifold, which obviuosly agree for $p = 1$ or $p = n - 1$.
The definition that has been preferred and that has been adopted here is the weakest one. The reason for this choice is that the compact odd-dimensional holomorphically parallelizable manifolds whose existence is shown in Theorem II.2.2 above carry a *strictly weakly positive* closed (p,p)-form, i.e. they are not Kähler but p-Kähler following the weakest definition. The point is that holomorphic parallelizability forces to this end since we have the following very special result:

II.2.3 PROPOSITION. *If M is a compact holomorphically parallelizable manifold of dimension n , and Ω is a closed strictly positive (p,p)-form on M, $1 < p < n - 1$, then M is Kähler.*

PROOF. We have seen in II.2.2 (ii) that if $\{\phi_\alpha\}_{1\leq\alpha\leq r}$ is a set of holomorphic 1-forms that give a basis for $\mathbf{T}^*_x(M)$ for every $x \in M$, $d\phi_A = 0$ for $A = (\alpha_1, \ldots \alpha_{n-p-1})$.
Now,

$$d\phi_\gamma = \frac{1}{2}\sum c^\gamma_{\beta\alpha}\,\phi_\alpha \wedge \phi_\beta.$$

Then

$$\sum_\alpha \phi_\alpha \wedge \overline{\phi}_\alpha$$

is the required Kähler form as shown by direct computation. □

However this result is false for a general compact manifold. In fact more meaningful examples will come out using the following result that shall not be proved here. In this situation in fact we shall found p-Kähler forms that are strictly positive. One has indeed the following way of constructing non trivial p-Kähler manifolds:

II.2.4 THEOREM. *Let*

$$f : M' \to M$$

be a smooth proper modification between compact complex manifolds with M Kahler. Let E be the exceptional set of f and let $Y = f(E)$.
Let us suppose that there exists an analytic discrete subset Z of Y such that:

(1) *$E \setminus f^{-1}(Z) \to Y \setminus Z$ is a holomorphic submersion,*

(2) *$M' \setminus f^{-1}(Z) \to M \setminus Z$ is a Kähler morphism (cfr.* [27, p. 24].
Then, for every p, $p > \dim f^{-1}(Z)$, M' is p-Kähler.

Moreover, if $p = \dim f^{-1}(Z)$, if we denote by $V_1, \dots, V_N$ the irreducible components of dimension p of $f^{-1}(Z)$ we have:

M' is p-Kähler $\Longleftrightarrow$ the following property holds:

if $\sum_{j=1}^{N} c_j [V_j]$, with $c_i \geq 0$, $\forall i$ represents the zero class in $H_{2p}(E)$, then

$$c_i = 0 \, \forall i.$$

The link between p-Kähler forms on M and M' is given by the following:

II.2.4 THEOREM. *Let M and M' be compact p-Kähler manifolds. Let*

$$f : M' \to M$$

be a modification with exceptional set E and let $Y = f(E)$, with $p > \dim Y$. Then for every p-Kähler form Ω on M, there exists a p-Kähler form $\hat{\Omega}$ on M', such that

$$\text{class } (\Omega) = \text{ class } (f_* \hat{\Omega})$$

in $\Lambda^{p,p}(M)_{\mathbb{R}}$.

PROOF. Let Ω and Ω' be p-Kähler forms on M and M' respectively.

Since $p > \dim Y$, arguing as in [26, pp. 251/2] we can find an open neighbourhood U of Y in M and a real current R on u such that:

$$f_* \Omega' = \sqrt{-1} \, \partial \bar{\partial} R.$$

Since $f_* \Omega'$ is smooth and $\partial\bar{\partial}$-exact in $U \setminus Y$, there exists a smooth real $\partial\bar{\partial}$-closed $(p-1, p-1)$-form β on $U \setminus Y$ such that

$$f_* \Omega'_{\restriction U \setminus Y} = \sqrt{-1} \, \partial \bar{\partial} \beta.$$

Moreover,

$$R - \beta = \gamma + \bar{\partial} C + \partial \overline{C}$$

in $U \setminus Y$ for some smooth real $\partial\bar{\partial}$-closed $(p-1, p-1)$-form γ and some current C.

Choose now an open set W such that

$$Y \subset W \subset\subset U$$

and a real function $g \in \mathcal{C}_0^\infty(U)$, with $g = 1$ in W. Set:

$$D = g(\beta + \gamma) + \bar{\partial} g C + \partial g \overline{C}.$$

Since $\sqrt{-1}\,\partial\bar{\partial} D$ is smooth in $M \setminus Y$,

$$\eta = (f_{\restriction_{M' \setminus E}})^* \sqrt{-1}\,\partial\bar{\partial} D$$

is a smooth (p,p) - form on the whole of M', whose support lies in $f^{-1}(U)$, and that it is positive on $f^{-1}(W)$.

Choose $\epsilon > 0$ such that

$$\hat{\Omega} = f^*\Omega + \epsilon\eta$$

is positive in M', hence:

$$\Omega - f_*\hat{\Omega} = \sqrt{-1}\,\partial\bar{\partial}\epsilon D. \quad \square$$

II.3 $(n-1)$-Kähler manifolds.

Let M be a compact complex manifold and

$$\mathcal{D} : \Gamma(\mathbf{T}M) \to \Gamma(\mathbf{T}^*M) \otimes \Gamma(\mathbf{T}M)$$

be a connection on M. The torsion tensor $T^{\mathcal{D}} \in \Gamma(\bigwedge^2(\mathbf{T}^*M \otimes \mathbf{T}M)$ associated with $\mathcal{D}$ is defined by

$$T^{\mathcal{D}}{}_{V,W} = \mathcal{D}_V W - \mathcal{D}_W V - [V, W],$$

for $V, W \in \Gamma(\mathbf{T}M)$. (We remark that all tensors considered are extended complex multilinearly to the complexification of $\mathbf{T}M$).

For a given hermitian metric $\mathbf{h}$ on M we can consider two natural connections preserving $\mathbf{h}$, the Riemannian connection and the $(1,0)$-connection or hermitian connection which in addition preserves the complex structure tensor J. In terms of the torsion tensor, the first is characterized by its vanishing and the second by the vanishing of its $(1,1)$ part, i.e.

$$-T^{\mathcal{D}}{}_{V,W} = T^{\mathcal{D}}{}_{JV,JW}.$$

When the Riemannian and the hermitian connections coincide, the metric is Kähler and this property is then equivalent with the condition that its torsion

tensor is identically zero.
Let $(z_1, \ldots, z_n)$ be a local coordinate system and in these coordinates let

$$\mathbf{h} = \sum h_{ij}\, dz_i \otimes d\bar{z}_j$$

and write the hermitian connection $\mathcal{D}$ as

$$\mathcal{D}(\partial_{z_j}) = \sum \omega_{jk} \otimes \partial_{z_k},$$

where

$$\partial_{z_k} = (\frac{\partial}{\partial z_k}).$$

Since $\mathcal{D}$ is (1,0), we get

$$\Omega = \partial H H^{-1}$$

where $\Omega = (\omega_{jk})$, and $H = (\mathbf{h}_{kl})$.
As a $\mathbf{T}M$-valued 2-form, $T^{\mathcal{D}}$ has no (1,1) component and its (2,0) and (0,2) components are J conjugates.

Again in local coordinates we have that

$$T^{\mathcal{D}} = \sum_{j,k,l} T^l{}_{jk} dz_j \wedge d\bar{z}_k \otimes \partial_{z_l},$$

with

$$T^l{}_{jk} = \sum_a (\partial_{z_j} h_{\bar{a}k} - \partial_{z_k} h_{\bar{a}j}) h^{l\bar{a}},$$

where (h^{kl}) denotes the inverse of H. We define now the *torsion 1-form* $\tau_{\mathbf{h}}$ or the *trace of the torsion tensor* as

$$\tau_{\mathbf{h}} = \sum \tau_k dz_k,$$

where $\tau_k = \sum_j T^j{}_{kj}$. We have:

II.3.1.THEOREM [Michelson, 20]. *Let M be a compact complex manifold of dimension n, $\mathbf{h}$ an hermitian metric on M with Kahler form ω . Then the following statements are equivalent:*

(1) $\tau_{\mathbf{h}} = 0$,
(2) $d\omega^{n-1} = 0$.

PROOF. By direct computation taking into account the facts that

$$\sqrt{-1}\,\tau_{\mathbf{h}} = \bar{\partial}^*\omega,$$

$$\bar{\partial}^* = -{}^*\partial^*,$$

$${}^*\omega = \frac{\omega^{n-1}}{(n-1)!}$$

and

$$P(\omega) = 0 \text{ if and only if } P^*(\omega) = 0$$

where P is any of the operators $\partial, \bar{\partial}, d, d^c$ and P^* denotes of course the formal adjoint of P with respect to $\mathbf{h}$. □

Then in view of II.1.3:

II.3.2. THEOREM. *A compact complex manifold of dimension n is (n-1)-Kähler if and only if M admits an hermitian metric* $\mathbf{h}$ *such that*

$$\tau_{\mathbf{h}} = 0.$$

PROOF. In view of II.3.1 we need only to show that if Ω is a $(n-1)$-Kähler form on M, there exists ω strictly positive $(1,1)$-form such that $\Omega = \omega^{n-1}$. By I.1.3 write

$$\Omega = \sigma_{n-1} \sum_j \eta_j \wedge \overline{\eta}_j$$

with non zero

$$\eta_j = a_{1_j} \wedge \ldots \wedge a_{n-1_j}.$$

Take then

$$\omega = \sqrt{-1}\sigma_1{}^{n-2} \sum_{jk} a_{j_k} \wedge \overline{a_j}_k. \quad \square$$

REMARK. Michelson calls a $(n-1)$-Kähler manifold a *balanced manifold.* Such manifolds are called semi-Kähler or co-symplectic somewhere in the literature.

An hermitian metric with the property about which in theorem II.3.2 will be called a *balanced metric*

II.2.4 becomes then:

II.3.3 THEOREM. *Let M and M' be compact balanced manifolds of dimension n. Let*

$$f : M' \to M$$

be a modification.

Then for every balanced metric **h** *on M, with Kähler form ω there exists a balaced metric* $\hat{\mathbf{h}}$ *on M' with Kähler form $\hat{\omega}$ such that*

$$\omega^{n-1} - f_*\hat{\omega}^{n-1}$$

is a $\partial\bar{\partial}$-exact current.

III

EXISTENCE OF SPECIAL METRICS ON COMPACT COMPLEX MANIFOLDS

III.1 Compact manifolds which are regular in the sense of Varouchas and $\partial\bar{\partial}$ - cohomology.

III.1.1 DEFINITION [25]. A compact complex manifold M is called *regular* if

$$\ker(\partial\bar{\partial} : \mathcal{E}^{p,q}(M) \to \mathcal{E}^{p+1,q+1}(M)) = \\ = \ker(\partial : \mathcal{E}^{p,q}(M) \to \mathcal{E}^{p+1,q}(M)) + +\overline{\partial}\mathcal{E}^{p,q-1}(M)$$

for every p, q.

This is equivalent with $Im\,\partial \cap \ker\,\overline{\partial} = Im\,\partial\bar{\partial}$.

The following two classes of compact complex manifolds obtained as images of Kähler manifolds are important in complex geometry:

III.1.2 DEFINITION. If

$$f : M' \to M$$

is a surjective meromorphic map between complex compact manifolds and M' is Kähler, M is called of class $\mathcal{C}$.

III.1.3 DEFINITION. If M is of class $\mathcal{C}$, f is a modification and M' is projective, M is called a Moišezon manifold.

III.1.4 PROPOSITION.

(1) *If M is compact Kähler, M is regular,*
(2) *If $f : M' \to M$ is a surjective holomorphic map between complex compact manifolds and M' is Kähler, then M is regular,*
(3) *If M is of class $\mathcal{C}$ is regular*
(4) *If M is Moišezon is regular*

(5) *If $f : M' \to M$ is a smooth proper modification and M is compact Kähler , then M' is regular.*

PROOF. For (1) and (2), see [25]. (3) and (4) follow directly from the definitions. As for (5), Varouchas has shown in [26] that a smooth modification with image of class $\mathcal{C}$ must have domain in the same class. Hence there are a compact Kähler manifold M'' and a holomorphic surjective map

$$M'' \to M',$$

so that M' is regular by (2). □

III.1.5 PROPOSITION. *Let M be a regular manifold. Let T be a d-closed real (p,p)-current on M, which is the (p,p)-component of a boundary. T is then d-exact and $\partial\bar{\partial}$ - exact.*

PROOF. There exists a real $(p, p+1)$ - current S such that

$$T = \partial\overline{S} + \bar{\partial}S.$$

T d - closed and real implies

$$\bar{\partial}T = 0,$$

hence

$$0 = -\partial\bar{\partial}\overline{S}.$$

Since M is regular,

$$\overline{S} = R + \bar{\partial}Q,\ \partial\overline{S} = \partial\bar{\partial}Q$$

and

$$T = \partial\bar{\partial}(Q - \overline{Q}) = d(\bar{\partial}Q - \overline{\partial Q}).\quad \square$$

If we consider the maps introduced in I.3.5 for $p = 1$:

$$\alpha : \Lambda^{1,1}(M)_{\mathbb{R}} \to H^{1,1}(M, \mathbb{R})$$

and

$$\beta : H^{1,1}(M, \mathbb{R}) \to \mathcal{V}^{1,1}(M)_{\mathbb{R}}$$

we obtain the following results whose proofs are evident:

III.1.6 PROPOSITION. *If the compact complex manifold M is regular, then $\beta \circ \alpha$ is an isomorphism.*

And

III.1.7 COROLLARY. *If the compact complex manifold M is regular, then it satisfies the following condition*

(B) *β is injective and $Im\,\beta$ contains all positive elements in $\mathcal{V}^{1,1}(M)_{\mathbb{R}}$.*

III.2 Balanced metrics on domains of modifications of complex manifolds with compact Kähler images

It is known since [Blanchard, 7] that if

$$f : M' \to M$$

is the blow up of M with center the submanifold Y, and M is compact Kähler, then M' is also Kähler . But if $f : M' \to M$ is just a smooth modification between compact complex manifolds with M Kähler M' is in general not Kähler as many examples show.
The basic one was given by [Hironaka, 16], enlightening the phenomenon particularly.

Let $M = \mathbb{P}_3(\mathbb{C})$, and Y be the irreducible curve:

$$\begin{cases} y^2 = x^2 + x^3 \\ z = 0 \end{cases}$$

with a unique singular point z with distinct tangents. There is an open neighbourhood U of z in M such that $Y \cap U = Y_1 \cup Y_2$ where Y_1 and Y_2 are smooth irreducible curves which intersect in z with distinct tangents.
There is then a smooth modification $f : M' \to M$ such that

$$f : M' \setminus f^{-1}(x) \to M \setminus \{z\}$$

is the blow up of $M \setminus \{z\}$ with center $Y \setminus \{z\}$ and

$$f : f^{-1}(U) \to U$$

is the composition of two blow up's: of U along Y_1, and after along Y'_2, where Y'_2, is the strict transform of Y_1 in the first blow up.

The exceptional set of f, $E = f^{-1}(Y)$, contains a curve $S = f^{-1}(z)$ which is the union of two components S_1 and S_2 each biholomorphic to $\mathbb{P}_1(\mathbb{C})$, which intersect in a single point with distinct tangents. By a Mayer-Vietoris argument, S_1 is homologous to zero hence M' cannot be Kähler.

But we have:

III.2.1 THEOREM. *M' is 2-Kähler.*

PROOF. By theorem II.1.3 it is enough to show that if T is a positive $(2,2)$-current on M' which is the $(2,2)$-component of a boundary, we have $T = 0$.

Let ω be a Kähler form on M. $f^*\omega$ is a closed real $(1,1)$-form which is positive semidefinite, so that from

$$0 = T(f^*\omega) = \int_{M'} (f^*\omega)_x \overrightarrow{T}_x ||T||$$

we get supp $T \subset E$.

By the Relative Support Theorem I.4.3 there exists then a non negative pluriharmonic function $h : Y \to \mathbb{R}$ such that

$$T = (h \circ f)[S].$$

Next we show that in this case h can be taken constant.

Indeed, the parametrisation of Y

$$\begin{cases} x = t^2 - 1 \\ y = t(t^2 - 1) \\ z = 0 \end{cases}$$

induces the biholomorphisms:

$$Y \setminus \{0\} \cong \mathbb{P}_1 \setminus \{a, b\} \cong \mathbb{C} \setminus \{0\}.$$

For every $x \in Y \setminus \{0\}$, $h \restriction_{f^{-1}(x)}$ is harmonic and therefore constant, so that we can project it to a non negative harmonic function

$$h' : Y \setminus \{0\} \to \mathbb{R}.$$

Taking into account the biholomorphisms above and a result in classical potential theory, there exists then a constant C and a holomorphic function

$$F : B(0, \epsilon) \to \mathbb{R}$$

such that

$$h'(z) = Re\, F(z) + C \log |z|, \ \forall z \in B(0, \epsilon) \setminus \{0\}.$$

Then F and C are independent on ϵ; moreover, replacing z by $1/w$, we see that F is a constant. Since h' is non negative, it must be $C = 0$, so that h' is a constant and h is a constant too.

This tells in particular that

$$dT = 0,$$

hence

$$T = \bar{\partial}S^{2,3} + \partial\overline{S^{2,3}}$$

for a $(2,3)$-current $S^{2,3}$ such that

$$\partial\bar{\partial}S^{2,3} = 0.$$

$\partial S^{2,3}$ is a holomorphic 2-form on M', hence d-closed, so that:

$$\int \partial S^{2,3} \wedge \overline{\partial S^{2,3}} \wedge f^*\omega = \int d(\partial S^{2,3} \wedge \overline{\partial S^{2,3}} \wedge f^*\omega) = 0,$$

that implies:

$$\partial S^{2,3} = 0,$$

so that:

$$T = d(S^{2,3} + \overline{S^{2,3}}),$$

hence T is also d - exact.

If h were different from zero, we would conclude that E is homologous to zero, which is not the case, as shown before.

We conclude then that $T = 0$. □

It makes sense to ask the following question:
given a smooth modification

$$f : M' \to M$$

between compact complex manifolds with M Kähler, is there a way of controlling the loss of Kählerianity on M', showing that M' has some intermediate Kähler degree and could be endowed of some special metric?

In fact, Alessandrini-Bassanelli have shown in [3] that the method utilized in III.2.1 could be generalized to obtain the following:

III.2.2 THEOREM. *Let*

$$f : M' \to M$$

be a smooth modification between compact complex manifolds of dimension n with M Kähler.

Then M' is (n-1)-Kähler (i.e. M' carries a balanced metric and for every Kähler metric $\mathbf{h}$ *on M, with Kähler form ω there exists a balanced metric* $\hat{\mathbf{h}}$ *on M' with Kähler form $\hat{\omega}$ such that ω^{n-1} and $f_*\hat{\omega}^{n-1}$ are cohomologous in $\Lambda^{n-1,n-1}(M)_{\mathbb{R}}$.)*

PROOF. By II.1.3 it is enough to show that if T is a positive real $(n-1, n-1)$-current on M' which is the $(n-1, n-1)$-component of a boundary, then T must be identically zero.

Let T be such a current, E be the exceptional set of f and ω the Kähler form of a Kähler metric on M.

$f^*\omega$ is a positive semidefinite d-closed $(1,1)$-form on M', positive outside E.

By I.1.5 we can write

$$T(f^*\omega^{n-1}) = \int_{M'} < T(x), f^*\omega^{n-1}(x) > ||T||,$$

which is equal to zero since T is the component of a boundary and since $f^*\omega^{n-1}$ is d-closed.

Further, since $f^*\omega^{n-1}$ is positive outside E, we get

$$\text{supp}\, T \subseteq E.$$

From this and the fact that T is $\partial\overline{\partial}$-closed, we can apply the Support Theorem I.2.5 for $p = n - 1$, and we then obtain that

$$0 = T - \sum_{j=1}^{s} c_j [E_j],$$

where $\{E_j\}_{1 \leq j \leq s}$, are the irreducible components of E and c_j are positive constants.

T is then a positive d-closed current on M' which is a regular manifold by III.1.2 (5), hence

$$T = \sqrt{-1}\partial\overline{\partial}Q.$$

By using a positive volume form on M', Q is identified with a distribution ϕ such that

$$\sqrt{-1}\partial\overline{\partial}\phi \geq 0,$$

i.e. ϕ is a plurisubharmonic function on M'. ϕ is therefore a constant, so that $T = 0$.

The last part of the statement will follow from II.3.3 □

We can do slightly better, by proving:

III.2.3 THEOREM. *Let*

$$f : M' \to M$$

be a smooth modification between compact complex manifolds of dimension n with M (n-1)-Kähler.

Then M' is (n-1)-Kähler (i.e. M' carries a balanced metric and for every balanced metric **h** *on M, with Kähler form ω there exists a balaced metric* $\hat{\mathbf{h}}$ *on M' with Kähler form $\hat{\omega}$ such that ω^{n-1} and $f_*\hat{\omega}^{n-1}$ are cohomologous in $\Lambda^{n-1,n-1}(M)_{\mathbb{R}}$.)*

PROOF. Again, by II.1.3 it is enough to show that if T is a positive real $(n-1, n-1)$-current on M' which is the $(n-1, n-1)$-component of a boundary, then T must be identically zero. We utilize the same argument as in the proof of Theorem III.2.2 replacing ω by the Kähler form of a balanced metric on M until we obtain $f_*T = 0$ hence $supp\, T \subseteq E$, and $T = \sum_{j=1}^{s} c_j[E_j]$ is d-closed.

We show next that T is d-exact. Since T is the component of a boundary,

$$T = \bar{\partial}S + \partial\overline{S}$$

for a suitable current S of bidegree $(1, 0)$. As

$$0 = \partial T = -\partial\bar{\partial}S,$$

∂S and ∂f_*S are holomorphic 2-forms on M' and M respectively.

Thus we can find a smooth $(1, 0)$-form ϕ and a distribution $\tau, = \alpha + \sqrt{-1}\,\beta$ on M such that

$$f_*S = \phi + \sqrt{-1}\,\partial\tau = \phi + \sqrt{-1}\,\partial\alpha - \partial\beta.$$

But

$$(*) \qquad 0 = f_*T = \bar{\partial}\phi + \partial\overline{\phi} - 2\sqrt{-1}\,\partial\bar{\partial}\beta,$$

so that β is a smooth function.

Hence the current:

$$S' = S - f^*(\phi + \sqrt{-1}\,\partial\beta)$$

is well defined on M' and, by (*),

$$T = \bar{\partial} S' + \partial \overline{S}'.$$

To conclude, set

$$Q = S' + \overline{S}'.$$

We have

$$\partial S' = \partial S - \partial f^*(\phi + \sqrt{-1}\,\partial\beta) = \partial S - f^* \partial f_* S$$

which vanishes on M' since it is a holomorphic form which is zero outside E. Then

$$dQ = d(S' + \overline{S}') = \bar{\partial} S' + \partial \overline{S}' = T.$$

The homology class of T is then zero.

The following homology sequences are exact $\forall i \geq 0$ [p.286]:

$$0 \longrightarrow H_i^*(E) \longrightarrow H_i(M') \xrightarrow{\alpha_i} H_i(M) \longrightarrow 0$$

$$0 \longrightarrow H_i^*(E) \longrightarrow H_i(E) \xrightarrow{\beta_i} H_i(Y) \longrightarrow 0$$

Since

$$\beta_{2(n-1)}(\text{class}(T)) \restriction_E = \sum_{j=1}^{s} c_j\, \beta_{2(n-1)}(\text{class}([E_j]) \restriction_E = 0$$

because codim $Y \geq 2$ and class $(T) = 0$ on M', we obtain (class $(T)) \restriction_E = 0$, thus

$$c_j = 0$$

for all j [8, th. 3.2].

The last part of the statement will follow from II.3.3. □

III.3 Balanced metrics on images of modifications of complex manifolds with compact Kähler domains

Conversely, let

$$f : M'' \to M$$

be a smooth modification of compact complex manifolds.

We suppose now that M'' has a Kähler metric.
In this case even when f is the blow up of M with smooth center Y, M is not necessarily Kähler : that is there is no converse, in the sense of putting a Kähler metric on the domain of a modification, of Blanchard's theorem.

Indeed, if we go back to the example of the preceding section, it can be seen that there are a projective manifold M'' and a holomorphic map

$$g : M'' \to \mathbb{P}_3(\mathbb{C}),$$

obtained by a finite number of blow-ups with smooth centers, which dominates f, i.e. there exists a holomorphic map

$$h : M \to \mathbb{P}_3(\mathbb{C})$$

such that

$$g = h \circ f.$$

Indeed h is the composition of two blow-ups with smooth centers:

$$m : M^* \to M',$$

the blow up of M' along S_1 ,

$$n : M'' \to M^*$$

otained by blowing up M^* along S'_2, which is the strict transform of S_2 in the blow up m.
Take then

$$h = n \circ m.$$

Since M'' is projective, and M' is not Kähler, if M^* is Kähler,

$$m : M^* \to M'$$

is the example we looked for; otherwise, if M^* is not Kähler , we look at

$$n : M'' \to M^*.$$

We may study then the converse problem to that of section III.2, by asking which degree of intermediate Kählerianity we can find on the image manifold. This problem has the following answers:

III.3.1 THEOREM. *Every manifold in the class $\mathcal{C}$ is (n-1)-Kähler.*

III.3.2 COROLLARY. *Moišezon manifolds are (n-1)-Kähler*

We shall obtain these results as corollaries to the following:

III.3.3 THEOREM. *Let M and M' be compact complex manifolds of dimension n.*

Let

$$f : M' \to M$$

be a modification.

Then if M' is (n-1)-Kähler and satisfies *(B) of III.1.7, M is (n-1)-Kähler and satisfies (B).*

Moreover, for every (n-1)-Kähler metric $\mathbf{h}$ *on M, with Kähler form ω there exists a balaced metric $\hat{\mathbf{h}}$ on M' with Kähler form $\hat{\omega}$ such that ω^{n-1} and $f_*\hat{\omega}^{n-1}$ are cohomologous in $\Lambda^{n-1,n-1}(M)_{\mathbb{R}}$.*

The proof of this theorem needs several preparatory results. We start with T a positive $\partial\bar{\partial}$-closed current of bidegree $(1,1)$ on M, and we try to find a "nice pull back of" T to M'.

If E is the exceptional set of f and $Y = f(E)$,

$$((f \restriction_{M' \setminus E})^{-1})_*(T \restriction_{M \setminus Y})$$

is a well defined positive $(1,1)$-current on $M' \setminus E$.

We are looking then for a positive $\partial\bar{\partial}$-closed positive extension to M', T' which also satisfies

$$(*) \quad \forall x \in M, \exists W, \text{ open neighbourhood of } x \text{ such that } T' \restriction f^{-1}(W)$$
$$\text{is limit of currents which are components of boundaries.}$$

But $((f \restriction_{M' \setminus E})^{-1})_*(T \restriction_{M \setminus Y})$ has an extension of order zero to M' if and only if it has locally finite mass across E, [18, p.10], i.e. $\forall x \in E$, $\exists V$ neighbourhood of x in M' such that

$$(**) \qquad \int_{V \setminus E} ((f \restriction_{M' \setminus E})^{-1})_*(T \restriction_{M \setminus Y}) \wedge \theta^{n-1} < +\infty,$$

where θ is a smooth strictly positive $(1,1)$-form on M'.

Since (**) is a local statement, we shall carry out computations in coordinates, starting with the case of a blow up with smooth center.

After having proved that $((f \restriction_{M'\setminus E})^{-1})_*(T \restriction_{M\setminus Y})$ admits extensions of order zero, we construct such an extension T' being also locally limit of currents which are components of boundaries and such that if condition (B) of III.1.7 holds, its class in the Aeppli group $\mathcal{V}^{1,1}(M')_{\mathbb{R}}$ coincides with that of f^*T. (It is remarkable that these properties are not enjoied in general by the simple extension defined as

$$T^0(\phi) = \int_{M'\setminus E} ((f \restriction_{M'\setminus E})^{-1})_*(T \restriction_{M\setminus Y}) \wedge \phi$$

for all compactly supported real $(n-1, n-1)$ - forms ϕ on M'.)

We start by recalling the following:

III.3.4 LEMMA. *Let U, G_1, G_2, be open subsets of $\mathbb{C}^n$ with*

$$G_2 \subset\subset G_1 \subset\subset U,$$

ϕ be the product of n-k smooth positive (1,1) - forms on U and $\{T_\epsilon\}$ be a sequence of positive currents on U of bidegree (k,k) converging to a current T on U in the weak topology.

Then

$$\limsup_\epsilon \int_{G_1} T_\epsilon \wedge \phi \leq \int_{G_2} T \wedge \phi$$

and

$$\int_{G_1} T \wedge \phi \leq \liminf_\epsilon T_\epsilon \wedge \phi.$$

Moreover, if L is a Borel subset relatively compact in U, then

$$\lim_\epsilon \int_L T_\epsilon \wedge \phi = \int_L T \wedge \phi,$$

if $||T||\,(bL) = O$, where bL is the topological boundary of L.

PROOF. [22, page 66]. □

Let B_k be the euclidean ball in $\mathbb{C}^k$ with center the origin and radius 1, set also $B_0 = \{0\}$. Set $U = B_m \times B_k$, $Y = B_m \times \{0\}$. Call $\pi : U' \to U$ the blow up of U with center Y. The natural Kähler form on U' is given then by

$$\omega' = \frac{\sqrt{-1}}{2}(\partial\bar{\partial}||t||^2 + \partial\bar{\partial}||z||^2 + \partial\bar{\partial}log||\zeta||^2)$$

with $t \in \mathbb{C}^m$, therefore:

$$\pi_* \omega' = \frac{\sqrt{-1}}{2}(\partial\bar{\partial}||t||^2 + \partial\bar{\partial}||z||^2 + \partial\bar{\partial}log||z||^2).$$

Let us take $y \in Y$ and identify the singular fibre $\pi^{-1}(y)$ with $\mathbb{P}_{n-1}$. Suppose finally that $n = m + k$. We have

III.3.5 PROPOSITION. *If $\{T_\epsilon\}$ is a sequence of positive $\partial\bar{\partial}$-closed forms on U of bidegree (1,1) converging to a current T on U in the weak topology then:*

for all $t^0 \in B_m$, there exists a neighbourhood V of $(t^0, 0)$ in U such that

$$\sup_\epsilon \int_{\pi^{-1}(V)} \pi^* T_\epsilon \wedge \omega'^{n-1} < \infty.$$

PROOF. Choose a unitary linear coordinate system

$$w = w(t, z) = (w_1, \ldots, w_n)$$

in $\mathbb{C}^n$ such that

$$(w_I, z) = (w_{i_1}, \ldots, w_{i_m}, z_1, \ldots, z_k)$$

form a coordinate system of $\mathbb{C}^n$ for every $I = (i_1, \ldots, i_m)$ with $1 \leq i_1 < \ldots < i_m \leq m$.

If we look at the form

$$\frac{\sqrt{-1}}{2\pi} \partial\bar{\partial} \log ||z||^2,$$

we see that its matrix is positive semidefinite: at $z \neq 0$ it has zero as a simple eigenvalue with eigendirection z, and

$$\frac{1}{\pi||z||^2}$$

as eigenvalue of multiplicity $k - 1$ with eigenspace $(z)^\perp$. Hence

$$(\frac{\sqrt{-1}}{2\pi} \partial\bar{\partial} \log ||z||^2)^h = 0$$

if $h \geq k$.

This implies that there exists a constant C, $C > 0$, such that

$$(\pi_*\omega')^{m+k-1} \leq C \sum_{j=m}^{m+k-1} \binom{m+k-1}{j} \sum_I (\frac{\sqrt{-1}}{2\pi}\partial\bar{\partial}\log||z||^2)^{m+k-1-j} \wedge$$
$$\wedge (\frac{\sqrt{-1}}{2}\partial\bar{\partial}||z||^2)^{j-m} \wedge (\frac{\sqrt{-1}}{2}\partial\bar{\partial}||w_I||^2)^m.$$

Let $t^0 \in B_m$ and let

$$\rho_I : \mathbb{C}^n \to \mathbb{C}^k$$

be defined by

$$\rho_I(t,z) = w_I(t,z).$$

Then there exists an open ball $\mathcal{B}_I$ with center $w_I(t^0, 0)$ in $\mathbb{C}^m$ and $r_I > 0$ such that

$$X_I = \rho_I^{-1}(\mathcal{B}_I) \cap (\mathbb{C}^m \times B_k(r_I)) \subset\subset U.$$

Thus, if we take

$$V = \cap_I X_I,$$

to get our contention we have only to prove:

$$(*) \qquad \sup_\epsilon \int_{X_I} T_\epsilon \wedge (\frac{\sqrt{-1}}{2\pi}\partial\bar{\partial}\log||z||^2)^{m+k-1-j} \wedge (\frac{\sqrt{-1}}{2}\partial\bar{\partial}||z||^2)^{j-m} \wedge$$
$$\wedge (\frac{\sqrt{-1}}{2}\partial\bar{\partial}||w_I||^2)^m < \infty.$$

The integrand in (*) is smaller or equal than:

$$\frac{1}{(\pi||z||^2)^{n-1-j}} T_\epsilon \wedge (\frac{\sqrt{-1}}{2}\partial\bar{\partial}||z||^2)^{n-1} \wedge (\frac{\sqrt{-1}}{2}\partial\bar{\partial}||w_I||^2)^m \in L_{loc}{}^1(U),$$

hence we may ignore the singularity of $\partial\bar{\partial}\log||z||^2$ in (*).

Since $\sqrt{-1}\,\partial\bar{\partial}T_\epsilon = 0$, there exists $(1,0)$-forms S_ϵ on U such that

$$T_\epsilon = \bar{\partial}S_\epsilon + \partial\overline{S}_\epsilon.$$

Thus if we denote by bX_I the topological boundary of X_I,

$$(**)\quad \int_{X_I} T_\epsilon \wedge (\frac{\sqrt{-1}}{2\pi}\partial\bar\partial \log ||z||^2)^{m+k-1-j}\wedge$$
$$\wedge (\frac{\sqrt{-1}}{2}\partial\bar\partial ||z||^2)^{j-m}\wedge(\frac{\sqrt{-1}}{2}\partial\bar\partial||w_I||^2)^m =$$
$$= \int_{bX_I}(S_\epsilon+\overline{S}_\epsilon)\wedge(\frac{\sqrt{-1}}{2\pi}\partial\bar\partial\log||z||^2)^{m+k-1-j}\wedge$$
$$\wedge (\frac{\sqrt{-1}}{2}\partial\bar\partial ||z||^2)^{j-m}\wedge(\frac{\sqrt{-1}}{2}\partial\bar\partial||w_I||^2)^m =$$
$$= \frac{1}{(\pi {r_I}^2)^{m+k-1-j}}\int_{bX_I}(S_\epsilon+\overline{S}_\epsilon)\wedge$$
$$\wedge (\frac{\sqrt{-1}}{2}\partial\bar\partial ||z||^2)^{k-1}\wedge(\frac{\sqrt{-1}}{2}\partial\bar\partial||w_I||^2)^m =$$
$$= \frac{1}{(\pi {r_I}^2)^{m+k-1-j}}\int_{X_I} T_\epsilon\wedge(\frac{\sqrt{-1}}{2}\partial\bar\partial ||z||^2)^{k-1}\wedge(\frac{\sqrt{-1}}{2}\partial\bar\partial||w_I||^2)^m.$$

The reason for the second equality in (**) is the following. Since

$$bX_I = [{\rho_I}^{-1}(b\mathcal{B}_I)\cap(\mathbb{C}^m\times B_k(r_I))]\cup[{\rho_I}^{-1}(\mathcal{B}_I)\cap(\mathbb{C}^m\times bB_k(r_I))] = Y_1\cup Y_2,$$

integration on Y_1 gives no contribution, since

$$(\frac{\sqrt{-1}}{2}\partial\bar\partial||w_I||^2)^m$$

is a $2m$ - form on the manifold $b\mathcal{B}_I$ of real dimension $2m-1$.
On the other end, we have on Y_2 that

$$\frac{\sqrt{-1}}{2\pi}\partial\bar\partial\log||z||^2 = \frac{1}{\pi {r_I}^2}\partial\bar\partial||z||^2,$$

by [22, p. 66].

Let us choose now an open relatively compact subset G of $\mathbb{C}^n$ such that

$$X_I \subset\subset G \subset\subset U,$$

then by (**) and Lemma III.3.4 we have

$$\limsup_{\epsilon} \int_{X_I} T_\epsilon \wedge (\frac{\sqrt{-1}}{2\pi}\partial\bar{\partial}\log||z||^2)^{m+k-1-j} \wedge$$
$$\wedge (\frac{\sqrt{-1}}{2}\partial\bar{\partial}||z||^2)^{j-m} \wedge \wedge (\frac{\sqrt{-1}}{2}\partial\bar{\partial}||w_I||^2)^m \leq$$
$$\leq \frac{1}{(\pi {r_I}^2)^{m+k-1-j}} \int_{X_I} T_\epsilon \wedge (\frac{\sqrt{-1}}{2}\partial\bar{\partial}||z||^2)^{k-1} \wedge (\frac{\sqrt{-1}}{2}\partial\bar{\partial}||w_I||^2)^m < \infty.$$

Thus also

$$\sup_{\epsilon} \int_{X_I} T_\epsilon \wedge (\frac{\sqrt{-1}}{2\pi}\partial\bar{\partial}\log||z||^2)^{m+k-1-j} \wedge (\frac{\sqrt{-1}}{2}\partial\bar{\partial}||z||^2)^{j-m} \wedge$$
$$\wedge (\frac{\sqrt{-1}}{2}\partial\bar{\partial}||w_I||^2)^m < \infty. \quad \square$$

III.3.6 PROPOSITION. *Let:*

$$f : M' \to M$$

the blow up of M along a submanifold Y, $E = f^{-1}(Y)$.

If T is a positive $\partial\bar{\partial}$ - closed current on M of bidegree (1,1), then the current

$$((f \restriction_{M' \setminus E})^{-1})_* (T \restriction_{M \setminus Y})$$

has locally finite mass across E, hence it extends to a current of order zero.

PROOF. We identify locally f with the blow up π and let $y = (t^0, y) \in Y$.

By smoothing T by convolution in a suitable open neighbourhood U of y in M, we get a family $\{T_\epsilon\}$ as in Proposition III.3.5.

Let us choose $r > 0$ and a sequence of real numbers $\{r_j\}_{j \geq 0}$ of positive real numbers with $r_j \to 0$ monotonously in such a way that

$$V_0 = B_m(t^0, r) \times B_k(r_0) \subset\subset U.$$

We have that

$$||T||(bV_j) = 0$$

where $V_j = B_m(t^0, r) \times B_k(r_j)$.

Let $U' = f^{-1}(U)$ and if $L \subset M'$, let χ_L be the characteristic function of L. Since

$$\lim_j \chi_{\pi^{-1}(V_0 \setminus V_j)} = \chi_{\pi^{-1}(V_0 \setminus Y)},$$

we obtain:

$$\lim_j \int_{\pi^{-1}(V_0 \setminus V_j)} ((\pi \restriction_{U' \setminus E})^{-1})_* (T \restriction_{U \setminus Y}) \wedge \omega'^{n-1} =$$

$$= \int_{\pi^{-1}(V_0 \setminus Y)} ((\pi \restriction_{U' \setminus E})^{-1})_* (T \restriction_{U \setminus Y}) \wedge \omega'^{n-1}.$$

But

$$\int_{\pi^{-1}(V_0 \setminus V_j)} ((\pi \restriction_{U' \setminus E})^{-1})_* (T \restriction_{U \setminus Y}) \wedge \omega'^{n-1} = \int_{V_0 \setminus V_j} T \wedge \pi_* \omega'^{n-1}.$$

Since $||T||(bV_j) = 0$, by the last part of Lemma III.3.4 we have:

$$\int_{V_0 \setminus V_j} T \wedge \pi_* \omega'^{n-1} = \lim_\epsilon \int_{V_0 \setminus V_j} T_\epsilon \wedge \pi_* \omega'^{n-1} \leq \sup_\epsilon \int_{V_0} T_\epsilon \wedge \pi_* \omega'^{n-1} < \infty.$$

Thus

$$\int_{\pi^{-1}(V_0 \setminus Y)} ((\pi \restriction_{U' \setminus E})^{-1})_* (T \restriction_{U \setminus Y}) \wedge \omega'^{n-1} < \infty. \quad \square$$

As we alrealy said we are interested in an extension T' which is also $\partial\bar{\partial}$-closed and satsfies property (*) after the statement of Theorem III.3.3. So we go on by recalling the following:

III.3.7 LEMMA. *Let Ω be an open subset of $\mathbb{C}^n$ and θ a smooth strictly positive (1,1) - form In Ω.*

Suppose that $\{T_\lambda\}$ is a sequence of smooth (k,k) - forms in Ω satisfying

$$\sup_\lambda \int_K T_\lambda \wedge \theta^{n-k} < \infty$$

for every compact subset K of Ω.

Then there exists a subsequence $\{T_{\lambda_\mu}\}$ of $\{T_\lambda\}$ converging in Ω in the weak topology.

PROOF. cfr. [22, p.69]. $\square$

III.3.8 COROLLARY. *In the hypotheses of Proposition III.3.5, there exists a subsequence* $\{\epsilon_\mu\}$, $\epsilon_\mu \to 0$, *such that*

$$\pi^* T_{\epsilon\mu} \text{ converges on } U' \text{ to a current } T'_U \text{ in the weak topology.}$$

T'_U *does not depend on the sequence* $\{\epsilon_\mu\}$, *but only on* T.

PROOF. Let us cover U' by coordinate patches Ω_j $j = 1, \dots n$. By Proposition III.3.5 Ω_j, $\pi^* T_\epsilon \restriction_{\Omega_j}$, and $\theta = \omega' \restriction_{\Omega_j}$ satisfy the assumptions of the preceding lemma, hence we can find a sequence $\{\epsilon_\mu\}$ such that $\pi^* T_{\epsilon_\mu}$ converges on each Ω_j, therefore on U', to a current T'_U. Let S_{ϵ_μ} be a sequence of currents with the same properties as $T_{\epsilon_m u}$ and set

$$\lim_\mu \pi^* S_{\epsilon_\mu} = S'_U.$$

Since $U = B_m \times B_n$, S_{ϵ_μ} and T_{ϵ_μ} are components of boundaries, we obtain

$$S'_U = T'_U$$

by applying the Relative support Theorem I.4.3 to the current $S'_U - T'_U$. □

By means of all this preliminary work we have obtained what we were looking for in the case of a blow-up. We may be summarize it in the following statement:

THEOREM III.3.9. *Let* M *and* M' *be complex manifolds of dimension* n *and*

$$f : M' \to M$$

be the blow-up of M *along the smooth submanifold* Y. *Let* E *be the exceptional set of* f *and* T *be a positive* $\partial\bar{\partial}$-*closed current on* M *of bidegree* $(1, 1)$.

$(f^{-1}_{\restriction_{M' \setminus E}})_* (T_{\restriction M \setminus Y})$ *can be extended to* M' *and there exists an extension* T' *which is positive,* $\partial\bar{\partial}$-*closed and satsfies the following property:*

$(*)$ $\forall x \in M$, $\exists$ *an open neighbourhood* W *of* x *such that* $T'_{\restriction_{f^{-1}(W)}}$ *is the limit in the weak topology of currents which are components of boundaries.*

In the case of a general modification, we prove now the last preparatory result:

III.3.10 THEOREM. *Let M and M' be complex manifolds of dimension n and $f : M' \to M$ be a proper modification and T be a positive $\partial\bar{\partial}$-closed current on M of bidegree $(1,1)$. Then the following hold:*

(1) *There exists a positive $\partial\bar{\partial}$-closed $(1,1)$-current T' on M' such that $f_*T' = T$ and $\forall x \in M$, $\exists$ an open neighbourhood W of x such that $T'_{\restriction_{f^{-1}(W)}}$ is the limit in the weak topology of currents which are components of boundaries.*

(2) *if M is compact the current T' is unique.*

If in addition M and M' satisfy (B) of III.1.7, and

$$f^* : \mathcal{V}^{1,1}(M)_{\mathbb{R}} \to \mathcal{V}^{1,1}(M')$$

is the induced map between Aeppli groups, then

$$f^*([T]) = [T'].$$

PROOF. (1) Let $x \in M$. By Lemma I.4.2 there exist an open neighbourhood V of x in M, a complex manifold Z and holomorphic maps $g : Z \to f^{-1}(V)$, $h : Z \to V$ such that $h = f \circ g$; $g : Z \to f^{-1}(V)$ is a blow up and h is obtained as a finite sequence of blow-ups with smooth centers. Because of this last property, by the preceding Theorem there exists a a positive $\partial\bar{\partial}$-closed current $\hat{T}$ on Z such that $h_*\hat{T} = T$.

After shrinking V if necessary, we may suppose that it is biholomorphic to an open ball by coordinate maps, and that there exists, for instance by smoothing by convolutions, a sequence $\{T_\epsilon\}$ of components of boundaries $T_\epsilon \to T$.

By construction, $\hat{T} = \lim_\epsilon T_\epsilon$ and it does not depend on the sequence $\{T_\epsilon\}$. Let us define $T'_V = g_*\hat{T}$. We have:

$$T'_V = \lim_\epsilon g_* h^* T_\epsilon = \lim_\epsilon (f \restriction_{f^{-1}(V)})^* T_\epsilon.$$

If we take: $T' \restriction_{f^{-1}(V)} = T'_V$, we obtain the required current, since T'_V does not depend on $\{T_\epsilon\}$ nor on the factorization $h = f \circ g$.

(2) Let T' and S' be currents on M' satisfying (1). Let $x \in M$ and Z, g, h as above. Applying (1) for the modification g, we obtain positive $\partial\bar{\partial}$-closed currents $\hat{T}$ and $\hat{S}$ on Z such that $g_*\hat{T} = T'$ and $g_*\hat{S} = S'$ on $f^{-1}(V)$. It follows

$$h_*\hat{T} = h_*\hat{S}$$

on V. Let $\{E'_\gamma\}$ and $\{F_\beta\}$ denote respectively the sets of irreducible components of $E \cap f^{-1}(V)$, where E is the exceptional set of f, and of F, the exceptional set of $g : Z \to f^{-1}(V)$; finally, let $\hat{E}'_\gamma$ denote the strict transform of $\{E'_\gamma\}$ under g. With all this in mind, the se of irreducible components of the exceptional set of h is given by $\{\hat{E}'_\gamma\} \cup \{F_\beta\}$, so that, by the Relative support theorem I.4.5 we obtain:

$$\hat{T} - \hat{S} = \sum_\gamma u_\gamma[\hat{E}'_\gamma] + \sum_\beta u_\beta[F_\beta]$$

where u_γ and u_β are pluriharmonic functions. Hence:

$$T' - S' = g_*(\hat{T} - \hat{S}) = \sum_\gamma u'_\gamma[E'_\gamma]$$

on $f^{-1}(V)$, where u'_γ are well defined pluriharmonic functions on $E'_\gamma \backslash g(\hat{E}'_\gamma \cap F)$ which are locally bounded on E'_γ. Thus u'_γ extends to E'_γ so that it holds on M':

$$T' - S' = \sum_\alpha u_\alpha[E_\alpha],$$

where $\{E_\alpha\}$ is the set of irreducible components of E, and u_α are pluriharmonic on E_α. But E_α is compact, so that u_α is a constant. The thesis will follow from the Relative support theorem I.4.6.

As for the last statement, let T' be a current satisfying property (1) in the theorem. By condition (B) of III.1.7 there exists d-closed real $(1,1)$-forms ϕ and ϕ' on M and M' respectively, such that $\beta([\phi]) = [T]$ and $\beta([\phi']) = [T']$ with $\beta : H^{1,1}(M, \mathbb{R}) \to \mathcal{V}^{1,1}(M)_\mathbb{R}$ defined in III.1.6. In other words there exists currents S and R on M and M' respectively, such that

$$T = \phi + \bar{\partial}S + \partial\bar{S}$$

and

$$T' = \phi' + \bar{\partial}R + \partial\bar{R}.$$

Therefore:

$$f_*\phi' - \phi = \bar{\partial}(S - f_*R) + \partial(\bar{S} - f_*\bar{R}).$$

The injectivity of β gives then that $f_*\phi' - \phi$ is d-exact, and because of the links between the cohomology rings of M, M', E and Y, see for instance [11, p. 285], we obtain:

$$f_*\phi' - \phi = \sum_\alpha c_\alpha[E_\alpha] + dQ$$

for a suitable current Q.

Let us choose V on which $T' \restriction_{f^{-1}(V)}$ is the limit of components of boundaries in such a way that $H^2(V, \mathbb{R}) = 0$ so that $\phi \restriction_V = d\psi$.

Therefore,

$$\sum_\alpha c_\alpha[E_\alpha] = \phi' - f^*\phi - dQ$$

is the limit of coimponents of boundaries on $f^{-1}(V)$. By the Relative support theorem I.4.6 $c_\alpha = 0 \quad \forall \alpha$, hence

$$f^*([T]) = f^*(\beta([\phi])) = \beta f^*([\phi]) = \beta([\phi']) = T'. \quad \square$$

We are now ready to prove the main result:

PROOF OF THEOREM III.3.3.

We observe first that if M' satisfies condition (B) of III.1.7, M also satisfies condition (B). Indeed, let us consider the commutative diagram:

$$\begin{array}{ccc}
H^{1,1}(M, \mathbb{R}) & \xrightarrow{\beta} & \mathcal{V}^{1,1}(M)_{\mathbb{R}} \\
f^* \downarrow & & \downarrow f^* \\
H^{1,1}(M', \mathbb{R}) & \xrightarrow{\beta} & \mathcal{V}^{1,1}(M')_{\mathbb{R}} \\
f_* \downarrow & & \downarrow f_* \\
H^{1,1}(M, \mathbb{R}) & \xrightarrow{\beta} & \mathcal{V}^{1,1}(M)_{\mathbb{R}}
\end{array}$$

By assumption, β at level M' is injective, hence β at the M level is injective too. Let T be a positive $\partial\bar{\partial}$-closed current of degree $(1,1)$ on M, and T' be the positive $\partial\bar{\partial}$-closed $(1,1)$ -current given by the preceding Theorem. Let ψ be a d-closed real $(1,1)$-form on M'such that $\beta([\psi]) = [T']$. Then:

$$\beta f_*([\psi]) = f_*\beta([\psi]) = [f_*T'] = [T].$$

Suppose now in addition that T is the $(n-1,n-1)$-component of a boundary. We prove our contention by showing that $T=0$ and applying Theorem II.1.3. Let T' as above, so that

$$f^*([T]) = T'.$$

T' is therefore a positive component of a boundary on a balanced manifold. This implies $T'=0$ by theorem II.1.3. Let E be the exceptional set of f and $Y \subset M$ such that $Y = f(E)$. We have then supp T $\subset Y$, but the codimension of Y is strictly greater than one, hence $T=0$, by the Support theorem I.2.4.

The last part of the statement is a direct consequence of Theorem II.2.4. □

ADDED IN PROOF. Alessandrini-Bassanelli have been able, in *The Class of compact Balanced Manifolds is invariant under Modifications*, these *Proceedings*, to get rid of the technical assumption that M' satisfy condition (B) of III.1.7. by showing that the unique current T' of Theorem III.3.9 actually belongs to the class $f^*([T]) \in \mathcal{V}^{1,1}(M')_{\mathbb{R}}$. The first part of the main Theorem III.3.3 becomes:

Let M and M' be compact complex manifolds of dimension n. Let

$$f : M' \to M$$

be a modification.

Then if M' is (n-1)-Kähler, M is (n-1)-Kähler.

References

1. L.Alessandrini and M.Andreatta, *Closed transverse (p,p)-forms on compact complex manifolds*, Compositio Math. **61** (1987), 181–200.
2. L.Alessandrini and G.Bassanelli, *A balanced proper modification of* $\mathbb{P}_3$, Comment. Math.Helvetici **66** (1991), 505–511.
3. ———, *Positive* $\partial\bar{\partial}$*-closed currents and non Kähler geometry*, to appear, J. of Geometrical Analysis (1992).
4. ———, *A metric characterization of manifolds bimeromorphic to compact Kähler spaces*, to appear, J. of Diff. Geometry (1992).
5. A.Andreotti and H.Grauert, *Théorèmes de finitude pour la cohomologie des espaces complexes*, Bull. Soc. Math. France **90** (1962), 193–259.
6. D.Barlet, *Convexite' au voisinage d'un cycle*, Seminaire F. Norguet 1979. Lecture Notes in Mathematics n°807, . Springer Verlag, 1980.
7. A.Blanchard, *Les variétés analytiques complexes*, Ann. Sc. Ecole Nor. Sup. **73** (1958), 157–202.
8. A.Borel and A.Haefliger, *La classe d'homologie fondamentale d'un espace analytique*, Bull. Soc. Math. France **89** (1961), 461–513.
9. H.Federer, *Geometric Measure Theory*, Springer Verlag, Berlin–New York, 1969.
10. H.Grauert and R.Remmert, *Plurisubharmonische Funktionen in komplexen Räumen*, Math. Z. **65** (1957), 175–194.
11. H.Grauert and O.Riemenschneider, *Verschwindungssätze für analytische Kohomologiegruppen auf komplexen Räumen*, Inv. Math. **11** (1970), 263-292.
12. A.Gray and L.M. Hervella, *The sixteen classes of almost hermitian manifolds and their linear invariants*, Annali Mat. Pura e Appl..
13. R.Harvey, *Holomorphic Chains and their boundaries*, Several Complex Variables, Proc AMS Symp. Pure Math,vol 30 part I, AMS, Providence R.I., 1977, pp. 309–382.
14. R.Harvey and A.W.Knapp, *Positive (p,p)-forms, Wirtinger's inequality and currents*, Proceedings Tulane University Program on Value Distribution Theory in Complex Analysis and related topics in Differential Geometry 1972/73, De kker, New York, 1974, pp. 43–62.
15. R.Harvey and J.B. Lawson, *An intrinsic characterization of Kähler manifolds*, Inv. Math. **74** (1983), 169–198.
16. H.Hironaka, *On the theory of birational blowing up*, Harvard Thesis (1960).
17. H.Hironaka and H.Rossi, *On the equivalence of embeddings of exceptional complex spaces*, Math. Ann. **156** (1964), 313–333.
18. P.Lelong, *Plurisubharmonic Functions and Positive Differential Forms*, Gordon and Breach, New York, 1969.
19. P.Libermann and C.M. Marle, *Symplectic Geometry and Analytical Mechanics*, Mathematics and its applications, D.Reidel Publishing Co., Dordrecht, 1987.
20. M.L. Michelson, *On the existence of special metrics in complex geometry*, Acta Math. **143** (1983), 261–295.
21. Y. Miyaoka, *Extension Theorems for Kähler metrics*, Proc. Japan Acad. **50** (1974), 407–410.

22. Y.T.Siu, *Analiticity of sets associated with Lelong Numbers and the extension of closed positive currents*, Inv.Math. **27** (1974), 53–156.
23. H.Skoda, *Prolongement des courants positifs, fermés de masse finie*, Inv. Math. **66** (1982), 361–376.
24. D.Sullivan, *Cycles for trhe dynamical study of foliated manifolds and complex manifolds*, Inv. Math. **36** (1976), 225–255.
25. J.Varouchas, *Propriété cohomologiques d'une classe de variétés analitiques complexes compactes*, Sem. d'Analyse Lelong-Dolbeault-Skoda 1983-84, Springer Lecture Notes,vol 1198, Springer Verlag, Berlin, 1985, pp. 233–243.
26. ———, *Sur l'image d'une variété Kählerienne compacte*, Sem. Norguet 1983-84, Springer Lecture Notes , vol. 1188, Springer Verlag, Berlin, 1985, pp. 245-259.
27. ———, *Kähler spaces and proper open morphisms*, Math. Ann. **283** (1989), 13-52.

Isocurved Deformations of Riemannian Homogeneous Metrics

Giuseppe Tomassini, Scuola Normale Superiore, Piazza dei Cavalieri 7, I-56126 Pisa, Italy. tomassini@vaxsns.sns.it

Franco Tricerri, † Dipartimento di Matematica, Università di Firenze, Viale Morgagni 67/A, I-50134 Firenze, Italy

0. Introduction

It is very simple to construct deformations of a Riemannian metric, but it is much more difficult to keep the curvature or some other Riemannian invariants of the deformed metrics under control. Often the expected result is the *rigidity* of the "special" Riemannian structure which we are concerned with, namely the impossibility to construct non-trivial deformations remaining in the same special class. More generally, some *finiteness* of the space of the non-trivial deformations is often conjectured, and sometimes proved. We refer to [BS] for a discussion of the case of Einstein metrics, and to [BR], [TV] for a conjecture of Gromov closely related to the subject developed in the present paper.

Here we construct and study deformations of Riemannian homogeneous metrics which preserve the Riemann curvature in a sense specified below. We were inspired by some examples constructed in [KTV], and motivated by the aim to complete the study started there.

In order to state our results in a precise form, it is convenient to consider the Riemannian curvature R as a map defined on the total space OM of the orthonormal frame bundle of (M, g) with values in the vector space $\mathrm{R}(V)$ of algebraic curvature tensors on $V = \mathbf{R}^n$. Such a map is defined by

$$R(u)(\xi_1, \xi_2, \xi_3, \xi_4) = (R_{\pi(u)})_{u\xi_1, u\xi_2, u\xi_3, u\xi_4} \ , \ \xi_i \in \mathbf{R}^n = V$$

where the elements u of OM are just the isometries between the Euclidean vector space V endowed with the standard inner product, and the tangent space T_pM at $p = \pi(u)$, π being the projection of OM onto M. The orthogonal group $O(n)$ is acting on the right on OM, and on the left on $\mathrm{R}(V)$. The map is *equivariant* w.r.t. such actions. In fact,

$$(a^{-1}R(u))(\xi_1, \xi_2, \xi_3, \xi_4) = R(u)(a\xi_1, a\xi_2, a\xi_3, a\xi_4) =$$

$$= (R_{\pi(u)})_{ua\xi_1, ua\xi_2, ua\xi_3, ua\xi_4} = R(ua)(\xi_1, \xi_2, \xi_3, \xi_4)$$

Supported by the Project 40 % M.U.R.S.T. "Geometria reale e complessa".
This paper is in final form and no part of it will be submitted elsewhere.

since $\pi(ua) = \pi(u)$. Therefore, R maps the fibres of OM (orbits of the action of $O(n)$ on OM) into the orbits of the orthogonal group in R(V). If (M, g) is *homogeneous*, $R(OM)$ is contained in one single orbit. The converse is not true, and the manifolds having this property are called *curvature homogeneous*. Thus, if (M, g) is curvature homogeneous, we have

$$R(OM) \subseteq O(n)K$$

where K is a fixed element of R(V). If K is the curvature tensor of some homogeneous space $G/H = M_0$ endowed with a G-invariant metric g_0, we say that (M, g) has the *same curvature* as the *model space* (M_0, g_0).

The model spaces considered in the present paper are special solvable Lie groups endowed with left invariant metrics. They are described and studied in Section 3. As Riemannian manifold, $M_0 = \mathbf{R}^p \times \mathbf{R}^q$ and $g_0 = g_C$ is a homogeneous metric depending on a linear map

$$C : \mathbf{R}^q \longrightarrow Sym(\mathbf{R}^p), \;\; x \longmapsto C(x),$$

of $\mathbf{R}^q$ into the space of the *symmetric* $p \times p$ matrices, such that $C(x)$ and $C(x')$ commute for any x and x'. This metric appears as a special case (i.e., $A = 0$) of a metric $g_{A,C}$ depending also on a second linear map A. This time

$$A : \mathbf{R}^p \longrightarrow \mathbf{so}(\mathbf{R}^q), \;\; w \longmapsto A(w),$$

maps $\mathbf{R}^p$ into the space of the *skew-symmetric* $q \times q$ matrices in such a way that $A(w)$ and $A(w')$ commute for any w and w'. If the following *compatibility condition*

$$C(A(w)x)w' = C(A(w')x)w$$

is satisfied, then all the metrics $g_{A,C}$ have the *same curvature* as g_C. In general, they are not locally homogeneous and therefore not isometric to the model metric. So, the family $g_{A,C}$ is a *non-trivial deformation* of the homogeneous metric g_C preserving the Riemann curvature.

The metrics $g_{A,C}$ can be deformed once more. This leads to a class of metrics $g_{A,C,h}$ depending also on a *diffeomorphism* h of $\mathbf{R}^p$ whose Jacobian matrix, at any point, is commuting with $C(x)$, for all $x \in \mathbf{R}^q$. Again, all these metrics have the same curvature as the model metric g_C. So we get non-trivial deformations of a homogeneous metric preserving the Riemann curvature and depending in an essential way on some arbitrary functions. In fact, we shall prove that, if some genericity assumptions are satisfied, the *isometry classes* of these metrics depend on a certain number of arbitrary functions. Namely, their *moduli space* is not *finite-dimensional* (see Section 9).

The metrics $g_{A,C}$ and $g_{A,C,h}$ are already defined in Section 1 as deformation of a *flat right invariant metric* on a Lie group. Section 2 is devoted to the computation of the Levi Civita connection and of the Riemann curvature of all these metrics. In particular, we prove that they are *curvature homogeneous* if A and C satisfy the compatibility condition stated above. Section 3 is devoted to the study of the model metric g_C. In Section 4 the norm of the covariant derivative is computed. This leads to the non-homogeneity of a *generic* metric $g_{A,C,h}$. Some linear algebra arguments give a useful canonical form in Section 5. This allows to construct some simple explicit examples satisfying the compatibility conditions and some important genericity assumptions which assure, in particular, the *irreducibility* and the *completeness* of the corresponding metrics. The irreducibility turns out to be a quite delicate problem which is treated in Section 6 and Section 7. Section 6 contains also a general condition which guarantees the completeness of the metrics $g_{A,C,h}$ (the metrics $g_{A,C}$ are always complete as it easily follows from this condition). The isometries between two metrics $g_{A,C,h}$ and $g_{A',C,h'}$ (with the same C, and therefore the same curvature) are studied and completely characterized in Section 8 under the assumption of *weakly generic Ricci curvature*. These results are employed in the last section in order to study the isometry classes of the metrics $g_{A,C}$ and $g_{A,C,h}$ arising from the special examples constructed in Section 5.

It is worthwhile to observe here that our method works because the model space $(M_0, g_0 = g_C)$ is *reducible*. This is equivalent to the requirement that the kernel of the map $x \longrightarrow C(x)$ is not trivial. Otherwise the compatibility condition forces A to be zero. In such a case, all the metrics $g_{0,A,h}$ are isometric to g_C, as proved in Section 2 and Section 3, and our deformations are *trivial*.

It would be interesting, in case there are any, to construct non-trivial isocurved deformations of an *irreducible* homogeneous Riemannian metric.

Finally we would like to thank L. Vanhecke for his constant interest and for his critical reading of this paper.

1. The metrics $g_{A,C}$ as deformation of a flat right-invariant metric on a Lie group

A Lie group admits a flat right (or left) invariant Riemannian metric if and only if it is the semidirect product of two Abelian Lie subgroups whose Lie algebras are mutually orthogonal and one factor is acting on the other by isometries ([ML] p. 298). Such a group can be realized as the semidirect product $\mathbf{R}^p \ltimes \mathbf{R}^q$, where $\mathbf{R}^p$ is acting on $\mathbf{R}^q$ as follows:

$$(w, x)(w_0, x_0) = (w + w_0, e^{-A(w)}x_0 + x). \tag{1.1}$$

In this formula, $e^{-A(w)}$ denotes the exponential of the operator $A(w)$, and
$A : \mathbf{R}^p \longrightarrow \mathbf{so}(q)$ is a linear map into the Lie algebra $\mathbf{so}(q)$ of the skew-symmetric operators on $\mathbf{R}^q$ such that

$$[A(w), A(w')] = 0 \tag{1.2}$$

for each $w, w' \in \mathbf{R}^p$. It is easily checked that the Maurer-Cartan form on $G = \mathbf{R}^p \ltimes \mathbf{R}^q$ is given by (dw, θ) where

$$\theta = dx + A(dw)x \ . \tag{1.3}$$

We adopt here an index-free matrix notation; so dw and dx denote the vector valued one-forms

$$dw = \begin{pmatrix} dw^1 \\ \vdots \\ dw^p \end{pmatrix} , \quad dx = \begin{pmatrix} dx^1 \\ \vdots \\ dx^q \end{pmatrix} \tag{1.4}$$

where $(w^1, \ldots, w^p)$ and $(x^1, \ldots, x^q)$ are, respectively, the coordinate functions of the vector spaces $\mathbf{R}^p$ and $\mathbf{R}^q$. dw is biinvariant and θ is right invariant. Therefore,

$$g_0 = {}^t dw \otimes dw + {}^t\theta \otimes \theta \tag{1.5}$$

defines a right invariant Riemannian metric on $\mathbf{R}^p \ltimes \mathbf{R}^q$, which turns out to be flat (see Section 2).

The subgroup $\mathbf{R}^p$ is acting on $\mathbf{R}^p \ltimes \mathbf{R}^q$ on the left by

$$w_0(w, x) = (w_0 + w, e^{-A(w_0)}x). \tag{1.6}$$

The orbit space of this action can be identified with $\mathbf{R}^q$. With this identification, the projection π of $\mathbf{R}^p \ltimes \mathbf{R}^q$ on $\mathbf{R}^q$ is given by

$$\pi(w, x) = e^{A(w)}x. \tag{1.7}$$

On the other hand, $\mathbf{R}^p$ can be identified with the orbit of (w_0, x_0) via the immersion

$$i_{(w_0,x_0)} : \mathbf{R}^p \longrightarrow \mathbf{R}^p \ltimes \mathbf{R}^q, \;\; w \longmapsto (w_0 + w, e^{-A(w)}x_0). \tag{1.8}$$

Then, we have

Proposition 1.1 *The metric g_0 is the unique Riemannian metric on $\mathbf{R}^p \ltimes \mathbf{R}^q$ such that the induced metric on the orbits is the Euclidean metric ${}^t dw \otimes dw$, and π is a Riemannian submersion on the Euclidean space $(\mathbf{R}^q, {}^t dx \otimes dx)$.*

Proof. From (1.8) we get at once

$$i^*_{(w_0,x_0)} dw = dw$$

and

$$i^*_{(w_0,x_0)} \theta = -e^{-A(w)}A(dw)x_0 + A(dw)e^{-A(w)}x_0 = 0.$$

Therefore,

$$i^*_{(w_0,x_0)} g_0 = {}^t dw \otimes dw.$$

On the other hand, (1.7) gives

$$\pi^* dx = e^{A(w)}(A(dw)x + dx) = e^{A(w)}\theta.$$

Hence, the vertical distribution spanned by the vectors tangent to the fibres of π (i.e., to the orbits of $\mathbf{R}^p$) is the kernel of θ. Moreover,

$$\pi^*({}^t dx \otimes dx) = {}^t\theta \otimes \theta.$$

So, π is a Riemannian submersion, as claimed.

Consider now a linear map

$$C : \mathbf{R}^q \longrightarrow Sym(\mathbf{R}^p), \;\; x \longmapsto C(x)$$

of $\mathbf{R}^q$ into the space $Sym(\mathbf{R}^p)$ of the *symmetric operators* $\mathbf{R}^q$ such that

$$[C(x), C(x')] = 0, \tag{1.9}$$

for each x, x' in $\mathbf{R}^q$. Deform the metric g_0 along the $\mathbf{R}^p$-orbits of $\mathbf{R}^p \ltimes \mathbf{R}^q$ by putting

$$g_{A,C} = {}^t\omega \otimes \omega + {}^t\theta \otimes \theta \tag{1.10}$$

where

$$\omega = e^{C(x)} dw. \tag{1.11}$$

Then, we obtain

Theorem 1.2 *If the maps A and C satisfy the following compatibility condition*

$$C(A(w)x)w' = C(A(w')x)w, \tag{1.12}$$

then all the metrics $g_{A,C}$ are curvature homogeneous with Riemann curvature depending only on C.

We postpone the proof of this theorem to the next section which will be devoted to the computation of the basic Riemannian invariants.

It is expected that the isometry classes of these metrics depend on the maps A and C, and therefore at most on a finite number of parameters (i.e., the components of the tensors A and C). If we want to introduce other degrees of freedom in such a way that these classes depend also on some arbitrary functions, we shall modify again the one-form ω as follows. Consider the subgroup $\mathcal{H}$ of the diffeomorphisms of $\mathbf{R}^p$ whose differentials are commuting with all operators $C(x)$, i.e.,

$$\mathcal{H} = \{h \in Diff(\mathbf{R}^p) \; : \; [dh_{|_w}, C(x)] = 0, \forall x \in \mathbf{R}^q, w \in \mathbf{R}^p\}. \tag{1.13}$$

Let h be an element of $\mathcal{H}$. Extend h to a diffeomorphism of $\mathbf{R}^p \ltimes \mathbf{R}^q$, denoted by the same symbol, by means of the formula

$$h(w, x) = (h(w), x), \tag{1.14}$$

and define the metric $g_{A,C,h}$ as follows:

$$g_{A,C,h} = {}^t h^*\omega \otimes h^*\omega + {}^t\theta \otimes \theta. \tag{1.15}$$

We recover $g_{A,C}$ by choosing h equal to the identity of $\mathbf{R}^p$. In general, we have

Theorem 1.3 *If the compatibility condition (1.12) is satisfied, then the metrics $g_{A,C,h}$ are curvature homogeneous with the same curvature as $g_{A,C}$.*

Also this theorem will follow from the computations in the next section.

2. Basic Riemannian invariants

We shall perform all the computations in the general case for the metrics $g_{A,C,h}$. We write down these metrics in the form

$$g_{A,C,h} = {}^t\omega \otimes \omega + {}^t\theta \otimes \theta, \tag{2.1}$$

where θ is given by (1.3), but

$$\omega = e^{C(x)} dh, \tag{2.2}$$

with $h \in \mathcal{H}$. Note that

$$dh_{|w} = J(w)dw \tag{2.3}$$

where $J(w)$ is the Jacobian matrix of h (evaluated at w). Then $J(w)$ is a smooth function with values in the subgroup H of $GL(p, \mathbf{R})$ of the non-singular linear operators of $\mathbf{R}^p$ commuting with all $C(x)$, for each $x \in \mathbf{R}^q$. Therefore, we can write

$$\omega = e^{C(x)} J(w)dw = J(w)e^{C(x)}dw.$$

Note that

$$d(J(w)dw) = 0.$$

Hence,

$$\begin{aligned} d\omega &= e^{C(x)}C(dx) \wedge J(w)dw = \\ &= C(dx) \wedge e^{C(x)}J(w)dw = \\ &= C(\theta) \wedge \omega - C(A(dw)x) \wedge e^{C(x)}J(w)dw = \\ &= C(\theta) \wedge \omega - e^{C(x)}J(w)C(A(dw)x) \wedge dw. \end{aligned}$$

But for all $\xi, \zeta \in \mathbf{R}^p$, we have

$$\begin{aligned} &2\ (C(A(dw))x) \wedge dw)(\xi, \zeta) = \\ &= C(A(dw(\xi))x)dw(\zeta) - C(A(dw(\zeta))x)dw(\xi) = \\ &= C(A(\xi)x)\zeta - C(A(\zeta)x)\xi = 0 \end{aligned}$$

because of the compatibility condition. Therefore,

$$d\omega = C(\theta) \wedge \omega. \tag{2.4}$$

A similar computation gives

$$\begin{aligned} d\theta &= -A(dw) \wedge dx \\ &= -A(dw) \wedge \theta + A(dw) \wedge A(dw)x. \end{aligned}$$

On the other hand, for all ξ, ζ in $\mathbf{R}^p$, we get

$$\begin{aligned} 2 \ & (A(dw) \wedge A(dw)x)(\xi,\zeta) = \\ = \ & A(\xi)A(\zeta)x - A(\zeta)A(\xi)x = 0, \end{aligned}$$

since $[A(\xi), A(\zeta)] = 0$. Therefore,

$$\begin{aligned} d\theta &= -A(dw) \wedge \theta = \\ &= -A(e^{-C(x)}J(w)^{-1}\omega) \wedge \theta. \end{aligned} \tag{2.5}$$

The structural equations for the Levi Civita connection can be written in the form

$$\begin{pmatrix} d\omega \\ d\theta \end{pmatrix} + \Lambda \wedge \begin{pmatrix} \omega \\ \theta \end{pmatrix} = 0, \ \Lambda +^t \Lambda = 0 \tag{2.6}$$

where the *connection forms*, in a matrix block notation, are given by

$$\Lambda = \begin{pmatrix} \lambda & \varphi \\ -^t\varphi & \psi \end{pmatrix}. \tag{2.7}$$

It is easy to prove that (2.6) is satisfied if and only if $\lambda = 0$ and the matrix valued 1-forms φ and ψ are given by

$$\psi = -A(dw), \tag{2.8}$$

$$\varphi = (\varphi^i_\alpha) \ , \ \varphi^i_\alpha = \langle C(u_\alpha)\omega, v_i\rangle. \tag{2.9}$$

In this formula, v_i, $i = 1, \ldots, p$ (respectively u_α, $\alpha = 1, \ldots, q$) are the vectors of the natural basis of $\mathbf{R}^p$ (respectively $\mathbf{R}^q$). Moreover, the bracket $\langle , \rangle$ denotes the standard inner product in $\mathbf{R}^m$. For the explicit computations it is useful to remark that ω and θ can be expressed as

$$\omega = \sum_i \omega^i v_i \ , \ \theta = \sum_\alpha \theta^\alpha u_\alpha, \tag{2.10}$$

where ω^i and θ^α are ordinary one-forms given by

$$\omega^i = \langle \omega, v_i\rangle \ , \ \theta^\alpha = \langle \theta, u_\alpha\rangle. \tag{2.11}$$

The *curvature forms* are obtained from the structural equation

$$\Omega = d\Lambda + \Lambda \wedge \Lambda. \tag{2.12}$$

So, if we put

$$\Omega = \begin{pmatrix} \mu & \nu \\ -^t\nu & \eta \end{pmatrix}, \tag{2.13}$$

we easily get

$$(2.14)\qquad \begin{cases} \mu = & -\varphi \wedge {}^t\varphi, \\ \nu = & d\varphi + \varphi \wedge \psi, \\ \eta = & d\psi - {}^t\varphi \wedge \varphi + \psi \wedge \psi. \end{cases}$$

Then, direct computations yield

$$(2.15)\qquad \mu = -\sum_\alpha C(u_\alpha)\omega \wedge C(u_\alpha)\omega,$$

$$(2.16)\qquad \nu = (\nu^i_\alpha)\ ,\ \nu^i_\alpha = \langle C(u_\alpha)v_i, C(\theta) \wedge \omega\rangle,$$

$$(2.17)\qquad \eta = 0.$$

We omit the computational details but recall that the compatibility condition of Theorem 1.2 and the commuting properties (1.2), (1.9), (1.13) play here an essential role just as in the proof of the formulae (2.4) and (2.5).

The *Riemann curvature* R is given by

$$(2.18)\qquad R = 2\sum_{i,j} \Omega^i_j \otimes \omega^i \wedge \omega^j + 4\sum_{i,\alpha} \Omega^\alpha_i \otimes \theta^\alpha \wedge \omega^i.$$

Since $\Omega^i_j = \mu^i_j$ and $\Omega^\alpha_i = -\nu^i_\alpha$, a direct computation leads to

Proposition 2.1 *The Riemann curvature of the metrics $g_{A,C,h}$ is given by*

$$(2.19)\qquad \begin{aligned} R = & - \sum_\alpha \langle C(u_\alpha)\omega, \omega\rangle \owedge \langle C(u_\alpha)\omega, \omega\rangle \\ & - 4\sum_\alpha \langle C(\theta) \wedge \omega, C(\theta) \wedge \omega\rangle\ , \end{aligned}$$

where $\owedge$ denotes the Kulkarni-Nomizu product of the symmetric bilinear forms

$$\rho_\alpha(\xi, \zeta) = \langle C(u_\alpha)\omega(\xi), \omega(\zeta)\rangle$$

(see [BS p. 42]).

It follows that the non-zero components of R are

$$(2.20)\qquad \begin{aligned} R_{hkij} = & - \sum_\alpha \langle C(u_\alpha)v_h, v_i\rangle\langle C(u_\alpha)v_k, v_j\rangle \\ & - \sum_\alpha \langle C(u_\alpha)v_k, v_i\rangle\langle C(u_\alpha)v_h, v_j\rangle, \end{aligned}$$

$$(2.21)\qquad R_{\beta j\alpha i} = - \langle C(u_\alpha)v_i, C(u_\beta v_j\rangle.$$

Since these components are constant and depend only on C, Theorem 1.2 and Theorem 1.3 are proved.

Remark 2.1 All the metrics $g_{A,0,h}$ are *flat*, and all the metrics $g_{0,C,h}$ are *isometric* to

$$(2.22)\qquad g_C = {}^t\omega_0 \otimes \omega_0 + {}^t dx \otimes dx,$$

where $\omega_0 = e^{C(x)}dw$. In fact, if $A = 0$, then $\theta = dx$, and

$$(2.23)\qquad h^* g_C = g_{0,C,h}.$$

Remark 2.2 Since all the metrics $g_{A,C,h}$ with $C \neq 0$ are curvature homogeneous, their *index of nullity* is constant. Therefore, the *nullity spaces* define a regular distribution E on $\mathbf{R}^p \times \mathbf{R}^q$, the so-called *nullity distribution* (see [KN]). It is easily proved from (2.19) that

$$E = \{\xi : \omega(\xi) = 0 \ , \ C(\theta(\xi)) = 0\}. \tag{2.24}$$

In fact, by definition, E is spanned by the vector fields ξ such that

$$R_{\xi,\xi_i,\xi_2,\xi_3} = 0, \text{for all } \xi_1, \xi_2 \,, \, \xi_3.$$

Therefore, if E is *zero-dimensional*, then the kernel of the map
$C : \mathbf{R}^q \longrightarrow Sym(\mathbf{R}^p)$ must be *trivial*. As we shall show in the next section (see Proposition 3.1) this implies that the metric is homogeneous. Hence, *each non-locally homogeneous Riemannian metric $g_{A,C,h}$ has non-zero index of nullity.*

3. The model space

For any choice of the map C the metrics g_C defined by (2.22) are *homogeneous*. In fact, g_C is a *left invariant metric* on the group $G = \mathbf{R}^p \rtimes \mathbf{R}^q$ with the product defined by

$$(w, x)(w_0, x_0) = (w_0 + e^{-C(x_0)}(w), x + x_0), \tag{3.1}$$

since dx is biinvariant and $\omega = e^{C(x)}(dw)$ is left invariant with respect to this product.

Moreover, $\mathbf{R}^p$ is a *normal subgroup* of G and $\mathbf{R}^q$ is acting on it via (3.1). It follows at once that G is *solvable*, but *not nilpotent*.

The Lie algebra $\mathbf{g}$ of G is the *semidirect sum* of two *Abelian* Lie algebras $\mathbf{h} = \mathbf{R}^p$ and $\mathbf{k} = \mathbf{R}^q$ endowed with the bracket

$$[w, x] = -C(x)w. \tag{3.2}$$

Then,

$$\mathbf{a} = \{x \in \mathbf{k} \ : \ C(x) = 0\} = KerC \tag{3.3}$$

is an *Abelian ideal* of $\mathbf{g}$, not only of $\mathbf{k}$. Therefore, $\mathbf{g}$ splits as direct sum of $\mathbf{a}$ and of its orthogonal complement $\mathbf{g}'$. Denote by $\mathbf{a}^\perp$ the orthogonal complement of $\mathbf{a}$ in the subalgebra $\mathbf{k} = \mathbf{R}^q$. Then, $\mathbf{g}'$ is a semidirect sum of $\mathbf{h}$ and $\mathbf{a}^\perp$. Moreover, if

$$x = x' + x''$$

is the decomposition of a vector x of $\mathbf{R}^q$ as sum of a vector of $\mathbf{a}^\perp$ and a vector of $\mathbf{a}$, from (3.1) we get

$$(w, x)(w_0, x_0) = (w, x')(w_0, x_0) + (0, x'' + x_0'').$$

Therefore, G splits as direct product of its normal subgroup $G' = \mathbf{R}^p \rtimes \mathbf{a}^\perp$ and of the additive normal subgroup $\mathbf{a}$.

Since

$$dx = dx' + dx''$$

and

$$\omega_0 = e^{C(x)}dw = e^{C(x')}dw,$$

the metric g_C can be written as

$$g_C = g'_C +^t dx'' \otimes dx'' ,$$

where

$$g'_C =^t \omega_0 \otimes \omega_0 +^t dx' \otimes dx' .$$

This shows that the Riemannian manifold (G, g_C) is the product of (G', g'_C) and of the Euclidean factor $(\mathbf{a},^t dx'' \otimes dx'')$.

The metric g'_C on G' is *irreducible* as we shall prove in Section 7.

As stated in Section 2, all the metrics $g_{A,C,h}$ have the same Riemann curvature as the *model space* (G, g_C). Therefore, by keeping C fixed, we have constructed a wide class of deformations of a homogeneous Riemannian metric which preserve the Riemann curvature. These metrics depend on a tensor A and on a diffeomorphism h of $\mathbf{R}^p$. If such "parameters" are chosen "generically", then $g_{A,C,h}$ is not homogeneous, even locally. But, in order to get these results, we are forced to consider *reducible homogeneous models* (G, g_C). In fact, we have

Proposition 3.1 *If $KerC = \{0\}$, then $A = 0$. Therefore, all the metrics $g_{0,C,h}$ are homogeneous since they are isometric to the homogeneous model g_C.*

Proof. The symmetric operators $C(x)$ are simultaneously diagonalizable because they commute. Take a basis $v_i, i = 1, \ldots, p$, which diagonalizes all the $C(x)$. Then,

$$C(x)v_i = \lambda_i(x)v_i, \tag{3.4}$$

where the λ_i are linear forms. We can write each λ_i as

$$\lambda_i(x) = \langle c_i, x\rangle, \tag{3.5}$$

for some suitable vector c_i in $\mathbf{R}^q$. It follows that

$$C(A(v_i)x)v_j = \langle c_j, A(v_i)x\rangle v_j = -\langle A(v_i)c_j, x\rangle v_j.$$

Then the compatibility condition (1.12) is satisfied if and only if

$$A(v_i)c_j = 0 \tag{3.6}$$

for all $i \neq j$. Note that the orthogonal complement of the vector subspace of $\mathbf{R}^q$ spanned by $c_1, \ldots, c_p$ is contained in the kernel of the map $x \longrightarrow C(x)$. From the hypothesis we obtain that $\mathbf{R}^q$ is spanned by $c_1, \ldots, c_p$. Then (3.6) implies that $A(v_i)$ is zero for all i. In fact, since $A(v_i)$ is skew-symmetric, also the vector

$$A(v_i)c_i = \sum_h \langle A(v_i)c_i, c_h\rangle c_h = \langle A(v_i)c_i, c_i\rangle c_i$$

is zero, and therefore $A = 0$ as claimed.

4. Non-homogeneity of a generic metric $g_{A,C,h}$

In general, a metric $g_{A,C,h}$ is not homogeneous, even locally. This is an implicit consequence of the results about the isometry classes (see Sections 8, 9). Nevertheless, this can also be proved directly by computing the norm of the covariant derivative of the Ricci tensor and by showing that it is not constant. First of all, from (2.5), (2.6) and (2.7) it is easy to compute the covariant derivatives of the forms ω and θ along a vector field ζ. We obtain

$$D_\zeta \omega = -C(u_\alpha)\omega(\zeta), \tag{4.1}$$

$$D_\zeta \theta = \sum_\alpha \langle C(u_\alpha)\omega(\zeta), \omega\rangle u_\alpha - A(dw(\zeta))\theta. \tag{4.2}$$

Then, the covariant differentials of ω and θ are given by

$${}^tD\omega = -{}^t\omega \otimes C(\theta), \tag{4.3}$$

$$D\theta = \sum_\alpha \langle C(u_\alpha)\omega, \omega\rangle u_\alpha - A(dw) \otimes \theta. \tag{4.4}$$

Remark 4.1 The components of $D\omega$ and $D\theta$ are constant if $A = 0$. In such a case, the metric is *infinitesimally homogeneous* in the sense of [SN], i.e., the Riemann curvature and all its covariant derivatives have constant components with respect to the dual frame of (ω, θ). It follows again that all the metrics $g_{0,C,h}$ are homogeneous.

By taking into account Proposition (2.1) (or, more explicitly, the formulae (2.20) and (2.21)), we find the following expression for the *Ricci curvature* tensor r of type (0,2) of the metrics $g_{A,C,h}$:

$$r = -\sum_\alpha trC(u_\alpha)\langle C(u_\alpha)\omega, \omega\rangle u_\alpha - trC(\theta)^2. \tag{4.5}$$

Here, $u_\alpha, \alpha = 1, \dots, q$, is the natural basis of $\mathbf{R}^q$ and $trC(u_\alpha)$ (respectively of $C(\theta)^2$) denotes the trace of the symmetric operators $C(u_\alpha)$ (respectively of $C(\theta)^2 = C(\theta) \circ C(\theta)$). Then, (4.1) and (4.2) give

$$\begin{aligned} D_\zeta r = &-\sum_i \sum_{\alpha\beta} trC(u_\alpha)\langle C(u_\alpha)v_i, C(u_\beta)\omega(\zeta)\rangle(\omega^i \otimes \theta^\beta + \theta^\beta \otimes \omega^i) + \\ &+\sum_{ik}\sum_\alpha trC(u_\alpha)\langle C(u_\beta)C(u_\alpha)v_k, v_k\rangle\langle C(u_\alpha)\omega(\zeta), v_i\rangle(\theta^\beta \otimes \omega^i + \omega^i \otimes \theta^\beta) + \\ &-\sum_k \sum_{\alpha\beta}\langle C(A(dw(\zeta))u_\alpha)u_\alpha C(u_\beta)v_k, v_k\rangle(\theta^\alpha \otimes \theta^\beta + \theta^\beta \otimes \theta^\alpha). \end{aligned} \tag{4.6}$$

It follows that the non-zero components of Dr with respect to the dual basis $\{E_i\ 1 \le i \le p,\ E_{p+\alpha},\ 1 \le \alpha \le q\}$ of (ω, θ) are given by

$$\begin{aligned} D_j r_{\beta i} &= D_j r_{i\beta} = -\sum_\alpha trC(u_\alpha)\langle C(u_\alpha)v_i, C(u_\beta)v_j\rangle + \\ &+ \sum_k \sum_\alpha \langle C(u_\beta)C(u_\alpha)v_k, v_k\rangle\rangle\langle C(u_\alpha)v_j, v_i\rangle\rangle, \end{aligned} \tag{4.7}$$

$$D_j r_{\alpha\beta} = -\sum_k \langle C(A(e^{-C(x)}J(w)^{-1}v_j)u_\alpha)C(u_\beta)v_k, v_k\rangle. \tag{4.8}$$

Hence,

$$\| Dr \|^2 = 2\sum_{\beta}\sum_{ij}(D_j r_{\beta i})^2 + \sum_{\alpha\beta}\sum_{j}(D_j r_{\alpha\beta})^2. \tag{4.9}$$

The first summation is constant. In fact, this term depends only on C. The second is a function of x and w depending on A (and C, of course). Put

$$F_A(x,w) = \sum_{\alpha\beta}\sum_{j}(D_j r_{\alpha\beta})^2. \tag{4.10}$$

From the compatibility condition, the symmetry of the operators $C(x)$ and from the relation (1.9), we get

$$\begin{aligned} D_j r_{\alpha\beta} &= -\sum_k \langle C(u_\beta)C(A(e^{-C(x)}J(w)^{-1}v_j)u_\alpha)v_k, v_k\rangle \\ &= -\sum_k \langle C(u_\beta)C(A(v_k)u_\alpha)e^{-C(x)}J(w)^{-1}v_j, v_k\rangle \\ &= -\sum_k \langle C(u_\beta)C(A(v_k)u_\alpha)v_k, e^{-C(x)}J(w)^{-1}v_j\rangle. \end{aligned} \tag{4.11}$$

Hence,

$$F_A(x,w) = \sum_{\alpha\beta}\sum_{j}\langle\sum_{k} C(u_\beta)C(A(v_k)u_\alpha)v_k, e^{-C(x)}J(w)^{-1}v_j\rangle^2. \tag{4.12}$$

This shows that F_A is not constant for a *generic choice* of the tensors A, i.e., such that

$$C(u_\beta)\sum_k C(A(v_k)u_\alpha)v_k \neq 0 \tag{4.13}$$

for some α and β.

With this hypothesis, the metric $g_{A,C,h}$ is not locally homogeneous.

5. The canonical form and the explicit example

It is always possible to put the tensors A and C in a *canonical form*. This will be very useful in order to construct explicit examples. First we prove two lemmas.

Lemma 5.1 *Let L be an element of $O(p)$, and*

$$C'(x) = LC(x)^tL,\ \ A'(w) = A(^tLw),\ \ h'(w) = Lh(^tLw), \tag{5.1}$$

for each x in $\mathbf{R}^q$, w in $\mathbf{R}^p$. Then the metrics $g_{A,C,h}$ and $g'_{A',C',h}$ are isometric.

Proof. Let f be the map

$$f(w,x) = (^tLw, x).$$

Then

$$\begin{aligned} f^*\omega &= f^*(e^{C(x)}dh) = \\ &= e^{C(x)}d(h\circ f) = e^{C(x)}\,{}^tLdh' = \\ &= {}^tLe^{C'(x)}dh' = {}^tL\omega'. \end{aligned}$$

Moreover,

$$\begin{aligned} f^*\theta &= f^*\omega(dx + A(dw)x) = \\ &= dx + A({}^tLdw)x = \\ &= dx + A'(dw)x = \\ &= \theta' \ . \end{aligned}$$

Therefore,

$$\begin{aligned} f^*g_{A,C,h} &= {}^t\omega' L \otimes^t L\omega' +^t \theta' \otimes^t \theta' = \\ &= {}^t\omega' \otimes^t \omega' +^t \theta' \otimes^t \theta' = \\ &= g'_{A',C',h} \end{aligned}$$

as claimed.

Lemma 5.2 *Let P be an element of $O(q)$, and*

$$C'(x) = C({}^tPx) \ , \ A'(w) = PA(w)^tP \ .$$

Then $g_{A,C,h}$ and $g'_{A',C',h}$ are isometric.

Proof. Define f by

$$f(w, x) = (w, {}^tPx).$$

Then we have

$$f^*\omega = e^{C({}^tPx)}dh = e^{C'(x)}dh = \omega'$$

and

$$\begin{aligned} f^*\theta &= {}^tPdx + A(dw)^tPx = \\ &= {}^tP(dx + A'(dw)x) = \\ &= {}^tP\theta'. \end{aligned}$$

As before, the metrics $g_{A,C,h}$ and $g'_{A',C',h}$ are isometric via f.

We have already remarked in the proof of Proposition 3.1 that there exists an orthonormal basis v_i' which diagonalizes simultaneously all the operators $C(x)$.

Let L be the orthogonal matrix associated to this change af basis. Then, Lemma 5.1 implies that our original metric $g_{A,C,h}$ is isometric to $g'_{A',C',h}$, where

$$A'(w) = A({}^tLw), \ h'(w) = Lh({}^tLw) \tag{5.2}$$

and $C'(x)$ is in diagonal form, i.e.,

$$\begin{pmatrix} \langle c_1, x\rangle & & & \\ & \ddots & & \\ & & \ddots & \\ & & & \langle c_p, x\rangle \end{pmatrix}. \tag{5.3}$$

Also the skew-symmetric operators $A'(w)$ commute and therefore they admit a common canonical form. More precisely, there exists an orthormal basis u'_α, $1 \le \alpha \le q$, of $\mathbf{R}^q$ such that the matrix associated to $A'(w)$ is of the form

$$\begin{pmatrix} 0 & \langle a_1, w\rangle & & & & & & \\ -\langle a_1, w\rangle & 0 & & & & & & \\ & & \ddots & & & & & \\ & & & 0 & \langle a_s, w\rangle & & & \\ & & & -\langle a_s, w\rangle & 0 & & & \\ & & & & & 0 & & \\ & & & & & & \ddots & \\ & & & & & & & 0 \end{pmatrix} \tag{5.4}$$

where the a_α are non-zero vectors of $\mathbf{R}^p$ and $2s \le q$. Denote by P the matrix of the corresponding change of basis. It follows from Lemma 5.2 that $g'_{A',C',h}$ is isometric to $g_{A'',C'',h'}$ where

$$C''(x) = C'({}^tPx)$$

and $A''(w)$ is associated to the matrix (5.4) in the new basis. Note that $C''(x)$ is still in diagonal form. Therefore, we have proved the following result:

Proposition 5.3 *Up to isometries, we can always suppose that the tensors C and A are given by (5.3) and (5.4), respectively.*

In such a case we say that $g_{A,C,h}$ is in *canonical form*. As proved in Section 3 (see the proof of Proposition 3.1), the compatibility condition

$$C(A(w)x)w' = C(A(w')x)w$$

for a metric $g_{A,C,h}$ in canonical form reads as follows:

$$A(v_i)c_j = 0 \tag{5.5}$$

for all $i \ne j$ (of course, for this we need only $C(x)$ in diagonal form). Then, the previous results permit the construction of explicit examples of operators $A(w)$ and $C(x)$ satisfying the compatibility condition, and therefore of metrics of type $g_{A,C,h}$. In particular, we shall extensively refer to the following example.

Example 5.1 Suppose $p = r + 1$ and $q = 2r + 1$. Let c_i be the vectors of $\mathbf{R}^q$ given by

$$c_i = \gamma_i u_{2i-1} + u_{2r+1} \tag{5.6}$$

if $i \leq r$, and

$$c_{r+1} = u_{2r+1}. \tag{5.7}$$

Choose r vectors a_i of $\mathbf{R}^p$, $1 \leq i \leq r$, as follows

$$a_i = \alpha_i v_i, \tag{5.8}$$

for some constant α_i. Let $A(w)$ and $C(x)$ be the operators associated to the matrices (5.4), (5.3). Then, the compatibility condition is satisfied. It follows that all the metrics

$$g_{\alpha,\gamma,h} = g_{A,C,h}, \tag{5.9}$$

where $\alpha = (\alpha_1, \ldots, \alpha_r)$, $\gamma = (\gamma_1, \ldots, \gamma_r)$, are curvature homogeneous as soon as the Jacobian matrix $J(w)$ of the diffeomorphism h of $\mathbf{R}^p$ commutes with the operators $C(x)$.

Since $C(x)$ is diagonal, from (5.3) and (5.6) we get at once that $[J(w), C(x)] = 0$ if and only if

$$\gamma_i J^i_j(w) = 0 \tag{5.10}$$

for each $i \leq r = p - 1$ and for any j.

Therefore, if all the coefficients γ_i are *non-zero*, the matrix $J(w)$ must be *diagonal*. In such a case h is given by

$$h(w) = (h^1(w^1), \ldots, h^p(w^p)), \tag{5.11}$$

where each function h^i depends only on the variable w^i and is *invertible*.

Hence, the metrics $g_{\alpha,\gamma,h} = g_{A,C,h}$ depend on $2r$ real parameters and $p = r+1$ functions $h^1(w^1), \ldots, h^p(w^p)$. Moreover, all these metrics are *curvature homogeneous* with curvature depending only on r real parameters $\gamma = (\gamma_1, \ldots, \gamma_r)$.

If all the parameters α_i are zero, then $g_{0,\gamma,h}$ is homogeneous and isometric to the *model space* $g_C = g_{0,\gamma,Id}$. But, in general, these metrics are *not locally homogeneous*, because (4.13) holds.

6. Irreducibility and completeness

In order to study the irreducibility of the metrics $g_{A,C,h}$ it is convenient to suppose that $C(x)$ is given in diagonal form (the canonical form is not necessary here). Then we have

$$C(x)v_i = \langle c_i, x\rangle v_i, \quad 1 \leq i \leq p, \tag{6.1}$$

where x is an element of $\mathbf{R}^q$, v_i, $1 \leq i \leq p$, the natural basis of $\mathbf{R}^p$ and c_i, $1 \leq i \leq p$, are suitable vectors in $\mathbf{R}^q$.

A priori some of these vectors could be zero, but this is an uninteresting case since we have

Proposition 6.1 *Suppose* $c_i \neq 0$ *for* $i \leq r$, *and* $c_{r+1} = \ldots = c_p = 0$, *for some* $r < p$; *then the metric* $g_{A,C,h}$ *is reducible.*

Proof. From (4.1) we get

$$D_\zeta \omega^i = -\langle c_i, \theta \rangle \omega^i(\zeta)$$

for all $i = 1, \ldots, p$. Therefore, $D_\zeta \omega^i = 0$ if $r + 1 \leq i$. This means that the distribution D defined by $\omega^i = 0$, $r + s \leq i$, is *parallel* and the metric is reducible by the de Rham Theorem.

Note that the induced metrics on the integral manifolds of the complementary distribution $\mathcal{D}^\perp$ are flat. Thus, the manifold has a $(p - r)$-dimensional Euclidean factor, if $c_i = 0$, for $r + 1 \leq i$.

Even if all the vectors c_i are non-zero, the metric can be reducible. In fact, we have

Proposition 6.2 *Let* $\mathcal{A}$ *be the vector subspace of* $\mathbf{R}^q$ *spanned by all the vectors of the form* $A(w_1) \circ \cdots \circ A(w_m)c_i$, *for* $1 \leq i \leq p, w_1, \ldots, w_m \in \mathbf{R}^p$, $m \geq 0$. *If* $\mathcal{A}$ *is a proper subspace, then the metric is reducible.*

Proof. Let $\mathcal{A}^\perp$ be the orthogonal complement of $\mathcal{A}$ in $\mathbf{R}^q$, and

$$\mathcal{D}^\perp = \{\xi : \omega(\xi) = 0, \ \ \theta(\xi) \in \mathcal{A}^\perp\}.$$

Then $\mathcal{D}^\perp$ is a *parallel* distribution. In fact, for each vector field ζ, we have from (4.1)

$$\begin{aligned} \omega(D_\zeta \xi) &= -(D_\zeta \omega)(\xi) + \zeta(\omega(\xi)) = \\ &= C(\theta(\xi))(\zeta) = 0 \end{aligned}$$

since $\omega(\xi) = 0$ and $\theta(\xi)$ is orthogonal to all vectors c_i (i.e., $\theta(\xi)$ belongs to the kernel of C). Moreover, from (4.2), we have

$$\begin{aligned} \theta(D_\zeta \xi) &= -(D_\zeta \theta)(\xi) + \zeta(\theta(\xi)) = \\ &= A(dw(\zeta))\theta(\xi) + \zeta(\theta(\xi)), \end{aligned}$$

because $\omega(\xi) = 0$. Note that $\zeta(\theta(\xi))$ is an element of $\mathcal{A}^\perp$. In fact,

$$\langle \theta(\xi), A(w_1) \circ \cdots \circ A(w_m)c_i \rangle = 0$$

for all $i, m \geq 0$ and for all (constant) vectors $w_1, \ldots, w_m$ in $\mathbf{R}^p$. By differentiating with respect to ζ, we obtain that $\zeta(\theta(\xi))$ also belongs to $\mathcal{A}^\perp$. Moreover,

$$\begin{aligned} &\langle A(dw(\zeta))\theta(\xi), A(w_1) \circ \cdots \circ A(w_m)c_i \rangle = \\ &= -\langle \theta(\xi), A(dw(\zeta))A(w_1) \circ \cdots \circ A(w_m)c_i \rangle = 0 \end{aligned}$$

for $\theta(\xi)$ is orthogonal to $\mathcal{A}$. Therefore, $A(dw(\zeta))\theta(\xi)$ is an element of $\mathcal{A}^\perp$ and the distribution $\mathcal{D}^\perp$ is parallel as claimed.

The subspace $\mathcal{A}$ of $\mathbf{R}^q$ plays a fundamental role in studying the reducibility or irreducibility of the metrics $g_{A,C,h}$. Recall that, from the compatibility condition (1.12), we have

$$A(v_i)c_j = 0$$

if $i \neq j$. Moreover, the operators $A(v_i), 1 \leq i \leq p$, commute. It follows that $\mathcal{A}$ is given by

$$\mathcal{A} = \mathrm{Span}\{A(v_i)^m c_i, \quad 1 \leq i \leq p, m \geq 0\}. \tag{6.2}$$

Note that the condition $\mathcal{A} = \mathbf{R}^q$ is necessary in order to get the irreducibility of $g_{A,C,h}$, but not sufficient. In fact we have

Proposition 6.3 *Suppose $\mathcal{A} = \mathbf{R}^q$ and that there exists a partition of the set $C = \{c_1, \ldots, c_p\}$ in two mutually orthogonal subsets. Then $g_{A,C,h}$ is reducible.*

Proof. Suppose that

$$\langle c_i, c_j \rangle = 0$$

for $1 \leq i \leq r$, $r+1 \leq j \leq p$. Let $\mathcal{A}_r$ be the subspace of $\mathcal{A}$ spanned by the vectors $A(v_i)^m c_i$, $1 \leq i \leq r, m \geq 0$. Note that its orthogonal complement $\mathcal{A}_r^{\perp}$ is spanned by $A(v_j)^{m'} c_j$, $r+1 \leq j$, because $A(v_i)^m c_i$ and $A(v_j)^{m'} c_j$ are orthogonal if $i \neq j$.

Let $\mathcal{D}_r$ be the distribution given by

$$\mathcal{D}_r = \{\xi \ : \ \omega^j(\xi) = 0, \ j \geq r+1, \ \theta(\xi) \in \mathcal{A}_r\} \ .$$

Then $\mathcal{D}_r$ is *parallel*. In fact, for each $j \geq r+1$ we have

$$\begin{aligned} \omega^j(D_\zeta \xi) &= -(D_\zeta \omega^j)(\xi) + \zeta(\omega^j(\xi)) = \\ &= \langle C(\theta(\xi))\omega(\zeta), v_j \rangle = \langle C(\theta(\xi))v_j, \omega(\zeta) \rangle = \\ &= \langle c_j, \theta(\xi) \rangle \langle v_j, \omega(\zeta) \rangle = \langle c_j, \theta(\xi) \rangle \omega^j(\zeta), \end{aligned}$$

and

$$\begin{aligned} \langle \theta(D_\zeta \xi), A(v_j)^m c_j \rangle &= -\langle (D_\zeta \theta)(\xi), A(v_j)^m c_j \rangle + \\ &+ \langle \zeta(\theta(\xi)), A(v_j)^m c_j \rangle = \zeta \langle \theta(\xi), A(v_j)^m c_j \rangle + \\ - \textstyle\sum_\alpha \langle C(u_\alpha)\omega(\zeta), \omega(\xi) \rangle \langle u_\alpha, A(v_j)^m c_j \rangle &+ \langle A(dw(\zeta))\theta(\xi), A(v_i)^m c_j \rangle = \\ = - \textstyle\sum_{i \leq r} \langle C(A(v_j)^m c_j)\omega(\zeta), v_i \rangle \omega^i(\xi) &- \langle \theta(\xi), A(dw(\zeta))A(v_j)^m c_j \rangle = \\ &= - \textstyle\sum_{i \leq r} \sum_h \langle C(A(v_j)^m c_j)v_h, v_i \rangle \omega^h(\zeta)\omega^i(\xi), \end{aligned}$$

because $A(dw(\zeta))A(v_j)^m c_j$ belongs to $\mathcal{A}_r^{\perp}$. On the other hand,

$$C(A(v_j)^m c_j)v_h = \langle c_h, A(v_j)^m c_j \rangle.$$

It follows that $\langle \theta(D_\zeta \xi), A(v_j)^m c_j \rangle = 0$ for each $j \geq r+1$, and $\mathcal{D}_r$ is parallel as claimed.

Therefore, if the metric $g_{A,C,h}$ is *irreducible* the following conditions must be satisfied:

i) $c_i \neq 0$, $1 \leq i \leq p$;

ii) $\mathbf{R}^q = \mathcal{A} = \mathrm{Span}\{A(v_i)^m c_i, 1 \le i \le p,\ m \ge 0\}$;

iii) there is no partition of $\mathcal{C}$ in mutually orthogonal subsets.

These necessary conditions are also *sufficient* if the diffeomorphism h of $\mathbf{R}^p$ reduces to the identity, i.e., for the metric $g_{A,C}$. In general we have

Theorem 6.4 *If* i), ii) *and* iii) *are satisfied and*

iv) $\langle J(w)^{-1}v_i, v_i\rangle \neq 0$

for all i, $1 \le i \le p$, *then* $g_{A,C,h}$ *is irreducible.*

The proof of this theorem requires the study of the *holonomy* and it will be postponed to the following section.

Remark 6.1 Condition iv) is automatically satisfied in the case of the metrics $g_{A,C,h} = g_{\alpha,\gamma,h}$, if all the parameters γ_i are different from zero, because $J(w)$ must be diagonal with non-zero entries. For all these examples the vectors c_i are different from zero, and there are no partitions of $\mathcal{C} = (c_i, \ldots, c_p)$ in mutually orthogonal subsets, because we have $\langle c_i, c_j\rangle \neq 0$, for all i and j. Moreover, if also all the parameters α_i are different from zero, then condition ii) holds as well. We conclude that, for any diffeomorphism $h \in Diff(\mathbf{R}^p)$ of type $h(w) = (h^1(w^1), \ldots, h^p(w^p))$, the metric $g_{\alpha,\gamma,h}$ is *irreducible* if

$$\alpha_i \neq 0 \tag{6.3}$$

and

$$\gamma_i \neq 0. \tag{6.4}$$

We finish this section by proving the following

Theorem 6.5 *Suppose* $C(x)$ *and* $J(w)$ *diagonal. Denote by* $a^i(w_i)$ *the (i,i)-entry of* $J(w)$ *(it depends only on* w^i*). If there exist two positive constants* a^i *and* b^i *such that*

$$a^i \le a^i(w^i) \le b^i,$$

then the metric $g_{A,C,h}$ *is complete.*

Proof. We follow the same method as in [KTV, Section 6c]. Let ϕ be the diffeomorphism of $\mathbf{R}^{p+q}$ given by

$$\phi(w, x) = (w, e^{-A(w)}x). \tag{6.5}$$

Then

$$\begin{aligned} \phi^*\theta &= e^{-A(w)}dx, \\ \phi^*\omega &= e^{C(A(w)x)}J(w)dw. \end{aligned} \tag{6.6}$$

So, we have

$$\phi^*g = {}^t dw {}^tJ(w)e^{2C(e^{-A(w)}x)}J(w) \otimes dw + {}^t dx \otimes dx. \tag{6.7}$$

The symmetric operator $e^{2C(e^{-A(w)}x)}$ is diagonal, because $C(x)$ is diagonal, for all x. If $C(x)v_i = \langle c_i, x\rangle$, then their diagonal entries are

$$e^{2\langle c_i, e^{-A(w)}x\rangle} = e^{2\langle e^{-A(w)}c_i, x\rangle} = e^{2\langle e^{w^i A(v_i)}c_i, x\rangle}.$$

Therefore, the (i, i)-entry depends only on w^i. It follows that

$$\phi^* g = \sum_i a^i(w^i)^2 e^{2\langle (e^{w^i A(v_i)}c_i, x\rangle} dw^i \otimes dw^i + \sum_\alpha dx^\alpha \otimes dx^\alpha. \tag{6.8}$$

On the other hand,

$$|\langle e^{w^i A(v_i)}c_i, x\rangle| \leq \| e^{w^i A(v_i)}c_i \| \| x \| = \| c_i \| \| x \|, \tag{6.9}$$

since $e^{w^i A(v_i)}$ is orthogonal. Hence,

$$(a^i)^2 e^{-2\|c_i\|\,\|x\|} \leq a^i(w^i)^2 e^{2\langle (e^{w^i A(v_i)}c_i, x\rangle} \leq (b_i)^2 e^{2\|c_i\|\,\|x\|}. \tag{6.10}$$

Let g_1 be the metric of $\mathbf{R} \times \mathbf{R}^q$ defined by

$$g_1 = a^1(w^1)^2 e^{2\langle (e^{w^1 A(v_1)}c_1, x\rangle} dw^1 \otimes dw^1 + \sum_\alpha dx^\alpha \otimes dx^\alpha. \tag{6.11}$$

It is a *generalized warped product metric* of two Euclidean metrics. By Proposition 2.1 of [SK] it is complete, since (6.10) holds. Then, a simple inductive argument shows that $g_{A,C,h}$ is complete.

7. Holonomy and irreducibility (continued)

In order to prove the irreducibility of the metric $g_{A,C,h}$ in the hypotheses of Theorem 6.4 it is enough to show that the *infinitesimal holonomy Lie algebra* **hol**(a) is acting irreducibly on T_aM (see [KN], vol. I).

This algebra is the Lie subalgebra of $\mathbf{so}(T_aM)$ spanned by the operators

$$(R_a)_{xy} : z \in T_aM \longmapsto (R_a)_{xy}z \in T_aM$$

and

$$(D^m R_a)_{x_1 \dots, x_m} : z \in T_aM \longmapsto (D^m R_a)_{x_1 \dots, x_m} z \in T_aM,$$

where $x_1, \dots, x_m$ are varying in T_aM. A priori, we have to compute all these operators. Actually, this will not be necessary, and the aim of the present section is to prove that already the operators of **hol**(a), provided by Lemmas 7.1, 7.2 and 7.3 span a Lie algebra which is acting irreducibly on T_aM.

We suppose, as in the previous section, that all the operators $C(x)$ are in diagonal form. Then, from (2.15) and (2.16) we get at once that the non-zero components of R are

$$R_{ijij} = -\langle c_i, c_j\rangle \tag{7.1}$$

and

$$R_{i\alpha i\beta} = -\langle c_i, u_\alpha \rangle \langle c_i, u_\beta \rangle. \tag{7.2}$$

Let $E_{AB}, 1 \leq A, B, \leq p+q$, the skew-symmetric operators defined by

$$E_{AB}(E_A) = E_B \ , \ E_{AB}(E_C) = 0 \ , \ C \neq A, B. \tag{7.3}$$

Here $E_i, 1 \leq i \leq p$, $E_{p+\alpha}, 1 \leq \alpha \leq q$ denote the dual vector fields of the one-forms ω^i, θ^i. Then we have

$$\begin{aligned} R_{E_i E_j} &= -\langle c_i, c_j \rangle E_{ij}, & (7.4)\\ R_{E_i E_{p+\alpha}} &= -\langle c_i, u_\alpha \rangle \sum_\beta \langle c_i, u_\beta \rangle E_{i\,p+\beta}. & (7.5) \end{aligned}$$

Since each c_i is non-zero (otherwise the metric is reducible as proved in Section 6), there exists at least one index α, such that $\langle c_i, u_\alpha \rangle \neq 0$. Therefore, we have

Lemma 7.1 *All the operators $\sum_\beta \langle c_i, u_\beta \rangle E_{i\,p+\beta}$, evaluated at a, belong to* $\mathbf{hol}(a)$.

The covariant derivatives of the vector fields $E_i, E_{i\,p+\alpha}$ are given by

$$D_\zeta E_i = \sum_\beta \varphi_i^\beta(\zeta) E_{p+\beta}$$

and by

$$D_\zeta E_{p+\alpha} = \sum_i \varphi_\alpha^i(\zeta) E_i + \sum_\beta \psi_\alpha^\beta(\zeta) E_{p+\beta}.$$

The connection one-forms φ_i^α and ψ_α^β are provided by (2.7) and (2.6), respectively. It follows that $D_\zeta E_i$ is zero, except when $\zeta = E_i$. Then we have

$$D_{E_i} E_i = -\sum_\beta \langle c_i, u_\beta \rangle E_{p+\beta}. \tag{7.6}$$

On the other hand, we have

$$D_{E_i} E_{p+\alpha} = \langle c_i, u_\alpha \rangle E_i + \sum_\beta \langle A(dw(E_i)) u_\alpha, u_\beta \rangle E_{p+\beta}.$$

Moreover,

$$\begin{aligned} A(dw(E_i)) &= A(e^{-C(x)} J(w)^{-1} \omega(E_i)) = A(e^{-C(x)} J(w)^{-1} v_i) = \\ &= A(J(w)^{-1} e^{-C(x)} v_i) = e^{-\langle c_i, x \rangle} A(J(w)^{-1} v_i) = \\ &= e^{-\langle c_i, x \rangle} \textstyle\sum_j \langle J(w)^{-1} v_i, v_j \rangle A(v_j). \end{aligned}$$

It follows that

$$D_{E_i} E_{p+\alpha} = \langle c_i, u_\alpha \rangle E_i + e^{-\langle c_i, x \rangle} \sum_j \langle J(w)^{-1} v_i, v_j \rangle A(v_j). \tag{7.7}$$

By taking these formulae into account, we get

Lemma 7.2 *If* $\langle J(w)^{-1}v_i, v_i\rangle \neq 0$, *then the operators*

$$\sum_\beta \langle A(v_i)c_i, u_\beta\rangle E_{i\ p+\beta},$$

evaluated at a, *belong to* $\mathbf{hol}(a)$.

Proof. First of all we have

$$(D_{E_i}R)_{E_i\ E_{p+\alpha}} = [D_{E_i}, R_{E_i\ E_{p+\alpha}}] - R_{E_i\ D_{E_i}E_{p+\alpha}}.$$

It follows that $[D_{E_i}, R_{E_i\ E_{p+\alpha}}]|_a$ belongs to $\mathbf{hol}(a)$. From (7.5) we get that

$$[D_{E_i}, \sum_\beta \langle c_i, u_\beta\rangle E_{i\, p+\beta}]|_a$$

is an element of $\mathbf{hol}(a)$ too. Then

$$\begin{aligned}[D_{E_i}, \textstyle\sum_\beta \langle c_i, u_\beta\rangle E_{i\, p+\beta}]|_a(E_i) = \textstyle\sum_\beta \langle c_i, u_\beta\rangle D_{E_i} E_{i\, p+\beta}(E_i) + \\ - \textstyle\sum_\beta \langle c_i, u_\beta\rangle E_{i\, p+\beta}(D_{E_i}E_i) = -e^{-\langle c_i, x\rangle}\langle J(w)^{-1}v_i, v_i\rangle \textstyle\sum_\beta \langle A(v_i)c_i, u_\beta\rangle E_\beta\end{aligned}$$

because $A(v_i)c_j = 0$ if $i \neq j$. Hence,

$$\begin{aligned}&[D_{E_i}, \textstyle\sum_\beta \langle c_i, u_\beta\rangle E_{i\, p+\beta}] = \\ &= -e^{-\langle c_i, x\rangle}\langle J(w)^{-1}v_i, v_i\rangle \textstyle\sum_\beta \langle A(v_i)c_i, u_\beta\rangle E_{i\,\beta}\end{aligned}$$

and the Lemma is proved.

Under the same hypothesis we have

Lemma 7.3 *For all* $m \geq 1$, *the operators*

$$\sum_\beta \langle A(v_i)^m c_i, u_\beta\rangle E_{i\ p+\beta},$$

evaluated at a, *are elements of* $\mathbf{hol}(a)$.

Proof. We proceed by induction on m. If $m = 1$, the lemma reduces to the previous one. Suppose that it is true for all integers $l \leq m-1$. A direct computation, as in Lemma 7.2, shows that

$$\begin{aligned}&(D^m_{E_i \ldots, E_i}R)_{E_i E_\alpha} = \langle c_i, u_\alpha\rangle e^{-m\langle c_i, x\rangle}\langle J(w)^{-1}v_i, v_i\rangle^m \\ &\sum_\beta \langle A(v_i)^m c_i, u_\beta\rangle E_{i\,\beta} + \ldots,\end{aligned}$$

where the omitted terms, evaluated at a, belong to $\mathbf{hol}(a)$ by the inductive hypothesis. Since $\langle c_i, u_\alpha\rangle \neq 0$ by hypothesis, the lemma is proved.

Now, we are able to prove the following

Theorem 7.4 *Let* $\mathbf{h}$ *be the Lie subalgebra of* $\mathbf{hol}(a)$ *spanned by all the operators of Lemma 7.1 and Lemma 7.3, evaluated at* a. *If the conditions of Theorem 6.4 are satisfied, then* $\mathbf{h}$ *acts irreducibly on* T_aM.

Proof. The Lie algebra $\mathbf{h}$ is a subalgebra of $\mathbf{so}(T_aM)$. Therefore, it is enough to prove that each bilinear form φ, which is invariant with respect to $\mathbf{h}$, is a multiple of the scalar product g_a induced by the metric.

Since all the involved operators have constant coefficients, it is sufficient to prove that φ is a multiple of g_a at the origin of $\mathbf{R}^p \times \mathbf{R}^q$. At this point $E_{i|_a} = v_i$ and $E_{p+\alpha|_a} = u_\alpha$. Moreover, g_a reduces to the standard inner product $\langle , \rangle$.

If φ is invariant with respect to $\mathbf{h}$, we have

$$\sum_\beta \langle A(v_h)^m c_h, u_\beta\rangle(\varphi(E_{h\,p+\beta}v_i, v_j) + \varphi(v_i, E_{h\,p+\beta}v_j)) = 0, \tag{7.8}$$

for all $m \geq 0, h, i, j = 1, \ldots, p$. For $h = i \neq j$ this relation reduces to

$$\sum_\beta \langle A(v_i)^m c_i, u_\beta\rangle \varphi(u_\beta, v_j) = 0,\ j \neq i.$$

By choosing $h = i = j$, we also get

$$\sum_\beta \langle A(v_i)^m c_i, u_\beta\rangle \varphi(u_\beta, v_i) = 0.$$

On the other hand, $\mathbf{R}^q$ is spanned by the vectors $A(v_i)^m c_i$, by hypothesis. It follows that

$$\varphi(u_\beta, v_i) = 0 \tag{7.9}$$

for all $\beta = 1, \ldots, q$ and $i = 1, \ldots, p$. Again, because of the invariance of φ, we have

$$\sum_\beta \langle A(v_h)^m c_h, u_\beta\rangle(\varphi(E_{h\,p+\beta}v_i, u_\alpha) + \varphi(v_i, E_{h\,p+\beta}u_\alpha)) = 0, \tag{7.10}$$

for all m, h, i, α. If we choose $h \neq i$ we get

$$\langle A(v_i)^m c_h, u_\alpha\rangle \varphi(v_i, v_h) = 0. \tag{7.11}$$

By choosing $h = 1$, we obtain

$$\sum_\beta \langle A(v_i)^m c_h, u_\beta\rangle \varphi(u_\beta, v_\beta) - \langle A(v_i)^m c_h, u_\alpha\rangle \varphi(v_i, v_i) = 0. \tag{7.12}$$

Recall that also the operators $R_{E_hE_k}$, evaluated at a, belong to $\mathbf{hol}(a)$. Thus from (7.4), it follows that

$$\langle c_h, c_k\rangle(\varphi(E_{h\,k}(v_i), v_j) + \varphi(v_i, E_{hk}(v_j)) = 0.$$

Therefore, if $k = i \neq h \neq j$, we get

$$\langle c_h, c_i\rangle \varphi(v_h, v_j) = 0 \tag{7.13}$$

and if $k = i \neq h = j$, then

$$\langle c_j, c_i\rangle(\varphi(v_i, v_i) - \varphi(v_j, v_j)) = 0. \tag{7.14}$$

Lemma 7.5 *We have* $\varphi(v_i, v_j) = 0$, *if* $i \neq j$, *and*

$$\varphi(v_i, v_i) = \lambda$$

for all $i = 1, \ldots, p$.

Proof. First of all, remark that there exists a vector c_j such that $\langle c_i, c_j \rangle \neq 0$, otherwise the condition of Theorem 6.4 is not satisfied. It is not restrictive to suppose $j = 2$. Then, from (7.13) we get

$$\langle c_1, c_2 \rangle \varphi(v_1, v_j) = 0$$

for all $j \neq 1$, and

$$\langle c_2, c_1 \rangle \varphi(v_2, v_j) = 0$$

for all $j \neq 2$. Therefore $\varphi(v_2, v_j) = 0$, and

$$\varphi(v_1, v_j) = \varphi(v_2, v_j) = 0$$

for all $j \neq 1, 2$. For the same reasons, there exists a third vector c_k such that $\langle c_1, c_k \rangle \neq 0$ or $\langle c_2, c_k \rangle \neq 0$. Suppose that $k = 3$. Then we have

$$\varphi(v_3, v_j) = 0$$

for all $j \neq 3$. By proceeding in the same way we get

$$\varphi(v_i, v_j) = 0$$

if $i \neq j$. The same arguments applied to (7.14) give

$$\varphi(v_i, v_i) = \varphi(v_j, v_j)$$

for all i and j, as claimed.

From (7.12), and the previous Lemma , we deduce that

$$\sum_{\beta} \langle A(v_i)^m c_i, u_\beta \rangle \varphi(u_\beta, u_\alpha) = \langle A(v_i)^m c_i, u_\alpha \rangle \lambda.$$

Thus,

$$\varphi(A(v_i)^m c_i, u_\alpha) = \lambda \langle A(v_i)^m c_i, u_\alpha \rangle$$

for all i and α. Since $\mathbf{R}^q$ is spanned by the vectors $A(v_i)^m c_i$, this means that φ and $\langle , \rangle$ are proportional, as claimed.

This achieves the proof of Theorem 7.4 and Theorem 6.4.

8. Isometries between metrics with weakly generic Ricci curvature

The purpose of this section is to study the isometry classes of the metrics $g_{A,C,h}$ when their Ricci curvature is "generic" in the sense specified below.

Recall that the *Ricci curvature* of $g_{A,C,h}$ is given by

$$r = -\sum_i trC(u_\alpha)\langle C(u_\alpha)\omega, \omega\rangle - trC(\theta)^2. \tag{8.1}$$

Therefore, if $C(x)$ is in diagonal form, we have

$$r = -\sum_i \langle c_i, \sum_m c_m\rangle \omega^i \otimes \omega^i - \sum_i \sum_{\alpha\beta} \langle c_i, u_\alpha\rangle\langle c_i, u_\beta\rangle \theta^\alpha \otimes \theta^\beta.$$

We can always suppose, up to a change of orthonormal basis in $\mathbf{R}^q$, that

$$\mathrm{Span}\{(c_1, \dots, c_p)\} = \mathrm{Span}\{(u_1, \dots, u_k\} \tag{8.2}$$

for some $k \leq q$. Hence,

$$r = -\sum_i \langle c_i, \sum_m c_m\rangle \omega^i \otimes \omega^i - \sum_{\alpha,\beta=1}^{k} \langle c_i, u_\alpha\rangle\langle c_i, u_\beta\rangle \theta^\alpha \otimes \theta^\beta. \tag{8.3}$$

It follows that we can put r in diagonal form just by performing an orthonormal change of basis in the space $\mathrm{Span}\{(c_1, \dots, c_p)\}$. After this change, we get

$$r = +\sum_i \lambda_i \omega^i \otimes \omega^i + \sum_{\alpha=1}^{k} \lambda_\alpha \tilde{\theta}^\alpha \otimes \tilde{\theta}^\alpha, \tag{8.4}$$

where

$$\tilde{\theta}^\alpha = \sum_{\beta=1}^{k} A^\alpha_\beta \theta^\beta \tag{8.5}$$

and (A^α_β) is a $k \times k$ orthogonal matrix. Note that the diagonal entries of r are given by

$$\lambda_i = \langle c_i, \sum_m c_m\rangle, \tag{8.6}$$

λ_α, $1 \leq \alpha \leq k$, and 0 with multiplicity $q-k$. This means that the *Ricci principal curvatures* are the scalars $\lambda_i (1 \leq i \leq p)$, $\lambda_\alpha (1 \leq \alpha \leq q)$, and $\lambda_0 = 0$ with multiplicity (at least) equal to $q-k$. Remark also that $q-k$ is just the *index* of *nullity* of the metric (see Section 2). We say that the tensor r is (*weakly*)*generic*, if the eigenvalues λ_i are non-zero, different from λ_α, and distinct, i.e., if

$$\lambda_i \neq 0, \lambda_i \neq \lambda_\alpha, \lambda_i \neq \lambda_j$$

when $i \neq j$. Of course, some of the eigenvalues λ_α could be zero or coincide.

Remark 8.1 The Ricci curvature of the metric $g_{A,C,h}$ of the example (5.4) is *weakly generic* if $\gamma_i \neq c_j$ for $i \neq j$, and the eigenvalues of the matrix

$$\begin{pmatrix} \gamma_1^2 & 0 & 0 & \dots & 0 & 0 & \gamma_1 \\ 0 & 0 & 0 & \dots & 0 & 0 & 0 \\ 0 & 0 & \gamma_2^2 & \dots & 0 & 0 & \gamma_2 \\ \vdots & \vdots & \vdots & \ddots & \vdots & \vdots & \vdots \\ \vdots & \vdots & \vdots & & \vdots & \vdots & \vdots \\ 0 & 0 & 0 & \dots & \gamma_r^2 & 0 & \gamma_r \\ 0 & 0 & 0 & \dots & 0 & 0 & 0 \\ \gamma_1 & 0 & \gamma_2 & \dots & \gamma_r & 0 & r+1 \end{pmatrix}$$

are different from $\lambda_i = \gamma_i^2 + (r+1)$. Generically, this is always satisfied.

Let $g = g_{A,C,h}$ and $g' = g_{A',C,h'}$ be two metrics with the *same* C, and therefore with the *same curvature*. Recall that these metrics are given by

$$g_{A,C,h} = {}^t\omega \otimes \omega + {}^t\theta \otimes \theta,$$
$$g_{A',C,h'} = {}^t\omega' \otimes \omega' + {}^t\theta' \otimes \theta'$$

where

$$\omega = e^{C(x)}dh \quad , \quad \theta = dx + A(dw)x,$$
$$(8.7) \qquad \omega' = e^{C(x)}dh' \quad , \quad \theta' = dx + A'(dw)x.$$

Moreover,

$$(8.8) \qquad dh = J(w)dw \,,\; dh' = J'(w)dw$$

and the Jacobian matrices $J(w)$ and $J'(w)$ commute with the operators $C(x)$.

We are now looking for the conditions to be satisfied by A, A', h and h' in order that there exists an *isometry* f between $g_{A,C,h}$ and $g_{A',C,h'}$, when their Ricci curvatures are *weakly generic*. Suppose that

$$f^*g = g' \;.$$

Then

$$f^*\omega' = L\omega + M\theta,$$
$$f^*\theta' = N\omega + P\theta,$$

where

$$\begin{pmatrix} L & M \\ N & P \end{pmatrix}$$

is an $O(p+q)$-valued function. Since f is an isometry, it preserves the Ricci curvatures, which are weakly generic. It follows from (8.4) and $\lambda_i \neq \lambda_j$, that M and N must be zero, and that L is a $p \times p$ *diagonal matrix*. The non-zero entries are $+1$ or -1, so it is *constant*. Thus, we get

$$f^*\omega' = L\omega, \tag{8.9}$$

where $L \in O(p)$ is diagonal, and

$$f^*\theta' = P\theta, \tag{8.10}$$

where P is an $O(q)$-valued function. Actually, P is *constant*. In order to prove this, recall that f is an *affine transformation* with respect to the Levi Civita connections of the metrics, i.e., for each one- form ϕ, it satisfies

$$D(f^*\phi) = f^*(D'\phi). \tag{8.11}$$

From (8.9) and (8.11) we get

$$f^*(D'\omega') = D(f^*\omega') = LD\omega, \tag{8.12}$$

since L is constant. The formulae (4.1) and (4.2) may be written as

$$\begin{aligned} {}^tD\omega &= -{}^t\omega \otimes C(\theta), \\ D\theta &= \textstyle\sum_\alpha \langle C(u_\alpha)\omega, \omega\rangle u_\alpha - A(dw) \otimes \theta. \end{aligned} \tag{8.13}$$

Of course, these formulae apply, mutatis mutandis, to both the metrics g and g'. By substituting in (8.12) we get

$$\begin{aligned} {}^t\omega \otimes C(\theta)^tL &= {}^tf^*\omega' \otimes C(f^*\theta') = \\ &= {}^t\omega^tL \otimes C(P\theta) = {}^t\omega \otimes {}^tLC(P\theta). \end{aligned}$$

It follows that

$${}^tLC(P\theta) = C(\theta)^tL.$$

Note that L commutes with all the operators $C(x)$ because all these operators are in diagonal form. Therefore, we get

$$C(P\theta) = C(\theta).$$

This can be written as

$$C(Px) = C(x) \tag{8.14}$$

for all $x \in \mathbf{R}^q$. On the other hand

$$C(x)v_i = \langle c_i, x\rangle$$

for $i = 1, \ldots, p$. Hence (8.14) is equivalent to

$$\langle c_i, x\rangle = \langle c_i, Px\rangle = \langle {}^tPc_i, x\rangle.$$

Therefore, we have proved the following Lemma

Lemma 8.1 *All the vectors c_i are invariant with respect to P, i.e., $Pc_i = c_i$, for $i \leq 1 \leq p$.*

Again from (8.11) and (8.10), we get

$$f^*(D'\theta') = D(f^*\theta') = D(P\theta), \tag{8.15}$$

but this time we have

$$D(P\theta) = DP \otimes \theta + PD\theta. \tag{8.16}$$

Therefore, from (8.13), we obtain

$$\begin{aligned} &\textstyle\sum_\alpha \langle C(u_\alpha L\omega, L\omega\rangle u_\alpha - A'(d(w \circ f)) \otimes P\theta = \\ &= dP \otimes \theta + \textstyle\sum_\alpha \langle C(u_\alpha \omega, \omega\rangle P u_\alpha - PA(dw) \otimes \theta. \end{aligned} \tag{8.17}$$

Note that

$$\sum_\alpha \langle C(u_\alpha)L\omega, L\omega\rangle = \sum_\alpha \langle C(u_\alpha)\omega, \omega\rangle$$

since L commutes with $C(u_\alpha)$ and is orthogonal. It follows at once that (8.17) is equivalent to the system

$$\begin{cases} \sum_\alpha \langle C(u_\alpha)\omega, \omega\rangle u_\alpha = \sum_\alpha \langle C(u_\alpha)\omega, \omega\rangle P u_\alpha, \\ dP - PA(dw) + A'(d(w \circ f))P = 0. \end{cases}$$

It is not difficult to prove that the first equation is equivalent to (8.14). Thus, we get only one new independent condition, summarized in the following Lemma.

Lemma 8.2 *The $O(q)$-valued function P satisfies the following differential equation*

$$dP - PA(dw) + A'(d(w \circ f))P = 0. \tag{8.18}$$

We are now ready to write down explicitely the differential equations coming from (8.9), (8.10), (8.7) and (8.8). First of all we have

$$f^*\omega' = e^{C(x \circ f)} d(h' \circ f) = Le^{C(x)} dh. \tag{8.19}$$

Then,

$$f^*\theta' = d(x \circ f) + A'(d(w \circ f))(x \circ f) = Pdx + PA(dw)x. \tag{8.20}$$

By taking into account the previous Lemma, this last equation can be rewritten as

$$\begin{aligned} d(x \circ f) - dP\,{}^tP(x \circ f) + PA(dw)^t P(x \circ f) \\ = Pdx + PA(dw)x. \end{aligned}$$

From this we get

$$\begin{aligned} d((x \circ f) - Px) = dP^t P((x \circ f) - Px) + \\ -PA(dw)^t P((x \circ f) - Px). \end{aligned}$$

This suggests we define a new function F by

$$F = x \circ f - Px \tag{8.21}$$

and rewrite the previous equation as

$$dF = dP\ {}^tPF - PA(dw)\ {}^tPF. \tag{8.22}$$

The equation (8.19), because of (8.8) and (8.14), is equivalent to

$$e^{C(F)}J'(w \circ f)d(w \circ f) = LJ(w)dw. \tag{8.23}$$

This equation contains an important information, namely *the function* $w \circ f$ *does not depend on* x. In fact, the second term contains only dw and it does not involve dx. It follows at once from (8.18) that *also* P *depends only on* w. Therefore, the second term of (8.22) involves only dw. This implies that *also* F *depends only on* w. Thus, we have

$$\begin{aligned} x \circ f &= P(w)x + F(w), \\ w \circ f &= G(w), \end{aligned} \tag{8.24}$$

for some function $F(w), G(w), P(w)$ satisfying (8.22), (8.23) and (8.18). Of course, this can be rewritten as

$$f(w, x) = (G(w), P(w)x + F(w)). \tag{8.25}$$

Now, we can prove our claim, namely

Lemma 8.3 *If the hypotheses of Theorem 6.4 are satisfied by* $C(x), A(w), A'(w), h(w)$ *and* $h'(w)$, *then the* $O(q)$*-valued function* $P(w)$ *is constant.*

Proof. From Lemma 8.1 we obtain for all i

$$Pc_i = c_i.$$

Therefore,

$$dPc_i = c_i.$$

Since P satisfies (8.18), we get

$$PA(dw)c_i + A'(dG)Pc_i = 0.$$

On the other hand, by the very definition of dw, we have $dw(v_i) = v_i$. It follows that

$$PA(v_i)c_i + A'(dG(v_i))c_i = 0.$$

Hence,

$$PA(v_i)c_i = \langle dG(v_i), v_i\rangle A'(v_i)c_i.$$

Recall that P is orthogonal and therefore

$$\| A(v_i)c_i \|^2 = \langle dG(v_i), v_i\rangle^2 \| A'(v_i)c_i \|^2 .$$

This implies that the function $\langle dG(v_i), v_i\rangle$ is constant. Thus, we have proved that

$$PA(v_i)c_i = \mu_i A'(v_i)c_i \tag{8.26}$$

for some constant μ_i. By an easy inductive argument, it follows that

$$PA(v_i)^m c_i = \mu_i^m A'(v_i)^m c_i \tag{8.27}$$

for any integer m. By hypothesis, $\mathbf{R}^q$ is spanned by the vectors $A(v_i)^m c_i$. This yields that $P(w)$ is constant, since it is constant for each generator of $\mathbf{R}^q$.

We summarize all the previous results in the following

Theorem 8.4 *Let f be an isometry between $g' = g_{A',C,h'}$ and $g = g_{A,C,h}$ with weakly generic Ricci curvature. Suppose that the genericity assumptions of Theorem 6.4, guaranteeing the irreducibility of both metrics, are satisfied. Then*

$$f(w,x) = (G(w), Px + F(w)), \tag{8.28}$$

where P is a constant element of $O(q)$ such that

$$C(Px) = C(x), \;\; x \in \mathbf{R}^q, \tag{8.29}$$

and $F(w)$, $G(w)$ satisfy the following equations:

$$\begin{aligned} dF(w) + PA(dw)\,{}^tPF(w) &= 0, \\ e^{C(F(w))}J'(G(w))dG(w) &= LJ(w)dw, \\ A'(dG(w)) &= PA(dw)\,{}^tP. \end{aligned} \tag{8.30}$$

We conclude this section by studying the differential equations (8.30) in detail.

First of all we have

Lemma 8.5 *The map F of $\mathbf{R}^p$ in $\mathbf{R}^q$ is uniquely determined by its value at 0. In fact, if $F(0) = x_0$, then*

$$F(w) = Pe^{-A(w)}x_0. \tag{8.31}$$

Proof. Put

$$\tilde{F}(w) = {}^tPF(w)\ .$$

Then

$$d\tilde{F} = {}^tPdF = -A(dw)\tilde{F}.$$

Therefore, $\tilde{F}(w) = e^{-A(w)}x_0$ and (8.31) follows at once.

The *diffeomorphism* G of $\mathbf{R}^p$ cannot be determined explicitly. Nevertheless, it exists and it is uniquely determined by its value w_0 at the origin 0 as well.

In order to prove this, rewrite (8.30) as

$$d(h' \circ G) = Le^{-C(F(w))}dh \tag{8.32}$$

where $F(w)$ is given by (8.31). Then, pull back the form $d(h' \circ G)$ via the diffeomorphism h^{-1} of $\mathbf{R}^p$. In this way we get

$$d(h' \circ G \circ h^{-1}) = Le^{-C((F \circ h^{-1})(w))}dw. \tag{8.33}$$

Put

(8.34) $$\tilde{G} = h' \circ G \circ h^{-1}.$$

Then we have

Lemma 8.6 *There exists a unique diffeomorphism* $\tilde{G}$ *of* $\mathbf{R}^p$ *satisfying*

$$d\tilde{G} = Le^{-C((F \circ h^{-1})(w))} dw,$$

where $F(w)$ *is given by (8.31), such that*

$$\tilde{G}(h(0)) = h'(w_0).$$

Proof. It is enough to show that the integrability condition $d^2\tilde{G} = 0$ is satisfied identically. Note that

$$d^2\tilde{G} = -Le^{-C(F(h^{-1}(w)))} C(d(F \circ h^{-1})) \wedge dw.$$

On the other hand,

$$\begin{aligned} h^*(C(d(F \circ h^{-1})) \wedge dw &= C(dF) \wedge dw = \\ &= C(Pe^{-A(w)} A(dw) x_0) \wedge J(w) dw = \\ &= J(w) C(A(dw) e^{A(w)} x_0) \wedge dw \end{aligned}$$

since $C(Px) = C(x)$ and $J(w)$ commutes with all the operators $C(x)$. Moreover, from the compatibility condition, we get at once that

$$C(A(dw)(e^{A(w)} x_0)) \wedge dw = 0$$

identically. Therefore, $d^2\tilde{G}$ vanishes identically and the Lemma follows.

From (8.30) we get at once that

(8.35) $$A'(G(w)) = PA(w)\,{}^tP + A'(w_0).$$

Let H be the diffeomorphism of $\mathbf{R}^p$, keeping the origin fixed and defined by

(8.36) $$H(w) = G(w) - w_0.$$

Then, (8.35) is verified if and only if H can be written as

(8.37) $$H(w) = Q(w + \tilde{H}(w)),$$

where Q is a constant non-singular $p \times p$ matrix such that

(8.38) $$A'(Qw) = PA(w)^tP,$$

and $\tilde{H}(w)$ is a $(Ker\ A)$-valued function. So, we get a more precise result. Namely,

Theorem 8.7 *Under the same hypotheses as in Theorem 8.4 the metrics* $g' = g_{A'C,h'}$ *and* $g = g_{A,C,h}$ *are isometric if and only if there exist*

i) *a* $p \times p$ *diagonal orthogonal matrix* L;

ii) *a* $q \times q$ *orthogonal matrix* P *such that* $C(Px) = C(x)$;

iii) *a non singular* $p \times p$ *matrix* Q *which verifies*

$$A'(w) = P(A(Q^{-1}w))^t P \,, \tag{8.39}$$

$$h'(w) = (\tilde{G} \circ h \circ G^{-1})(w). \tag{8.40}$$

Moreover, the diffeomorphism $\tilde{G}$ *is uniquely determined by Lemma 8.6 and*

$$G(w) = w_0 + Q(w + \tilde{H}(w)), \tag{8.41}$$

where $\tilde{H}$ *is a* $(Ker\ A)$*-valued function.*

9. The isometry classes of the metrics $g_{\alpha,\gamma,h}$

We apply the results of the previous sections to the metrics of the family described in (5.1).

First of all we fix C, i.e., $\gamma = (\gamma_1, \ldots, \gamma_r)$, in such a way that the homogeneous *model metric* $g_0 = g_C = g_\gamma$ has *weakly generic* Ricci curvature. Then, we deform this metric by introducing the operators A depending on r parameters $\alpha = (\alpha_1, \ldots, \alpha_r)$. We get the metrics

$$g_{A,C} = g_{\alpha,\gamma}.$$

If all the parameters α_i are different from zero, these metrics are *irreducible*. Otherwise, if $\alpha_i \neq 0$ only for $i \leq r' < r$, $g_{\alpha,\gamma}$ splits as the product of $g_{\alpha',\gamma}$ where $\alpha' = (\alpha_1, \ldots, \alpha'_r)$, and of the Euclidean metric of $\mathbf{R}^{r-r'}$ (see Section 6). Of course, $g_{\alpha',\gamma}$ is irreducible. We recover the model metric g_0 by choosing $\alpha = (0, \ldots, 0)$. This shows again that g_0 is *reducible*(see also Section 3), and that g is the product of an irreducible metric g'_C and of the Euclidean metric of $\mathbf{R}^r$. Note that r is just the dimension of the kernel of the linear map $x \longrightarrow C(x)$, according to the results of Section 3.

The *isometry classes* of such metrics, supposed to be irreducible, are parametrized by the points of the space

$$(\mathbf{R}^+)^r = \{(\alpha_1, \ldots, \alpha_r) \in \mathbf{R}^r \mid \alpha_i > 0\}.$$

In fact we have

Theorem 9.1 *Two irreducible metrics* $g_{\alpha,\gamma}$ *and* $g_{\alpha',\gamma}$ *are isometric if and only if* $\alpha'_i = \pm\alpha_i$ *for each index* i.

Proof. From Theorem 8.7 we have that $g_{\alpha,\gamma}$ and $g_{\alpha',\gamma}$ are isometric if and only if there exist a diagonal matrix L in $O(p)$, a $q \times q$ orthogonal matrix P such that $C(Px) = C(x)$, and if

$$e^{C(F(w))} dG = L dw. \tag{9.1}$$

Moreover, we must have

$$A'(dG) = PA(dw)\ {}^t P, \tag{9.2}$$

where $F(w)$ is given by

$$F(w) = Pe^{-A(w)}x_0. \tag{9.3}$$

Recall that $p = r+1$, $q = 2r+1$, and

$$C(x)v_i = \langle c_i, x\rangle v_i, \quad i \leq r+1 \tag{9.4}$$

where c_i is given by (5.6) or (5.7), i.e.,

$$c_i \gamma_i u_{2i-1} + u_{2r+1}, \quad i \leq 2r; \quad c_{r+1} = u_{2r+1}. \tag{9.5}$$

On the other hand, the skew-symmetric operators $A(w)$ and $A'(w)$ are defined by

$$\begin{cases} A(w)u_{2i-1} &= \alpha_i \langle w, v_i\rangle u_{2i}, \quad i \leq r, \\ A(w)u_{2i-1} &= \alpha_i' \langle w, v_i\rangle u_{2i}, \quad i \leq r, \\ A(w)u_{2r+1} &= A'(w)u_{2r+1} = 0. \end{cases} \tag{9.6}$$

Since $C(Px) = C(x)$ is equivalent to $Pc_i = c_i$, for all i, (9.5) gives

$$P(u_{2r+1}) = u_{2r+1}\ , \quad P(u_{2i-1}) = u_{2i-1}, \ i \leq 2r. \tag{9.7}$$

But P is orthogonal and therefore P must be *diagonal*, and

$$P(u_{2i}) = \epsilon_{2i} u_{2i} \tag{9.8}$$

where $\epsilon_{2i} = \pm 1$. From (9.1), (9.3), (9.2) and $dw(v_i) = v_i$ we obtain, by a simple computation,

$$l_i e^{\langle e^{w^i A(w^i)} c_i, x_0\rangle} A'(v_i) = PA(v_i)\ {}^tP \tag{9.9}$$

where l_i denotes the (i,i)-entry of L. Then, (9.6) and (9.7) imply

$$l_i e^{\langle e^{w^i A(w^i)} c_i, x_0\rangle} \alpha_i' = \alpha_i \epsilon_i \tag{9.10}$$

where $\epsilon_{2i-1} = 1$. Therefore, the function $e^{\langle e^{w^i A(w^i)} c_i, x_0\rangle}$ of the variable w^i alone, must be constant. By differentiating once, we get

$$\langle e^{w^i A(v^i)} c_i, x_0\rangle = 0. \tag{9.11}$$

From this equation, by differentiating m times, we obtain

$$\langle e^{w^i A(v^i)} A(v_i)^m c_i, x_0\rangle = 0. \tag{9.12}$$

Therefore, by putting $w^i = 0$, we get

$$\langle A(v_i)^m c_i\, ,\, x_0\rangle = 0$$

for all i and m. Since $\mathbf{R}^q$ is spanned by the vectors $A(v_i)^m c_i$, we must have

$$x_0 = 0. \tag{9.13}$$

Then, $F(w)$ is identically zero and $G(w)$ is given by

$$G(w) = Lw + w_0. \tag{9.14}$$

The equation (9.2) reduces to

$$A'(Lw) = PA(w)^t P. \tag{9.15}$$

By taking into account (9.6) and (9.8), we find that $\alpha'_i = \pm\alpha_i$ as claimed.

Remark 9.1 All the metrics $g_{\alpha,\gamma}$ are *complete* because the assumptions of Theorem 6.5 are trivially verified.

We can deform again each metric $g_{\alpha,\gamma}$ by means of the diffeomorphism h of $\mathbf{R}^p$ whose Jacobian matrix $J(w)$ commutes with $C(x)$. In this way we get the metric

$$g_{\alpha,\gamma,h} = g_{A,C,h}.$$

All these metrics are *curvature homogeneous*, with the same curvature as the model metric g_0. Therefore, they have weakly generic Ricci curvature. Moreover, if $\alpha \neq 0$, they are *irreducible* and *complete* as soon as the entries of the Jacobian matrix of h (which is diagonal) are bounded by two positive constants. In any case, we can again apply Theorems 8.4 and 8.7 to study the isometry classes of these metrics.

If two such metrics $g_{\alpha,\gamma,h}$ and $g_{\alpha,\gamma,h'}$ (same α and γ) are isometric, then

$$h'(w) = \tilde{G}(h(G^{-1}(w))). \tag{9.16}$$

From Lemma 8.6 we know that $\tilde{G}$ is *uniquely determined up to the choice of* p=2r *arbitrary constants*, i.e., the components of $h'(w_0)$. The diffeomorphism G is determined up to the choice of the matrix Q of $GL(p, \mathbf{R})$, and up to the $(Ker\ A)$-valued function $\tilde{H}(w)$. Remark that $Q\tilde{H}(w)$ is in the kernel of the operator $x \longrightarrow A(x)$ as well. But $Ker\ A$ is one-dimensional and it is spanned by the vector $v_{r+1} = v_p$. Therefore, we have

$$\tilde{H}(w) = f(w)v_p \tag{9.17}$$

and

$$Q^i_p = 0 \tag{9.18}$$

if $i < p$. On the other hand, (8.32) implies that the Jacobian matrix of the map $h' \circ G$ is *diagonal*, as well as that of h'. From (8.41) we get

$$\begin{aligned}\frac{\partial (h' \circ G)^i}{\partial w^j} &= \sum_m \frac{\partial h'^i}{\partial w^m}\frac{\partial}{\partial w^j}(\textstyle\sum_k Q^m_k w^k + f(w)Q^m_p) \\ &= \frac{\partial h'^i}{\partial w^i}(Q^i_j + \frac{\partial f}{\partial w^j}Q^i_p).\end{aligned}$$

Then, if $i \neq j$, the following relation holds:

$$Q^i_j + \frac{\partial f}{\partial w^j}Q^i_p = 0. \tag{9.19}$$

From (9.18) we obtain that all non-diagonal entries of Q are zero, except those of type Q^p_j. In other terms, we have

$$Q = \begin{pmatrix} Q^1_1 & 0 & \cdots & 0 \\ 0 & Q^2_2 & \cdots & 0 \\ \vdots & \vdots & \ddots & 0 \\ Q^p_1 & Q^p_2 & \cdots & Q^p_p \end{pmatrix}. \tag{9.20}$$

Thus (9.19) reduces to

$$Q^p_j + \frac{\partial f}{\partial w^j} Q^p_p = 0 \ , \ j \neq p. \tag{9.21}$$

Hence, f is given by

$$f(w) = -(Q^p_p)^{-1} \sum_{j<p} Q^p_j w^j + f_p(w^p) \tag{9.22}$$

where f_p is an arbitrary function of the last variable w^p only. It follows that

$$G(w) = w_0 + Qw + f_p(w^p) v_p\,. \tag{9.23}$$

Therefore, if Q is given, G *is uniquely determined up to a function* $f_p(w_p)$ *of one variable* (f_p is a diffeomorphism of $\mathbf{R}$).

In conclusion we obtain that the diffeomorphism h' is *uniquely determined* up to

i) a vector $h'(w_0)$ of $\mathbf{R}^q$;

ii) a matrix Q of $GL(p, \mathbf{R})$ of type (9.20);

iii) a diagonal matrix L of $O(p)$;

iv) a diagonal matrix P of $O(q)$ given by (9.7) and(9.8);

v) a function $f_p(w^p)$ of the variable w^p only that is invertible with C^∞ inverse.

This, roughly speaking, gives an implicit *description of the orbit of* $g_{\alpha,\gamma,h}$ *under the action of the diffeomorphism group of* $\mathbf{R}^{p+q}$.

Now, recall that the diffeomorphism h' of $\mathbf{R}^p$, involved in the definition of the metric $g_{\alpha,\gamma,h'}$, has the form

$$h'(w) = (h'^1(w^1), \ldots, h'^{p-1}(w^{p-1}), h'^p(w^p)),$$

where the ith component depends only on the variable w^i. It follows that the *isometry classes* of the metrics $g_{\alpha,\gamma,h}$, as h is varying in $Diff\,(\mathbf{R}^p)$, depend at least on $(p-1)$ arbitrary functions $h'^1(w^1), \ldots, h'^{p-1}(w^{p-1})$, modulo some finite-dimensional parameter space. In other words, the *moduli space* of the metrics $g_{\alpha,\gamma,h}$, $h \in Diff\,(\mathbf{R}^p)$, is *not finite-dimensional*.

References

[BR] M. Berger, L'oeuvre de André Lichnerowicz en géométrie riemannienne, *Physique Quantique et Géométrie*, (eds. D. Bernard and Y. Choquet-Bruhat), Colloque Géométrie et Physique 1986 en l'honneur d'André Lichnerowicz, Travaux en Cours, Hermann, Paris, 1988, pp. 11-24.

[BS] A. L. Besse, *Einstein manifolds*, Springer-Verlag, Berlin, Heidelberg, New York, 1987.

[KN] S. Kobayashi and K. Nomizu, *Foundations of Differential Geometry*, Vol. I and II, Interscience Publishers, New York, 1963, 1969.

[KTV] O. Kowalski, F. Tricerri, L. Vanhecke, Curvature homogeneous spaces with a solvable Lie group as homogeneous model, *J. Math. Soc. Japan* **44** (1992), 461-484.

[ML] J. Milnor, Curvature of left invariant metrics on Lie group, *Adv. in Math.* **21** (1976), 293-329.

[SK] K. Sekigawa, On the Riemannian manifolds of the form $B \times_f F$, *Kōdai Math. Sem. Rep.* **26** (1975), 343-347.

[TV] F. Tricerri and L. Vanhecke, Curvature homogeneous Riemannian manifolds, *Ann. Sci. Ec. Norm. Sup.* **22** (1989), 535-554.

On the Propagation of Extendibility of *CR* Functions

A. TUMANOV*
University of Illinois, Urbana, Illinois 61801

Abstract. We obtain a new simple proof of the propagation of extendibility of CR functions along CR orbits. In addition, we give a sharp result on regularity of Bishop's equation, which governs analytic discs attached to a real manifold in complex space.

Introduction

In this paper we give a new simple proof of the main result of [T3] on the propagation of extendibility of CR functions.

Let M be a smooth real generic manifold in $\mathbf{C}^N$. Hanges and Treves [HT] show that *analyticity* of CR functions on M propagates along complex curves in M (see also [HS]). Trépreau [Tr] shows that *wedge-extendibility* of CR functions propagates along *CR orbits* in M, a CR orbit being the set of all points in M that can be reached from a given point by *CR curves*, which are real curves that run in complex tangential directions to M. Moreover, Trépreau [Tr] describes the variation of the direction of extendibility along CR orbits by showing that the wave-front set of a CR function is a union of CR orbits in the conormal bundle with respect to a natural CR structure there.

In [T3] we study propagation of *CR extendibility* of CR functions. We say that a CR function on M is CR-extendible at $p \in M$ if it extends to be CR on some manifold with boundary attached to M near p. We show that CR extendibility propagates along CR orbits the same way that wedge-extendibility does. We describe the variation of the direction of CR extendibility in terms of a certain differential geometric (partial) connection and the corresponding parallel displacement in the normal bundle. This description is dual to that of Trépreau. The result of [T3] has already found an interesting application.

Supported by NSF Grant DMS-9401652.

Merker [M] uses this result in his proof of Trépreau's [Tr] conjecture on simultaneous wedge-extendibility of CR functions on globally minimal manifolds. (Jöricke [J] has obtained an independent proof of the same result.) Merker [M] also gives a new simplified description of the connection defined in [T3].

Although the result of [T3] is formulated in natural geometric terms, the proof in [T3] consists of computations in local coordinates. We believe this result deserves a simple coordinate-free proof. Furthermore, we provide an improved version of our earlier use of Bishop's equation. We obtain extendibility of CR function by the method of analytic discs. In particular, propagation of extendibility is first obtained along boundaries of analytic discs attached to M (Theorem 3.2 here). Given a CR curve, we approximate it by a sequence of analytic discs and then apply Theorem 3.2 to each disc, which results in propagation along CR curves whence CR orbits. However, we lose some amount of Lipschitz smoothness after each step. Since the number of discs in the sequence is unbounded, we should be able to solve Bishop's equation with arbitrarily small loss of smoothness. Jöricke has observed that the result of [T2] on Bishop's equation that we presumably use in [T3] does not suffice for our needs, namely the size of the discs would depend on the allowed loss of smoothness. We fix this difficulty here.

The paper is organized as follows. In Section 1 we study solvability and regularity of Bishop's equation. In Section 2 we recall some definitions and facts on analytic discs and extendibility of CR functions. In Section 3 we formulate the main results, and in Section 4 we give the proofs. Although we occasionally refer to [T3] and [Tr], this exposition is independent of these papers.

I wish to thank Burglind Jöricke for drawing my attention to the regularity issue in the main result of [T3].

1. Bishop's equation

Let Δ and $b\Delta$ denote respectively the unit disc and the unit circle on the complex plane $\mathbf{C}$. Let T denote the Hilbert transform or the harmonic conjugation operator on the unit circle. Recall that for a function ϕ on $b\Delta$, $T\phi$ is such a function on $b\Delta$ that $\phi + iT\phi$ extends holomorphically into Δ and $P_0(T\phi) = 0$, where

$$P_0\phi = \frac{1}{2\pi}\int_0^{2\pi} \phi(e^{i\theta})\, d\theta.$$

We also use the Hilbert transform T_1 normalized by the condition $(T_1\phi)(1) = 0$, that is

$$T_1\phi = T\phi - (T\phi)(1).$$

The operators T and T_1 are bounded in $C^{k,\alpha}$ (where $k \geq 0$ and $0 < \alpha < 1$), which is the space of functions with derivatives through order k satisfying a Lipschitz condition with exponent α. The operators T and T_1 are bounded in L^p $(1 < p < \infty)$, also. We denote the norm in $C^{k,\alpha}$ by $\|\,.\,\|_{k,\alpha}$. The sup-norm is denoted by $\|\,.\,\|_0$.

The matter of this section is the following equation that arises (Bishop [B]) in constructing analytic discs with boundaries in a given manifold:

$$y = T_1 H(y, ., t) + y_0, \tag{1.1}$$

where H is a given $\mathbf{R}^m$-valued function of $y \in \mathbf{R}^m$, $\zeta \in b\Delta$, and $t \in \mathbf{R}^l$. The solution $y = y(\zeta, t, y_0)$ to (1.1) is a function of $\zeta \in b\Delta$, the variables $t \in \mathbf{R}^l$ and $y_0 \in \mathbf{R}^m$ being parameters.

We state some properties of the Lipschitz spaces $C^{k,\alpha}$. Let $u = u(t, s)$, $u \in C^{0,\alpha}(\mathbf{R}^p \times \mathbf{R}^q)$ and $0 < \beta < \alpha, \epsilon = \alpha - \beta$. Then

$$\| u(t_1, .) - u(t_2, .) \|_{0,\epsilon} \leq C \| u \|_{0,\alpha} |t_1 - t_2|^{\beta}, \tag{1.2}$$

where C depends on α and β only. If $u \in C^{1,\alpha}$,

$$\| u(t_1, .) - u(t_2, .) \|_{0,\alpha} \leq C \| u \|_{1,\alpha} |t_1 - t_2|, \tag{1.3}$$

where C depends on α only.

A little less obvious statement is as follows.

LEMMA 1.1. *Let $H \in C^{1,\alpha}(\mathbf{R}^m)$ and $y_1, y_2 \in C^{0,1}(b\Delta, \mathbf{R}^m)$, $\| y_j \|_{0,1} \leq C$, $j = 1, 2$. Then*

$$\| H(y_1) - H(y_2) \|_{0,\alpha} \leq C_1 \| H \|_{1,\alpha} \| y_1 - y_2 \|_{0,\alpha}, \tag{1.4}$$

where C_1 depends on C and α only.

A proof of this lemma is given in [T2]. Another proof is easily obtained by using the characterization of the Lipschitz spaces in terms of estimates of derivatives of the harmonic extention. The author does not know whether the lemma still holds if merely $y_1, y_2 \in C^{0,\alpha}$, but we do not need that result here.

We now state the main result on solvability and regularity of the equation (1.1). Let $B_r^m \subset \mathbf{R}^m$ denote the ball of radius r centered at the origin.

THEOREM 1.2. *Let $H \in C^{k,\alpha}(B_1^m \times b\Delta \times B_1^l, \mathbf{R}^m), k \geq 1, 0 < \alpha < 1$. For every constant $C > 0$ there is $c > 0$ such that if $\| H \|_{k,\alpha} < C$, $\| H \|_0 < c$, $\| H_y \|_0 < c$, and $\| H_\zeta \|_0 < c$, then:*

(i) *(1.1) has a unique solution $\zeta \mapsto y(\zeta, t, y_0)$ in $L^2(b\Delta)$;*

(ii) *$y \in C^{k,\alpha}(b\Delta)$, and $\| y \|_{k,\alpha}$ is uniformly bounded with respect to the parameters;*

(iii) *$y \in C^{k,\beta}(b\Delta \times B_1^l \times B_c^m, \mathbf{R}^m)$ for all $0 < \beta < \alpha$.*

Note that the statement (iii) does not hold for $\alpha = \beta$ because the Hilbert transform does not preserve $C^{k,\alpha}$ smoothness with respect to parameters — see the example in the end of this section. A close analogue of this theorem is given in [T2], but in that version c would depend on $\alpha - \beta$. The standard approach to solving (1.1) uses the implicit function theorem in Banach spaces (cf. [Bo]), but seemingly, some regularity of the solution is lost this way.

PROOF. We follow [T2] until Lemma 1.3 where we deviate from [T2]. There exists a unique solution $y \in L^2(b\Delta)$ because for sufficiently small c the mapping

$$y \mapsto F(y) = T_1 H(y, ., t) + y_0$$

is a contraction in $L^2(b\Delta)$. To study further properties of the solution, we consider the successive approximations:

$$y_0(\zeta) = y_0, \quad y_{\nu+1} = F(y_\nu).$$

The sequence y_ν is bounded in $C^{1,\delta}$ for $\delta = \alpha/2$. Indeed,

$$\| F(y) \|_{1,\delta} \leq |y_0| + \| T_1 \|_{1,\delta} \| H(y(.), ., t) \|_{1,\delta} .$$

We choose c so small that $\| H(., ., t) \|_{1,\delta}$ is small. Thus we obtain that $\| y_{\nu+1} \|_{1,\delta} \leq c_1 + c_2 \| y_\nu \|_{1,\delta}$, where c_1 and c_2 are small, whence it follows that $\| y_\nu \|_{1,\delta}$ is bounded and even small.

The sequence converges in $C^{0,\delta}$ for $\delta = \alpha/2$. This follows because in the norm of $C^{0,\delta}$ the mapping F is a contraction on functions bounded in $C^{1,\delta}$. Indeed, by Lemma 1.1,

$$\| F(y_1) - F(y_2) \|_{0,\delta} \leq C_1 \| T_1 \|_{0,\delta} \| H \|_{1,\delta} \| y_1 - y_2 \|_{0,\delta} \leq C_2 \| y_1 - y_2 \|_{0,\delta},$$

where C_2 is small. Thus, the limit of the sequence is the unique solution $y = y(\zeta)$ to (1.1) for fixed values of the parameters t and y_0.

The solution y belongs to $C^{1,\delta}$ (where $\delta = \alpha/2$) because it is posible, by the Arzela lemma, to extract from the sequence y_ν a subsequence converging in $C^{1,\delta}$.

We now show that the solution $y = y(\zeta, t, y_0)$ satisfies a Lipschitz condition with exponent 1 with respect to all the variables. We estimate the increment with respect to t. We set $y_j(\zeta) = y(\zeta, t_j, y_0)$ for $j = 1, 2$. Then

$$\begin{aligned} y_1 - y_2 &= T_1 H(y_1, ., t_1) - T_1 H(y_2, ., t_2) \\ &= T_1(H(y_1, ., t_1) - H(y_2, ., t_1)) + T_1(H(y_2, ., t_1) - H(y_2, ., t_2)). \end{aligned}$$

The last expression consist of two terms, which are estimated respectively by (1.4) and (1.3) as follows:

$$\| y_1 - y_2 \|_{0,\delta} \leq C_1 \| H(.., ., t_1) \|_{1,\delta} \| y_1 - y_2 \|_{0,\delta} + C_2 \| H(y_2, \ldots) \|_{1,\delta} |t_1 - t_2|.$$

Since the coefficient of $\| y_1 - y_2 \|_{0,\delta}$ on the right-hand side is small, we obtain

$$\| y_1 - y_2 \|_0 \leq \| y_1 - y_2 \|_{0,\delta} \leq C_3 |t_1 - t_2|,$$

that is $y(\zeta, t, y_0)$ satisfies a Lipschitz condition with exponent 1 with respect to t. Similarly, we come to the same conclusion regarding the parameter y_0. Thus, the first derivatives of y exist in L^∞.

We now differentiate the equation (1.1) with respect to all the variables. For a function ϕ on $b\Delta$, we set $\phi'(e^{i\theta}) = d\phi(e^{i\theta})/d\theta$. For differentiating with respect to θ, we use the fact that the Hilbert transform T commutes with $d/d\theta$. The validity of differentiating with respect to t and y_0 follows because T is bounded in L^2. Thus we obtain

$$\begin{aligned} y' &= T(H_y y' + i\zeta H_\zeta), \\ y_t &= T_1(H_y y_t + H_t), \\ y_{y_0} &= T_1(H_y y_{y_0}) + \mathbf{1}, \end{aligned} \tag{1.5}$$

where $\mathbf{1}$ is the identity matrix. These equations are viewed as equations in the derivatives of y at a certain y. The following lemma is applicable to these equations.

LEMMA 1.3. *Let $\zeta \mapsto u(\zeta, t)$ be a L^∞ function on $b\Delta$ depending on a parameter $t \in B_1^l \subset \mathbf{R}^l$. Let u satisfy the equation*

$$u = T_*(pu) + q,$$

where T_* is either T or T_1, $p, q \in C^{k,\alpha}(b\Delta \times B_1^l), k \geq 0, 0 < \alpha < 1$. For every $C > 0$ there exists a $c > 0$ such that if $\| p \|_{k,\alpha} < C$, $\| p \|_0 < c$, then the solution $u \in C^{k,\beta}(b\Delta \times B_1^l, \mathbf{R}^m)$ for all $0 < \beta < \alpha$.

The theorem immediately follows by applying Lemma 1.3 to (1.5) successively. In proving the lemma, we make use of the Cauchy type integral. Let $K\phi$ denote the inner limiting values of the Cauchy type integral of a function ϕ on the unit circle:

$$(K\phi)(z) = \lim_{r \to 1-0} \frac{1}{2\pi i} \int_{b\Delta} \frac{\phi(\zeta)\, d\zeta}{\zeta - rz}, \quad z \in b\Delta.$$

The operator K is bounded in $C^{k,\alpha}$ $(k \geq 0,\ 0 < \alpha < 1)$.

Let $f \in C^{k,\alpha}(b\Delta)$ and $\phi \in H^\infty(\Delta)$, that is ϕ is holomorphic and bounded in Δ. We use the following estimate [T4, Proposition 1.1]:

$$\| K(f\bar{\phi}) \|_{k,\alpha} \leq \text{const} \| f \|_{k,\alpha} \| \phi \|_{L^\infty}. \tag{1.6}$$

PROOF OF LEMMA 1.3. We consider the equation

$$u = T(pu) + q, \tag{1.7}$$

the case $T_* = T_1$ being similar.

It suffices to restrict to the case in which $P_0 u = 0$. Indeed, otherwise $v = u - P_0 u$ satisfies the equation $v = T(pv) + \tilde{q}$, where $\tilde{q} = q + (P_0 q)(Tp - 1)$. Then $P_0 v = 0$ and if the lemma holds for v, it will hold for $u = v + P_0 q$ as well.

Note that $T^2 f = -f + P_0 f$. Applying T to (1.7), we get

$$Tu = -pu + P_0(pu) + Tq. \tag{1.8}$$

We set $\phi = [u + i(Tu - P_0(pu))]/2$. Then ϕ extends holomorphically into Δ with respect to ζ, and $u = \phi + \bar{\phi}$, $Tu - P_0(pu) = -i(\phi - \bar{\phi})$. Using ϕ, we rewrite (1.8) in the form

$$\phi = \bar{\phi} + Q\bar{\phi} + R, \tag{1.9}$$

where

$$Q = -2ip(1+ip)^{-1},$$
$$R = i(1+ip)^{-1}Tq.$$

Since ϕ is holomorphic,

$$K\phi = \phi, \quad K\bar{\phi} = P_0\bar{\phi}.$$

Applying K to (1.9), we get

$$\phi = K(Q\bar{\phi} + R) + P_0\bar{\phi} = K(Q\bar{\phi} + R) + \frac{i}{2}P_0(pu). \tag{1.10}$$

Owing to (1.6), the right-hand side of (1.10) belongs to $C^{k,\alpha}$ with respect to ζ. Therefore ϕ whence u belongs to $C^{k,\alpha}$ with respect to ζ.

Let $k = 0$. We now study the behavior of the solution with respect to the parameter t. Let $u_j = u(\zeta, t_j)$, $j = 1, 2$. Similarly, we introduce ϕ_j, p_j, Q_j, R_j, $j = 1, 2$. Using (1.10), we get

$$\begin{aligned} &\phi_1 - \phi_2 = E_1 + E_2, \\ &\frac{1}{2} \| u_1 - u_2 \|_0 \le \| \phi_1 - \phi_2 \|_0 \le ||E_1||_0 + ||E_2||_0, \end{aligned} \tag{1.11}$$

where

$$E_1 = K((Q_1 - Q_2)\bar{\phi}_1 + R_1 - R_2) + \frac{i}{2}P_0((p_1 - p_2)u_1),$$
$$E_2 = K(Q_2(\bar{\phi}_1 - \bar{\phi}_2)) + \frac{i}{2}P_0(p_2(u_1 - u_2)).$$

Let $0 < \beta < \alpha$. Applying (1.2), we get

$$\| E_1 \|_0 \le \| E_1 \|_{0,\alpha-\beta} \le \text{const}\, |t_1 - t_2|^\beta, \tag{1.12}$$

where the constant depends only on α, β, c, and C. Applying (1.6), we get

$$\| E_2 \|_0 \le C_1(\| Q_2 \|_{0,\alpha/2} \| \phi_1 - \phi_2 \|_0 + \| p_2 \|_0 \| u_1 - u_2 \|_0). \tag{1.13}$$

Note this estimate is independent of β. We choose c in the statement of the lemma so small that p whence Q and therefore the coefficients of $\| \phi_1 - \phi_2 \|_0$ and $\| u_1 - u_2 \|_0$ in (1.13) are small. Now putting together (1.11) through (1.13), we get

$$\| u_1 - u_2 \|_0 \le \text{const}\, |t_1 - t_2|^\beta,$$

where the constant depends only on α, β, c, and C. This completes the case $k = 0$.

For $k \ge 1$, the lemma is proved by induction. The derivatives of u with respect to ζ and t are solutions to linear equations obtained by differentiating (1.7). The validity of differentiating is justified the same way as in the proof of Theorem 1.2. We apply this lemma to the resulting equations. The lemma whence Theorem 1.2 is now proved.

We conclude this section with a simple example showing that the Hilbert transform does not preserve $C^{k,\alpha}$ smoothness with respect to parameters. As a result, the statement (iii) in Theorem 1.2 does not hold for $\alpha = \beta$. We consider the Hilbert transform on the real line $\mathbf{R}$:

$$(T\phi)(x) = \frac{1}{\pi} \int_{-\infty}^{+\infty} \frac{\phi(\xi)\, d\xi}{x - \xi}.$$

Let $\phi(x) = x$ for $x \geq 0$ and $\phi(x) = x^2$ for $x \leq 0$. Let $\psi(x,t) = \min(\phi(x)^\alpha, |t|^\alpha)$, $0 < \alpha < 1$. Then $\psi \in C^{0,\alpha}$ with respect to both x and t, but $T\psi|_{x=0}$ involves $|t|^\alpha \ln |t|$, therefore $T\psi$ does not belong to $C^{0,\alpha}$.

2. CR extendibility and analytic discs

Let M be a smooth real manifold in $\mathbf{C}^N$. Let $T_p^c(M) = T_p(M) \cap JT_p(M)$ be the maximal complex subspace of the tangent space $T_p(M)$ at the point $p \in M$, where J is the operator of multiplication by the imaginary unit in $\mathbf{C}^N$. Recall that the manifold M is called a *CR manifold* if all spaces $T_p^c(M)$, $p \in M$ have the same dimension. This dimension is called the *CR dimension* of M and it is denoted by $\mathrm{CRdim}(M)$ here. The spaces $T_p^c(M)$ form the *complex tangent bundle* $T^c(M)$.

Recall also that the manifold M is called *generic* if $T_p(M) + JT_p(M) = T_p(\mathbf{C}^N) \simeq \mathbf{C}^N$, $p \in M$. A generic manifold $M \in \mathbf{C}^N$ is always a CR manifold; it can be defined by a local equation

$$x = h(y, w), \tag{2.1}$$

where $z = x + iy \in \mathbf{C}^m, w \in \mathbf{C}^n, m + n = N, n = \mathrm{CRdim}(M)$, $h(y,w)$ is an $\mathbf{R}^m$-valued function such that h and the partial derivatives h_y, h_w are small.

Recall that a smooth complex valued function on M is called a *CR function* if its differential is $\mathbf{C}$-linear on $T^c(M)$. A continuous function is called CR if the last condition holds in the sense of distribution theory. We denote by $\mathrm{CR}(M)$ the set of all continuous CR functions on M.

An *analytic disc* in $\mathbf{C}^N$ is a continuous mapping $A : \bar{\Delta} \to \mathbf{C}^N$ that is holomorphic in the unit disc Δ. We say that A is *attached* to M if $A(b\Delta) \subset M$.

According to the approximation theorem by Baouendi and Treves [BT], every CR function is locally a uniform limit of complex polynomials. This implies that a CR function on M extends holomorphically into sufficiently small analytic discs attached to M.

Moreover, if such discs cover a domain or a CR manifold then the function extends to be holomorphic or CR there.

There are many analytic discs attached to a CR manifold which means the following (cf. [Bo]).

PROPOSITION 2.1. *Let M be a CR manifold of class $C^{k,\alpha}(k \geq 0, 0 < \alpha < 1)$ in $\mathbf{C}^N$ and $p \in M$. Let $w : \mathbf{C}^N \to \mathbf{C}^n$, $w(p) = 0$, be a complex affine mapping such that the restriction $w|_{T_p^c(M)}$ is an isomorphism. For any $q \in M$ close to p and any small $\mathbf{C}^n$-valued function $\phi \in C^{k,\alpha}(\Delta) \cap H(\bar{\Delta})$ with $\phi(1) = w(q)$, there is a unique $C^{k,\alpha}$ smooth analytic disc A attached to M such that $A(1) = q$, $w \circ A = \phi$.*

For completeness, we include a proof.

PROOF. We take w as a coordinate function and complete it to a coordinate system (z, w) in $\mathbf{C}^N$.

First assume that M is generic and is given by (2.1). We seek an analytic disc $\zeta \mapsto A(\zeta) = (x(\zeta) + iy(\zeta), \phi(\zeta))$ such that $x(\zeta) = h(y(\zeta), \phi(\zeta))$ for $|\zeta| = 1$. Since the functions $x(\zeta)$ and $y(\zeta)$ are harmonic conjugates, they must satisfy the Bishop equation

$$y = T_1 h(y, \phi) + y_0,$$

where $y_0 = \mathrm{Im} z(q)$. By Theorem 1.2, this equation has a unique solution, which defines the needed disc A.

If M is not generic, we split the coordinates $z = (z', z'')$ in such a way that the projection M' of M onto the space of the (z', w) coordinates is generic. Then due to the previous argument, there exists a unique analytic disc A' such that $A'(1) = (z'(q), w(q))$, $w \circ A = \phi$. The remaining coordinate function z'' is CR on M hence also on M'. By the Baouendi-Treves approximation theorem, it extends into the disc A' to form the needed disk A. The proof is complete.

In this paper, we will also study CR functions on manifolds with boundaries. We say that a manifold M' with boundary is *attached* to M at $p \in M$ if $bM' \cap U = M \cap U$ for some neighborhood U of p. In fact, $\dim(M') = \dim(M) + 1$. If M' is attached to M, one can consider continuous CR functions on $M \cup M'$. We denote the set of such functions by $\mathrm{CR}(M \cup M')$. The approximation theorem by Baouendi-Treves still holds for $\mathrm{CR}(M \cup M')$.

PROPOSITION 2.2. *Let M be a generic manifold in $\mathbf{C}^N$. Then, for every point $p \in M$, there exists a neighborhood $U \subset M$ of p with the following property. For every point*

$q \in U$ and every manifold M' attached to M at q, there exists such a neighborhood U', $q \in U' \subset M'$, that every CR function on $M \cup M'$ is a uniform limit of polynomials on $U \cup U'$.

The proof is actually a repetition of the one given in [BT]. For instance, this proposition shows that a CR function on $M \cup M'$ extends holomorphically into sufficiently small analytic discs attached to $M \cup M'$.

Let M be a generic manifold in $\mathbf{C}^N$. We denote by $N(M)$ the normal bundle $T(\mathbf{C}^N)/T(M)$. Let M' be attached to M at $p \in M$. We say that M' is attached to M at (p, ξ) if $\xi \in N_p(M)$ is represented by a vector $v \in T_p(M')$ directed inside M'.

Let f be a CR function on M, i. e. $f \in$CR(M). We say that f is *CR-extendible* at (p, ξ) if it extends continuously to be CR on some M' attached to M at (p, ξ). We also say that f is CR-extendible at p in the direction of ξ in this case.

The next simple proposition shows that CR extendibility may arise due to just one analytic disc attached to $M \cup M'$, because we can deform the disc to cover another small manifold attached to M.

PROPOSITION 2.3. [T3] *Let M, M', U, U' be the same as in Proposition 2.2. Let A be a sufficiently small analytic disc attached to $U \cup U'$ such that $A(\zeta) \in U$ if $\zeta \in b\Delta$ and $|\zeta - 1|$ is small enough. Assume, in addition, that $v = -\partial A(1)/\partial\zeta \notin T_p(M)$, where $p = A(1)$. Then all CR functions on $M \cup M'$ extend to be CR on some manifold M'' attached to M at (p, ξ), where $\xi \in N_p(M)$ is represented by v. If M, M' and A belong to $C^{k,\alpha}(k \geq 1, 0 < \alpha < 1)$, then M'' is $C^{k,\beta}$ for all $\beta < \alpha$.*

PROOF. We construct M'' to be a union of a family of analytic discs attached to $U \cup U'$ containing the given disc A. Then M'' will be automatically attached to M at (p, ξ). According to Proposition 1.1, every CR function on $M \cup M'$ is a uniform limit of polynomials on $U \cup U'$. By maximum principle, those polynomials will then converge to a CR function on M'', completing the proof.

Let $p = 0$. Since the disc A is small, we can assume that $M, M' \subset \mathbf{C}^N$ are defined by the parametric equation

$$x = h(y, w, t), \tag{2.2}$$

where $z = x + iy \in \mathbf{C}^m$, $w \in \mathbf{C}^n$, $m + n = N$, $t \geq 0$, h is a $C^{k,\alpha}$-smooth $\mathbf{R}^m$-valued function with sufficiently small partial derivatives h_y, h_w in a neighborhood of $(0, 0, 0)$. This means that M is given by $x = h(y, w, 0)$, and M' is a part of the manifold with boundary defined by (2.2) for $t \geq 0$.

We set $p = 0$. The given disc $\zeta \mapsto A(\zeta) = (x(\zeta) + iy(\zeta), w(\zeta))$ must satisfy the Bishop equation

$$y = T_1 h(y, w, \tau)$$

for some function $\tau \geq 0$ on the circle; $\tau = 0$ in a neighborhood of $\zeta = 1$. We perturb this equation by including the initial condition $y(1) = y_0$ and adding an arbitrary constant w_0 to the given function w. We get

$$y = T_1 h(y, w + w_0, \tau) + y_0,$$

where τ is the same as for the given disc. The solution to this equation gives the needed family of discs. The proposition is proved.

3. Propagation of CR extendibility

Let $M \subset \mathbf{C}^N$ be a generic manifold in $\mathbf{C}^N$. We study the propagation of extendibility of CR functions along piecewise smooth real curves ($\gamma : [0, 1] \to \mathbf{C}^N$) running in complex tangential directions ($d\gamma(t)/dt \in T^c_{\gamma(t)}(M)$). We call such curves *CR curves*. Let $S \subset M$ be a submanifold with the property $\text{CRdim}(S) = \text{CRdim}(M)$. Then a CR curve that begins at $p \in S$ must lie in S entirely. Conversely, the CR orbit S of a point $p \in M$, which is the set of all points in M that can be reached from p by CR curves, is a CR manifold with the property $\text{CRdim}(S) = \text{CRdim}(M)$. So, we assume that a submanifold S like that is given initially.

We can expect that all CR functions on M automatically extend in some of the directions of $JT_p(S)$. For instance, if the manifold S is *minimal* at a point $p \in S$ (see [T1], [T2]), then all CR functions on M simultaneously extend at p in directions that span $JT_p(S)$. On the contrary, we are interested in obtaining extendibility of CR functions in directions other than directions in which CR functions extend automatically. This leads to the following definition.

$$E_p = T_p(\mathbf{C}^N)/\left(T_p(M) + JT_p(S)\right), \quad p \in S.$$

The spaces E_p form a bundle E over S. It is a quotient bundle of the normal bundle $N(M)$ restricted to S.

Let $p \in S$, and let M' be attached to M at (p, u), $u \in N_p(M)$. We also say that M' is attached to M at (p, ξ) if u represents $\xi \in E_p$, $\xi \neq 0$. Thus, it makes sense to consider CR-extendibility at (p, ξ), $p \in S$, $\xi \in E_p$.

To state the main results, we need the real dual bundle E^* of the bundle E. The fiber E_p^* consists of all real forms on $\mathbf{C}^N$ that vanish on $T_p(M)$ and $JT_p(S)$. Let $T^*(\mathbf{C}^N)$ be the bundle of *holomorphic* ($\mathbf{C}$-linear) forms on $\mathbf{C}^N$. Then $T^*(\mathbf{C}^N)$ is a *complex* manifold. We can identify real and holomorphic forms by $\phi \leftrightarrow \mathrm{Re}\phi$. In this identification

$$E_p^* = \{\phi \in T_p^*(\mathbf{C}^N) : \mathrm{Re}\phi|_{T_p(M)} = 0,\ \phi|_{T_p(S)} = 0\}.$$

The bundle E^* is a subset in the conormal bundle

$$N_p^*(M) = \{\phi \in T_p^*(\mathbf{C}^N) : \mathrm{Re}\phi|_{T_p(M)} = 0\}.$$

Since M is generic, the fiber $N_p^*(M)$ is *totally real* in $T_p^*(\mathbf{C}^N) \simeq \mathbf{C}^N$, that is it does not contain complex subspaces. Therefore, under the natural projection $N^*(M) \to M$, the space $T_\phi^c(N^*(M))$ at $\phi \in T_p(M)$ maps injectively to $T_p^c(M)$, and $\dim_{\mathbf{C}} T_\phi^c(N^*(M)) \leq \mathrm{CRdim}(M)$.

Trépreau [Tr] observes that although $N^*(M)$ itself generally is not a CR manifold, the submanifold E^* is.

PROPOSITION 3.1. *E^* is a CR manifold and* $\mathrm{CRdim}(E^*) = \mathrm{CRdim}(M)$. *The isomorphism between $T_\phi E^*$ and $T_p^c(S)$, $\phi \in E_p^*$, is given by the differential of the natural projection $E^* \to S$.*

For completness, we provide a proof.

PROOF. Let M be defined by a local equation $r(z) = 0$, where $r = (r^1, \ldots, r^m)$ is a smooth $\mathbf{R}^m$-valued function of $z \in \mathbf{C}^N$ such that the holomorphic differentials $\partial r^j = \sum r_k^j dz_k$, $r_k^j = \partial r^j / \partial z_k$ are linearly independent.

We first compute the complex tangent space to $N^*(M)$. We introduce the coordinates (z, w) in $T^*(\mathbf{C}^N)$ by setting

$$T^*(\mathbf{C}^N) = \left\{\sum w_k\, dz_k : w_k \in \mathbf{C}, 1 \leq k \leq N\right\}.$$

A vector $X \in T(\mathbf{C}^N)$ is tangent to M if and only if

$$\langle dr, X\rangle = 2\mathrm{Re}\langle \partial r, X\rangle = 0.$$

We recall this relation implies that ∂r is purely imaginary on M. Therefore, using the identification $\phi \leftrightarrow \mathrm{Re}\phi$, we get

$$N_z^*(M) = \left\{\lambda\partial r = \sum \lambda_j \partial r^j(z) : \lambda = (\lambda_1, \ldots, \lambda_m) \in \mathbf{R}^m\right\}.$$

A real vector $\tilde{X} \in T(T^*(\mathbf{C}^N))$ has the form

$$\tilde{X} = 2\text{Re}\left(\sum \dot{w}_k \frac{\partial}{\partial w_k}\right) + X, \quad X = 2\text{Re}\left(\sum X^k \frac{\partial}{\partial z_k}\right),$$

where X is the projection of $\tilde{X}$ on $\mathbf{C}^N$.

Let $\tilde{X} \in T(N^*(M))$. Then $X \in T(M)$, and differentiating $\lambda\partial r = \sum \lambda_j r^j_k dz_k$ yields

$$\dot{w}_k = \sum \dot{\lambda}_j r^j_k + \sum \lambda_j (r^j_{kl} X^l + r^j_{k\bar{l}} \bar{X}^l). \tag{3.1}$$

where $r^j_{kl} = \partial^2 r^j / \partial z_k \partial z_l$, $r^j_{k\bar{l}} = \partial^2 r^j / \partial z_k \partial \bar{z}_l$, $\dot{\lambda}_j \in \mathbf{R}$.

Let $\tilde{X} \in T^c(N^*(M))$. This happens if and only if $J\tilde{X} \in T(N^*(M))$. Therefore, X must belong to $T^c(M)$, and

$$i\dot{w}_k = \sum \dot{\mu}_j r^j_k + \sum \lambda_j (r^j_{kl} iX^l + r^j_{k\bar{l}} \overline{iX^l}) \tag{3.2}$$

must hold for some $\dot{\mu}_j \in \mathbf{R}$. The relations (3.1) and (3.2) yield

$$\sum (\dot{\lambda}_j + i\dot{\mu}_j) r^j_k + 2 \sum \lambda_j r^j_{k\bar{l}} \bar{X}^l = 0. \tag{3.3}$$

Let $\lambda\partial r \in E^*$. We need to show that for every $X \in T^c(M)$ the equation (3.3) has a unique solution $\dot{\lambda}_j$. The uniqueness follows because ∂r^j are linearly independent. The existence follows because there are $n = \text{CRdim}(M)$ linearly independent relations among the $N = n + m$ equations (3.3). Indeed, let $Y \in T^c(M)$, that is $\langle \partial r, Y \rangle = 0$. Let $\xi = \sum X^k \partial/\partial z_k$ and $\eta = \sum Y^k \partial/\partial z_k$ be the (1,0) components of X and Y, that is $X = 2\text{Re}\xi$, $Y = 2\text{Re}\eta$. Since $X, Y \in T^c(M)$, we have $\xi, \eta \in T^c(M) \otimes \mathbf{C}$. We multiply (3.3) by Y_k and sum as k runs from 1 to N. We have

$$\sum (\dot{\lambda}_j + i\dot{\mu}_j) r^j_k Y^k + 2 \sum \lambda_j r^j_{k\bar{l}} \bar{X}^l Y^k = 0. \tag{3.4}$$

The first term is zero because $Y \in T^c(M)$. We now view our vectors as vector fields. By Cartan's formula,

$$2 \sum \lambda_j r^j_{k\bar{l}} \bar{X}^l Y^k = 2\lambda\bar{\partial}\partial r(\xi, \eta) = \bar{\xi}\langle \lambda\partial r, \eta \rangle - \eta \langle \lambda\partial r, \bar{\xi} \rangle - \langle \lambda\partial r, [\bar{\xi}, \eta] \rangle. \tag{3.5}$$

In this expression, the first two terms are zero because $X, Y \in T^c(M)$. We now use that $T^c(S) = T^c(M)|_S$. Since $\bar{\xi}, \eta \in T^c(M) \otimes \mathbf{C}$, we have $\bar{\xi}, \eta \in T(S) \otimes \mathbf{C}$, whence $[\bar{\xi}, \eta] \in T(S) \otimes \mathbf{C}$. Since $\lambda\partial r \in E^*$, the last term in (3.5) is zero also. Thus, (3.4) is zero identically for every $Y \in T^c(M)$. Hence, the solution to (3.3) exists for every $X \in T^c(M)$.

We need to show that the lift $\tilde{X} \in T^c(N^*(M))$ of $X \in T^c(M)$ obtaned by solving (3.3) belongs to $T(E^*)$. Since λ forms a set of coordinates in the fiber of $N^*(M)$, we have $\tilde{X} = \sum \dot{\lambda}_j \, \partial/\partial\lambda_j + X$. Note that $\tilde{X} \in T(E^*)$ if and only if $\tilde{X}\langle\lambda\partial r, Y\rangle = 0$ for every vector field $Y = 2\mathrm{Re}\left(\sum Y^k \, \partial/\partial z_k\right)$ on S. Expanding this relation yields

$$\langle\dot{\lambda}\partial r, Y\rangle + X\langle\lambda\partial r, Y\rangle = 0.$$

By Cartan's formula,

$$2i\lambda_j \mathrm{Im}\left(\sum r^j_{k\bar{l}} \bar{X}^l Y^k\right) = 2\lambda\bar{\partial}\partial r(X, Y) = X\langle\lambda\partial r, Y\rangle - Y\langle\lambda\partial r, X\rangle - \langle\lambda\partial r, [X, Y]\rangle.$$

The last two terms are zero because $X \in T^c(M)$ and $[X, Y] \in T(S)$. Thus, the condition that $\tilde{X} \in T(E^*)$ takes the form

$$\sum\left(\dot{\lambda}_j r^j_k Y^k + 2i\lambda_j \mathrm{Im}\left(\sum r^j_{k\bar{l}} \bar{X}^l Y^k\right)\right) = 0. \tag{3.6}$$

Note again, (3.3) implies (3.4) for our X and Y. Since $r^j_k Y^k$ is purely imaginary, we immediately obtain (3.6) as the imaginary part of (3.4). Hence, $\tilde{X} \in T^c(E^*)$, what we need. The proof is now complete.

We now study analytic discs in $T^*(\mathbf{C}^N)$ attached to E^*. If A^* is such a disc, then the projection of A^* on $\mathbf{C}^N$ forms a disc A attached to S. Let M be given by a local equation (2.1) so that the projection of $T^c_p(M), p \in M$ onto the w space is bijective. Because the fibers of E^* are totally real and $\mathrm{CRdim}(E^*) = \mathrm{CRdim}(M)$, according to Proposition 2.1, a small disc A^* attached to E^* is uniquely defined by $A^*(1)$ and the w component of the projection of A^* on $\mathbf{C}^N$.

Let $p \in S$ and let A be a small analytic disc attached to S with $A(1) = q$, $A(-1) = p$. We define a linear isomorphism $R^*_A : E^*_q \to E^*_p$. Let $\phi \in E^*_q$. There exists a unique analytic disc A^* attached to E^* such that $A^*(1) = \phi$ and $A^*(\zeta) \in E^*_{A(\zeta)}$ for $\zeta \in b\Delta$. Indeed, let $\zeta \mapsto w(\zeta)$ be the w component of the disc A. By Proposition 2.1, we can construct the disc A^*, $A^*(1) = \phi$, so that the projection A' of A^* on $\mathbf{C}^N$ has w component $\zeta \mapsto w(\zeta)$. Then A and A' are the same because they have the same w components. We say that A^* is a *lift* of A to E^*. We set

$$R^*_A(\phi) = A^*(-1), \text{ where } A^*(\zeta) \text{ is the lift of } A \text{ to } E^* \text{ with } A^*(1) = \phi.$$

The mapping R^*_A is linear because the set of all lifts of A to E^* is a vector space. The mapping R^*_A is an isomorphism because the mapping $R^*_{A'}$, where $A'(\zeta) = A(-\zeta)$, is the inverse of R^*_A.

For the given analytic disc A we consider also the operator $R_A : E_p \to E_q$ the adjoint operator to R_A^*.

THEOREM 3.2. *Let M be a generic manifold in $\mathbf{C}^N$ and let $S \subset M$ be a CR submanifold such that* $\mathrm{CRdim}(S) = \mathrm{CRdim}(M)$. *Let A be a sufficiently small analytic disc attached to S; $A(1) = q$, $A(-1) = p$. Assume M, S, and A are $C^{k,\alpha}, k \geq 2, 0 < \alpha < 1$. For any $u \in E_p$, any $C^{k,\alpha}$ manifold M' attached to M at (p, u), and any $\epsilon > 0$, there exist $v \in E_q$ and a manifold M'' attached to M at (q, v) such that*

(i) $|v - R_A u| < \epsilon$;

(ii) *M'' is $C^{k,\beta}$ for every $0 < \beta < \alpha$;*

(iii) *if $f \in \mathrm{CR}(M \cup M')$, f extends to be CR on M''.*

Let M be a generic manifold in $\mathbf{C}^N$ and let $S \subset M$ be a CR submanifold such that $\mathrm{CRdim}(S) = \mathrm{CRdim}(M)$. Let $\gamma : [0, 1] \to S$ be a CR curve in S connecting two points $p, q \in S$, that is $\gamma(0) = p$, $\gamma(1) = q$. Since E^* has a CR structure, we consider CR curves in it. We say that a CR curve $\gamma^* : [0, 1] \to E^*$ is a (horizontal) lift of γ to E^* if $\gamma^*(t) \in E^*_{\gamma(t)}$, $t \in [0, 1]$. The set of complex tangent subspaces $T^c_\phi(E^*)$ forms the set of horizontal subspaces of a $T^c(S)$-partial connection in the vector bundle $E^* \to S$ (see [M], [Tr], [T3]). This means that for every CR curve γ in S and every $\phi \in E^*_p$ there exists a unique lift γ^* with $\gamma^*(0) = \phi$. Moreover, all lifts of γ form a vector space. Indeed, the equation (3.3) that defines the lift of $X \in T^c_p(M)$ to $\tilde{X} \in T^c_\phi(E^*)$, where $\phi = \lambda \partial r \in E^*$, is linear in λ. Therefore, the lifts of a CR curve exist and form a vector space because they satisfy a linear differential equation. Hence, the *parallel displacement* operator $\Pi^*_\gamma : E^*_p \to E^*_q$ that takes $\gamma^*(0) = \phi$ to $\gamma^*(1)$ is a linear isomorphism. Let $\Pi_\gamma : E_p \to E_q$ be the dual isomorphism (the adjoint of Π^{*-1}_γ), the parallel displacement operator along γ for the dual connection in E. (Merker [M] gives a natural description of Π_γ without referring to E^*.)

The substance of the paper is the following

THEOREM 3.3. *Let M be a generic manifold in $\mathbf{C}^N$ and let $S \subset M$ be a CR submanifold such that* $\mathrm{CRdim}(S) = \mathrm{CRdim}(M)$. *Let M and S be $C^{k,\alpha}, k \geq 2, 0 < \alpha < 1$. Let $\gamma : [0, 1] \to S$ be a piecewise C^1-smooth CR curve in S such that $\gamma(0) = p$, $\gamma(1) = q$. For any $u \in E_p$, any $C^{k,\alpha}$ manifold M' attached to M at (p, u), any $\epsilon > 0$, and any $0 < \beta < \alpha$ there exist $v \in E_q$ and a manifold M'' attached to M at (q, v) such that*

(i) $|v - \Pi_\gamma u| < \epsilon$;

(ii) *M'' is $C^{k,\beta}$;*

(iii) *if $f \in \mathrm{CR}(M \cup M')$, f extends to be CR on M''.*

We note that in contrast to Theorem 3.2, M'' and even the germ of M'' at q depend on β here. In general, the size of M'' tends to zero as either $\epsilon \to 0$ or $\beta \to \alpha$.

4. Proof of the main results

We introduce the following notation. For a $C^{1,\alpha}$ function ϕ on the unit circle with $\phi(1) = 0$, we write

$$\mathcal{J}(\phi) = \frac{1}{\pi} \int_0^{2\pi} \frac{\phi(e^{i\theta})\, d\theta}{|e^{i\theta} - 1|^2},$$

where the integral is understood in the sense of principal value. Note that for any function $\phi \in C^{1,\alpha}(\bar{\Delta})$ holomorphic in Δ with $\phi(1) = 0$, we have

$$\mathcal{J}(\phi) = -\frac{\partial \phi(1)}{\partial \zeta} = i \frac{d}{d\theta}\Big|_{\theta=0} \phi(e^{i\theta}). \tag{4.1}$$

Let M be a generic manifold in $\mathbf{C}^N$ and let A be an analytic disc attached to M. A *(infinitesimal) deformation* of the disc A is a continuous mapping $\dot{A} : \bar{\Delta} \to T(\mathbf{C}^N) \simeq \mathbf{C}^N \times \mathbf{C}^N$ holomorphic in Δ such that $\dot{A}(\zeta) \in T_{A(\zeta)}(\mathbf{C}^N), \zeta \in b\Delta$, that is $\dot{A}$ is a lift of A to $T(\mathbf{C}^N)$.

The following simple lemma now replaces most of the computations in [T3].

LEMMA 4.1. *Let $\dot{A}$ be a deformation of the disc A such that $\dot{A}(1) = 0$. Let A^* be a lift of A to $T^*(\mathbf{C}^N)$. Then*

$$\mathrm{Re}\langle A^*(1), -\frac{\partial \dot{A}(1)}{\partial \zeta}\rangle = \mathcal{J}(\mathrm{Re}\langle A^*, \dot{A}\rangle). \tag{4.2}$$

PROOF. For brevity, we use the following notation: $\phi' = \frac{d}{d\theta}\big|_{\theta=0} \phi(e^{i\theta})$. Since $\dot{A}$ satisfies the Cauchy-Riemann equations at $\zeta = 1$, we have

$$\mathrm{Re}\langle A^*(1), -\frac{\partial \dot{A}(1)}{\partial \zeta}\rangle = \mathrm{Re}\langle A^*(1), i\dot{A}'\rangle = -\mathrm{Im}\langle A^*(1), \dot{A}'\rangle. \tag{4.3}$$

Since $\dot{A}(1) = 0$, we have

$$\langle A^*(1), \dot{A}'\rangle = \langle A^*, \dot{A}\rangle'. \tag{4.4}$$

Since the function $\langle A^*, \dot{A}\rangle$ is holomorphic, (4.1) yields

$$\langle A^*, \dot{A}\rangle' = -i\mathcal{J}(\langle A^*, \dot{A}\rangle). \tag{4.5}$$

Note that $\mathcal{J}$ is real. Therefore, combining (4.3) through (4.5), we get (4.2). The lemma is proved.

PROOF OF THEOREM 3.2. We deform the given disc A with the fixed point $A(1) = q$ to get a new one attached to $M \cup M'$. According to Proposition 2.3, the deformed disc will produce a manifold M'' attached to M at q such that CR functions on $M \cup M'$ extend to M''. We will show that the direction v of the deformed disc at p can be arbitrarily close to $R_A u$.

Let $A(1) = q = 0$. Since the disc A is small, we can assume that $M, M' \subset \mathbf{C}^N$ are defined by the parametric equation (2.2):

$$x = h(y, w, t), \quad t \geq 0. \tag{4.6}$$

As before, M is defined by $x = h(y, w, 0)$ while M' is given by (4.6) in a neighborhood of $A(-1) = p = (x_0 + iy_0, w_0)$, $x_0 = h(y_0, w_0, 0)$. The fact that M' is attached to M at (p, u), $u \in E_p$ means that u is represented by the vector $(h_t, 0) = (h_t(y_0, w_0, 0), 0) \in T_p(\mathbf{C}^N)$.

We construct the deformation of the disc A by solving the Bishop equation

$$y = T_1 h(y, w, t\chi),$$

where $\zeta \mapsto w(\zeta)$ is the w-component of the given disc A, $t \geq 0$, and $\chi \geq 0$ is a smooth function on the circle supported in a small neighborhood of $\zeta = -1$ that we will choose soon. The solution to the equation defines a family of discs $\zeta \mapsto A(\zeta, t)$ attached to $M \cup M'$. The initial disc A is obtained at $t = 0$. The direction of the disc $\zeta \mapsto A(\zeta, t)$ at q is given by $v(t) = -\partial A(1, t)/\partial\zeta \bmod (T_q(M) + JT_q(S)) \in E_q$. Since the given disc A is attached to S, $v(0) = 0 \in E_q$. We denote by a dot the derivative with respect to t at $t = 0$. Then $\dot{A}$ is an infinitesimal deformation of A. On $b\Delta$, we have $\dot{A} = (h_y\dot{y} + h_t\chi + i\dot{y}, 0) = (h_t\chi, 0) \bmod T(M)$ because $(h_y\dot{y} + i\dot{y}, 0) \in T(M), \zeta \in b\Delta$. Let A^* be a lift of A to E^*, $A^*(1) = \phi$. Then by Lemma 4.1

$$\mathrm{Re}\langle\phi, \dot{v}\rangle = \mathcal{J}(\mathrm{Re}\langle A^*, \dot{A}\rangle). \tag{4.7}$$

Let χ approach $4\pi\delta_{-1}$, where δ_{-1} is Dirac's delta-function on $b\Delta$ supported at $\zeta = -1$. Then $\dot{A}|_{b\Delta} \bmod (T(M) + JT(S))$ approaches $4\pi\delta_{-1}u$, and the right-hand side of (4.7) approaches $\mathrm{Re}\langle A^*(-1), u\rangle = \mathrm{Re}\langle R_A^*\phi, u\rangle = \mathrm{Re}\langle \phi, R_A u\rangle$. Since this holds for all $\phi \in E_q^*$, $\dot{v} \to R_A u$ as $\chi \to 4\pi\delta_{-1}$. For small $t > 0$, $v(t)/t \approx \dot{v} \approx R_A u$, what we need.

The proof of Theorem 3.2 is complete.

Theorem 3.3 is proved by reduction to Theorem 3.2 in a very similar way as in [T3]. To complete the reduction, we use the well known fact that $T^c(M)$ can be viewed as the set of all "infinitely small" analytic discs attached to M.

PROPOSITION 4.2. *Let M be a C^1-smooth (not necessarily CR) manifold in $\mathbf{C}^N$ and $\zeta \mapsto A(\zeta, t)$ a C^1 one parameter family of discs attached to M such that $A(\zeta, 0) = A(1, t) = p \in M$. Then $\partial A(\zeta, 0)/\partial t \in T_p^c(M)$. Conversely, let M be a $C^{k,\alpha}$ $(k \geq 1, 0 < \alpha < 1)$ CR manifold and let K be a compact set in M. There exists a family of discs $\zeta \mapsto A(\zeta, t, p, u)$, where $t \in \mathbf{R}$, $p \in K$, and $u \in T_p^c(M)$, such that*

(i) *A is $C^{k,\beta}$ for all $0 < \beta < \alpha$, and A is $C^{k,\alpha}$ in ζ;*

(ii) *$A|_{t=0} = A|_{\zeta=1} = p$;*

(iii) *$\partial A(-1, 0, p, u)/\partial t = u$.*

For completeness, we include a proof.

PROOF. Since the question is local, we assume that M is defined by the equation $r = 0$, where r is a C^1 vector valued function in a neighborhood of $p \in \mathbf{C}^N$. Then $r(A(\zeta, t)) = 0$ on the unit circle $b\Delta$. Differentiating this equation with respect to t yields:

$$\mathrm{Re}\langle \partial r(A(\zeta, t)), \partial A(\zeta, t)/\partial t\rangle = 0, \quad \zeta \in b\Delta.$$

Setting $t = 0$, we get

$$\mathrm{Re}\langle \partial r(p), \partial A(\zeta, 0)/\partial t\rangle = 0, \quad \zeta \in b\Delta.$$

Note that the function after "Re" is holomorphic in Δ and vanishes at $\zeta = 1$. Therefore, it vanishes identically in $\bar{\Delta}$, which means precisely that $\partial A(\zeta, 0)/\partial t \in T_p^c(M)$, the first assertion of the proposition.

Conversely, let M be defined by the equation $x = h(y, w)$ as in (2.1). According to Proposition 2.1, there exists a unique disc $\zeta \mapsto A(\zeta, t, p, u)$, with the w-component equal to $p_w + t u_w(1-\zeta)/2$ and $A|_{\zeta=1} = p$, where p_w and u_w are the w-components of p and u. Then (i) follows from Theorem 1.2; by the uniqueness argument in Proposition 2.1, at $t = 0$ we obtain the constant disc $\{p\}$ which yields (ii). Further, according to the construction, (iii) holds for w-components of its sides. Since we have proved the left side belongs to $T_p^c(M)$, (iii) holds automatically for z-components, also. The proposition is proved.

PROOF OF THEOREM 3.3. Without a loss of generality, we restrict to the case in which $\gamma : [0, 1] \to S$ is small and consists of a single smooth piece. Let $\gamma(0) = p$, $\gamma(1) = q$. Following [T3], we approximate γ by a sequence of discs A_j, $j = 1, \ldots, k$, such that $A_j(1) = A_{j+1}(-1)$. For brevity, we call a sequence of discs with this property a *chain*. The parallel displacement operator along γ will then be approximated by the composition of the mappings R_{A_j} corresponding to the discs. As a result, Theorem 3.3 will follow from Theorem 3.2.

Let X be a complex tangential vector field on S having γ among its integral curves, that is $X(p) \in T_p^c(S)$, $p \in S$, and $d\gamma(t)/dt = X(\gamma(t))$, $0 \le t \le 1$.

The discs A_j, $j = 1, \ldots, k$, are constructed by Proposition 4.2 as follows. We set

$$q_k = q, \quad A_j(\zeta) = A(\zeta, 1/k, q_j, -X(q_j)), \quad q_{j-1} = A_j(-1),$$

where j runs from k to 1. So, we are moving backwards along X. According to Euler's method of solving ordinary differential equations, the sequence q_j approximates the integral curve γ of the field X. Generally, we do not get to the point p, but $q_0 \to p$ as $k \to \infty$.

Let $\phi \in E_q^*$. Let A_j^* be lifts of A_j to E^* such that they form a chain and $A_k(1) = \phi$. The discs A_j^* get arbitrarily small as $k \to \infty$. By Proposition 4.2, they approach complex tangential directions to E^*. Therefore, due to the same argument as above, they approximate the curve γ^* the lift of γ to E^* with $\gamma^*(1) = \phi$. Hence, by definition of $R_{A_j}^*$ and Π_γ^*, $(R^*A_1)^{-1} \circ \cdots \circ (R^*A_k)^{-1}(\phi)$ approaches $(\Pi_\gamma^*)^{-1}(\phi)$ though they may belong to different fibers of E^*. Since ϕ is arbitrary, we conclude by duality that $R_{A_k} \circ \cdots \circ R_{A_1}$ approaches Π_γ as $k \to \infty$.

Since q_0 approaches p, it gets to a neighborhood of p where M' is attached to M in some direction $v_0 \in E_{q_0}$ close to $u \in E_p$. Let $M_0 = M'$. We apply Theorem 3.2 successively to get a sequence of manifolds M_j attached to M at (q_j, v_j) so that u_j is close to $R_{A_j} u_{j-1}$, $1 \le j \le k$, and every $f \in \mathrm{CR}(M \cup M')$ extends to be CR on every M_j. Hence, the manifold $M'' = M_k$, that is attached to M at (q, v), where $v = v_k$ is arbitrarily close to $\Pi_\gamma u$, is what we need. The proof is complete.

References

[BT] M. S. Baouendi and F. Treves, *A property of the functions and distributions annihilated by a locally integrable system of complex vector fields*, Ann. of Math. **114** (1981), 387–421.

[B] E. Bishop, *Differentiable manifolds in complex Euclidean space*, Duke Math. J. **32** (1965), 1–21.

[Bo] A. Boggess, *CR manifolds and the tangential Cauchy-Riemann complex*, CRC Press, 1991.

[HS] N. Hanges and J. Sjöstrand, *Propagation of analyticity for a class of non-micro-characteristic operators*, Ann. Math. **116** (1982), 559–577.

[HT] N. Hanges and F. Treves, *Propagation of holomorphic extendibility of CR function*, Math. Ann. **263** (1983), 157–177.

[J] B. Jöricke, *Deformation of CR manifolds, minimal poits and CR manifolds with the microlocal analytic extention property*, to appear.

[M] J. Merker, *Global minimality of generic manifolds and holomorphic extendibility of CR functions*, Duke Math. J., to appear.

[Tr] J.-M. Trépreau, *Sur la propagation des singularites dans les varietes CR*, Bull. Soc. Math. France **118** (1990), 403–450.

[T1] A. E. Tumanov, *Extending CR functions on a manifold of finite type over a wedge*, Mat. Sbornik **136** (1988), 129–140.

[T2] ———, *Extending CR functions into a wedge*, Mat. Sbornik **181** (1990), 951–964.

[T3] ———, *Connections and propagation of analyticity for CR functions*, Duke Math. J. **73** (1994), 1–24.

[T4] ———, *Analytic discs and the regularity of CR mappings in higher codimension*, Duke Math. J., to appear.

Completions of Instantons Moduli Space and Control Theory

Giorgio Valli, Dipartimento di Matematica, Università di Roma Tor Vergata, Via della Ricerca Scientifica, 00133 Roma, Italy

Introduction.

Let $\mathcal{M}_{d,N}$ be the moduli space of framed SU(N)-instantons over the 4-sphere, with instanton number d; or, equivalently (Donaldson, 1984), the moduli space of algebraic vector bundles E over $P^2(\mathbb{C})$, with a fixed trivialization at the $\mathbb{P}^1$ at infinity, of rank N, and $c_2(E) = d$.

In this note we will describe two different "completions" of $\mathcal{M}_{d,N}$, which we will call $\overline{\mathcal{M}}^P{}_{d,N}$ and $\overline{\mathcal{M}}_{d,N}$. For moduli spaces of (usually unframed) instantons, two compactifications have generally been studied: the "Uhlenbeck compactification", defined in terms of "ideal connections" with δ-function singularities in the curvatures; and the "Gieseker compactification", defined in term of equivalence classes of semistable sheaves: and the relations between these two constructions have been recently studied (Li, 1993, and Morgan, 1993).

For framed instantons, the right notion is of a completion. Anyway, $\overline{\mathcal{M}}^P{}_{d,N}$ is very much related to the Uhlenbeck compactification, while we conjecture that $\overline{\mathcal{M}}_{d,N}$ is actually the natural restriction of the Gieseker compactification. Another difference between the two is that $\overline{\mathcal{M}}^P{}_{d,N}$ is a contractible affine variety, while $\overline{\mathcal{M}}_{d,N}$ is non-singular, but with non trivial topology.

The advantage of $\overline{\mathcal{M}}^P{}_{d,N}$ and $\overline{\mathcal{M}}_{d,N}$ on more elaborate constructions, is that they are defined in terms of the Barth-Donaldson monads, i.e. they are linear algebraic in nature. This allows many constructions and computations, but one has to use many techniques and results from control theory literature, which do not appear to be well-known to pure mathematicians.

Let us sketch the construction of $\overline{\mathcal{M}}^P{}_{d,N}$ and $\overline{\mathcal{M}}_{d,N}$. As a first step, (partially following Sontag (1987)) we define the group $\mathcal{R}_N$ of rational functions on $P^2(\mathbb{C})$, with values in $SL(N,\mathbb{C})$, which are the identity at the $\mathbb{P}^1$ at infinity, and which are non singular outside a "grid" $\{z=\alpha_i \text{ or } w=\beta_j\}$ (with z,w inhomogeneous coordinates). There is a natural 1 to 1 correspondance between $\mathcal{R}_N$

and the disjoint union (over $d \in \mathbb{N}$) of the $\mathcal{M}_{d,N}$'s. This may be proved in many ways (Sontag, 1987, and Valli, 1992); for example by seeing any R(z,w) as a "canonical transition function" of an associated holomorphic $\mathbb{C}^N$-bundle over $\mathbb{P}^2$. This yields a formula (due to A.D. King).

$$R(z,w) = I + C(z-A_1)^{-1}(w-A_2)^{-1}B$$

where A_1, A_2, B,C are monad matrices defining E. By multiplying elements in $\mathcal{R}_N$ by scalar polynomials, we get an algebraic embedding of $\mathcal{M}_{d,N}$ into a corresponding affine space of polynomials: and we define $\overline{\mathcal{M}^P}_{d,N}$ as the closure in such space of the quasi-projective $\mathcal{M}_{d,N}$. We have a natural algebraic stratification $\overline{\mathcal{M}^P}_{d,N} = \Sigma_{0 \le k \le d} (\mathcal{M}_{d-k,N} \times \mathbb{C}^{2k})$.

The completion $\overline{\mathcal{M}}_{d,N}$ has a simpler definition, obtained by relaxing the non-degeneracy condition for the monads. Its properties are, on the contrary, slightly subtler to prove. For example, $\overline{\mathcal{M}}_{d,N}$ is smooth, has a natural holomorphic and proper map $\overline{\mathcal{M}}_{d,N} \longrightarrow \overline{\mathcal{M}^P}_{d,N}$ (with generic fibre at the boundary $\mathbb{P}^{N-1}$), and it stratifies accordingly. Moreover, $\mathcal{M}_{d,N}$ is homotopy equivalent to a singular compact algebraic variety, a Grassmannian of invariant subspaces of $\mathbb{C}^{d(d+1)N}$ under two commuting nilpotent transformations (shifts). This opens the way to topological and geometrical applications (see theorem 10).

Note. This paper is in final form, and no part of it will be published elsewhere.

1 Instantons as rational matrix functions on $P^2(\mathbb{C})$.

Let $\mathcal{M}_{d,N}$ be the moduli space of self dual Yang-Mills SU(N)-connections over S^4, of instanton number d, modulo gauge transformations which are the identity at the point at infinity. The space $\mathcal{M}_{d,N}$ is often referred to as the moduli space of "framed" instantons. It is a smooth, complex, connected, non compact manifold of complex dimension 8dN ($N \ge 2$). By a theorem of Donaldson (1984), $\mathcal{M}_{d,N}$ is diffeomorphic to the moduli space of holomorphic vector bundles E of rank N and second Chern class d on $P^2(\mathbb{C})$, with a fixed holomorphic trivialization at $\mathbb{P}^1$ at infinity. Let us introduce inhomogeneous coordinates (z,w) on $\mathbb{P}^2$. $\mathcal{M}_{d,N}$ can be described as the moduli space of matrices (Barth-Donaldson monads) $A_1:\mathbb{C}^d \longrightarrow \mathbb{C}^d$, $A_2:\mathbb{C}^d \longrightarrow \mathbb{C}^d$, $B:\mathbb{C}^N \longrightarrow \mathbb{C}^d$, $C:\mathbb{C}^d \longrightarrow \mathbb{C}^N$, satisfying the following

$$\forall (z,w) \in \mathbb{C}^2, (A_1-z, A_2-w, B):\mathbb{C}^{2d+N} \longrightarrow \mathbb{C}^d \quad \text{is surjective;} \qquad \textbf{(1A)}$$

$$\forall (z,w) \in \mathbb{C}^2, \begin{pmatrix} A_1-z \\ A_2-w \\ C \end{pmatrix} :\mathbb{C}^d \longrightarrow \mathbb{C}^{2d+N} \quad \text{is injective;} \qquad \textbf{(1B)}$$

$$[A_1, A_2] + BC = 0 \qquad \textbf{(2)}$$

modulo the $GL(d,\mathbb{C})$ action

$$A_i \longmapsto S^{-1}A_iS, \; B \longmapsto S^{-1}B, \; C \longmapsto CS. \qquad \textbf{(3)}$$

Let us define a "grid" (or "net") in $P^2(\mathbb{C})$ as the union of a finite number of "vertical lines" $\{z=\alpha_i\}_{1 \le i \le h}$, and of a finite number of "horizontal lines" $\{w=\beta_j\}_{1 \le j \le k}$. Let $\mathcal{S}_N$ be the ring of rational functions $R(z,w)$ on $P^2(\mathbb{C})$, with values in N×N matrices, with poles contained in some grid. Partially following Sontag (1987), we define the group $\mathcal{R}_N$ of those rational matrix functions $R \in \mathcal{S}_N$ satisfying

1) $R = I$ at P^1 at infinity;

2) det R= 1 almost everywhere.

Let us denote $\mathcal{M}_N$ the disjoint union (over $d \in \mathbb{N}$) of the $\mathcal{M}_{d,N}$'s.

Proposition 1. There is a natural bijective correspondence between $\mathcal{M}_N$ and $\mathcal{R}_N$.

Proof.

A detailed proof is in Valli (1992), see also Sontag (1987).

We can give an interpretation of the elements as "canonical transition functions" for the bundles E in $\mathcal{M}_{d,N}$, arguing as follows. Outside a finite number of "vertical jumping lines" $\{z=\alpha_i\}_{1 \le i \le h}$, the trivialization at infinity induces a trivialization of E

$$\Phi_1: (P^2(\mathbb{C}) - \cup_{1 \le i \le h} \{z=\alpha_i\}) \times \mathbb{C}^N \longrightarrow E_{|(P^2(\mathbb{C}) - \cup_{1 \le i \le h} \{z=\alpha_i\})}$$

Similarly, we have a trivialization Φ_2 of E, outside the union of a finite number of "horizontal jumping lines" $\{w=\beta_j\}_{1 \le j \le k}$.

The map $R = (\Phi_2)^{-1}\Phi_1$ is an element of $\mathcal{R}_N$. Indeed, it is an element of $\mathcal{S}_N$, satisfying 1). Moreover, its inverse has the same properties, therefore R is invertible outside the grid. The determinant det R is a scalar rational function satisfying the same properties; and it is easy to show that it is constant, and therefore equal to 1, being 1 at infinity.

Conversely, given $R(z,w) \in \mathcal{R}_N$, we can use it as a transition function to construct an algebraic vector bundle over $P^2(\mathbb{C})$-the grid, which extends uniquely to an algebraic vector bundle over $P^2(\mathbb{C})$, with the desired properties. ♦

We have also an explicit formula, due to A.D. King, for the element $R(z,w)$ corresponding to the vector bundle $E \in \mathcal{M}_{d,N}$, in terms of the monad matrices of E; its proof is by constructing the trivializations in terms of the monads.

Proposition 2. (A.D. King) The rational matrix function $R(z,w)$ associated to the monad data (A_1,A_2,B,C) is

$$R(z,w) = I + C(z-A_1)^{-1}(w-A_2)^{-1}B \tag{4}$$

and we have

$$R(z,w)^{-1} = I - C(w-A_2)^{-1}(z-A_1)^{-1}B \tag{5}$$

Remarks. These formulas show how the coefficients in the power series expansion of each $R(z,w)$ are the "invariants" (in the sense of classical invariant theory) of the $GL(d,\mathbb{C})$-action .

Formulas above do not appear to be much known; to our knowledge, their only brief appearance in the mathematical literature is in (Sontag, 1987). In (Valli, 1992) there is a proof of proposition 1 in terms of loop group theory (which is related to monads on $\mathbb{P}^1$, see Valli (1994)). It is also easy to show how (5) follows from (4) by using equation (2).

We remark that proposition 1 has no topological content.

2 Instantons as matrix-valued polynomials in two complex variables.

In order to find a holomorphic (and topological) version of proposition 1, we multiply elements in $\mathcal{R}_N$ by scalar polynomials in the two complex variables z,w: we obtain matrix polynomials; and for spaces of polynomials, the topology and the complex structures are clearly defined.

Let P(z,w) be a polynomial in the two complex variables z,w, with coefficients in N×N complex matrices. We call P ***monic***, of bidegree (d,d), if there exist monic scalar polynomials in one complex variable r,s, such that

$$P(z,w) = r(z)s(w) + \Sigma_{0\leq i,j\leq d-1} M_{ij}z^i w^j \qquad (6)$$

We call a monic polynomial P(z,w) in (5) ***based*** if $\det(P)= r(z)^N s(w)^N$. This implies that P is an invertible matrix outside a grid in $\mathbb{C}^2$. Let us denote

$$\mathcal{P}_{d,N} = \{\text{based monic polynomials in (z,w), of bidegree (d,d)}\}$$

We remark that each space $\mathcal{P}_{d,N}$ is an affine algebraic variety.

Let us denote $\mathcal{P}_N$ as the disjoint union (over d´N) of the $\mathcal{P}_{d,N}$'s. Let ≅ be the equivalence relation over $\mathcal{P}_{d,N}$, generated by multiplication by scalar monomials $(z-\alpha)(w-\beta)$.

Lemma 3. We have a natural bijective correspondence G between $\mathcal{P}_N/\cong$ and $\mathcal{R}_N$.

Proof.

If $P = r(z)s(w) + \Sigma_{0\leq i,j\leq d-1} M_{ij}\, z^i w^j \in \mathcal{P}_N$, we define $G(P) = P/_{r(z)s(w)}$. We claim that G establishes a bijective correspondence between $\mathcal{P}_N/\cong$ and $\mathcal{R}_N$.

It is obvious that G is invariant under ≅. Moreover $P/_{r(z)s(w)} \in \mathcal{R}_N$, because P is a based monic polynomial. G is obviously injective; and every rational function $R(z,w) \in \mathcal{R}_N$ comes from a polynomial, which must be based and monic because det(R) = 1, and R= I at infinity. ♦

Proposition 4. For any integer d, there is an algebraic affine embedding

$$\Psi_d: \mathcal{M}_{d,N} \longrightarrow \mathcal{P}_{d,N}$$

covering the bijection between $\mathcal{M}_N$ and $\mathcal{R}_N$ given by proposition 1.

Proof.

Using King's formula, we define P(z,w) in terms of the monad matrices (A_1,A_2,B,C) by

$$P(z,w) = \det(A_1-z)\det(A_2-w) + C\,(z-A_1)^{\#}(w-A_2)^{\#}B \qquad (7)$$

(where, if X is a square matrix,we denote $X^{\#}= \det(X)X^{-1}$).

This formula clearly shows that Ψ_d is algebraic. It is proved in (Valli, 1992) that it has maximal rank, because it is obtained by composing each element in $\mathcal{M}_{d,N}$ (seen as a moduli space of holomorphic maps) with a fixed holomorphic map. ♦

We can now define the completion $\overline{\mathcal{M}}^P{}_{d,N}$ as the closure of $\Psi_d(\mathcal{M}_{d,N})$ in $\mathcal{P}_{d,N}$.

Proposition 5.

1) $\overline{\mathcal{M}}^P{}_{d,N}$ is a contractible algebraic affine variety.

2) $\overline{\mathcal{M}}^P{}_{d,N}$ has a stratification

$$\overline{\mathcal{M}}^P{}_{d,N} = \cup_{0\leq k\leq d} \mathcal{M}_{k,N} \times \mathbb{C}^{2(d-k)} \qquad (8)$$

Proof.

1) Consider the scaling action of $\mathbb{R}_+$ on $\mathcal{P}_{d,N}$ given by $P_t(z,w) = 1/_{t^{2d}}\, P(tz,tw)$. It

corresponds to an action on $\mathcal{M}_{d,N}$ given by

$$(A_1,A_2,B,C) \longmapsto (1/_t A_1, 1/_t A_2, 1/_t B, 1/_t C).$$

Taking the limit for $t \longrightarrow \infty$, we get a retraction of $\overline{\mathcal{M}^P}_{d,N}$ into the point (i.e. the polynomial) $z^d w^d$.

2) Let $P(z,w) = r(z)s(w) + \Sigma_{0\leq i,j\leq d-1} M_{ij} z^i w^j \in \mathcal{P}_{d,N}$. Then $P = r(z)s(w)R(z,w)$ with $R \in \mathcal{R}_N$ coming from an $E \in \mathcal{M}_{d,N}$.

We can write uniquely $R(z,w) = P^*(z,w)/_{r^*(z)s^*(w)}$, with $P^*(z,w) \in \Psi_d (\mathcal{M}_{d,N})$ and $r^*(z)$, $s^*(w)$ polynomials of degree k. When $P \in \mathcal{M}_{d,N}$, then $P=P^*$, $d=d'$, $r=r^*$, $s=s^*$. By semicontinuity, we see that $P \in \overline{\mathcal{M}^P}_{d,N}$ if and only if $k \leq d$, and the polynomials r^*, s^* divide r, s (respectively). The integer k stratifies $\overline{\mathcal{M}^P}_{d,N}$, and the unique corresponding stratum is uniquely described by R, and by the monic polynomials, of degree d-k, $r/_{r^*}$, $s/_{s^*}$. ♦

Remark. One can construct an "Uhlenbeck completion" of $\mathcal{M}_{d,N}$, as a restriction of the standard Uhlenbeck compactification (Li (1993) and Morgan (1993)). It is contractible, and it has a stratification with strata $\mathcal{M}_{k,N} \times S^{(d-k)}(\mathbb{C}^2)$; i.e. with the symmetric product $S^{(d-k)}(\mathbb{C}^2)$ in place of $S^{(d-k)}(\mathbb{C}) \times S^{(d-k)}(\mathbb{C})$.

3 The smooth completion $\mathcal{M}_{d,N}$.

Let us define the moduli space $\overline{\mathcal{M}}_{d,N}$ of matrices (A_1,A_2,B,C) satisfying (1A) and (2), i.e.

$\forall\ (z,w) \in \mathbb{C}^2$, $(A_1-z, A_2-w, B):\mathbb{C}^{2d+N} \longrightarrow \mathbb{C}^d$ is surjective; **(1A)**

$[A_1, A_2] + BC = 0$ **(2)**

modulo the $GL(d,\mathbb{C})$ action (3).

Proposition 6.

1) $\overline{\mathcal{M}}_{d,N}$ has a natural stucture of a smooth complex manifold of complex dimension 2dN.

2) $\mathcal{M}_{d,N}$ sits inside $\overline{\mathcal{M}}_{d,N}$ as an open dense subset.

3) There is a proper holomorphic map $\pi: \overline{\mathcal{M}}_{d,N} \longrightarrow \overline{\mathcal{M}^P}_{d,N}$ which is the identity on $\mathcal{M}_{d,N}$, and has generic fibre $\mathbb{P}^{N-1}$ at the boundary $\overline{\mathcal{M}}_{d,N} - \mathcal{M}_{d,N}$.

Proof.

1) It is not difficult to prove that, because of condition (1A), the linearization of the map $(A_1,A_2,B,C) \longmapsto [A_1, A_2] + BC$ has maximal rank d^2. Similarly, again by (1A), the $GL(d,\mathbb{C})$-action on the space of matrices (A_1,A_2,B) satisfying (1A) is free and with a closed graph (see Helmke (1986 and 1993)); therefore the same is true for the action (3) on the complete space of matrices (A_1,A_2,B,C) satisfying (1A) and (2). By general principles, this proves 1).

2) is obvious.

3) Define $\pi(A_1,A_2,B,C)$ to be the polynomial in $\mathcal{P}_{d,N}$ given by King's formula (7) . The map π is clearly the identity on $\mathcal{M}_{d,N}$, therefore its image is actually contained in $\overline{\mathcal{M}^P}_{d,N}$.

To prove properness is slightly more substantial. Let us take a sequence $(A_1,A_2,B,C)(n)$ of monad matrices satisfying (1A) and (2); and suppose the corresponding sequence of matrix polynomials $P_n(z,w)$ converges. We want to prove that there exist a sequence $g_n \in GL(d,\mathbb{C})$ such that the sequence of monads obtained by applying g_n to $(A_1,A_2,B,C)(n)$ has a convergent subsequence .

Let us recall from control theory the definition of the space $\Sigma_{d,k}$; it is defined as the moduli space of matrices (A,B), $A:\mathbb{C}^d \longrightarrow \mathbb{C}^d$, $B:\mathbb{C}^k \longrightarrow \mathbb{C}^d$, satisfying

$$\forall z \in \mathbb{C}, (A-z, B):\mathbb{C}^{d+k} \longrightarrow \mathbb{C}^d \quad \text{is surjective} \qquad \textbf{(9)}$$

modulo the usual $GL(d,\mathbb{C})$-action (3).

The space $\Sigma_{d,k}$ has been widely studied; in particular (Helmke, 1986 and 1993) it has the homology of the Grassmannian $G_d(\mathbb{C}^{d+k-1})$, and the map $\Sigma_{d,k} \longrightarrow \mathbb{C}^d$ which assign to any (A,B) the coefficients of the characteristic polynomial of A is proper.

We divide the proof into a number of steps, each being sufficiently elementary.

Step 1) For any (A_1,A_2,B,C) satisfying 1A) and 2), let us consider the block d×dN matrix

$$B^* = \{B, A_2B, A_2^2B, \;, A_2^{d-1}B\} \qquad \textbf{(10)}$$

Then the pair (A_2,B^*) satisfies (9), and therefore represents an element of $\Sigma_{d,dN}$.

This follows immediately from (1A), (2), and the definitions.

Step 2) The convergence of $P_n(z,w) = \pi(A_1,A_2,B,C)(n)$ implies a uniform bound on the characteristic polynomials of A_1,A_2; which therefore subconverge. In particular, the sequence $(A_1,B^*)(n) \in \Sigma_{d,dN}$ subconverges in $\Sigma_{d,dN}$.

This follows immediately from the compactness properties of $\Sigma_{d,k}$ quoted above. This shows that, up to the $GL(d,\mathbb{C})$-action, A_1 and B^* subconverge. It is standard to see that this is equivalent to the subconvergence of $(A_1^iA_2^j B)$ for any (i,j), $0 \le i,j \le d-1$.

Step 3) Up to subsequence, each matrix $(CA_1^iA_2^j B)$ $0 \le i,j \le d-1$, converges.

This is independent from the second step. Observe that the $(CA_1^iA_2^j B)$'s are the coefficients of the power series expansion in z^{-1}, w^{-1} of the associated rational function $R(z,w) \in \mathcal{R}_N$. The convergence of $P(z,w) = r(z)s(w) + \sum_{0 \le i,j \le d-1} M_{ij} z^i w^j = r(z)s(w)R(z,w)$ and of the characteristic polynomials r, s, prove the statement.

Step 4) The sequence of C's also converges.

Indeed, let $\mathcal{S} = \mathcal{S}(A_1,A_2,B)$ be the $d \times d^2N$ matrix having the matrices $A_1^iA_2^j B$ as column blocks. We define similarly the $N \times d^2N$ matrix $\mathcal{C} = \mathcal{C}(A_1,A_2,B) = \{CA_1^iA_2^j B\}$. We have $C\mathcal{S} = \mathcal{C}$. But it is standard to see that any $\mathcal{S}$ must have rank d, as a consequence of the non-degeneracy condition (1A), and of step 2. Therefore $\mathcal{S}$ has a right inverse, and the sequence of C's converges.

We omit the proof of the last statement, but, as a special case, see the forthcoming examples (the proof in the general case is not much harder). ♦

Proposition 7.

1)The completion $\overline{\mathcal{M}}_{d,N}$ has a stratification

$$\overline{\mathcal{M}}_{d,N} = \cup_{0 \le k \le d} \mathcal{M}^k{}_{d,N} \qquad \textbf{(11)}$$

where $\mathcal{M}^k{}_{d,N}$ is a complex analytic variety of complex dimension $2kN + (d-k)(N+1)$.

2) There exist algebraic maps

$$p_k: \mathcal{M}^k{}_{d,N} \longrightarrow \mathcal{M}_{k,N} \qquad \textbf{(12)}$$

which are stratified submersions with respect to a natural stratification of $\mathcal{M}^k{}_{d,N}$ by smooth complex submanifolds.

Proof (sketch).

We omit the complete proof. The integer k can be easily defined as measuring the amount by which (1B) is not satisfied. Then one has to decompose (A_1,A_2,B,C) into block matrices (of dimension depending on k). One block represents a genuine element in $\mathcal{M}_{k,N}$, and gives the map p_k. What is left (the fiber) has singularities modelled on the space of pairs of commuting square matrices of size (d-k), and therefore constant on $\mathcal{M}_{k,N}$. ♦

One can formalize the above remarks as follows. Despite the fact that the fibers of p_k are not compact, due to the possibility of the eigenvalues going to infinity, one can restrict to the case (equivalent from a topological point of view) of having the spectrum of A_1,A_2 contained in a ball. Then the restriction of each map p_k is a locally trivial fiber bundle, by Thom's first isotopy lemma for manifolds with faces (see Verona, 1984). This has important topological applications (see theorem 10).

Examples.

1) Let us take d=1, N= 2. Then the rational functions associated to elements of $\mathcal{M}_{1,2}$ are exactly those of the form

$$R(z,w) = I + M/_{(z-\alpha)(w-\beta)}$$

where M is a nilpotent *non zero* 2×2 matrix.

Therefore $\mathcal{M}_{1,2}$ is $\mathbb{C}^2 \times \mathcal{N}^*$, where $\mathcal{N}^*$ is the space of such matrices. The map which associates to any $M\in \mathcal{N}^*$ its kernel describes $\mathcal{N}^*$ as a principal $\mathbb{C}^*$-bundle over $\mathbb{P}^1$; and one can prove that it is topologically the natural homogeneous map $SO(3)\longrightarrow S^2$.

The completion $\overline{\mathcal{M}}^P{}_{1,2}$ is $\mathbb{C}^2 \times \mathcal{N}$, where $\mathcal{N}$ is the space of nilpotent 2×2 matrices, obtained by adding one point (the zero matrix) to $\mathcal{N}^*$.

The completion $\overline{\mathcal{M}}_{1,2}$ is $\mathbb{C}^2 \times \tilde{\mathcal{N}}$, which is obtained by adding the zero section to the principal $\mathbb{C}^*$ - bundle $\mathcal{N}^* \longrightarrow \mathbb{P}^1$, obtaining a $\mathbb{C}$-bundle. In other words,

$$\tilde{\mathcal{N}} = \{ (M,v) \in \mathcal{N} \times \mathbb{P}^1 \mid v \in \ker M \}.$$

We remark that $\overline{\mathcal{M}}_{1,2}$ has a simpler topology than $\mathcal{M}_{1,2}$ (as expected).

2) In the stratification given by proposition 7, the lower dimensional stratum is

$$\mathcal{M}^0{}_{d,N} = \{(A_1,A_2,B), \text{ satisfying (1A) and } [A_1,A_2]= 0 \} / GL(d,\mathbb{C})$$

It is actually homotopy equivalent to the whole completion $\overline{\mathcal{M}}_{d,N}$ (see lemma 8).

Generically in $\mathcal{M}^0{}_{d,N}$, either A_1 or A_2 have d distinct eigenvalues. It follows that $\mathcal{M}^0{}_{d,N}$ has an open dense subset diffeomorphic to $C_4(d,\mathbb{P}^{N-1})$, the configuration space of d distinct points in $\mathbb{R}^4$ (the joint common spectrum (α_i,β_i) of A_1, A_2), with "labels" in $\mathbb{P}^{N-1}$ (the corresponding row vector in B).

3) The space $\mathcal{M}_{d,1}$ is empty. Nevertheless one can define the moduli spaces $\overline{\mathcal{M}}_{d,1}$ and $\mathcal{M}^0{}_{d,1}$.

It is easily checked that $\mathcal{M}^0{}_{d,1}$ is not singular. The map $\mathcal{M}^0{}_{d,1}\longrightarrow S^d(\mathbb{R}^4)$ which associates to $(A_1,A_2,b)\in \mathcal{M}^0{}_{d,1}$ the joint spectrum of (A_1,A_2), describes $\mathcal{M}^0{}_{d,1}$ as a **desingularization** of the symmetric product $S^d(\mathbb{R}^4)$. ♦

Let us conclude this paper with a suggestion of possible topological applications of the constructions above.

Lemma 8. The completion $\overline{\mathcal{M}_{d,N}}$ is homotopically equivalent to the moduli space

$\mathcal{H}_{d,N} = \{(A_1,A_2,B), A_1, A_2$ nilpotent, satisfying (1A) and $[A_1,A_2]= 0 \} / GL(d,\mathbb{C})$

Proof.

Let us consider the vector valued function F on $\overline{\mathcal{M}_{d,N}}$, whose components are the coefficients of the characteristic polynomials of A_1, A_2; and the matrices $\{CA_1^{\,i}A_2^{\,j}B\}$; and let $f = |F|^2$. The function f has $\mathcal{H}_{d,N}$ as (degenerate) set of minima, and it has no other critical submanifold. By following the gradient flow of f, one gets a continuous (but not even C^1, because $\overline{\mathcal{M}_{d,N}}$ is smooth, while $\mathcal{H}_{d,N}$ is singular) retraction of $\overline{\mathcal{M}_{d,N}}$ onto $\mathcal{H}_{d,N}$.

Let us consider the map (see Helmke, 1993).

$$\phi: \mathcal{H}_{d,N} \longrightarrow G_d(\mathbb{C}^{d(d+1)N}) \qquad (13)$$

defined by $(A_1,A_2,B) \longmapsto$ Image of the rows of $(A_1^{\,i}A_2^{\,j}B)$. The map ϕ describes an algebraic embedding of the compact complex analytic variety $\mathcal{H}_{d,N}$. We can describe the image of ϕ as follows.

Let I be the set of indexes $I = \{(i,j) \mid 0 \le i,j \le d-1, i+j \le d-1\}$. Let us consider $V = \mathbb{C}^I \otimes \mathbb{C}^N$. Let us take shift operators T, S acting on the full space of indexes $\mathbb{N}\times\mathbb{N}$ by $T(i,j) = (i-1,j)$, $S(i,j)= (i,j-1)$. The operators T, S define nilpotent commuting operators on V, by agreeing that no vector corresponds in V to negative indices.

Let us consider the Grassmannian $\mathcal{G}_{d,N}$ of those d-dimensional complex subspaces of V, invariant under T,S. It is a compact singular algebraic variety of dimension dN -1. The following is an easy generalization of lemma 1.9 in (Helmke, 1993).

Proposition 9. The map ϕ gives a biholomorphism between $\mathcal{H}_{d,N}$ and $\mathcal{G}_{d,N}$. ♦

Remark. By intersecting Schubert cells in $G_d(\mathbb{C}^{d(d+1)N})$ with $\mathcal{H}_{d,N}$, one can get, in the lower dimensional cases, a cell decomposition of $\mathcal{H}_{d,N}$, thus computing the homology. One gets (after the trivial case $\mathcal{H}_{1,2}= \mathbb{P}^1$)

$$H_0(\mathcal{H}_{2,2},\mathbb{Z}) = \mathbb{Z} \quad H_2(\mathcal{H}_{2,2},\mathbb{Z}) = \mathbb{Z} \quad H_4(\mathcal{H}_{2,2},\mathbb{Z}) = \mathbb{Z}^2 \quad H_6(\mathcal{H}_{2,2},\mathbb{Z}) = \mathbb{Z}$$

$$H_0(\mathcal{H}_{3,2},\mathbb{Z}) = \mathbb{Z}, \ H_2(\mathcal{H}_{3,2},\mathbb{Z}) = \mathbb{Z}, \ H_4(\mathcal{H}_{3,2},\mathbb{Z}) = \mathbb{Z}^2,$$

$$H_6(\mathcal{H}_{3,2},\mathbb{Z}) = \mathbb{Z}^3, \ H_8(\mathcal{H}_{3,2},\mathbb{Z}) = \mathbb{Z}^2, \ H_{10}(\mathcal{H}_{3,2},\mathbb{Z}) = \mathbb{Z}$$

while all the odd dimensional groups are 0.

(By a similar method, we have computed the homology also in the cases N=2 and $d \le 6$; and N=3, $d \le 3$, finding similar patterns).

Let BU(d) be the "classifying space" of d-dimensional complex subspaces of a separable Hilbert space. The map ϕ gives an embedding $\mathcal{H}_{d,N} \longrightarrow BU(d)$. We make the following conjecture (which we have checked in the cases N=2 , $d \le 6$; and N=3, $d \le 3$).

Conjecture. The embedding $\mathcal{H}_{d,N} \longrightarrow BU(d)$ induces isomorphisms in homology and homotopy groups up to dimension 2d+1. ♦

It is easy, from the monad description, to construct a map $\mathcal{M}_{d,N} \longrightarrow \mathcal{M}_{d+1,N}$ (see Kirwan, 1992). It is the so-called "Taubes map" (gluing of one instanton); one can see that it extends to a map between the completions. Let us take N=2 for simplicity. By a relative standard inductive argument going back to Arnold (see §5 in Segal, 1979), and by using the remarks after proposition 7, we get the following (note that the hypotheses are much weaker that the above conjecture).

Theorem 10. Suppose that Taubes map $\overline{\mathcal{M}}_{d,2} \longrightarrow \overline{\mathcal{M}}_{d+1,2}$ induces an isomorphism in homology groups up to range d. Then the induced maps in homology

$$H_i(\mathcal{M}_{d,2},\mathbb{Z}) \longrightarrow H_i(\mathcal{M}_{d+1,2},\mathbb{Z})$$

are isomorphisms for i<d, and surjections for i=d. ♦

The above theorem is noteworthy because it settles the so-called Atiyah-Jones conjecture in a homotopy range that it is widely expected, in terms of the topology of a simpler object.

For example, by the remarks after proposition 9, we have that the Taubes maps $\mathcal{M}_{d,2} \longrightarrow \mathcal{M}_{d+1,2}$, for $d \le 5$, induces maps $H_i(\mathcal{M}_{d,2},\mathbb{Z}) \longrightarrow H_i(\mathcal{M}_{d+1,2},\mathbb{Z})$ which are isomorphisms for $i < d$, and surjections for $i \le d$.

Remark. R. Cohen and J. Jones have recently proved, by using the Uhlenbeck completion, that the maps $H_i(\mathcal{M}_{d,2},\mathbb{Q}) \longrightarrow H_i(\mathcal{M}_{d+1,2},\mathbb{Q})$ are isomorphisms for $i < 3d$, and surjections for $i=3d$.

References.

Donaldson S. K.(1984). *Instantons and geometric invariant theory,* Comm. Math. Phys. **93** , 453-460.

Helmke U. (1986). *Topology of the moduli space for reachable dynamical systems: the complex case,* Math. Systems Theory **19**, 155-187.

Helmke U. (1993). *The cohomology of Moduli Spaces of Linear Dynamical Systems,* Regensburg.

Kirwan F. (1992), *Geometric invariant theory and the Atiyah-Jones conjecture,* Preprint Oxford .

Jun Li (1993). *Algebraic Geometric interpretation of Donaldson's polynomial invariants,* J. Differ. Geometry **37** , 417-466.

Morgan J.W. (1993). *Comparison of the Donaldson polynomial invariants with their algebro-geometric analogue,* Topology **32** , 449-488.

Sontag E. (1987). *A remark on bilinear systems and moduli spaces of instantons,* Systems Control Letters **9** , 361-367.

Segal G. (1979). *The topology of spaces of rational functions,* Acta Mat., **193**, 39-72.

Valli G. (1992). *Instantons as matrix-valued rational functions on CP(2),* Preprint University of Rome 2.

Valli G. (1994). *Interpolation theory, loop groups and instantons,* J. reine angew. Math. **446**, 137-163.

Verona A. (1984). *Stratified Mappings - Structure and Triangulability,* Springer LNM **1102**.

Itérées et Points Fixes d'Applications Holomorphes

Jean-Pierre VIGUÉ ; Mathématiques ; URA CNRS D1322 Groupes de Lie et Géométrie ; Université de Poitiers ; 40, avenue du Recteur Pineau ; F-86022 POITIERS CEDEX ; FRANCE

1. INTRODUCTION

J'ai déjà consacré un certain nombre de travaux à l'étude des points fixes des applications holomorphes [16, 17, 18 et 19]. En particulier dans [19], étant donnés un domaine borné D de $\mathbb{C}^n$ et une application holomorphe $f : D \longrightarrow D$, j'ai considéré comme H. Cartan [4] et E. Bedford [3] la suite des itérées $f^n = f \circ \ldots \circ f$ (n fois) de f. Ceci m'a permis de montrer en particulier que l'ensemble Fix f des points fixes de f est une sous-variété fermée (non nécessairement connexe) de D.

Une partie de ces travaux ont été généralisés à la dimension infinie par P. Mazet et J.-P. Vigué [10 et 11]. Là aussi, les itérées de certaines applications holomorphes jouent un rôle fondamental. D'autre part, M. Abate [1 et 2] a bien montré, au moins dans le cas convexe, la relation entre l'existence de points fixes de f et le fait que la suite des itérées f^n de f n'est pas compactement divergente.

Dans cet article, nous allons considérer un domaine borné taut de $\mathbb{C}^n$ ou une variété complexe hyperbolique taut de dimension finie X. Soit M une variété complexe. Nous allons chercher et étudier les variétés-limites et les points fixes d'une application holomorphe $f : M \times X \longrightarrow X$. Nous allons montrer qu'en fait, pour tout $m \in M$, les ensembles Fix $f(m, .)$ sont deux à deux isomorphes. Ces résultats étendent et précisent des résultats de [18].

Dans une dernière partie, nous utiliserons ces résultats pour étudier les points fixes d'une application holomorphe

$$f : X_1 \times X_2 \longrightarrow X_1 \times X_2,$$

où $X_1 \times X_2$ est un produit de variétés hyperboliques taut. En particulier, nous étudierons les rapports entre l'existence de points fixes pour f et pour chacune des applications partielles $f_2(x_1, .)$ et $f_1(., x_2)$.

Alors que nous avions achevé la rédaction de cet article, M. Abate nous a communiqué son preprint Iteration of holomorphic families dans lequel il montre, de manière indépendante, certains résultats sur les variétés-limites d'une application holomorphe dépendant holomorphiquement d'un paramètre. Il étudie aussi les suites compactement divergentes.

[C'est pour moi un grand plaisir que ces résultats soient publiés dans les Actes du congrès Complex Analysis and Geometry organisé par MM. Ancona et Silva, à Trento. J'ai eu très souvent l'occasion de participer à ce congrès et je remercie chaleureusement les organisateurs qui m'y ont invité.]

2. RAPPELS

Si X_1 et X_2 sont deux variétés complexes de dimension finie, on munit, comme d'habitude, l'ensemble $H(X_1, X_2)$ des applications holomorphes de X_1 dans X_2 de la topologie compacte-ouverte.

Rappelons maintenant les définitions suivantes.

DÉFINITION 2.1. - Une suite $(f_n)_{n\in\mathbb{N}}$ d'applications holomorphes d'une variété complexe de dimension finie X_1 dans une autre X_2 est dite compactement divergente si, pour tout compact K_1 de X_1 et tout compact K_2 de X_2, il existe un entier $n_0 \in \mathbb{N}$ tel que, pour tout $n \geq n_0$, $f_n(K_1) \cap K_2 = \emptyset$.

DÉFINITION 2.2. - On dit qu'une variété complexe X est taut si, pour toute suite d'applications holomorphes $(f_n)_{n\in\mathbb{N}}$ du disque-unité Δ dans X, on peut extraire de la suite $(f_n)_{n\in\mathbb{N}}$ une sous-suite qui est convergente dans $H(\Delta, X)$ ou une sous-suite compactement divergente.

D'après les résultats classiques (voir par exemple H. Wu [20] ou S. Kobayashi [9]), si X est taut, le même résultat reste vrai pour l'ensemble $H(Y, X)$ des applications holomorphes d'une variété connexe Y dans X.

Je ne rappellerai pas la définition et les propriétés des pseudodistances de Carathéodory c_X et de Kobayashi k_X sur une variété analytique complexe X pour lesquelles je renvoie le lecteur à [6], [7], [8] et [9].

Rappelons enfin qu'une variété X est dite hyperbolique si la pseudodistance de Kobayashi k_X sur X est une distance qui définit la topologie de X.

Pour une variété complexe de dimension finie, M. Abate [1] (voir aussi [2]) a montré le résultat suivant.

THÉORÈME 2.3. - *Soit X une variété complexe taut de dimension finie. Soit $f : X \longrightarrow X$ une application holomorphe. Alors les conditions suivantes sont équivalentes :*

(i) *la suite des itérées $(f^n)_{n\in\mathbb{N}}$ de f n'est pas compactement divergente ;*

(ii) *la suite des itérées $(f^n)_{n\in\mathbb{N}}$ de f n'a pas de sous-suite compactement divergente ;*

(iii) *pour tout $x \in X$, $\bigcup_{n\in\mathbb{N}}\{f^n(x)\}$ est relativement compact dans X ;*

(iv) *il existe $x \in X$, tel que $\bigcup_{n\in\mathbb{N}}\{f^n(x)\}$ soit relativement compact dans X.*

Si on suppose de plus que X est un domaine borné convexe de $\mathbb{C}^n$, ces conditions sont aussi équivalentes à la condition suivante :

(v) *f admet au moins un point fixe dans X.*

Remarquons au sujet de la démonstration que le fait que X est un domaine borné convexe de $\mathbb{C}^n$ entraîne que les boules pour la distance de Carathéodory c_X et de Kobayashi

k_X sont relativement compactes dans X (voir par exemple L. Harris [8]). On en déduit que X est taut.

Pour un domaine convexe borné X, l'existence d'un point fixe de f entraîne que l'image par f de la boule $B_c(a,r)$ de centre a et de rayon r pour la distance de Carathéodory c_X est contenue dans $B_c(a,r)$. Ainsi, la suite des itérées f^n n'est pas compactement divergente. La réciproque est un peu plus délicate et utilise le fait que le centre asymptotique de la suite des itérées $f^n(x)$ d'un point x est un convexe compact de X. Le résultat se déduit alors du théorème de Brouwer.

Dans cet article, nous allons donc étudier la suite des itérées f^n d'une application holomorphe $f : X \longrightarrow X$ d'une variété complexe taut X dans elle-même. Si la suite f^n n'est pas compactement divergente, on peut en extraire (voir E. Bedford [3], J.-P. Vigué [19] ou M. Abate [2]) une sous-suite f^{n_j} qui converge uniformément sur tout compact de X vers une application holomorphe

$$F : X \longrightarrow X.$$

Quitte à supposer que $n_{j+1} - n_j$ et que $n_{j+1} - 2n_j$ tendent vers $+\infty$ et à extraire un certain nombre de sous-suites, la suite $f^{n_{j+1}-n_j}$ converge uniformément sur tout compact de X vers une application holomorphe ρ, et ρ est une rétraction holomorphe de X sur une sous-variété fermée V de X appelée sous-variété limite de f et que nous noterons $\Lambda(f)$. Rappelons que, d'après H. Cartan [5], une rétraction holomorphe est linéarisable dans une carte locale, ce qui entraîne en particulier que l'image d'une rétraction holomorphe est une sous-variété complexe fermée de X. De plus, f est un automorphisme analytique de $\Lambda(f)$; tout point adhérent à la suite des itérées f^n s'écrit $\gamma o \rho$, où γ est un automorphisme analytique de $\Lambda(f)$. Par suite, la variété-limite $\Lambda(f)$ est unique.

D'autre part, nous utiliserons le résultat suivant dont la démonstration m'a été suggérée par H. Cartan (voir aussi T. Franzoni and E. Vesentini [7]).

THÉORÈME 2.4. - *Soit X une variété complexe hyperbolique, soit M une variété complexe connexe et soit $f : M \times X \longrightarrow X$ une application holomorphe. S'il existe $m_0 \in M$ tel que $f(m_0,.)$ soit un automorphisme analytique de X, alors, pour tout $m \in M$, $f(m,.) = f(m_0,.)$. [Ainsi, f est indépendant de $m \in M$].*

Démonstration. - Soit $\tau = f(m_0,.)$, et considérons l'application holomorphe

$$g : M \times X \longrightarrow M \times X$$

définie par $g(m,x) = (m, \tau^{-1}(f(m,x)))$. Alors, $g(m_0,x) = (m_0,x)$, et il nous faut montrer que, pour tout $m \in M$, $g(m,.) = \mathrm{id}|_X$. D'après le théorème de prolongement analytique, il suffit de le montrer pour m suffisamment proche de m_0. Soit N un voisinage de m_0 dans M analytiquement isomorphe à un domaine borné. Soit $B_k((m_0,x_0),r)$ une boule pour la distance de Kobayashi de centre (m_0,x_0) et de rayon r dans $N \times X$. Comme $g(m_0,x_0) = (m_0,x_0)$, g laisse stable $B_k((m_0,x_0),r)$, et si r est choisi suffisamment petit, $B_k((m_0,x_0),r)$ est isomorphe à un domaine borné de $\mathbb{C}^n$. On sait que $g(m_0,x_0) = (m_0,x_0)$, montrons que $g'(m_0,x_0) = \mathrm{id}$. Or,

$$g'(m_0,x_0).(u,v) = (u, v + h(u)),$$

où h est une application linéaire. En itérant, on trouve

$$(g'(m_0,x_0))^k.(u,v) = (u, v + kh(u)).$$

D'après les inégalités de Cauchy, il existe une constante K telle que $\|kh\| \leq K$ pour tout $k \in \mathbb{N}$, ce qui entraîne que $h = 0$. Par suite, $g'(m_0, x_0) = \text{id}$. D'après le théorème d'unicité de H. Cartan, $g(m, x) = (m, x)$, ce qui démontre le théorème.

3. APPLICATIONS DÉPENDANT HOLOMORPHIQUEMENT D'UN PARAMÈTRE

On considère une variété complexe hyperbolique taut X et une variété complexe connexe M. Soit $f : M \times X \longrightarrow X$ une application holomorphe. Pour tout $m \in M$, on peut considérer la suite des itérées $f^n(m,.)$ définie par récurrence par $f^0(m, x) = x$, et pour $n > 0$,

$$f^n(m, x) = f(m, f^{n-1}(m, x)).$$

On a d'abord le lemme suivant.

LEMME 3.1. - *S'il existe $m_0 \in M$, tel que la suite des itérées $f^n(m_0,.)$ ne soit pas compactement divergente, alors, pour tout $m \in M$, la suite $f^n(m,.)$ n'est pas compactement divergente.*

Démonstration. - D'après l'hypothèse, f^n est une suite d'applications holomorphes de $M \times X$ dans X, et il existe un point x_0 de X tel que la suite $f^n(m_0, x_0)$ soit relativement compacte dans X. Comme X est taut, il en est de même pour tout $m \in M$ et pour tout $x \in X$.

Nous supposerons dans tout ce paragraphe qu'il existe au moins un $m_0 \in M$ tel que la suite des itérées $f^n(m_0,.)$ ne soit pas compactement divergente. Il en est alors de même pour tout $m \in M$. Nous pouvons alors énoncer et montrer le théorème suivant.

THÉORÈME 3.2. - *Il existe une application holomorphe $\rho : M \times X \longrightarrow X$ telle que, pour tout $m \in M$, $\rho(m,.)$ soit une rétraction holomorphe sur la variété-limite $\Lambda(f(m,.))$. De plus, toutes les variétés-limites $\Lambda(f(m,.))$ sont deux à deux isomorphes. Plus précisément, pour tout $m \in M$, pour tout $m_0 \in M$, la restriction de $\rho(m_0,.)$ à la variété-limite $\Lambda(f(m,.))$ est un isomorphisme analytique de $\Lambda(f(m,.))$ sur $\Lambda(f(m_0,.))$, et on a :*

$$\rho(m_0,.) \circ f(m,.) \circ \rho(m,.)|_{\Lambda(f(m_0,.))} = f(m_0,.)|_{\Lambda(f(m_0,.))}.$$

Démonstration. - La construction de ρ se fait en reprenant la construction rappelée au paragraphe précédent dans le cas avec paramètre : on considère f^n, puis une suite extraite f^{n_j} qui converge uniformément sur tout compact vers $F : M \times X \longrightarrow X$. Ensuite, on montre, quitte à extraire encore des sous-suites, que $f^{n_{j+1}-n_j}$ converge vers l'application holomorphe ρ cherchée. Il est clair que, pour tout $m \in M$, $\rho(m,.)$ est une rétraction holomorphe sur $\Lambda(f(m,.))$. Ceci suffit déjà à montrer que toutes les variétés-limites ont la même dimension.

Soit $m_0 \in M$ et considérons l'application

$$G : M \times \Lambda(f(m_0,.)) \longrightarrow \Lambda(f(m_0,.))$$

définie par

$$G(m, x) = \rho(m_0, \rho(m, x)).$$

Il est clair que G est bien une application holomorphe de $M \times \Lambda(f(m_0,.))$ dans $\Lambda(f(m_0,.))$. Par construction, pour tout $x \in \Lambda(f(m_0,.))$,

$$G(m_0, x) = \rho(m_0, \rho(m_0, x)) = \rho(m_0, x) = x.$$

D'après le théorème 2.3, pour tout $m \in M$, pour tout $x \in \Lambda(f(m_0,.))$, $G(m,x) = x$. Ceci montre que $\rho(m,.)$ est un isomorphisme analytique de $\Lambda(f(m_0,.))$ sur son image dans $\Lambda(f(m,.))$. De même, $\rho(m_0,.)$ est une application surjective de $\Lambda(f(m,.))$ sur $\Lambda(f(m_0,.))$. Il suffit d'échanger les rôles de m et m_0 pour montrer que $\rho(m_0,.)$ est un isomorphisme analytique de $\Lambda(f(m,.))$ sur $\Lambda(f(m_0,.))$, l'isomorphisme réciproque étant donné par $\rho(m,.)$.

Montrons maintenant la dernière formule du théorème. Pour cela, considérons l'application holomorphe

$$H : M \times \Lambda(f(m_0,.)) \longrightarrow \Lambda(f(m_0,.))$$

définie par

$$H(m,x) = \rho(m_0,.) \circ f(m,.) \circ \rho(m,x).$$

Il est clair que, pour tout $x \in \Lambda(f(m_0,.))$,

$$H(m_0,x) = f(m_0,x),$$

et $f(m_0,.)$ est un automorphisme analytique de $\Lambda(f(m_0,.))$. D'après le théorème 2.4, $H(m,x)$ est indépendant de m, ce qui montre que

$$H(m,x) = H(m_0,x),$$

et ceci achève la démonstration du théorème.

Ainsi donc, pour tout $m \in M$, l'automorphisme analytique $f(m,.)|_{\Lambda(f(m,.))}$ s'obtient en composant $f(m_0,.)|_{\Lambda(f(m_0,.))}$ avec $\rho(m_0,.)$ et $\rho(m_0,.)^{-1}$. Comme par construction, les points fixes de $f(m,.)$ sont contenus dans $\Lambda(f(m,.))$, on en déduit facilement que $\rho(m_0,.)$ induit un isomorphisme analytique de l'ensemble Fix $f(m,.)$ des points fixes de $f(m,.)$ sur Fix $f(m_0,.)$.

Nous avons donc montré le théorème 3.3.

THÉORÈME 3.3. - *Soit X une variété complexe hyperbolique taut, soit M une variété complexe connexe, et soit $f : M \times X \longrightarrow X$ une application holomorphe. Alors, pour tout $m \in M$, pour tout $m_0 \in M$, $\rho(m_0,.)$ induit un isomorphisme analytique de l'ensemble* Fix *$f(m,.)$ des points fixes de $f(m,.)$ sur l'ensemble* Fix $f(m_0,.)$.

Bien sûr, il ne faut pas croire que l'ensemble des points fixes de $f(m,.)$ est indépendant de $m \in M$. Considérons par exemple $X = \Delta \times \Delta$ dans $\mathbb{C}^2$. Soit M une variété analytique complexe quelconque, et soit

$$g : M \times \Delta \longrightarrow \Delta$$

une application holomorphe. Soit

$$f(m,(x_1,x_2)) = (x_1, g(m,x_1)).$$

Alors l'ensemble des points fixes de $f(m,.)$ est exactement l'ensemble des points

$$\{(x_1,x_2) \in M | x_2 = g(m,x_1)\}.$$

C'est donc le graphe de l'application $g(m,.)$, et il est clair que cet ensemble dépend de m en général. On peut aussi remarquer que Fix $f(m,.)$ est toujours isomorphe au disque-unité Δ.

Cependant, pour certains domaines particuliers, on peut montrer que l'ensemble Fix $f(m,.)$ est indépendant de m. Ainsi, nous avons le théorème suivant déduit des travaux de E. Vesentini [13, 14 et 15]. (Je donne le résultat dans le cadre un peu plus général des domaines bornés d'espaces de Banach complexe).

THÉORÈME 3.4. - *Soit B la boule-unité ouverte d'un espace de Banach complexe E. Soit M une variété complexe connexe. Supposons que tous les points de la frontière de B soient des points complexe-extrémaux de $\overline{B}$. Soit $f : M \times B \longrightarrow B$ une application holomorphe telle que, pour tout $m \in M$, $f(m, 0) = 0$. Alors, il existe un sous-espace vectoriel fermé F de E tel que, pour tout $m \in M$,*

$$\text{Fix } f(m,.) = B \cap F.$$

Démonstration. - D'après E. Vesentini [13, 14 et 15], on sait que si $m_0 \in M$, l'ensemble Fix $f(m_0,.)$ est de la forme $B \cap F$, où F est un sous-espace vectoriel fermé de E. Soit $v \in F$ un vecteur de norme 1. Considérons l'application holomorphe $g : M \times \Delta \longrightarrow B$ définie par

$$g(m, \zeta) = f(m, \zeta v).$$

Alors, $\frac{\partial g}{\partial \zeta}(m_0, 0) = v$, et l'application

$$m \longrightarrow \frac{\partial g}{\partial \zeta}(m, 0)$$

envoie M dans $\overline{B}$. Le principe du maximum montre que g envoie M dans la frontière ∂B de B. Comme tous les points de la frontière de B sont des points complexe-extrémaux de $\overline{B}$, le principe du maximum fort de E. Thorp et R. Whitley [12] montre que $\frac{\partial g}{\partial \zeta}(m_0, 0)$ est constante et égale à v. D'après E. Vesentini [13, 14 et 15], ceci montre que $g(m, \zeta) = \zeta v$, pour tout $m \in M$. Ainsi donc,

$$B \cap F \subset \text{Fix } f(m,.).$$

En échangeant les rôles de m et m_0, on montre l'égalité

$$\text{Fix } f(m,.) = B \cap F,$$

ce qui démontre le théorème.

On peut se demander ce qui se passe si on suppose seulement qu'il existe $m_0 \in M$ tel que $f(m_0, 0) = 0$. Alors, il faut distinguer deux cas : si Fix $f(m_0,.) = \{0\}$, alors Fix $f(m,.)$ dépend de m en général. Si on suppose en revanche que Fix $f(m_0,.)$ est de dimension au moins 1, on a la proposition suivante.

PROPOSITION 3.5. - *Soit B la boule-unité ouverte d'un espace de Banach complexe E. Soit M une variété complexe connexe. Supposons que tous les points de la frontière de B soient des points complexe-extrémaux de $\overline{B}$. Soit $f : M \times B \longrightarrow B$ une application holomorphe telle qu'il existe $m_0 \in M$, $f(m_0, 0) = 0$. Alors, il existe un sous-espace vectoriel fermé F de E tel que, pour tout $m \in M$,*

$$\text{Fix } f(m,.) = B \cap F.$$

Démonstration. - Soit

$$F = \{v \in E | \frac{\partial f}{\partial x}(m_0, 0).v = v\}.$$

Par hypothèse, F est non réduit à $\{0\}$. Soit $v \in F$ un vecteur de norme 1. Quitte à considérer un voisinage suffisamment petit de m_0 dans M, on peut supposer que M est un ouvert d'un espace de Banach et que $m_0 = 0$. Soit $h : \Delta \longrightarrow M$ une application holomorphe telle que $h(0) = 0, h'(0) = 0$. Soit $\varphi : \Delta \longrightarrow B$ définie par

$$\varphi(\zeta) = f(h(\zeta), \zeta v).$$

On a

$$\varphi(0) = 0, \varphi'(0) = \frac{\partial f}{\partial m}(0,0).h'(0) + \frac{\partial f}{\partial x}(0,0).v = v.$$

D'après le principe du maximum fort [12],

$$\varphi(\zeta) = \zeta v.$$

Ceci entraîne que, pour tout m suffisamment proche de 0, $f(m, \zeta v) = \zeta v$. Par continuité, $f(m, 0) = 0$, et on peut appliquer le théorème 3.4.

Rappelons que, si X est un domaine borné convexe de $\mathbb{C}^n$, et si $f : X \longrightarrow X$ est une application holomorphe, il existe d'après [16 et 17] une rétraction holomorphe $\rho : X \longrightarrow \text{Fix } f$. Si maintenant on considère une variété complexe connexe M et une application $f : M \times X \longrightarrow X$, on peut construire, en utilisant les méthodes de [17] ou celles de P. Mazet et J.-P. Vigué [11], une application $\rho : M \times X \longrightarrow X$ telle que, pour tout $m \in M$, $\rho(m, .)$ soit une rétraction holomorphe de X sur Fix $f(m, .)$. On obtient ainsi, dans le cas d'un domaine borné convexe X, une démonstration directe du fait que toutes les sous-variétés Fix $f(m, .)$ sont deux à deux isomorphes.

4. POINTS FIXES SUR UN PRODUIT DE VARIÉTÉS COMPLEXES

Soient X_1 et X_2 deux variétés complexes connexes de dimension finie, et supposons que X_1 et X_2 soient hyperboliques et taut. Soit $f = (f_1, f_2) : X_1 \times X_2 \longrightarrow X_1 \times X_2$ une application holomorphe. Alors $f_1 : X_1 \times X_2 \longrightarrow X_1$ est une application holomorphe de X_1 dans X_1 qui dépend d'un paramètre appartenant à X_2. Nous allons donc étudier, en nous aidant des résultats précédents, les relations entre les variétés Fix f, Fix $f_2(x_1, .)$ et Fix $f_1(., x_2)$.

Montrons d'abord la proposition suivante.

PROPOSITION 4.1. - *Si* Fix f *est non vide, alors, pour tout* $x_1 \in X_1$, Fix $f_2(x_1, .)$ *est non vide. De même, pour tout* $x_2 \in X_2$, Fix $f_1(., x_2)$ *est non vide. De plus, si* $x = (x_1, x_2) \in$ Fix f, *on a :*

$$\dim_{x_2} \text{Fix } f_2(x_1, .) + \dim_{x_1} \text{Fix } f_1(., x_2) \leq \dim_x \text{Fix } f.$$

Démonstration. - Supposons que Fix f soit non vide. Il existe donc un point $x^0 = (x_1^0, x_2^0)$ de $X_1 \times X_2$ tel que $f(x_1^0, x_2^0) = (x_1^0, x_2^0)$. Ceci entraîne bien sûr que Fix $f_2(x_1^0, .)$ et Fix $f_1(., x_2^0)$ sont non vides. D'après le théorème 3.3, pour tout $x_1 \in X_1$, Fix $f_2(x_1, .)$ est non vide. De même, pour tout $x_2 \in X_2$, Fix $f_1(., x_2)$ est non vide.

Si on considère un point $x = (x_1, x_2)$ appartenant à Fix f, il est clair que $\{x_1\}\times$ Fix $f_2(x_1, .)$ et que Fix $f_1(., x_2) \times \{x_2\}$ sont contenus dans Fix f. On en déduit tout de suite l'inégalité annoncée.

Remarquons que cette inégalité n'est pas une égalité en général. En effet, considérons comme produit $X_1 \times X_2$ le bidisque $\Delta \times \Delta$, et soit

$$f : \Delta \times \Delta \longrightarrow \Delta \times \Delta$$

définie par $f(x_1, x_2) = (1/2(x_1 + x_2), 1/2(x_1 + x_2))$. Il est clair que, pour tout $x_1 \in \Delta$, $f_2(x_1, .)$ a un point fixe unique, ce qui entraîne que

$$\dim \text{Fix } f_2(x_1, .) = 0.$$

De même,

$$\dim \text{Fix } f_1(., x_2) = 0.$$

Cependant, l'ensemble des points fixes de f est égal à

$$\{(x_1, x_2) \in \Delta \times \Delta | x_1 = x_2\},$$

qui est de dimension 1.

Si on cherche une majoration de la dimension de l'ensemble des points fixes de f, on a par exemple

$$\dim_x \text{Fix } f \leq \inf (\dim X_2 + \dim_{x_1} \text{Fix } f_1(., x_2), \dim X_1 + \dim_{x_2} \text{Fix } f_2(x_1, .)).$$

Cette majoration provient simplement du fait que

$$\text{Fix } f \cap (\{x_1\} \times X_2) = \{x_1\} \times \text{ Fix } f_2(x_1, .).$$

Enfin, on peut se demander si l'existence d'un point fixe pour $f_1(., x_2)$ et pour $f_2(x_1, .)$ entraîne que f a au moins un point fixe dans $X_1 \times X_2$. Nous avons la proposition suivante.

PROPOSITION 4.2. - *Soit $f : X_1 \times X_2 \longrightarrow X_1 \times X_2$ une application holomorphe. Supposons que $f_2(x_1, .)$ et $f_1(., x_2)$ aient au moins un point fixe dans X_2 et X_1 respectivement. Soient $\rho_2 : X_1 \times X_2 \longrightarrow X_2$ (resp. $\rho_1 : X_1 \times X_2 \longrightarrow X_1$) la rétraction holomorphe sur* Fix $f_2(x_1, .)$ *(resp.* Fix $f_1(., x_2)$*). Alors l'ensemble des points fixes de f est égal à l'ensemble des points fixes d'une des deux applications suivantes :*

$$g : (x_1, x_2) \longrightarrow (\rho_1(x_1, \rho_2(x_1, x_2)), \rho_2(x_1, x_2)),$$

$$h : (x_1, x_2) \longrightarrow (\rho_1(x_1, x_2), \rho_2(\rho_1(x_1, x_2), x_2)).$$

Démonstration. - Si $x^0 = (x_1^0, x_2^0)$ est un point fixe de f, il est clair que

$$\rho_1(x_1^0, x_2^0) = x_1^0, \ \rho_2(x_1^0, x_2^0) = x_2^0,$$

ce qui montre que

$$g(x_1^0, x_2^0) = (x_1^0, x_2^0), \ h(x_1^0, x_2^0) = (x_1^0, x_2^0).$$

Réciproquement, si $x^0 = (x_1^0, x_2^0)$ est un point fixe de g, alors on a $\rho_2(x_1^0, x_2^0) = x_2^0$, ce qui montre que $f_2(x_1^0, x_2^0) = x_2^0$. On remarque alors que $\rho_1(x_1^0, x_2^0) = x_1^0$, ce qui entraîne que $f_1(x_1^0, x_2^0) = x_1^0$, et la proposition est démontrée.

Un des intérêts de cette proposition réside dans l'application suivante : Soit $f : X_1 \times X_2 \longrightarrow X_1 \times X_2$ une application holomophe et supposons que $f_2(x_1, .)$ admette un point fixe unique $\varphi_2(x_1)$. De même, supposons que $f_1(., x_2)$ admette un point fixe unique $\varphi_1(x_2)$. Alors

$$\rho_1(x_1, x_2) = \varphi_1(x_2);$$

$$\rho_2(x_1, x_2) = \varphi_2(x_1).$$

Pour que f admette un point fixe dans $X_1 \times X_2$, il faut et il suffit que l'application

$$\varphi_2 o \varphi_1 : X_2 \longrightarrow X_2 \text{ (resp. } \varphi_1 o \varphi_2 : X_1 \longrightarrow X_1 \text{)}$$

admette un point fixe dans X_2 (resp. X_1).

Dans le cas où X_2 est convexe, d'après le théorème 2.3 (voir M. Abate [1]), il faut et il suffit que la suite des itérées $(\varphi_2 o \varphi_1)^n$ de $\varphi_2 o \varphi_1$ ne soit pas compactement divergente.

Si on suppose maintenant que X_1 et X_2 sont égaux au disque-unité Δ, étant donnés deux applications holomorphes $\varphi_1 : X_2 \longrightarrow X_1$ et $\varphi_2 : X_1 \longrightarrow X_2$, il existe une application holomorphe

$$f = (f_1, f_2) : X_1 \times X_2 \longrightarrow X_1 \times X_2$$

telle que

$$\text{Fix } f_1(., x_2) = \{x_1 \in X_1 | x_1 = \varphi_1(x_2)\} ;$$
$$\text{Fix } f_2(x_1, .) = \{x_2 \in X_2 | x_2 = \varphi_2(x_1)\}.$$

Il suffit par exemple de prendre

$$f(x_1, x_2) = (1/2(x_1 + \varphi_1(x_2)), 1/2(x_2 + \varphi_2(x_1))) .$$

L'existence de points fixes pour f ne dépend donc pas de la forme précise de f, mais seulement des applications φ_1 et φ_2.

Si on prend par exemple

$$\varphi_1(x_2) = x_2, \ \varphi_2(x_1) = \frac{x_1 + a}{1 + \bar{a}x_1} (a \neq 0),$$

la suite des itérées $(\varphi_2 o \varphi_1)^n$ est compactement divergente, et f n'a pas de point fixe dans $X_1 \times X_2$, bien que $f_1(., x_2)$ et $f_2(x_1, .)$ en aient dans X_1 et X_2 respectivement.

BIBLIOGRAPHIE

1. M. Abate. Iteration theory of holomorphic maps on taut manifolds. Mediterranean Press, Rende, Cosenza, 1990.

2. M. Abate. *Iteration theory, compactly divergent sequences and commuting holomorphic maps*. Ann. Scuola Norm. Sup. Pisa Cl. Sci. (4), **18** (1991), p. 167-191.

3. E. Bedford. *On the automorphism group of a Stein manifold.* Math. Ann., **266** (1983), p. 215-227.

4. H. Cartan. *Sur les fonctions de plusieurs variables complexes : l'itération des transformations intérieures d'un domaine borné*. Math. Z., **35** (1932), p. 760-773.

5. H. Cartan. *Sur les rétractions d'une variété*. C. R. Acad. Sc. Paris Série I Math., **303** (1986), p. 715-716.

6. S. Dineen. The Schwarz Lemma. Oxford Math. Monographs, Clarendon Press, Oxford, 1989.

7. T. Franzoni and E. Vesentini. Holomorphic maps and invariant distances. Math. Studies **40**, North-Holland, Amsterdam, 1980.

8. L. Harris. *Schwarz-Pick systems of pseudometrics for domains in normed linear spaces*. In Advances in Holomorphy, Mathematical Studies **34**, North-Holland, Amsterdam, 1979, p. 345-406.

9. S. Kobayashi. *Intrinsic distances, measures and geometric function theory.* Bull. Amer. Math. Soc., **82** (1976), p. 357-416.

10. P. Mazet et J.-P. Vigué. *Points fixes d'une application holomorphe d'un domaine borné dans lui-même.* Acta Mathematica, **166** (1991), p. 1-26.

11. P. Mazet et J.-P. Vigué. *Convexité de la distance de Carathéodory et points fixes d'applications holomorphes.* Bull. Sc. Math. (2), **116** (1992), p. 285-305.

12. E. Thorp and R. Whitley. *The strong maximum modulus theorem for analytic functions into a Banach space.* Proc. Amer. Math. Soc., **18** (1967), p. 640-646.

13. E. Vesentini. *Complex geodesics.* Compositio Math., **44** (1981), p. 375- 394.

14. E. Vesentini. *Complex geodesics and holomorphic mappings.* Symposia M*ath.*, **26** (1982), p. 211-230.

15. E. Vesentini. *Invariant distances and invariant differential metrics in locally convex spaces*, in Spectral theory, Banach Center Publications, Warsaw, **8** (1982), p. 493-512.

16. J.-P. Vigué. *Géodésiques complexes et points fixes d'applications holomorphes.* Adv. in Math., **52** (1984), p. 241-247.

17. J.-P. Vigué. *Points fixes d'applications holomorphes dans un domaine borné convexe de* $\mathbb{C}^n$. Trans. Amer. Math. Soc., **289** (1985), p. 345-353.

18. J.-P. Vigué. *Points fixes d'une limite d'applications holomorphes.* Bull. Sc. Math. (2), **110** (1986), p. 411-424.

19. J.-P. Vigué. *Sur les points fixes d'applications holomorphes.* C. R. Acad. Sc. Paris Série I Math., **303** (1986), p. 927-930.

20. H. Wu. *Normal families of holomorphic mappings.* Acta Math., **119** (1967), p. 194-233.

Holomorphic Functions on an Algebraic Group Invariant Under Zariski-Dense Subgroups

Jörg Winkelmann, Mathematisches Institut, Ruhr-Universität Bochum, Universitätstrasse 150, D-44780 Bochum, Germany. winkelmann@ruba.rz.ruhr-uni-bochum.de

1. Introduction

Let G be a reductive complex linear-algebraic group, Γ a subgroup, $\mathcal{O}(G)^\Gamma$ the algebra of Γ-invariant holomorphic functions on G. It is known [1] that $\mathcal{O}(G)^\Gamma = \mathbb{C}$ if Γ is dense in G with respect to the algebraic Zariski-topology.

We are interested in similar results for non-reductive groups. If G is a complex linear-algebraic group with G/G' non-reductive (where G' denotes the commutator group), then there exists a surjective group morphism $\tau : G \to (\mathbb{C}, +)$ and $\Gamma = \tau^{-1}(\mathbb{Z})$ is a Zariski-dense subgroup of G with $\mathcal{O}(G)^\Gamma \simeq \mathcal{O}(\mathbb{C}^*) \neq \mathbb{C}$. Hence we are led to the question whether the following two properties are equivalent:

Let G be a connected complex linear-algebraic group.

(i) G/G' is reductive.

(ii) $\mathcal{O}(G)^\Gamma = \mathbb{C}$ for every Zariski-dense subgroup Γ.

The above argument gave us $(ii) \Rightarrow (i)$ and the result of Barth and Otte [1] implies the equivalence of (i) and (ii) for G reductive.

We will prove that (i) and (ii) are likewise equivalent in the following two cases:

a) G is solvable.

b) The adjoint representation of S on $\mathcal{L}ie(U)$ has no zero weight, where S denotes a maximal connected semisimple subgroup of G and U the unipotent radical of G.

Case b) is equivalent to each of the following two conditions

b') G/G' is reductive and the semisimple elements are dense in G'.

b") G/G' is reductive and $N_{G'}(T)/T$ is finite, where T is a maximal torus in G' and $N_{G'}(T)$ denotes the normalizer of T in G'.

For instance, if we take G to be a semi-direct product $SL_2(\mathbb{C}) \ltimes_\rho (\mathbb{C}^n, +)$ with $\rho : SL_2(\mathbb{C}) \to GL_n(\mathbb{C})$ irreducible, then G fulfills the condition of case b) if and only if n is an even number.

The proof for case a) is based on the usual solvable group methods and the structure theorem on holomorphically separable solvmanifolds by Huckleberry and E. Oeljeklaus.

The proof for case b) relies on the discussion of semisimple elements of infinite order in such a Γ. For this reason we conclude the paper with an example of Margulis which implies that $G = SL_2(\mathbb{C}) \ltimes_\rho (\mathbb{C}^3, +)$ (ρ irreducible) admits a Zariski-dense discrete subgroup Γ such that no element of Γ is semisimple. Thus condition b) is really needed in order to find semisimple elements in Zariski-dense subgroups. In consequence, our method does not work for the example of Margulis. However, this only means that we can not prove $\mathcal{O}(G)^{\Gamma} = \mathbb{C}$ for Margulis' example. We have no knowledge whether there actually exist non-constant holomorphic functions in this case.

Finally we discuss invariant meromorphic and plurisubharmonic functions on certain groups.

2. Solvable groups

Here we will discuss solvable groups. First we will develop some auxiliary lemmata.

Lemma 1. *Let G be a connected complex linear-algebraic group such that G/G' is reductive.*

Then $[G, G'] = G'$.

Proof. By taking the appropriate quotient, we may assume $[G, G'] = \{e\}$. We have to show that this implies $G' = \{e\}$. Now $[G, G'] = \{e\}$ means that G' is central, hence $Ad(G)$ factors through G/G'. But G/G' is reductive and acts trivially (by conjugation) on both G/G' and G'. Due to complete reducibility of representations of reductive groups it follows that $Ad(G)$ is trivial, i.e. G is abelian, i.e. $G' = \{e\}$. □

Lemma 2. *Let G be a connected complex linear-algebraic group, $H \subset G'$ a connected complex Lie subgroup which is normal in G'.*

Then H is algebraic.

Proof. Let U denote the unipotent radical of G'. Then G'/U is semisimple. Now $A = (H \cap U)^0$ is algebraic, because every connected complex Lie subgroup of a unipotent group is algebraic. Normality of H implies that $H/(H \cap U)$ is semisimple. Hence H/A is semisimple, too. It follows that H/A is an algebraic subgroup of G'/A. Thus H has to be algebraic. □

Lemma 3. *Let G be a connected topological group, H a normal subgroup, such that $H \cap G'$ is totally disconnected.*

Then H is central.

Proof. For each $h \in H$ the set $S_h = \{ghg^{-1}h^{-1} : g \in G\}$ is both totally disconnected and connected and therefore reduces to $\{e\}$. □

Lemma 4. *Let G be a connected complex linear-algebraic group, $A \subset G'$ a complex Lie subgroup which is normal in G and Zariski-dense in G'. Assume moreover that $[G, G'] = G'$.*

Then $A = G'$.

Proof. The connected component A^0 of A is algebraic (Lemma 2). Thus G/A^0 is again algebraic. Moreover G'/A^0 is the commutator group of G/A^0. Therefore, by replacing G with G/A^0 we may assume that A is totally disconnected. But totally disconnected normal subgroups of connected Lie groups are central (Lemma 3). Since A is Zariski-dense in G', and $[G,G'] = G'$, this may occur only for $A = G' = \{e\}$. Thus $A = G'$. □

Theorem 1. *Let G be a connected solvable complex linear-algebraic group, Γ a subgroup which is dense in the algebraic Zariski-topology. Assume that G/G' is reductive.*

Then $\mathcal{O}(G)^\Gamma = \mathbb{C}$.

Proof. Let $G/\Gamma \to G/H$ denote the holomorphic reduction, i.e.

$$H = \{g \in G : f(g) = f(e) \forall f \in \mathcal{O}(G)^\Gamma\}.$$

Now $\Gamma \subset H$ normalizes H^0. Since Γ is Zariski-dense in G and the normalizer of a connected Lie subgroup is necessarily algebraic, it follows that H^0 is normal in G. Let $A = H^0 \cap G'$. This is again a closed normal subgroup in G. By a result of Huckleberry and E. Oeljeklaus [3] H/H^0 is almost nilpotent (i.e. admits a subgroup of finite index which is nilpotent). Let Γ_0 be a subgroup of finite index in Γ with $\Gamma_0/(\Gamma_0 \cap H^0)$ nilpotent. By definition this means there exists a number k such that $C^k\Gamma_0 \subset H^0$ where C^k denotes the central series. Now $[G,G'] = G'$ implies $C^kG = G'$ for all $k \geq 1$. Therefore $C^k\Gamma_0$ is Zariski-dense in G'. It follows that $A = H^0 \cap G'$ is a closed normal Lie subgroup of G which is Zariski-dense in G'. By the preceding lemma it follows that $A = G'$, i.e. $G' \subset H$. Now G/G' is assumed to be reductive. Thus the statement of the theorem now follows from the result for reductive groups ([1]). □

3. Groups with many semisimple elements

Here we will prove the following theorem.

Theorem 2. *Let G be a connected complex linear-algebraic group. Assume that G/G' is reductive and that furthermore one (hence all) of the following equivalent conditions is fulfilled.*

(1) G' contains a dense open subset Ω such that each element in Ω is semisimple.

(2) For any maximal torus T in G' the quotient $N_{G'}(T)/T$ is finite.

(3) Let S denote a maximal connected semisimple subgroup of G and U the unipotent radical of G. Let $\rho : S \to GL(\mathcal{L}ie\, U)$ denote the representation obtained by restriction from the adjoint representation $Ad : G \to GL(\mathcal{L}ie\, G)$. The condition is that all weights of ρ are non-zero.

Under these assumptions $\mathcal{O}(G)^\Gamma = \mathbb{C}$ for any Zariski-dense subgroup $\Gamma \subset G$.

Examples.

(a) Let G be a reductive group. Then G/G' is reductive and G' semisimple, hence $N_{G'}(T)/T$ finite for any maximal torus $T \subset G'$. Therefore this theorem is a generalization of the result of Barth and Otte [1] on redutive groups.

(b) Let G be a parabolic subgroup of a semisimple group S. G/G' is obviously reductive. Furthermore a maximal torus T in G is already a maximal torus in S. Hence $N_S(T)/T$ is finite. Consequently $N_G(T)/T$ is finite and G fulfills the assumptions of the theorem.

(c) Let G be a semi-direct product of $SL_2(\mathbb{C})$ with a unipotent group $U \simeq \mathbb{C}^n$ induced by an irreducible representation $\xi : SL_2(\mathbb{C}) \to GL(U)$. Then G fulfills the assumptions of the theorem if and only if n is even.

Now we will demonstrate that (1), (2) and (3) are indeed equivalent. The equivalence of (2) and (3) is rather obvious from standard results on algebraic groups. For the equivalence of (1) and (2) we need some elementary facts on semisimple elements in a connected algebraic group G. Let G_s denote the set of all semisimple elements in G and T be a maximal torus in G. Now $g \in G_s$ iff g is conjugate to an element in T. It follows that G_s is the image of the map $\zeta : G \times T \to G_s$ given by $\zeta(g,t) = gtg^{-1}$. In particular G_s is a constructible set. Now a torus contains only countably many algebraic subgroups, hence a generic element $h \in T$ generates a Zariski-dense subgroup of T. It follows that for a generic element $h \in T$ the assumption $g \in G$ with $ghg^{-1} \in T$ implies $gTg^{-1} = T$. From this it follows that a generic fiber of ζ has the dimension $\dim N_G(T)$. Therefore the dimension of $G_s = Image(\zeta)$ equals $\dim G - \dim N_G(T)$. Thus we obtained the following lemma, which implies the equivalence of (1) and (2).

Lemma 5. *Let G be a connected linear-algebraic group, T a maximal torus and G_s the set of semisimple elements in T.*

Then G_s is dense in G if and only if $\dim N_G(T) = \dim T$.

Next we state some easy consequences of the assumptions of Theorem 2.

Lemma 6. *Let G be an algebraic group fulfilling the assumptions of Theorem 2 and $\tau : G \to H$ a surjective morphism of algebraic groups.*

Then H likewise fulfills the assumptions of Theorem 2.

Proof. Surjectivity of τ gives a surjective morphism of algebraic groups from G/G' onto H/H'. Therefore H/H' is reductive. The surjectivity of τ furthermore implies $\tau(G') = H'$. Since morphisms of algebraic groups map semisimple elements to semisimple elements, it follows that H fulfills condition (1). □

Lemma 7. *Let G be an algebraic group fulfilling the assumptions of Theorem 2. Then the center Z of G must be reductive.*

Proof. Condition (2) implies that $(Z \cap G')^0$ is contained in a maximal torus of G'. Since G/G' is reductive, this implies that Z is reductive. □

The following lemma illuminates why semisimple elements are important for our purposes.

Lemma 8. *Let G be a complex linear-algebraic group, $g \in G$ an element of infinite order, Γ the subgroup generated by g and H the Zariski-closure of Γ.*

Then $Z = H/\Gamma$ is a Cousin group (hence in particular $\mathcal{O}(Z) = \mathbb{C}$) if g is semisimple; but Z is biholomorphic to some $(\mathbb{C}^*)^n$ (hence holomorphically separable) if g is not semisimple.

Proof. Note that $\bar{\Gamma} = H$ implies $H = H^0\Gamma$. Hence $H/\Gamma = H^0/(H^0 \cap \Gamma)$ is connected. If g is semisimple, the Zariski-closure of Γ is reductive and the statement follows from [1]. If g is not semisimple then $H \simeq (\mathbb{C}^*)^{n-1} \times \mathbb{C}$ for some $n \geq 1$ and g is not contained in the maximal torus of H. This implies $H/\Gamma \simeq (\mathbb{C}^*)^n$. □

Lemma 9. *Let G be a connected real Lie group, Γ a subgroup such that each element $\gamma \in \Gamma$ is of finite order.*

Then Γ is almost abelian and relatively compact in G.

Proof. If G is abelian, then $G \simeq \mathbb{R}^k \times (S^1)^n$. In this case $\Gamma \subset (S^1)^n$ and the statement is immediate.

Now let us assume that G may be embedded into a complex linear-algebraic group $\tilde{G}$. Let H denote the (complex-algebraic) Zariski-closure of Γ in $\tilde{G}$. By the theorem of Tits [8] Γ is almost solvable, hence H^0 is solvable. Now the commutator group of H^0 is unipotent and therefore contains no non-trivial element of finite order. Hence $\Gamma \cap H^0$ is abelian, which completes the proof for this case, since we discussed already the abelian case.

Finally let us discuss the general case. By the above considerations $Ad(\Gamma_0)$ is contained in an abelian connected compact subgroup K of $Ad(G)$ for some subgroup Γ_0 of finite index in Γ. Now $N = (Ad)^{-1}(K)$ is a central extension $1 \to Z \to N \to K \to 1$. (where Z is the center of G). But complete reducibility of the representations of compact groups implies that this sequence splits on the Lie algebra level. Hence N is abelian and we can complete the proof as before. □

Lemma 10. *Let G be a complex linear-algebraic group, Γ a Zariski-dense subgroup.*

Then Γ contains a finitely generated subgroup Γ_0 such that the Zariski-closure of Γ_0 contains G'.

Proof. Consider all finitely generated subgroups of Γ and their Zariski-closures in G. There is one such group Γ_0 for which the dimension of the Zariski-closure A is maximal. Clearly A must contain the connected component of the Zariski-closure for any finitely generated subgroup of Γ. This implies that A^0 is normal in G. Furthermore maximality implies that the group Γ/A^0 contains no element of infinite order. Hence Γ/A^0 is almost abelian, which implies $G' \subset A$ (Γ/A^0 is Zariski-dense in G/A^0). □

Caveat: There is no hope for $A = G$, e.g. take $G = \mathbb{C}^*$ and let Γ denote the subgroup which consists of all roots of unity.

Theorem 2 follows by induction on $dim(G)$ using the following lemma.

Lemma 11. *Let G be a positive-dimensional complex linear-algebraic group fulfilling the assumptions of Theorem 2, Γ a Zariski-dense subgroup.*

Then there exists a positive-dimensional normal algebraic subgroup A with $\mathcal{O}(G)^\Gamma \subset \mathcal{O}(G)^A$.

Proof. If G is abelian, the assumptions imply that G is reductive and $\mathcal{O}(G)^\Gamma = \mathbb{C}$.

Otherwise let $H = \{g : f(g) = f(e) \forall f \in \mathcal{O}(G)^\Gamma\}$. Now $\Gamma \subset H$, hence Γ normalizes H^0. The normalizer of a connected Lie subgroup is algebraic, thus H^0 is a normal subgroup of G. It follows that $(H \cap G')^0$ is a normal algebraic subgroup (Lemma 2). This completes the proof unless $H \cap G'$ is discrete.

This leaves the case where $(H \cap G')$ is discrete. Then H^0 is contained in the center Z of G (Lemma 3). Let A denote the Zariski-closure of H^0. The center is reductive (Lemma 7). It follows that for each Z-orbit every H^0-invariant functions is already A-invariant.

Therefore we can restrict to the case where H is discrete. Now Γ is discrete and contains a subgroup Γ_0 which is finitely generated and whose Zariski-closure contains G'. By a theorem of Selberg Γ_0 contains a subgroup of finite index Γ_1 which is torsion-free. Now let $\Gamma_2 = \Gamma_1 \cap G'$. Then being Zariski-dense, Γ_2 must contain a semisimple element of infinite order. Using Lemma 8, this yields a contradiction to the assumption that H is discrete. □

4. An example

At a first glance, it seems to be obvious that a Zariski-dense subgroup should contain enough elements of infinite order to generate a subgroup which is still Zariski-dense. However, one has to be careful.

Lemma 12. *Let $G = \mathbb{C}^* \ltimes \mathbb{C}$ with group law $(\lambda, z) \cdot (\mu, w) = (\lambda\mu, z + \lambda w)$ and Γ the subgroup generated by the elements $a_n = (e^{2\pi i/n}, 0)$ $(n \in \mathbb{N})$ and $a_0 = (1,1)$. Then*

$\gamma \in G' = \{(1,x) : x \in \mathbb{C}\}$ *for any element* $\gamma \in \Gamma$ *of infinite order, although* Γ *is Zariski-dense in* G.

Proof. It is clear that Γ is Zariski-dense in G. The other assertion follows from the fact that any element in G is either unipotent or semisimple. Hence every element $g \in G \setminus G'$ is conjugate to an element in $\mathbb{C}^* \times \{0\}$. □

5. Margulis' example

We will use an example of Margulis to demonstrate the following.

Proposition 1. *There exists a discrete Zariski-dense subgroup* Γ *in* $G = SL_2(\mathbb{C}) \ltimes_\rho (\mathbb{C}^3, +)$ *with* ρ *irreducible such that* Γ *contains no semisimple element.*

Thus the condition G/G' reductive is not sufficient to guarantee the existence of semisimple elements in Zariski-dense subgroups.

Margulis [5,M2] constructed his example in order to prove that there exist free non-commutative groups acting on $\mathbb{R}^n$ properly discontinuous and by affin-linear transformations, thereby contradicting a conjecture of Milnor [7].

We will now start with the description of Margulis' example. Let B denote the bilinear form on $\mathbb{R}^3$ given by $B(x_1, x_2, x_3) = x_1^2 + x_2^2 - x_3^2$, $W = \{x \in \mathbb{R}^3 : B(x,x) = 0\}$ the zero cone and $W^+ = \{x \in W : x_3 > 0\}$ the positive part. Let $S = \{x \in W^+ : |x| = 1\}$. Let H be the connected component of the isometry group $O(2,1)$ of B. (As a Lie group H is isomorphic to $PSL_2(\mathbb{R})$.) Let $G_{\mathbb{R}} = H \ltimes (\mathbb{R}^3, +)$ the group of affine-linear transformations on $\mathbb{R}^3$ whose linear part is in H.

The following is easy to verify: Let x^+, x^- be different vectors in S. Then there exists a unique vector x^0 such that $B(x^0, x^0) = 1$, $B(x^+, x^0) = 0 = B(x^-, x^0)$ and x^0, x^-, x^+ form a positively oriented basis of the vector space $\mathbb{R}^3$. Furthermore for any $\lambda \in]0,1[$ there is an element $g \in H$ (depending on $x^+, x^- \in S$ and λ) defined as follows:

$$g : ax^0 + bx^- + cx^+ \mapsto ax^0 + \frac{b}{\lambda}x^- + \lambda c x^+$$

Conversely any non-trivial diagonalizable element $g \in H$ is given in such a way and x^+, x^- and λ are uniquely determined by g.

The result of Margulis is the following:

Let $x^+, x^-, \tilde{x}^+, \tilde{x}^-$ be four different points in S, $\lambda, \tilde{\lambda} \in]0,1[$, $v, \tilde{v} \in \mathbb{R}^3$ such that v, x^-, x^+ resp. $\tilde{v}, \tilde{x}^-, \tilde{x}^+$ forms a positively oriented basis of $\mathbb{R}^3$. Let $h, \tilde{h} \in H$ be the elements corresponding to x^-, x^+, λ resp. $\tilde{x}^-, \tilde{x}^+, \tilde{\lambda}$ and $g, \tilde{g} \in G_{\mathbb{R}} = H \ltimes \mathbb{R}^3$ given by $g = (h, v)$, $\tilde{g} = (\tilde{h}, \tilde{v})$.

Then there exists a number $N = N(g, \tilde{g})$ such that the elements g^N, $\tilde{g}^N$ generate a (non-commutative) free discrete subgroup $\Gamma \subset G_{\mathbb{R}}$ such that the action on $\mathbb{R}^3$ is properly discontinuous and free.

Now an element $g \in G_{\mathbb{R}}$ is conjugate to an element in H if and only if $g(w) = w$ for some $w \in \mathbb{R}^3$. Hence no element in Γ is conjugate to an element in H. In particular no element in semisimple. Furthermore it is clear that Γ is Zariski-dense in the complexification $G = SL_2(\mathbb{C}) \ltimes \mathbb{C}^3$ of $G_{\mathbb{R}}$.

6. Meromorphic functions

Proposition 2. *Let G be a connected complex linear-algebraic group with $G = G'$ and an open subset Ω such that each element in Ω is semisimple. Let Γ be a Zariski-dense subgroup.*

Then any Γ-invariant plurisubharmonic or meromorphic function is constant and there exist no Γ-invariant hypersurface.

Proof. We may assume that Γ is closed (in the Hausdorff topology). Since $G = G'$, it follows that H^0 is a normal algebraic subgroup for each Zariski-dense subgroup H. Therefore we may assume that Γ is discrete and furthermore it suffices (by induction on $\dim(G)$ to demonstrate that the functions resp. hypersurfaces are invariant under a positive-dimensional subgroup. Now $G = G'$ implies that Γ admits a finitely generated subgroup Γ_0 which is still Zariski-dense. By the theorem of Selberg Γ_0 admits a subgroup of finite index Γ_1 which is torsion-free. Thus Γ_1 contains a semisimple element of infinite order γ which generates a subgroup I whose Zariski-closure $\bar{I}$ is a torus. $G = G'$ implies that this torus is contained in a connected semisimple subgroup S of G. Now known results on subgroups in semisimple groups [4][2] imply that the functions resp. hypersurfaces are invariant under $\bar{I}$, which is positive-dimensional. □

For this result it is essential to require $G = G'$ and not only G/G' reductive.

Lemma 13. *Let $G = \mathbb{C}^* \times \mathbb{C}^*$ and $\Gamma \simeq \mathbb{Z}$ a (possibly Zariski-dense) discrete subgroup.*

Then G admits Γ-invariant non-constant plurisubharmonic and meromorphic functions.

Proof. $G/\Gamma \simeq \mathbb{C}^2/\Lambda$ with $\Lambda \simeq \mathbb{Z}^3$. Let $V = < \Lambda >_{\mathbb{R}}$ the real subvector space of $\mathbb{C}^2$ spanned by Λ and $t : \mathbb{C}^2 \to \mathbb{C}^2/V \simeq \mathbb{R}$ a $\mathbb{R}$-linear map. Then t^2 yields a Γ-invariant plurisubharmonic function on G.

Let $L = V \cap iV$ and $\gamma \in \Lambda \setminus L$. Let $H = < \gamma >_{\mathbb{C}}$. Then $H \neq L$, hence $H + L = \mathbb{C}^2$. It follows that the H-orbits in G/Γ are closed and induce a fibration $G/\Gamma \to G/H\Gamma$ onto a one-dimensional torus. One-dimensional tori are projective and therefore admit non-constant meromorphic functions. □

REFERENCES

1. BARTH, W.; OTTE, M.: Invariante Holomorphe Funktionen auf reduktiven Liegruppen. *Math. Ann.* **201**, 91–112 (1973)
2. BERTELOOT, F.; OELJEKLAUS, K.: Invariant Plurisubharmonic Functions and Hypersurfaces on Semisimple Complex Lie Groups. *Math. Ann.* **281**, 513–530 (1988)
3. HUCKLEBERRY, A.T.; OELJEKLAUS, E.: On holomorphically separable complex solvmanifolds. *Ann. Inst. Fourier* **XXXVI 3**, 57–65 (1986)
4. HUCKLEBERRY, A.T.; MARGULIS, G.A.: Invariant analytic hypersurfaces. *Invent. Math.* **71**, 235–240 (1983)
5. MARGULIS, G.A.: Free totally discontinuous groups of affine transformations. *Soviet Math. Doklady* **28**, 435–439 (1983)
6. MARGULIS, G.A.: Complete affine locally flat manifolds with free fundamental group. *Zapiski Naučn. Sem. Leningrad Otd. Mat. Inst. Steklov* **134**, 190-205 (1984) [in Russian]
7. MILNOR, J.: On fundamental groups of completely affinely flat manifolds. *Adv. Math.* **25**, 178–187 (1977)
8. TITS, J.: Free subgroups in linear groups. *J. Algebra* **20**, 250–270 (1972)

Pseudoconvexity and Pseudoconcavity of Dihedrons of C^n

Giuseppe Zampieri, Dipartimento di Matematica Pura ed Applicata, Università di Padova, Via Belzoni 7, I-35131 Padova, Italy. zampieri@pdmat1.unipd.it

ABSTRACT. Let X be a complex manifold of dimension n, $\mathcal{O}_X$ the sheaf of holomorphic functions on X, W a domain of X, z_o a point of ∂W. It is classical that, when $\partial W \in C^2$, then $\varinjlim_{B \ni z_o} H^j(B \cap W, \mathcal{O}_X) = 0\, \forall j \notin [s^-(z_o), n-(s^+(z_o)+1)]$, where $s^{+,-,0}(z_o)$ are respectively the numbers of positive, negative, and null eigenvalues of the Levi form of W at z_o. It is also classical that $s^-(z) \equiv 0\, \forall z$ close to z_o, if and only if all the non-trivial cohomology groups of $\mathcal{O}_X$ on W vanish. More generally (see [11]), it is known that the upper bound $n-(s^+(z_o)+1)$ should be replaced by $\sup_z s^-(z)$. In this paper we generalize all of the above results to the case of a dihedron with transversal faces (see also [15] as for the dihedron with convex tangent cone). The two tools we use are rather different from eachother: algebraic analysis in §2, and the Hörmander L^2-estimates in §3.

1. STATEMENTS

Let X be a complex manifold, z_o a point of X, ϕ a real C^2-function on X at z_o. For a local system of complex coordinates $z = (z_j)$, we denote by $L_\phi(z_o)$ the Hermitian form with matrix $(\frac{\partial^2 \phi}{\partial z_i \partial \bar{z}_j}(z_o))_{ij}$. Let M be a real C^2-submanifold of X at z_o, denote by $T^*_M X$ the conormal bundle to M in X, and fix $p \in \dot{T}^*_M X \overset{\text{def.}}{=} T^*_M X \setminus \{0\}$ with $\pi(p) = z_o$. Choose an equation $\phi = 0$ for M with $\partial\phi(z_o) = p$. One sees that the restriction of $L_\phi(z_o)$ to $T^{\mathbb{C}}_{z_o} M = T_{z_o} M \cap \sqrt{-1} T_{z_o} M$ does not depend on the choice of ϕ (provided that $\partial\phi(z_o) = p$). One then defines $L_M(p) = L_\phi(z_o)|_{T^{\mathbb{C}}_{z_o} M}$ and denotes by $s_M^{+,-,0}(p)$ the numbers of respectively positive, negative, and null eigenvalues for $L_M(p)$.

Let W be a dihedron of X at z_o with transversal faces M_1^+, M_2^+. Define $M_3 = M_1^+ \cap M_2^+$, and denote by $N(W)$ (resp. TW, resp. $N(W)^{oa}$) the normal (resp. tangent, resp. exterior conormal) cone to W. Choose extensions M_i, $i = 1, 2$ of M_i^+, set $T^*_{M_i} X^+ = T^*_{M_i} X \cap N(W)^{oa}$, and define

$$(1.1) \qquad s_i^- = \inf_{\forall p' \in T^*_{M_i} X^+ \cap \pi^{-1}(B)} s^-_{M_i}(p'), \; S_i^- = \sup_{\forall p' \in T^*_{M_i} X^+ \cap \pi^{-1}(B)} s^-_{M_i}(p'),$$

where B is a neighborhood of z_o. If $T_{z_o}W$ is non-convex, define

$$(1.2) \qquad s^- = s_3^- + 1, \quad S^- = \sup(S_1^-, S_2^-, S_3^- + 1).$$

If $T_{z_o}W$ is convex, define

$$(1.3) \qquad s^- = s_3^-, \quad S^- = \sup(S_1^-, S_2^-, S_3^-),$$

and assume also, in this case, that for an extension M_i of M_i^+, $i = 1, 2$, one has $s_{M_i}^-(p') \leq S^- \, \forall p' \in T_{M_i}^*X \cap \pi^{-1}(B)$.

Theorem 1.1. *Assume that M_3 is generic (i.e. $T_{M_3}^*X_{z_o} \cap \sqrt{-1}T_{M_3}^*X_{z_o} = \{0\}$). Then*

$$(1.4) \qquad \varinjlim_B \mathrm{H}^j(W \cap B, \mathcal{O}_X) = 0 \quad \forall j \notin [s^-, S^-].$$

Moreover if $s^- \geq 1$, then $\varinjlim_B \Gamma(B, \mathcal{O}_X) \twoheadrightarrow \varinjlim_B \Gamma(W \cap B, \mathcal{O}_X)$ is surjective.

To prove Th. 1.1 we shall adopt two different methods: algebraic analysis and L^2-estimates. Let us point out that the first appears more appropriate for proving the vanishing of (1.4) for $j < s^-$, and the second for $j > S^-$.

§2. The method of algebraic analysis

§2.1. Real Symplectic geometry. Let X be a real manifold, T^*X be the cotangent bundle to X, p be a point of $\dot{T}^*X$. Let Λ_1^+, Λ_2^+ be two conic Lagrangian C^1-submanifolds of T^*X in a neighborhood of p with boundary Σ, which intersect along Σ with clean intersection. Define a "dihedral Lagrangian" by $\Lambda = \Lambda_1^+ \cup \Lambda_2^+$.

Theorem 2.1. *There exists a homogeneous contact transformation χ on T^*X at p which interchanges Λ with the conormal bundle T_Y^*X to a C^1-hypersurface Y.*

Proof. **(a)** First we extend Λ_i^+ to Λ_i, across Σ and transform $\chi(\Lambda_i) = T_{M_i}^*X$ with $\operatorname{codim} M_i = 1$.

(b) The cleaness of the intersection $T_{M_1}^*X \cap T_{M_2}^*X$ is equivalent to the fact that the M_i's intersect with order 2. In fact let $S \overset{\text{def.}}{=} \pi(T_{M_1}^*X \cap T_{M_2}^*X)$. This is a submanifold of codim 1 in M_i. Take real coordinates $t = (t_1, t_2, t')$ such that

$$M_2 = \{t_1 = 0\}, \, M_1 = \{t_1 = g(t_2, t')\}, \, S = \{t_1 = t_2 = 0\},$$

(with $g = o(|(t_2, t')|)$, $p = (0; dt_1)$). We have $\operatorname{Hess} g \cdot u = 0 \, \forall u \in S$ since $g|_S \equiv 0$, and $dg|_S \equiv 0$. We also have

$$T_p T_{M_2}^*X = \{(u; t dt_1); u_1 = 0, t \in \mathbb{R}\}, \; T_p T_{M_1}^*X = \{(u; t dt_1 + \operatorname{Hess} g \cdot u); u_1 = 0, t \in \mathbb{R}\}.$$

Thus cleanness is equivalent to the implication $\operatorname{Hess} g \cdot u = 0 \Rightarrow u_2 = 0$, i.e. $\operatorname{Hess} g \cdot u = cu_2^2$, $c \neq 0$. It follows that M_1, M_2 intersect with order 2 along S.

(c) Write

$$\mathring{\Lambda}_i^+ = \Lambda_i^+ \setminus \Sigma,\ \Lambda_i^- = \Lambda_i \setminus \mathring{\Lambda}_i^+,\ \mathring{\Lambda}_i^- = \Lambda_i \setminus \Lambda_i^+,$$
$$M_i^\pm = \pi(\chi(\Lambda_i^\pm)),\ \mathring{M}_i^\pm = \pi(\chi(\mathring{\Lambda}_i^\pm)).$$

Either $M_1^+ \cup M_2^+$ or $M_1^+ \cup M_2^-$ is a C^1–hypersurface Y. In the first case $\chi(\Lambda) = T_Y^* X$ and the proof is complete.

In the second case, we recall the notations of (b). Assume $g > ct_2^2$ (resp. $g < -ct_2^2$). Take $r > (2c)^{-1}$, and consider the contact transformation $\tilde{\chi}$ which, in coordinates $(t, \theta) \in T^*X$, is given by:

$$\tilde{\chi} : \begin{cases} t \mapsto t \pm r\theta/|\theta|, \\ \theta \mapsto \theta \end{cases}$$

where we take $+$ (resp. $-$) if $g > ct_2^2$ (resp. $g < -ct_2^2$). Let

$$\Omega_0 = \{t; \pm[(t_1 \mp \frac{1}{2c})^2 + t_2^2] > \pm\frac{1}{(2c)^2}\},\ \Omega_1 = \{t; \pm[(t_1 \mp \frac{1}{2c})^2 + t_2^2] < \pm(r - \frac{1}{2c})^2\}$$

and denote by $N(\Omega_i)^{oa}$, $i = 1, 2$, the exterior conormal bundle to Ω_i We have $\tilde{\chi}(N(\Omega_0)^{oa}) = N(\Omega_1)^{oa}$. Write $\chi(T_{M_i}^* X) = T_{M_i'}^* X$, $\chi(S \times_{M_1} T_{M_1}^* X) = S' \times_{M_1'} T_{M_1'}^* X$. We also have $M'_1 \subset \{t_1' \leq r\} \setminus \Omega_1$, $M'_2 = \{t_1' = r\}$. Thus $M'^+_1 \cup M'^-_2 \notin C^1$. Hence $M'^+_1 \cup M'^+_2 \in C^1$, since $TM_1'|_{S'} = TM_2'|_{S'}$, $(S' = M_1' \cap M_2')$.

Q.E.D.

Remark 2.2. Let $\Lambda_2^- = \Lambda_2 \setminus \Lambda_2^+$. It is easy to see that in order to transform both $\Lambda_1^+ \cup \Lambda_2^+$ and $\Lambda_1^+ \cup \Lambda_2^-$ into conormal bundles to (different) hypersurfaces, the intersection $\Lambda_1 \cap \Lambda_2$ must be clean.

Let $D^b(X)$ be the derived category of the category of complexes of sheaves of groups with bounded cohomology. Let SS denote the microsupport in the sense of [12], and let $D^b(X; p)$, $(p \in T^*X)$ be the localization of $D^b(X)$ by the null–system $\{\mathcal{F}; \mathrm{SS}(\mathcal{F}) \not\ni p\}$. Let $\mathcal{F}, \mathcal{G} \in D^b(X; p)$.

Theorem 2.3. *Let $\Lambda = \Lambda_1^+ \cup \Lambda_2^+$ be a dihedral Lagrangian at p, and let $\mathrm{SS}(\mathcal{F}) \subset \Lambda$, $\mathrm{SS}(\mathcal{G}) \subset \Lambda$. If $\mathcal{F} \simeq \mathcal{G}$ microlocally at a regular point $p' \in \Lambda \setminus \Sigma$, then $\mathcal{F} \simeq \mathcal{G}$ at p.*

Proof. By a contact transformation χ, we transform $\chi(\Lambda) = T_Y^* X$. By choosing a quantization Φ_K above χ, we get, for suitable complexes of groups $M^\cdot$, $N^\cdot$:

$$\Phi_K(\mathcal{F}) \simeq M_Y^\cdot, \quad \Phi_K(\mathcal{G}) \simeq N^\cdot,$$
$$\mathrm{Hom}_{D^b(X;p')}(M_Y^\cdot, N_Y^\cdot) \xrightarrow{\sim} \mathrm{H}^0\mu\mathrm{hom}(M_Y^\cdot, N_Y^\cdot)_{p'} \simeq \mathrm{Hom}_{\mathbb{Z}}(M^\cdot, N^\cdot).$$

Thus $M_Y^\cdot \xrightarrow{\sim} N_Y^\cdot$ at p' iff $M^\cdot \simeq N^\cdot$.

Q.E.D.

In particular there is an unique $\mathcal{F}$ at p with $\mathrm{SS}(\mathcal{F}) \subset \Lambda$ and simple with a prescribed shift at a regular point $p' \in \Lambda$.

§2.2. Vanishing of cohomology for microfunctions at the boundary. Let X be a complex manifold, $S \subset M$ two real C^2–submanifolds of X with $\operatorname{codim}_X M = l_M$, $\operatorname{codim}_M S = 1$. Set $\Lambda_1 = T^*_M X, \Lambda_2 = T^*_S X$. Clearly $\Lambda_1 \cap \Lambda_2$ is clean. Denote by $\bar\Omega^{\pm}$ the closed components of $M \setminus S$ and put $\Lambda_1^{\pm} = \bar\Omega^{\pm} \times_M T^*_M X$. This choice of sign induces a choice for the components $\Lambda_2^{\pm} = \rho^{-1}(N(\bar\Omega^{\pm})^o)$ of $\Lambda_2 \setminus \Sigma$ (with $N(\bar\Omega^{\pm})^o$ denoting the conormal cones and $\rho : M \times_X T^*X \to T^*M$ the projection). We also put $\mathring{\Lambda}_i^{\pm} = \Lambda_i^{\pm} \setminus \Sigma$ and write Ω instead of Ω^+. We have $\mathrm{SS}(\mathbb{Z}_{\bar\Omega}) \subset \Lambda_1^+ \cup \Lambda_2^+$, and $\mathrm{SS}(\mathbb{Z}_\Omega) \subset \Lambda_1^+ \cup \Lambda_2^-$ with both sheaves simple with shift $\frac{1}{2}\operatorname{codim} M$ in $\mathring{\Lambda}_1^+$. By § 2.1 they are completely described by the mentioned properties. Recall the Levi form $L_M(p)$, $p \in \dot{T}^*_M X$ (resp. $L_S(p)$, $p \in \dot{T}^*_S X$) and denote by $s_M^{+,-,0}(p)$ (resp. $s_S^{+,-,0}(p)$) the numbers of its respectively positive, negative, and null eigenvalues. We also define $\gamma_M(z) = \dim(T^*_M X_z \cap \sqrt{-1}T^*_M X_z)$, $\gamma_S(z) = \dim(T^*_S X_z \cap \sqrt{-1}T^*_S X_z)$.

Let $\mathcal{O}_X$ be the sheaf of holomorphic functions on X and, for a locally closed subset $A \subset X$, let $\mathbb{Z}_A$ be the sheaf which is 0 on $X \setminus A$, and constant with stalk $\mathbb{Z}$ on A. Recall the complex $\mu_A(\mathcal{O}_X) \overset{\text{def.}}{=} \mu\mathrm{hom}(\mathbb{Z}_A, \mathcal{O}_X)$ by [K-S 1]. Let $S \subset M$ with $\operatorname{codim}_M S = 1$. Denote by Ω an open component of $M \setminus S$. Fix $p \in S \times_M T^*_M X, \pi(p) = z_o$, and assume $\sqrt{-1}p \notin T^*_S X$. The next statement is in part contained in [6].

Theorem 2.4. *We have*

$$\begin{aligned}(2.1)\qquad & \mathrm{H}^j\mu_\Omega(\mathcal{O}_X) = 0 \quad \forall j < l_M + s_M^-(p) - \gamma_M(z_o), \\ & \mathrm{H}^j\mu_{\bar\Omega}(\mathcal{O}_X) = 0 \quad \forall j < l_M + 1 + s_S^-(p) - \gamma_S(z_o).\end{aligned}$$

Proof. (a) We first assume $s_M^-(p) - \gamma_M(z_o) = 1 + s_S^-(p) - \gamma_S(z_o)$. We make a complex homogeneous contact transformation χ on T^*X between neighborhoods of p and $q \overset{\text{def.}}{=} \chi(p)$:

$$(2.2)\qquad \begin{cases} \chi(T^*_M X) = T^*_{\tilde M} X & \operatorname{codim} \tilde M = 1, s_{\tilde M}^-(q) = 0, \\ \chi(T^*_S X) = T^*_{\tilde S} X & \operatorname{codim} \tilde S = 1. \end{cases}$$

cf. [14]. (Note that the only assumption we use for (2.2) is $\sqrt{-1}p \notin T^*_S X$.) According to the proof of Th. 2.1 (b), $\pi(T^*_{\tilde M} X \cap T^*_{\tilde S} X) = \tilde M \cap \tilde S$, and $\Sigma_{\tilde M} \supset \Sigma_{\tilde S}$ or $\Sigma_{\tilde M} \subset \Sigma_{\tilde S}$, where $\Sigma_{\tilde M}$, $\Sigma_{\tilde S}$ are the closed half–spaces with boundaries $\tilde M$, $\tilde S$ and interior conormal q. Thus, either $\chi(\Lambda_1^+ \cup \Lambda_2^+) = T^*_Y X$, or $\chi(\Lambda_1^+ \cup \Lambda_2^-) = T^*_Y X$ for $Y \in C^1$. So, either $\Phi_K(\mathbb{Z}_{\bar\Omega})$, or $\Phi_K(\mathbb{Z}_\Omega)$ is microlocally isomorphic to a constant sheaf along Y. This implies that

$$(2.3)\qquad s_{\tilde S}^-(q) = 0,\ \Sigma_{\tilde M} \supset \Sigma_Y \supset \Sigma_{\tilde S},$$

and that

$$\begin{aligned}& \Phi_K(\mathbb{Z}_{\bar\Omega}) \simeq \mathbb{Z}_{\Sigma_Y}[s_M^-(p) + l_M - \gamma_M(z_o) - 1], \\ (2.4)\qquad & \Phi_K(\mathbb{Z}_\Omega) \simeq \mathbb{Z}_{\Sigma_Y \setminus \Sigma_{\tilde S}}[s_M^-(p) + l_M - \gamma_M(z_o) - 1].\end{aligned}$$

It follows that

$$(2.5)\qquad \chi_*\mu_{\tilde{\Omega}}(\mathcal{O}_X) = R\Gamma_{\Sigma_Y}(\mathcal{O}_X)[-s^-_M(p) - l_M + \gamma_M(z_o) + 1]$$
$$\chi_*\mu_{\Omega}(\mathcal{O}_X) = R\Gamma_{(\Sigma_Y\setminus\Sigma_{\tilde{S}})}(\mathcal{O}_X)[-s^-_M(p) - l_M + \gamma_M(z_o) + 1].$$

The conclusion for the first of (2.5) is immediate. As for the second, one has to remark that $\mathcal{H}^1_{\Sigma_{\tilde{S}}}(\mathcal{O}_X) \hookrightarrow \mathcal{H}^1_{\Sigma_Y}(\mathcal{O}_X)$ is injective.

(b) Let $s^-_M(p) - \gamma_M(z_o) = s^-_S(p) - \gamma_S(z_o)$. In this case we make a contact transformation χ as in (2.2), but we now require that $s^-_{\tilde{S}}(q) = 0$ instead of $s^-_{\tilde{M}}(q) = 0$. The proof is then similar to the above one.

Q.E.D.

§2.3. Proof of the vanishing of (1.4) for $j < s^-$. We assume in the beginning that $T_{z_o}W$ is convex (i.e. $c = 0$), and focus the attention on p_1, the conormal to M_1 at z_o exterior to W. We set $(M_1^+ \times_{M_1} T^*_{M_1}X) \sqcup \rho^{-1}(N(M_1^+)^{oa}) = \Lambda_1^+ \sqcup \Lambda_2^+$. We have

$$(2.6)\qquad N(W)^{oa} = \Lambda_1^+ \sqcup \Lambda_2^- \text{ at } p_1,$$

and therefore

$$(2.7)\qquad \mathrm{SS}(\mathbb{Z}_W) = \mathrm{SS}(\mathbb{Z}_{\overset{\circ}{M}{}_1^+}) \text{ at } p_1.$$

Thus by the results of § 2.1,

$$(2.8)\qquad \mathbb{Z}_W \simeq \mathbb{Z}_{\overset{\circ}{M}{}_1^+}[-1], \text{ in } D^b(X;p_1)$$

and by those of § 2.2,

$$(2.9)\qquad \mu_W(\mathcal{O}_X)_{p_1} \simeq \mu_{\overset{\circ}{M}{}_1^+}(\mathcal{O}_X)_{p_1}[+1] \text{ is concentrated in degree} \geq s^-_{M_1}(p_1).$$

By a similar argument, we may prove (2.9) with p_1, M_1 replaced by p_2, M_2, and also show that

$$(2.10)\qquad \mu_W(\mathcal{O}_X)_{p'} \simeq \mu_{M_3}(\mathcal{O}_X)_{p'}[2] \text{ is concentrated in degree} \geq s^-_{M_3}(p')\forall p' \in T^*_{M_3}X^+,\ p' \neq p_1, p_2.$$

Recall the triangle in $D^b(X)$

$$(2.11)\qquad (\mathcal{O}_X)_{\bar{W}} \to R\Gamma_W(\mathcal{O}_X) \to R\dot{\pi}_*\mu_W(\mathcal{O}_X) \overset{+1}{\to}.$$

Then $\mathrm{H}^j R\Gamma_W(\mathcal{O}_X)_{z_o} = 0,\ 0 < {}^\forall j < \inf_{i\,p'} s^-_{M_i}(p') = \inf_{p'} s^-_{M_3}(p') = s^-$.

We consider now the case $c = 1$ (i.e. $T_{z_o}W$ non–convex). Then

$$(2.12)\qquad N(W)^{oa} = \Lambda_1^+ \sqcup \Lambda_2^+ \text{ at } p_1.$$

It follows that

$$(2.13)\qquad \mathbb{Z}_W \simeq \mathbb{Z}_{M_1^+}[-1] \text{ in } D^b(X;p_1),$$

and therefore, by § 2.2,

$$(2.14)\qquad \mu_W(\mathcal{O}_X)_{p_1} \simeq \mu_{M_1^+}(\mathcal{O}_X)_{p_1}[1] \text{ is concentrated in degree } \geq 1 + s^-_{M_3}(p_1).$$

(Similarly for p_2, M_2.) We also have, by [12],

$$(2.15)\qquad \mu_W(\mathcal{O}_X)_{p'} \simeq \mu_{M_3}(\mathcal{O}_X)_{p'}[+1] \text{ is concentrated in degree } \geq 1 + s^-_{M_3}(p').)$$

We then conclude as in the first case, using (2.11).

Q.E.D.

3. The method of L^2–estimates

§3.1. Let W be a domain of $X = \mathbb{C}^n$ endowed with an exhaustion function ϕ (i.e. a function such that $\{\phi \le c\} \subset\subset W \, \forall c$), which verifies

(1) ϕ is C^1 everywhere in W and C^2 outside a finite disjoint family of C^2–hypersurfaces $N_i \subset W$

(2) ϕ has locally bounded second derivatives

(3) For a complex plane $\mathcal{R} = \mathcal{R}_z \subset T_z X$ of dimension b, depending continuously on z, and for $v \mapsto v'$ denoting the orthogonal projection $T_z X \to \mathcal{R}_z$, one has

$$\partial\bar\partial\phi v^t \bar v - \partial\bar\partial\phi v'^t \bar v' \ge \lambda |v''|^2, \; (v'' = v - v', \, \lambda > 0).$$

Proposition 3.1. *In the above situation one has*

$$\mathrm{H}^j(W, \mathcal{O}_X) = 0 \quad \forall j > b. \tag{3.1}$$

Proof. We recall here, with suitable modifications, the results by Hörmander in [10, ch. 4, 5] and in [11]. We make use of the norm $\|\cdot\|_\phi^2 = \int |\cdot|^2 e^{-\phi} \mathrm{d}V$ where ϕ is the function satisfying (1)–(3), and $\mathrm{d}V$ is the element of volume defined by a Hermitian metric on W. Let $\omega_1, \dots, \omega_n$ (resp. $\bar\omega_1, \dots, \bar\omega_n$) be a basis of holomorphic (resp. antiholomorphic) orthonormal 1–forms, and let $L^2_{i\,j}(W, \phi)$ denote the space of alternate $i\,j$–forms $f = \sum_{|H|=i\,|J|=j} f_{H\,J} \omega_H \wedge \bar\omega_J$ which have alternate coefficients $f_{H\,J}$ with finite $\|\cdot\|_\phi$ norm. (Here $\omega_H = \omega_{h_1} \wedge \dots \wedge \omega_{h_i}$, $\bar\omega_J = \bar\omega_{j_1} \wedge \dots \wedge \bar\omega_{j_j}$.) Let T, S be the closed densely defined operators

$$L^2_{i\,j-1}(W, \phi) \xrightarrow{T} L^2_{i\,j}(W, \phi) \xrightarrow{S} L^2_{i\,j+1}(W, \phi), \tag{3.2}$$

which induce $\bar\partial$ on C^∞_c–forms. According to [10 ch. 4 and 5], the vanishing of (3.1) may be reduced to the exactness of (3.2). By Banach's Theorem, and by the approximation result of [10, Lemma 5.2.1],the latter should follow from the inequality

$$\|f\|_\phi \le c(\|T^* f\|_\phi + \|Sf\|_\phi) \quad \text{for any } i, j\text{–form } f \text{ with } C^\infty_c\text{–coefficients.} \tag{3.3}$$

We prove now (3.3). We define the operator δ^α by $\delta^\alpha(w) = e^\phi \frac{\partial(we^{-\phi})}{\partial\omega_\alpha}$. Thus δ^α is the adjoint of $-\frac{\partial}{\partial\bar\omega_\alpha}$ with respect to the scalar product underlying the $\|\cdot\|_\phi$–norm, apart from an error (cf. the formula before [10, (5.2.3)]) which occurs because $\bar\partial\omega^i$ and $\bar\partial\bar\omega^i$ are not necessarily 0. One has also the commutation relations

$$\int_W \left(\delta^\alpha f_{H\,\alpha K} \overline{\delta^\beta f_{H\,\beta K}} - \frac{\partial f_{H\,\alpha K}}{\partial\bar\omega_\beta} \overline{\frac{\partial f_{H\,\beta K}}{\partial\bar\omega_\alpha}} \right) e^{-\phi}\,\mathrm{d}V = \tag{3.4}$$

$$\int_W f_{H\,\alpha K} \bar f_{H\,\beta K} \frac{\partial^2\phi}{\partial\omega_\alpha \partial\bar\omega_\beta} e^{-\phi}\,\mathrm{d}V + \sum_i \int_{N_i} f_{H\,\alpha K} J(\frac{\partial\phi}{\partial\omega_\alpha}) \bar f_{H\,\beta K} \bar n^i_\beta \,\mathrm{d}S + R.$$

Here R is an error like in [10, (5.2.8)], $J(\partial\phi)$ is the jump of $\partial\phi$ from the two sides of N_i, n^i is the normal to N_i (from the negative to the positive side), and $\mathrm{d}S$ is the element of area in N_i. But the above jump is 0 due to the hypothesis "$\phi \in C^1$". We choose $f = \sum'_{|H|=i\,|J|=j} f_{H\,J}\omega_H \wedge \bar{\omega}_J$ where $\sum'$ denotes the summation over strictly increasing multiindices. From (3.4) and from [10, (5.2.4) and (5.2.5)], we get at once

$$(3.5)\qquad \sum'_{|H|=i,\,|K|=j-1}\sum_{\alpha,\beta}\int_W f_{H\,\alpha K}\bar{f}_{H\,\beta K}\frac{\partial^2\phi}{\partial\omega_\alpha\partial\bar{\omega}_\beta}e^{-\phi}\,dV$$
$$+\left(\sum'_{|H|=i,\,|J|=j}\sum_{\alpha}\int_W\left|\frac{\partial f_{H\,J}}{\partial\bar{\omega}_\alpha}\right|^2 e^{-\phi}\,dV\right)\le 2(\|T^*f\|^2_\phi+\|Sf\|^2_\phi)+C\|f\|^2_\phi + R,$$

where the error R is given by integrals which involve products $f_{H\,J}\bar{f}_{H\,J'}$, $f_{H\,J}\overline{\frac{\partial f_{H\,J'}}{\partial\bar{\omega}_\alpha}}$, and $f_{H\,J}\overline{\delta^\alpha f_{H\,J'}}$. We pick up in the second entry on the left hand side of (3.5) those terms with $1 \le \alpha \le b$ and $\alpha \in J$, and rewrite them, after rearranging indices $J = \alpha K$, by the aid of (3.4) for $\alpha = \beta \le b$. We keep unchanged the remaining terms.

Thus the second entry in the left of (3.5) can be rewritten as

$$(3.6)\qquad -\sum'_{|H|=i\,|K|=j-1}\sum_{\alpha=1}^{b}\int_W |f_{H\,\alpha K}|^2\frac{\partial^2\phi}{\partial\omega_\alpha\partial\bar{\omega}_\alpha}e^{-\phi}\,dV +$$
$$\Bigg[\sum'_{|H|=i,|K|=j-1}\sum_{\alpha=1}^{b}\int_W|\delta^\alpha f_{H\,\alpha K}|^2e^{-\phi}\,dV+\sum'_{|H|=i\,|J|=j}\sum_{\alpha\notin J,\alpha\le b}\int_W\left|\frac{\partial f_{H\,J}}{\partial\bar{\omega}_\alpha}\right|^2e^{-\phi}\,dV+$$
$$\sum'_{|H|=i\,|J|=j}\sum_{\alpha=b+1}^{n}\int_W\left|\frac{\partial f_{H\,J}}{\partial\bar{\omega}_\alpha}\right|^2e^{-\phi}\,dV\Bigg] + R.$$

We denote by Q the term between brackets in (3.6). One easily checks (cf. the argument which leads to [10, (5.2.10)]) that all the above errors R verify

$$|R| \le C\|f\|^2_\phi + Q.$$

Thus from (3.5) we get:

$$(3.7)\qquad \sum'_{|H|=i,\,|K|=j-1}\sum_{\alpha\,\beta}\int_W f_{H\,\alpha K}\bar{f}_{H\,\beta K}\frac{\partial^2\phi}{\partial\omega_\alpha\partial\bar{\omega}_\beta}e^{-\phi}\,dV$$
$$-\sum'_{|H|=i,\,|K|=j-1}\sum_{\alpha=1}^{b}\int_W|f_{H\,\alpha K}|^2\frac{\partial^2\phi}{\partial\omega_\alpha\partial\bar{\omega}_\alpha}e^{-\phi}\,dV\le 2(\|T^*f\|^2_\phi+\|Sf\|^2_\phi)+C'\|f\|^2_\phi.$$

We remark that (3.7) holds for any choice of an orthonormal basis of 1–forms $\omega_1 \dots \omega_n$. In particular we choose coordinates so that $\mathcal{R}$ is the complex plane of the first b coordinates, and assume that $\left(\frac{\partial^2\phi}{\partial\omega_\alpha\partial\bar{\omega}_\beta}\right)_{\alpha\,\beta\leq b}$ is diagonal. Denote by $v_{H\,K}$ the vector $(f_{H\,\alpha K})_\alpha$, and denote by $v'_{H\,K}$ and $v''_{H\,K}$ the components with $\alpha < b$ and $\alpha \geq b+1$ respectively. With these notations the term on the left hand side of (3.7) can be rewritten as:

$$\sum'_{|H|=i\,|K|=j-1}\int_W\left(\partial\bar{\partial}\phi v_{H\,K}{}^t\overline{v_{H\,K}} - \partial\bar{\partial}\phi v'_{H\,K}{}^t\overline{v'_{H\,K}}\right)e^{-\phi}\,\mathrm{d}V.$$

By (3) the latter can be estimated from below by $\lambda \sum'_{|H|=i\,|K|=j-1}\int_W |v''|^2 e^{-\phi}\,\mathrm{d}V$. But we have

$$\sum'_{|H|=i\,|K|=j-1}|v''|^2(=\sum'_{|H|=i\,|K|=j-1}\sum_{\alpha>b}|f_{H\,\alpha K}|^2)\geq\sum'_{|H|=i\,|J|=j}|f_{H\,J}|^2,$$

because $|J| > b$ and hence J can be written, up to the order, as $J = \alpha\, K$ for suitable $\alpha > b$ and K.

In conclusion the term on the left side of (3.7) can be estimated from below by $\lambda\|f\|^2_\phi$, and hence (3.7) gives

$$\lambda\|f\|^2_\phi \leq 2(\|f\|^2_\phi + \|Sf\|^2_\phi) + C'\|f\|^2_\phi. \tag{3.8}$$

We also observe that λ and C' can be taken constant when the coefficients of f have support in a prescribed compact set of W. But in general, they should be replaced by $\lambda(z), C'(z)$. In such case, one should also replace ϕ by $\chi(\phi)$, where χ is a convex function of an 1–dimensional real argument t, with such an increasing rate at ∞ that:

$$\chi'(t) \geq \sup_{\{z;\phi(z)\leq t\}}\frac{2+C'(z)}{\lambda(z)}.$$

Then (3.3) follows, with ϕ replaced by $\chi(\phi)$.

Q.E.D.

Remark 3.2. Fix $z_o \in \partial W$ and let ϕ be an exhaustion function which satisfies (1)–(3) of Proposition 3.1 in a neighborhood of z_o. Then

$$\lim_{\overrightarrow{B}}\mathrm{H}^j(W\cap B, \mathcal{O}_X) = 0 \quad \forall j > b. \tag{3.9}$$

In fact it is easy to find an open set $\tilde{W} \subset W \cap B$ with $\tilde{W} \cap B' = W \cap B'$ $(B' \subset B)$ and which is endowed with an exhaustion function ϕ satisfying (1)–(3) in the whole $\tilde{W}$ (cf. [2]).

§3.2. Proof of the vanishing of (1.4) for $\mathbf{j > S^-}$. We first treat the case of $T_{z_o}W$ non–convex. We write any point $z \in W$ close to ∂W as $z = z^* - r\frac{\zeta}{(\sum\zeta_i^2)^{\frac{1}{2}}}$ for $(z^*; \frac{\zeta}{|\zeta|})(\overset{\text{def.}}{=}: p^*) \in N(W)^{oa}/\mathbb{R}^+$ and for $r \in R^+$. (Since M_3 is generic, we may assume that $(\sum \zeta_i^2)^{\frac{1}{2}} = |\zeta| > 0 \forall \zeta \in (T^*_{M_3}X)_{z_o}$.) We define $\delta(z) \overset{\text{def.}}{:=} r$. We decompose $\bar{W} = \bar{W}_1 \sqcup \bar{W}_2 \sqcup \bar{W}_3$ in such a way that $\forall z \in W_i$ the support point $z^* \in \partial W$ belongs to M_i. Clearly $N_i \overset{\text{def.}}{:=} (\partial W_i \cap \partial W_3) \cap W$, $i = 1, 2$ are two hypersurfaces of W. We define $\phi(z) \overset{\text{def.}}{:=} -\log\delta(z)$. We also write $\delta_i = \delta|_{W_i}$, $\phi_i = \phi|_{W_i}$. We shall prove in what follows that ϕ fulfills (3) for a space $\mathcal{R}$ of dimension $b = S^-(= \sup(S_1^-, S_2^-, S_3^- + 1))$ (cf. (1.1) and (1.2)).

First step. We prove that

$$s^-_{\phi_3}(z) = s^-_{M_3}(p^*) + 1 \ \forall z \in W_3$$

$$s^-_{\phi_i}(z) = s^-_{M_i}(p^*) \ \forall z \in W_i,\ i = 1, 2 \tag{3.10}$$

In fact we first remark that W is foliated by the C^1-hypersurfaces:

$$M_\epsilon = \{z \in X; \delta(z) = \epsilon\}(= \{z^* - \epsilon \frac{\zeta}{(\sum \zeta_i^2)^{\frac{1}{2}}}; p^* = (z^*; \frac{\zeta}{|\zeta|}) \in N(W)^{oa}/\mathbb{R}^+\}).$$

Let $\chi = \chi_\epsilon$ be the complex symplectic transformation $(z;\zeta) \mapsto (z + \epsilon \frac{\zeta}{(\sum \zeta_i^2)^{\frac{1}{2}}}; \zeta)$. Then

$$\chi_\epsilon(T^*_{M_\epsilon}X) = N(W)^{oa}, \tag{3.11}$$

where we consider on the left side of (3.11) the conormals which are exterior to $\{\delta > \epsilon\}$. On each W_i, M_ϵ is C^2, the correspondence χ interchanges $T^*_{M_\epsilon}X$ with $T^*_{M_i}X$, and χ' interchanges:

$$T_pT^*_{M_\epsilon}X \cap \sqrt{-1}T_pT^*_{M_\epsilon}X \overset{\chi'}{\simeq} T_qT^*_{M_i}X \cap \sqrt{-1}T_qT^*_{M_i}X, \tag{3.12}$$

(with $q = \chi(p)(= tp^*$ for some $t \in \mathbb{R}^+)$). Moreover on W_3 (resp. W_i, $i = 1, 2$) one has $\dim T^{\mathbb{C}}M_\epsilon = \dim T^{\mathbb{C}}M_3 + 1$ (resp. $\dim T^{\mathbb{C}}M_\epsilon = \dim T^{\mathbb{C}}M_i$). Thus $\operatorname{rank} L_{M_\epsilon} = \operatorname{rank} L_{M_3} + 1$ (resp. $\operatorname{rank} L_{M_\epsilon} = \operatorname{rank} L_{M_i}$) on W_3 (resp. W_i). It follows easily

$$s^-_{M_\epsilon}(z) = s^-_{M_3}(p^*) + 1 \quad (\text{resp. } s^-_{M_\epsilon}(z) = s^-_{M_i}(p^*)) \text{ in } W_3 \text{ (resp. } W_i). \tag{3.13}$$

Remark that

$$\partial\bar{\partial}(-\log\delta)w^t\bar{w} = \delta^{-2}(\partial\delta w \bar{\partial}\delta\bar{w}) - \delta^{-1}\partial\bar{\partial}\delta w^t\bar{w}. \tag{3.14}$$

From (3.13), (3.14) (and since $M_\epsilon = \{\delta = \epsilon\}$), we get at once the conclusion.

Second step. We prove now that there is a plane $\mathcal{R}_z \subset T_zX$ of dimension $b = S^-$ depending continuously on z and containing a plane $\mathcal{S}^-_z$ (possibly discontinuous) such that

$$L_{\phi_i}(z)|_{\mathcal{S}^-_z} < 0, \quad \dim(\mathcal{S}^-_z) \geq s^-_{\phi_i}(z). \tag{3.15}$$

This is obvious in each W_i for ϕ_i is C^2 there and one can take as $\mathcal{S}^-_z$ a negative eigenspace for $L_{\phi_i}(z)$.

As for the points $z \in N_i$, choose a negative eigenspace $\mathcal{S}^-_{M_3}(p)$ for $L_{M_3}(p)$ ($p \in M_3 \times_{M_i} T^*_{M_i}X$, $\pi(p) = z$). Then a choice of a negative eigenspace for $L_{\phi_i}(z)$, $i = 1, 2$, is $\mathcal{S}^-_{M_3}(p) \oplus \mathbb{C}v^i_z$ for a suitable $v^i_z \in T^{\mathbb{C}}_{z^*}M_i$, whereas for $L_{\phi_3}(z)$ is $\mathcal{S}^-_{M_3}(p) \oplus \mathbb{C}w_z \forall w_z \in T^{\mathbb{C}}_{z^*}M_i$ transversal to $T^{\mathbb{C}}_{z^*}M_3$. Thus by choosing $w_z = v^i_z$ at N_i one can make a continuous choice of $\mathcal{R}_z \supset \mathcal{S}^-_z$ (with $\dim \mathcal{R}_z \leq S^-$ due to the conclusion of the first step).

Third step. We denote by $v \mapsto v'$ the orthogonal projection $T_zX \to \mathcal{R}_z$, set $v'' = v - v'$, and prove (3). In fact we first replace ϕ by $\phi + c|z|^2$ so that $L_\phi(z)$ is non–degenerate $\forall z$. It is easy to prove that $L_\phi(z)|_{\mathcal{R}_z}$ is also non–degenerate. In fact otherwise one finds $u \notin \mathcal{S}_z^-$ with $L_\phi(z)|_{\mathcal{S}_z^- \oplus \mathbb{C}u} < 0$.

Thus we may choose coordinates so that $L_\phi(z)$ is diagonal with eigenvalues $\lambda_\alpha(z)$, and $\mathcal{R}_z$ is the complex plane of $v' = (v_1, \ldots, v_b)\,(b = S^-)$. Since $\mathcal{R}_z \supset \mathcal{S}_z^-$, then $\lambda_\alpha(z) > 0\,\forall\alpha > b$. It follows (with $v'' = (v_{b+1}, \ldots, v_n)$):

$$L_\phi(z)v^t\bar{v} - L_\phi(z)v'^t\bar{v}' = \sum_{\alpha=b+1}^{n} \lambda_\alpha |v_\alpha|^2 \geq \lambda |v''|^2. \tag{3.16}$$

Thus ϕ fulfills (1)–(3) and Proposition 3.1 may be applied. This concludes the proof of the vanishing of (1.4) for $j > S^-$ when $T_{z_o}W$ is non–convex.

When $T_{z_o}W$ is convex we recall that we assume $s^-_{M_i}(p') \leq S^- \,\forall p' \in T^*_{M_i}X \cap \pi^{-1}(B)$ for suitable extensions M_i of M_i^+, $i = 1, 2$. Let W_i, $i = 1, 2$ be the open domains with boundary M_i such that $W = W_1 \cap W_2$. One has a long exact sequence

$$\cdots \to \mathrm{H}^j(W_1, \mathcal{O}_X) \oplus \mathrm{H}^j(W_2, \mathcal{O}_X) \to \mathrm{H}^j(W, \mathcal{O}_X) \to \mathrm{H}^{j+1}(W_1 \cup W_2, \mathcal{O}_X) \to \ldots$$

The terms in the direct sum are 0 $\forall j > S^-$. (In this case one only needs to use the "first step" above and Proposition 3.1.) The last term is 0 $\forall j > S^-$ by (1.4) applied to the non–convex domain $W_1 \cup W_2$. So is the second term too.

Q.E.D.

References

[1] A. Andreotti, H. Grauert, *Théorèmes de finitude pour la cohomologie des éspaces complexes*, Bull. Soc. Math. France **90** (1962), 193–259.

[2] A. Andreotti, C.D. Hill, *E.E. Levi convexity and the Hans Lewy problem. Part II: Vanishing theorems*, Ann. Sc. Norm. Sup. Pisa (1972), 747–806.

[3] M.S. Baouendi, L.P. Rothschild, F. Treves, *CR Structures with group action and extendability of CR functions*, Invent. Math. **82** (1985), 359–396.

[4] A. D'Agnolo, G. Zampieri, *Generalized Levi's form for microdifferential systems, D-modules* and microlocal geometry Walter de Gruyter and Co., Berlin New–York (1992), 25–35.

[5] ———, *A vanishing theorem for sheaves of microfunctions at the boundary on CR manifolds*, Comm. in P.D.E. **17 (5, 6)** (1992), 989–999.

[6] ———, *On microfunctions at the boundary along CR–manifolds*, Preprint (1991).

[7] H. Grauert, *Kantenkohomologie.*, Compositio Math. **44** (1981), 79–101.

[8] G.M. Henkin, *H. Lewy's equation and analysis on pseudoconvex manifolds (Russian)*, I. Uspehi Mat. Nauk. **32 (3)** (1977), 57–118.

[9] G.M. Henkin, J. Leiterer, *Andreotti–Grauert theory by integral formulas*, Birkhauser Progress in Math. **74** (1988).

[10] L. Hörmander, *An introduction to complex analysis in several complex variables*, Van Nostrand, Princeton N.J. (1966).

[11] ———, *L^2 estimates and existence theorems for the $\bar{\partial}$ operator*, Acta Math. **113** (1965), 89–152.

[12] M. Kashiwara, P. Schapira, *Microlocal study of sheaves*, Astérisque **128** (1985.).

[13] M. Sato, M. Kashiwara, T. Kawaï, *Hyperfunctions and pseudodifferential equations*, Springer Lecture Notes in Math. **287** (1973), 265–529.

[14] J. M. Trépreau, *Systèmes differentiels à caractéristiques simples et structures réelles-complexes (d'après Baouendi-Trèves et Sato-Kashiwara-Kawaï)*, Sém. Bourbaki **595** (1981–82).

[15] G. Zampieri, *The Andreotti–Grauert vanishing theorem for dihedrons of* $\mathbb{C}^n$, To appear (1991).
[16] ———, *Simple sheaves along dihedral Lagrangeans*, To appear (1994).

Canonical Symplectic Structure of a Levi Foliation

Giuseppe Zampieri, Dipartimento di Matematica Pura ed Applicata, Università di Padova, Via Belzoni 7, I-35131 Padova, Italy. zampieri@pdmat1.unipd.it

INTRODUCTION

The aim of this paper is to state the microlocal structure of a Levi foliation and to provide the link with the classical local structure by C. Rea [R], M. Freeman [F] and F. Sommer [S]. The basic method is the analysis of CR structures in homogeneous symplectic spaces.

Let X be a complex manifold, M a real submanifold of X, T^*X the cotangent bundle to X, T^*_MX the conormal bundle to M in X, α and σ respectively the canonical 1 and 2 form on T^*X. We fix $p \in \dot{T}^*_MX$, and assume that $\alpha|_{\dot{T}^*_MX} \neq 0$ and $\operatorname{rank}\sigma|_{T^*_MX} \equiv$ const in a neighborhood of p. By the first (resp. second) T^*_MX is " regular" (resp. " CR") in $\dot{T}^*X$ at p. We may then interchange $\dot{T}^*_MX$ with the conormal bundle to a hypersurface, the boundary of a pseudoconvex domain whose Levi form has constant rank. By [R] and [F] there exists a local foliation (the " Levi foliation") of such hypersurface by the (complex) integral leaves of the Levi null-space. (We give here a new proof and even an improvement of the results by [R], [F].) This induces a foliation of $\dot{T}^*_MX$ with complex leaves (by a suitable (complex) identification of the Levi null-space into TT^*_MX). We note that the similar results of [B-F] can not be applied here for the dimension of leaves is possibly > 1. Thus we provide a new proof which is confined, for the seek of simplicity, to the case of real analytic M. Finally we show that $\dot{T}^*_MX$, (still real analytic), can be defined as the zero-set of real and/or imaginary part of holomorphic symplectic coordinates of $\dot{T}^*X$ which makes the leaves of $\dot{T}^*_MX$ to be complex planes. It is well known, on the other hand, that the similar structure for the Levi leaves of M in complex coordinates of X, holds only if M is " Levi flat". As an application we get a result on decomposition of analytic wave front set for CR hyperfunctions, and a generalization of the celebrated " edge of the wedge" Theorem.

CH. 1. CR MANIFOLDS AND LEVI FOLIATIONS

§**1.1. CR Manifolds.** Let X be a complex manifold, $\bar{X}$ the complex conjugate, $X^{\mathbb{R}}$ the real (C^2–)submanifold underlying X. The diagonal identification $j: X^{\mathbb{R}} \xrightarrow{\sim} X \times_X \bar{X}$ induces the following identification for $TX^{\mathbb{R}}$ and its complexification $\mathbb{C} \otimes_{\mathbb{R}} TX^{\mathbb{R}}$:

$$\begin{array}{ccc} T(X^{\mathbb{R}}) & \overset{j'}{\xrightarrow{\sim}} & TX \times_{TX} T\bar{X} \simeq (TX)^{\mathbb{R}} \\ \downarrow & & \downarrow \\ \mathbb{C} \otimes_{\mathbb{R}} T(X^{\mathbb{R}}) & \underset{j'^{\mathbb{C}}}{\xrightarrow{\sim}} & TX \times_X T\bar{X}. \end{array} \tag{1.1}$$

Let M be a C^2–submanifold of $X^{\mathbb{R}}$, TM the tangent bundle to M, $l = \operatorname{codim} M$, and take a system $(r_j) = 0$, $j = 1, ..., l$ of independent equations for M. By (1.1) we get the identifications:

$$\begin{aligned} TM &\simeq \{Z + \bar{Z} \in TX \times_{TX} T\bar{X}|_M; (Z + \bar{Z})(r_j) = 0\} \\ \mathbb{C} \otimes_{\mathbb{R}} TM &\simeq \{Z + \bar{S} \in TX \times_X T\bar{X}|_M; (Z + \bar{S})(r_j) = 0\}. \end{aligned} \tag{1.2}$$

Let p_1 and p_2 be the two projections from $TX \times_{TX} T\bar{X}$ to TX and $T\bar{X}$ resp., and put $j_1 = p_1 \circ j'$, $j_2 = p_2 \circ j'$. Clearly

$$j_1^{-1}(Z) = \Re e Z, \quad j_2^{-1}(\bar{Z}) = \Re e \bar{Z}.$$

Let J be the authomorphism of $TX^{\mathbb{R}}$ which underlies the multiplication by $\sqrt{-1}$ in TX:

$$J : \Re e Z \mapsto \Re e(\sqrt{-1} j_1(Z)).$$

Definition 1.1.

$$\begin{aligned} T^{\mathbb{C}}M &= TM \cap J(TM) \hookrightarrow M \times_X TX^{\mathbb{R}} \\ T^{1,0}M &= j_1(T^{\mathbb{C}}M) \hookrightarrow M \times_X TX \\ T^{0,1}M &= j_2(T^{C}M) \hookrightarrow M \times_X T\bar{X}. \end{aligned}$$

A vector field $\Re e Z = \frac{1}{2}(Z + \bar{Z})$ (resp. Z resp. $\bar{Z}$) belongs to $T^{\mathbb{C}}M$ (resp. $T^{1,0}M$, resp. $T^{0,1}M$) if and only if $Z(r_j) = 0$.

Definition 1.2. *M is said to be CR when* $\operatorname{rank} T_z^{\mathbb{C}}M \equiv \text{const}\ \forall z \in M$.
M is said to be generic when $\operatorname{rank} T_z^{\mathbb{C}}M \equiv n - l$.

If M is CR, then $\operatorname{rank} T^{\mathbb{C}}M$ is called the CR–dimension of M.

Remark 1.3. Let $\gamma_M(z) = \dim(T_M^*X_z \cap {}^tJ(T_M^*X)_z)$. Then $\dim(T_z^{\mathbb{C}}M) = n - l + \gamma_M(z)$ and $\dim(T_zM + J(T_zM)) = n - \gamma_M(z)$. Thus M is generic if and only if $\gamma_M(z) = 0$ or else $T_zM + J(T_zM) = T_zX$.

§**1.2. Levi foliations.** Let X be a complex manifold and M a CR submanifold of $X^{\mathbb{R}}$. We recall the subbundles $T^{\mathbb{C}}M$, $T^{1,0}M$, and $T^{0,1}M$ of $M \times_{X^{\mathbb{R}}} TX^{\mathbb{R}}$, $M \times_X TX$, and $M \times_X T\bar{X}$ resp., and remind that they have the same (constant) dimension ($=$ CR–dim(M)). We observe that $T^{1,0}M$ and $T^{0,1}M$ are integrable ($=$ closed under Lie–brackets); it is well known that $T^{\mathbb{C}}M$ is not integrable in general. Let $\pi : T^*X \to X$ be the cotangent bundle to X, $\dot{T}^*X$ the bundle T^*X with the 0–section removed, T_M^*X the conormal bundle to M in X.

Definition 1.4. Let $p \in \dot{T}^*_M X$, and let r be a function which satisfies $r|_M \equiv 0$ and $\partial r = p$. We define the Levi form of M with respect to p by:

$$L_M(p) = \partial\bar{\partial} r(z)|_{T^{\mathbb{C}}_z M} \quad (z = \pi(p)). \tag{1.3}$$

We shall see in Ch. 2 that $L_M(p)$ can be intrisically defined by the aid of the canonical 2–form σ on T^*X. In particular (1.3) only depends, in signature and rank, on M and p and not on the choice of r.(But this can also be checked directly.) Let $s_M^{+,-,0}(p)$ denote the numbers of resp. positive, negative, and null eigenvalues of $L_M(p)$. We take a system of equations $(r_j)_j = 0$ for M, put $p_j = \partial r_j(z)$, denote by $\operatorname{Ker} L_M(p_j)$ the totally isotropic space of $L_M(p)$, and define:

$$N_M(z) = \bigcap_{j=1,\dots,l} \operatorname{Ker} L_M(p_j), \tag{1.4}$$

the Levi null–space of M at z. Thus a vector field $Z + \bar{Z} \in M \times_{X^{\mathbb{R}}} TX^{\mathbb{R}}$ belongs to N_M if and only if

$$Z(r_j) = 0, \quad L_M(p_j)(Z(z), \cdot) = 0 \ \forall j.$$

Lemma 1.5. *let M be CR and C^3, let $m = n - l + \gamma_M$ be the CR dim of M, let r be a C^3–function s.t. $r|_M \equiv 0$, $\partial r(z_o) \neq 0$, and assume*

$$s_M^-(\partial r(z)) \, (\text{or } s_M^+(\partial r(z))) \ \equiv \ \text{const} \ \forall z \in M.$$

Then we may find complex coordinates $z = (z', z'') \in X \simeq \mathbb{C}^n$ s.t. $T^{\mathbb{C}}M = \{z; z'' = 0\}$, $\partial r(z_o) = dy_n$, and

$$r|_M \ = \ y_n + \left[\sum_{j=1}^{s^+} |z'_j|^2 - \sum_{j=s^++1}^{s^++s^-} |z'_j|^2\right] \ + \ Q(z,\bar{z}) \ + o(|z|)^3, \tag{1.5}$$

where $Q = 0(|z|)^3$ and

$$\partial_{z_i}\partial_{\bar{z}_j} Q \equiv 0 \quad i,j = s^+ + s^- + 1, \dots, m. \tag{1.6}$$

Proof. We change notations and write $z = (z', z'', z''')$, $z = x + \sqrt{-1}y$ with $\partial r(z_o) = dy_n$, $(T^*_M X)_{z_o} = \operatorname{Vect}\{dy'', dy''', dx''\}, T^{\mathbb{C}}_{z_o} M = \{z; z'' = z''' = 0\}$. By the implicit function Th. y'', y''', x'' are functions of $x''', z', \bar{z}'$ on M. Substitution in r gives, over M:

$$r \ = \ y_n \ + \ a(z', \bar{z}') + \sum_{ij} \Re e\, b_{ij} z'_i x'''_j \ + \sum_{ij} c_{ij} x'''_i x'''_j \ + \ 0(|z|)^3.$$

By the coordinate change:

$$\left(z_n \mapsto z_n - \sqrt{-1} \left(\sum_{ij} b_{ij} z'_i z'''_j + \sum_{ij} c_{ij} z'''_i z'''_j \right), \ z_{n-1} \mapsto z_{n-1} \dots \right),$$

we get on M:

$$r \;=\; y_n + a(z',\bar{z}') + 0(|y'''|)0(|z|) + 0(|z|)^3.$$

Thus again by the implicit function Th.:

$$r \;=\; y_n + a(z',\bar{z}') + Q(z,\bar{z}) + o(|z|)^3 \;\;(Q = 0(|z|)^3).$$

It is easy to prove, by a coordinate change in $\mathbb{C}_{z_n} \times \mathbb{C}^m_{z'}$ that a can be transformed into the quadratic term of (1.5). Finally if $\left(\partial^2_{z'_i\bar{z}'_j}Q\right)_{i\,j=s^++s^-+1,\ldots,m} \neq 0$, then both $s^{\pm}_M(\partial r)$ should be non-constant.
Q.E.D.

In the following we give a new proof of the classical result by [R], [F], (and [S]), and also enlarge the class of manifolds M to which the statement applies.

Theorem 1.6. *Let M be CR and C^3, and assume* $\dim(N_M(z)) \equiv \text{const}\ \forall z \in M$. *Then N_M is integrable.*

Proof. We have to prove that for any couple of vector fields $Z+\bar{Z}$ and $S+\bar{S}$ of N_M, we have $[Z+\bar{Z}\,,\,S+\bar{S}] \in N_M$. Let $Z = \sum_j a_j\partial_{z_j}$, $S = \sum_j b_j\partial_{z_j}$, and set $R = \sum_j c_j\partial_{z_j}$ with $c_j :\overset{\text{def.}}{=} \sum_h(\bar{a}_h\partial_{\bar{z}_h}(b_j) - \bar{b}_h\partial_{\bar{z}_h}(a_j))$. We have

$$[Z+\bar{Z}\,,\,S+\bar{S}] \;=\; [Z\,,\,S] + \overline{[Z\,,\,S]} + R + \bar{R}. \tag{1.7}$$

Thus we have in fact to prove that $R+\bar{R} \in N_M$. But first $\forall r$ with $r|_M = 0$, $\partial r|_M \neq 0$, we have:

$$\begin{aligned} R(r) &= -\partial\bar{\partial}r(\bar{a},b) + \partial\bar{\partial}r(\bar{b},a) \\ &= -2\sqrt{-1}\Im m(\partial_{z_i}\partial_{\bar{z}_j}r)(\bar{a},b) \;=\; 0, \end{aligned}$$

due to $Z \in N_M$ or $S \in N_M$. (Here $\partial\bar{\partial}r$ denotes the matrix $(\partial_{z_i}\partial_{\bar{z}_j}r)_{i\,j}$ and a,b the vectors $(a_j),(b_j)$ resp.) Second we have $(\partial\bar{\partial}r)(c,\cdot) = 0$. In fact we fix $z_o = 0$ and choose coordinates $z = (z',z'')$ s.t. r takes the form of Lemma 1.5. Then

(1.8)

$$\begin{aligned} (\partial\bar{\partial}r\;c)_j|_{z=z_o} &= \sum_i \partial_{z_i}\partial_{\bar{z}_j}r(\sum_h \bar{a}_h\partial_{\bar{z}_h}b_i - \bar{b}_h\partial_{\bar{z}_h}a_i)|_{z=z_o} \\ &= \left(\sum_h \bar{a}_h\partial_{\bar{z}_h}(\sum_i \partial_{z_i}\partial_{\bar{z}_j}rb_i) - \sum_h \bar{b}_h\partial_{\bar{z}_h}(\sum_i \partial_{z_i}\partial_{\bar{z}_j}ra_i) - \right. \\ &\quad \left. \sum_{h\,i} \bar{a}_h b_i\partial_{z_i}\partial_{\bar{z}_j}\partial_{\bar{z}_h}r + \sum_{h\,i} \bar{b}_h a_i\partial_{z_i}\partial_{\bar{z}_j}\partial_{\bar{z}_h}r\right)|_{z=z_o}. \end{aligned}$$

But $a_i(z_o) = 0, b_h(z_o) = 0 \forall i, h \neq s^+ + s^- + 1, \ldots, m$. Besides $\partial_{z_i}\partial_{\bar{z}_j}\partial_{\bar{z}_h}r(z_o) = \partial_{z_i}\partial_{\bar{z}_j}\partial_{\bar{z}_h}\,Q(z_o) = 0\,\forall h, i = s^+ + s^- + 1, \ldots m$ and $\forall j = 1, \ldots, m$ due to (1.6). Then (1.8) is 0. The proof is complete.
Q.E.D.

A local foliation of M by complex leaves of dimension d is equivalent to a projection $\rho : M \to M'$ where M' is a real manifold of dimension $2n - l - 2d$ and the fibers of ρ are complex manifolds of dimension d. This is the same as to give a system of coordinates $(t, w) \in M \simeq \mathbb{R}^{2n-l-2d} \times \mathbb{C}^d$ modulo the transformations $\tilde{t} = \tilde{t}(t), \tilde{w} = \tilde{w}(t, w)$ with $\tilde{w}$ depending holomorphically on w. Or a system of charts $M' \times \mathbb{C}^d \xrightarrow{\Phi} M$ with Φ holomorphic in w (cf. [F]).

Corollary 1.7. Let *M be CR and C^3, and assume* $\dim N_M \equiv$ const *in a neighborhood of a point z_o. Then M is foliated near z_o by the complex integral leaves of N_M.*

Proof. By Th. 1.6, we have the hypotheses of Frobenius' Theorem. Thus M is locally foliated by the integral leaves of N_M. These leaves are complex manifolds since their tangent planes are complex planes.

Q.E.D.

Remark 1.8. The proof of Th. 1.6 shows that $T^{\mathbb{C}}M$ is integrable (if and) only if $T^{\mathbb{C}}M = N_M$ (i.e. M is "Levi flat"). This comes also from an easier argument. In fact let Γ be an integral leaf of $T^{\mathbb{C}}M$, assume Γ is defined by $z' = 0$, take an equation $r = 0$ for M, and apply $\partial_{z''}\partial_{\bar{z}''}$ to the identity $r(z'', \bar{z}'') \equiv 0$. It follows $\partial_z \partial_{\bar{z}} r|_{T^{\mathbb{C}}M} = \partial_{z''}\partial_{\bar{z}''} r = 0$.

§1.3. Real analytic CR manifolds. The partial complexification. Let X be a complex manifold, M a real analytic submanifold of $X^{\mathbb{R}}$, $\mathcal{O}_X$ the sheaf of holomorphic functions on X, $\mathcal{A}_M (\overset{\text{def.}}{=} \mathcal{O}_X|_M)$ the sheaf of real analytic functions on M. The following is known as the "Tomassini Theorem".

Proposition 1.9. *Let M be real analytic and generic. For $f \in \mathcal{A}_M$ there is $g \in \mathcal{O}_X$ with $g|_M = f$ if and only if $\bar{Z}f = 0\ \forall \bar{Z} \in T^{0,1}M$. Such g is unique.*

Proof. Let $j : M \hookrightarrow X^{\mathbb{R}}$ with complexification $j^{\mathbb{C}} : M^{\mathbb{C}} \hookrightarrow X \times \bar{X}$, and set $h = f \circ j$ with complexification $h^{\mathbb{C}} = f^{\mathbb{C}} \circ j^{\mathbb{C}}$. Our hypothesis $\bar{Z}f^{\mathbb{C}} = 0$ is the condition for $h^{\mathbb{C}}$ to be constant on the fibers of $p_1 \circ j^{\mathbb{C}}$ or else to define a holomorphic function g on $p_1 \circ j^{\mathbb{C}}(M^{\mathbb{C}}) = X$ (where the last equality follows from the genericity of M in X); such g satisfies $g|_M = f$.

Any other g_1 must coincide with g on $p_1 \circ j^{\mathbb{C}}(M^{\mathbb{C}}) = X$.

Q.E.D.

Proposition 1.10. *Let M be a real analytic CR manifold. Then there exists a complex manifold Y, $M \subset Y \subset X$ which contains M as a generic submanifold. Such Y is unique upto biholomorphic transformations.*

Proof. We set $Y = p_1 \circ j^{\mathbb{C}}(M^{\mathbb{C}})$ and note that $\operatorname{rank}(p_1 \circ j^{\mathbb{C}}) = \dim(TM + J(TM)) = n - \gamma_M$. Thus Y is a complex manifold and M is generic in Y. Moreover if $f \in \mathcal{O}_X$ and $f|_M = 0$, then $f|_Y = 0$ by Prop. 1.9. Thus any complex manifold $Y_1 \subset X$ which contains M must contain Y.

Q.E.D.

We shall refer to Y as to the " partial complexification" of M. We remark that if $f \in \mathcal{A}_M$ verifies $\bar{Z}f = 0 \forall \bar{Z} \in T^{0,1}M$, then one still has existence of a holomorphic function g on the partial complexification Y of X (and hence a fortiori in X) with $g|_M = f$. The unicity in Y (not in X) still holds.

Remark 1.11. Define $\bar{\partial}^b \stackrel{\text{def.}}{:=} j^{\mathbb{C}'^{-1}}(T^{0,1}X)$. It is easy to check that $\bar{\partial}^b$ is the orthogonal to ${}^t j^{\mathbb{C}'}(T^*X \times \{0\})$ (in the duality between $T^*M^{\mathbb{C}}$ and $TM^{\mathbb{C}}$). It is also immediate to see that the above definition and property remain unchanged by replacing the embedding $M \overset{j}{\hookrightarrow} X$ by $M \overset{j}{\hookrightarrow} Y$. In particular it is not restrictive to assume M generic in X. This is equivalent as to require that the embedding $M^{\mathbb{C}} \hookrightarrow X \times \bar{X}$ is non-characteritic for the system $\bar{\partial}$ on $X \times \bar{X}$. Thus Cauchy–Kowalevsky theorem can be applied and gives:

$$\mathcal{A}_M^{\bar{\partial}^b} \simeq \mathcal{O}_X|_M,$$

where $\mathcal{A}_M^{\bar{\partial}^b}$ denotes the $\mathcal{A}_M$–solutions of $\bar{\partial}^b$. This is another way of stating Propositions 1.9 and 1.10.

§1.4. Real analytic CR manifolds M: foliations of M and T^*_MX by complex leaves. Let M be a real analytic submanifold of $X^{\mathbb{R}}$, $\pi : T^*_MX \to M$ the conormal bundle to M in X, $L_M(p)$, $p \in \dot{T}^*_MX$ the Levi form of M. Assume, for a while, that M is a hypersurface, the boundary of a weakly pseudoconvex (or pseudoconcave) domain Ω. (This is equivalent as to require that $L_M(p)$ is semidefinite with constant sign when p ranges through $\dot{T}^*_MX^+$ (or $\dot{T}^*_MX^-$) the exterior (or interior) conormal bundle to Ω. In this situation it is proved by [B-F] that any germ of complex curve Γ on M is the image, via π, of one in $\dot{T}^*_MX$. We prove here that if one considers a foliation instead of a single leaf, then the above result holds true even for $\dim\Gamma > 1$.

Let us come back to $l = \operatorname{codim} M \geq 1$, and fix $p_o \in \dot{T}^*_MX$ with $\pi(p_o) = z_o$.

Theorem 1.12. *Let M be a real analytic CR submanifold of $X^{\mathbb{R}}$, S a hypersurface of $X^{\mathbb{R}}$ which contains M, and assume*

$$s_S^-(p) \equiv 0 \quad \forall p \in M \times_S T^*_SX,\ p \text{ close to } p_o. \tag{1.9}$$

Then any local foliation of M at z_o by complex leaves is induced by one of $M \times_S T^*_SX$ at p_o.

Proof. Let Γ be a complex leaf on M. If Γ is defined, in complex coordinates $z = (z_1, z', z'')$ on X by $z_1 = z' = 0$, and if S is given by $r = 0$ with $\partial r \neq 0$, then application of $\partial_{z''}\partial_{\bar{z}''}$ to the identity $r(z'', \bar{z}'') \equiv 0$, gives $L_S(p)(w, \bar{w}) = 0\, \forall w \in T\Gamma$ (with $p = \partial r(z), z \in M$). Moreover one has:

$$L_S(p)(w, \cdot) = 0 \quad \text{for } p = \partial r(z)\, \forall z \in M. \tag{1.10}$$

In fact if there existed $v \in T_z^{\mathbb{C}}S$ s.t. $\Re L_S(p)(w, \bar{v}) > 0$, we would get :

$$\begin{aligned} L_S(p)(w - tv, \overline{w - tv}) &= -2tL_S(p)(w, \bar{v}) + t^2 L_S(p)(v, \bar{v}) \\ &< 0 \text{ (resp. } > 0) \text{ for } t > 0 \text{ (resp. } t < 0\text{) and } t \text{ small,} \end{aligned}$$

which is a contradiction to (1.9). We denote by $g : M \to M' \simeq \mathbb{R}^{2n-l-2d}$ the foliation of M with complex leaves of $\dim d$ (≥ 1). We assume $\partial r(z_o) = dy_1$ and define $R = g^{-1}(M \cap \mathbb{C}_{z_1})$. Clearly R is Levi–flat (i.e. $N_R = T^{\mathbb{C}}R$). Thus we may find complex coordinates $z = (z_1, z', z'')$, $z = x + \sqrt{-1}y$:

$$R = \{z; y_1 = 0,\ z' = 0\}. \tag{1.11}$$

In fact, since $(Z + \bar{Z})g = 0 \, \forall Z \in T^{\mathbb{C}}R$ (due to $\operatorname{Ker} g' = T^{\mathbb{C}}R$), then g extends to a holomorphic function $\tilde{g} : Y \to \mathbb{C}_{z_1}$ where Y is the partial complexification of R in X. Thus, in complex coordinates in which $\tilde{g}$ is $(z_1, z'') \mapsto z_1$, R takes the form (1.11).

Since $S \supset R$, we get from (1.11): $r = y_1 + 0(|z'|)0(|(z_1, z'')|) + o(|z|)^2$. Let $\Gamma = \mathbb{C}^d_{z''}$; then:

$$\partial_{\bar{z}}\left(\partial_z r | \Gamma\right) \equiv \left(\partial_{\bar{z}''}\partial_z r\right)|_\Gamma \equiv 0 \text{ i.e. } \partial r|_\Gamma \text{ is holomorphic.} \tag{1.12}$$

(In fact $\partial_{z_1} r|_\Gamma \equiv -\sqrt{-1}$ and $\partial_{\bar{z}''}\partial_{z_i} r|_\Gamma \equiv 0 \, \forall i \neq 1$ due to (1.10).) Thus the foliation of M by the complex leaves Γ, gives a foliation of $M \times_S T^*_S X$ by the complex leaves $\Gamma_t = \{(z; t\partial_z r(z)); z \in \Gamma\}$, $t \in \mathbb{R}$.

Q.E.D.

Remark 1.13. Let M be real analytic CR and verify $s_{\bar{M}}(p) \equiv 0 \, \forall p \in \dot{T}^*_M X$. Then $N_M \equiv T^{\mathbb{C}}M$ whence M is (locally) foliated by the integral leaves of N_M. By applying the argument of Theorem 1.12 to a system of independent equations $(r_j)_j = 0, j = 1, \ldots, l$ for M one gets a foliation of $T^*_M X$ by the holomorphic leaves $\Gamma_{(t_j)} = \{(z; \sum_{j=1}^l t_j \partial_{z_j} r\ (z)); z \in \Gamma\}$, $(t_j) \in \mathbb{R}^l$. As the proof of Th. 1.12 explains, this is obtained by a suitable identification of N_M to a complex subbundle of $M \times_M TT^*_M X$.

CH. 2. CANONICAL FORM OF $\mathbb{R}$-LAGRANGIAN REGULAR SUBMANIFOLDS WITH CONSTANT LEVI–RANK

§2.1. Complex homogeneous symplectic structures and Levi forms. Let X be a complex manifold of dim n, $\pi : T^*X \to X$ the cotangent bundle to X, $\dot{T}^*X$ the bundle T^*X with the 0–section removed, $\alpha = \alpha^{\mathbb{R}} + \sqrt{-1}\alpha^{\mathbb{I}}$ the canonical 1–form, $\sigma = \sigma^{\mathbb{R}} + \sqrt{-1}\sigma^{\mathbb{I}}$ the canonical 2–form, H (resp. $H^{\mathbb{R}}$, $H^{\mathbb{I}}$) the Hamiltonian isomorphism $T^*T^*X \to TT^*X$ associated to σ (resp. $\sigma^{\mathbb{R}}$, $\sigma^{\mathbb{I}}$). Let $X^{\mathbb{R}}$ (resp. $(T^*X)^{\mathbb{R}}$) be the real manifold underlying X (resp. T^*X). The diagonal identification (1.1) gives an identification $T^*X^{\mathbb{R}} \overset{{}^t j'}{\underset{\leftarrow}{\sim}} (T^*X)^{\mathbb{R}}$. We fix $p \in \dot{T}^*X$ and define

$$\begin{aligned} e(p) &= T_p T^*X \\ \lambda_0(p) &= T_p(\pi^{-1}\pi(p)) \\ \nu(p) &= \mathbb{C}H(\alpha(p)). \end{aligned} \tag{2.1}$$

We often drop p in (2.1). A plane $v \subset e$ is called ($\mathbb{C}$–)involutive (resp. Lagrangian, resp. isotropic) if and only if $v^\perp \subset v$ (resp. $v^\perp = v$ resp. $v^\perp \supset v$) with $\cdot^\perp$ denoting the σ–orthogonal. A (complex analytic) submanifold $V \subset T^*X$ is ($\mathbb{C}$–)involutive (rep. Lagrangian, resp. isotropic) if at each $p \in V$ the tangent plane $v = T_p V$ has the corresponding property in e.

V is said to be regular when $\alpha|_V \neq 0$.

A (C^1)–submanifold $\Lambda \subset T^*X^{\mathbb{R}}$ is called $\mathbb{R}$–Lagrangian if $\lambda(p) \overset{\text{def.}}{:=} T_p\Lambda$ is Lagrangian for $\sigma^{\mathbb{R}}(p)$. Λ is called I–symplectic when $\sigma^I(p)$ is non-degenerate on $\lambda(p)$. All manifolds of T^*X ($T^*X^{\mathbb{R}}$) will be conic i.e. locally invariant under ν (resp. $\Re e\, \nu$).

Let λ be an $\mathbb{R}$–Lagrangian plane of e; we set

$$(2.2)\qquad \begin{aligned} \mu &= \lambda \cap \sqrt{-1}\lambda \\ d &= \dim(\mu) \\ \gamma &= \dim(\lambda \cap \sqrt{-1}\lambda \cap \lambda_0) \\ l &= \dim(\lambda \cap \lambda_0). \end{aligned}$$

Let σ^μ be the symplectic form induced by σ on $\mu^\perp/\mu$. We define the Levi form $L = L_{\lambda/\lambda_0}$ by:

$$(2.3)\qquad L_{\lambda/\lambda_0} = \sigma^\mu(u, v^c)|_{u,v\in\lambda_0^\mu},$$

(where $\lambda_0^\mu \overset{\text{def.}}{:=} ((\lambda_0 \cap (\lambda + \sqrt{-1}\lambda)) + (\lambda \cap \sqrt{-1}\lambda)) \,/\, (\lambda \cap \sqrt{-1}\lambda)$ and $\cdot^c$ denotes the conjugate in the sum $\lambda + \sqrt{-1}\lambda \,/\, \lambda \cap \sqrt{-1}\lambda$). This is a Hermitian form in λ_0^μ whose kernel is $\left(\lambda \cap \lambda_0 \,/\, \lambda \cap \sqrt{-1}\lambda \cap \lambda_0\right)^{\mathbb{C}}$ (where $\cdot^{\mathbb{C}}$ denotes the complexification). Remark that $\dim_{\mathbb{C}}(\lambda_0^\mu) = n - d$, and $\dim_{\mathbb{C}}\left(\lambda \cap \lambda_0 \,/\, \lambda \cap \sqrt{-1}\lambda \cap \lambda_0\right)^{\mathbb{C}} = l - 2\gamma$. It follows

$$(2.4)\qquad \operatorname{rank} L_{\lambda/\lambda_0} = n - d - l + 2\gamma.$$

One may also check that

$$(2.5)\qquad \operatorname{sign} L_{\lambda/\lambda_0} = \frac{1}{2}\tau(\lambda, \sqrt{-1}\lambda, \lambda_0),$$

where τ is the inertia index (cf. [K-S]).

Let M be a (C^2–)submanifold of $X^{\mathbb{R}}$, and let $T^*_M X$ be the cotangent bundle to M in X identified to an $\mathbb{R}$–Lagrangian submanifold of $T^*X^{\mathbb{R}}$. We fix $p \in T^*_M X$, $z = \pi(p)$, and put

$$(2.6)\qquad \begin{aligned} \lambda_M(p) &= T_p T^*_M X \\ T_z^{\mathbb{C}} M &= T_z M \cap \sqrt{-1} T_z M. \end{aligned}$$

We also define μ_M, d_M, γ_M in the obvious way. We take a system of independent equations $r_j = 0$ for M with $\partial r_1(z) = p$ and define:

$$\psi : M \times \mathbb{R}^l \overset{\sim}{\to} T^*_M X \quad (z; (t_j)) \mapsto (z; \sum_j t_j \partial r_j(z)),$$

which induces:

$$\begin{array}{ccc} TM \times \mathbb{R}^l & \overset{\psi'}{\underset{\sim}{\to}} & \lambda_M \\ \downarrow & & \downarrow \\ (\mathbb{C} \otimes_{\mathbb{R}} TM) \times \mathbb{C}^l & \overset{\psi_*}{\underset{\sim}{\to}} & \lambda_M + \sqrt{-1}\lambda_M \hookrightarrow T_p T^* X. \end{array}$$

We recall the identifications (1.1) and (1.2) of §1.1. They give

$$\lambda_M(p) = \{(u; \sum_j t_j \partial r_j + \partial\bar{\partial} r_1(z)u + \partial\bar{\partial} r_1(z)\bar{u}; (t_j) \in \mathbb{R}^l, \quad \partial r_1(z)u + \bar{\partial} r_1(z)\bar{u} = 0\}$$

$$\lambda_M(p) + \sqrt{-1}\lambda_M(p) = \{(u; \sum_j \tau_j \partial r_j + \partial\bar{\partial} r_1(z)u + \partial\bar{\partial} r_1(z)\bar{w}; (\tau_j) \in \mathbb{C}^l, \quad \partial r_1(z)u + \bar{\partial} r_1(z)\bar{w} = 0\}$$

$$\lambda_M(p) \cap \sqrt{-1}\lambda_M(p) = \{(u; {}^t\partial\bar{\partial} r_1(z)u + \partial\bar{\partial} r_1(z)u); \partial r_1(z)u = 0, \partial\bar{\partial} r_1(z)\bar{u} \in T^*_S X_z + \sqrt{-1}T^*_S X_z\} \oplus \{(0;v); v \in T^*_M X_z \cap \sqrt{-1}T^*_M X_z\}.$$

Recall the Levi form $L_M(p) = \partial\bar{\partial} r_1|_{T^{\mathbb{C}}_z M}$; then

$$\lambda_M(p) \cap \sqrt{-1}\lambda_M(p) \simeq \operatorname{Ker}(L_M(p)) \oplus (\lambda_M(p) \cap \sqrt{-1}\lambda_M(p) \cap \lambda_0(p)). \tag{2.7}$$

Observe that

$$\begin{aligned} \operatorname{rank} L_M(p) &= \dim T^{\mathbb{C}}_z M - \dim \operatorname{Ker}(L_M(p)) \\ &= (n - l + \gamma_M) - (d_M - \gamma_M) = \operatorname{rank} L_{\lambda_M/\lambda_0}(p). \end{aligned} \tag{2.8}$$

(One similarly proves that $\operatorname{sign} L_M(p) = \operatorname{sign} L_{\lambda_M/\lambda_0}(p)$ (cf. [D'A-Z]).)

§2.2. Statement of the main result. Let X be a complex manifold of $\dim n$, M a real analytic submanifold of $X^{\mathbb{R}}$ of $\operatorname{codim} l$, p_o a point of $\dot{T}^*_M X$. We assume that $T^*_M X$ is regular at p_o; this is equivalent as to require $\sqrt{-1}p_o \notin T^*_M X$. We also assume

$$d_M(p) \equiv \text{const} \quad \forall p \in \dot{T}^*_M X \text{ in a neighborhood of } p_o. \tag{2.9}$$

Let V be a regular involutive submanifold of $\dot{T}^*X$. It is well known that V has a (local) foliation (= the " Hamiltonian foliation") by means of the integral leaves of its Hamiltonian flow $v^\perp$ (= the "bicharacteristic leaves" of V). We define an equivalence relation $\sim$ on V by identifying all points in a same bicharacteristic leaf. We set $V' = V/\sim$ and denote by $\rho : V \to V'$ the canonical projection. The form σ induces a non–degenerate form σ' (denoted sometimes $\sigma^{v^\perp}$) on $v' = v/v^\perp (\simeq TV')$. Our main result is:

Theorem 2.1. *Let $T^*_M X$ be real analytic, regular at p_o, and satisfy (2.9). Then there exists a germ V at p_o of regular involutive complex submanifold of $\dot{T}^*X$, such that if $\rho : V \to V'$ is the Hamiltonian foliation of V, then*

$$\dot{T}^*_M X = \rho^{-1}(\Lambda') \quad \text{for a germ } \Lambda' \text{ of } \mathbb{R}\text{–Lagrangian and } \mathbb{I}\text{–symplectic submanifold of } V'. \tag{2.10}$$

Remark 2.2. Once Th. 1.1 is established, we can find complex symplectic coordinates $(z,\zeta) \in T^*X$, $z = (z', z'')$, $\zeta = (\zeta', \zeta'')$, $z = x + \sqrt{-1}y$, $\zeta = \xi + \sqrt{-1}\eta$ such that $V = \{(z;\zeta); \zeta'' = 0\}$, $V' = T^*X'$, $X' \simeq \mathbb{C}^{n-d}$. On the other hand any $\mathbb{R}$–Lagrangian $\mathbb{I}$–symplectic submanifold $\Lambda' \subset T^*\mathbb{C}^{n-d}$ can be interchanged, by a complex symplectic transformation of $T^*\mathbb{C}^{n-d}$ with $T^*_{\mathbb{R}^{n-d}}\mathbb{C}^{n-d}$. Thus by a complex symplectic transformation χ in $\dot{T}^*X$, we get

$$\dot{T}^*_M X \overset{\chi}{\simeq} \dot{T}^*_{\mathbb{R}^{n-d}}\mathbb{C}^{n-d} \times \mathbb{C}^d. \tag{2.11}$$

Remark 2.3. When M is CR (i.e. $\gamma_M \equiv$ const at $z_o = \pi(p_o)$), then (2.9) is equivalent to:

$$\operatorname{rank} L_M(p) \equiv \text{const} \quad \forall p \in T^*_M X \text{ close to } p_o, \tag{2.12}$$

(due to (2.8)).

Remark 2.4. Remark 2.2 shows that it is possible to straighten the integral leaves of μ_M by a complex symplectic transformation of $\dot{T}^*X$. This transformation is not induced by a change of complex coordinates $z = (z', z'') \in X$ for otherwise M should take the form $M = \{z;\ y'' = 0\}$ which is never true unless M is Levi flat i.e. $N_M = T^{\mathbb{C}}M$ (cf. [R]).

§2.3. **Proof of Theorem 2.1.** We put $\Lambda_M = \dot{T}^*_M X$. By the regularity assumption, we may interchange Λ_M, by a complex symplectic transformation χ, to the conormal bundle to a hypersurface. And we may also require that $s^- = 0$ in $\chi(p)$ for such hypersurface. But we have indeed $s^- \equiv 0$ in a neighborhood of $\chi(p)$ because of (2.9) and since the constancy of $s^\pm - \gamma$ is a symplectic invariant. Thus this hypersurface is in fact the boundary of a pseudoconvex domain. By the same reason $s^+ \equiv$ const at $\chi(p)$.

Thus it is not restrictive to assume from the beginning that M is the boundary of a pseudoconvex domain with dim $N_M \equiv$ const (say d). By Th. 1.7 M is foliated by the integral leaves of N_M, and by Th. 1.12 this foliation is induced, via π, by a foliation of Λ_M whose leaves are tangent to $\mu_M = \lambda_M \cap \sqrt{-1}\lambda_M$ (with $\lambda_M = T\Lambda_M$). Thus we have a projection:

$$\rho : \Lambda_M \to \Lambda', \quad \operatorname{Ker} \rho' = \mu_M.$$

(Λ' is only a space of (real) parameters and does not carry, for the moment, any symplectic structure.) Let $\Lambda'^{\mathbb{C}}$ be a complexification of Λ', and V the partial complexification of Λ_M in T^*X. The latter exists because dim $\mu_M \equiv$ const (cf. §1.3). Let $\tilde{\rho} : V \to \Lambda'^{\mathbb{C}}$ be the (unique) extension of ρ; this exists because ρ is the projection along the integral leaves of μ_M whence $(Z + \bar{Z})\rho = 0 \, \forall Z + \bar{Z} \in T^{\mathbb{C}}\Lambda_M = \mu_M$ (cf. Prop. 1.9). We claim that V is a regular involutive complex submanifold of $\dot{T}^*X$, and $\tilde{\rho}$ is the projection along its bicharacteristic leaves. (Thus in particular $V' \simeq \Lambda'^{\mathbb{C}}$.) In fact

$$\dim_{\mathbb{C}} v = \dim_{\mathbb{C}}(\lambda_M + \sqrt{-1}\lambda_M) = 2n - d, \quad \dim_{\mathbb{C}} \lambda'^{\mathbb{C}} = 2(n-d).$$

Thus $\mathrm{Ker}\,\tilde{\rho}'$ is a complex bundle on V whose dimension equals $\dim_{\mathbb{C}} v - \dim_{\mathbb{C}} \lambda'^{\mathbb{C}} = d$. The same is obviously true for $v^{\perp}$ (since $v^{\perp} = \mathrm{Vect}\{H_{s_1}, ..., H_{s_d}\}$ for a system of d independent equations $s_j = 0$ for V in $\dot{T}^*X$). Moreover:

$$v^{\perp} = \mu_M(p) = \mathrm{Ker}\,\tilde{\rho}'(p) \quad \forall p \in \Lambda_M.$$

Hence

$$v^{\perp} = \mathrm{Ker}\,\tilde{\rho}'(p) \quad \forall p \in V,$$

because two holomorphic bundles on V which coincide over Λ_M, must coincide everywhere. This proves our claim.

Last we have $\lambda' = \lambda_M/v^{\perp} = \lambda_M/\mu_M$ (for $v^{\perp} = \mu_M$ on Λ_M); in particular $\Lambda' \subset V'$ is $\mathbb{R}$–Lagrangian (due to $\lambda'^{\perp} = \lambda_M^{\perp}/\mu_M = \lambda'$) and $\mathbb{I}$–symplectic (due to $\mathrm{Ker}\,\sigma|_{\lambda_M} = \mu_M$). The proof is complete.

Q.E.D.

CH. 3. APPLICATIONS: DECOMPOSITION OF WAVE FRONT SET OF CR HYPERFUNCTIONS AND THE " EDGE OF THE WEDGE" THEOREM

§3.1. Let X be a complex manifold of $\dim n$, M a real analytic CR submanifold of $\mathrm{codim}\, l$, U an open cone of $\dot{T}^*_M X$. We assume $\sqrt{-1}p \notin T^*_M X\ \forall p \in U$, and:

$$\text{(3.1)} \qquad \mathrm{rank}\, L_M(p) \equiv \mathrm{const} \quad \forall p \in U.$$

Let $\mathcal{O}_X$ be the sheaf of complex analytic functions on X and $\mathbb{Z}_M$ the sheaf which is 0 in $X \setminus M$ and the constant sheaf with stalk $\mathbb{Z}$ in M. We assume M oriented and define the complexes of resp. Sato's microfunctions and hyperfunctions along M by:

$$\text{(3.2)} \qquad \begin{cases} \mathcal{C}_{M|X} = \mu\mathrm{hom}(\mathbb{Z}_M, \mathcal{O}_X)[l] \\ \mathcal{B}_{M|X} = \mathbf{R}\mathcal{H}\mathrm{om}(\mathbb{Z}_M, \mathcal{O}_X)[l](= \mathbb{R}\pi_*(\mathcal{C}_{M|X}), \end{cases}$$

where $\mu\mathrm{hom}(\cdot,\cdot)$ is the bifunctor of [K–S, ch. 4]. $\mathcal{B}_{M|X}$ turns out to coincide with the complex of CR hyperfunctions along M (= tangential $\bar{\partial}$–complex over forms whose coefficients are (usual) hyperfunctions $\mathcal{B}_M$). Let $\mathrm{sp} : \pi^{-1}\mathcal{B}_{M|X} \to \mathcal{C}_{M|X}$ be the (spectral) morphism obtained from the second term of (3.2) by adjunction. For $f \in H^j(\mathcal{B}_{M|X})$ we set

$$\text{(3.3)} \qquad WF^j(f) = \mathrm{supp}(\mathrm{sp}(f)).$$

(Sometimes we drop the superscript j.) For M generic and $s_M^- = 0$, WF^j coincides with the classical analytic wave front set (cf. [B-C-T]). According to [K-S] the geometric correspondence $\chi : T^*_M X \xrightarrow{\sim} T^*_{\mathbb{R}^{n-d}}\mathbb{C}^{n-d} \times \mathbb{C}^d$, ($d = \mathrm{corank}\, L_M$), stated in §2.2, can be " quantized" to an isomorphism

$$\text{(3.4)} \qquad \mathcal{C}_{M|X} \simeq \mathcal{C}_{\mathbb{R}^{n-d}\times\mathbb{C}^d|X}[-s_M^- + \gamma_M].$$

Thus $\mathcal{C}_{M|X}$ is isomorphic, upto a shift, to the usual sheaf of microfunctions with holomorphic parameters. Then the theory by [S-K-K] applies. In particular $\mathcal{C}_{M|X}$ is concentrated in degree $s_M^- - \gamma_M$ and:

Proposition 3.1. *$H^{s_M^- - \gamma_M}(\mathcal{C}_{M|X})|_U$ satisfies the principle of the analytic continuation along the integral leaves of $\mu_M = \lambda_M \cap \sqrt{-1}\lambda_M$ and is transversally conically flabby with respect to such a foliation. The first means that if $\Omega' \subset \Omega \subset U$ are open cones with $\Omega = \rho^{-1}\rho(\Omega)$ and $\rho(\Omega) = \rho(\Omega')$, then:*

$$f \in H^{s_M^- - \gamma_M}\left(\mathcal{C}_{M|X}\right)(\Omega), f|_{\Omega'} = 0 \quad \text{implies} \quad f|_\Omega = 0.$$

The second means that if we fix $U' \times K$ in U with U' open conic in $\dot{T}^*_{\mathbb{R}^{n-d}}\mathbb{C}^{n-d}$, K compact Stein in $\mathbb{C}^d$, and denote by ρ_1 the restriction of ρ to $U' \times K$, then $\rho_{1*}\left(H^{s_M^- - \gamma_M}\left(\mathcal{C}_{M|X}\right)\right)$ is a conically flabby sheaf on U'.

Remark 3.2. As a consequence of the transversal conical flabbiness we get the following statement. Let η_j, $j = 1, \ldots, N$ be closed cones of U verifying $\eta_j = \rho^{-1}\rho(\eta_j)$. Then any $f \in H^{s_M^- - \gamma_M}\left(\mathcal{C}_{M|X}\right)$ s.t. $WF(f) \subset \cup_j \eta_j$ can be decomposed as $f = \sum_{j=1,\ldots,N} f_j$ with $WF(f_j) \cap (U' \times K) \subset \eta_j$.

We shall assume henceforth that M is generic i.e. $\gamma_M \equiv 0$. This is equivalent as to assume that

$$\text{(3.5)} \qquad \text{the fibers of } \pi \text{ and } \rho \text{ are transversal.}$$

Let δ be an open convex cone of $T_M X = M \times_X TX \,/\, TM$. A domain $W \subset X$ is said to be a wedge with profile δ when

$$T_M X \setminus C_M(X \setminus W) \supset \delta,$$

where $C_M(\cdot)$ denotes the Whitney normal cone along M. (These wedges form the smallest C^1–invariant family which contains, in the identification $X \simeq \mathbb{R}^{2n-l} \times \mathbb{R}^l \simeq M \times \mathbb{R}^l$, the domains $W = M \times \Delta$ for Δ open convex cone of $\mathbb{R}^l$.) Let η be a closed convex proper cone of $T^*_M X$ with $\eta \supset M$. We have

$$\text{(3.6)} \qquad H^j_\eta(T^*_M X, \mathcal{C}_{M|X}) \overset{b}{\sim} \underset{\substack{\longleftarrow \longrightarrow \\ W}}{\lim} H^j(W, \mathcal{O}_X),$$

where W ranges through the family of wedges with profile $\delta = \operatorname{int} \eta^{oa}$ (the interior of the antipodal to η^o, the polar cone to η). The isomorphism (3.6) is called the " boundary value" morphism. Thus $f \in H^{s_M}\left(\mathcal{B}_{M|X}\right)_z$ verifies $WF(f) \subset \eta$ if and only if f is the boundary value of a cohomology class of degree s_M^- on $W \cap B$ where B is a neighborhood of z and W a wedge of profile $\operatorname{int} \eta^{oa}$.

The following extends to the generic real analytic submanifolds $M \subset X$ with constant Levi–rank the celebrated Martineau's "edge of the wedge" Theorem which corresponds to the case $M = \mathbb{R}^n \subset X = \mathbb{C}^n$. (Cf. also [S-T].) Fix $z \in M$, $z \in \pi(U)$.

Proposition 3.3. *Let M be a real analytic generic submanifold of $X^{\mathbb{R}}$ which verifies (3.1). Let $\eta_j = M \times Z_j$, $j = 1, ..., N$ be closed convex proper cones of U, and let $F_j \in H^{s_M^-}(W_j \cap B, \mathcal{O}_X)$ where B ranges through the family of neighborhoods of z, and W_j of wedges with profile $\delta_j = M \times \operatorname{int} Z_j^{oa}$. Assume that $\sum_j b(F_j) = 0$.*

Then there exist $F_{ij} \in H^{s_M^-}(W_{ij} \cap B, \mathcal{O}_X)$ where W_{ij} are wedges with profile the convex hull δ_{ij} of δ_i, δ_j, such that:

$$F_{ij} = -F_{ji} \forall i j, \quad F_j = \sum_i F_{ij} \forall j.$$

Proof. Let $f_j = b(F_j)|_U$. We have

$$\operatorname{supp}(f_j) \subset \bigcup_{i \neq j} (\eta_i \cap \eta_j) = M \times \underset{i \neq j}{\cup} (Z_i \cap Z_j).$$

By applying Remark 3.2, we may find sections $f_{ij} \in \mathcal{C}_{M|X}(U \cap \pi^{-1}(B))$:

$$\left(\operatorname{supp} f_{ij} \subset \eta_i \cap \eta_j = \eta_{ij}, \; f_{ij} = -f_{ji} \; f_j = \sum_i f_{ij} \right).$$

Finally by the surjectivity of

$$\mathcal{H}^{s_M^-}_{\eta_{ij}}(\mathcal{C}_{M|X})_z \twoheadrightarrow \dot{\pi}_*^{-1} \mathcal{H}^{s_M^-}_{\dot{\eta}_{ij}}(\mathcal{C}_{M|X})_z,$$

(due to $H^{s_M^- + 1}(\mathbb{R}\Gamma_{T^*_X X}(\mathcal{C}_{M|X})) = 0$), and by (3.6), we get $f_{ij} = b(F_{ij})$ for suitable $F_{ij} \in H^{s_M^-}(B \cap W_{ij}, \mathcal{O}_X)$ where W_{ij} are wedges with profile $\delta_{ij} = \operatorname{int} \eta_{ij}^{oa}$.
Q.E.D.

Let $f \in H^{s_M^-}(\mathcal{B}_{M|X})_z$, $z \in \pi(U)$. By similar argument we get:

Proposition 3.4. *(i) Assume $WF(f) \subset U$, and let $p \in U$, $\pi(p) = z$. Then $p \notin WF(f)$ if and only if $f = \sum_{j=1}^N b(F_j)$ for $F_j \in H^{s_M^-}(B \cap W_j, \mathcal{O}_X)$, where W_j are wedges whose profile δ_j verifies $p \notin \operatorname{int} \delta_j^{oa} \subset \eta$.*

(ii) Let $WF(f) \subset \underset{j=1,\dots,N}{\cup} \eta_j$ with $\eta_j = M \times Z_j \subset U$. Then we may decompose

$$f = \sum_{j=1}^N b(F_j), \quad F_j \in H^{s_M^-}(B \cap W_j, \mathcal{O}_X),$$

where W_j are wedges with profile $\delta_j = \operatorname{int} \eta_j^{oa}$.

References

[B-F] E. Bedford, J.E. Fornaess, *Complex manifolds in pseudoconvex boundaries*, Duke Math. J. **48** (1981), 279–287.

[B-C-T] M.S. Baouendi, C.H. Chang, F. Treves, *Microlocal hypo-analyticity and extension of C.R. functions*, J. of Diff. Geom. **18** (1983), 331–391.

[B-R-T] M.S. Baouendi, L.P. Rothschild, F. Treves, *CR Structures with group action and extendability of CR functions*, Invent. Math. **82** (1985), 359–396.

[D'A-Z] A. D'Agnolo, G. Zampieri, *Generalized Levi's form for microdifferential systems*, *$\mathcal{D}$-modules* and microlocal geometry Walter de Gruyter and Co., Berlin New-York (1992), 25–35.

[F] M. Freeman, *Local complex foliation of real submanifolds*, Math. Ann. **209** (1974), 1–30.

[H] L. Hörmander, *An introduction to complex analysis in several complex variables*, Van Nostrand, Princeton N.J. (1966).

[K-S] M. Kashiwara, P. Schapira, *Microlocal study of sheaves*, Astérisque **128** (1985).

[R] C. Rea, *Levi–flat submanifolds and holomorphic extension of foliations*, Ann. SNS Pisa **26** (1972), 664–681.

[S-K-K] M. Sato, M. Kashiwara, T. Kawai, *hyperfunctions and pseudodifferential equations*, Springer Lecture Note in Math. **287** (1973), 265–529.

[S-T] P. Schapira, J.M. Trepreau, *Microlocal pseudoconveexity and "edge of the wedge" theorem*, Duke Math. J. **61 1** (1990), 105–118.

[S] F. Sommer, *Komplex-analytishe Blätterung reeller hyperflächen in* $\mathbb{C}^n$, Math. Ann. **137** (1959), 392–411.

[Tr 1] J.-M. Trépreau, *Sur la propagation des singularités dans les varietés CR*, Bull. Soc. math. de France **118** (1990), 129–140.

[Tr 2] J. M. Trépreau, *Systèmes differentiels à caractéristiques simples et structures réelles-complexes (d'après Baouendi-Trèves et Sato-Kawai-Kashiwara)*, Sémin. Bourbaki **595** (1981-82).

[Tu 1] A. Tumanov, *Extending CR functions on a manifold of finite type over a wedge*, Mat. Sb. **136** (1988), 129–140.

[Tu 2] A. Tumanov, *Connections and propagation of analyticity for CR functions*, Duke Math. Jour. **73 1** (1994), 1–24.